A TEXTBOOK OF FLUID MECHANICS

A TEXTBOOK OF
FLUID MECHANICS

[In S.I. Units]

By

Dr. R.K. BANSAL

B.Sc. Engg. (Mech.), M.Tech., Hons. (I.I.T., Delhi)
Ph.D., M.I.E. (India)
Formerly
Professor, Department of Mechanical Engineering,
Dean (U.G. Studies), Delhi College of Engineering, Delhi.

LAXMI PUBLICATIONS (P) LTD

(An ISO 9001:2008 Company)

BENGALURU • CHENNAI • COCHIN • GUWAHATI • HYDERABAD
JALANDHAR • KOLKATA • LUCKNOW • MUMBAI • RANCHI • NEW DELHI
BOSTON (USA) • ACCRA (GHANA) • NAIROBI (KENYA)

A TEXTBOOK OF FLUID MECHANICS

Printed and bound in India
Typeset at : Shubham Composer, Delhi.
First Edition : 2005, *Reprint* : 2007, 2008, 2009, 2010, 2011, 2012, 2014, 2015,
New Edition : 2017, Second Edition : 2020
ISBN : 978-81-318-0294-6

PUBLISHED IN INDIA BY

(An ISO 9001:2008 Company)
113, GOLDEN HOUSE, DARYAGANJ,
NEW DELHI - 110002, INDIA
Telephone : 91-11-4353 2500, 4353 2501
Fax : 91-11-2325 2572, 4353 2528
www.laxmipublications.com info@laxmipublications.com

Branches

✆ **Bengaluru**	080-26 75 69 30
✆ **Chennai**	044-24 34 47 26
✆ **Cochin**	0484-237 70 04, 405 13 03
✆ **Guwahati**	0361-254 36 69, 251 38 81
✆ **Hyderabad**	040-27 55 53 83, 27 55 53 93
✆ **Jalandhar**	0181-222 12 72
✆ **Kolkata**	033-22 27 43 84
✆ **Lucknow**	0522-220 99 16
✆ **Mumbai**	022-24 91 54 15, 24 92 78 69
✆ **Ranchi**	0651-220 44 64

C—R/017/01
Printed at : Ajit Printers, Delhi

PREFACE TO THE FIRST EDITION

I am glad to present the book entitled, 'Fluid Mechanics' to the engineering students of mechanical of Visveswaraiah Technology University (V.T.U.) RGPV, Bhopal and other Universities. The course contents have been planned in such a way that the general requirements of all engineering students namely civil, aeronautical, automobile manufacturing, industrial production and industrial management, and chemical are fulfilled.

During my long experience of teaching this subject to undergraduate and postgraduate engineering students, I have observed that the students face difficulty in understanding clearly the basic principles, fundamental concepts and theory without adequate solved problems along with the text. To meet this very basic requirement to the students, a large number of the questions taken from different B.E. Degree Examinations of various University of India have been solved along with the text in S.I. Units.

The book is written in a simple and easy-to-follow language, so that even an average students can grasp the subject by self-study. At the end of each chapter highlights, theoretical questions and many unsolved numerical problems with answer are given for the students to solve them.

In the chapter of Flow Past Immersed Bodies, more details of Boundary Layers are given for the benefit of the students.

I express my appreciation and gratefulness to my publisher Shri R.K. Gupta (as Mechanical Engineer) for his most co-operative, painstaking attitude and untiring efforts for bringing out the book in a short period.

Mrs. Nirmal Bansal deserves special credit as she not only provided an ideal atmosphere at home for book writing but also gave inspiration and valuable suggestions.

Though every care has been taken in checking the manuscripts and proof reading, yet claiming perfection is very difficult. I shall be very grateful to the readers and users of this book for pointing any mistakes that might have crept in. Suggestions for improvement are most welcome and would be incorporated in the next edition with a view to make the book more useful.

R.K. BANSAL

SYLLABUS

FLUID MACHINES

1. **Properties of Fluids:** Introductory concepts and definitions, properties of fluids, classification of fluids and regimes of flow. 6 Hrs.
2. **Fluid Statics:** Pascal's law of pressure, pressure variation in a static fluid, manometers, hydrostatic force on submerged plane surfaces and curved surfaces. Buoyancy and stability criteria. 8 Hrs.
3. **Fluid Kinematics:** Fluid flow concepts, lines of flow, types of function and velocity potential function for 2-D flow. Relationship between them and flow nets. 6 Hrs.
4. **Dimensional Analysis :** Dimensions of physical quantities, dimensional homogeneity Buckingham's π theorem, the Raleigh's method, important dimensionless numbers, similitude. 5 Hrs.
5. **Fluid Dynamics:** General energy and momentum equations, Euler's equation of motion along a stream line, Bernoulli's equation derived from Fundamentals and Euler's equation, Bernoulli's equation for real fluids. 5 Hrs.
6. **Fluid Flow Measurements:** Venturimeter, orifice meter, pitot tube and V-notch. 4 Hrs.
7. **Flow Through Pipes:** Friction loss in pipe flow — Minor losses in pipe flow — Energy line and hydraulic gradient line, Drezy and Chezy equations. 5 Hrs.
8. **Laminar Flow and Viscous Effects:** Reynold's number, critical Reynold's number, Laminar flow through a round pipe—Hagen poisuille's equation, Laminar flow between parallel stationary plates. 5 Hrs.
9. **Flow Past Immersed Bodies:** Lift and drag skin friction and form drag, Boundary layer concept, calculation of laminar boundary layers thickness, displacement and momentum thickness. 5 Hrs.
10. **Introduction to Compressible Flow:** Sonic velocity, Mach number, isentropic flow, speed of sound wave. 3 Hrs.

CONTENTS

Properties of Fluids

▶ 1.1 INTRODUCTION

Fluid mechanics is that branch of science which deals with the behaviour of the fluids (liquids or gases) at rest as well as in motion. Thus this branch of science deals with the static, kinematics and dynamic aspects of fluids. The study of fluids at rest is called fluid statics. The study of fluids in motion, where pressure forces are not considered, is called fluid kinematics and if the pressure forces are also considered for the fluids in motion, that branch of science is called fluid dynamics.

▶ 1.2 PROPERTIES OF FLUIDS

1.2.1 Density or Mass Density. Density or mass density of a fluid is defined as the ratio of the mass of a fluid to its volume. Thus mass per unit volume of a fluid is called density. It is denoted the symbol ρ (rho). The unit of mass density in SI unit is kg per cubic metre, *i.e.*, kg/m^3. The density of liquids may be considered as constant while that of gases changes with the variation of pressure and temperature.

Mathematically, mass density is written as

$$\rho = \frac{\text{Mass of fluid}}{\text{Volume of fluid}}.$$

The value of density of water is 1 gm/cm^3 or 1000 kg/m^3.

1.2.2 Specific Weight or Weight Density. Specific weight or weight density of a fluid is the ratio between the weight of a fluid to its volume. Thus weight per unit volume of a fluid is called weight density and it is denoted by the symbol *w*.

$$\text{Thus mathematically, } w = \frac{\text{Weight of fluid}}{\text{Volume of fluid}} = \frac{\text{(Mass of fluid)} \times \text{Acceleration due to gravity}}{\text{Volume of fluid}}$$

$$= \frac{\text{Mass of fluid} \times g}{\text{Volume of fluid}}$$

$$= \rho \times g \qquad \left\{\because \frac{\text{Mass of fluid}}{\text{Volume of fluid}} = \rho\right\}$$

$$\therefore \quad w = \rho g \qquad ...(1.1)$$

The value of specific weight or weight density (w) for water is 9.81×1000 Newton/m^3 in SI units.

1.2.3 Specific Volume. Specific volume of a fluid is defined as the volume of a fluid occupied by a unit mass or volume per unit mass of a fluid is called specific volume. Mathematically, it is expressed as

$$\text{Specific volume} = \frac{\text{Volume of fluid}}{\text{Mass of fluid}} = \frac{1}{\left(\dfrac{\text{Mass of fluid}}{\text{Volume}}\right)} = \frac{1}{\rho}$$

Thus specific volume is the reciprocal of mass density. It is expressed as m^3/kg. It is commonly applied to gases.

1.2.4 Specific Gravity. Specific gravity is defined as the ratio of the weight density (or density) of a fluid to the weight density (or density) of a standard fluid. For liquids, the standard fluid is taken water and for gases, the standard fluid is taken air. Specific gravity is also called relative density. It is dimensionless quantity and is denoted by the symbol S.

$$\text{Mathematically,} \quad S(\text{for liquids}) = \frac{\text{Weight density (density) of liquid}}{\text{Weight density (density) of water}}$$

$$S(\text{for gases}) = \frac{\text{Weight density (density) of gas}}{\text{Weight density (density) of air}}$$

Thus weight density of a liquid $= S \times$ Weight density of water

$= S \times 1000 \times 9.81$ N/m^3

The density of a liquid $= S \times$ Density of water

$= S \times 1000$ kg/m^3. ...(1.1A)

If the specific gravity of a fluid is known, then the density of the fluid will be equal to specific gravity of fluid multiplied by the density of water. For example the specific gravity of mercury is 13.6, hence density of mercury = 13.6 × 1000 = 13600 kg/m^3.

Problem 1.1 *Calculate the specific weight, density and specific gravity of one litre of a liquid which weighs 7 N.*

Solution. Given :

$$\text{Volume} = 1 \text{ litre} = \frac{1}{1000} \text{ m}^3 \quad \left(\because \ 1 \text{ litre} = \frac{1}{1000} \text{ m}^3 \text{ or } 1 \text{ litre} = 1000 \text{ cm}^3\right)$$

Weight = 7 N

(*i*) Specific weight (w) $= \dfrac{\text{Weight}}{\text{Volume}} = \dfrac{7\text{N}}{\left(\dfrac{1}{1000}\right)\text{m}^3}$ = **7000 N/m³. Ans.**

(*ii*) Density (ρ) $= \dfrac{w}{g} = \dfrac{7000}{9.81}$ kg/m^3 = **713.5 kg/m³. Ans.**

(*iii*) Specific gravity $= \dfrac{\text{Density of liquid}}{\text{Density of water}} = \dfrac{713.5}{1000}$ $\{\because$ Density of water = 1000 kg/m$^3\}$

= **0.7135. Ans.**

Problem 1.2 *Calculate the density, specific weight and weight of one litre of petrol of specific gravity = 0.7*

Solution. Given : Volume = 1 litre = $1 \times 1000 \text{ cm}^3 = \frac{1000}{10^6} \text{ m}^3 = 0.001 \text{ m}^3$

Sp. gravity $S = 0.7$

(i) Density (ρ)

Using equation (1.1 *A*),

Density (ρ) $= S \times 1000 \text{ kg/m}^3 = 0.7 \times 1000 =$ **700 kg/m³. Ans.**

(ii) Specific weight (w)

Using equation (1.1), $w = \rho \times g = 700 \times 9.81 \text{ N/m}^3 =$ **6867 N/m³. Ans.**

(iii) Weight (W)

We know that specific weight $= \frac{\text{Weight}}{\text{Volume}}$

or $w = \frac{W}{0.001}$ or $6867 = \frac{W}{0.001}$

$\therefore$ $W = 6867 \times 0.001 =$ **6.867 N. Ans.**

Problem 1.2A. *10 m³ of mercury weighs 136 × 10⁴ N. Calculate its specific weight mass density, specific volume and specific gravity.* (VTU Aug., 2002)

Solution. Given :

Volume = 10 m³ ; Weight = 136×10^4 N

(i) Specific weight (w) $= \frac{\text{Weight}}{\text{Volume}} = \frac{136 \times 10^4}{10} =$ **136000 N/m³. Ans.**

(ii) Mass density (ρ) $= \frac{w}{g} = \frac{136000}{9.81} =$ **13863.4 kg/m³. Ans.**

(iii) Specific volume $= \frac{1}{\rho} = \frac{1}{13863.4} =$ **72.13×10^{-6} m³/kg. Ans.**

(iv) Specific gravity $= \frac{\text{Density of mercury}}{\text{Density of water}} = \frac{13863.4}{1000} =$ **13.86. Ans.**

▶ 1.3 VISCOSITY

Viscosity is defined as the property of a fluid which offers resistance to the movement of one layer of fluid over another adjacent layer of the fluid. When two layers of a fluid, a distance '*dy*' apart, move one over the other at different velocities, say *u* and *u* + *du* as shown in Fig. 1.1, the viscosity together with relative velocity causes a shear stress acting between the fluid layers.

The top layer causes a shear stress on the adjacent lower layer while the lower layer causes a shear stress on the adjacent top layer. This shear

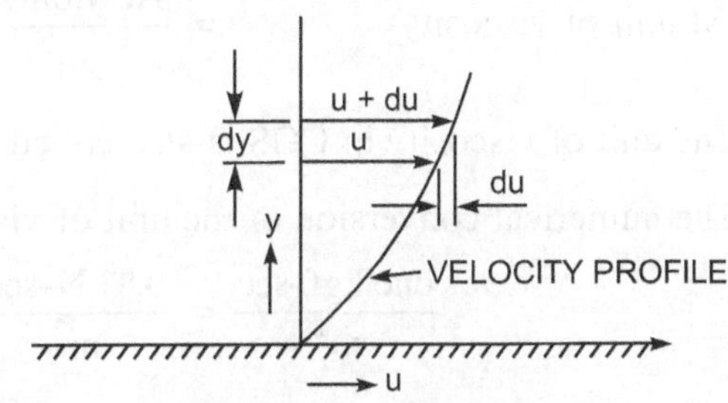

Fig. 1.1 *Velocity variation near a solid boundary.*

stress is proportional to the rate of change of velocity with respect to y. It is denoted by symbol τ called Tau.

Mathematically, $$\tau \propto \frac{du}{dy}$$

or $$\tau = \mu \frac{du}{dy} \quad \text{...(1.2)}$$

where μ (called *mu*) is the constant of proportionality and is known as the co-efficient of dynamic viscosity or only viscosity. $\frac{du}{dy}$ represents the rate of shear strain or rate of shear deformation or velocity gradient.

From equation (1.2), we have $$\mu = \frac{\tau}{\left(\frac{du}{dy}\right)} \quad \text{...(1.3)}$$

Thus viscosity is also defined as the shear stress required to produce unit rate of shear strain.

1.3.1 Units of Viscosity. The units of viscosity is obtained by putting the dimensions of the quantities in equation (1.3)

$$\mu = \frac{\text{Shear stress}}{\dfrac{\text{Change of velocity}}{\text{Change of distance}}} = \frac{\text{Force/Area}}{\left(\dfrac{\text{Length}}{\text{Time}}\right) \times \dfrac{1}{\text{Length}}}$$

$$= \frac{\text{Force}/(\text{length})^2}{\dfrac{1}{\text{Time}}} = \frac{\text{Force} \times \text{Time}}{(\text{Length})^2}$$

In MKS system, force is represented by kgf and length by metre (m), in CGS system, force is represented by dyne and length by cm and in SI system force is represented by Newton (N) and length by metre (m).

$\therefore$ MKS unit of viscosity $= \frac{\text{kgf-sec}}{\text{m}^2}$

CGS unit of viscosity $= \frac{\text{dyne-sec}}{\text{cm}^2}$

In the above expression N/m^2 is also known as Pascal which is represented by Pa. Hence N/m^2 = Pa = Pascal

$\therefore$ SI unit of viscosity = Ns/m^2 = Pa s.

SI unit of viscosity $= \frac{\text{Newton-sec}}{\text{m}^2} = \frac{\text{Ns}}{\text{m}^2}$

The unit of viscosity in CGS is also called Poise which is equal to $\frac{\text{dyne-sec}}{\text{cm}^2}$.

The numerical conversion of the unit of viscosity from MKS unit to CGS unit is given below :

$$\frac{\text{one kgf-sec}}{\text{m}^2} = \frac{9.81\ \text{N-sec}}{\text{m}^2} \qquad \{\because 1\ \text{kgf} = 9.81\ \text{Newton}\}$$

But one Newton = one kg (mass) × one $\left(\frac{\text{m}}{\text{sec}^2}\right)$ (acceleration)

$$= \frac{(1000 \text{ gm}) \times (100 \text{ cm})}{\text{sec}^2} = 1000 \times 100 \frac{\text{gm-cm}}{\text{sec}^2}$$

$$= 1000 \times 100 \text{ dyne} \qquad \left\{\because \text{ dyne} = \text{gm} \times \frac{\text{cm}}{\text{sec}^2}\right\}$$

$$\therefore \quad \frac{\text{one kgf-sec}}{\text{m}^2} = 9.81 \times 100000 \frac{\text{dyne-sec}}{\text{cm}^2} = 9.81 \times 100000 \frac{\text{dyne-sec}}{100 \times 100 \times \text{cm}^2}$$

$$= 98.1 \frac{\text{dyne-sec}}{\text{cm}^2} = 98.1 \text{ poise} \qquad \left\{\because \frac{\text{dyne-sec}}{\text{cm}^2} = \text{Poise}\right\}$$

Thus for solving numerical problems, if viscosity is given in poise, it must be divided by 98.1 to get its equivalent numerical value in MKS.

But $$\frac{\text{one kgf-sec}}{\text{m}^2} = \frac{9.81 \text{ Ns}}{\text{m}^2} = 98.1 \text{ poise}$$

$$\therefore \quad \frac{\text{one Ns}}{\text{m}^2} = \frac{98.1}{9.81} \text{ poise} = 10 \text{ poise} \quad \text{or} \quad \text{One poise} = \frac{1}{10} \frac{\text{Ns}}{\text{m}^2}.$$

Alternate Method. One poise $= \dfrac{\text{dyne} \times \text{s}}{\text{cm}^2} = \left(\dfrac{1 \text{ gm} \times 1 \text{ cm}}{\text{s}^2}\right) \times \dfrac{\text{s}}{\text{cm}^2}$

But dyne $$= 1 \text{ gm} \times \frac{1 \text{ cm}}{\text{s}^2}$$

$$\therefore \text{ One poise} = \frac{1 \text{ gm}}{\text{s cm}} = \frac{\frac{1}{1000} \text{ kg}}{\text{s} \frac{1}{100} \text{ m}}$$

$$= \frac{1}{1000} \times 100 \frac{\text{kg}}{\text{s m}} = \frac{1}{10} \frac{\text{kg}}{\text{s m}} \quad \text{or} \quad 1 \frac{\text{kg}}{\text{s m}} = 10 \text{ poise.}$$

Note. (*i*) In SI units second is represented by 's' and not by 'sec'.

(*ii*) If viscosity is given in poise, it must be divided by 10 to get its equivalent numerical value in SI units.

Sometimes a unit of viscosity as centipoise is used where

$$1 \text{ centipoise} = \frac{1}{100} \text{ poise} \quad \text{or} \quad 1 \text{ cP} = \frac{1}{100} \text{ P} \qquad [\text{cP} = \text{Centipoise, P} = \text{Poise}]$$

The viscosity of water at 20°C is 0.01 poise or 1.0 centipoise.

1.3.2 Kinematic Viscosity.

1.3.2 Kinematic Viscosity. It is defined as the ratio between the dynamic viscosity and density of fluid. It is denoted by the Greek symbol (ν) called 'nu'. Thus, mathematically,

$$\nu = \frac{\text{Viscosity}}{\text{Density}} = \frac{\mu}{\rho} \qquad \ldots(1.4)$$

The units of kinematic viscosity is obtained as

$$\nu = \frac{\text{Units of } \mu}{\text{Units of } \rho} = \frac{\text{Force} \times \text{Time}}{(\text{Length})^2 \times \frac{\text{Mass}}{(\text{Length})^3}} = \frac{\text{Force} \times \text{Time}}{\frac{\text{Mass}}{\text{Length}}}$$

$$= \frac{\text{Mass} \times \dfrac{\text{Length}}{(\text{Time})^2} \times \text{Time}}{\left(\dfrac{\text{Mass}}{\text{Length}}\right)} \qquad \left\{\because \text{ Force} = \text{Mass} \times \text{Acc.} = \text{Mass} \times \frac{\text{Length}}{\text{Time}^2}\right\}$$

$$= \frac{(\text{Length})^2}{\text{Time}}.$$

In MKS and SI, the unit of kinematic viscosity is metre^2/sec or m^2/sec while in CGS units it is written as cm^2/s. In CGS units, kinematic viscosity is also known stoke.

Thus, one stoke $= \text{cm}^2/\text{s} = \left(\dfrac{1}{100}\right)^2 \text{m}^2/\text{s} = 10^{-4}\ \text{m}^2/\text{s}$

Centistoke means $= \dfrac{1}{100}$ stoke.

1.3.3 Newton's Law of Viscosity. It states that the shear stress (τ) on a fluid element layer is directly proportional to the rate of shear strain. The constant of proportionality is called the co-efficient of viscosity. Mathematically, it is expressed as given by equation (1.2) or as

$$\tau = \mu \frac{du}{dy}.$$

Fluids which obey the above relation are known as **Newtonian** fluids and the fluids which do not obey the above relation are called **Non-newtonian** fluids.

1.3.4 Variation of Viscosity with Temperature. Temperature affects the viscosity. The viscosity of liquids decreases with the increase of temperature while the viscosity of gases increases with the increase of temperature. This is due to reason that the viscous forces in a fluid are due to cohesive forces and molecular momentum transfer. In liquids the cohesive forces predominates the molecular momentum transfer, due to closely packed molecules and with the increase in temperature, the cohesive forces decreases with the result of decreasing viscosity. But in case of gases the cohesive force are small and molecular momentum transfer predominates. With the increase in temperature, molecular momentum transfer increases and hence viscosity increases. The relation between viscosity and temperature for liquids and gases are:

(*i*) **For liquids,** $\mu = \mu_0 \left(\dfrac{1}{1 + \alpha t + \beta t^2}\right)$ ...(1.4*A*)

where μ = Viscosity of liquid at t°C, in poise

μ_0 = Viscosity of liquid at 0°C, in poise

α, β = are constants for the liquid

For water, $\mu_0 = 1.79 \times 10^{-3}$ poise, $\alpha = 0.03368$ and $\beta = 0.000221$.

The equation (1.4 *A*) shows that with the increase of temperature, the viscosity decreases.

(*ii*) **For a gas,** $\mu = \mu_0 + \alpha t - \beta t^2$ ...(1.4*B*)

where for air $\mu_0 = 0.000017$, $\alpha = 0.000000056$, $\beta = 0.1189 \times 10^{-9}$.

The equation (1.4*B*) shows that with the increase of temperature, the viscosity increases.

1.3.5 Classification of Fluids. The fluids may be classified into the following five types :

1. Ideal fluid,
2. Real fluid,
3. Newtonian fluid,
4. Non-Newtonian fluid, and
5. Ideal plastic fluid.

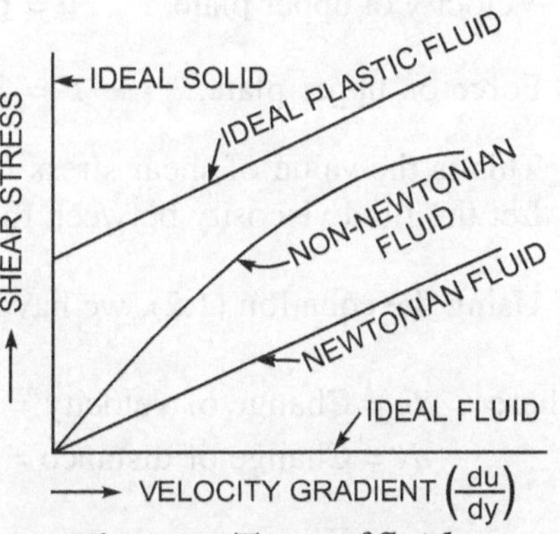

Fig. 1.2 *Types of fluids.*

1. Ideal Fluid. A fluid, which is incompressible and is having no viscosity, is known as an ideal fluid. Ideal fluid is only an imaginary fluid as all the fluids, which exist, have some viscosity.

2. Real Fluid. A fluid, which possesses viscosity, is known as real fluid. All the fluids, in actual practice, are real fluids.

3. Newtonian Fluid. A real fluid, in which the shear stress is directly, proportional to the rate of shear strain (or velocity gradient), is known as a Newtonian fluid.

4. Non-Newtonian Fluid. A real fluid, in which the shear stress is not proportional to the rate of shear strain (or velocity gradient), known as a Non-Newtonian fluid.

5. Ideal Plastic Fluid. A fluid, in which shear stress is more than the yield value and shear stress is proportional to the rate of shear strain (or velocity gradient), is known as ideal plastic fluid.

Problem 1.3 *If the velocity distribution over a plate is given by* $u = \frac{2}{3}y - y^2$ *in which u is the velocity in metre per second at a distance y metre above the plate, determine the shear stress at* $y = 0$ *and* $y = 0.15$ *m. Take dynamic viscosity of fluid as 8.63 poises.*

Solution. Given : $u = \frac{2}{3}y - y^2 \quad \therefore \quad \frac{du}{dy} = \frac{2}{3} - 2y$

$$\left(\frac{du}{dy}\right)_{\text{at } y=0} \text{ or } \left(\frac{du}{dy}\right)_{y=0} = \frac{2}{3} - 2(0) = \frac{2}{3} = 0.667$$

Also

$$\left(\frac{du}{dy}\right)_{\text{at } y=0.15} \text{ or } \left(\frac{du}{dy}\right)_{y=0.15} = \frac{2}{3} - 2\times.15 = .667 - .30 = 0.367$$

Value of $\mu = 8.63$ poise $= \frac{8.63}{10}$ SI units $= 0.863$ N s/m^2

Now shear stress is given by equation (1.2) as $\tau = \mu \frac{du}{dy}$.

(*i*) Shear stress at $y = 0$ is given by

$$\tau_0 = \mu\left(\frac{du}{dy}\right)_{y=0} = 0.863 \times 0.667 = \mathbf{0.5756\ N/m^2.\ Ans.}$$

(*ii*) Shear stress at $y = 0.15$ m is given by

$$(\tau)_{y=0.15} = \mu\left(\frac{du}{dy}\right)_{y=0.15} = 0.863 \times 0.367 = \mathbf{0.3167\ N/m^2.\ Ans.}$$

Problem 1.4 *A plate, 0.025 mm distant from a fixed plate, moves at 60 cm/s and requires a force of 2 N per unit area i.e., 2 N/m*^{2} *to maintain this speed. Determine the fluid viscosity between the plates.*

Solution. Given :

Distance between plates, $dy = .025$ mm

$= .025 \times 10^{-3}$ m

Velocity of upper plate, $u = 60$ cm/s = 0.6 m/s

Force on upper plate, $F = 2.0 \dfrac{N}{m^2}$.

This is the value of shear stress *i.e.*, τ

Let the fluid viscosity between the plates is μ.

F

dy = .025 mm

u = 60 cm/s

FIXED PLATE

Fig. 1.3

Using the equation (1.2), we have $\tau = \mu \dfrac{du}{dy}$.

where du = Change of velocity = $u - 0 = u = 0.60$ m/s

dy = Change of distance = $.025 \times 10^{-3}$ m

τ = Force per unit area = $2.0 \dfrac{N}{m^2}$

$$\therefore \quad 2.0 = \mu \frac{0.60}{.025 \times 10^{-3}} \quad \therefore \quad \mu = \frac{2.0 \times .025 \times 10^{-3}}{0.60} = 8.33 \times 10^{-5} \frac{Ns}{m^2}$$

$= 8.33 \times 10^{-5} \times 10$ poise = $\mathbf{8.33 \times 10^{-4}}$ **poise. Ans.**

Problem 1.5 *A flat plate of area 1.5×10^6 mm² is pulled with a speed of 0.4 m/s relative to another plate located at a distance of 0.15 mm from it. Find the force and power required to maintain this speed, if the fluid separating them is having viscosity as 1 poise.*

Solution. Given :

Area of the plate, $A = 1.5 \times 10^6 \text{ mm}^2 = 1.5 \text{ m}^2$

Speed of plate relative to another plate, $du = 0.4$ m/s

Distance between the plates, $dy = 0.15 \text{ mm} = 0.15 \times 10^{-3}$ m

Viscosity $\mu = 1 \text{ poise} = \dfrac{1}{10} \dfrac{Ns}{m^2}$.

Using equation (1.2) we have $\tau = \mu \dfrac{du}{dy} = \dfrac{1}{10} \times \dfrac{0.4}{.15 \times 10^{-3}} = 266.66 \dfrac{N}{m^2}$

(*i*) $\therefore$ Shear force, $F = \tau \times \text{area} = 266.66 \times 1.5 =$ **400 N. Ans.**

(*ii*) Power* required to move the plate at the speed 0.4 m/sec

$= F \times u = 400 \times 0.4 =$ **160 W. Ans.**

Problem 1.6 *Determine the intensity of shear of an oil having viscosity = 1 poise. The oil is used for lubricating the clearance between a shaft of diameter 10 cm and its journal bearing. The clearance is 1.5 mm and the shaft rotates at 150 r.p.m.*

Solution. Given : $\mu = 1 \text{ poise} = \dfrac{1}{10} \dfrac{Ns}{m^2}$

Dia. of shaft, $D = 10 \text{ cm} = 0.1$ m

Distance between shaft and journal bearing,

$dy = 1.5 \text{ mm} = 1.5 \times 10^{-3}$ m

Speed of shaft, $N = 150$ r.p.m.

---

* Power = $F \times u$ N m/s = $F \times u$ W ($\because$ Nm/s = Watt)

Tangential speed of shaft is given by

$$u = \frac{\pi DN}{60} = \frac{\pi \times 0.1 \times 150}{60} = 0.785 \text{ m/s}$$

Using equation (1.2), $\tau = \mu \frac{du}{dy}$,

where du = change of velocity between shaft and bearing = $u - 0 = u$

$$= \frac{1}{10} \times \frac{0.785}{1.5 \times 10^{-3}} = \mathbf{52.33\ N/m^2.\ Ans.}$$

Problem 1.7 *Calculate the dynamic viscosity of an oil, which is used for lubrication between a square plate of size 0.8 m × 0.8 m and an inclined plane with angle of inclination 30° as shown in Fig. 1.4. The weight of the square plate is 300 N and it slides down the inclined plane with a uniform velocity of 0.3 m/s. The thickness of oil film is 1.5 mm.*

Solution. Given :

Area of plate, $A = 0.8 \times 0.8 = 0.64 \text{ m}^2$

Angle of plane, $\theta = 30°$

Weight of plate, $W = 300$ N

Velocity of plate, $u = 0.3$ m/s

Thickness of oil film, $t = dy = 1.5 \text{ mm} = 1.5 \times 10^{-3}$ m

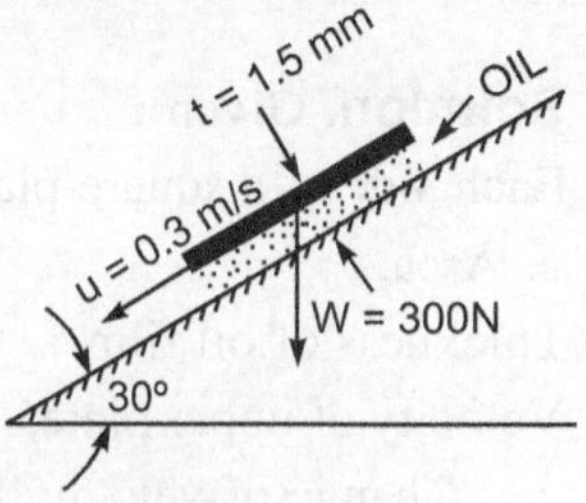

t = 1.5 mm
OIL
u = 0.3 m/s
W = 300N
30°

Fig. 1.4

Let the viscosity of fluid between plate and inclined plane is μ.

Component of weight W, along the plane = $W \cos 60° = 300 \cos 60° = 150$ N

Thus the shear force, F, on the bottom surface of the plate = 150 N

and shear stress, $\tau = \frac{F}{\text{Area}} = \frac{150}{0.64} \text{ N/m}^2$

Now using equation (1.2), we have

$$\tau = \mu \frac{du}{dy}$$

where du = change of velocity = $u - 0 = u = 0.3$ m/s

$$dy = t = 1.5 \times 10^{-3} \text{ m}$$

$$\therefore \quad \frac{150}{0.64} = \mu \frac{0.3}{1.5 \times 10^{-3}}$$

$$\therefore \quad \mu = \frac{150 \times 1.5 \times 10^{-3}}{0.64 \times 0.3} = 1.17 \text{ N s/m}^2 = 1.17 \times 10 = \mathbf{11.7\ poise.\ Ans.}$$

Problem 1.8 *Two horizontal plates are placed 1.25 cm apart, the space between them being filled with oil of viscosity 14 poises. Calculate the shear stress in oil if upper plate is moved with a velocity of 2.5 m/s.* (A.M.I.E., 1972)

Solution. Given :

Distance between plates, $dy = 1.25 \text{ cm} = 0.0125$ m

Viscosity, $\mu = 14 \text{ poise} = \frac{14}{10} \text{ N s/m}^2$

Velocity of upper plate, $u = 2.5$ m/sec.

Shear stress is given by equation (1.2) as, $\tau = \mu \dfrac{du}{dy}$

where du = Change of velocity between plates = $u - 0 = u = 2.5$ m/sec.

$dy = 0.0125$ m.

$$\therefore \quad \tau = \frac{14}{10} \times \frac{2.5}{.0125} = \mathbf{280\ N/m^2.\ Ans.}$$

Problem 1.9 *The space between two square flat parallel plates is filled with oil. Each side of the plate is 60 cm. The thickness of the oil film is 12.5 mm. The upper plate, which moves at 2.5 metre per sec requires a force of 98.1 N to maintain the speed. Determine :*

(i) the dynamic viscosity of the oil in poise, and

(ii) the kinematic viscosity of the oil in stokes if the specific gravity of the oil is 0.95.

(A.M.I.E., Winter 1977)

Solution. Given :

Each side of a square plate = 60 cm = 0.60 m

∴ Area, $A = 0.6 \times 0.6 = 0.36\ m^2$

Thickness of oil film, $dy = 12.5\ mm = 12.5 \times 10^{-3}$ m

Velocity of upper plate, $u = 2.5$ m/sec

∴ Change of velocity between plates, $du = 2.5$ m/sec

Force required on upper plate, $F = 98.1$ N

$$\therefore \text{ Shear stress,} \quad \tau = \frac{\text{Force}}{\text{Area}} = \frac{F}{A} = \frac{98.1\ \text{N}}{0.36\ \text{m}^2}$$

(*i*) Let μ = Dynamic viscosity of oil

$$\text{Using equation (1.2),} \quad \tau = \mu \frac{du}{dy} \quad \text{or} \quad \frac{98.1}{0.36} = \mu \times \frac{2.5}{12.5 \times 10^{-3}}$$

$$\therefore \quad \mu = \frac{98.1}{0.36} \times \frac{12.5 \times 10^{-3}}{2.5} = 1.3635\ \frac{\text{Ns}}{\text{m}^2} \qquad \left(\because \frac{1\ \text{Ns}}{\text{m}^2} = 10 \text{ poise}\right)$$

$$= 1.3635 \times 10 = \mathbf{13.635\ poise.\ Ans.}$$

(*ii*) Sp. gr. of oil, S = 0.95

Let ν = kinematic viscosity of oil

Using equation (1.1 A),

Mass density of oil, $\rho = S \times 1000 = 0.95 \times 1000 = 950\ kg/m^3$

$$\text{Using the relation, } \nu = \frac{\mu}{\rho}, \text{ we get } \nu = \frac{1.3635\left(\frac{\text{Ns}}{\text{m}^2}\right)}{950} = .001435\ \text{m}^2/\text{sec} = .001435 \times 10^4\ \text{cm}^2/\text{s}$$

$$= \mathbf{14.35\ stokes.\ Ans.} \qquad (\because \text{cm}^2/\text{s} = \text{stoke})$$

Problem 1.10 *Find the kinematic viscosity of an oil having density 981 kg/m^3. The shear stress at a point in oil is 0.2452 N/m^2 and velocity gradient at that point is 0.2 per second.*

Solution. Given :

Mass density, $\rho = 981\ kg/m^3$

Shear stress, $\tau = 0.2452\ N/m^2$

Velocity gradient, $\frac{du}{dy} = 0.2$ s

Using the equation (1.2), $\tau = \mu \frac{du}{dy}$ or $0.2452 = \mu \times 0.2$

$\therefore$ $\mu = \frac{0.2452}{0.200} = 1.226$ Ns/m^2

Kinematic viscosity ν is given by

$$\therefore \quad \nu = \frac{\mu}{\rho} = \frac{1.226}{981} = .125 \times 10^{-2} \text{ m}^2\text{/sec}$$

$$= 0.125 \times 10^{-2} \times 10^4 \text{ cm}^2\text{/s} = 0.125 \times 10^2 \text{ cm}^2\text{/s}$$

$= 12.5$ cm^2/s $=$ **12.5 stoke. Ans.** $(\because$ cm^2/s = stoke)

Problem 1.11. *Determine the specific gravity of a fluid having viscosity 0.05 poise and kinematic viscosity 0.035 stokes.*

Solution. Given :

Viscosity, $\mu = 0.05$ poise $= \frac{0.05}{10}$ N s/m^2

Kinematic viscosity, $\nu = 0.035$ stokes

$= 0.035$ cm^2/s $\{\because$ Stoke = cm^2/s$\}$

$= 0.035 \times 10^{-4}$ m^2/s

Using the relation $\nu = \frac{\mu}{\rho}$, we get $0.035 \times 10^{-4} = \frac{0.05}{10} \times \frac{1}{\rho}$

$$\therefore \quad \rho = \frac{0.05}{10} \times \frac{1}{.035 \times 10^{-4}} = 1428.5 \text{ kg/m}^3$$

$\therefore$ Sp. gr. of liquid $= \frac{\text{Density of liquid}}{\text{Density of water}} = \frac{1428.5}{1000} = 1.4285 \simeq$ **1.43. Ans.**

Problem 1.12 *Determine the viscosity of a liquid having kinematic viscosity 6 stokes and specific gravity 1.9.*

Solution. Given :

Kinematic viscosity $\nu = 6$ stokes $= 6$ cm^2/s $= 6 \times 10^{-4}$ m^2/s

Sp. gr. of liquid $= 1.9$

Let the viscosity of liquid $= \mu$

Now sp. gr. of a liquid $= \frac{\text{Density of the liquid}}{\text{Density of water}}$

or $1.9 = \frac{\text{Density of liquid}}{1000}$

$\therefore$ Density of liquid $= 1000 \times 1.9 = 1900 \frac{\text{kg}}{\text{m}^3}$

$\therefore$ Using the relation $\nu = \frac{\mu}{\rho}$, we get

$$6 \times 10^{-4} = \frac{\mu}{1900}$$

$\because$ $\mu = 6 \times 10^{-4} \times 1900 = 1.14 \text{ Ns/m}^2$

$= 1.14 \times 10 =$ **11.40 poise. Ans.**

Problem 1.13 *The velocity distribution for flow over a flat plate is given by $u = \frac{3}{4}y - y^2$ in which u is the velocity in metre per second at a distance y metre above the plate. Determine the shear stress at y = 0.15 m. Take dynamic viscosity of fluid as 8.6 poise.*

Solution. Given : $u = \frac{3}{4}y - y^2$

$\therefore$ $\frac{du}{dy} = \frac{3}{4} - 2y$

At $y = 0.15$, $\frac{du}{dy} = \frac{3}{4} - 2 \times 0.15 = 0.75 - 0.30 = 0.45$

Viscosity, $\mu = 8.5 \text{ poise} = \frac{8.5}{10} \frac{\text{Ns}}{\text{m}^2}$ $\left(\because 10 \text{ poise} = 1 \frac{\text{Ns}}{\text{m}^2}\right)$

Using equation (1.2), $\tau = \mu \frac{du}{dy} = \frac{8.5}{10} \times 0.45 \frac{\text{N}}{\text{m}^2} = \mathbf{0.3825 \frac{N}{m^2}}$. **Ans.**

Problem 1.14 *The dynamic viscosity of an oil, used for lubrication between a shaft and sleeve is 6 poise. The shaft is of diameter 0.4 m and rotates at 190 r.p.m. Calculate the power lost in the bearing for a sleeve length of 90 mm. The thickness of the oil film is 1.5 mm.* (VTU, 2000)

Solution. Given :

Viscosity $\mu = 6$ poise

$= \frac{6}{10} \frac{\text{Ns}}{\text{m}^2} = 0.6 \frac{\text{Ns}}{\text{m}^2}$

Dia. of shaft, $D = 0.4$ m

Speed of shaft, $N = 190$ r.p.m

Sleeve length, $L = 90 \text{ mm} = 90 \times 10^{-3}$ m

Thickness of oil film, $t = 1.5 \text{ mm} = 1.5 \times 10^{-3}$ m

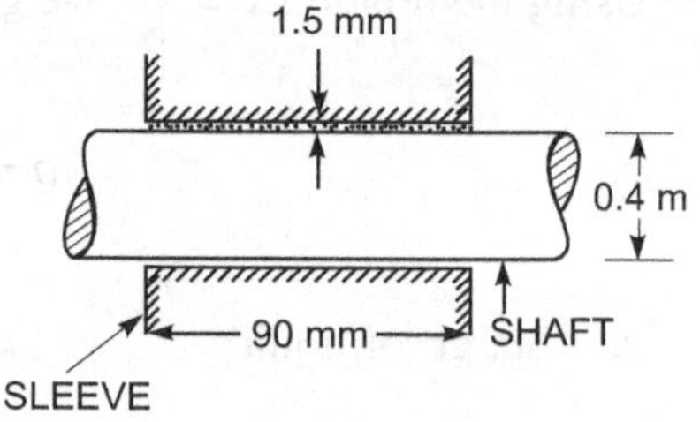

Fig. 1.5

Tangential velocity of shaft, $u = \frac{\pi DN}{60} = \frac{\pi \times 0.4 \times 190}{60} = 3.98$ m/s

Using the relation $\tau = \mu \frac{du}{dy}$

where du = Change of velocity $= u - 0 = u = 3.98$ m/s

dy = Change of distance $= t = 1.5 \times 10^{-3}$ m

$$\tau = 10 \times \frac{3.98}{1.5 \times 10^{-3}} = 1592 \text{ N/m}^2$$

This is shear stress on shaft

$\therefore$ Shear force on the shaft, F = Shear stress × Area

$$= 1592 \times \pi D \times L = 1592 \times \pi \times .4 \times 90 \times 10^{-3} = 180.05 \text{ N}$$

Torque on the shaft, $T = \text{Force} \times \dfrac{D}{2} = 180.05 \times \dfrac{0.4}{2} = 36.01 \text{ Nm}$

$\therefore$ *Power lost $= \dfrac{2\pi NT}{60} = \dfrac{2\pi \times 190 \times 36.01}{60} =$ **716.48 W. Ans.**

Problem 1.15 *If the velocity profile of a fluid over a plate is a parabolic with the vertex 20 cm from the plate, where the velocity is 120 cm/sec. Calculate the velocity gradients and shear stresses at a distance of 0, 10 and 20 cm from the plate, if the viscosity of the fluid is 8.5 poise.*

Solution. Given :

Distance of vertex from plate = 20 cm

Velocity at vertex, u = 120 cm/sec

Viscosity, $\mu = 8.5 \text{ poise} = \dfrac{8.5}{10} \dfrac{\text{N s}}{\text{m}^2} = 0.85.$

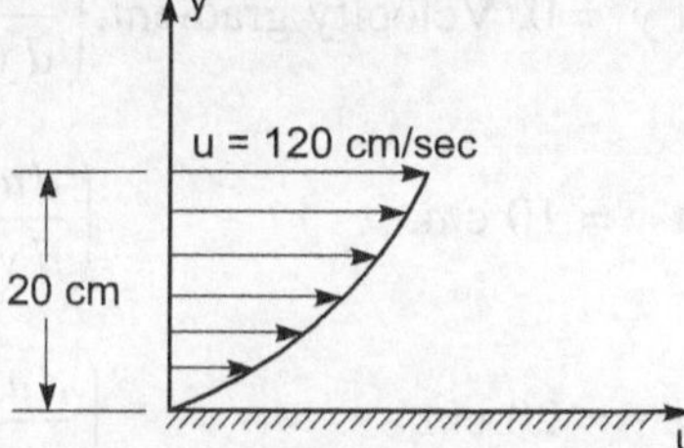

Fig. 1.6

The velocity profile is given parabolic and equation of velocity profile is

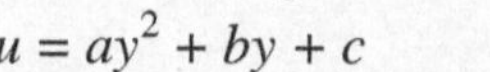

$$u = ay^2 + by + c \qquad \ldots(i)$$

where a, b and c are constants. Their values are determined from boundary conditions as :

(*a*) at $y = 0$, $u = 0$

(*b*) at $y = 20$ cm, $u = 120$ cm/sec

(*c*) at $y = 20$ cm, $\dfrac{du}{dy} = 0$.

Substituting boundary condition (*a*) in equation (*i*), we get

$$c = 0.$$

Boundary condition (*b*) on substitution in (*i*) gives

$$120 = a(20)^2 + b(20) = 400a + 20b \qquad \ldots(ii)$$

Boundary condition (*c*) on substitution in equation (*i*) gives

$$\frac{du}{dy} = 2ay + b \qquad \ldots(iii)$$

or $$0 = 2 \times a \times 20 + b = 40a + b$$

Solving equations (*ii*) and (*iii*) for a and b

From equation (*iii*), $b = -40a$

Substituting this value in equation (*ii*), we get

$$120 = 400a + 20 \times (-40a) = 400a - 800a = -400a$$

---

* Power in S.I. unit = T * ω = T × $\dfrac{2\pi N}{60}$ Watt = $\dfrac{2\pi NT}{60}$ Watt

$$\therefore \quad a = \frac{120}{-400} = -\frac{3}{10} = -0.3$$

$$\therefore \quad b = -40 \times (-0.3) = 12.0$$

Substituting the values of a, b and c in equation (i),

$$u = -0.3y^2 + 12y.$$

Velocity Gradient

$$\frac{du}{dy} = -0.3 \times 2y + 12 = -0.6y + 12$$

at $y = 0$, Velocity gradient, $\left(\frac{du}{dy}\right)_{y=0} = -0.6 \times 0 + 12 =$ **12/s. Ans.**

at $y = 10$ cm, $\left(\frac{du}{dy}\right)_{y=10} = -0.6 \times 10 + 12 = -6 + 12 =$ **6/s. Ans.**

at $y = 20$ cm, $\left(\frac{du}{dy}\right)_{y=20} = -0.6 \times 20 + 12 = -12 + 12 =$ **0. Ans.**

Shear Stresses

Shear stress is given by, $\tau = \mu \frac{du}{dy}$

(i) Shear stress at $y = 0$, $\tau = \mu \left(\frac{du}{dy}\right)_{y=0} = 0.85 \times 12.0 = 10.2 \text{ N/m}^2.$

(ii) Shear stress at $y = 10$, $\tau = \mu \left(\frac{du}{dy}\right)_{y=10} = 0.85 \times 6.0 = 5.1 \text{ N/m}^2.$

(iii) Shear stress at $y = 20$, $\tau = \mu \left(\frac{du}{dy}\right)_{y=20} = 0.85 \times 0 =$ **0. Ans.**

Problem 1.16 *A Newtonian fluid is filled in the clearance between a shaft and a concentric sleeve. The sleeve attains a speed of 50 cm/s, when a force of 40 N is applied to the sleeve parallel to the shaft. Determine the speed if a force of 200 N is applied.* (A.M.I.E., Summer 1980)

Solution. Given : Speed of sleeve, $u_1 = 50$ cm/s

when force, $F_1 = 40$ N.

Let speed of sleeve is u_2 when force, $F_2 = 200$ N.

Using relation $\tau = \mu \frac{du}{dy}$

where τ = Shear stress $= \frac{\text{Force}}{\text{Area}} = \frac{F}{A}$

du = Change of velocity = $u - 0 = u$

dy = Clearance = y

$$\therefore \quad \frac{F}{A} = \mu \frac{u}{y}$$

$$\therefore \quad F = \frac{A\mu u}{y} \propto u \qquad \{\because \text{ A, } \mu \text{ and } y \text{ are constant}\}$$

$$\therefore \quad \frac{F_1}{u_1} = \frac{F_2}{u_2}$$

Substituting values, we get $\frac{40}{50} = \frac{200}{u_2}$

$$\therefore \quad u_2 = \frac{50 \times 200}{40} = 50 \times 5 = \mathbf{250\ cm/s.\ Ans.}$$

Problem 1.17 *A 15 cm diameter vertical cylinder rotates concentrically inside another cylinder of diameter 15.10 cm. Both cylinders are 25 cm high. The space between the cylinders is filled with a liquid whose viscosity is unknown. If a torque of 12.0 Nm is required to rotate the inner cylinder at 100 r.p.m., determine the viscosity of the fluid.* (A.M.I.E., Winter 1979)

Solution. Given :

Diameter of cylinder = 15 cm = 0.15 m

Dia. of outer cylinder = 15.10 cm = 0.151 m

Length of cylinders, L = 25 cm = 0.25 m

Torque, T = 12.0 Nm

Speed, N = 100 r.p.m.

Let the viscosity = μ

Tangential velocity of cylinder, $u = \frac{\pi DN}{60} = \frac{\pi \times 0.15 \times 100}{60} = 0.7854$ m/s

Surface area of cylinder, $A = \pi D \times L = \pi \times 0.15 \times 0.25 = .1178\ \text{m}^2$

Now using relation $\tau = \mu \frac{du}{dy}$

where $du = u - 0 = u = .7854$ m/s

$$dy = \frac{0.151 - 0.150}{2} \text{ m} = .0005 \text{ m}$$

Substituting these values in equation (*i*), we get

$$\tau = \frac{\mu \times .7854}{.0005}$$

$\therefore$ Shear force, F = Shear stress $\times$ Area $= \frac{\mu \times .7854}{.0005} \times .1178$

$\therefore$ Torque, $T = F \times \frac{D}{2}$

$$12.0 = \frac{\mu \times .7854}{.0005} \times .1178 \times \frac{.15}{2}$$

$\therefore$ $$\mu = \frac{12.0 \times .0005 \times 2}{.7854 \times .1178 \times .15} = 0.864 \text{ N s/m}^2$$

$$= 0.864 \times 10 = \mathbf{8.64 \text{ poise. Ans.}}$$

Problem 1.17A. *A vertical cylinder of diameter 180 mm rotates concentrically inside another cylinder of diameter 181.2 mm. Boththe cylinders are 300 mm high. The space between the cylinders isfilled with a liquid whose viscosity is unknown. Determine the viscosity of the fluid if a torque of 20 Nm is required to rotate the inner cylinder at 120 r.p.m.* (VTU, Aug, 2003)

Solution. Given:

Dia. of inner cylinder, $D_i = 180$ mm $= 0.180$ m ; Dia. of outer cylinder

$D_o = 181.2$ mm $= 0.1812$ m ; height of cylinders, $h = 300$ mm $= 0.3$

m,

Torque, $T = 20$ Nm ; speed, $N = 120$ r.p.m.

Tangential velocity of rotating cylinder,

$$\mu = \frac{\pi DN}{60} = \frac{\pi D_i N}{60}$$

$$= \frac{\pi \times 0.180 \times 120}{60} = 1.13 \text{ m/s}$$

(Here $D = D_i$ as inner cylinder is rotating)

Surface area of rotating cylinder $= \pi D_i \times h$

$$= \pi \times 0.18 \times 0.3 = 0.1696 \text{ m}^2$$

Now using the rotation, $\tau = \mu \dfrac{du}{dy}$

where $du = u - 0 = u = 1.13$ m/s ; $dy = \dfrac{D_o - D_i}{2}$

$$= \frac{0.1812 - 0.180}{2} = 0.0006 \text{ m, we get}$$

$$\tau = \mu \times \frac{1.13}{0.0006}$$

$\therefore$ Shear force, $$F = \tau \times \text{Area} = \left(\frac{\mu \times 1.13}{0.0006}\right) \times 0.1696$$

$\therefore$ Torque, $$T = F \times \frac{D}{2} = \frac{F \times D_i}{2}$$

or $$20 = \left(\frac{\mu \times 1.13}{0.0006} \times 0.1696\right) \times \frac{0.180}{2}$$

$\therefore$ $$\mu = \frac{20 \times 0.0006 \times 2}{1.13 \times 0.1696 \times 0.180} = 0.695 \text{ N s/m}^2$$

$$= 0.695 \times 10 = \mathbf{6.95 \text{ poise. Ans.}}$$

Problem 1.18 *Two large plane surfaces are 2.4 cm apart. The space between the surfaces is filled with glycerine. What force is required to drag a very thin plate of surface area 0.5 square metre between the two large plane surfaces at a speed of 0.6 m/s, if :*

(i) the thin plate is in the middle of the two plane surfaces, and

(ii) the thin plate is at a distance of 0.8 cm from one of the plane surfaces ? Take the dynamic viscosity of glycerine = 8.10×10^{-1} N s/m^2. (VTU, Aug. 2003)

Solution. Given :

Distance between two large surfaces = 2.4 cm

Area of thin plate, $A = 0.5 \text{ m}^2$

Velocity of thin plate, $u = 0.6$ m/s

Viscosity of glycerine, $\mu = 8.10 \times 10^{-1}$ N s/m^2

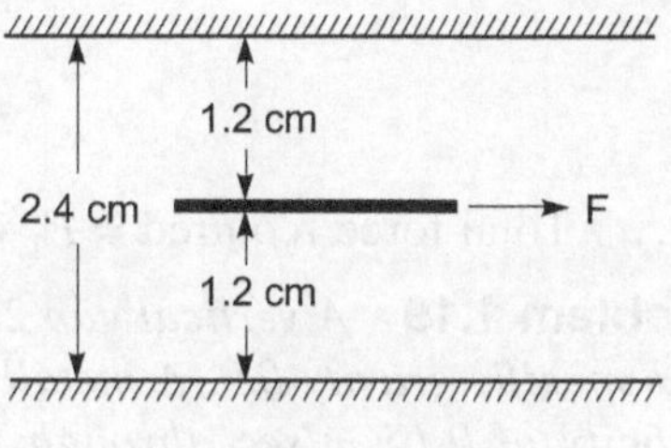

Fig. 1.7 (*a*)

Case I. When the thin plate is in the middle of the two plane surfaces [Refer to Fig. 1.7 (*a*)]

Let F_1 = Shear force on the upper side of the thin plate

F_2 = Shear force on the lower side of the thin plate

F = Total force required to drag the plate

Then $F = F_1 + F_2$

The shear stress (τ_2) on the upper side of the thin plate is given by equation,

$$\tau_1 = \mu \left(\frac{du}{dv} \right)_1$$

where du = Relative velocity between thin plate and upper large plane surface
= 0.6 m/sec

dy = Distance between thin plate and upper large plane surface
= 1.2 cm = 0.012 m (plate is a thin one and hence thickness of plate is neglected)

$$\therefore \quad \tau_1 = 8.10 \times 10^{-1} \times \left(\frac{0.6}{.012} \right) = 40.5 \text{ N/m}^2$$

Now shear force, F_1 = Shear stress × Area
$= \tau_1 \times A = 40.5 \times 0.5 = 20.25$ N

Similarly shear stress (τ_2) on the lower side of the thin plate is given by

$$\tau_2 = \mu \left(\frac{du}{dy} \right)_2 = 8.10 \times 10^{-1} \times \left(\frac{0.6}{0.012} \right) = 40.5 \text{ N/m}^2$$

∴ Shear force, $F_2 = \tau_2 \times A = 40.5 \times 0.5 = 20.25$ N

∴ Total force, $F = F_1 + F_2 = 20.25 + 20.25 =$ **40.5 N. Ans.**

Case II. When the thin plate is at a distance of 0.8 cm from one of the plane surfaces [Refer to Fig. 1.7 (*b*)].

Let the thin plate is a distance 0.8 cm from the lower plane surface.

Then distance of the plate from the upper plane surface
= 2.4 – 0.8 = 1.6 cm = .016 m
(Neglecting thickness of the plate)

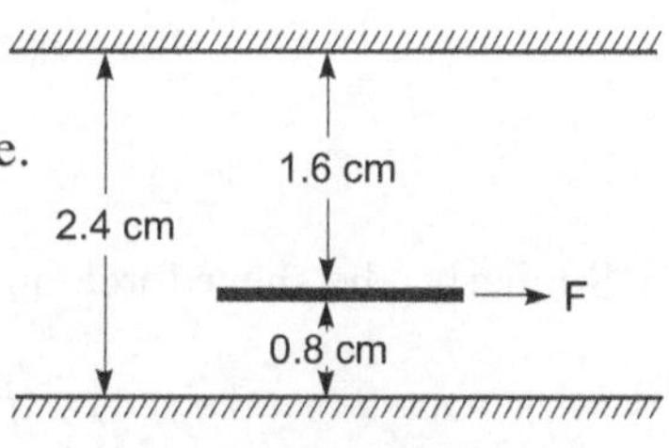

Fig. 1.7 (*b*)

The shear force on the upper side of the thin plate,

F_1 = Shear stress × Area = $\tau_1 \times A$

$$= \mu \left(\frac{du}{dy}\right)_1 \times A = 8.10 \times 10^{-1} \times \left(\frac{0.6}{0.016}\right) \times 0.5 = 15.18 \text{ N}$$

The shear force on the lower side of the thin plate,

$$F_2 = \tau_2 \times A = \mu \left(\frac{du}{dy}\right)_2 \times A$$

$$= 8.10 \times 10^{-1} \times \left(\frac{0.6}{0.8/100}\right) \times 0.5 = 30.36 \text{ N}$$

∴ Total force required = $F_1 + F_2$ = 15.18 + 30.36 = **45.54 N. Ans.**

Problem 1.19 *A vertical gap 2.2 cm wide of infinite extent contains a fluid of viscosity 2.0 N s/m² and specific gravity 0.9. A metallic plate 1.2 m × 1.2 m × 0.2 cm is to be lifted up with a constant velocity of 0.15 m/sec, through the gap. If the plate is in the middle of the gap, find the force required. The weight of the plate is 40 N.*

Solution. Given :

Width of gap = 2.2 cm, viscosity, μ = 2.0 N s/m²

Sq. gr. of fluid = 0.9

∴ Weight density of fluid

$$= 0.9 \times 1000 = 900 \text{ kgf/m}^3 = 900 \times 9.81 \text{ N/m}^3$$

(∵ 1 kgf = 9.81 N)

Volume of plate = 1.2 m × 1.2 m × 0.2 cm

$= 1.2 \times 1.2 \times .002 \text{ m}^3 = .00288 \text{ m}^3$

Thickness of plate = 0.2 cm

Velocity of plate = 0.15 m/sec

Weight of plate = 40 N.

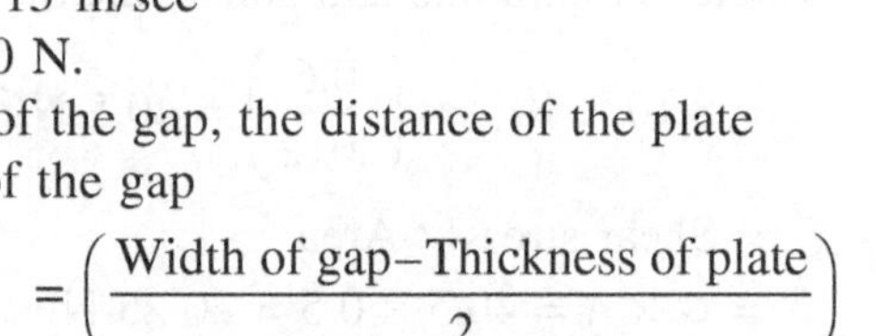

Fig. 1.8

When plate is in the middle of the gap, the distance of the plate of plate from vertical surface, of the gap

$$= \left(\frac{\text{Width of gap} - \text{Thickness of plate}}{2}\right)$$

$$= \frac{(2.2 - 0.2)}{2} = 1 \text{ cm} = .01 \text{ m}.$$

Now the shear force on the left side of the metallic plate,

$$F_1 = \text{Shear stress} \times \text{Area}$$

$$= \mu \left(\frac{du}{dy}\right)_1 \times \text{Area} = 2.0 \times \left(\frac{0.15}{.01}\right) \times 1.2 \times 1.2 \text{ N}$$

(∵ Area = 1.2 × 1.2 m²)

= 43.2 N.

Similarly, the shear force on the right side of the metallic plate,

$$F_2 = \text{Shear stress} \times \text{Area} = 2.0 \times \left(\frac{0.15}{.01}\right) \times 1.2 \times 1.2 = 43.2 \text{ N}$$

$\therefore$ Total shear force $= F_1 + F_2 = 43.2 + 43.2 = 86.4$ N.

In this case the weight of plate (which is acting vertically downward) and upward thrust is also to be taken into account.

The upward thrust = Weight of fluid displaced

= (Weight density of fluid) × Volume of fluid displaced

$= 9.81 \times 900 \times .00288$ N

($\because$ Volume of fluid displaced = Volume of plate = .00288)

= 25.43 N.

The net force acting in the downward direction due to weight of the plate and upward thrust

= Weight of plate – Upward thrust = 40 – 25.43 = 14.57 N

$\therefore$ Total force required to lift the plate up

= Total shear force + 14.57 = 86.4 + 14.57 = **100.97 N. Ans.**

▶ 1.4 THERMODYNAMIC PROPERTIES

Fluids consist of liquids or gases. But gases are compressible fluids and hence thermodynamic properties play an important role. With the change of pressure and temperature, the gases undergo l a r g e variation in density. The relationship between pressure (absolute), specific volume and temperature (absolute) of a gas is given by the equation of state as

$$p \forall = RT \text{ or } \frac{p}{\rho} = RT \qquad \text{...(1.5)}$$

where p = Absolute pressure of a gas in N/m^2

$\forall$ = Specific volume $= \frac{1}{\rho}$

R = Gas constant

T = Absolute temperature in °K

ρ = Density of a gas.

1.4.1 Dimension of R. The gas constant, R, depends upon the particular gas. The dimension of R is obtained from equation (1.5) as

$$R = \frac{p}{\rho T}$$

(*i*) In MKS units
$$R = \frac{\text{kgf/m}^2}{\left(\frac{\text{kg}}{\text{m}^3}\right)\text{°K}} = \frac{\text{kgf-m}}{\text{kg °K}}$$

(*ii*) In SI units, p is expressed in Newton/m^2 or N/m^2.

$$\therefore \quad R = \frac{\text{N/m}^2}{\frac{\text{kg}}{\text{m}^3} \times \text{K}} = \frac{\text{N-m}}{\text{kg-K}} = \frac{\text{Joule}}{\text{kg-K}} \qquad \text{[Joule = N-m]}$$

$$= \frac{\text{J}}{\text{kg-K}}$$

For air, R in MKS $= 29.3 \frac{\text{kgf-m}}{\text{kg °K}}$

$$R \text{ in SI} = 29.3 \times 9.81 \frac{\text{N-m}}{\text{kg°K}} = 287 \frac{\text{J}}{\text{kg-K}}$$

1.4.2 Isothermal Process. If the changes in density occurs at constant temperature, then the process is called isothermal and relationship between pressure (p) and density (ρ) is given by

$$\frac{p}{\rho} = \text{Constant} \quad \text{...(1.6)}$$

1.4.3 Adiabatic Process. If the change in density occurs with no heat exchange to and from the gas, the process is called adiabatic. And if no heat is generated within the gas due to friction, the relationship between pressure and density is given by

$$\frac{p}{\rho^k} = \text{Constant} \quad \text{...(1.7)}$$

where k = Ratio of specific heat of a gas at constant pressure and constant volume.
= 1.4 for air.

1.4.4 Universal Gas Constant.

Let m = Mass of a gas in kg
$\forall$ = Volume of gas of mass m
p = Absolute pressure
T = Absolute temperature

Then, we have $p\forall = mRT$ ...(1.8)

where R = Gas constant.

Equation (1.8) can be made universal, *i.e.*, applicable to all gases if it is expressed in **mole-basis**.

Let n = Number of moles in volume of a gas
$\forall$ = Volume of the gas

$$M = \frac{\text{Mass of the gas molecules}}{\text{Mass of a hydrogen atom}}$$

m = Mass of a gas in kg

Then, we have $n \times M = m$.

Substituting the value of m in equation (1.8), we get

$$p\forall = n \times M \times RT \quad \text{...(1.9)}$$

The product $M \times R$ is called universal gas constant and is equal to 848 $\frac{\text{kgf-m}}{\text{kg-mole °K}}$ in MKS units and 8314 J/kg-mole K in SI units.

One kilogram mole is defined as the product of one kilogram mass of the gas and its molecular weight.

Problem 1.20 *A gas weighs 16 N/m^3 at 25°C and at an absolute pressure of 0.25 N/mm^2. Determine the gas constant and density of the gas.*

Solution. Given :

Weight density, $w = 16 \text{ N/m}^2$

Temperature, $t = 25°\text{C}$

$\therefore$ $T = 273 + t = 273 + 25 = 288°\text{K}$

$$p = 0.25 \text{ N/mm}^2 \text{ (abs.)} = 0.25 \times 10^6 \text{ N/m}^2 = 25 \times 10^4 \text{ N/m}^2$$

(*i*) Using relation $w = \rho g$, density is obtained as

$$\rho = \frac{w}{g} = \frac{16}{9.81} = \mathbf{1.63 \text{ kg/m}^3\text{. Ans.}}$$

(*ii*) Using equation (1.5), $\frac{p}{\rho} = RT$

$$\therefore \quad R = \frac{p}{\rho T} = \frac{25 \times 10^4}{1.63 \times 288} = \mathbf{532.55 \frac{Nm}{kg\ K}\text{. Ans.}}$$

Problem 1.21 *A cylinder of 0.6 m³ in volume contains air at 50°C and 0.3 N/mm² absolute pressure. The air is compressed to 0.3 m³. Find (i) pressure inside the cylinder assuming isothermal process and (ii) pressure and temperature assuming adiabatic process. Take k = 1.4.*

Solution. Given :

Initial volume, $\forall_1 = 0.6 \text{ m}^3$

Temperature $t_1 = 50°\text{C}$

$\therefore \quad T_1 = 273 + 50 = 323°\text{K}$

Pressure $p_1 = 0.3 \text{ N/mm}^2 = 0.3 \times 10^6 \text{ N/m}^2 = 30 \times 10^4 \text{ N/m}^2$

Final volume $\forall_2 = 0.3 \text{ m}^3$

$k = 1.4$

(*i*) Isothermal process :

Using equation (1.6), $\frac{p}{\rho} = \text{Constant or } p\forall = \text{Constant.}$

$$\therefore \quad p_1\forall_1 = p_2\forall_2$$

$$\therefore \quad p_2 = \frac{p_1\forall_1}{\forall_2} = \frac{30 \times 10^4 \times 0.6}{0.3} = 0.6 \times 10^6 \text{ N/m}^2 = \mathbf{0.6 \text{ N/mm}^2\text{. Ans.}}$$

(*ii*) Adiabatic process :

Using equation (1.7), $\frac{p}{\rho^k} = \text{Constant or } p\ \forall^k = \text{Constant}$

$$\therefore \quad p_1\forall_1^k = p_2\forall_2^k.$$

$$\therefore \quad p_2 = p_1\frac{\forall_1^k}{\forall_2^k} = 30 \times 10^4 \times \left(\frac{0.6}{0.3}\right)^{1.4} = 30 \times 10^4 \times 2^{1.4}$$

$$= 0.791 \times 10^6 \text{ N/m}^2 = \mathbf{0.791 \text{ N/mm}^2\text{. Ans.}}$$

For temperature, using equation (1.5), we get

$$p\forall = RT \text{ and also } p\ \forall^k = \text{Constant}$$

$$\therefore \quad p = \frac{RT}{\forall} \text{ and } \frac{RT}{\forall} \times \forall^k = \text{Constant}$$

or $\quad RT\forall^{k-1} = \text{Constant}$

or $\quad T\ \forall^{k-1} = \text{Constant} \qquad \{\because R \text{ is also constant}\}$

$\therefore \qquad T_1 \forall_1^{k-1} = T_2 \forall_2^{k-1}$

$$\therefore \qquad T_2 = T_1 \left(\frac{\forall_1}{\forall_2}\right)^{k-1} = 323 \left(\frac{0.6}{0.3}\right)^{1.4-1.0} = 323 \times 2^{0.4} = 426.2°\text{K}$$

$$\therefore \qquad t_2 = 426.2 - 273 = \mathbf{153.2°C.\ Ans.}$$

Problem 1.22 *Calculate the pressure exerted by 5 kg of nitrogen gas at a temperature of 10°C if the volume is 0.4 m³. Molecular weight of nitrogen is 28. Assume, ideal gas laws are applicable.*

Solution. Given :

Mass of nitrogen $= 5$ kg

Temperature, $t = 10°\text{C}$

$\therefore \qquad T = 273 + 10 = 283°\text{K}$

Volume of nitrogen, $\forall = 0.4\ \text{m}^3$

Molecular weight $= 28$

Using equation (1.9), we have $p\forall = n \times M \times RT$

where $M \times R$ = Universal gas constant = $8314 \ \dfrac{\text{N-m}}{\text{kg-mole °K}}$

and one kg-mole = (kg-mass) × Molecular weight = (kg-mass) × 28

$$\therefore \qquad \text{R for nitrogen} = \frac{8314}{28} = 296.9 \ \frac{\text{N-m}}{\text{kg °K}}$$

The gas laws for nitrogen is $p\forall = mRT$, where R = Characteristic gas constant

or $$p \times 0.4 = 5 \times 296.9 \times 283$$

$$\therefore \qquad p = \frac{5 \times 296.9 \times 283}{0.4} = 1050283.7\ \text{N/m}^2 = \mathbf{1.05\ N/mm^2.\ Ans.}$$

▶ 1.5 COMPRESSIBILITY AND BULK MODULUS

Compressibility is the reciprocal of the bulk modulus of elasticity, K which is defined as the ratio of compressive stress to volumetric strain.

Consider a cylinder fitted with a piston as shown in Fig. 1.9.

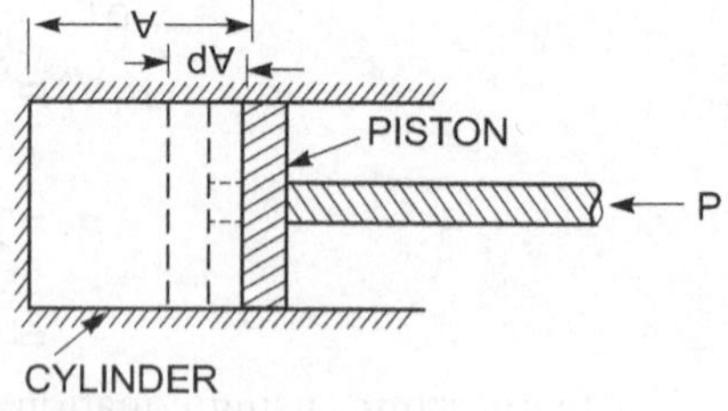

Fig. 1.9

Let $\forall$ = Volume of a gas enclosed in the cylinder

p = Pressure of gas when volume is $\forall$

Let the pressure is increased to $p + dp$, the volume of gas decreases from $\forall$ to $\forall - d\forall$.

Then increase in pressure $= dp$ kgf/m²

Decrease in volume $= d\forall$

$$\therefore \quad \text{Volumetric strain} = -\frac{d\forall}{\forall}$$

– ve sign means the volume decreases with increase of pressure.

$\therefore$ Bulk modulus $\quad K = \dfrac{\text{Increase of pressure}}{\text{Volumetric strain}}$

$$= \frac{dp}{\frac{-d\forall}{\forall}} = \frac{-dp}{d\forall}\forall \qquad ...(1.10)$$

Compressibility is given by $\quad = \dfrac{1}{K}$ ...(1.11)

Relationship between Bulk Modulus (K) and Pressure (p) for a Gas

The relationship between bulk modulus of elasticity (K) and pressure for a gas for two different processes of compression are as :

(*i*) **For Isothermal Process.** Equation (1.6) gives the relationship between pressure (p) and density (ρ) of a gas as

$$\frac{p}{\rho} = \text{Constant}$$

or

$$p\forall = \text{Constant} \qquad \left\{\because \forall = \frac{1}{\rho}\right\}$$

Differentiating this equation, we get (p and $\forall$ both are variables)

$$pd\forall + \forall dp = 0 \quad \text{or} \quad pd\forall = -\forall dp \quad \text{or} \quad p = \frac{-\forall dp}{d\forall}$$

Substituting this value in equation (1.10), we get

$$K = p \qquad ...(1.12)$$

(*ii*) **For Adiabatic Process.** Using equation (1.7) for adiabatic process

$$\frac{p}{\rho^k} = \text{Constant or } p\,\forall^k = \text{Constant}$$

Differentiating, we get $pd(\forall^k) + \forall^k(dp) = 0$

or $\quad p \times k \times \forall^{k-1}\, d\forall + \forall^k\, dp = 0$

or $\quad pkd\forall + \forall dp = 0 \qquad$ [Cancelling $\forall^{k-1}$ to both sides]

or $\quad pkd\forall = -\forall dp \quad$ or $\quad pk = -\dfrac{\forall dp}{d\forall}$

Hence from equation (1.10), we have

$$K = pk \qquad ...(1.13)$$

where K = Bulk modulus and k = Ratio of specific heats.

Problem 1.23 *Determine the bulk modulus of elasticity of a liquid, if the pressure of the liquid is increased from 70 N/cm² to 130 N/cm². The volume of the liquid decreases by 0.15 per cent.*

(VTU, Jan. 2004)

Solution. Given :

Initial pressure $= 70\ \text{N/cm}^2$

Final pressure $= 130\ \text{N/cm}^2$

$\therefore$ dp = Increase in pressure = 130 – 70 = 60 N/cm^2

Decrease in volume = 0.15%

$$\therefore \quad -\frac{d\forall}{\forall} = +\frac{0.15}{100}$$

Bulk modulus, K is given by equation (1.10) as

$$K = \frac{dp}{-\frac{d\forall}{\forall}} = \frac{60 \text{ N/cm}^2}{\frac{.15}{100}} = \frac{60 \times 100}{.15} = \mathbf{4 \times 10^4 \text{ N/cm}^2. \text{ Ans.}}$$

Problem 1.24 *What is the bulk modulus of elasticity of a liquid which is compressed in a cylinder from a volume of 0.0125 m^3 at 80 N/cm^2 pressure to a volume of 0.0124 m^3 at 150 N/cm^2 pressure ?*

Solution. Given :

Initial volume, $\forall = 0.0125 \text{ m}^3$

Final volume $= 0.0124 \text{ m}^3$

$\therefore$ Decrease in volume, $d\forall = .0125 - .0124 = .0001 \text{ m}^3$

$$\therefore \quad -\frac{d\forall}{\forall} = \frac{.0001}{.0125}$$

Initial pressure = 80 N/cm^2

Final pressure = 150 N/cm^2

$\therefore$ Increase in pressure, $dp = (150 - 80) = 70 \text{ N/cm}^2$

Bulk modulus is given by equation (1.10) as

$$K = \frac{dp}{-\frac{d\forall}{\forall}} = \frac{70}{\frac{.0001}{.0125}} = 70 \times 125 \text{ N/cm}^2$$

$$= \mathbf{8.75 \times 10^3 \text{ N/cm}^2. \text{ Ans.}}$$

▶ 1.6 SURFACE TENSION AND CAPILLARITY

Surface tension is defined as the tensile force acting on the surface of a liquid in contact with a gas or on the surface between two immiscible liquids such that the contact surface behaves like a membrance under tension. The magnitude of this force per unit length of the free surface will have the same value as the surface energy per unit area. It is denoted by Greek letter σ (called sigma). In MKS units, it is expressed as kgf/m while in SI units as N/m.

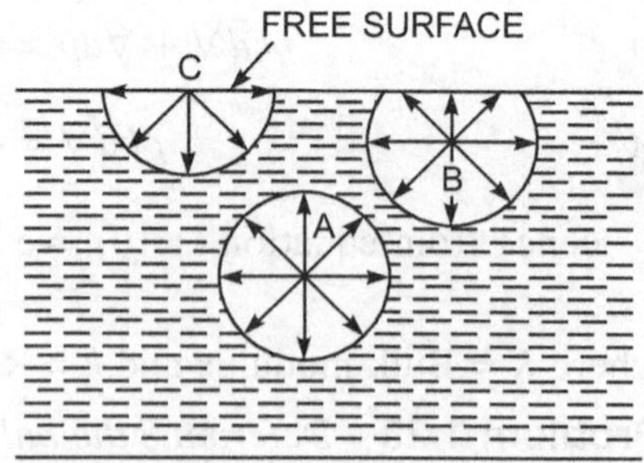

Fig. 1.10 *Surface tension.*

The phenomenon of surface tension is explained by Fig. 1.10. Consider three molecules A, B, C of a liquid in a mass of liquid. The molecule A is attracted in all directions equally by the surrounding molecules of the liquid. Thus the resultant force acting on the molecule A is zero. But the molecule B, which is situated near the free surface, is acted upon by upward and downward forces which are unbalanced. Thus a net resultant force on molecule B is acting in the downward direction. The molecule C, situated on the free surface of liquid, does experience a resultant downward force. All the molecules on the free surface experience a downward force. Thus the free surface of the liquid acts like a very thin film under tension of the surface of the liquid act as though it is an elastic membrance under tension.

1.6.1 Surface Tension on Liquid Droplet.

Consider a small spherical droplet of a liquid of radius 'r'. On the entire surface of the droplet, the tensile force due to surface tension will be acting.

Let σ = Surface tension of the liquid

p = Pressure intensity inside the droplet (in excess of the outside pressure intensity)

d = Dia. of droplet.

Let the droplet is cut into two halves. The forces acting on one half (say left half) will be

(*i*) tensile force due to surface tension acting around the circumference of the cut portion as shown in Fig. 1.11 (*b*) and this is equal to

$$= \sigma \times \text{Circumference}$$

$$= \sigma \times \pi d$$

(*ii*) pressure force on the area $\frac{\pi}{4} d^2$ and $= p \times \frac{\pi}{4} d^2$ as shown in Fig. 1.11 (*c*). These two forces will be equal and opposite under equilibrium conditions, *i.e.*,

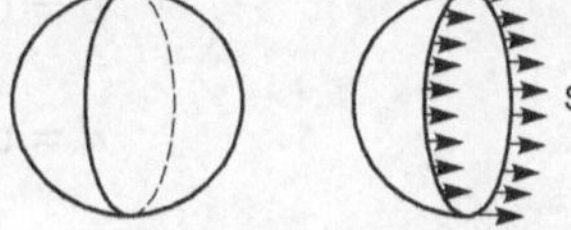

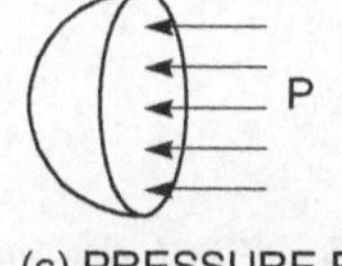

Fig. 1.11 *Forces on droplet.*

$$p \times \frac{\pi}{4} d^2 = \sigma \times \pi d$$

or
$$p = \frac{\sigma \times \pi d}{\frac{\pi}{4} \times d^2} = \frac{4\sigma}{d} \quad \text{...(1.14)}$$

Equation (1.14) shows that with the decrease of diameter of the droplet, pressure intensity inside the droplet increases.

1.6.2 Surface Tension on a Hollow Bubble.

A hollow bubble like a soap bubble in air has two surfaces in contact with air, one inside and other outside. Thus two surfaces are subjected to surface tension. In such case, we have

$$p \times \frac{\pi}{4} d^2 = 2 \times (\sigma \times \pi d)$$

$$\therefore \quad p = \frac{2\sigma \pi d}{\frac{\pi}{4} d^2} = \frac{8\sigma}{d} \quad \text{...(1.15)}$$

1.6.3 Surface Tension on a Liquid Jet.

Consider a liquid jet of diameter 'd' and length 'L' as shown in Fig. 1.12.

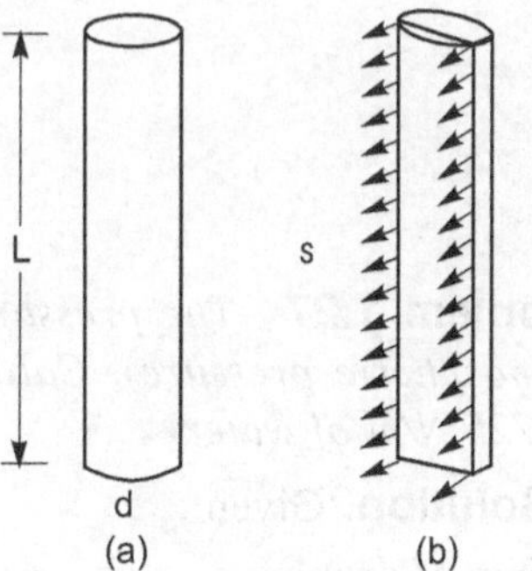

Fig. 1.12 *Forces on liquid jet.*

Let p = Pressure intensity inside the liquid jet above the outside pressure

σ = Surface tension of the liquid.

Consider the equilibrium of the semi jet, we have

Force due to pressure $= p \times$ area of semi jet

$= p \times L \times d$

Force due to surface tension $= \sigma \times 2L$.

Equating the forces, we have

$$p \times L \times d = \sigma \times 2L$$

$$\therefore \qquad p = \frac{\sigma \times 2L}{L \times d} \qquad \text{...(1.16)}$$

Problem 1.25 *The surface tension of water in contact with air at 20°C is 0.0725 N/m. The pressure inside a droplet of water is to be 0.02 N/cm² greater than the outside pressure. Calculate the diameter of the droplet of water.*

Solution. Given :

Surface tension, $\sigma = 0.0725$ N/m

Pressure intensity, p in excess of outside pressure is

$$p = 0.02 \text{ N/cm}^2 = 0.02 \times 10^4 \ \frac{\text{N}}{\text{m}^2}$$

Let d = dia. of the droplet

Using equation (1.14), we get $p = \dfrac{4\sigma}{d}$ or $0.02 \times 10^4 = \dfrac{4 \times 0.0725}{d}$

$$\therefore \qquad d = \frac{4 \times 0.0725}{0.02 \times (10)^4} = .00145 \text{ m} = .00145 \times 1000 = \textbf{1.45 mm. Ans.}$$

Problem 1.26 *Find the surface tension in a soap bubble of 40 mm diameter when the inside pressure is 2.5 N/m² above atmospheric pressure.*

Solution. Given :

Dia. of bubble, $d = 40 \text{ mm} = 40 \times 10^{-3}$ m

Pressure in excess of outside, $p = 2.5$ N/m²

For a soap bubble, using equation (1.15), we get

$$p = \frac{8\sigma}{d} \quad \text{or} \quad 2.5 = \frac{8 \times \sigma}{40 \times 10^{-3}}$$

$$\sigma = \frac{2.5 \times 40 \times 10^{-3}}{8} \text{ N/m} = \textbf{0.0125 N/m. Ans.}$$

Problem 1.27 *The pressure outside the droplet of water of diameter 0.04 mm is 10.32 N/cm² (atmospheric pressure). Calculate the pressure within the droplet if surface tension is given as 0.0725 N/m of water.*

Solution. Given :

Dia. of droplet, $d = 0.04 \text{ mm} = .04 \times 10^{-3}$ m

Pressure outside the droplet $= 10.32 \text{ N/cm}^2 = 10.32 \times 10^4 \text{ N/m}^2$

Surface tension, $\sigma = 0.0725$ N/m

The pressure inside the droplet, in excess of outside pressure is given by equation (1.14)

$$\text{or} \qquad p = \frac{4\sigma}{d} = \frac{4 \times 0.0725}{.04 \times 10^{-3}} = 7250 \text{ N/m}^2 = \frac{7250 \text{ N}}{10^4 \text{ cm}^2} = 0.725 \text{ N/cm}^2$$

∴ Pressure inside the droplet = p + Pressure outside the droplet

$$= 0.725 + 10.32 = \mathbf{11.045\ N/cm^2}.\ \textbf{Ans.}$$

1.6.4 Capillarity. Capillarity is defined as a phenomenon of rise or fall of a liquid surface in a small tube relative to the adjacent general level of liquid when the tube is held vertically in the liquid. The rise of liquid surface is known as capillary rise while the fall of the liquid surface is known as capillary depression. It is expressed in terms of cm or mm of liquid. Its value depends upon the specific weight of the liquid, diameter of the tube and surface tension of the liquid.

Expression for Capillary Rise. Consider a glass tube of small diameter 'd' opened at both ends and is inserted in a liquid, say water. The liquid will rise in the tube above the level of the liquid.

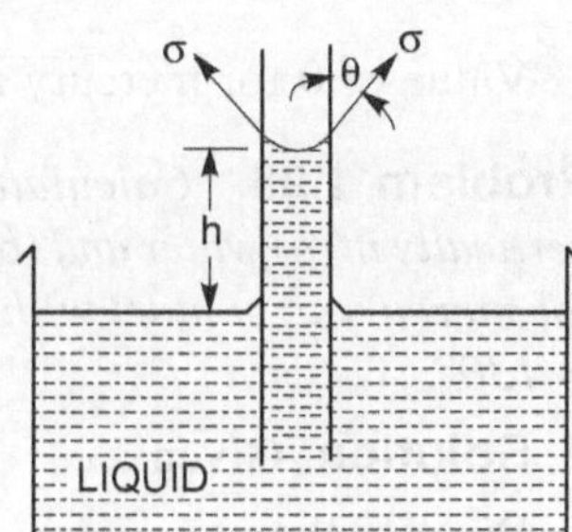

Fig. 1.13 *Capillary rise.*

Let h = height of the liquid in the tube. Under a state of equilibrium, the weight of liquid of height h is balanced by the force at the surface of the liquid in the tube. But the force at the surface of the liquid in the tube is due to surface tension.

Let σ = Surface tension of liquid

θ = Angle of contact between liquid and glass tube.

The weight of liquid of height h in the tube = (Area of tube × h) × ρ × g

$$= \frac{\pi}{4} d^2 \times h \times \rho \times g \qquad \text{...(1.17)}$$

where ρ = Density of liquid

Vertical component of the surface tensile force

$$= (\sigma \times \text{Circumference}) \times \cos\theta$$

$$= \sigma \times \pi d \times \cos\theta \qquad \text{...(1.18)}$$

For equilibrium, equating (1.17) and (1.18), we get

$$\frac{\pi}{4} d^2 \times h \times \rho \times g = \sigma \times \pi d \times \cos\theta$$

or

$$h = \frac{\sigma \times \pi d \times \cos\theta}{\frac{\pi}{4} d^2 \times \rho \times g} = \frac{4\sigma\cos\theta}{\rho \times g \times d} \qquad \text{...(1.19)}$$

The value of θ between water and clean glass tube is approximately equal to zero and hence $\cos\theta$ is equal to unity. Then rise of water is given by

$$h = \frac{4\sigma}{\rho \times g \times d} \qquad \text{... (1.20)}$$

Expression for Capillary Fall. If the glass tube is dipped in mercury, the level of mercury in the tube will be lower than the general level of the outside liquid as shown in Fig. 1.14.

Let h = Height of depression in tube.

Then in equilibrium, two forces are acting on the mercury inside the tube. First one is due to surface tension acting in the downward direction and is equal to $\sigma \times \pi d \times \cos\theta$.

Second force is due to hydrostatic force acting upward and is equal to intensity of pressure at a depth '*h*' × Area

$$= p \times \frac{\pi}{4} d^2 = \rho g \times h \times \frac{\pi}{4} d^2 \ \{\because p = \rho g h\}$$

Equating the two, we get

$$\sigma \times \pi d \times \cos\theta = \rho g h \times \frac{\pi}{4} d^2$$

$$\therefore \qquad h = \frac{4\sigma\cos\theta}{\rho g d} \qquad \text{...(1.21)}$$

Value of θ for mercury and glass tube is 128°.

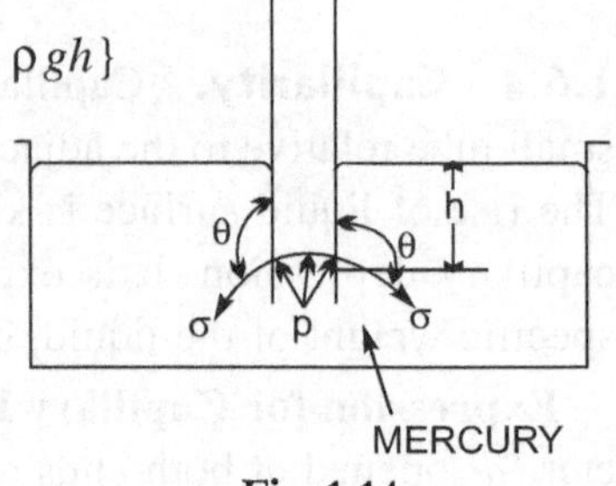

Fig. 1.14

Problem 1.28 *Calculate the capillary rise in a glass tube of 2.5 mm diameter when immersed vertically in (a) water and (b) mercury. Take surface tensions σ = 0.0725 N/m for water and σ = 0.52 N/m for mercury in contact with air. The specific gravity for mercury is given as 13.6 and angle of contact = 130°.*

Solution. Given :

Dia. of tube, $d = 2.5 \text{ mm} = 2.5 \times 10^{-3}$ m

Surface tenstion, σ for water = 0.0725 N/m

σ for mercury = 0.52 N/m

Sp. gr. of mercury = 13.6

∴ Density = $13.6 \times 1000 \text{ kg/m}^3$.

(*a*) **Capillary rise for water (θ = 0)**

Using equation (1.20), we get $h = \frac{4\sigma}{\rho \times g \times d} = \frac{4 \times 0.0725}{1000 \times 9.81 \times 2.5 \times 10^{-3}}$

= .0118 m = **1.18 cm. Ans.**

(*b*) **For mercury**

Angle of constant between mercury and glass tube, θ = 130°

Using equation (1.21), we get $h = \frac{4\sigma\cos\theta}{\rho \times g \times d} = \frac{4 \times 0.52 \times \cos 130°}{13.6 \times 1000 \times 9.81 \times 2.5 \times 10^{-3}}$

= –.004 m = **– 0.4 cm. Ans.**

The negative sign indicates the capillary depression.

Problem 1.29 *Calculate the capillary effect in millimetres in a glass tube of 4 mm diameter, when immersed in (i) water, and (ii) mercury. The temperature of the liquid is 20°C and the values of the surface tension of water and mercury at 20°C in contact with air are 0.073575 N/m and 0.51 N/m respectively. The angle of contact for water is zero that for mercury 1.30°. Take density of water at 20°C as equal to 998 kg/m³.* (U.P.S.C. Engg. Exam., 1974)

Solution. Given :

Dia of tube, $d = 4 \text{ mm} = 4 \times 10^{-3}$ m

The capillary effect (*i.e.*, capillary rise or depression) is given by equation (1.20) as

$$h = \frac{4\sigma\cos\theta}{\rho \times g \times d}$$

where σ = surface tenstion in kgf/m
θ = angle of contact, and ρ = density

(*i*) **Capillary effect for water**

$$\sigma = 0.073575 \text{ N/m}, \theta = 0°$$
$$\rho = 998 \text{ kg/m}^3 \text{ at } 20°\text{C}$$

$$\therefore \quad h = \frac{4 \times 0.073575 \times \cos 0°}{998 \times 9.81 \times 4 \times 10^{-3}} = 7.51 \times 10^{-3} \text{ m} = \mathbf{7.51 \text{ mm. Ans.}}$$

(*ii*) **Capillary effect for mercury**

$$\sigma = 0.51 \text{ N/m}, \theta = 130° \text{ and}$$
$$\rho = \text{sp. gr.} \times 1000 = 13.6 \times 1000 = 13600 \text{ kg/m}^2$$

$$\therefore \quad h = \frac{4 \times 0.51 \times \cos 130°}{13600 \times 9.81 \times 4 \times 10^{-3}} = -2.46 \times 10^{-3} \text{ m} = \mathbf{-2.46 \text{ mm. Ans.}}$$

The negative sign indicates the capillary depression.

Problem 1.30. *The capillary rise in the glass tube is not to exceed 0.2 mm of water. Determine its minimum size, given that surface tension for water in contact with air = 0.0725 N/m.*

Solution. Given :

Capillary rise, $h = 0.2 \text{ mm} = 0.2 \times 10^3 \text{ m}$
Surface tension, $\sigma = 0.0725 \text{ N/m}$
Let dia. of tube $= d$
The angle θ for water $= 0$
Density (ρ) for water $= 1000 \text{ kg/m}^3$

Using equation (1.20), we get

$$h = \frac{4\sigma}{\rho \times g \times d} \text{ or } 0.2 \times 10^{-3} = \frac{4 \times 0.0725}{1000 \times 9.81 \times d}$$

$$\therefore \quad d = \frac{4 \times 0.0725}{1000 \times 9.81 \times 2 \times 10^{-3}} = 0.148 \text{ m} = \mathbf{14.8 \text{ cm. Ans.}}$$

Thus minimum diameter of the tube should be 14.8 cm.

Problem 1.31 *Find out the minimum size of glass tube that can be used to measure water level if the capillary rise in the tube is to be restricted to 2 mm. Consider surface tension of water in contact with air as 0.073575 N/m.* (Converted to SI Units, A.M.I.E., Summer, 1985)

Solution. Given :

Capillary rise, $h = 2.0 \text{ mm} = 2.0 \times 10^{-3} \text{ m}$
Surface tension, $\sigma = 0.073575 \text{ N/m}$
Let dia. of tube $= d$
The angle θ for water $= 0$
The density for water, $\rho = 1000 \text{ kg/m}^3$

Using equation (1.20), we get

$$h = \frac{4\sigma}{\rho \times g \times d} \text{ or } 2.0 \times 10^{-3} = \frac{4 \times 0.073575}{1000 \times 9.81 \times d}$$

$$\therefore \qquad d = \frac{4 \times 0.073575}{1000 \times 9.81 \times 2 \times 10^{-3}} = 0.015 \text{ m} = \textbf{1.5 cm. Ans.}$$

Thus minimum diameter of the tube should be 1.5 cm.

Problem 1.32 *An oil of viscosity 5 poise is used for lubrication between a shaft and sleeve. The diameter of the shaft is 0.5 m and it rotates at 200 r.p.m. Calculate the power lost in oil for a sleeve length of 100 mm. The thickness of oil film is 1.0 mm.* [Delhi University, December, 1992 (NS)]

Solution. Given :

Viscosity, $\mu = 5$ poise

$$= \frac{5}{10} = 0.5 \text{ N s/m}^2$$

Dia. of shaft, $D = 0.5$ m

Speed of shaft, $N = 200$ r.p.m.

Sleeve length, $L = 100 \text{ mm} = 100 \times 10^{-3} \text{ m} = 0.1 \text{ m}$

Thickness of oil film, $t = 1.0 \text{ mm} = 1 \times 10^{-3}$ m

Tangential velocity of shaft, $u = \dfrac{\pi DN}{60} = \dfrac{\pi \times 0.5 \times 200}{60} = 5.235$ m/s

Using the relation, $\tau = \mu \dfrac{du}{dy}$

where, du = Change of velocity $= u - 0 = u = 5.235$ m/s

dy = Change of distance $= t = 1 \times 10^{-3}$ m

$$\therefore \qquad \tau = \frac{0.5 \times 5.235}{1 \times 10^{-3}} = 2617.5 \text{ N/m}^2$$

This is the shear stress on the shaft

$\therefore$ Shear force on the shaft, F = Shear stress × Area $= 2617.5 \times \pi D \times L$ $\quad (\because \quad \text{Area} = \pi D \times L)$

$$= 2617.5 \times \pi \times 0.5 \times 0.1 = 410.95 \text{ N}$$

Torque on the shaft, $T = \text{Force} \times \dfrac{D}{2} = 410.95 \times \dfrac{0.5}{2} = 102.74$ Nm

$$\therefore \qquad \text{Power* lost} = T \times \omega \text{ Watts} = T \times \frac{2\pi N}{60} \text{ W}$$

$$= 102.74 \times \frac{2\pi \times 200}{60} = 2150 \text{ W} = \textbf{2.15 kW. Ans.}$$

▶1.7 VAPOUR PRESSURE AND CAVITATION

A change from the liquid state to the gaseous state is known as vaporization. The vaporization (which depends upon the prevailing pressure and temperature condition) occurs because of continuous escaping of the molecules through the free liquid surface.

---

* Power in case of S.I. Unit $= T \times \omega$ or $\dfrac{2\pi NT}{60}$ Watts or $\dfrac{2\pi NT}{60{,}000}$ kW. The angular velocity $\omega = \dfrac{2\pi N}{60}$.

Consider a liquid (say water) which is confined in a closed vessel. Let the temperature of liquid is 20°C and pressure is atmospheric. This liquid will vaporise at 100°C. When vaporization takes place, the molecules escapes from the free surface of the liquid. These vapour molecules get accumulated in the space between the free liquid surface and top of the vessel. These accumulated vapours exert a pressure on the liquid surface. This pressure is known as **vapour pressure** of the liquid. Or this is the pressure at which the liquid is converted into vapours.

Again consider the same liquid at 20°C at atmospheric pressure in the closed vessel. If the pressure above the liquid surface is reduced by some means, the boiling temperature will also reduce. If the pressure is reduced to such an extent that it becomes equal to or less than the vapour pressure, the boiling of the liquid will start, though the temperature of the liquid is 20°C. Thus a liquid may boil even at ordinary temperature, if the pressure above the liquid surface is reduced so as to be equal or less than the vapour pressure of the liquid at that temperature.

Now consider a flowing liquid in a system. If the pressure at any point in this flowing liquid becomes equal to or less than the vapour pressure, the vaporization of the liquid starts. The bubbles of these vapours are carried by the flowing liquid into the region of high pressure where they collapse, giving rise to high impact pressure. The pressure developed by the collapsing bubbles is so high that the material from the adjoining boundaries gets eroded and cavities are formed on them. This phenomenon is known as **cavitation.**

Hence the cavitation is the phenomenon of formation of vapour bubbles of a flowing liquid in a region where the pressure of the liquid falls below the vapour pressure and sudden collapsing of these vapour bubbles in a region of higher pressure. When the vapour bubbles collapse, a very high pressure is created. The metallic surfaces, above which the liquid is flowing, is subjected to these high pressures, which cause pitting action on the surface. Thus cavities are formed on the metallic surface and hence the name is cavitation.

HIGHLIGHTS

1. The weight density or specific weight of a fluid is equal to weight per unit volume. It is also equal to,
$$w = \rho \times g.$$
2. Specific volume is the reciprocal of mass density.
3. The shear stress is proportional to the velocity gradient $\frac{du}{dy}$. Mathematically, $\tau = \mu \frac{du}{dy}$.
4. Kinematic viscosity ν is given by $\nu = \frac{\mu}{\rho}$.
5. Poise and stokes are the units of viscosity and kinematic viscosity respectively.
6. To convert the unit of viscosity from poise to MKS units, poise should be divided by 98.1 and to convert poise into SI units, the poise should be divided by 10. SI unit of viscosity is Ns/m^2 or Pa s, where N/m^2 = Pa = Pascal.
7. For a perfect gas, the equation of state is $\frac{p}{\rho} = RT$

   where R = gas constant and for air = $29.3 \frac{\text{kgf-m}}{\text{kg}^\circ\text{K}}$ = 287 J/kg °K.
8. For isothermal process, $\frac{p}{\rho}$ = Constant whereas for adiabatic process, $\frac{p}{\rho^k}$ = constant.

9. Bulk modulus of elasticity is given as $K = \dfrac{-dp}{\left(\dfrac{d\forall}{\forall}\right)}$.

10. Compressibility is the reciprocal of bulk modulus of elasticity or $= \dfrac{1}{K}$.

11. Surface tension is expressed in N/m or dyne/cm. The relation between surface tension (σ) and difference of pressure (p) between the inside and outside of a liquid drop is given as $p = \dfrac{4\sigma}{d}$

For a soap bubble, $p = \dfrac{8\sigma}{d}$.

For a liquid jet, $p = \dfrac{2\sigma}{d}$.

12. Capillary rise or fall of a liquid is given by $h = \dfrac{4\sigma \cos\theta}{wd}$.

The value of θ for water is taken equal to zero and for mercury equal to 128°.

EXERCISE 1

(A) THEORETICAL PROBLEMS

1. Define the following fluid properties :
   Density, weight density, specific volume and specific gravity of a fluid.
2. Differentiate between : (*i*) Liquids and gases, (*ii*) Real fluids and ideal fluids, (*iii*) Specific weight and specific volume of a fluid.
3. What is the difference between dynamic viscosity and kinematic viscosity ? State their units of measurements.
4. Explain the terms : (*i*) Dynamic viscosity, and (*ii*) Kinematic viscosity. Give their dimensions. *(A.M.I.E., Summer 1988)*
5. State the Newton's law of viscosity and give examples of its application. *(Delhi University, June 1996)*
6. Enunciate Newton's law of viscosity. Explain the importance of viscosity in fluid motion. What is the effect of temperature on viscosity of water and that of air? *(A.M.I.E., Winter 1987)*
7. Define Newtonian and Non-Newtonian fluids.
8. What do you understand by terms : (*i*) Isothermal process, (*ii*) Adiabatic process, and (*iii*) Universal-gas constant.
9. Define compressibility. Prove that compressibility for a perfect gas undergoing isothermal compression is $\dfrac{1}{p}$ while for a perfect gas undergoing isentropic compression is $\dfrac{1}{wp}$.
10. Define surface tension. Prove that the relationship between surface tension and pressure inside a droplet of liquid in excess of outside pressure is given by $p = \dfrac{4\sigma}{d}$.
11. Explain the phenomenon of capillarity. Obtain an expression for capillary rise of a liquid.
12. (*a*) Distinguish between ideal fluids and real fluids. Explain the importance of compressibility in fluid flow. *(A.M.I.E., Summer 1988)*
    (*b*) Define the terms : density, specific volume, specific gravity, vacuum pressure, compressible and incompressible fluids. *(R.G.P. Vishwavidyalaya, Bhopal S 2002)*
13. Define and explain Newton's law of viscosity. *(Delhi University, April 1992)*
14. Convert 1 kg/s-m dynamic viscosity in poise. *(A.M.I.E., Winter 1991)*

15. Why does the viscosity of a gas increases with the increase in temperature while that of a liquid decreases with increase in temperature ? *(A.M.I.E., Winter 1990)*

16. (*a*) How does viscosity of a fluid vary with temperature ?
(*b*) Cite examples where surface tension effects play a prominent role. *(J.N.T.U., Hyderabad S 2002)*

17. (*i*) Develop the expression for the relation between gauge pressure P inside a droplet of liquid and the surface tension.
(*ii*) Explain the following :
Newtonian and Non-Newtonian fluids, vapour pressure, and compressibility.
(R.G.P.V., Bhopal S 2001)

(B) NUMERICAL PROBLEMS

1. One litre of crude oil weighs 9.6 N. Calculate its specific weight, density and specific gravity.
(A.M.I.E., Summer 1986) [**Ans.** 9600 N/m^3, 978.6 kg/m^3, 0.978]

2. The velocity distribution for flow over a flat plate is given by $u = \frac{3}{2} y - y^{3/2}$, where u is the point velocity in metre per second at a distance y metre above the plate. Determine the shear stress at $y = 9$ cm. Assume dynamic viscosity as 8 poise. *(Nagpur University)* [**Ans.** 0.839 N/m^2]

3. A plate, 0.025 mm distant form a fixed plate, moves at 50 cm/s and requires a force of 1.471 N/m^2 to maintain this speed. Determine the fluid viscosity between the plates in the poise. [**Ans.** 7.357×10^{-4}]

4. Determine the intensity of shear of an oil having viscosity = 1.2 poise and is used for lubrication in the clearance between a 10 cm diameter shaft and its journal bearing. The clearance is 1.0 mm and shaft rotates at 200 r.p.m. [**Ans.** 125.56 N/m^2]

5. Two plates are placed at a distance of 0.15 mm apart. The lower plate is fixed while the upper plate having surface area 1.0 m^2 is pulled at 0.3 m/s. Find the force and power required to maintain this speed, if the fluid separating them is having viscosity 1.5 poise. [**Ans.** 300 N, 89.8 W]

6. An oil film of thickness 1.5 mm is used for lubrication between a square plate of size 0.9 m × 0.9 m and an inclined plane having an angle of inclination 20°. The weight of the square is 392.4 N and it slides down the plane with a uniform velocity of 0.2 m/s. Find the dynamic viscosity of the oil.
[**Ans.** 12.42 poise] *(VTU, Jan. 2004)*

7. In a stream of glycerine in motion, at a certain point the velocity gradient is 0.25 metre per sec per metre. The mass density of fluid is 1268.4 kg per cubic metre and kinematic viscosity is 6.30×10^{-4} square metre per second. Calculate the shear stress at the point. *(U.P.S.C., 1975)* [**Ans.** 0.2 N/m^2]

8. Find the kinematic viscosity of an oil having density 980 kg/m^2 when at a certain point in the oil, the shear stress is 0.25 N/m^2 and velocity gradient 0.3/s. $\left[\textbf{Ans.}\ 0.000849 \frac{\text{m}^2}{\text{sec}} \text{ or } 8.49 \text{ stokes}\right]$

9. Determine the specific gravity of a fluid having viscosity 0.07 poise and kinematic viscosity 0.042 stokes. [**Ans.** 1.667]

10. Determine the viscosity of a liquid having kinematic viscosity 6 stokes and specific gravity 2.0.
[**Ans.** 11.99 poise]

11. If the velocity distribution of a fluid over a plate is given by $u = (3/4)\, y - y^2$, where u is the velocity in metre per second at a distance of y metres above the plate, determine the shear stress at $y = 0.15$ metre. Take dynamic viscosity of the fluid as 8.5×10^{-5} kg-sec/m^2.
(A.M.I.E., Winter 1974) [**Ans.** 3.825×10^{-5} kgf/m^2]

12. An oil of viscosity 5 poise is used for lubrication between a shaft and sleeve. The diameter of shaft is 0.5 m and it rotates at 200 r.p.m. Calculate the power lost in the oil for a sleeve length of 100 mm. The thickness of the oil film is 1.0 mm. [**Ans.** 2.15 kW]

13. The velocity distribution over a plate is given by $u = \frac{2}{3} y - y^2$ in which u is the velocity in m/sec at a distance of y m above the plate. Determine the shear stress at $y = 0$, 0.1 and 0.2 m. Take $\mu = 6$ poise.
[**Ans.** 0.4, 0.028 and 0.159 N/m^2]

14. In question 13, find the distance in metres above the plate, at which the shear stress is zero.
[**Ans.** 0.333 m]

15. The velocity profile of a viscous fluid over a plate is parabolic with vertex 20 cm from the plate, where the velocity is 120 cm/s. Calculate the velocity gradient and shear stress at distances of 0, 5 and 15 cm from the plate, given the viscosity of the fluid = 6 poise.
[**Ans.** 12/s, 7.18 N/m^2; 9/s, 5.385 N/m^2 ; 3/s, 1.795 N/m^2]

16. The weight of a gas is given as 17.658 N/m^3 at 30°C and at an absolute pressure of 29.43 N/cm^2. Determine the gas constant and also the density of the gas.
$\left[\textbf{Ans.} \frac{1.8 \text{ kg}}{\text{m}^3}, \frac{539.55 \text{ N-m}}{\text{kg}^\circ \text{K}}\right]$

17. A cylinder of 0.9 m^3 in volume contains air at 0°C and 39.24 N/cm^2 absolute pressure. The air is compressed to 0.45 m^3. Find (*i*) the pressure inside the cylinder assuming isothermal process, (*ii*) pressure and temperature assuming adiabatic process. Take $k = 1.4$ for air.
[**Ans.** (*i*) 78.48 N/cm^2, (*ii*) 103.5 N/m^2, 140°C]

18. Calculate the pressure exerted by 4 kg mass of nitrogen gas at a temperature of 15°C if the volume is 0.35 m^3. Molecular weight of nitrogen is 28. [**Ans.** 97.8 N/cm^2]

19. The pressure of a liquid is increased from 60 N/cm^2 to 100 N/cm^2 and volume decreases by 0.2 per cent. Determine the bulk modulus of elasticity. [**Ans.** 2×10^4 N/cm^2]

20. Determine the bulk modulus of elasticity of a fluid which is compressed in a cylinder from a volume of 0.009 m^3 at 70 N/cm^2 pressure to a volume of 0.0085 m^3 at 270 N/cm^2 pressure. [**Ans.** 3.6×10^3 N/cm^2]

21. The surface tension of water in contact with air at 20°C is given as 0.0716 N/m. The pressure inside a droplet of water is to be 0.0147 N/cm^2 greater than the outside pressure, calculate the diameter of the droplet of water. [**Ans.** 1.94 mm]

22. Find the surface tension in a soap bubble of 30 mm diameter when the inside pressure is 1.962 N/m^2 above atmosphere. [**Ans.** 0.00735 N/m]

23. The surface tension of water in contact with air is given as 0.0725 N/m. The pressure outside the droplet of water of diameter 0.02 mm is atmospheric $\left(10.32 \frac{\text{N}}{\text{cm}^2}\right)$. Calculate the pressure within the droplet of water. [**Ans.** 11.77 N/cm^2]

24. Calculate the capillary rise in a glass tube of 3.0 mm diameter when immersed vertically in (*a*) water, and (*b*) mercury. Take surface tensions for mercury and water as 0.0725 N/m and 0.52 N/m respectively in contact with air. Specific gravity for mercury is given as 13.6. [**Ans.** 0.966 cm, 0.3275 cm]

25. The capillary rise in the glass tube used for measuring water level is not to exceed 0.5 mm. Determine its minimum size, given that surface tension for water in contact with air = 0.07112 N/m. [**Ans.** 5.8 cm]

26. (**SI Units**). One litre of crude oil weighs 9.6 N. Calculate its specific weight, density and specific gravity. (*Converted to SI units, A.M.I.E., Summer 1986*) [**Ans.** 9600 N/m^3; 979.6 kg/m^3 ; 0.9786]

27. (**SI Units**). A piston 796 mm diameter and 200 mm long works in a cylinder of 800 mm diameter. If the annular space is filled with a lubricating oil of viscosity 5 cp (centi-poise), calculate the speed of descent of the piston in vertical position. The weight of the piston and axial load are 9.81 N.
[**Ans.** 7.84 m/s]

28. (**SI Units**). Find the capillary rise of water in a tube 0.03 cm diameter. The surface tension of water is 0.0735 N/m. [**Ans.** 9.99 cm]

29. Calculate the specific weight, density and specific gravity of two litres of a liquid which weight 15 N.
(*Delhi University, April 1992*) [**Ans.** 7500 N/m^3, 764.5 kg/m^3, 0.764]

30. A 150 mm diameter vertical cylinder rotates concentrically inside another cylinder of diameter 151 mm. Both the cylinders are of 250 mm height. The space between the cylinders is filled with a liquid of viscosity 10 poise. Determine the torque required to rotate the inner cylinder at 100 r.p.m.
(*Delhi University, April 1992*) [**Ans.** 13.87 Nm]

31. A shaft of diameter 120 mm is rotating inside a journal bearing of diameter 122 mm at a speed of 360 r.p.m. The space between the shaft and the bearing is filled with a lubricating oil of viscosity 6 poise. Find the power absorbed in oil if the length of bearing is 100 mm.(*Delhi University, May 1998*) [**Ans.** 115.73 W]

32. A shaft of diameter 100 mm is rotating inside a journal bearing of diameter 102 mm at a space of 360 r.p.m. The space between the shaft and bearing is filled with a lubricating oil of viscosity 5 poise. The length of the bearing is 200 mm. Find the power absorbed in the lubricating oil.

(*Delhi University, June 1996*) [**Ans.** 111.58 W]

33. Assuming that the bulk modulus of elasticity of water is 2.07×10^6 kN/m^2 at standard atmospheric conditions, determine the increase of pressure necessary to produce 1% reduction in volume at the same temperature. (*Delhi University, June 1997*)

[**Hint.** $K = 2.07 \times 10^6$ kN/m^2; $\frac{-d\forall}{\forall} = \frac{1}{100} = 0.01.$

Increase in pressure (dp) $= K \times \left(\frac{-d\forall}{\forall}\right) = 2.07 \times 10^6 \times 0.01 =$ **2.07×10^4 kN/m^2. Ans.**]

34. A square plate of size 1 m × 1 m and weighing 350 N slides down an inclined plane with a uniform velocity of 1.5 m/s. The inclined plane is laid on a slope of 5 vertical to 12 horizontal and has an oil film of 1 mm thickness. Calculate the dynamic viscosity of oil. (*J.N.T.U., Hyderabad, S 2002*)

[**Hint.** $A = 1 \times 1 = 1$ m^2, W = 350 N, $u = 1.5$ m/s, $\tan\theta = \frac{5}{12} = \frac{BC}{AB}$

Component of weight along the plane = W × sin θ

where $\sin\theta = \frac{BC}{AC} = \frac{5}{13}$ $\quad\left(\because AC = \sqrt{AB^2 + BC^2} = \sqrt{12^2 + 5^2} = 13\right)$

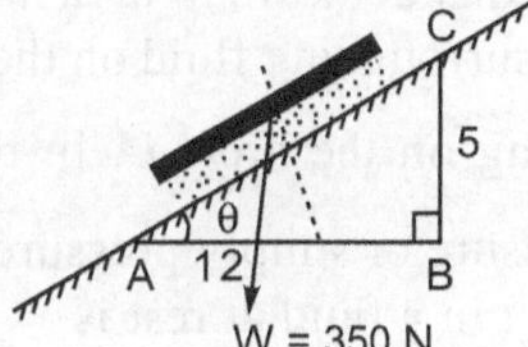

Fig. 1.15

$$\therefore \quad F = W\sin\theta = 350 \times \frac{5}{13} = 134.615$$

Now $\tau = \mu \frac{du}{dy}$, where $du = u - 0 = u = 1.5$ m/s and

$dy = 1 \text{ mm} = 1 \times 10^{-3}$ m

or $\frac{F}{A} = \mu \frac{du}{dy}$, $\therefore \mu = \frac{F}{A} \times \frac{dy}{du} = \frac{134.615}{1} \times \frac{1\times10^{-3}}{1.5} = 0.0897 \frac{Ns}{m^2}$ = **0.897 poise. Ans.**]

Pressure and its Measurement

▶ 2.1 FLUID PRESSURE AT A POINT

Consider a small area dA in large mass of fluid. If the fluid is stationary, then the force exerted by the surrounding fluid on the area dA will always be perpendicular to the surface dA. Let dF is the force acting on the area dA in the normal direction. Then the ratio of $\frac{dF}{dA}$ is known as the intensity of pressure or simply pressure and this ratio is represented by p. Hence mathematically the pressure at a point in a fluid at rest is

$$p = \frac{dF}{dA}.$$

If the force (F) is uniformly distributed over the area (A), then pressure at any point is given by

$$p = \frac{F}{A} = \frac{\text{Force}}{\text{Area}}.$$

∴ Force or pressure force, $F = p \times A$.

The units of pressure are : (*i*) kgf/m^2 and kgf/cm^2 in MKS units, (*ii*) Newton/m^2 or N/m^2 and N/mm^2 in SI units. N/m^2 is known as Pascal and is represented by Pa. Other commonly used units of pressure are :

$$\text{kPa} = \text{kilo pascal} = 1000 \text{ N/m}^2$$
$$\text{bar} = 100 \text{ kPa} = 10^5 \text{ N/m}^2.$$

▶ 2.2 PASCAL'S LAW

It states that the pressure or intensity of pressure at a point in a static fluid is equal in all directions. This is proved as :

The fluid element is of very small dimensions *i.e.*, dx, dy and ds.

Consider an **arbitrary** fluid element of wedge shape in a fluid mass at rest as shown in Fig. 2.1. Let the width of the element perpendicular to the plane of paper is unity and p_x,

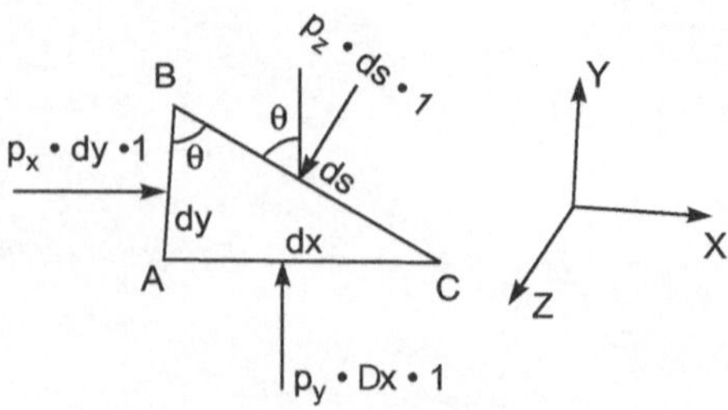

Fig. 2.1 *Forces on a fluid element.*

p_y and p_z are the pressures or intensity of pressure acting on the face AB, AC and BC respectively. Let $\angle ABC = \theta$. Then the forces acting on the element are :

1. Pressure forces normal to the surfaces.
2. Weight of element in the vertical direction.

The forces on the faces are :

Force on the face AB $= p_x \times$ Area of face AB

$= p_x \times dy \times 1$

Similarly force on the face $AC = p_y \times dx \times 1$

Force on the face BC $= p_z \times ds \times 1$

Weight of element $=$ (Mass of element) $\times g$

$$= (\text{Volume} \times \rho) \times g = \left(\frac{AB \times AC}{2} \times 1\right) \times \rho \times g,$$

where ρ = density of fluid.

Resolving the forces in x-direction, we have

$$p_x \times dy \times 1 - p\,(ds \times 1) \sin (90° - \theta) = 0$$

or

$$p_x \times dy \times 1 - p_z\, ds \times 1 \cos \theta = 0.$$

But from Fig. 2.1, $ds \cos \theta = AB = dy$

$\therefore$

$$p_x \times dy \times 1 - p_z \times dy \times 1 = 0$$

or

$$p_x = p_z \qquad \text{...(2.1)}$$

Similarly, resolving the forces in y-direction, we get

$$p_y \times dx \times 1 - p_z \times ds \times 1 \cos (90° - \theta) - \frac{dx \times dy}{2} \times 1 \times \rho \times g = 0$$

or

$$p_y \times dx - p_z\, ds \sin \theta - \frac{dxdy}{2} \times \rho \times g = 0.$$

But $ds \sin \theta = dx$ and also the element is very small and hence weight is negligible.

$\therefore$

$$p_y dx - p_z \times dx = 0$$

or

$$p_y = p_z \qquad \text{...(2.2)}$$

From equations (2.1) and (2.2), we have

$$p_x = p_y = p_z \qquad \text{...(2.3)}$$

The above equation shows that the pressure at any point in x, y and z directions is equal.

Since the choice of fluid element was completely arbitrary, which means the pressure at any point is the same in all directions.

▶ 2.3 PRESSURE VARIATION IN A FLUID AT REST

The pressure at any point in a fluid at rest is obtained by the Hydrostatic Law which states that the rate of increase of pressure in a vertically downward direction must be equal to the specific weight of the fluid at that point. This is proved as :

Consider a small fluid element as shown in Fig. 2.2

Let ΔA = Cross-sectional area of element

ΔZ = Height of fluid element

p = Pressure on face AB

Z = Distance of fluid element from free surface.

The forces acting on the fluid element are :

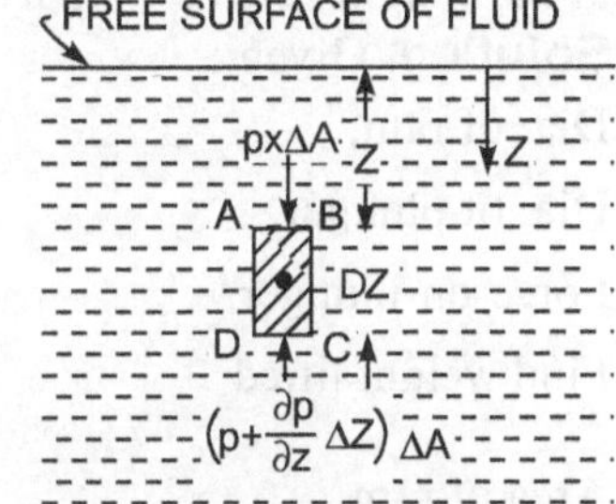

Fig. 2.2 *Forces on a fluid element.*

1. Pressure force on $AB = p \times \Delta A$ and acting perpendicular to face AB in the downward direction.

2. Pressure force on $CD = \left(p + \frac{\partial p}{\partial Z}\Delta Z\right) \times \Delta A$, acting perpendicular to face CD, vertically upward direction.

3. Weight of fluid element = Density $\times g \times$ Volume $= \rho \times g \times (\Delta A \times \Delta Z)$.

4. Pressure forces on surfaces BC and AD are equal and opposite. For equilibrium of fluid element, we have

$$p\Delta A - \left(p + \frac{\partial p}{\partial Z}\Delta Z\right)\Delta A + \rho \times g \times (\Delta A \times \Delta Z) = 0$$

or
$$p\Delta A - p\Delta A - \frac{\partial p}{\partial Z}\Delta Z\Delta A + \rho \times g \times \Delta A \times Z = 0$$

or
$$-\frac{\partial p}{\partial Z}\Delta Z\Delta A + \rho \times g \times \Delta A\Delta Z = 0$$

or
$$\frac{\partial p}{\partial Z}\Delta Z\Delta A = \rho \times g \times \Delta A\Delta Z \quad \text{or} \quad \frac{\partial p}{\partial Z} = \rho \times g \text{ [cancelling } \Delta A\Delta Z \text{ on both sides]}$$

$$\therefore \quad \frac{\partial p}{\partial Z} = \rho \times g = w \qquad (\because \rho \times g = w) \qquad \text{...(2.4)}$$

where w = Weight density of fluid.

Equation (2.4) states that rate of increase of pressure in a vertical direction is equal to weight density of the fluid at that point. This is **Hydrostatic Law.**

By integrating the above equation (2.4) for liquids, we get

$$\int dp = \int \rho g Z$$

or
$$p = \rho g Z \qquad \text{...(2.5)}$$

where p is the pressure above atmospheric pressure and Z is the height of the point from free surfaces.

From equation (2.5), we have
$$Z = \frac{p}{\rho \times g} \qquad \text{...(2.6)}$$

Here Z is called **pressure head.**

Problem 2.1 *A hydraulic press has a ram of 30 cm diameter and a plunger of 4.5 cm diameter. Find the weight lifted by the hydraulic press when the force applied at the plunger is 500 N.*

Solution. Given :

Dia. of ram, $D = 30 \text{ cm} = 0.3 \text{ m}$

Dia. of plunger, $d = 4.5 \text{ cm} = 0.045 \text{ m}$

Force on plunger, $F = 500 \text{ N}$

Find weight lifted $= W$

Area of ram, $A = \frac{\pi}{4}D^2 = \frac{\pi}{4}(0.3)^2 = 0.07068 \text{ m}^2$

Area of plunger, $a = \frac{\pi}{4}d^2 = \frac{\pi}{4}(0.045)^2 = .00159 \text{ m}^2$

Pressure intensity due to plunger

$$= \frac{\text{Force on plunger}}{\text{Area of plunger}} = \frac{F}{a} = \frac{500}{.00159} \text{ N/m}^2.$$

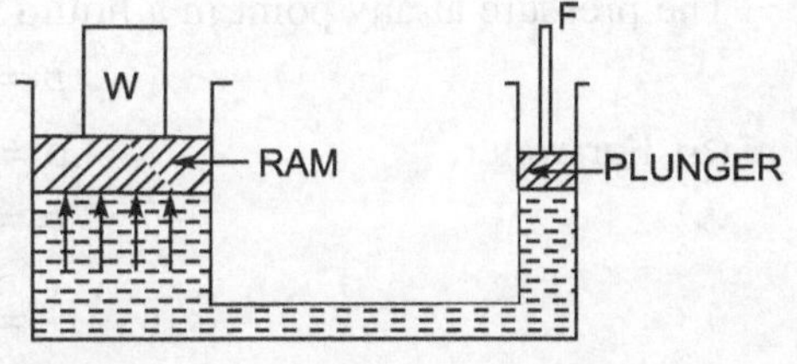

Fig. 2.3

Due to Pascal's law, the intensity of pressure will be equally transmitted in all directions. Hence the pressure intensity at the ram

$$= \frac{500}{.00159} = 314465.4 \text{ N/m}^2$$

But pressure intensity at ram $= \dfrac{\text{Weight}}{\text{Area of ram}} = \dfrac{W}{A} = \dfrac{W}{.07068}$ N/m^2

$$\frac{W}{.07068} = 314465.4$$

$\therefore$ Weight $= 314465.4 \times .07068 = 22222$ N $=$ **22.222 kN. Ans.**

Problem 2.2 *A hydraulic press has a ram of 20 cm diameter and a plunger of 3 cm diameter. It is used for lifting a weight of 30 kN. Find the force required at the plunger.*

Solution. Given :

Dia. of ram, $D = 20$ cm $= 0.2$ m

$\therefore$ Area of ram, $A = \dfrac{\pi}{4} D^2 = \dfrac{\pi}{4}(.2)^2 = 0.0314 \text{ m}^2$

Dia. of plunger $d = 3$ cm $= 0.03$ m

$\therefore$ Area of plunger, $a = \dfrac{\pi}{4}(.03)^2 = 7.068 \times 10^{-4} \text{ m}^2$

Weight lifted, $W = 30$ kN $= 30 \times 1000$ N $= 30000$ N.

See Fig. 2.3.

Pressure intensity developed due to plunger $= \dfrac{\text{Force}}{\text{Area}} = \dfrac{F}{a}$.

By Pascal's Law, this pressure is transmitted equally in all directions

Hence pressure transmitted at the ram $= \dfrac{F}{a}$

$\therefore$ Force acting on ram = Pressure intensity $\times$ Area of ram

$$= \frac{F}{a} \times A = \frac{F \times .0314}{7.068 \times 10^{-4}} \text{ N}$$

But force acting on ram = Weight lifted = 30000 N

$$\therefore \quad 30000 = \frac{F \times .0314}{7.068 \times 10^{-4}}$$

$$\therefore \quad F = \frac{30000 \times 7.068 \times 10^{-4}}{.0314} = \textbf{675.2 N. Ans.}$$

Problem 2.3 *Calculate the pressure due to a column of 0.3 of (a) water, (b) an oil of sp. gr. 0.8, and (c) mercury of sp. gr. 13.6. Take density of water, $\rho = 1000$ kg/m^3.*

Solution. Given :

Height of liquid column, $Z = 0.3$ m.

The pressure at any point in a liquid is given by equation (2.5) as

$$p = \rho g Z$$

(*a*) For water, $\rho = 1000 \text{ kg/m}^3$

$\therefore$ $p = \rho g Z = 1000 \times 9.81 \times 0.3 = 2943 \text{ N/m}^2$

$$= \frac{2943}{10^4} \text{ N/cm}^2 = \mathbf{0.2943 \text{ N/cm}^2. \text{ Ans.}}$$

(*b*) For oil of sp. gr. 0.8,

From equation (1.1*A*), we know that the density of a fluid is equal to specific gravity of fluid multiplied by density of water.

$\therefore$ Density of oil, $\rho_0 = \text{Sp. gr. of oil} \times \text{Density of water}$ (ρ_0 = Density of oil)

$$= 0.8 \times \rho = 0.8 \times 1000 = 800 \text{ kg/m}^3$$

Now pressure, $p = \rho_0 \times g \times Z$

$$= 800 \times 9.81 \times 0.3 = 2354.4 \frac{\text{N}}{\text{m}^2} = \frac{2354.4}{10^4} \frac{\text{N}}{\text{cm}^2}.$$

$$= \mathbf{0.2354 \frac{N}{cm^2}. \text{ Ans.}}$$

(*c*) For mercury, sp. gr. $= 13.6$

From equation (1.1*A*) we know that the density of a fluid is equal to specific gravity of fluid multiplied by density of water

$\therefore$ Density of mercury, $\rho_s = \text{Specific gravity of mercury} \times \text{Density of water}$

$$= 13.6 \times 1000 = 13600 \text{ kg/m}^3$$

$\therefore$ $p = \rho_s \times g \times Z$

$$= 13600 \times 9.81 \times 0.3 = 40025 \frac{\text{N}}{\text{m}^2}$$

$$= \frac{40025}{10^4} = \mathbf{4.002 \frac{N}{cm^2}. \text{ Ans}}$$

Problem 2.4 *The pressure intensity at a point in a fluid is given 3.924 N/cm². Find the corresponding height of fluid when the fluid is : (a) water, and (b) oil of sp. gr. 0.9.*

Solution. Given :

Pressure intensity, $p = 3.924 \frac{\text{N}}{\text{cm}^2} = 3.924 \times 10^4 \frac{\text{N}}{\text{m}^2}.$

The corresponding height, *Z*, of the fluid is given by equation (2.6) as

$$Z = \frac{p}{\rho \times g}$$

(*a*) For water, $\rho = 1000 \text{ kg/m}^3$

$\therefore$ $Z = \frac{p}{\rho \times g} = \frac{3.924 \times 10^4}{1000 \times 9.81} = \mathbf{4 \text{ m of water. Ans.}}$

(*b*) For oil, sp. gr. $= 0.9$

$\therefore$ Density of oil $\rho_0 = 0.9 \times 1000 = 900 \text{ kg/m}^3$

$\therefore$ $Z = \frac{p}{\rho_0 \times g} = \frac{3.924 \times 10^4}{900 \times 9.81} = \mathbf{4.44 \text{ m of oil. Ans.}}$

Problem 2.5 *An oil of sp. gr. 0.9 is contained in a vessel. At a point the height of oil is 40 m. Find the corresponding height of water at the point.*

Solution. Given :

Sp. gr. of oil, $S_0 = 0.9$

Height of oil, $Z_0 = 40$ m

Density of oil, $\rho_0 =$ Sp. gr. of oil × Density of water = 0.9 × 1000 = 900 kg/m^3

Intensity of pressure, $p = \rho_0 \times g \times Z_0 = 900 \times 9.81 \times 40 \frac{\text{N}}{\text{m}^2}$

$$\therefore \text{ Corresponding height of water} = \frac{p}{\text{Density of water} \times g}$$

$$= \frac{900 \times 9.81 \times 40}{1000 \times 9.81} = 0.9 \times 40 = \textbf{36 m of water. Ans.}$$

Problem 2.6 *An open tank contains water upto a depth of 2 m and above it an oil of sp. gr. 0.9 for a depth of 1 m. Find the pressure intensity (i) at the interface of the two liquids, and (ii) at the bottom of the tank.*

Solution. Given :

Height of water, $Z_1 = 2$ m

Height of oil, $Z_2 = 1$ m

Sp. gr. of oil, $S_0 = 0.9$

Density of water, $\rho_1 = 1000$ kg/m^3

Density of oil, $\rho_2 =$ Sp. gr. of oil × Density of water $= 0.9 \times 1000 = 900$ kg/m^3

Fig. 2.4

Pressure intensity at any point is given by

$$p = \rho \times g \times Z.$$

(*i*) At interface, *i.e.*, at *A*

$$p = \rho_2 \times g \times 1.0$$
$$= 900 \times 9.81 \times 1.0$$
$$= 8829 \frac{\text{N}}{\text{m}^2} = \frac{8829}{10^4} = \textbf{0.8829 N/cm}^2\textbf{. Ans.}$$

(*ii*) At the bottom, *i.e.*, at *B*

$$p = \rho_2 \times gZ_2 + \rho_1 \times g \times Z_1 = 900 \times 9.81 \times 1.0 + 1000 \times 9.81 \times 2.0$$

$$= 8829 + 19620 = 28449 \text{ N/m}^2 = \frac{28449}{10^4} \text{ N/cm}^2 = \textbf{2.8449 N/cm}^2\textbf{. Ans.}$$

Problem 2.7 *The diameters of a small piston and a large piston of a hydraulic jack ate 3 cm and 10 cm respectively. A force of 80 N is applied on the small piston. Find the load lifted by the large piston when :*

(a) the pistons are at the same level.

(b) small piston is 40 cm above the large piston.

The density of the liquid in the jack is given as 1000 kg/m^3.

Solution. Given :

Dia. of small piston, $d = 3$ cm

$\therefore$ Area of small piston, $a = \frac{\pi}{4} d^2 = \frac{\pi}{4} \times (3)^2 = 7.068 \text{ cm}^2$

Dia. of large piston, $D = 10$ cm

$\therefore$ Area of larger piston, $A = \frac{p}{4} \times (10)^2 = 78.54 \text{ cm}^2$

Force on small piston, $F = 80$ N

Let the load lifted $= W.$

(*a*) **When the pistons are at the same level**

Pressure intensity on small piston

$$\frac{F}{a} = \frac{80}{7.068} \text{ N/cm}^2$$

This is transmitted equally on the large piston.

$\therefore$ Pressure intensity on the large piston

$$= \frac{80}{7.068}$$

$\therefore$ Force on the large piston = Pressure × Area

$$= \frac{80}{7.068} \times 78.54 \text{ N} = \mathbf{888.96 \text{ N. Ans.}}$$

Fig. 2.5

(*b*) **When the small piston is 40 cm above the large piston**

Pressure intensity on the small piston

$$= \frac{F}{a} = \frac{80}{7.068} \frac{\text{N}}{\text{cm}^2}$$

$\therefore$ Pressure intensity at section $A - A$

$$= \frac{F}{a} + \text{Pressure intensity due to height of 40 cm of liquid.}$$

But pressure intensity due to 40 cm of liquid

$$= \rho \times g \times h = 1000 \times 9.81 \times 0.4 \text{ N/m}^2$$

$$= \frac{1000 \times 9.81 \times .40}{10^4} \text{ N/cm}^2 = 0.3924 \text{ N/cm}^2$$

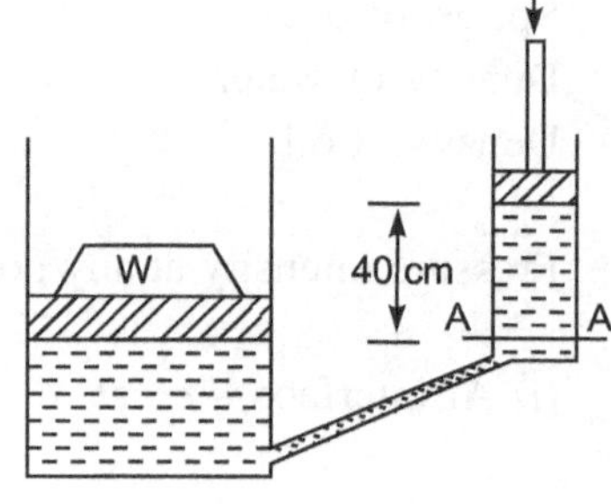

Fig. 2.6

$\therefore$ Pressure intensity at section

$$A - A = \frac{80}{7.068} + 0.3924$$

$$= 11.32 + 0.3924 = 11.71 \text{ N/cm}^2$$

$\therefore$ Pressure intensity transmitted to the large piston = 11.71 N/cm^2

$\therefore$ Force on the large piston = Pressure × Area of the large piston

$$= 11.71 \times A = 11.71 \times 78.54 = 919.7 \text{ N.}$$

▶ 2.4 ABSOLUTE, GAUGE, ATMOSPHERIC AND VACUUM PRESSURES

The pressure on a fluid is measured in two different systems. In one system, it is measured above the absolute zero or complete vacuum and it is called the absolute pressure and in other system, pressure is measured above the atmospheric pressure and it is called gauge pressure. Thus :

1. Absolute pressure is defined as the pressure which is measured with reference to absolute vacuum pressure.

2. Gauge pressure is defined as the pressure which is measured with the help of a pressure measuring instrument, in which the atmospheric pressure is taken as datum. The atmospheric pressure on the scale is marked as zero.

3. Vacuum pressure is defined as the pressure below the atmospheric pressure.

The relationship between the absolute pressure, gauge pressure and vacuum pressure are shown in Fig. 2.7.

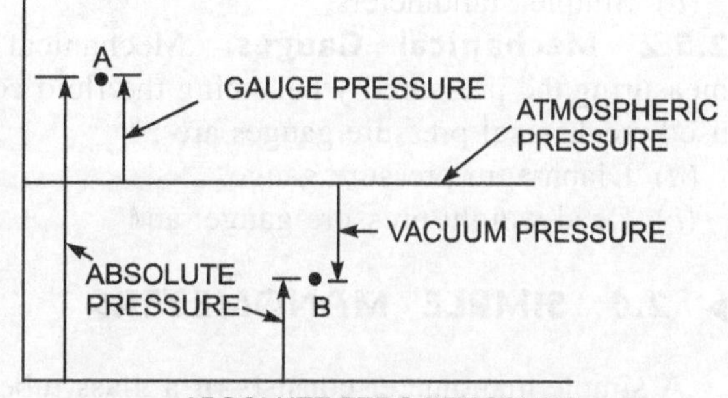

Fig. 2.7 *Relationship between pressures.*

Mathematically :

(*i*) Absolute pressure

= Atmospheric pressure + Gauge pressure

or $$p_{ab} = p_{atm} + p_{gauge}$$

(*ii*) Vacuum pressure

= Atmospheric pressure – Absolute pressure.

Note. (*i*) The atmospheric pressure at sea level at 15°C is 101.3 kN/m^2 or 10.13 N/cm^2 in SI unit. In case of MKS units, it is equal to 1.033 kgf/cm^2.

(*ii*) The atmospheric pressure head is 760 mm of mercury or 10.33 m of water.

Problem 2.8 *What are the gauge pressure and absolute pressure at a point 3 m below the free surface of a liquid having a density of 1.53 × 10^3 kg/m^3 if the atmospheric pressure is equivalent to 750 mm of mercury ? The specific gravity of mercury is 13.6 and density of water = 1000 kg/m^3.*

(A.M.I.E., Summer 1986)

Solution. Given :

Depth of liquid, $Z_1 = 3$ m

Density of liquid, $\rho_1 = 1.53 \times 10^3$ kg/m^3

Atmospheric pressure head, $Z_0 = 750$ mm of Hg

$$= \frac{750}{1000} = 0.75 \text{ m of Hg}$$

∴ Atmospheric pressure, $p_{atm} = \rho_0 \times g \times Z_0$

where ρ_0 = Density of Hg = Sp. gr. of mercury × Density of water = 13.6 × 1000 kg/m^3

and Z_0 = Pressure head in terms of mercury.

∴ $$p_{atm} = (13.6 \times 1000) \times 9.81 \times 0.75 \text{ N/m}^2 \quad (\because \; Z_0 = 0.75)$$

$$= 100062 \text{ N/m}^2$$

Pressure at a point, which is at a depth of 3 m from the free surface of the liquid is given by,

$$p = \rho_1 \times g \times Z_1$$

$$= (1.53 \times 1000) \times 9.81 \times 3 = 45028 \text{ N/m}^2$$

∴ Gauge pressure, p = **45028 N/m^2. Ans**

Now absolute pressure = Gauge pressure + Atmospheric pressure

= 45028 + 100062 = **145090 N/m^2. Ans.**

▶ 2.5 MEASUREMENT OF PRESSURE

The pressure of a fluid is measured by the following devices :

1. Manometers 2. Mechanical Gauges.

2.5.1 Manometers. Manometers are defined as the devices used for measuring the pressure at a point in a fluid by balancing the column of fluid by the same or another column of the fluid. They are classified as :

(*a*) Simple Manometers, (*b*) Differential Manometers.

2.5.2 Mechanical Gauges. Mechanical gauges are defined as the devices used for measuring the pressure by balancing the fluid column by the spring or dead weight. The commonly used mechanical pressure gauges are :

(*a*) Diaphragm pressure gauge, (*b*) Bourdon tube pressure gauge,
(*c*) Dead-weight pressure gauge, and (*d*) Bellows pressure gauge.

▶ 2.6 SIMPLE MANOMETERS

A simple manometer consists of a glass tube having one of its ends connected to a point where pressure is to be measured and other end remains open to atmosphere. Common types of simple manometers are :

1. Piezometer,
2. U-tube Manometer, and
3. Single Column Manometer.

2.6.1 Piezometer. It is the simplest form of manometer used for measuring gauge pressures. One end of this manometer is connected to the point where pressure is to be measured and other end is open to the atmosphere as shown in Fig. 2.8. The rise of liquid gives the pressure head at that point. If at a point *A*, the height of liquid say water is *h* in piezometer tube, then pressure at *A*

$$= \rho \times g \times h \ \frac{\text{N}}{\text{m}^2}.$$

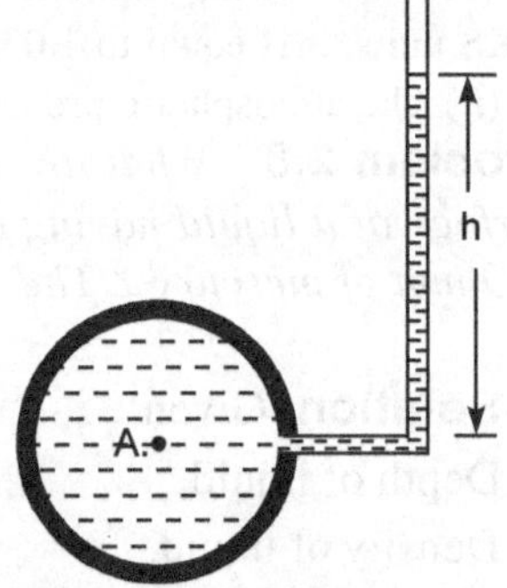

Fig. 2.8 *Piezometer.*

2.6.2 U-tube Manometer. It consists of glass tube bent in *U*-shape, one end of which is connected to a point at which pressure is to be measured and other end remains open to the atmosphere as shown in Fig. 2.9. The tube generally contains mercury or any other liquid whose specific gravity is greater than the specific gravity of the liquid whose pressure is to be measured.

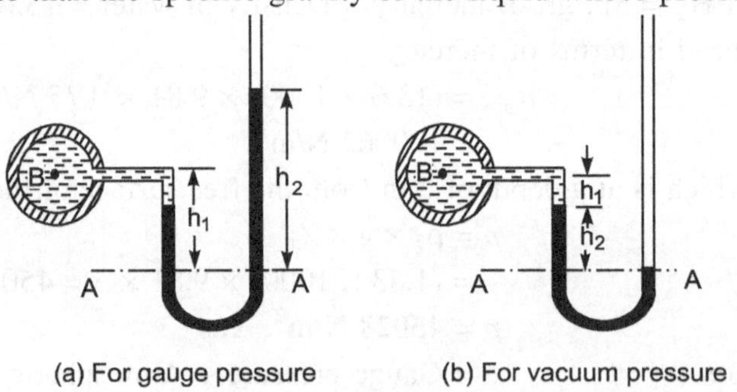

Fig. 2.9 *U-tube Manometer.*

(*a*) **For Gauge Pressure.** Let *B* is the point at which pressure is to be measured, whose value is *p*. The datum line is *A-A*.

Let h_1 = Height of light liquid above the datum line
h_2 = Height of heavy liquid above the datum line
S_1 = Sp. gr. of light liquid
ρ_1 = Density of light liquid = $1000 \times S_1$
S_2 = Sp. gr. of heavy liquid
ρ_2 = Density of heavy liquid = $1000 \times S_2$

As the pressure is the same for the horizontal surface. Hence pressure above the horizontal datum line *A-A* in the left column and in the right column of *U*-tube manometer should be same.

Pressure above *A-A* in the left column $= p + \rho_1 \times g \times h_1$

Pressure above *A-A* in the right column $= \rho_2 \times g \times h_2$

Hence equating the two pressures $p + \rho_1 g h_1 = \rho_2 g h_2$

$$\therefore \quad p = (\rho_2 g h_2 - \rho_1 \times g \times h_1). \quad ...(2.7)$$

(*b*) **For Vacuum Pressure.** For measuring vacuum pressure, the level of the heavy liquid in the manometer will be as shown in Fig. 2.9 (*b*). Then

Pressure above *A-A* in the left column $= \rho_2 g h_2 + \rho_1 g h_1 + p$

Pressure head in the right column above $A\text{-}A = 0$

$$\therefore \quad \rho_2 g h_2 + \rho_1 g h_1 + p = 0$$

$$\therefore \quad p = -(\rho_2 g h_2 + \rho_1 g h_1). \quad ...(2.8)$$

Problem 2.9 *The right limb of a simple U-tube manometer containing mercury is open to the atmosphere while the left limb is connected to a pipe in which a fluid of sp. gr. 0.9 is flowing. The centre of the pipe is 12 cm below the level of mercury in the right limb. Find the pressure of fluid in the pipe if the difference of mercury level in the two limbs is 20 cm.*

Solution. Given :

Sp. gr. of fluid, $S_1 = 0.9$

$\therefore$ Density of fluid, $\rho_1 = S_1 \times 1000 = 0.9 \times 1000 = 900 \text{ kg/m}^3$

Sp. gr. of mercury, $S_2 = 13.6$

$\therefore$ Density of mercury, $\rho_2 = 13.6 \times 1000 \text{ kg/m}^3$

Difference of mercury level $h_2 = 20 \text{ cm} = 0.2 \text{ m}$

Height of fluid from *A-A*, $h_1 = 20 - 12 = 8 \text{ cm} = 0.08 \text{ m}$

Let p = Pressure of fluid in pipe

Equating the pressure above *A-A*, we get

$$p + \rho_1 g h_1 = \rho_2 g h_2$$

or $p + 900 \times 9.81 \times 0.08 = 13.6 \times 1000 \times 9.81 \times .2$

$$p = 13.6 \times 1000 \times 9.81 \times .2 - 900 \times 9.81 \times 0.08$$

$$= 26683 - 706 = 25977 \text{ N/m}^2 = \mathbf{2.597 \text{ N/cm}^2}. \textbf{ Ans.}$$

Fig. 2.10

Problem 2.10 *A simple U-tube manometer containing mercury is connected to a pipe in which a fluid of sp. gr. 0.8 and having vacuum pressure is flowing. The other end of the manometer is open to atmosphere. Find the vacuum pressure in pipe, if the difference of mercury level in the two limbs is 40 cm and the height of fluid in the left from the centre of pipe is 15 cm below.*

Solution. Given :

Sp. gr. of fluid, $S_1 = 0.8$

Sp. gr. of mercury, $S_2 = 13.6$

Density of fluid, $\rho_1 = 800$

Density of mercury, $\rho_2 = 13.6 \times 1000$

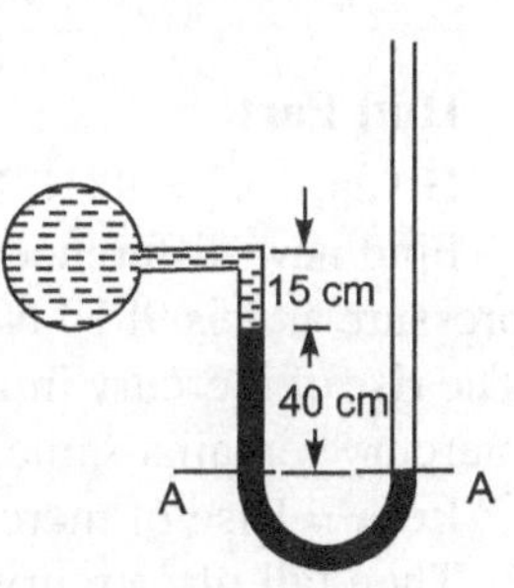

Fig. 2.11

Difference of mercury level, $h_2 = 40 \text{ cm} = 0.4 \text{ m}$. Height of liquid in left limb, $h_1 = 15 \text{ cm} = 0.15 \text{ m}$. Let the pressure in pipe $= p$. Equating pressure above datum line *A-A*, we get

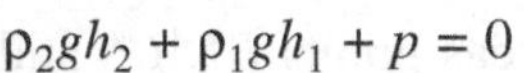

$$\rho_2 g h_2 + \rho_1 g h_1 + p = 0$$

$\therefore \quad p = -[\rho_2 g h_2 + \rho_1 g h_1]$

$= -[13.6 \times 1000 \times 9.81 \times 0.4 + 800 \times 9.81 \times 0.15]$

$= -[53366.4 + 1177.2] = -54543.6 \text{ N/m}^2 = \mathbf{-5.454 \text{ N/cm}^2}$. **Ans.**

Problem 2.11 *A U-Tube manometer is used to measure the pressure of water in a pipe line, which is in excess of atmospheric pressure. The right limb of the manometer contains mercury and is open to atmosphere. The contact between water and mercury is in the left limb. Determine the pressure of water in the main line, if the difference in level of mercury in the limbs of U-tube is 10 cm and the free surface of mercury is in level with the centre of the pipe. If the pressure of water in pipe line is reduced to 9810 N/m², calculate the new difference in the level of mercury. Sketch the arrangements in both cases.* (A.M.I.E., Winter 1989)

Solution. Given :

Difference of mercury = 10 cm = 0.1 m

The arrangement is shown in Fig. 2.11 (*a*)

Let p_A = (pressure of water in pipe line (*i.e.*, at point *A*)

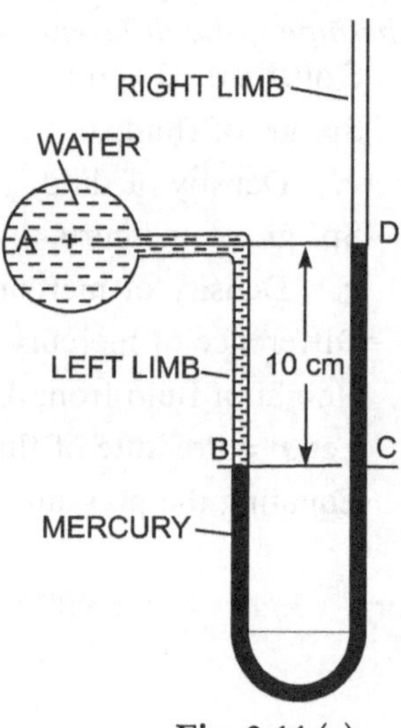

Fig. 2.11 (*a*)

The points *B* and *C* lie on the same horizontal line. Hence pressure at *B* should be equal to pressure at *C*. But pressure at *B*

= Pressure at *A* + Pressure due to 10 cm (or 0.1 m) of water

$= p_A + \rho \times g \times h$

where $\rho = 1000 \text{ kg/m}^3$ and $h = 0.1$ m

$= p_A + 1000 \times 9.81 \times 0.1$

$= p_A + 981 \text{ N/m}^2$ ...(*i*)

Pressure at *C* = Pressure at *D* + Pressure due to 10 cm of mercury

$= 0 + \rho_0 \times g \times h_0$

where ρ_0 for mercury $= 13.6 \times 1000 \text{ kg/m}^3$

and $h_0 = 10 \text{ cm} = 0.1 \text{ m}$

$\therefore$ Pressure at $C = 0 + (13.6 \times 1000) \times 9.81 \times 0.1$

$= 13341.6$ N ...(*ii*)

But pressure at *B* is equal to pressure at *C*. Hence equating the equations (*i*) and (*ii*), we get

$$p_A + 981 = 13341.6$$

$$\therefore \quad p_A = 13341.6 - 981$$

$$= \mathbf{12360.6 \frac{N}{m^2}}\text{. } \textbf{Ans.}$$

IInd Part

Given, $p_A = 9810 \text{ N/m}^2$

Find new difference of mercury level. The arrangement is shown in Fig. 2.11 (*b*). In this case the pressure at *A* is 9810 N/m² which is less than the 12360.6 N/m². Hence mercury in left limb will rise. The rise of mercury in left limb will be equal to the fall of mercury in right limb as the total volume of mercury remains same.

Let *x* = Rise of mercury in left limb in cm

Then fall of mercury in right limb = *x* cm

The points *B*, *C* and *D* show the initial conditions whereas points *B**, *C** and *D** show the final conditions.

The pressure at B^* = Pressure at C^*

or Pressure at A + Pressure due to $(10 - x)$ cm of water

= Pressure at D^* + Pressure due to $(10 - 2x)$ cm of mercury

or $p_A + \rho_1 \times g \times h_1 = p_D{}^* + \rho_2 \times g \times h_2$

or $1910 + 1000 \times 9.81 \times \left(\dfrac{10 - x}{100}\right)$

$= 0 + (13.6 \times 1000) \times 9.81 \times \left(\dfrac{10 - 2x}{100}\right)$

Fig. 2.11 (*b*)

Dividing by 9.81, we get

or $1000 + 100 - 10x = 1360 - 272x$

or $272x - 10x = 1360 - 1100$

or $262x = 260$

$\therefore \quad x = \dfrac{260}{262} = 0.992$ cm

$\therefore$ New difference of mercury $= 10 - 2x$ cm $= 10 - 2 \times 0.992$

$=$ **8.016 cm. Ans.**

Problem 2.12 *Fig. 2.12 shows a conical vessel having its outlet at A to which a U-tube manometer is connected. The reading of the manometer given in the figure shows when the vessel is empty. Find the reading of the manometer when the vessel is completely filled with water.* (A.M.I.E., Winter 1975)

Solution. Vessel is empty. Given :

Difference of mercury level $h_2 = 20$ cm

Let h_1 = Height of water above X-X

Sp. gr. of mercury, $S_2 = 13.6$

Sp. gr. of water, $S_1 = 1.0$

Density of mercury, $\rho_2 = 13.6 \times 1000$

Density of water, $\rho_1 = 1000$

Equating the pressure above datum line X-X, we have

$$\rho_2 \times g \times h_2 = \rho_1 \times g \times h_1$$

or $13.6 \times 1000 \times 9.81 \times 0.2 = 1000 \times 9.81 \times h_1$

$h_1 = 2.72$ m of water.

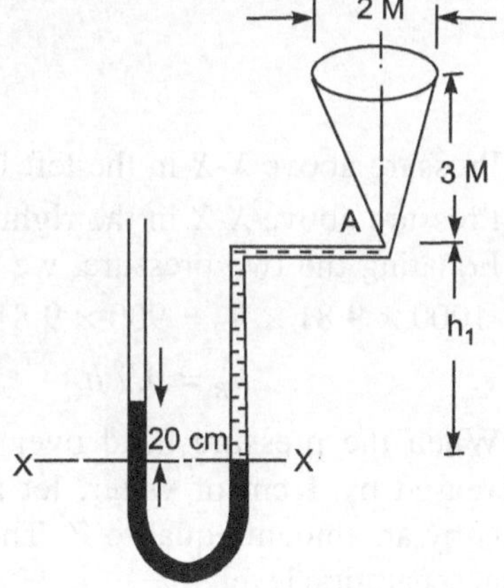

Fig. 2.12

Vessel is full of water. When vessel is full of water, the pressure in the right limb will increase and mercury level in the right limb will go down. Let the distance through which mercury goes down in the right limb be, y cm as shown in Fig. 2.13. The mercury will rise in the left by a distance of y cm. Now the datum line is Z-Z. Equating the pressure above the datum line Z-Z.

Pressure in left limb = Pressure in right limb

$13.6 \times 1000 \times 9.81 \times (0.2 + 2y/100)$

$= 1000 \times 9.81 \times (3 + h_1 + y/100)$

or $\quad 13.6 \times (0.2 + 2y/100) = (3 + 2.72 + y/100) \quad (\because \quad h_1 = 2.72 \text{ cm})$

or $\quad 2.72 + 27.2y/100 = 3 + 2.72 + y/100$

or $\quad (27.2y - y)/100 = 3.0$

or $\quad 26.2y = 3 \times 100 = 300$

$$\therefore \quad y = \frac{300}{26.2} = 11.45 \text{ cm}$$

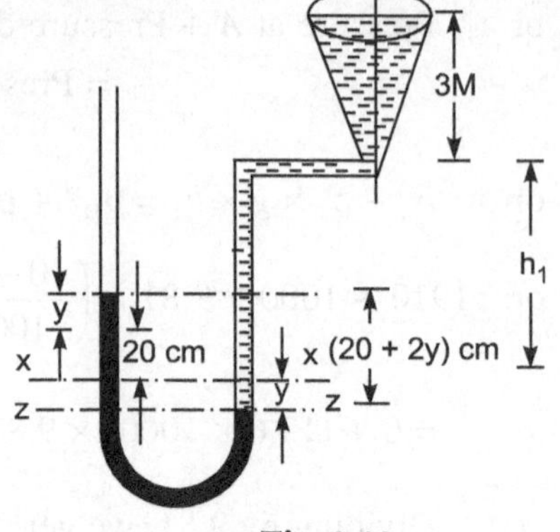

The difference of mercury level in two limbs

$= (20 + 2y)$ cm of mercury

$= 20 + 2 \times 11.45 = 20 + 22.90$

$= 42.90$ cm of mercury

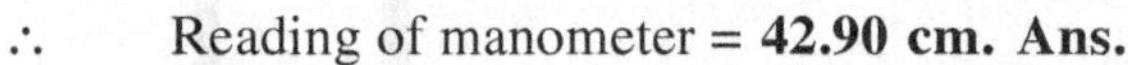

$\therefore$ Reading of manometer = **42.90 cm. Ans.**

Fig. 2.13

Problem 2.13 *A pressure gauge consists of two cylindrical bulbs B and C each of 10 sq. cm cross-sectional area, which are connected by a U-tube with vertical limbs each of 0.25 sq. cm cross-sectional area. A red liquid of specific gravity 0.9 is filled into C and clear water is filled into B, the surface of separation being in the limb attached to C. Find the displacement of the surface of separation when the pressure on the surface in C is greater than that in B by an amount equal to 1 cm head of water.* (A.M.I.E., Summer, 1978)

Solution. Given :

Area of each bulb B and C, $\quad A = 10 \text{ cm}^2$

Area of each vertical limb, $\quad a = 0.25 \text{ cm}^2$

Sp. gr. of red liquid $= 0.9 \quad \therefore$ Its density $= 900 \text{ kg/m}^3$

Let $\quad X\text{-}X$ = Initial separation level

h_C = Height of red liquid above $X\text{-}X$

h_B = Height of water above $X\text{-}X$

Pressure above $X\text{-}X$ in the left limb $= 1000 \times 9.81 \times h_B$

Pressure above $X\text{-}X$ in the right limb $= 900 \times 9.81 \times h_C$

Equating the two pressure, we get

$$1000 \times 9.81 \times h_B = 900 \times 9.81 \times h_C$$

$$\therefore \quad h_B = 0.9\, h_C \quad \ldots(i)$$

When the pressure head over the surface in C is increased by 1 cm of water, let the separation level falls by an amount equal to Z. Then $Y\text{-}Y$ becomes the final separation level.

Now fall in surface level of C multiplied by cross-sectional area of bulb C must be equal to the fall in separation level multiplied by cross-sectional area of limb.

$\therefore$ Fall in surface level of C

$$= \frac{\text{Fall in separation level} \times a}{A}$$

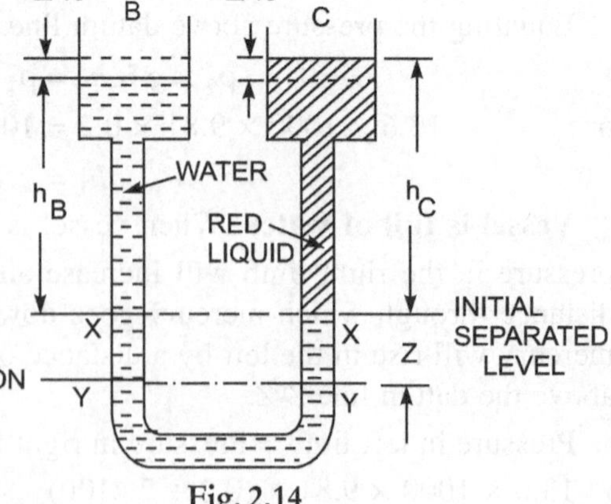

Fig. 2.14

$$= \frac{Z \times a}{A} = \frac{Z \times 0.25}{10} = \frac{Z}{40}.$$

Also fall in surface level of C

$$= \text{Rise in surface level of } B$$

$$= \frac{Z}{40}$$

The pressure of 1 cm (or 0.01 m) of water = ρgh = 1000 × 9.81 × 0.01= 98.1 N/m^2

Consider final separation level Y-Y

Pressure above Y-Y in the left limb $= 1000 \times 9.81 \left(Z + h_B + \frac{Z}{40} \right)$

Pressure above Y-Y in the right limb $= 900 \times 9.81 \left(Z + h_C - \frac{Z}{40} \right) + 98.1$

Equating the two pressure, we get

$$1000 \times 9.81 \left(Z + h_B + \frac{Z}{40} \right) = \left(Z + h_C - \frac{Z}{40} \right) 900 \times 9.81 + 98.1$$

Dividing by 9.81, we get

$$1000 \left(Z + h_B + \frac{Z}{40} \right) = 900 \left(Z + h_C - \frac{Z}{40} \right) + 10$$

Dividing by 1000, we get $Z + h_B + \frac{Z}{40} = 0.9 \left(Z + h_C - \frac{Z}{40} \right) + 0.01$

But from equation (*i*), $h_B = 0.9\, h_C$

$$\therefore \quad Z + 0.9\, h_C + \frac{Z}{40} = \frac{39Z}{40} \times 0.9 + 0.9\, h_C + 0.01$$

or

$$\frac{41Z}{40} = \frac{39}{40} \times .9Z + .01$$

or

$$Z\left(\frac{41}{40} - \frac{39 \times .9}{40} \right) = .01 \quad \text{or} \quad Z\left(\frac{41 - 35.1}{40} \right) = .01$$

$$\therefore \quad Z = \frac{40 \times 0.01}{5.9} = \mathbf{0.0678\ m = 6.78\ cm.\ Ans.}$$

2.6.3 Single Column Manometer. Single column manomter is a modified form of a *U*-tube manometer in which a reservoir, having a large cross-sectional area (about 100 times) as compared to the area of the tube is connected to one of the limbs (say left limb) of the manometer as shown in Fig. 2.15. Due to large cross-sectional area of the reservoir, for any variation in pressure, the change in the liquid level in the reservoir will be very small which may be neglected and hence the pressure is given by the height of liquid in the other limb. The other limb may be vertical or inclined. Thus there are two types of single column manometer as :

1. Vertical Single Column Manometer.
2. Inclined Single Column Manometer.

1. Vertical Single Column Manometer

Fig. 2.15 shows the vertical single column manometer. Let *X-X* be the datum line in the reservoir and in the right limb of the manometer, when it is not connected to the pipe. When the manometer is connected to the pipe, due to high pressure at A, the heavy liquid in the reservoir will be pushed downward and will rise in the right limb.

Let Δh = Fall of heavy liquid in reservoir

h_2 = Rise of heavy liquid in right limb

h_1 = Height of centre of pipe above X-X

p_A = Pressure at A, which is to be measured

A = Cross-sectional area of the reservoir

a = Cross-sectional area of the right limb

S_1 = Sp. gr. of liquid in pipe

S_2 = Sp. gr. of heavy liquid in reservoir and right limb

ρ_1 = Density of liquid in pipe

ρ_2 = Density of liquid in reservoir

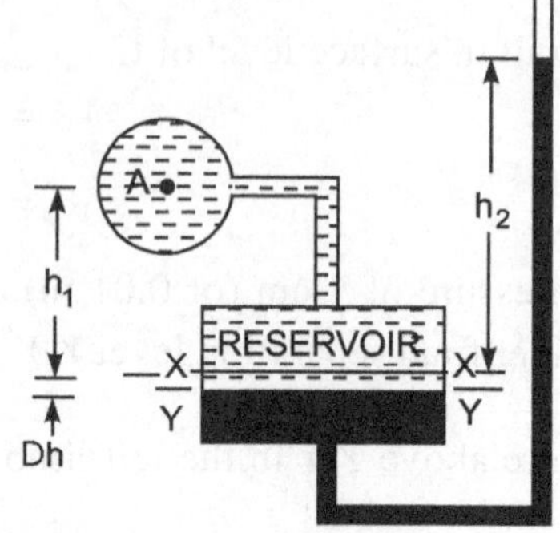

Fig. 2.15 *Vertical single column manometer.*

Fall of heavy liquid in reservoir will cause a rise of heavy liquid level in the right limb.

$$\therefore \quad A \times \Delta h = a \times h_2$$

$$\therefore \quad \Delta h = \frac{a \times h_2}{A} \qquad \ldots(i)$$

Now consider the datum line Y-Y as shown in Fig. 2.15. Then pressure in the right limb above Y-Y.

$$= \rho_2 \times g \times (\Delta h + h_2)$$

Pressure in the left limb above Y-Y = $\rho_1 \times g \times (\Delta h + h_1) + p_A$

Equating these pressures, we have

$$\rho_2 \times g \times (\Delta h + h_2) = \rho_1 \times g \times (\Delta h + h_1) + p_A$$

or

$$p_A = \rho_2 g\,(\Delta h + h_2) - \rho_1 g(\Delta h + h_1)$$

$$= \Delta h[\rho_2 g - \rho_1 g] + h_2\rho_2 g - h_1\rho_1 g$$

But from equation (*i*), $\quad \Delta h = \dfrac{a \times h_2}{A}$

$$\therefore \quad p_A = \frac{a \times h_2}{A}\,[\rho_2 g - \rho_1 g] + h_2\rho_2 g - h_1\rho_1 g \qquad \ldots(2.9)$$

As the area A is very large as compared to a, hence ratio $\dfrac{a}{A}$ becomes very small and can be neglected.

Then $p_A = h_2\rho_2 g - h_1\rho_1 g \qquad \ldots(2.10)$

From equation (2.10), it is clear that as h_1 is known and hence by knowing h_2 or rise of heavy liquid in the right limb, the pressure at A can be calculated.

2. Inclined Single Column Manometer

Fig. 2.16 shows the inclined single column manometer. This manometer is more sensitive. Due to inclination the distance moved by the heavy liquid in the right limb will be more.

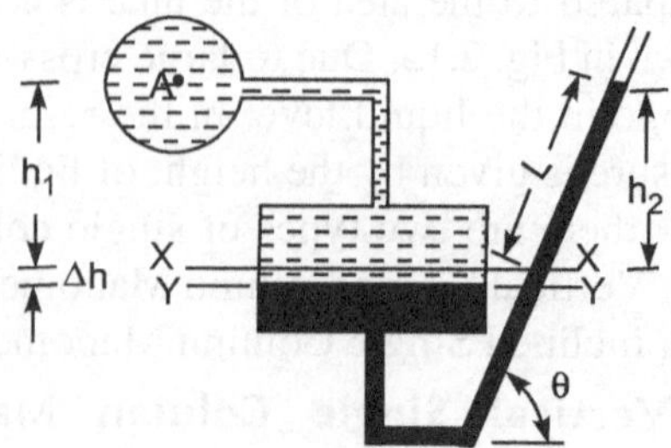

Fig. 2.16 *Inclined single column manometer.*

Let L = Length of heavy liquid moved in right limb from X-X

θ = Inclination of right limb with horizontal

h_2 = Vertical rise of heavy liquid in right limb from X-$X = L \times \sin\theta$

From equation (2.10), the pressure at A is

$$p_A = h_2 \rho_2 g - h_1 \rho_1 g.$$

Substituting the value of h_2, we get

$$p_A = \sin\theta \times \rho_2 g - h_1 \rho_1 g. \quad \text{...(2.11)}$$

Problem 2.14 *A single column manometer is connected to a pipe containing a liquid of sp. gr. 0.9 as shown in Fig. 2.17. Find the pressure in the pipe if the area of the reservoir is 100 times the area of the tube for the manometer reading shown in Fig. 2.17. The specific gravity of mercury is 13.6.*

Solution. Given :

Sp. gr. of liquid in pipe, $S_1 = 0.9$

$\therefore$ Density $\rho_1 = 900 \text{ kg/m}^3$

Sp. gr. of heavy liquid, $S_2 = 13.6$

Density, $\rho_2 = 13.6 \times 1000$

$$\frac{\text{Area of reservoir}}{\text{Area of right limb}} = \frac{A}{a} = 100$$

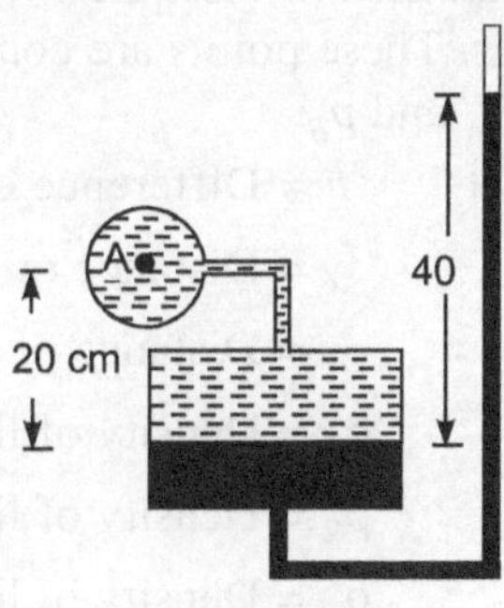

Fig. 2.17

Height of liquid, $h_1 = 20 \text{ cm} = 0.2 \text{ m}$

Rise of mercury in right limb,

$h_2 = 40 \text{ cm} = 0.4 \text{ m}$

Let p_A = Pressure in pipe

Using equation (2.9), we get

$$p_A = \frac{a}{A} h_2[\rho_2 g - \rho_1 g] + h_2 \rho_2 g - h_1 \rho_1 g$$

$$= \frac{1}{100} \times 0.4[13.6 \times 1000 \times 9.81 - 900 \times 9.81] + 0.4 \times 13.6 \times 1000 \times 9.81 - 0.2 \times 900 \times 9.81$$

$$= \frac{0.4}{100} [133416 - 8829] + 53366.4 - 1765.8$$

$= 533.664 + 53366.4 - 1765.8 \text{ N/m}^2 = 52134 \text{ N/m}^2 =$ **5.21 N/cm². Ans.**

▶ 2.7 DIFFERENTIAL MANOMETERS

Differential manometers are the devices used for measuring the difference of pressures between two points in a pipe or in two different pipes. A differential manometer consists of a *U*-tube, containing a heavy liquid, whose two ends are connected to the points, whose difference of pressure is to be measured. Most commonly types of differential manometers are :

1. *U*-tube differential manometer and
2. Inverted *U*-tube differential manometer.

2.7.1 U-tube Differential Manometer. Fig. 2.18 shows the differential manometers of *U*-tube type.

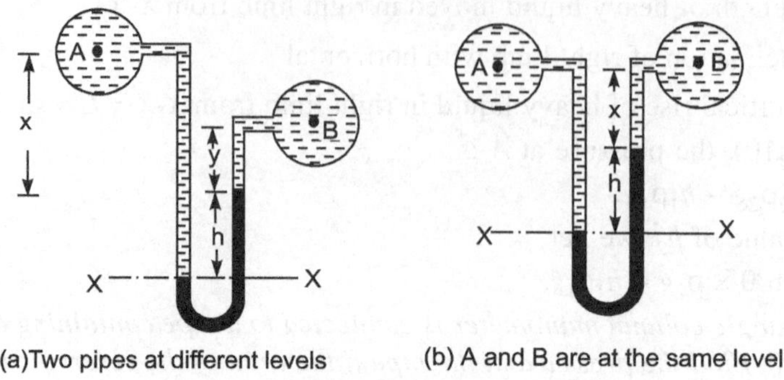

(a)Two pipes at different levels (b) A and B are at the same level

Fig. 2.18 *U-tube differential manometers.*

Fig. 2.18 (*a*). Let the two points *A* and *B* are at different level and also contains liquids of different sp. gr. These points are connected to the *U*-tube differential manometer. Let the pressure at *A* and *B* are p_A and p_B.

Let h = Difference of mercury level in the *U*-tube.

y = Distance of the centre of *B*, from the mercury level in the right limb.

x = Distance of the centre of *A*, from the mercury level in the right limb.

ρ_1 = Density of liquid at *A*.

ρ_2 = Density of liquid at *B*.

ρ_g = Density of heavy liquid or mercury.

Taking datum line at *X-X*.

Pressure above *X-X* in the left limb $= \rho_1 g(h + x) + p_A$

where p_A = pressure at *A*.

Pressure above *X-X* in the right limb $= \rho_g \times g \times h + \rho_2 \times g \times y + p_B$

where p_B = Pressure at *B*.

Equating the two pressure, we have

$$\rho_1 g(h + x) + p_A = \rho_g \times g \times h + \rho_2 g y + p_B$$

$$\therefore \quad p_A - p_B = \rho_g \times g \times h + \rho_2 g y - \rho_1 g(h + x)$$

$$= h \times g(\rho_g - \rho_1) + \rho_2 g y - \rho_1 g x \qquad \text{...(2.12)}$$

$\therefore$ Difference of pressure at *A* and $B = h \times g(\rho_g - \rho_1) + \rho_2 g y - \rho_1 g x$

Fig. 2.18 (*b*). *A* and *B* are at the same level and contains the same liquid of density ρ_1. Then

Pressure above *X-X* in right limb $= \rho_g \times g \times h + \rho_1 \times g \times x + p_B$

Pressure above *X-X* in left limb $= \rho_1 \times g \times (h + x) + p_A$

Equating the two pressure

$$\rho_g \times g \times h + \rho_1 g x + p_B = \rho_1 \times g \times (h + x) + p_A$$

$$\therefore \quad p_A - p_B = \rho_g \times g \times h + \rho_1 g x - \rho_1 g(h + x)$$

$$= g \times h(\rho_g - \rho_1). \qquad \text{...(2.13)}$$

Problem 2.15 *A pipe contains an oil of sp. gr. 0.9. A differential manometer connected at the two points A and B shows a difference in mercury level as 15 cm. Find the difference of pressure at the two points.*

Solution. Given :

Sp. gr. of oil, $S_1 = 0.9$ ∴ Density, $\rho_1 = 0.9 \times 1000 = 900$ kg/m^3

Difference in mercury level, $h = 15$ cm $= 0.15$ m

Sp. gr. of mercury, $S_g = 13.6$ ∴ Density, $\rho_g = 13.6 \times 1000$ kg/m^3

The difference of pressure is given by equation (2.13)

or $$p_A - p_B = g \times h(\rho_g - \rho_1)$$

$$= 9.81 \times 0.15\ (13600 - 900) = \mathbf{18688\ N/m^2.\ Ans.}$$

Problem 2.16 *A differential manometer is connected at the two points A and B of two pipes as shown in Fig. 2.19. The pipe A contains a liquid of sp. gr. = 1.5 while pipe B contains a liquid of sp. gr. = 0.9. The pressures at A and B are 1 kgf/cm² and 1.80 kgf/cm² respectively. Find the difference in mercury level in the differential manometer.*

Fig. 2.19

Solution. Given :

Sp. gr. of liquid at A, $S_1 = 1.5$ ∴ $\rho_1 = 1500$

Sp. gr. of liquid at B, $S_2 = 0.9$ ∴ $\rho_2 = 900$

Pressure at A, $p_A = 1$ kgf/cm^2 $= 1 \times 10^4$ kgf/m^2

$= 10^4 \times 9.81$ N/m^2 (∵ 1 kgf = 9.81 N)

Pressure at B, $p_B = 1.8$ kgf/cm^2

$= 1.8 \times 10^4$ kgf/m^2

$= 1.8 \times 10^4 \times 9.81$ N/m^2 (∵ 1 kgf = 9.81 N)

Density of mercury $= 13.6 \times 1000$ kg/m^3

Taking X-X as datum line.

Pressure above X-X in the left limb

$$= 13.6 \times 1000 \times 9.81 \times h + 1500 \times 9.81 \times (2 + 3) + p_A$$

$$= 13.6 \times 1000 \times 9.81 \times h + 7500 \times 9.81 + 9.81 \times 10^4$$

Pressure above X-X in the right limb $= 900 \times 9.81 \times (h + 2) + p_B$

$$= 900 \times 9.81 \times (h + 2) + 1.8 \times 10^4 \times 9.81$$

Equating the two pressure, we get

$$13.6 \times 1000 \times 9.81h + 7500 \times 9.81 + 9.81 \times 10^4$$

$$= 900 \times 9.81 \times (h + 2) + 1.8 \times 10^4 \times 9.81$$

Dividing by 1000×9.81, we get

$$13.6h + 7.5 + 10 = (h + 2.0) \times .9 + 18$$

or $$13.6h + 17.5 = 0.9h + 1.8 + 18 = 0.9h + 19.8$$

or $$(13.6 - 0.9)h = 19.8 - 17.5 \text{ or } 12.7h = 2.3$$

∴ $$h = \frac{2.3}{12.7} = 0.181 \text{ m} = \mathbf{18.1\ cm.\ Ans.}$$

Problem 2.17 *A differential manometer is connected at the two points A and B as shown in Fig. 2.20. At B air pressure is 9.81 N/cm² (abs), find the absolute pressure at A.*

Solution. Given :

Air pressure at $B = 9.81$ N/cm^2

or $p_B = 9.81 \times 10^4$ N/m^2

Density of oil $= 0.9 \times 1000 = 900 \text{ kg/m}^3$

Density of mercury $= 13.6 \times 1000 \text{ kg/m}^3$

Let the pressure at A is p_A

Taking datum line at X-X

Pressure above X-X in the right limb

$$= 1000 \times 9.81 \times 0.6 + p_B$$
$$= 5886 + 98100 = 103986$$

Pressure above X-X in the left limb

$$= 13.6 \times 1000 \times 9.81 \times 0.1 + 900 \times 9.81 \times 0.2 + p_A$$
$$= 13341.6 + 1765.8 + p_A$$

Equating the two pressure head

$$103986 = 13341.6 + 1765.8 + p_A$$

$$\therefore \quad p_A = 103986 - 15107.4 = 88876.8$$

$$\therefore \quad p_A = 88876.8 \text{ N/m}^2 = \frac{88876.8 \text{N}}{10000 \text{ cm}^2} = 8.887 \frac{\text{N}}{\text{cm}^2}.$$

$\therefore$ Absolute pressure at A = **8.887 N/cm². Ans.**

Fig. 2.20

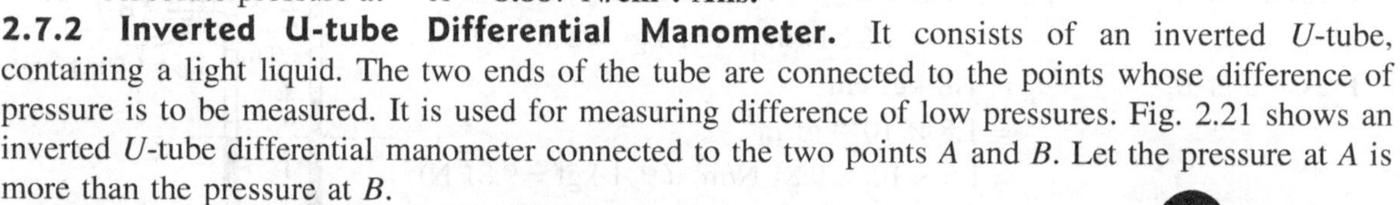

2.7.2 Inverted U-tube Differential Manometer. It consists of an inverted U-tube, containing a light liquid. The two ends of the tube are connected to the points whose difference of pressure is to be measured. It is used for measuring difference of low pressures. Fig. 2.21 shows an inverted U-tube differential manometer connected to the two points A and B. Let the pressure at A is more than the pressure at B.

Let h_1 = Height of liquid in left limb below the datum line X-X

h_2 = Height of liquid in right limb

h = Difference of light liquid

ρ_1 = Density of liquid at A

ρ_2 = Density of liquid at B

ρ_s = Density of light liquid

p_A = Pressure at A

p_B = Pressure at B.

Taking X-X as datum line. Then pressure in the left limb below X-X

$$= p_A - \rho_1 \times g \times h_1.$$

Fig. 2.21

Pressure in the right limb below X-X

$$= p_B - \rho_2 \times g \times h_2 - \rho_s \times g \times h$$

Equating the two pressure

$$p_A - \rho_1 \times g \times h_1 = p_B - \rho_2 \times g \times h_2 - \rho_s \times g \times h$$

or
$$p_A - p_B = \rho_1 \times g \times h_1 - \rho_2 \times g \times h_2 - \rho_s \times g \times h. \quad \text{...(2.14)}$$

Problem 2.18 *Water is flowing through two different pipes to which an inverted differential manometer having an oil of sp. gr. 0.8 is connected. The pressure head in the pipe A is 2 m of water, find the pressure in the pipe B for the manometer readings as shown in Fig. 2.22.*

Solution. Given :

Pressure head at $A = \frac{p_A}{\rho g} = 2$ m of water

$$\therefore \quad p_A = \rho \times g \times 2 = 1000 \times 9.81 \times 2 = 19620 \text{ N/m}^2$$

Fig. 2.22 shows the arrangement. Taking X-X as datum line.

Pressure below X-X in the left limb $= p_A - \rho_1 \times g \times h_1$

$= 19620 - 1000 \times 9.81 \times 0.3 = 16677$ N/m^2.

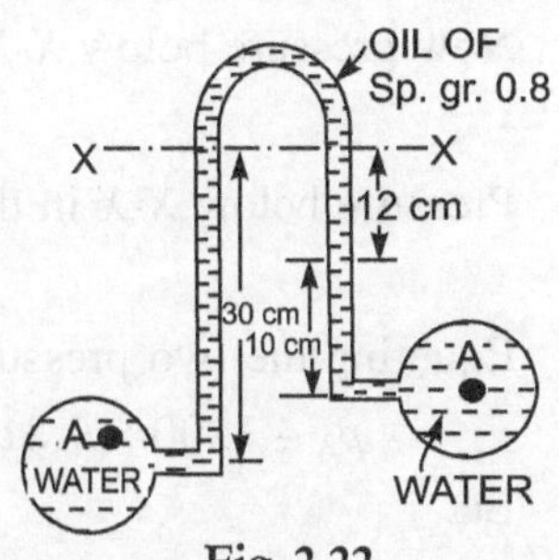

Fig. 2.22

Pressure below *X-X* in the right limb

$$= p_B - 1000 \times 9.81 \times 0.1 - 800 \times 9.81 \times 0.12$$
$$= p_B - 981 - 941.76 = p_B - 1922.76$$

Equating the two pressure, we get

$$16677 = p_B - 1922.76$$

or $$p_B = 16677 + 1922.76 = 18599.76 \text{ N/m}^2$$

or $$p_B = \mathbf{1.8599 \text{ N/cm}^2.\ Ans.}$$

Problem 2.19 *In Fig. 2.23, an inverted differential manometer is connected to two pipes A and B which convey water. The fluid in manometer is oil of sp. gr. 0.8. For the manometer readings shown in the figure, find the pressure difference between A and B.*

Solution. Given :

Sp. gr. of oil $= 0.8 \quad \therefore \quad \rho_s = 800$ kg/m^3

Difference of oil in the two limbs

$$= (30 + 20) - 30 = 20 \text{ cm}$$

Taking datum line at *X-X*

Pressure in the left limb below *X-X*

$$= p_A - 1000 \times 9.81 \times 0$$
$$= p_A - 2943$$

Fig. 2.23

Pressure in the right limb below *X-X*

$$= p_B - 1000 \times 9.81 \times 0.3 - 800 \times 9.81 \times 0.2$$
$$= p_B - 2943 - 1569.6 = p_B - 4512.6$$

Equating the two pressure $p_A - 2943 = p_B - 4512.6$

$$\therefore \quad p_B - p_A = 4512.6 - 2943 = \mathbf{1569.6 \text{ N/m}^2.\ Ans.}$$

Problem 2.20 *Find out the differential reading 'h' of an inverted U-tube manometer containing oil of specific gravity 0.7 as the manometric fluid when connected across pipes A and B as shown in Fig. 2.24 below, conveying liquids of specific gravities 1.2 and 1.0 and immiscible with manometric fluid. Pipes A and B are located at the same level and assume the pressures at A and B to be equal.*

(A.M.I.E., Winter, 1985)

Solution. Given :

Fig. 2.24 shows the arrangement. Taking *X-X* as datum line.

Let p_A = Pressure at *A*

p_B = Pressure at *B*

Density of liquid in pipe *A* = Sp. gr. × 1000

$= 1.2 \times 1000$

$= 1200$ kg/m^2

Density of liquid in pipe *B* $= 1 \times 1000 = 1000$ kg/m^3

Density of oil $= 0.7 \times 1000 = 700$ kg/m^3

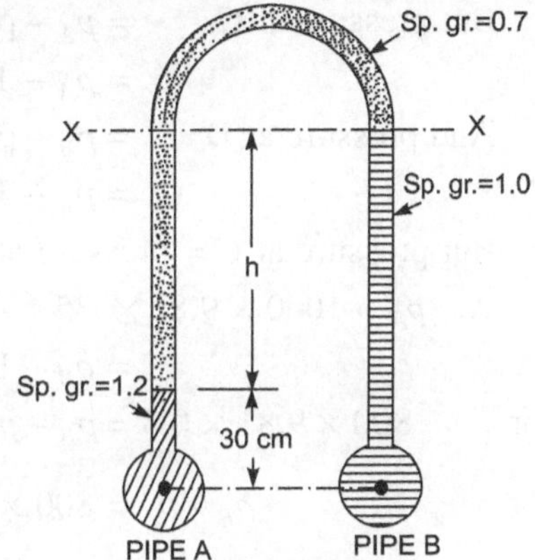

Fig. 2.24

Now pressure below X-X in the left limb

$$= p_A - 1200 \times 9.81 \times 0.3 - 700 \times 9.81 \times h$$

Pressure below X-X in the right limb

$$= p_B - 1000 \times 9.81 \times (h + 0.3)$$

Equating the two pressure, we get

$$p_A - 1200 \times 9.81 \times 0.3 - 700 \times 9.81 \times h = p_B - 1000 \times 9.81\ (h + 0.3)$$

But $\quad p_A = p_B \quad$ (given)

$$\therefore \quad -1200 \times 9.81 \times 0.3 - 700 \times 9.81 \times h = -1000 \times 9.81\ (h + 0.3)$$

Dividing by 1000×9.81

$$-1.2 \times 0.3 - 0.7h = -(h + 0.3)$$

or $\quad 0.3 \times 1.2 + 0.7h = h + 0.3$ or $0.36 - 0.3 = h - 0.7h = 0.3h$

$$\therefore \quad h = \frac{0.36 - 0.30}{0.30} = \frac{0.06}{0.30}\text{ m}$$

$$= \frac{1}{5}\text{ m} = \frac{1}{5} \times 100 = \textbf{20 cm. Ans.}$$

Problem 2.21 *An inverted U-tube manometer is connected to two horizontal pipes A and B through which water is flowing. The vertical distance between the axes of these pipes is 30 cm. When an oil of specific gravity 0.8 is used as a gauge fluid, the vertical heights of water columns in the two limbs of the inverted manometer (when measured from tlhe respective centre lines of the pipes) are found to be same and equal to 35 cm. Determine the difference of pressure between the pipes.*

(A.M.I.E., Summer 1990)

Solution. Given :

Specific gravity of measuring liquid = 0.8

The arrangement is shown in Fig. 2.24 (*a*).

Let $\quad p_A$ = pressure at A

$\quad p_B$ = pressure at B.

The points C and D lie on the same horizontal line.

Hence pressure at C should be equal to pressure at D.

Fig. 2.24 (*a*)

But pressure at C $\quad = p_A - \rho g h$

$$= p_A - 1000 \times 9.81 \times (0.35)$$

And pressure at D $\quad = p_B - \rho_1 g h_1 - \rho_2 g h_2$

$$= p_B - 1000 \times 9.81 \times (0.35) - 800 \times 9.81 \times 0.3$$

But pressure at C = pressure at D

$$\therefore \quad p_A - 1000 \times 9.81 \times .35 = p_B - 1000 \times 9.81 \times 0.35 - 800 \times 9.81 \times 0.3$$

or $\quad 800 \times 9.81 \times 0.3 = p_B - p_A$

or $\quad p_B - p_A = 800 \times 9.81 \times 0.3 = \textbf{2354.4}\ \mathbf{\frac{N}{m^2}}$**. Ans.**

▶ 2.8 PRESSURE AT A POINT IN COMPRESSIBLE FLUID

For compressible fluids, density (ρ) changes with the change of pressure and temperature. Such problems are encountered in aeronautics, oceanography and meteorology where we are concerned with atmospheric* air where density, pressure and temperature changes with elevation. Thus for fluids with variable density, equation (2.4) cannot be integrated, unless the relationship between p and ρ is known. For gases the equation of state is

$$\frac{p}{\rho} = RT \qquad \text{...(2.15)}$$

or

$$\rho = \frac{p}{RT}$$

Now equation (2.4) is

$$\frac{dp}{dZ} = w = \rho g = \frac{p}{RT} \times g$$

$$\therefore \qquad \frac{dp}{p} = \frac{g}{RT} dZ \qquad \text{...(2.16)}$$

In equation (2.4), Z is measured vertically downward. But if Z is measured vertically up, then $\frac{dp}{dZ} = -\rho g$ and hence equation (2.16) becomes

$$\frac{dp}{p} = \frac{-g}{RT} dZ \qquad \text{...(2.17)}$$

2.8.1 Isothermal Process. **Case I.** If temperature T is constant which is true for **isothermal process**, equation (2.17) can be integrated as

$$\int_{p_0}^{p} \frac{dp}{p} = -\int_{Z_0}^{Z} \frac{g}{RT} dz = -\frac{g}{RT}\int_{Z_0}^{Z} dz$$

or

$$\log \frac{p}{p_0} = \frac{-g}{RT} [Z - Z_0]$$

where p_0 is the pressure where height is Z_0. If the datum line is taken at Z_0, then $Z_0 = 0$ and p_0 becomes the pressure at datum line.

$$\therefore \qquad \log \frac{p}{p_0} = \frac{-g}{RT} Z$$

$$\frac{p}{p_0} = e^{-gZ/RT}$$

or pressure at a height Z is given by $p = p_0 e^{-gZ/RT}$ ...(2.18)

2.8.2 Adiabatic Process. If temperature T is not constant but the process follows adiabatic law then the relation between pressure and density is given by

$$\frac{p}{\rho^k} = \text{Constant} = C \qquad \text{...}(i)$$

---

* The standard atmospheric pressure, temperature and density referred to STP at the sea-level are : Pressure = 101.325 kN/m^2 ; Temperature = 15°C and Density = 1.225 kg/m^3.

where k is ratio of specific constant.

$$\therefore \qquad \rho^k = \frac{p}{C}$$

or
$$\rho = \left(\frac{p}{C}\right) \qquad \text{...(ii)}$$

Then equation (2.4) for Z measured vertically up becomes,

$$\frac{dp}{dZ} = -\rho g = -\left(\frac{p}{C}\right)^{1/k} g$$

or
$$\frac{dp}{\left(\frac{p}{C}\right)^{1/k}} = -gdZ \text{ or } C^{1/k}\frac{dp}{\frac{1}{p^k}} = -gdZ$$

Integrating, we get $\int_{p_0}^{p} C^{1/k}\ p^{-1/k}\ dp = \int_{Z_0}^{Z} -gdZ$

or
$$C^{1/k}\left[\frac{p^{-1/k+1}}{-\frac{1}{k}+1}\right]_{p_0}^{p} = -g\ [Z]_{Z_0}^{Z}$$

or
$$\left[\frac{C^{1/k}\ p^{-1/k+1}}{-\frac{1}{k}+1}\right]_{p_0}^{p} = -g\ [Z]_{Z_0}^{Z} \qquad \text{[}C\text{ is a constant, can be taken inside]}$$

But from equation (i), $\qquad C^{1/k} = \left(\frac{p}{\rho^k}\right)^{1/k} = \frac{p^{1/k}}{\rho}$

Substituting this value of $C^{1/k}$ above, we get

$$\left[\frac{p^{1/k}}{\rho}\frac{p^{-1/k+1}}{-\frac{1}{k}+1}\right]_{p_0}^{p} = -g[Z-Z_0]$$

or
$$\left[\frac{p^{\frac{1-1+k}{k}}}{\rho^{\frac{k-1}{k}}}\right]_{p_0}^{p} = -g[Z-Z_0] \text{ or } \left[\frac{k}{k-1}\frac{p}{\rho}\right]_{p_0}^{p} = -g[Z-Z_0]$$

or
$$\frac{k}{k-1}\left[\frac{p}{\rho}-\frac{p_0}{\rho_0}\right] = -g[Z-Z_0]$$

If datum line is taken at Z_0, where pressure, temperature and density are p_0, T_0 and ρ_0, then $Z_0 = 0$.

$$\therefore \qquad \frac{k}{k-1}\left[\frac{p}{\rho}-\frac{p_0}{\rho_0}\right] = -gZ \quad \text{or} \quad \frac{p}{\rho}-\frac{p_0}{\rho_0} = -gZ\,\frac{(k-1)}{k}$$

or
$$\frac{p}{\rho} = \frac{p_0}{\rho_0} - gZ\frac{(k-1)}{k} = \frac{p_0}{\rho_0}\left[1-\frac{k-1}{k}\,gZ\,\frac{\rho_0}{p_0}\right]$$

or
$$\frac{p}{\rho}\times\frac{\rho_0}{p_0}=\left[1+\frac{k-1}{k}gZ\frac{\rho_0}{p_0}\right] \qquad ...(iii)$$

But from equation (*i*),
$$\frac{p}{\rho^k}=\frac{p_0}{\rho_0^k} \text{ or } \left(\frac{\rho_0}{\rho}\right)^k=\frac{p_0}{p} \text{ or } \frac{\rho_0}{\rho}=\left(\frac{p_0}{p}\right)^{1/k}$$

Substituting the value of $\frac{\rho_0}{\rho}$ in equation (*iii*), we get

$$\frac{p}{p_0}\times\left(\frac{p_0}{p}\right)^{1/k}=\left[1-\frac{k-1}{k}gZ\frac{\rho_0}{p_0}\right]$$

or
$$\frac{p}{p_0}\times\left(\frac{p}{p_0}\right)^{-1/k}=\left[1-\frac{k-1}{k}gZ\frac{\rho_0}{p_0}\right]$$

or
$$\left(\frac{p}{p_0}\right)^{1-\frac{1}{k}}=\left(\frac{p}{p_0}\right)^{\frac{k-1}{k}}=\left[1-\frac{k-1}{k}gZ\frac{\rho_0}{p_0}\right]$$

$\therefore$
$$\frac{p}{p_0}=\left[1-\frac{k-1}{k}gZ\frac{\rho_0}{p_0}\right]^{\frac{k}{k-1}}$$

$\therefore$ Pressure at a height Z from ground level is given by

$$p=p_0\left[1-\frac{k-1}{k}gZ\frac{\rho_0}{p_0}\right]^{\frac{k}{k-1}} \qquad ...(2.19)$$

in equation (2.19), p_0 = pressure at ground level, where $Z_0 = 0$

ρ_0 = density of air at ground level

Equation of state is
$$\frac{p_0}{\rho_0}=RT_0 \text{ or } \frac{\rho_0}{p_0}=\frac{1}{RT_0}$$

Substituting the values of $\frac{\rho_0}{p_0}$ in equation (2.19), we get

$$p=p_0\left[1-\frac{k-1}{k}\frac{gZ}{RT_0}\right]^{\frac{k}{k-1}} \qquad ...(2.20)$$

2.8.3 Temperature at any Point in Compressible Fluid. For the adiabatic process, the temperature at any height in air is calculated as :

Equation of state at ground level and at a height Z from ground level is written as

$$\frac{p_0}{\rho_0}=RT_0 \text{ and } \frac{p}{\rho}=RT$$

Dividing these equations, we get

$$\left(\frac{p_0}{\rho_0}\right)\div\frac{p}{\rho}=\frac{RT_0}{RT}=\frac{T_0}{T} \quad \text{or} \quad \frac{p_0}{\rho_0}\times\frac{\rho}{p}=\frac{T_0}{T}$$

or
$$\frac{T}{T_0}=\frac{\rho_0}{p_0}\times\frac{p}{\rho}=\frac{p}{p_0}\times\frac{\rho_0}{\rho} \qquad ...(i)$$

But $\frac{p}{p_0}$ from equation (2.20) is given by

$$\frac{p}{p_0} = \left[1 - \frac{k-1}{k}\frac{gZ}{RT_0}\right]^{\frac{k}{k-1}}$$

Also for adiabatic process $\frac{p}{\rho^k} = \frac{p_0}{\rho_0^k}$ or $\left(\frac{\rho_0}{\rho}\right)^k = \frac{p_0}{p}$

or
$$\frac{\rho_0}{\rho} = \left(\frac{p_0}{p}\right)^{\frac{1}{k}} = \left(\frac{p}{p_0}\right)^{-\frac{1}{k}}$$

$$= \left[1 - \frac{k-1}{k}\frac{gZ}{RT_0}\right]^{\left(\frac{k}{k-1}\right)\times\left(-\frac{1}{k}\right)} = \left[1 - \frac{k-1}{k}\frac{gZ}{RT_0}\right]^{-\frac{1}{k-1}}$$

Substituting the values of $\frac{p}{p_0}$ and $\frac{\rho_0}{\rho}$ in equation (*i*), we get

$$\frac{T}{T_0} = \left[1 - \frac{k-1}{k}\frac{gZ}{RT_0}\right]^{\frac{k}{k-1}} \times \left[1 - \frac{k-1}{k}\frac{gZ}{RT_0}\right]^{-\frac{1}{k-1}}$$

$$= \left[1 - \frac{k-1}{k}\frac{gZ}{RT_0}\right]^{\frac{k}{k-1}-\frac{1}{k-1}} = \left[1 - \frac{k-1}{k}\frac{gZ}{RT_0}\right]$$

$\therefore$
$$T = T_0\left[1 - \frac{k-1}{k}\frac{gZ}{RT_0}\right] \qquad \text{...(2.21)}$$

2.8.4 Temperature Lapse-Rate (L). It is defined as the rate at which the temperature changes with elevation. To obtain an expression for the temperature lapse-rate, the temperature given by equation (2.21) is differentiated with respect to Z as

$$\frac{dT}{dZ} = \frac{d}{dZ}\left[T_0\left(1 - \frac{k-1}{k}\frac{gZ}{RT_0}\right)\right]$$

where T_0, K, g and R are constant

$\therefore$
$$\frac{dT}{dZ} = -\frac{k-1}{k} \times \frac{g}{RT_0} \times T_0 = \frac{-g}{R}\left(\frac{k-1}{k}\right)$$

The temperature lapse-rate is denoted by L and hence

$$L = \frac{dT}{dZ} = \frac{-g}{R}\left(\frac{k-1}{k}\right) \qquad \text{...(2.22)}$$

In equation (2.22), if (*i*) $k = 1$ which means isothermal process, $\frac{dT}{dZ} = 0$, which means temperature is constant with height.

(*ii*) If $k > 1$, the lapse-rate is negative which means temperature decreases with the increase of height.

In atmosphere, the value of k varies with height and hence the value of temperature lapse-rate also varies. From the sea-level upto an elevation of about 11000 m (or 11 km), the temperature of air decreases uniformly at the rate of 0.0065°C/m from 11000 m to 32000 m, the temperature remains constant at – 56.5°C and hence in this range lapse-rate is zero. Temperature rises again after 32000 m in air.

Problem 2.22 (SI Units) *If the atmosphere pressure at sea level is 10.143 N/cm², determine the pressure at a height of 2500 m assuming the pressure variation follows (i) Hydrostatic law, and (ii) isothermal law. The density of air is given as 1.208 kg/m³.*

Solution. Given :

Pressure at sea-level, $p_0 = 10.143 \text{ N/cm}^2 = 10.143 \times 10^4 \text{ N/m}^2$

Height, $Z = 2500$ m

Density of air, $\rho_0 = 1.208 \text{ kg/m}^3$

(*i*) **Pressure by hydrostatic law.** For hydrostatic law, ρ is assumed constant and hence p is given by equation $\frac{dp}{dZ} = -\rho g$

Integrating, we get $\int_{p_0}^{p} dp = \int -\rho g\, dZ = -\rho g \int_{Z_0}^{Z} dZ$

or $p - p_0 = -\rho g\, [Z - Z_0]$

For datum line at sea-level, $Z_0 = 0$

$$\therefore \quad p - p_0 = -\rho g Z \quad \text{or} \quad p = p_0 - \rho g Z$$

$$= 10.143 \times 10^4 - 1.208 \times 9.81 \times 2500 \; [\because \; \rho = \rho_0 = 1.208]$$

$$= 101430 - 29626 = 71804 \frac{\text{N}}{\text{m}^2} \text{ or } \frac{71804}{10^4} \text{ N/cm}^2$$

$$= \mathbf{7.18 \text{ N/cm}^2. \text{ Ans.}}$$

(*ii*) **Pressure by Isothermal Law.** Pressure at any height Z by isothermal law is given by equation (2.18) as

$$p = p_0\, e^{-gZ/RT}$$

$$= 10.143 \times 10^4\, e^{-\frac{gZ \times \rho_0}{p_0}} \qquad \left[\because \; \frac{p_0}{\rho_0} = RT \text{ and } \rho_0 g = w_0\right]$$

$$= 10.143 \times 10^4\, e^{-\frac{Z\rho_0 \times g}{p_0}}$$

$$= 10.143 \times 10^4\, e^{(-2500 \times 1.208 \times 9.81)/10.143 \times 10^4}$$

$$= 101430 \times e^{-.292} = 101430 \times \frac{1}{1.3391} = 75743 \text{ N/m}^2$$

$$= \frac{75743}{10^4} \text{ N/cm}^2 = \mathbf{7.574 \text{ N/cm}^2. \text{ Ans.}}$$

Problem 2.23 *The barometric pressure at sea level is 760 mm of mercury while that on a mountain top is 735 mm, if the density of air is assumed constant at 1.2 kg/m³, what is the elevation of the mountain top.* (A.M.I.E., Summer, 1988)

Solution. Given :

Pressure* at sea, $p_0 = 760$ mm of Hg

$$= \frac{760}{1000} \times 13.6 \times 1000 \times 9.81 \text{ N/m}^2 = 101396 \text{ N/m}^2$$

---

* Here pressure head (Z) is given as 760 mm of Hg. Hence $(p/\rho g) = 760$ mm of Hg. The density (ρ) for mercury $= 13.6 \times 1000 \text{ kg/m}^3$. Hence pressure ($p$) will be equal to $\rho \times g \times Z$ *i.e.*, $13.6 \times 1000 \times 9.81 \times \frac{760}{1000} \text{ N/m}^2$.

Pressure at mountain, $p = 735$ mm of Hg

$$= \frac{735}{1000} \times 13.6 \times 1000 \times 9.81 = 98060 \text{ N/m}^2$$

Density of air, $\rho = 1.2 \text{ kg/m}^3$

Let h = Height of the mountain from sea-level.

We know that as the elevation above the sea-level increases, the atmospheric pressure decreases. Here the density of air is given constant, hence the pressure at any height 'h' above the sea-level is given by the equation,

$$p = p_0 - \rho \times g \times h$$

or

$$h = \frac{p_0 - p}{\rho \times g} = \frac{101396 - 98060}{1.2 \times 9.81} = \mathbf{283.33 \text{ m. Ans.}}$$

Problem 2.24 *Calculate the pressure at a height of 7500 m above sea level if the atmospheric pressure is 10.143 N/cm² and temperature is 15°C at the sea-level, assuming (i) air is incompressible, (ii) pressure variation follows isothermal law, and (iii) pressure variation follows adiabatic law. Take the density of air at the sea-level as equal to 1.285 kg/m³. Neglect variation of g with altitude.*

Solution. Given :

Height above sea-level, $Z = 7500$ m

Pressure at sea-level, $p_0 = 10.143 \text{ N/cm}^2 = 10.143 \times 10^4 \text{ N/m}^2$

Temperature at sea-level, $t_0 = 15°\text{C}$

$\therefore$ $T_0 = 273 + 15 = 288°\text{K}$

Density of air, $\rho = \rho_0 = 1.285 \text{ kg/m}^3$

(*i*) Pressure when air is incompressible :

$$\frac{dp}{dZ} = -\rho g$$

$\therefore$

$$\int_{p_0}^{p} dp = -\int_{Z_0}^{Z} \rho g dz \quad \text{or} \quad p - p_0 = -\rho g[Z - Z_0]$$

or

$$p = p_0 - \rho g Z \qquad \{\because Z_0 = \text{datum line} = 0\}$$

$$= 10.143 \times 10^4 - 1.285 \times 9.81 \times 7500$$

$$= 101430 - 94543 = 6887 \text{ N/m}^2 = \mathbf{0.688 \frac{N}{cm^2}. \text{ Ans.}}$$

(*ii*) Pressure variation follows isothermal law :

Using equation (2.18), we have $p = p_0 e^{-gZ/RT}$

$$= p_0 e^{-gZ\rho_0/p_0} \qquad \left\{\because \frac{p_0}{\rho_0} = RT \therefore \frac{\rho_0}{p_0} = \frac{1}{RT}\right\}$$

$$= 101430\, e^{-gZ\rho_0/p_0} = 101430\, e^{-7500 \times 1.285 \times 9.81/101430}$$

$$= 101430\, e^{-.9320} = 101430 \times .39376$$

$$= \mathbf{39939 \text{ N/m}^2 \text{ or } 3.993 \text{ N/cm}^2. \text{ Ans.}}$$

(*iii*) Pressure variation follows adiabatic law : [$k = 1.4$]

Using equation (2.19), we have

$$p = p_0 \left[1 - \frac{k-1}{k} gZ \frac{\rho_0}{p_0}\right]^{k/(k-1)}, \text{ where } \rho_0 = 1.285 \text{ kg/m}^3$$

$$\therefore \quad p = 101430\left[1-\frac{(1.4-1.0)}{1.4}\times 9.81\times\frac{(7500\times 1.285)}{101430}\right]^{\frac{1.4}{1.4-1.0}}$$

$$= 101430\ [1 - .2662]^{1.4/.4} = 101430 \times (.7337)^{3.5}$$

$$= \mathbf{34310\ N/m^2\ or\ 3.431\ \frac{N}{cm^2}.\ Ans.}$$

Problem 2.25 *Calculate the pressure and density of air at a height of 4000 m from sea-level where pressure and temperature of the air are 10.143 N/cm² and 15°C respectively. The temperature lapse rate is given as 0.0065°C/m. Take density of air at sea-level equal to 1.285 kg/m³.*

Solution. Given :

Height, $Z = 4000$ m

Pressure at sea-level, $p_0 = 10.143\ \text{N/cm}^2 = 10.143 \times 10^4 = 101430\ \frac{\text{N}}{\text{m}^2}$

Temperature at sea-level, $t_0 = 15°\text{C}$

$\therefore \quad T_0 = 273 + 15 = 288°\text{K}$

Temperature lapse-rate, $L = \frac{dT}{dZ} = -\ 0.0065°\text{K/m}$

$\rho_0 = 1.285\ \text{kg/m}^3$

Using equation (2.22), we have $L = \frac{dT}{dZ} = -\frac{g}{R}\left(\frac{k-1}{k}\right)$

or $$-\ 0.0065 = -\frac{9.81}{R}\left(\frac{k-1}{k}\right),\ \text{where } R = \frac{p_0}{\rho_0 T_0} = \frac{101430}{1.285\times 288} = 274.09$$

$$\therefore \quad -\ 0.0065 = \frac{-9.81}{274.09}\times\left(\frac{k-1}{k}\right)$$

$$\therefore \quad \frac{k-1}{k} = \frac{0.0065\times 274.09}{9.81} = 0.1815$$

$$\therefore \quad k[1 - .1815] = 1$$

$$\therefore \quad k = \frac{1}{1-.1815} = \frac{1}{.8184} = 1.222$$

This means that the value of power index $k = 1.222$.

(*i*) **Pressure** at 4000 m height is given by equation (2.19) as

$$p = p_0\left[1-\frac{k-1}{k}gZ\frac{\rho_0}{p_0}\right]^{\frac{k}{k-1}},\ \text{where } k = 1.222 \text{ and } \rho_0 = 1.285$$

$$\therefore \quad p = 101430\left[1-\left(\frac{1.222-1.0}{1.222}\right)\times 9.81\times\frac{4000\times 1.285}{101430}\right]^{\frac{1.222}{1.222-1.0}}$$

$$= 101430\ [1 - 0.09]^{5.50} = 101430 \times .595$$

$$= 60350\ \text{N/m}^2 = \mathbf{6.035\ \frac{N}{cm^2}.\ Ans.}$$

(*ii*) **Density.** Using equation of state, we get

$$\frac{p}{\rho} = RT$$

where p = Pressure at 4000 m height
ρ = Density at 4000 m height
T = Temperature at 4000 m height

Now T is calculated from temperature lapse-rate as

$$t \text{ at } 4000 \text{ m} = t_0 + \frac{dT}{dZ} \times 4000 = 15 - .0065 \times 4000 = 15 - 26 = -11°\text{C}$$

$$\therefore \quad T = 273 + t = 273 - 11 = 262°\text{K}$$

$\therefore$ Density is given by $\rho = \frac{p}{RT} = \frac{60350}{274.09 \times 262}$ kg/m^3 = **0.84 kg/m^3. Ans.**

Problem 2.26 *An aeroplane is flying at an altitude of 5000 m. Calculate the pressure around the aeroplane, given the lapse-rate in the atmosphere as 0.0065°K/m. Neglect variation of g with altitude. Take pressure and temperature at ground level as 10.143 N/cm^2 and 15°C and density of air as 1.285 kg/cm^3.*

Solution. Given :

Height, $Z = 5000$ m

Lapse-rate, $L = \frac{dT}{dZ} = -.0065°\text{K/m}$

Pressure at ground level, $p_0 = 10.143 \times 10^4 \text{ N/m}^2$

$t_0 = 15°\text{C}$

$\therefore$ $T_0 = 273 + 15 = 288°\text{K}$

Density, $\rho_0 = 1.285 \text{ kg/m}^3$

$\therefore$ Temperature at 5000 m height $= T_0 + \frac{dT}{dZ} \times \text{Height} = 288 - .0065 \times 5000$

$$= 288 - 32.5 = 255.5°\text{K}$$

First find the value of power index k as

From equation (2.22), we have $L = \frac{dT}{dZ} = -\frac{g}{R}\left(\frac{k-1}{k}\right)$

or
$$-.0065 = -\frac{9.81}{R}\left(\frac{k-1}{k}\right)$$

where $R = \frac{p_0}{\rho_0 T_0} = \frac{101430}{1.285 \times 288} = 274.09$

$$\therefore \quad -.0065 = -\frac{9.81}{274.09}\left(\frac{k-1}{k}\right)$$

$$\therefore \quad k = 1.222$$

The pressure is given by equation (2.19) as

$$p = p_0 \left[1 - \frac{k-1}{k} gZ \frac{\rho_0}{p_0}\right]^{\left(\frac{k}{k-1}\right)}$$

$$= 101430\left[1-\left(\frac{1.222-1.0}{1.222}\right)\times 9.81\times\frac{5000\times 1.285}{101430}\right]^{\frac{1.222}{1.222-1.0}}$$

$$= 101430\left[1-\frac{.222}{1.222}\times 9.81\times\frac{5000\times 1.285}{101430}\right]^{\frac{1.222}{.222}}$$

$$= 101430\,[1-0.11288]^{5.50} = 101430\times 0.5175 = 52490 \text{ N/m}^3$$

$$= \mathbf{5.249\ N/cm^2.\ Ans.}$$

HIGHLIGHTS

1. The pressure at any point in a fluid is defined as the force per unit area.
2. The Pascal's law states that intensity of pressure for a fluid at rest is equal in all directions.
3. Pressure variation at a point in a fluid at rest is given by the hydrostatic law which states that the rate of increase of pressure in the vertically downward direction is equal the specific weight of the fluid, $\frac{dp}{dZ} = w = \rho \times g.$
4. The pressure at any point in a incompressible fluid (liquid) is equal to the product of density of fluid at that point, acceleration due to gravity and vertical height from free surface of fluid, $p = \rho \times g \times Z.$
5. Absolute pressure is the pressure in which absolute vacuum pressure is taken as datum while gauge pressure is the pressure in which the atmospheric pressure is taken as datum,
$$p_{abs.} = p_{atm} + p_{gauge.}$$
6. Manometer is a device used for measuring pressure at a point in a fluid.
7. Manometers are classified as *(a)* Simple manometers and *(b)* Differential manometers.
8. Simple manometers are used for measuring pressure at a point while differential manometers are used for measuring the difference of pressures between the two points in a pipe, or two different pipes.
9. A single column manometer (or micrometer) is used for measuring small pressures, where accuracy is required.
10. The pressure at a point in static compressible fluid is obtained by combining two equations, *i.e.*, equation of state for a gas and equation given by hydrostatic law.
11. The pressure at a height Z in a static compressible fluid (gas) under going isothermal compression $\left(\frac{p}{\rho} = \text{const.}\right)$
$$p = p_0\, e^{-gZ/RT}$$
where p_0 = Absolute pressure at sea-level or at ground level
Z = Height from sea or ground level
R = Gas constant
T = Absolute temperature.
12. The pressure and temperature at a height Z in a static compressible fluid (gas) undergoing adiabatic compression (p/ρ^k = const.)
$$p = p_0\left[1-\frac{k-1}{k}\,gZ\,\frac{\rho_0}{p_0}\right]^{\frac{k}{k-1}} = p_0\left[1-\frac{k-1}{k}\,\frac{gZ}{RT_0}\right]^{\frac{k}{k-1}}$$

and temperature, $$T = T_0\left[1 - \frac{k-1}{k}\frac{gZ}{RT_0}\right]$$

where p_0, T_0 are pressure and temperature at sea-level $k = 1.4$ for air.

13. The rate at which the temperature changes with elevation is known as Temperature Lapse-Rate. It is given by

$$L = \frac{-g}{R}\left(\frac{k-1}{k}\right)$$

if (*i*) $k = 1$, temperature is zero.

(*ii*) $k > 1$, temperature decreases with the increase of height.

EXERCISE 2

(A) THEORETICAL PROBLEMS

1. Define pressure. Obtain an expression for the pressure intensity at a point in a fluid.
2. State and prove the Pascal's law.
3. What do you understand by Hydrostatic Law ?
4. Differentiate between : (*i*) Absolute and gauge pressure, (*ii*) Simple manometer and differential manometer, and (*iii*) Piezometer and pressure gauges.
5. What do you mean by vacuum pressure ?
6. What is a manometer ? How are they classified ?
7. What do you mean by single column manometers ? How are they used for the measurement of pressure ?
8. What is the difference between U-tube differential manometers and inverted U-tube differential manometers ? Where are they used ?
9. Distinguish between manometers and mechanical gauges. What are the different types of mechanical pressure gauges ?
10. Derive an expression for the pressure at a height Z from sea-level for a static air when the compression of the air is assumed isothermal. The pressure and temperature at sea-levels are p_0 and T_0 respectively.
11. Prove that the pressure and temperature for an adiabatic process at a height Z from sea-level for a static air are :

$$p_0 = p_0\left[1 - \frac{k-1}{k}\frac{gZ}{RT_0}\right]^{\frac{k}{k-1}} \text{ and } T = T_0\left[1 - \frac{k-1}{k}\frac{gZ}{RT_0}\right]$$

where p_0 and T_0 are the pressure and temperature at sea-level.

12. What do you understand by the term, 'Temperature Lapse-Rate' ? Obtain an expression for the temperature Lapse-Rate.
13. What is hydrostatic pressure distribution ? Give one example where pressure distribution is non-hydrostatic. (*A.M.I.E., Winter 1990*)
14. Explain briefly the working principle of Bourdon Pressure Gauge with a neat sketch. (*J.N.T.U., Hyderabad, S 2002*)

(B) NUMERICAL PROBLEMS

1. A hydraulic press has a ram of 30 cm diameter and a plunger of 5 cm diameter. Find the weight lifted by the hydraulic press when the force applied at the plunger is 400 N. [**Ans.** 14.4 kN]
2. A hydraulic press has a ram of 20 cm diameter and a plunger of 4 cm diameter. It is used for lifting a weight of 20 kN. Find the force required at the plunger. [**Ans.** 800 N]

3. Calculate the pressure due to a column of 0.4 m of (*a*) water, (*b*) an oil of sp. gr. 0.9, and (*c*) mercury of sp. gr. 13.6. Take density of water, $\rho = 1000 \frac{kg}{m^3}$. **[Ans.** (*a*) 0.3924 N/cm^2, (*b*) 0.353 N/cm^2, (*c*) 5.33 N/cm^2]

4. The pressure intensity at a point in a fluid is given 4.9 N/cm^2. Find the corresponding height of fluid when it is : (*a*) water, and (*b*) an oil of sp. gr. 0.8. **[Ans.** (*a*) 5 m of water, (*b*) 6.25 m of oil]

5. An oil of sp. gr. 0.8 is contained in a vessel. At a point the height of oil is 20 m. Find the corresponding height of water at that point. **[Ans.** 16 m]

6. An open tank contains water upto a depth of 1.5 m and above it an oil of sp. gr. 0.8 for a depth of 2 m. Find the pressure intensity : (*i*) at the interface of the two liquids, and (*ii*) at the bottom of the tank. **[Ans.** (*i*) 1.57 N/cm^2, (*ii*) 3.04 N/cm^2]

7. The diameters of a small piston and a large piston of a hydraulic jack are 2 cm and 10 cm respectively. A force of 60 N is applied on the small piston. Find the load lifted by the large piston, when : (*a*) the pistons are at the same level, and (*b*) small piston is 20 cm above the large piston. The density of the liquid in the jack is given as $1000 \frac{kg}{m^3}$. **[Ans.** (*a*) 1500 N, (*b*) 1520.5 N]

8. Determine the gauge and absolute pressure at a point which is 2.0 m below the free surface of water. Take atmospheric pressure as 10.1043 N/cm^2. **[Ans.** 1.962 N/cm^2 (gauge), 12.066 N/cm^2 (abs.)]

9. A simple manometer is used to measure the pressure of oil (sp. gr. = 0.8) flowing in a pipe line. Its right limb is open to the atmosphere and left limb is connected to the pipe. The centre of the pipe is 9 cm below the level of mercury (sp. gr. 13.6) in the right limb. If the difference of mercury level in the two limbs is 15 cm, determine the absolute pressure of the oil in the pipe in N/cm^2. (*A.M.I.E., Winter, 1977*) **[Ans.** 12.058 N/cm^2]

10. A simple manometer (*U*-tube) containing mercury is connected to a pipe in which an oil of sp. gr. 0.8 is flowing. The pressure in the pipe is vacuum. The other end of the manometer is open to the atmosphere. Find the vacuum, pressure in pipe, if the difference of mercury level in the two limbs is 20 cm and height of oil in the left limb from the centre of the pipe is 15 cm below. **[Ans.** – 27.86 N/cm^2]

11. A single column vertical manometer (*i.e.,* micrometer) is connected to a pipe containing oil of sp. gr. 0.9. The area of the reservoir is 80 times the area of the manometer tube. The reservoir contains mercury of sp. gr. 13.6. The level of mercury in the reservoir is at a height of 30 cm below the centre of the pipe and difference of mercury levels in the reservoir and right limb is 50 cm. Find the pressure in the pipe. **[Ans.** 6.474 N/cm^2]

12. A pipe contains an oil of sp. gr. 0.8. A differential manometer connected at the two points *A* and *B* of the pipe shows a difference in mercury level as 20 cm. Find the difference of pressure at the two points. **[Ans.** 25113.6 N/m^2]

13. A *U*-tube differential manometer connects two pressure pipes *A* and *B*. Pipe *A* contains carbon tetrachloride having a specific gravity 1.594 under a pressure of 11.772 N/cm^2 and pipe *B* contains oil of sp. gr. 0.8 under a pressure of 11.772 N/cm^2. The pipe *A* lies 2.5 m above pipe *B*. Find the difference of pressure measured by mercury as fluid filling *U*-tube. (*A.M.I.E. December, 1974*) **[Ans.** 31.36 cm of mercury]

14. A differential manometer is connected at the two points *A* and *B* as shown in Fig. 2.25. At *B* air pressure is 7.848 N/cm^2 (abs.), find the absolute pressure at *A*. **[Ans.** 6.91 N/cm^2]

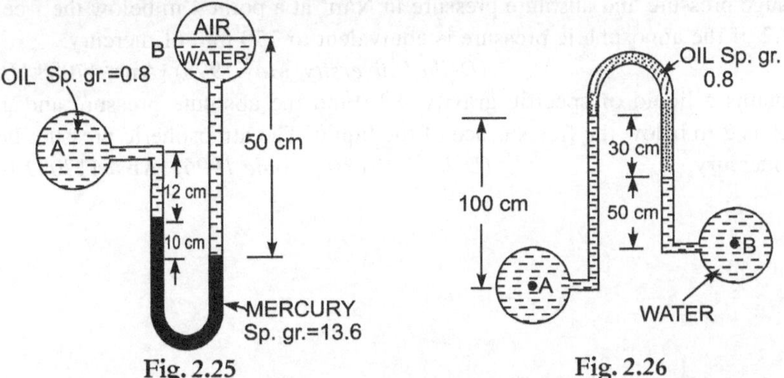

Fig. 2.25

Fig. 2.26

15. An inverted differential manometer containing an oil of sp. gr. 0.9 is connected to find the difference of pressures at two points of a pipe containing water. If the manometer reading is 40 cm, find the difference of pressures. [**Ans.** 392.4 N/m^2]

16. In above Fig. 2.26 shows an inverted differential manometer connected to two pipes *A* and *B* containing water. The fluid in manometer is oil of sp. gr. 0.8. For the manometer readings shown in the figure, find the difference of pressure head between *A* and *B*. [**Ans.** 0.26 m of water]

17. If the atmospheric pressure at sea-level is 10.143 N/cm^2, determine the pressure at a height of 2000 m assuming that the pressure variation follows : (*i*) Hydrostatic law, and (*ii*) Isothermal law. The density of air is given as 1.208 kg/m^3. [**Ans.** (*i*) 7.77 N/cm^2, (*ii*) 8.03 N/cm^3]

18. Calculate the pressure at a height of 8000 m above sea-level if the atmospheric pressure is 101.3 kN/m^2 and temperature is 15°C at the sea-level assuming (*i*) air is incompressible, (*ii*) pressure variation follows adiabatic law, and (*iii*) pressure variation follows isothermal law. Take the density of air at the sea-level as equal to 1.285 kg/m^3. Neglect variation of *g* with altitude.

[**Ans.** (*i*) 607.5 N/m^2, (*ii*) 31.5 kN/m^2 (*iii*) 37.45 kN/m^2]

19. Calculate the pressure and density of air at a height of 3000 m above sea-level where pressure and temperature of the air are 10.143 N/cm^2 and 15°C respectively. The temperature lapse-rate is given as 0.0065° K/m. Take density of air at sea-level equal to 1.285 kg/m^3. [**Ans.** 6.896 N/cm^2, 0.937 kg/m^3]

20. An aeroplane is flying at an altitude of 4000 m. Calculate the pressure around the aeroplane, given the lapse-rate in the atmosphere as 0.0065° K/m. Neglect variation of *g* with altitude. Take pressure and temperature at ground level as 10.143 N/cm^2 and 15°C respectively. The density of air at ground level is given as 1.285 kg/m^3. [**Ans.** 6.038 N/cm^2]

21. The atmosphere pressure at the sea-level is 101.3 kN/m^2 and the temperature is 15°C. Calculate the pressure 8000 m above sea-level, assuming (*i*) air is in compressible, (*ii*) isothermal variation of pressure and density, and (*iii*) adiabatic variation of pressure and density. Assume density of air at sea-level as 1.285 kg/m^3. Neglect variation of '*g*' with altitude.

[**Ans.** (*i*) 501.3 N/m^2, (*ii*) 37.45 kN/m^2, (*iii*) 31.5 kN/m^2]

22. An oil of sp. gr. is 0.8 under a pressure of 137.2 kN/m^2

(*i*) What is the pressure head expressed in metre of water ?

(*ii*) What is the pressure head expressed in metre of oil ? [**Ans.** (*i*) 14 m, (*ii*) 17.5 m]

23. The atmospheric pressure at the sea-level is 101.3 kN/m^2 and temperature is 15°C. Calculate the pressure 8000 m above sea-level, assuming : (*i*) isothermal variation of pressure and density, and (*ii*) adiabatic variation of pressure and density. Assume density of air at sea-level as 1.285 kg/m^3. Neglect variation of '*g*' with altitude.

Derive the formula that you may use. (*Delhi University, 1992*) [**Ans.** (*i*) 37.45 kN/m^2 (*ii*) 31.5 kN/m^2]

24. What are the gauge pressure and absolute pressure at a point 4 m below the free surface of a liquid of specific gravity 1.53, if atmospheric pressure is equivalent to 750 mm of mercury.

(*Delhi University, 1992*) [**Ans.** 60037 N/m^2 and 160099 N/m^2]

25. Find the gauge pressure and absolute pressure in N/m^2 at a point 4 m below the free surface of a liquid of sp. gr. 1.2, if the atmospheric pressure is equivalent to 750 mm of mercury.

(*Delhi University, May 1998*) [**Ans.** 47088 N/m^2 ; 147150 N/m^2]

26. A tank contains a liquid of specific gravity 0.8. Find the absolute pressure and gauge pressure at a point, which is 2 m below the free surface of the liquid. The atmospheric pressure head is equivalent to 760 mm of mercury. (*Delhi University, June 1996*) [**Ans.** 117092 N/m^2 ; 15696 N/m^2]

Hydrostatic Forces on Submerged Surfaces

▶ 3.1 INTRODUCTION

This chapter deals with the fluids (*i.e.,* liquids and gases) at rest. This means that there will be no relative motion between adjacent or neighbouring fluid layers. The velocity gradient, which is equal to the change of velocity between two adjacent fluid layers divided by the distance between the layers, will be zero or $\frac{du}{dy} = 0$. The shear stress which is equal to $\mu \frac{\partial u}{\partial y}$ will also be zero. Then the forces acting on the fluid particles will be :

1. due to pressure of fluid normal to the surface,
2. due to gravity (or self-weight of fluid particles).

▶ 3.2 TOTAL PRESSURE AND CENTRE OF PRESSURE

Total pressure is defined as the force exerted by a static fluid on a surface either plane or curved when the fluid comes in contact with the surfaces. This force always acts normal to the surface.

Centre of pressure is defined as the point of application of the total pressure on the surface. There are four cases of submerged surfaces on which the total pressure force and centre of pressure is to be determined. The submerged surfaces may be :

1. Vertical plane surface,
2. Horizontal plane surface,
3. Inclined plane surface, and
4. Curved surface.

▶ 3.3 VERTICAL PLANE SURFACE SUBMERGED IN LIQUID

Consider a plane vertical surface of arbitrary shape immersed in a liquid as shown in Fig. 3.1.

Let A = Total area of the surface

$\bar{h}$ = Distance of C.G. of the area from free surface of liquid

G = Centre of gravity of plane surface

P = Centre of pressure

h^* = Distance of centre of pressure from free surface of liquid.

(*a*) **Total Pressure (F).** The total pressure on the surface may be determined by dividing the entire surface into a number of small parallel strips. The force on small strip is then calculated and the total pressure force on the whole area is calculated by integrating the force on small strip.

Consider a strip of thickness dh and width b at a depth of h from free surface of liquid as shown in Fig. 3.1

FREE SURFACE OF LIQUID

Fig. 3.1

Pressure intensity on the strip, $p = \rho g h$

(See equation 2.5)

Area of the strip, $dA = b \times dh$

Total pressure force on strip, $dF = p \times \text{Area}$

$= \rho g h \times b \times dh$

$\therefore$ Total pressure force on the whole surface,

$$F = \int dF = \int \rho g h \times b \times dh = \rho g \int b \times h \times dh$$

But $\int b \times h \times dh = \int h \times dA$

= Moment of surface area about the free surface of liquid

= Area of surface × Distance of C.G. from free surface

$= A \times \bar{h}$

$$\therefore \quad F = \rho g A \bar{h} \qquad \text{...(3.1)}$$

For water the value of $\rho = 1000$ kg/m^3 and $g = 9.81$ m/s^2. The force will be in Newton.

(*b*) **Centre of Pressure (h*).** Centre of pressure is calculated by using the "Principle of Moments", which states that the moment of the resultant force about an axis is equal to the sum of moments of the components about the same axis.

The resultant force F is acting at P, at a distance h^* from free surface of the liquid as shown in Fig. 3.1. Hence moment of the force F about free surface of the liquid $= F \times h^*$ ...(3.2)

Moment of force dF, acting on a strip about free surface of liquid

$= dF \times h$ $\{\because \; dF = \rho g h \times b \times dh\}$

$= \rho g h \times b \times dh \times h$

Sum of moments of all such forces about free surface of liquid

$= \int \rho g h \times b \times dh \times h = \rho g \int b \times h \times h dh$

$= \rho g \int b h^2 \, dh = \rho g \int h^2 \, dA$ $(\because \; b dh = dA)$

But $\int h^2 dA = \int b h^2 dh$

= Moment of Inertia of the surface about free surface of liquid

$= I_0$

$\therefore$ Sum of moments about free surface

$$= \rho g I_0 \qquad \text{...(3.3)}$$

Equating (3.2) and (3.3), we get

$$F \times h^* = \rho g I_0$$

But $$F = \rho g A \bar{h}$$

$\therefore$ $$\rho g A \bar{h} \times h^* = \rho g I_0$$

or $$h^* = \frac{\rho g I_0}{\rho g A \bar{h}} = \frac{I_0}{A\bar{h}} \qquad ...(3.4)$$

By the theorem of parallel axis, we have

$$I_0 = I_G + A \times \overline{h^2}$$

where I_G = Moment of Inertia of area about an axis passing through the C.G. of the area and parallel to the free surface of liquid.

Substituting I_G in equation (3.4), we get

$$h^* = \frac{I_G + A\overline{h^2}}{Ah} = \frac{I_G}{A\bar{h}} + \bar{h} \qquad ...(3.5)$$

In equation (3.5), $\bar{h}$ is the distance of C.G. of the area of the vertical surface from free surface of the liquid. Hence from equation (3.5), it is clear that :

(*i*) Centre of pressure (*i.e.*, h^*) lies below the centre of gravity of the vertical surface.

(*ii*) The distance of centre of pressure from free surface of liquid is independent of the density of the liquid.

Table 3.1 The moments of inertia and other geometric properties of some important plane surfaces

Plane surface	*C.G. from the base*	*Area*	*Moment of inertia about an axis passing through C.G. and parallel to base* (I_G)	*Moment of inertia about base* (I_0)
1. Rectangle (G, a, x, b)	$x = \frac{d}{2}$	bd	$\frac{bd^3}{12}$	$\frac{bd^3}{3}$
2. Triangle (G, h, x, b)	$x = \frac{h}{3}$	$\frac{bh}{2}$	$\frac{bh^3}{36}$	$\frac{bh^3}{12}$

Plane surface	C.G. from the base	Area	Moment of inertia about an axis passing through C.G. and parallel to base (I_G)	Moment of inertia about base (I_0)
3. Circle	$x = \dfrac{d}{2}$	$\dfrac{\pi d^2}{4}$	$\dfrac{\pi d^4}{64}$	—
4. Trapezium	$x = \left(\dfrac{2a+b}{a+b}\right)\dfrac{h}{3}$	$\dfrac{(a+b)}{2} \times h$	$\left(\dfrac{a^2+4ab+b^2}{36(a+b)}\right) \times h^3$	—

Problem 3.1 *A rectangular plane surface is 2 m wide and 3 m deep. It lies in vertical plane in water. Determine the total pressure and position of centre of pressure on the plane surface when its upper edge is horizontal and (a) coincides with water surface, (b) 2.5 m below the free water surface.*

Solution. Given :

Width of plane surface, $b = 2$ m

Depth of plane surface, $d = 3$ m

(*a*) **Upper edge coincides with water surface (Fig. 3.2).** Total pressure is given by equation (3.1) as

$$F = \rho g A \bar{h}$$

where $\rho = 1000$ kg/m^3, $g = 9.81$ m/s^2

$$A = 3 \times 2 = 6 \text{ m}^2,\ \bar{h} = \frac{1}{2}(3) = 1.5 \text{ m.}$$

$$\therefore \quad F = 1000 \times 9.81 \times 6 \times 1.5$$

$$= \mathbf{88290 \text{ N. Ans.}}$$

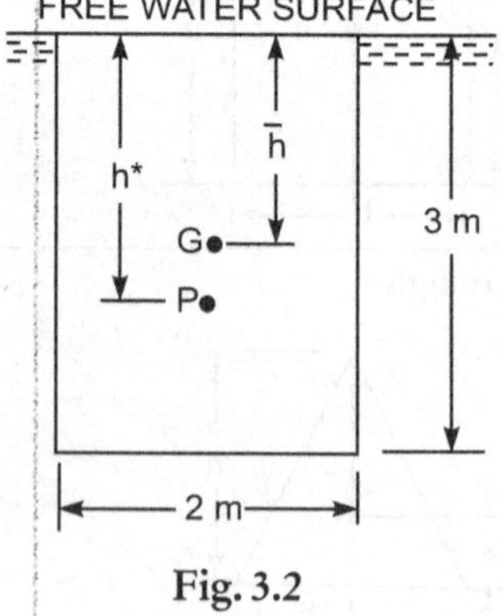

Fig. 3.2

Depth of centre of pressure is given by equation (3.5) as

$$h^* = \frac{I_G}{A\bar{h}} + \bar{h}$$

where I_G = M.O.I. about C.G. of the area of surface

$$= \frac{bd^3}{12} = \frac{2 \times 3^3}{12} = 4.5 \text{ m}^4$$

$$\therefore \qquad h^* = \frac{4.5}{6 \times 1.5} + 1.5 = 0.5 + 1.5 = \mathbf{2.0\ m.\ Ans.}$$

(*b*) **Upper edge is 2.5 m below water surface (Fig. 3.3).** Total pressure (F) is given by (3.1)

$$F = \rho g A \bar{h}$$

where $\bar{h}$ = Distance of C.G. from free surface of water

$$= 2.5 + \frac{3}{2} = 4.0 \text{ m}$$

$$\therefore \qquad F = 1000 \times 9.81 \times 6 \times 4.0$$

$$= \mathbf{235440\ N.\ Ans.}$$

Centre of pressure is given by $h^* = \dfrac{I_G}{A\bar{h}} + \bar{h}$

where $I_G = 4.5$, $A = 6.0$, $\bar{h} = 4.0$

$$\therefore \qquad h^* = \frac{4.5}{6.0 \times 4.0} + 4.0$$

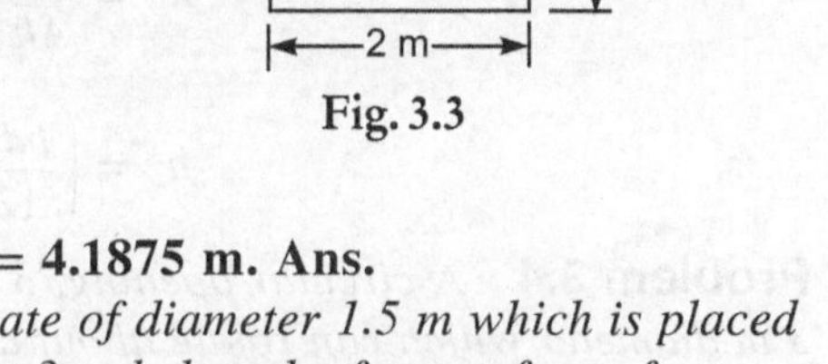

Fig. 3.3

$$= 0.1875 + 4.0 = 4.1875 = \mathbf{4.1875\ m.\ Ans.}$$

Problem 3.2 *Determine the total pressure on a circular plate of diameter 1.5 m which is placed vertically in water in such a way that the centre of the plate is 3 m below the free surface of water. Find the position of centre of pressure also.*

Solution. Given : Dia. of plate, $d = 1.5$ m

$$\therefore \text{ Area,} \qquad A = \frac{\pi}{4}(1.5)^2 = 1.767 \text{ m}^2$$

$$\bar{h} = 3.0 \text{ m}$$

Total pressure is given by equation (3.1),

$$F = \rho g A \bar{h}$$

$$= 1000 \times 9.81 \times 1.767 \times 3.0 \text{ N}$$

$$= \mathbf{52002.81\ N.\ Ans.}$$

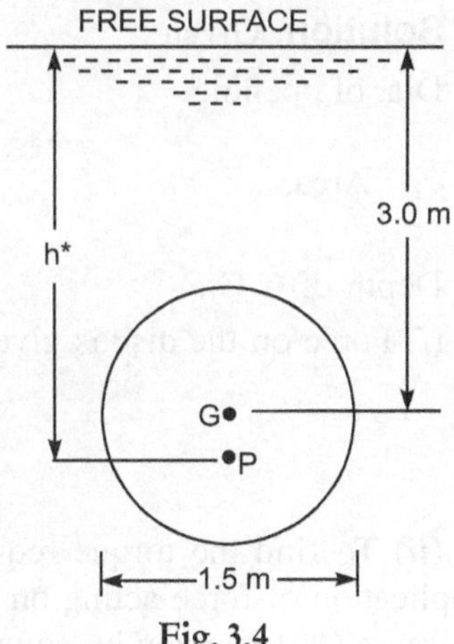

Fig. 3.4

Position of centre of pressure (h^*) is given by equation (3.5)

$$h^* = \frac{I_G}{A\bar{h}} + \bar{h}$$

where $I_G = \dfrac{\pi d^4}{64} = \dfrac{\pi \times 1.5^4}{64} = 0.2485 \text{ m}^4$

$$\therefore \qquad h^* = \frac{0.2485}{1.767 \times 3.0} + 3.0 = 0.0468 + 3.0$$

$$= \mathbf{3.0468\ m.\ Ans.}$$

Problem 3.3 *A rectangular sluice gate is situated on the vertical wall of a lock. The vertical side of the sluice is 'd' metres in length and depth of centroid of the area is 'p' m below the water surface. Prove that the depth of pressure is equal to* $\left(p+\frac{d^2}{12p}\right)$.

Solution. Given :

Depth of vertical gate $= d$ m

Let the width of gate $= b$ m

∴ Area, $A = b \times d \text{ m}^2$

Depth of C.G. from free surface

$$\bar{h} = p \text{ m}.$$

Fig. 3.5

Let h^* is the depth of centre of pressure from free surface, which is given by equation (3.5) as

$$h^* = \frac{I_G}{A\bar{h}} + \bar{h}, \text{ where } I_G = \frac{bd^2}{12}$$

$$\therefore \qquad h^* = \left(\frac{bd^3}{12} \Big/ b \times d \times p\right) + p = \frac{d^2}{12p} + p \quad \text{or} \quad \mathbf{p + \frac{d^2}{12}}. \textbf{ Ans.}$$

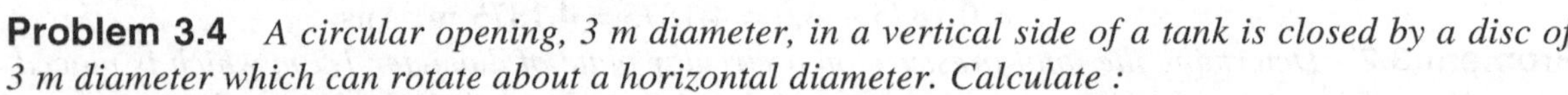

Problem 3.4 *A circular opening, 3 m diameter, in a vertical side of a tank is closed by a disc of 3 m diameter which can rotate about a horizontal diameter. Calculate :*

(i) the force on the disc, and

(ii) the torque required to maintain the disc in equilibrium in the vertical position when the head of water above the horizontal diameter is 4 m. (A.M.I.E., Winter, 1977)

Solution. Given :

Dia. of opening, $d = 3$ m

∴ Area, $A = \frac{\pi}{4} \times 3^2 = 7.0685 \text{ m}^2.$

Depth of C.G., $\bar{h} = 4$ m

(*i*) Force on the disc is given by equation (3.1) as

$$F = \rho g A\bar{h} = 1000 \times 9.81 \times 7.0685 \times 4.0$$

$$= \mathbf{277368 \text{ N} = 277.368 \text{ kN}.} \textbf{ Ans.}$$

(*ii*) To find the torque required to maintain the disc in equilibrium, first calculate the point of application of force acting on the disc, *i.e.*, centre of pressure of the force F. The depth of centre of pressure (h^*) is given by equation (3.5) as

$$h^* = \frac{I_G}{A\bar{h}} + \bar{h} = \frac{\frac{\pi}{64} d^4}{\frac{\pi}{4} d^2 \times 4.0} + 4.0 \qquad \left\{\because I_G = \frac{\pi}{64} d^4\right\}$$

$$= \frac{d^2}{16 \times 4.0} + 4.0 = \frac{3^2}{16 \times 4.0} + 4.0 = 0.14 + 4.0 = 4.14 \text{ m}$$

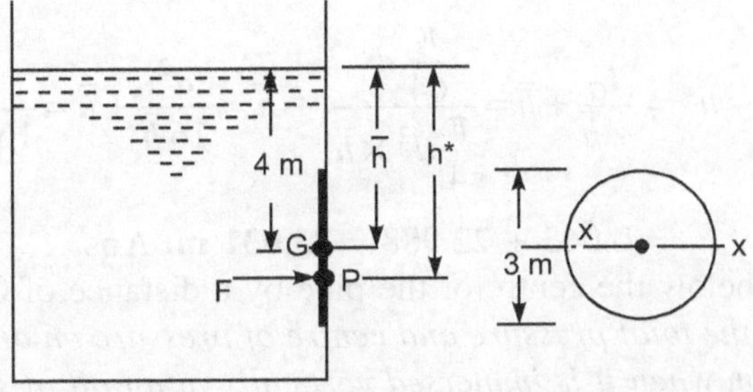

Fig. 3.6

The force F is acting at a distance of 4.14 m from free surface. Moment of this force about horizontal diameter X-X

$$= F \times \left(h^* - \bar{h}\right) = 277368\ (4.14 - 4.0) = \mathbf{38831\ Nm.\ Ans.}$$

Hence a torque of 38831 Nm must be applied on the disc in the clockwise direction.

Problem 3.5 *A pipe line which is 4 m in diameter contains a gate valve. The pressure at the centre of the pipe is 19.6 N/cm². If the pipe is filled with oil of sp. gr. 0.87, find the force exerted by the oil upon the gate and position of centre of pressure.* (Converted to SI Units, A.M.I.E., Winter, 1975)

Solution. Given :

Dia. of pipe, $d = 4$ m

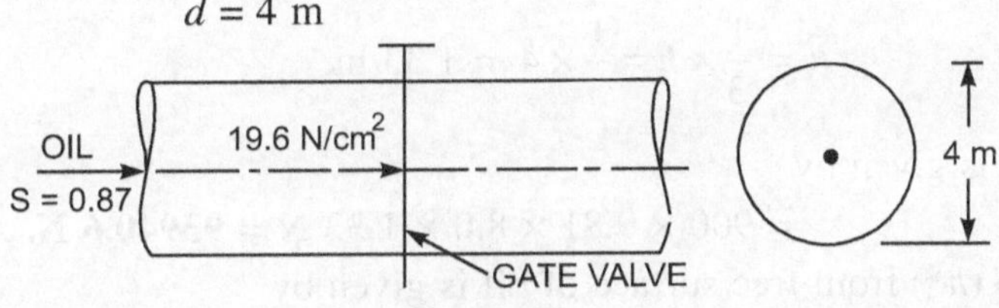

Fig. 3.7

∴ Area, $A = \dfrac{\pi}{4} \times 4^2 = 4\pi\ m^2$

Sp. gr. of oil, $S = 0.87$

∴ Density of oil, $\rho_0 = 0.87 \times 1000 = 870\ kg/m^3$

∴ Weight density of oil, $w_0 = \rho_0 \times g = 870 \times 9.81\ N/m^3$

Pressure at the centre of pipe, $p = 19.6\ N/cm^2 = 19.6 \times 10^4\ N/m^2$

∴ Pressure head at the centre $= \dfrac{p}{w_0} = \dfrac{19.6 \times 10^4}{870 \times 9.81} = 22.988$ m

∴ The height of equivalent free oil surface from the centre of pipe = 22.988 m.

The depth of C.G. of the gate valve from free oil surface $\bar{h} = 22.988$ m.

(*i*) Now the force exerted by the oil on the gate is given by

$$F = \rho g A \bar{h}$$

where ρ = density of oil = 870 kg/m³

$$F = 870 \times 9.81 \times 4\pi \times 22.988 = \mathbf{2465500\ N = 2.465\ MN.\ Ans.}$$

(*ii*) Position of centre of pressure (h^*) is given by (3.5) as

$$h^* = \frac{I_G}{A\bar{h}} + \bar{h} = \frac{\frac{\pi}{64}d^4}{\frac{\pi}{4}d^2 \times \bar{h}} + \bar{h} = \frac{d^2}{16\bar{h}} + \bar{h} = \frac{4^2}{16 \times 22.988} + 22.988$$

$$= 0.043 + 22.988 = \mathbf{23.031\ m.\ Ans.}$$

Or centre of pressure is below the centre of the pipe by a distance of 0.043 m. **Ans.**

Problem 3.6 *Determine the total pressure and centre of pressure on an isosceles triangular plate of base 4 m and altitude 4 m when it is immersed vertically in an oil of sp. gr. 0.9. The base of the plate coincides with the free surface of oil.*

FREE OIL SURFACE — 4 m — G — P — 4 m

Fig. 3.8

Solution. Given :

Base of plate, $b = 4$ m

Height of plate, $h = 4$ m

∴ Area, $A = \frac{b \times h}{2} = \frac{4 \times 4}{2} = 8.0\ \text{m}^2$

Sp. gr. of oil, $S = 0.9$

∴ Density of oil, $\rho = 900\ \text{kg/m}^3$.

The distance of C.G. from free surface of oil,

$$\bar{h} = \frac{1}{3} \times h = \frac{1}{3} \times 4 = 1.33\ \text{m}.$$

Total pressure (F) is given by $F = \rho g A \bar{h}$

$$= 900 \times 9.81 \times 8.0 \times 1.33\ \text{N} = \mathbf{93940.6\ N.\ Ans.}$$

Centre of pressure (h^*) from free surface of oil is given by

$$h^* = \frac{I_G}{A\bar{h}} + \bar{h}$$

where I_G = M.O.I. of triangular section about its C.G.

$$= \frac{bh^3}{36} = \frac{4 \times 4^3}{36} = 7.11\ \text{m}^4$$

∴ $$h^* = \frac{7.11}{8.0 \times 1.33} + 1.33 = 0.6667 + 1.33 = \mathbf{1.99\ m.\ Ans.}$$

Problem 3.7 *A vertical sluice gate is used to cover an opening in a dam. The opening is 2 m wide and 1.2 m high. On the upstream of the gate, the liquid of sp. gr. 1.45, lies upto a height of 1.5 m above the top of the gate, whereas on the downstream side the water is available upto a height touching the top of the gate. Find the resultant force acting on the gate and position of centre of pressure. Find also the force acting horizontally at the top of the gate which is capable of opening it. Assume that the gate is hinged at the bottom.* (A.M.I.E., May 1975)

Solution. Given :

Width of gate, $b = 2$ m

Depth of gate, $d = 1.2$ m

∴ Area, $A = b \times d = 2 \times 1.2 = 2.4\ \text{m}^2$

Sp. gr. of liquid $= 1.45$

∴ Density of liquid, $\rho_1 = 1.45 \times 1000 = 1450 \text{ kg/m}^3$

Let F_1 = Force exerted by the fluid of sp. gr. 1.45 on gate

F_2 = Force exerted by water on the gate.

The force F_1 is given by $F_1 = \rho_1 g \times A \times \bar{h}_1$

where $\rho_1 = 1.45 \times 1000 = 1450 \text{ kg/m}^2$

$\bar{h}_1$ = Depth of C.G. of gate from free surface of liquid

$$= 1.5 + \frac{1.2}{2} = 2.1 \text{ m.}$$

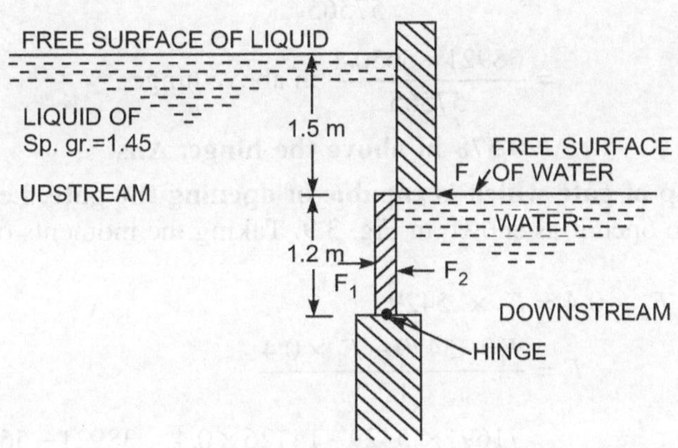

Fig. 3.9

∴ $F_1 = 1450 \times 9.81 \times 2.4 \times 2.1 = 71691 \text{ N}$

Similarly, $F_2 = \rho_2 g . A\bar{h}_2$

where $\rho_2 = 1{,}000 \text{ kg/m}^3$

$\bar{h}_2$ = Depth of C.G. of gate from free surface of water

$$= \frac{1}{2} \times 1.2 = 0.6 \text{ m}$$

∴ $F_2 = 1000 \times 9.81 \times 2.4 \times 0.6 = 14126 \text{ N}$

(i) **Resultant force on the gate** = $F_1 - F_2 = 71691 - 14126 =$ **57565 N. Ans.**

(ii) **Position of centre of pressure of resultant force.** The force F_1 will be acting at a depth of $h_1{}^*$ from free surface of liquid, given by the relation

$$h_1{}^* = \frac{I_G}{A\bar{h}_1} + \bar{h}_1$$

where $$I_G = \frac{bd^3}{12} = \frac{2 \times 1.2^3}{12} = 0.288 \text{ m}^4$$

∴ $$h_1{}^* = \frac{.288}{2.4 \times 2.1} + 2.1 = 0.0571 + 2.1 = 2.1571 \text{ m}$$

∴ Distance of F_1 from hinge

$$= (1.5 + 1.2) - h_1{}^* = 2.7 - 2.1571 = 0.5429 \text{ m}$$

The force F_2 will be acting at a depth of $h_2{}^*$ from free surface of water and is given by

$$h_2^* = \frac{I_G}{A\bar{h}_2} + \bar{h}_2$$

where $I_G = 0.288$ m^4, $\bar{h}_2 = 0.6$ m, $A = 2.4$ m^2,

$$h_2^* = \frac{0.288}{2.4 \times 0.6} + 0.6 = 0.2 + 0.6 = 0.8 \text{ m}$$

Distance of F_2 from hinge $= 1.2 - 0.8 = 0.4$ m

The resultant force 57565 N will be acting at a distance given by

$$= \frac{71691 \times .5429 - 14126 \times 0.4}{57565}$$

$$= \frac{38921 - 5650.4}{57565} \text{ m above hinge}$$

$= \mathbf{0.578}$ **m above the hinge. Ans.**

(*iii*) **Force at the top of gate which is capable of opening the gate.** Let F is the force required on the top of the gate to open it as shown in Fig. 3.9. Taking the moments of F, F_1 and F_2 about the hinge, we get

$$F \times 1.2 + F_2 \times 0.4 = F_1 \times .5429$$

or

$$F = \frac{F_1 \times .5429 - F_2 \times 0.4}{1.2}$$

$$= \frac{71691 \times .5429 - 14126 \times 0.4}{1.2} = \frac{38921 - 5650.4}{1.2}$$

$= \mathbf{27725.5}$ **N. Ans.**

Problem 3.8 *A caisson for closing the entrance to a dry dock is of trapezoidal form 16 m wide at the top and 10 m wide at the bottom and 6 m deep. Find the total pressure and centre of pressure on the caisson if the water on the outside is just level with the top and dock is empty.*

Solution. Given :

Width at top $= 16$ m

Width at bottom $= 10$ m

Depth, $d = 6$ m

Area of trapezoidal $ABCD$,

$$A = \frac{(BC + AD)}{2} \times d$$

$$= \frac{(10 + 16)}{2} \times 6 = 78 \text{ m}^2$$

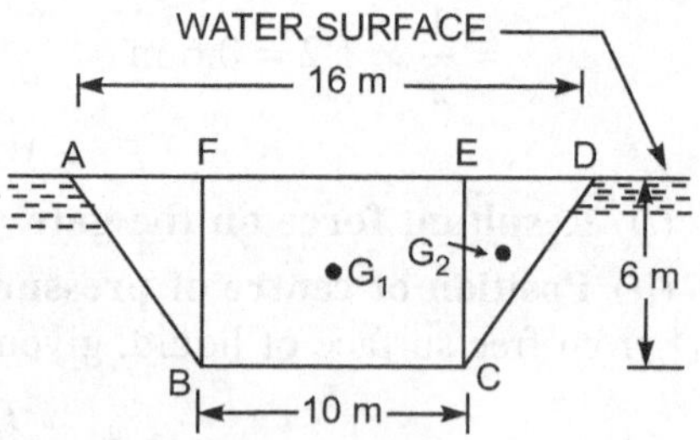

Fig. 3.10

Depth of C.G. of trapezoidal area $ABCD$ from free surface of water,

$$\bar{h} = \frac{10 \times 6 \times 3 + \frac{(16 - 10)}{2} \times 6 \times \frac{1}{3} \times 6}{78}$$

$$= \frac{180 + 36}{78} = 2.769 \text{ m from water surface.}$$

(*i*) **Total Pressure (F).** Total pressure, F is given by

$$F = \rho g A\bar{h} = 1000 \times 9.81 \times 78 \times 2.769 \text{ N}$$
$$= 2118783 \text{ N} = \mathbf{2.118783 \text{ MN. Ans.}}$$

(*ii*) **Centre of Pressure (h*).** Centre of pressure is given by equation (3.5) as

$$h^* = \frac{I_G}{A\bar{h}} + \bar{h}$$

where I_G = M.O.I. of trapezoidal *ABCD* about its C.G.

Let I_{G_1} = M.O.I. of rectangle *FBCE* about its C.G.

I_{G_2} = M.O.I. of two Δs *ABF* and *ECD* about its C.G.

Then $$I_{G_1} = \frac{bd^3}{12} = \frac{10 \times 6^3}{12} = 180 \text{ m}^4$$

I_{G_1} is the M.O.I. of the rectangle about the axis passing through G_1.

∴ M.O.I. of the rectangle about the axis passing through the C.G. of the trapezoidal I_{G_1} + Area of rectangle × x_1^2

where x_1 is distance between the C.G. of rectangle and C.G. of trapezoidal

$$= (3.0 - 2.769) = 0.231 \text{ m}$$

∴ M.O.I. of *FBCE* passing through C.G. of trapezoidal

$$= 180 + 10 \times 6 \times (0.231)^2 = 180 + 3.20 = 183.20 \text{ m}^4$$

Now $$I_{G_2} = \text{M.O.I. of } \Delta ABD \text{ in Fig. 3.11 about } G_2 = \frac{bd^3}{36}$$

$$= \frac{(16-10)\times 6^3}{36} = 36 \text{ m}^4$$

The distance between the C.G. of triangle and C.G. of trapezoidal

$$= (2.769 - 2.0) = 0.769$$

∴ M.O.I. of the two Δs about an axis passing through C.G. of trapezoidal

$$= I_{G_2} + \text{Area of triangles} \times (.769)^2$$

$$= 36.0 + \frac{6 \times 6}{2} \times (.769)^2$$

$$= 36.0 + 10.64 = 46.64$$

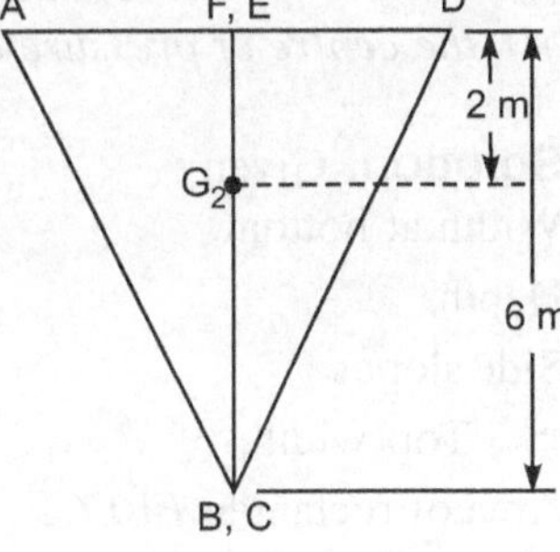

Fig. 3.11

∴ I_G = M.O.I. of trapezoidal about its C.G.

= M.O.I. of rectangle about the C.G. of trapezoidal

+ M.O.I. of triangles about the C.G. of the trapezoidal

$$= 183.20 + 46.64 = 229.84 \text{ m}^4$$

∴ $$h^* = \frac{I_G}{A\bar{h}} + \bar{h}$$

where $A = 78$, $\bar{h} = 2.769$

$$h^* = \frac{229.84}{78 \times 2.769} + 2.769 = 1.064 + 2.769 = \mathbf{3.833 \text{ m. Ans.}}$$

Alternate Method

The distance of the C.G. of the trapezoidal channel from surface *AD* is given by (refer to Table 3.1 on page 71)

$$x = \frac{(2a+b)}{(a+b)} \times \frac{h}{3}$$

$$= \frac{(2 \times 10 + 16)}{(10+16)} \times \frac{6}{3} \qquad (\because \quad a = 10, b = 16 \text{ and } h = 6)$$

$$= \frac{36}{26} \times 2 = 2.769 \text{ m}$$

This is also equal to the distance of the C.G. of the trapezoidal from free surface of water.

$$\bar{h} = 2.769 \text{ m}$$

$\therefore$ Total pressure, $\quad F = \rho g A \bar{h} \qquad (\because \quad A = 78)$

$$= 1000 \times 9.81 \times 78 \times 2.769 \text{ N} = \mathbf{2118783 \text{ N. Ans.}}$$

Centre of Pressure

$$(h^*) = \frac{I_G}{A\bar{h}} + \bar{h}$$

Now I_G from Table 3.1 is given by,

$$I_G = \frac{\left(a^2 + 4ab + b^2\right)}{36(a+b)} \times h^3 = \frac{\left(10^2 + 4 \times 10 \times 16 + 16^2\right)}{36(10+16)} \times 6^3$$

$$= \frac{(100 + 640 + 256)}{36 \times 26} \times 216 = 229.846 \text{ m}^4$$

$$\therefore \quad h^* = \frac{229.846}{78 \times 2.769} + 2.769 \qquad (\because \quad A = 78 \text{ m}^2)$$

$$= \mathbf{3.833 \text{ m. Ans.}}$$

Problem 3.9 *A trapezoidal channel 2 m wide at the bottom and 1 m deep has side slopes 1 : 1. Determine :*

(i) the total pressure, and

(ii) the centre of pressure on the vertical gate closing the channel when it is full of water.

(A.M.I.E., Summer, 1978)

Solution. Given :

Width at bottom $= 2$ m

Depth, $d = 1$ m

Side slopes $= 1 : 1$

$\therefore$ Top width, $AD = 2 + 1 + 1 = 4$ m

Area of rectangle *FBEC*, $A_1 = 2 \times 1 = 2 \text{ m}^2$

Area of two triangles *ABF* and *ECD*, $A_2 = \frac{(4-2)}{2} \times 1 = 1 \text{ m}^2$

Fig. 3.12

$\therefore$ Area of trapezoidal *ABCD*, $A = A_1 + A_2 = 2 + 1 = 3 \text{ m}^2$

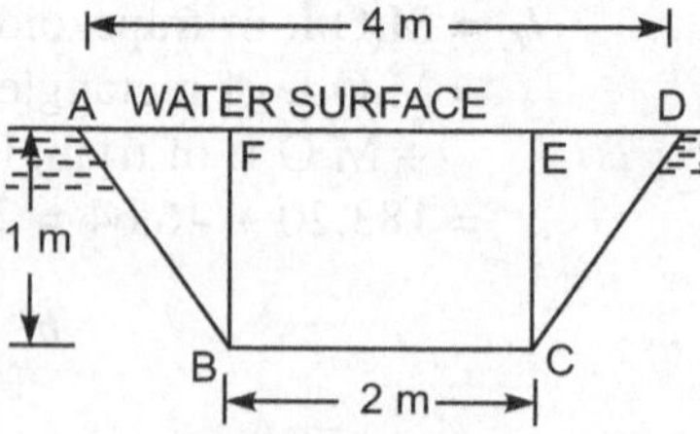

Depth of C.G. of rectangle *FBEC* from water surface,

$$\bar{h}_1 = \frac{1}{2} = 0.5 \text{ m}$$

Depth of C.G. of two triangles *ABF* and *ECD* from water surface,

$$\bar{h}_2 = \frac{1}{3} \times 1 = \frac{1}{3} \text{ m}$$

∴ Depth of C.G. of trapezoidal *ABCD* from free surface of water

$$\bar{h} = \frac{A_1 \times \bar{h}_1 + A_2 \times \bar{h}_2}{(A_1 + A_2)} = \frac{2 \times 0.5 + 1 \times 0.33333}{(2+1)} = .44444$$

(*i*) **Total Pressure (F).** Total pressure *F* is given by

$$F = \rho g A \bar{h}$$

$$= 1000 \times 9.81 \times 3.0 \times 0.44444 = \textbf{13079.9 N. Ans.}$$

(*ii*) **Centre of Pressure (h*).** M.O.I. of rectangle *FBCE* about its C.G.,

$$I_{G_1} = \frac{bd^3}{12} = \frac{2 \times 1^3}{12} = \frac{1}{6} \text{ m}^4$$

M.O.I. of *FBCE* about an axis passing through the C.G. of trapezoidal

or $$I_{G_1}{}^* = I_{G_1} + A_1 \times [\text{Distance between C.G. of rectangle and C.G. of trapezoidal}]^2$$

$$= \frac{1}{6} + 2 \times [\bar{h}_1 - \bar{h}]^2$$

$$= \frac{1}{6} + 2 \times [0.5 - .4444]^2 = .1666 + .006182 = 0.1727$$

M.O.I. of the two triangles *ABF* and *ECD* about their C.G.,

$$I_{G_2} = \frac{bd^3}{36} = \frac{(1+1) \times 1^3}{36} = \frac{2}{36} = \frac{1}{18} \text{ m}^4.$$

M.O.I. of the two triangles about the C.G. of trapezoidal,

$$I_{G_2}{}^* = I_{G_1} + A_2 \times [\text{Distance between C.G. of triangles and C.G. of trapezoidal}]^2$$

$$= \frac{1}{18} + 1 \times [\bar{h} - \bar{h}_2]^2 = \frac{1}{18} + 1 \times \left[.4444 - \frac{1}{3}\right]^2$$

$$= \frac{1}{18} + (.1111)^2 = 0.0555 + (.1111)^2$$

$$= .0555 + 0.01234 = 0.06789 \text{ m}^4$$

∴ M.O.I. of the trapezoidal about its C.G.

$$I_G = I_{G_1}{}^* + I_{G_2}{}^* = .1727 + .06789 = 0.24059 \text{ m}^4$$

Then centre of pressure (h^*) on the vertical trapezoidal,

$$h^* = \frac{I_G}{A\bar{h}} + \bar{h} = \frac{0.24059}{3 \times .4444} + .4444 = 0.18046 + .4444 = 0.6248$$

$$\simeq \textbf{0.625 m. Ans.}$$

Alternate Method

The distance of the C.G. of the trapezoidal channel from surface *AD* is given by (refer to Table 3.1 on page 71).

$$x = \frac{(2a+b)}{(a+b)} \times \frac{h}{3} = \frac{(2\times2+4)}{(2+4)} \times \frac{1}{3} \qquad (\because \quad a = 2,\ b = 4 \text{ and } h = 1)$$

$$= 0.444 \text{ m}$$

$\therefore$ $\bar{h} = x = 0.444$ m

$\therefore$ Total pressure, $F = \rho g A\bar{h} = 1000 \times 9.81 \times 3.0 \times .444 \qquad (\because \quad A = 3.0)$

$= \mathbf{13079}$ **N. Ans.**

Centre of pressure, $h^* = \dfrac{I_G}{A\bar{h}} + \bar{h}$

where I_G from Table 3.1 is given by

$$I_G = \frac{(a^2+4ab+b^2)}{36(a+b)} \times h^3 = \frac{(2^2+4\times2\times4+4^2)}{36(2+4)} \times 1^3 = \frac{52}{36\times6} = 0.2407 \text{ m}^4$$

$$\therefore \quad h^* = \frac{0.2407}{3.0\times.444} + .444 = \mathbf{0.625 \text{ m. Ans.}}$$

Problem 3.10 *A square aperture in the vertical side of a tank has one diagonal vertical and is completely covered by a plane plate hinged along one of the upper sides of the aperture. The diagonals of the aperture are 2 m long and the tank contains a liquid of specific gravity 1.15. The centre of aperture is 1.5 m below the free surface. Calculate the thrust exerted on the plate by the liquid and position of its centre of pressure.* (A.M.I.E., Summer, 1986)

Solution. Given : Diagonals of aperture, $AC = BD = 2$ m

$\therefore$ Area of square aperture, A = Area of ΔACB + Area of ΔACD

$$= \frac{AC\times BO}{2} + \frac{AC\times OD}{2} = \frac{2\times1}{2} + \frac{2\times1}{2} = 1 + 1 = 2.0 \text{ m}^2$$

Sp. gr. of liquid $= 1.15$

$\therefore$ Density of liquid, $\rho = 1.15 \times 1000 = 1150$ kg/m^3

Depth of centre of aperture from free surface,

$\bar{h} = 1.5$ m.

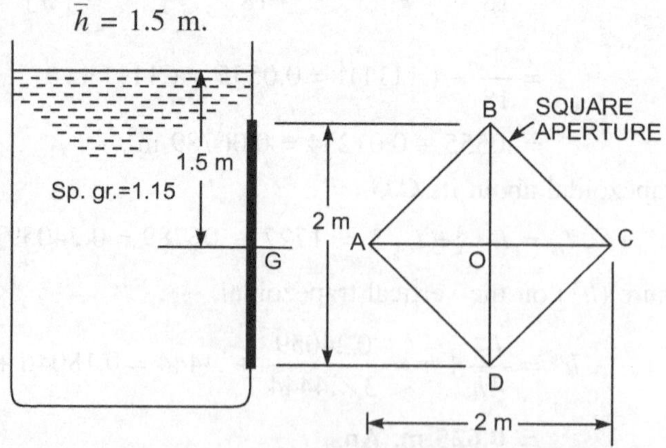

Fig. 3.13

(*i*) The thrust on the plate is given by

$$F = \rho g A\bar{h} = 1150 \times 9.81 \times 2 \times 1.5 = \mathbf{33844.5.\ Ans.}$$

(*ii*) Centre of pressure (h^*) is given by

$$h^* = \frac{I_G}{A\bar{h}} + \bar{h}$$

where I_G = M.O.I. of *ABCD* about diagonal *AC*

= M.O.I. of triangle *ABO* about *AC* + M.O.I. of triangle *ACD* about *AC*

$$= \frac{AC \times OB^3}{12} + \frac{AC \times OD^3}{12} \qquad \left(\because \text{M.O.I. of a triangle about its base} = \frac{bh^3}{12}\right)$$

$$= \frac{2 \times 1^3}{12} + \frac{2 \times 1^3}{12} = \frac{1}{6} + \frac{1}{6} = \frac{1}{3}\ \text{m}^4$$

$$\therefore \qquad h^* = \frac{\frac{1}{3}}{2 \times 1.5} + 1.5 = \frac{1}{3 \times 2 \times 1.5} + 1.5 = \mathbf{1.611\ m.\ Ans.}$$

Problem 3.11 *A tank contains water upto a height of 0.5 m above the base. An immiscible liquid of sp. gr. 0.8 is filled on the top of water upto 1 m height. Calculate :*

(i) total pressure on one side of the tank,

(ii) the position of centre of pressure for one side of the tank, which is 2 m wide.

Solution. Given :

Depth of water = 0.5 m

Depth of liquid = 1 m

Sp. gr. of liquid = 0.8

Density of liquid, $\rho_1 = 0.8 \times 1000 = 800$ kg/m^3

Density of water, $\rho_2 = 1000$ kg/m^3

Width of tank = 2 m

(*i*) **Total pressure on one side** is calculated by drawing pressure diagram, which is shown in Fig. 3.14.

Intensity of pressure on top, $p_A = 0$

Intensity of pressure on *D* (or *DE*), $p_D = \rho_1 g.h_1$

$= 800 \times 9.81 \times 1.0 = 7848$ N/m^2

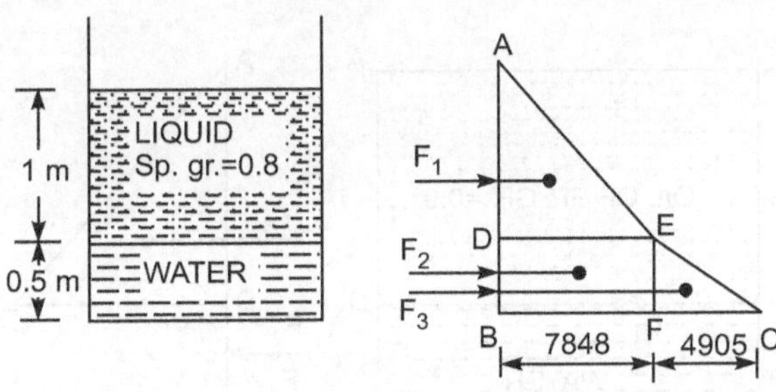

Fig. 3.14

Intensity of pressure on base (or *BC*), $p_B = \rho_1 g h_1 + \rho_2 g \times 0.5$

$= 7848 + 1000 \times 9.81 \times 0.5 = 7848 + 4905 = 12753$ N/m^2

Now force $\qquad F_1$ = Area of ΔADE × Width of tank

$$= \frac{1}{2} \times AD \times DE \times 2.0 = \frac{1}{2} \times 1 \times 7848 \times 2.0 = 7848\ \text{N}$$

Force F_2 = Area of rectangle $DBFE$ × Width of tank

$= 0.5 \times 7848 \times 2 = 7848$ N

F_3 = Area of ΔEFC × Width of tank

$$= \frac{1}{2} \times EF \times FC \times 2.0 = \frac{1}{2} \times 0.5 \times 4905 \times 2.0 = 2452.5 \text{ N}$$

∴ Total pressure, $F = F_1 + F_2 + F_3$

$= 7848 + 7848 + 2452.5 =$ **18148.5 N. Ans.**

(*ii*) **Centre of Pressure (h*).** Taking the moments of all force about *A*, we get

$$F \times h^* = F_1 \times \frac{2}{3} AD + F_2 (AD + \frac{1}{2} BD) + F_3[AD + \frac{2}{3} BD]$$

$$18148.5 \times h^* = 7848 \times \frac{2}{3} \times 1 + 7848 \left(1.0 + \frac{0.5}{2}\right) + 2452.5 \left(1.0 + \frac{2}{3} \times .5\right)$$

$$= 5232 + 9810 + 3270 = 18312$$

$$\therefore \quad h^* = \frac{18312}{18148.5} = \textbf{1.009 m from top. Ans.}$$

Problem 3.12 *A cubical tank has sides of 1.5 m. It contains water for the lower 0.6 m depth. The upper remaining part is filled with oil of specific gravity 0.9. Calculate for one vertical side of the tank :*

(*a*) *total pressure, and*

(*b*) *position of centre of pressure.* (A.M.I.E., Winter, 1987)

Solution. Given :

Cubical tank of sides 1.5 m means the dimensions of the tank are 1.5 m × 1.5 m × 1.5 m.

Depth of water = 0.6 m

Depth of liquid = 1.5 – 0.6 = 0.9 m

Sp. gr. of liquid = 0.9

Density of liquid, $\rho_1 = 0.9 \times 1000 = 900$ kg/m^3

Density of water, $\rho_2 = 1000$ kg/m^3

(*a*) **Total pressure** on one vertical side is calculated by drawing pressure diagram, which is shown in Fig. 3.15.

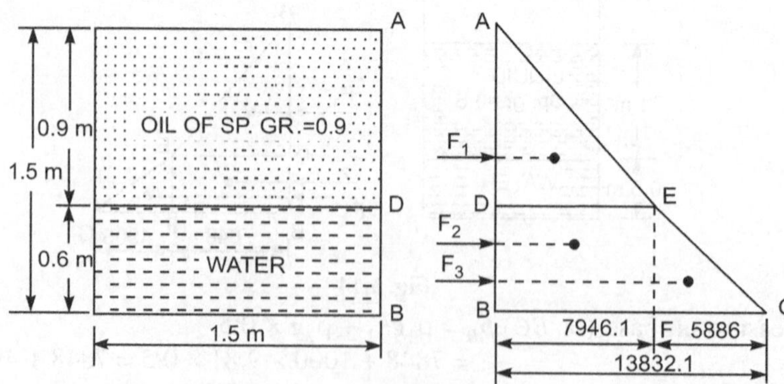

Fig. 3.15

Intensity of pressure at A, $p_A = 0$

Intensity of pressure at $D, p_D = \rho_1 g \times h = 900 \times 9.81 \times 0.9 = 7946.1 \text{ N/m}^2$

Intensity of pressure at B, $p_B = \rho_1 g h_1 + \rho_2 g h_2 = 900 \times 9.81 \times 0.9 + 1000 \times 9.81 \times 0.6$

$= 7946.1 + 5886 = 13832.1 \text{ N/m}^2$

Hence in pressure diagram :

$$DE = 7946.1 \text{ N/m}^2, BC = 13832.1 \text{ N/m}^2, FC = 5886 \text{ N/m}^2$$

The pressure diagram is split into triangle ADE, rectangle $BDEF$ and triangle EFC. The total pressure force consists of the following components :

(*i*) Force $F_1 = \text{Area of triangle } ADE \times \text{Width of tank}$

$= \left(\frac{1}{2} \times AD \times DE\right) \times 1.5$ $\quad (\because \text{ Width} = 1.5 \text{ m})$

$= \left(\frac{1}{2} \times 0.9 \times 7946.1\right) \times 1.5 \text{ N} = 5363.6 \text{ N}$

This force will be acting at the C.G. of the triangle ADE, *i.e.*, at a distance of $\frac{2}{3} \times 0.9 = 0.6$ m below A

(*ii*) Force $F_2 = \text{Area of rectangle } BDEF \times \text{Width of tank}$

$= (BD \times DE) \times 1.5 = (0.6 \times 7946.1) \times 1.5 = 7151.5$

This force will be acting at the C.G. of the rectangle $BDEF$ *i.e.*, at a distance of $0.9 + \frac{0.6}{2} = 1.2$ m below A.

(*iii*) Force $F_3 = \text{Area of triangle } EFC \times \text{Width of tank}$

$= \left(\frac{1}{2} \times EF \times FC\right) \times 1.5 = \left(\frac{1}{2} \times 0.6 \times 5886\right) \times 1.5 = 2648.7 \text{ N}$

This force will be acting at the C.G. of the triangle EFC, *i.e.*, at a distance of $0.9 + \frac{2}{3} \times 0.6 = 1.30$ m below A.

$\therefore$ Total pressure force on one vertical face of the tank,

$$F = F_1 + F_2 + F_3$$
$= 5363.6 + 7151.5 + 2648.7 =$ **15163.8 N. Ans.**

(*b*) **Position of centre of pressure**

Let the total force F is acting at a depth of h^* from the free surface of liquid, *i.e.*, from A. Taking the moments of all forces about A, we get

$$F \times h^* = F_1 \times 0.6 + F_2 \times 1.2 + F_3 \times 1.3$$

or
$$h^* = \frac{F_1 \times 0.6 + F_2 \times 1.2 + F_3 \times 1.3}{F}$$

$$= \frac{5363.6 \times 0.6 + 7151.5 \times 1.2 + 2648.7 \times 1.3}{15163.8}$$

= **1.005 m from A. Ans.**

▶ 3.4 HORIZONTAL PLANE SURFACE SUBMERGED IN LIQUID

Consider a plane horizontal surface immersed in a static fluid. As every point of the surface is at the same depth from the free surface of the liquid, the pressure intensity will be equal on the entire surface and equal to, $p = \rho g h$, where h is depth of surface.

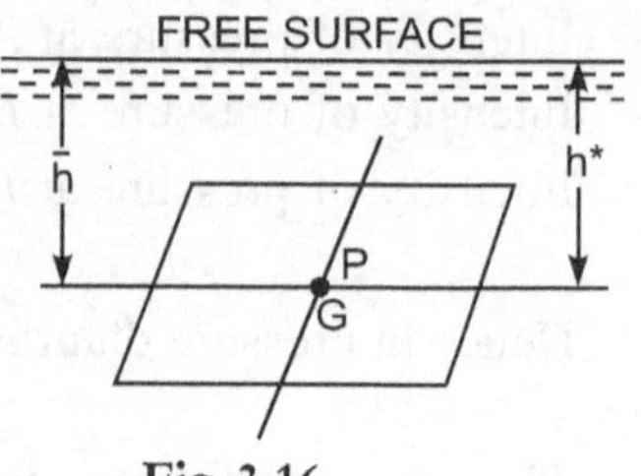

Fig. 3.16

Let $\qquad A$ = Total area of surface

Then total force, F, on the surface

$$= p \times \text{Area} = \rho g \times h \times A = \rho g A\bar{h}$$

where $\bar{h}$ = Depth of C.G. from free surface of liquid = h

also $\quad h^*$ = Depth of centre of pressure from free surface = h.

Problem 3.13 *Fig. 3.17 shows a tank full of water. Find :*

(i) Total pressure on the bottom of tank.

(ii) Weight of water in the tank.

(iii) Hydrostatic paradox between the results of (i) and (ii). Width of tank is 2 m.

Solution. Given :

Depth of water on bottom of tank

$$h_1 = 3 + 0.6 = 3.6 \text{ m}$$

Width of tank $\qquad = 2$ m

Length of tank at bottom $\qquad = 4$ m

$\therefore$ Area at the bottom, $\quad A = 4 \times 2 = 8 \text{ m}^2$

(*i*) Total pressure F, on the bottom is

$$F = \rho g A\bar{h} = 1000 \times 9.81 \times 8 \times 3.6$$

$$= \mathbf{282528 \text{ N. Ans.}}$$

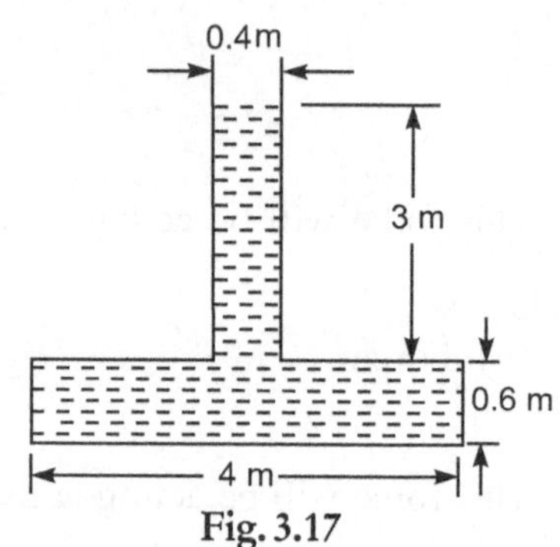

Fig. 3.17

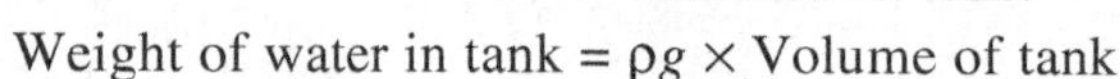

(*ii*) Weight of water in tank = ρg × Volume of tank

$$= 1000 \times 9.81 \times [3 \times 0.4 \times 2 + 4 \times .6 \times 2]$$

$$= 1000 \times 9.81\ [2.4 + 4.8] = \mathbf{70632 \text{ N. Ans.}}$$

(*iii*) From the results of (*i*) and (*ii*), it is observed that the total weight of water in the tank is much less than the total pressure at the bottom of the tank. This is known as **Hydrostatic paradox.**

▶ 3.5 INCLINED PLANE SURFACE SUBMERGED IN LIQUID

Consider a plane surface of arbitrary shape immersed in a liquid in such a way that the plane of the surface makes an angle θ with the free surface of the liquid as shown in Fig. 3.18.

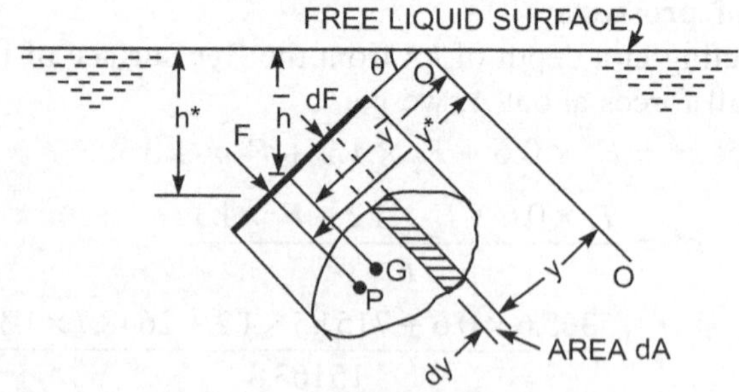

Fig. 3.18 *Inclined immersed surface.*

Let $\quad A$ = Total area of inclined surface

$\bar{h}$ = Depth of C.G. of inclined area from free surface

h^* = Distance of centre of pressure from free surface of liquid

θ = Angle made by the plane of the surface with free liquid surface.

Let the plane of the surface, if produced meet the free liquid surface at *O*. Then *O-O* is the axis perpendicular to the plane of the surface.

Let $\bar{y}$ = distance of the C.G. of the inclined surface from *O-O*

y^* = distance of the centre of pressure from *O-O*.

Consider a small strip of area *dA* at a depth '*h*' from free surface and at a distance *y* from the axis *O-O* as shown in Fig. 3.18.

Pressure intensity on the strip, $p = \rho g h$

$\therefore$ Pressure force, *dF*, on the strip, $dF = p \times \text{Area of strip} = \rho g h \times dA$

Total pressure force on the whole area, $F = \int dF = \int \rho g h dA$

But from Fig. 3.18, $\dfrac{h}{y} = \dfrac{\bar{h}}{\bar{y}} = \dfrac{h^*}{y^*} = \sin\theta$

$$\therefore \quad h = y \sin\theta$$

$$\therefore \quad F = \int \rho g \times y \times \sin\theta \times dA = \rho g \sin\theta \int y dA$$

But $\int y dA = A\bar{y}$

where $\bar{y}$ = Distance of C.G. from axis *O-O*

$$\therefore \quad F = \rho g \sin\theta\, \bar{y} \times A$$

$$= \rho g A \bar{h} \qquad (\because\ \bar{h} = \bar{y} \sin\theta) \quad \text{...(3.6)}$$

Centre of Pressure (h*)

Pressure force on the strip, $dF = \rho g h dA$

$= \rho g y \sin\theta\, dA$ $\quad [h = y \sin\theta]$

Moment of the force, *dF*, about axis *O-O*

$$= dF \times y = \rho g y \sin\theta\, dA \times y = \rho g \sin\theta\, y^2 dA$$

Sum of moments of all such forces about *O-O*

$$= \int \rho g \sin\theta\, y^2\, dA = \rho g \sin\theta \int y^2\, dA$$

But $\int y^2\, dA$ = M.O.I. of the surface about *O-O* = I_0

$\therefore$ Sum of moments of all forces about *O-O* = $\rho g \sin\theta\, I_0$ ...(3.7)

Moment of the total force, *F*, about *O-O* is also given by

$= F \times y^*$ ...(3.8)

where y^* = Distance of centre of pressure from *O-O*.

Equating the two values given by equations (3.7) and (3.8)

$$F \times y^* = \rho g \sin\theta\, I_0$$

or $$y^* = \frac{\rho g \sin\theta\, I_0}{F} \quad \text{...(3.9)}$$

Now $y^* = \dfrac{h^*}{\sin\theta}$, $F = \rho g A \bar{h}$

and I_0 by the theorem of parallel axis $= I_G + A\bar{y}^2$.

Substituting these values in equation (3.9), we get

$$\frac{h^*}{\sin\theta} = \frac{\rho g \sin\theta}{\rho g A\bar{h}}\left[I_G + A\bar{y}^2\right]$$

$$\therefore \quad h^* = \frac{\sin^2\theta}{A\bar{h}}\left[I_G + A\bar{y}^2\right]$$

But $\quad \frac{\bar{h}}{\bar{y}} = \sin\theta \quad \text{or} \quad \bar{y} = \frac{\bar{h}}{\sin\theta}$

$$\therefore \quad h^* = \frac{\sin^2\theta}{A\bar{h}}\left[I_G + A \times \frac{\bar{h}^2}{\sin^2\theta}\right]$$

or $$h^* = \frac{I_G \sin^2\theta}{A\bar{h}} + \bar{h} \qquad ...(3.10)$$

If $\theta = 90°$, equation (3.10) becomes same as equation (3.5) which is applicable to vertically plane submerged surfaces.

In equation (3.10), I_G = M.O.I. of inclined surfaces about an axis passing through G and parallel to O-O.

Problem 3.14 (*a*) *A rectangular plane surface 2 m wide and 3 m deep lies in water in such a way that its plane makes an angle of 30° with the free surface of water. Determine the total pressure and position of centre of pressure when the upper edge is 1.5 m below the free water surface.*

Solution. Given :

Width of plane surface, $b = 2$ m

Depth, $d = 3$ m

Angle, $\theta = 30°$

Distance of upper edge from free water surface = 1.5 m

(*i*) **Total pressure** force is given by equation (3.6) as

$$F = \rho g A\bar{h}$$

where $\rho = 1000$ kg/m^3

$A = b \times d = 3 \times 2 = 6$ m^2

$\therefore \quad \bar{h}$ = Depth of C.G. from free water surface
$= 1.5 + 1.5 \sin 30°$

$\{\because \quad \bar{h} = AE + EB = 1.5 + BC \sin 30° = 1.5 + 1.5 \sin 30°\}$

$= 1.5 + 1.5 \times \frac{1}{2} = 2.25$ m

$\therefore \quad F = 1000 \times 9.81 \times 6 \times 2.25 =$ **132435 N. Ans.**

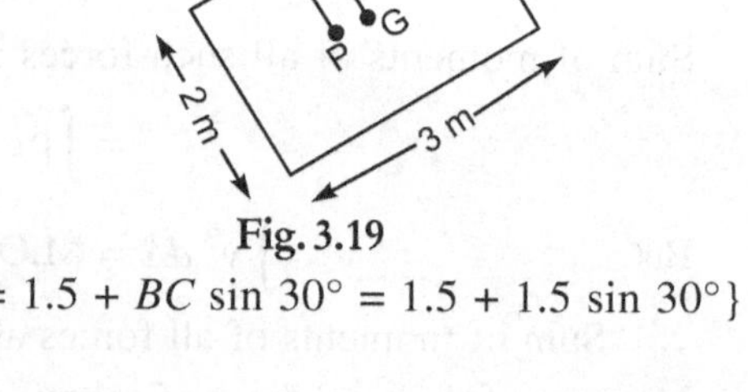

Fig. 3.19

(*ii*) **Centre of pressure (h*)**

Using equation (3.10), we have

$$h^* = \frac{I_G \sin^2\theta}{A\bar{h}} + \bar{h}, \quad \text{where } I_G = \frac{bd^3}{12} = \frac{2 \times 3^3}{12} = 4.5 \text{ m}^4$$

$$\therefore \quad h^* = \frac{4.5 \times \sin^2 30°}{6 \times 2.25} + 2.25 = \frac{4.5 \times \frac{1}{4}}{6 \times 2.25} + 2.25$$

$= 0.0833 + 2.25 =$ **2.3333 m. Ans.**

Problem 3.14 (*b*) *A rectangular plane surface 3 m wide and 4 m deep lies in water in such a way that its plane makes an angle of 30° with the free surface of water. Determine the total pressure force and position of centre of pressure, when the upper edge is 2 m below the free surface.*

(Delhi University, Dec. 2002)

Solution. Given :

$$b = 3 \text{ m}, \; d = 4 \text{ m}, \; \theta = 30°$$

Distance of upper edge from free surface of water = 2 m

(*i*) **Total pressure force** is given by equation (3.6) as

$$F = \rho g A \bar{h}$$

where $\rho = 1000 \text{ kg/m}^3$,

$A = b \times d = 3 \times 4 = 12 \text{ m}^2$

and $\bar{h}$ = Depth of C.G. of plate from free water surface

$= 2 + \text{BE} = 2 + \text{BC} \sin\theta$

$= 2 + 2 \sin 30° = 2 + 2 \times \frac{1}{2} = 3 \text{ m}$

Fig. 3.19 (*a*)

$\therefore$ $F = 1000 \times 9.81 \times 12 \times 3 = 353167 \text{ N} = \mathbf{353.167 \text{ kN. Ans.}}$

(*ii*) **Centre of pressure (h*)**

Using equation (3.10), we have $h^* = \dfrac{I_G \sin^2\theta}{A\bar{h}} + \bar{h}$

where $I_G = \dfrac{bd^3}{12} = \dfrac{3 \times 4^3}{12} = 16 \text{ m}^4$

$$\therefore \quad h^* = \frac{16 \times \sin^2 30°}{12 \times 3} + 3 = \frac{16 \times \frac{1}{4}}{36} + 3 = \mathbf{3.111 \text{ m. Ans.}}$$

Problem 3.15 (*a*) *A circular plate 3.0 m diameter is immersed in water in such a way that its greatest and least depth below the free surface are 4 m and 1.5 m respectively. Determine the total pressure on one face of the plate and position of the centre of pressure.*

Solution. Given :

Dia. of plate, $d = 3.0 \text{ m}$

$\therefore$ Area, $A = \dfrac{\pi}{4} d^2 = \dfrac{\pi}{4} (3.0)^2 = 7.0685 \text{ m}^2$

Distance $DC = 1.5 \text{ m}, \; BE = 4 \text{ m}$

Distance of C.G. from free surface

$$= \bar{h} = CD + GC \sin\theta = 1.5 + 1.5 \sin\theta$$

But $\sin\theta = \dfrac{AB}{BC} = \dfrac{BE - AE}{BC} = \dfrac{4.0 - DC}{3.0} = \dfrac{4.0 - 1.5}{3.0}$

$= \dfrac{2.5}{3.0} = 0.8333$

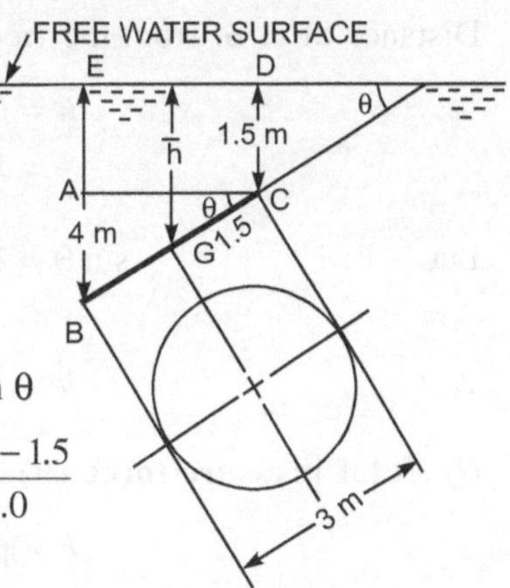

Fig. 3.20

$\therefore$ $\bar{h} = 1.5 + 1.5 \times .8333 = 1.5 + 1.249 = 2.749 \text{ m}$

(*i*) **Total pressure (F)**

$$F = \rho g A \bar{h}$$

$$= 1000 \times 9.81 \times 7.0685 \times 2.749 = \textbf{190621 N. Ans.}$$

(*ii*) **Centre of pressure (h*)**

Using equation (3.10), we have $h^* = \dfrac{I_G \sin^2 \theta}{A\bar{h}} + \bar{h}$

where $I_G = \dfrac{\pi}{64} d^4 = \dfrac{\pi}{64} (3)^4 = 3.976 \text{ m}^4$

$$h^* = \frac{3.976 \times (.8333) \times .8333}{7.0685 \times 2.749} + 2.749 = 0.1420 + 2.749$$

$$= \textbf{2.891 m. Ans.}$$

Problem 3.15 (*b*) *If in the above problem, the given circular plate is having a concentric circular hole of diameter 1.5 m, then calculate the total pressure and position of the centre of pressure on one face of the plate.*

Solution. Given : [Refer to Fig. 3.20 (*a*)]

Dia. of plate, $d = 3.0$ m

$\therefore$ Area of solid plate $= \dfrac{\pi}{4} d^2 = \dfrac{\pi}{4} (3)^2 = 7.0685 \text{ m}^2$

Dia. of hole in the plate, $d_0 = 1.5$ m

$\therefore$ Area of hole $= \dfrac{\pi}{4} d_0^2 = \dfrac{\pi}{4} \cdot (1.5)^2 = 1.7671 \text{ m}^2$

$\therefore$ Area of the given plate, A = Area of solid plate – Area of hole

$= 7.0685 - 1.7671 = 5.3014 \text{ m}^2$

Distance $CD = 1.5$, $BE = 4$ m

Fig. 3.20 (*a*)

Distance of C.G. from the free surface,

$$\bar{h} = CD + GC \sin \theta$$

$$= 1.5 + 1.5 \sin \theta$$

But $$\sin \theta = \frac{AB}{BC} = \frac{BE - AE}{BC} = \frac{4 - 1.5}{3} = \frac{2.5}{3}$$

$\therefore$ $$\bar{h} = 1.5 + 1.5 \times \frac{2.5}{3} = 1.5 + 1.25 = 2.75 \text{ m}$$

(*i*) **Total pressure force (F)**

$$F = \rho g A \bar{h}$$

$$= 1000 \times 9.81 \times 5.3014 \times 2.75$$

$$= 143018 \text{ N} = \textbf{143.018 kN. Ans.}$$

(*ii*) **Position of centre of pressure (h*)**

Using equation (3.10), we have

$$h^* = \frac{I_G \sin^2 \theta}{A\bar{h}} + \bar{h}$$

where $I_G = \frac{\pi}{64}[d^4 - d_0^4] = \frac{\pi}{64}[3^4 - 1.5^4] \text{ m}^4$

$$A = \frac{\pi}{4}[d^2 - d_0^2] = \frac{\pi}{4}[3^2 - 1.5^2] \text{ m}^2$$

$$\sin\theta = \frac{2.5}{3} \text{ and } \bar{h} = 2.75$$

$$\therefore \quad h^* = \frac{\frac{\pi}{64}\left[3^4 - 1.5^4\right] \times \left(\frac{2.5}{3}\right)^2}{\frac{\pi}{4}\left[3^2 - 1.5^2\right] \times 2.75} + 2.75$$

$$= \frac{\frac{1}{16}\left[3^2 + 1.5^2\right] \times \left(\frac{2.5}{3}\right)^2}{2.75} + 2.75 = \frac{1 \times 11.25 \times 6.25}{16 \times 2.75 \times 9} + 2.75$$

$$= 0.177 + 2.75 = \mathbf{2.927 \text{ m. Ans.}}$$

Problem 3.16 *A circular plate 3 metre diameter is submerged in water as shown in Fig. 3.21. Its greatest and least depths are below the surfaces being 2 metre and 1 metre respectively. Find : (i) the total pressure on front face of the plate, and (ii) the position of centre of pressure.*

(A.M.I.E., Winter, 1983)

Solution. Given :

Dia. of plate, $d = 3.0$ m

$\therefore$ Area, $A = \frac{\pi}{4}(3.0)^2 = 7.0685 \text{ m}^2$

Distance, $DC = 1$ m, $BE = 2$ m

In ΔABC, $\sin\theta = \frac{AB}{AC} = \frac{BE - AE}{BC} = \frac{BE - DC}{BC} = \frac{2.0 - 1.0}{3.0} = \frac{1}{3}$

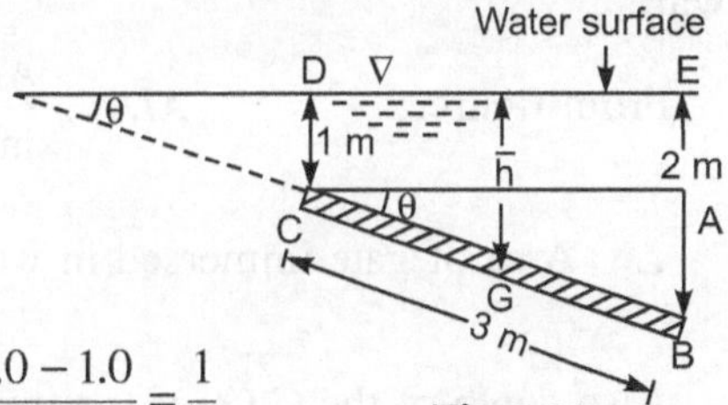

Fig. 3.21

The centre of gravity of the plate is at the middle of BC, *i.e.*, at a distance 1.5 m from C. The distance of centre of gravity from the free surface of the water is given by

$$\bar{h} = CD + CG \sin\theta = 1.0 + 1.5 \times \tfrac{1}{3} \qquad (\because \sin\theta = \tfrac{1}{3})$$

$$= 1.5 \text{ m.}$$

(*i*) Total pressure on the front face of the plate is given by

$$F = \rho g A \bar{h}$$

$$= 1000 \times 9.81 \times 7.0685 \times 1.5 = \mathbf{104013 \text{ N. Ans.}}$$

(*ii*) Let the distance of the centre of pressure from the free surface of the water be h^*. Then using equation (3.10), we have

$$h^* = \frac{I_G \sin^2 \theta}{A\bar{h}} + \bar{h}$$

where $I_G = \frac{\pi}{64} d^4 = \frac{\pi}{64}(3)^4$, $A = \frac{\pi}{4} d^2$, $\bar{h} = 1.5$ m and $\sin\theta = \frac{1}{3}$

Substituting the values, we get

$$h^* = \frac{\frac{\pi}{64} d^4 \times \left(\frac{1}{3}\right)^2}{\frac{\pi}{4} d^2 \times 1.5} + 1.5 = \frac{d^2}{16} \times \frac{1}{9 \times 1.5} + 1.5$$

$$= \frac{3^2}{16 \times 9 \times 1.5} + 1.5 = .0416 + 1.5 = \mathbf{1.5416\ m.\ Ans.}$$

Problem 3.17 *A rectangular gate 5 m × 2 m is hinged at its base and inclined at 60° to the horizontal as shown in Fig. 3.22. To keep the gate in a stable position, a counter weight of 5000 kgf is attached at the upper end of the gate as shown in figure. Find the depth of water at which the gate begins to fall. Neglect the weight of the gate and friction at the hinge and pulley.*

Solution. Given :

Length of gate $= 5$ m

Width of gate $= 2$ m

$\theta = 60°$

Weight, $W = 5000$ kgf

$= 5000 \times 9.81$ N

$= 49050$ N $\quad (\because 1 \text{ kgf} = 9.81 \text{ N})$

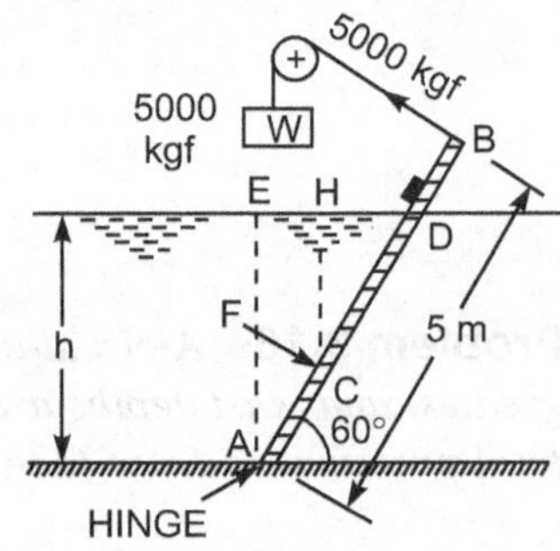

Fig. 3.22

As the pulley is frictionless, the force acting at $B = 49050$ N. First find the total force F acting on the gate AB for a given depth of water.

From figure, $AD = \frac{AE}{\sin\theta} = \frac{h}{\sin 60} = \frac{5}{\sqrt{3}/2} = \frac{2h}{\sqrt{3}}$

$\therefore$ Area of gate immersed in water, $A = AD \times \text{Width} \times \frac{2h}{\sqrt{3}} \times 2 = \frac{4h}{\sqrt{3}}\ \text{m}^2$

Also depth of the C.G. of the immersed area $= \bar{h} = \frac{h}{2} = 0.5\,h$

$\therefore$ Total force F is given by $F = \rho g A \bar{h} = 1000 \times 9.81 \times \frac{4h}{\sqrt{3}} \times \frac{h}{\sqrt{2}} = \frac{19620}{\sqrt{3}} h^2 \text{N}$

The centre of pressure of the immersed surface, h^* is given by

$$h^* = \frac{I_G \sin^2\theta}{A\bar{h}} + \bar{h}$$

where I_G = M.O.I. of the immersed area

$$= \frac{b \times (AD)^3}{12} = \frac{2}{12} \times \left(\frac{2h}{\sqrt{3}}\right) \qquad \left\{\because AD = \frac{2h}{\sqrt{3}}\right\}$$

$$= \frac{16h^3}{12 \times 3 \times \sqrt{3}} = \frac{4h^3}{9 \times \sqrt{3}}\ \text{m}^4$$

$$\therefore \quad h^* = \frac{4h^3}{9 \times \sqrt{3}} \times \frac{\left(\frac{\sqrt{3}}{2}\right)^2}{\frac{4h}{\sqrt{3}} \times \frac{h}{2}} + \frac{h}{2} = \frac{3h^3}{18h^2} \times \frac{h}{2} = \frac{h}{6} + \frac{h}{2} = \frac{h + 3h}{6} = \frac{2h}{3}$$

Now in the ΔCHD, $\quad CH = h^* = \dfrac{2h}{3}, \angle CDH = 60°$

$$\therefore \quad \frac{CH}{CD} = \sin 60°$$

$$\therefore \quad CD = \frac{CH}{\sin 60°} = \frac{h^*}{\sin 60°} = \frac{2h}{3 \times \frac{\sqrt{3}}{2}} = \frac{4h}{3 \times \sqrt{3}}$$

$$\therefore \quad AC = AD - CD = \frac{2h}{\sqrt{3}} - \frac{4h}{3\sqrt{3}} = \frac{6h - 4h}{3\sqrt{3}} = \frac{2h}{3\sqrt{3}} \text{ m}$$

Taking the moments about hinge, we get

$$49050 \times 5.0 = F \times AC = \frac{19620}{\sqrt{3}} h^2 \times \frac{2h}{3\sqrt{3}}$$

or

$$.245250 = \frac{39240\, h^3}{3 \times 3}$$

$$\therefore \quad h^3 = \frac{9 \times 245250}{39240} = 56.25$$

$$\therefore \quad h = (56.25)^{1/3} = \mathbf{3.83\ m.\ Ans.}$$

Problem 3.18 *An inclined rectangular sluice gate AB 1.2 m by 5 m size as shown in Fig. 3.23 is installed to controlled the discharge of water. The end A is hinged. Determine the force normal to the gate applied at B to open it.* (A.M.I.E., Summer, 1980)

Solution. Given :

$$A = \text{Area of gate} = 1.2 \times 5.0 = 6.0 \text{ m}^2$$

Depth of C.G. of the gate from free surface of the water $= \bar{h}$

$$= DG = BC - BE$$

$$= 5.0 - BG \sin 45°$$

$$= 5.0 - 0.6 \times \frac{1}{\sqrt{2}} = 4.576 \text{ m}$$

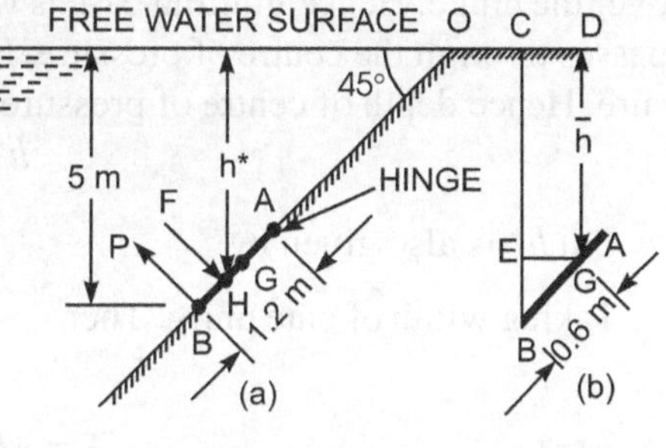

Fig. 3.23

The total pressure force (F) acting on the gate,

$$F = \rho g A \bar{h}$$

$$= 1000 \times 9.81 \times 6.0 \times 4.576$$

$$= 269343 \text{ N}$$

This force is acting at H, where the depth of H from free surface is given by

$$h^* = \frac{I_G \sin^2 \theta}{A\bar{h}} + \bar{h}$$

where I_G = M.O.I. of gate $= \frac{bd^3}{12} = \frac{5.0 \times 1.2^3}{12} = 0.72$ m

$\therefore$ Depth of centre of pressure $h^* = \frac{0.72 \times \sin^2 45°}{6 \times 4.576} + 4.576 = 0.13 + 4.576 = 4.589$ m

But from Fig. 3.23 (*a*), $\frac{h^*}{OH} = \sin 45°$

$\therefore$ Distance, $OH = \frac{h^*}{\sin 45°} = \frac{4.589}{\frac{1}{\sqrt{2}}} = 4.589 \times \sqrt{2} = 6.489$ m

Distance, $BO = \frac{5}{\sin 45°} = 5 \times \sqrt{2} = 7.071$ m

Distance, $BH = BO - OH = 7.071 - 6.489 = 0.582$ m

$\therefore$ Distance $AH = AB - BH = 1.2 - 0.582 = 0.618$ m

Taking the moments about the hinge *A*

$$P \times AB = F \times (AH)$$

where *P* is the force normal to the gate applied at *B*

$$\therefore \quad P \times 1.2 = 269343 \times 0.618$$

$$\therefore \quad P = \frac{269343 \times 0.618}{1.2} = \mathbf{138708\ N.\ Ans.}$$

Problem 3.19 *A gate supporting water is shown in Fig. 3.24. Find the height h of the water so that the gate tips about the hinge. Take the width of the gate as unity.*

Solution. Given : $\theta = 60°$

Distance, $AC = \frac{h}{\sin 60°} = \frac{2h}{\sqrt{3}}$

where h = Depth of water.

The gate will start tipping about hinge *B* if the resultant pressure force acts at *B*. If the resultant pressure force passes through a point which is lying from *B* to *C* anywhere on the gate, the gate will tip over the hinge. Hence limiting case is when the resultant force passes through *B*. But the resultant force passes through the centre of pressure. Hence for the given position point *B* becomes the centre of pressure. Hence depth of centre of pressure,

$$h^* = (h - 3) \text{ m}$$

But h^* is also given by $= \frac{I_G \sin^2 \theta}{Ah} + \bar{h}$

Taking width of gate unity. Then

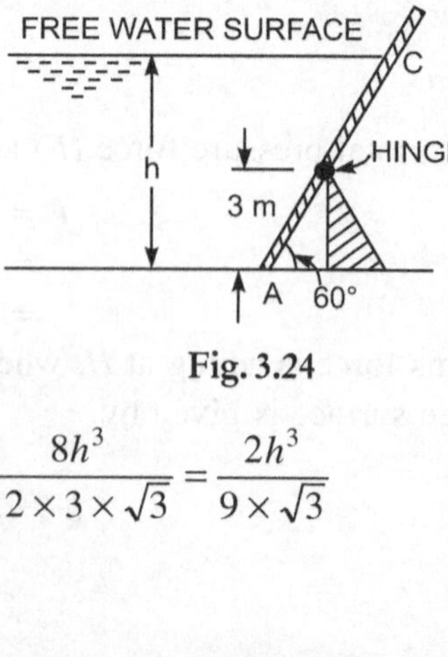

Fig. 3.24

Area, $A = AC \times 1 = \frac{2h}{\sqrt{3}} \times 1 \; ; \; \bar{h} = \frac{h}{2}$

$$I_G = \frac{bd^3}{12} = \frac{1 \times AC^3}{12} = \frac{1 \times \left(\frac{2h}{\sqrt{3}}\right)^3}{12} = \frac{8h^3}{12 \times 3 \times \sqrt{3}} = \frac{2h^3}{9 \times \sqrt{3}}$$

$$\therefore \qquad h^* = \frac{2h^3}{9\times\sqrt{3}} \times \frac{\sin^2 60°}{\frac{2h}{\sqrt{3}} \times \frac{h}{2}} + \frac{h}{2} = \frac{2h^3 \times \frac{3}{4}}{9h^2} + \frac{h}{2} = \frac{h}{6} + \frac{h}{2} = \frac{2h}{3}$$

Equating the two values of h^*,

$$h - 3 = \frac{2h}{3} \quad \text{or} \quad h - \frac{2h}{3} = 3 \quad \text{or} \quad \frac{h}{3} = 3$$

$$\therefore \qquad h = 3 \times 3 = 9 \text{ m}$$

$\therefore$ Height of water for tipping the gate = **9 m. Ans.**

Problem 3.20 *A rectangular sluice gate AB, 2 m wide and 3 m long is hinged at A as shown in Fig. 3.25. It is kept closed by a weight fixed to the gate. The total weight of the gate and weight fixed to the gate is 343350 N. Find the height of the water 'h' which will just cause the gate to open. The centre of gravity of the weight and gate is at G.*

Solution. Given :

Width of gate, $b = 2$ m ; Length of gate $L = 3$ m

$\therefore$ Area, $A = 2 \times 3 = 6 \text{ m}^2$

Weight of gate and $W = 343350$ N

Angle of inclination, $\theta = 45°$

Let h is the required height of water.

Depth of C.G. of the gate and weight = $\bar{h}$

From Fig. 3.25 (*a*),

$$\bar{h} = h - ED = h - (AD - AE)$$

$$= h - (AB \sin\theta - EG \tan\theta) \qquad \left\{\because \tan\theta = \frac{AE}{EG} \therefore AE = EG \tan\theta\right\}$$

$$= h - (3 \sin 45° - 0.6 \tan 45°)$$

$$= h - (2.121 - 0.6) = (h - 1.521) \text{ m}$$

The total pressure force, F is given by

$$F = \rho g A \bar{h} = 1000 \times 9.81 \times 6 \times (h - 1.521)$$

$$= 58860\,(h - 1.521) \text{ N.}$$

The total force F is acting at the centre of pressure as shown in Fig. 3.25 (*b*) at H. The depth of H from free surface is given by h^* which is equal to

$$h^* = \frac{I_G \sin^2\theta}{A\bar{h}} + \bar{h}, \text{ where } I_G = \frac{bd^3}{12} = \frac{2\times 3^3}{12} = \frac{54}{12} = 4.5 \text{ m}^4$$

$$\therefore \qquad h^* = \frac{4.5 \times \sin^2 45°}{6 \times (h - 1.521)} + (h - 1.521) = \frac{0.375}{(h - 1.521)} + (h - 1.521) \text{ m}$$

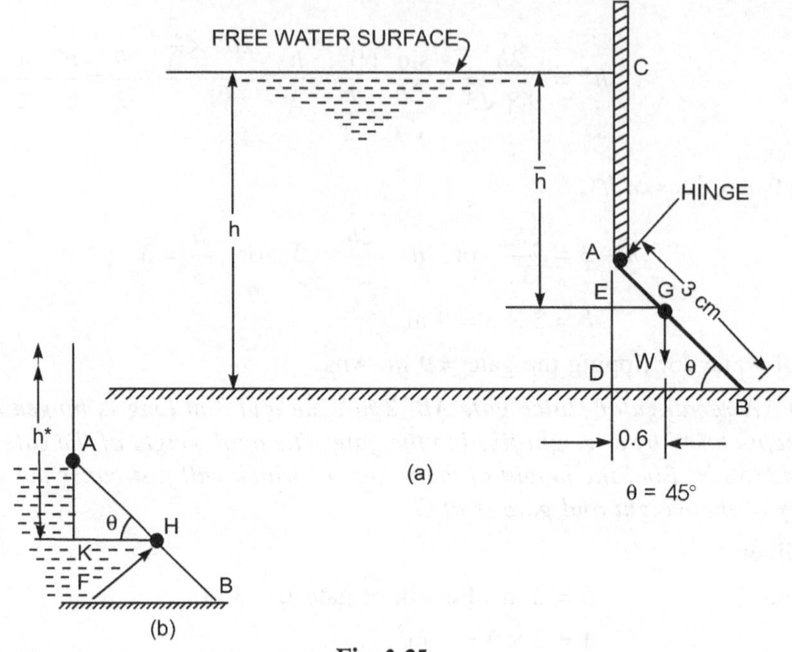

Fig. 3.25

Now taking moments about hinge A, we get

$$343350 \times EG = F \times AH$$

or
$$343350 \times 0.6 = F \times \frac{AK}{\sin 45^\circ}$$

$$\left[\text{From } \Delta AKH, \text{Fig. 3.24 } (b) \; AK = AH \sin\theta = AH \sin 45^\circ \;\therefore\; AH = \frac{AK}{\sin 45^\circ}\right]$$

$$= \frac{58860\,(h - 1.521) \times AK}{\sin 45^\circ}$$

$\therefore$
$$AK = \frac{343350 \times 0.6 \times \sin 45^\circ}{58860\,(h - 1.521)} = \frac{0.3535 \times 7}{(h - 1.521)} \quad \text{...}(i)$$

But
$$AK = h^* - AC = \frac{.375}{(h - 1.521)} + (h - 1.521) - AC \quad \text{...}(ii)$$

But
$$AC = CD - AD = h - AB \sin 45^\circ = h - 3 \times \sin 45^\circ = h - 2.121$$

$\therefore$ Substituting this value in (*ii*), we get

$$AK = \frac{.375}{h - 1.521} + (h - 1.521) - (h - 2.121)$$

$$= \frac{.375}{h - 1.521} + 2.121 - 1.521 = \frac{.375}{h - 1.521} + 0.6 \quad \text{...}(iii)$$

Equating the two values of AK from (*i*) and (*iii*)

$$\frac{0.3535 \times 7}{h - 1.521} = \frac{0.375}{h - 1.521} + 0.6$$

or $$0.3535 \times 7 = 0.375 + 0.6\,(h - 1.521) = 0.375 + 0.6\,h - 0.6 \times 1.521$$

or $$0.6h = 2.4745 - .375 + 0.6 \times 1.521 = 2.0995 + 0.9126 = 3.0121$$

$\therefore$ $$h = \frac{3.0121}{0.6} = \mathbf{5.02\ m.\ Ans.}$$

Problem 3.21 *Find the total pressure and position of centre of pressure on a triangular plate of base 2 m and height 3 m which is immersed in water in such a way that the plan of the plate makes an angle of 60° with the free surface of the water. The base of the plate is parallel to water surface and at a depth of 2.5 m from water surface.*

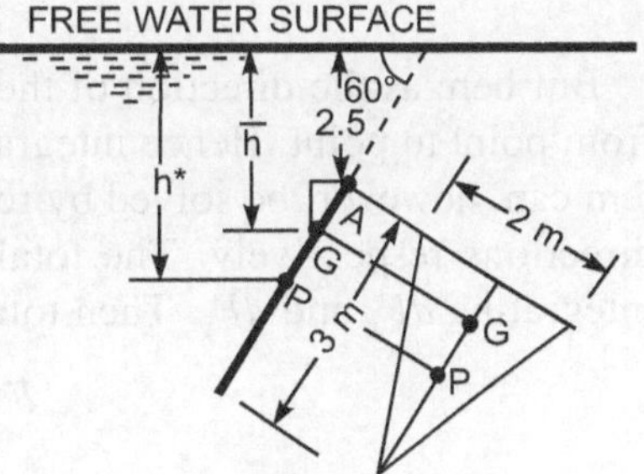

Fig. 3.26

Solution. Given :

Base of plate, $b = 2$ m

Height of plate, $h = 3$ m

$\therefore$ Area, $$A = \frac{b \times h}{2} = \frac{2 \times 3}{2} = 3\ \text{m}^2$$

Inclination, $\theta = 60°$

Depth of centre of gravity from free surface of water,

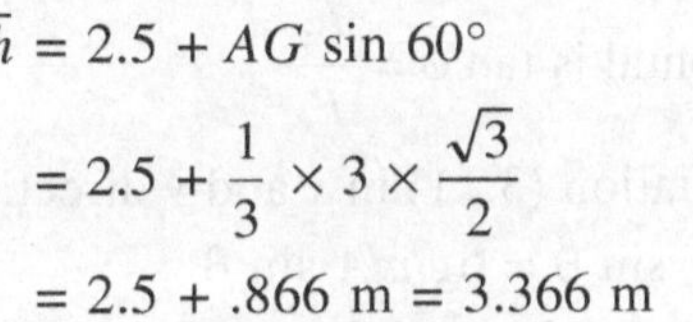

$$\bar{h} = 2.5 + AG \sin 60°$$

$$= 2.5 + \frac{1}{3} \times 3 \times \frac{\sqrt{3}}{2} \qquad \left\{\because AG = \frac{1}{3} \text{ of height of triangle}\right\}$$

$$= 2.5 + .866 \text{ m} = 3.366 \text{ m}$$

(*i*) **Total pressure force (F)**

$$F = \rho g A \bar{h} = 1000 \times 9.81 \times 3 \times 3.366 = \mathbf{99061.38\ N.\ Ans.}$$

(*ii*) **Centre of pressure (h*).** Depth of centre of pressure from free surface of water is given by

$$h^* = \frac{I_G \sin^2 \theta}{A\bar{h}} + \bar{h}$$

where $$I_G = \frac{bh^3}{36} = \frac{2 \times 3^3}{36} = \frac{3}{2} = 1.5\ \text{m}^4$$

$\therefore$ $$h^* = \frac{1.5 \times \sin^2 60°}{3 \times 3.366} + 3.366 = 0.1114 + 3.366 = \mathbf{3.477\ m.\ Ans.}$$

▶ 3.6 CURVED SURFACE SUBMERGED IN LIQUID

Consider a curved surface *AB*, submerged in a static fluid as shown in Fig. 3.27. Let dA is the area of a small strip at a depth of h from water surface.

Then pressure intensity on the area dA is $= \rho g h$

and pressure force, $$dF = p \times \text{Area} = \rho g h \times dA \qquad \ldots(3.11)$$

This force dF acts normal to the surface.

Hence total pressure force on the curved surface should be

$$F = \int \rho g h dA \qquad \ldots(3.12)$$

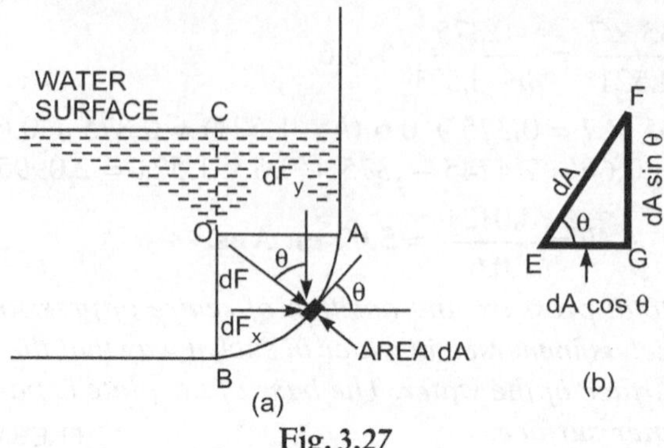

Fig. 3.27

But here as the direction of the forces on the small areas are not in the same direction, but varies from point to point. Hence integration of equation (3.11) for curved surface is impossible. The problem can, however, be solved by resolving the force dF in two components dF_x and dF_y in the x and y directions respectively. The total force in the x and y directions, *i.e.*, F_x and F_y are obtained by integrating dF_x and dF_y. Then total force on the curved surface is

$$F = \sqrt{F_x^2 + F_y^2} \quad ...(3.13)$$

and inclination of resultant with horizontal is $\tan \phi = \dfrac{F_y}{F_x}$ ...(3.14)

Resolving the force dF given by equation (3.11) in x and y directions :

$$dF_x = dF \sin \theta = \rho g h dA \sin \theta \qquad \{\because dF = \rho g h dA\}$$

and

$$dF_y = dF \cos \theta = \rho g h dA \cos \theta$$

Total forces in the x and y direction are :

$$F_x = \int dF_x = \int \rho g h dA \sin \theta = \rho g \int h dA \sin \theta \quad ...(3.15)$$

and

$$F_y = \int dF_y = \int \rho g h dA \cos \theta = \rho g \int h dA \cos \theta \quad ...(3.16)$$

Fig. 3.27 (*b*) shows the enlarged area dA. From this figure, *i.e.*, ΔEFG,

$$EF = dA$$
$$FG = dA \sin \theta$$
$$EG = dA \cos \theta$$

Thus in equation (3.15) $dA \sin \theta = FG$ = Vertical projection of the area dA and hence the expression $\rho g \int h dA \sin \theta$ represents the total pressure force on the projected area of the curved surface on the vertical plane. Thus

F_x = Total pressure force on the projected area of the curved surface on vertical plane. ...(3.17)

Also $dA \cos \theta = EG$ = horizontal projection of dA and hence $h dA \cos \theta$ is the volume of the liquid contained in the elementary area dA up to free surface of the liquid. Thus $\int h dA \cos \theta$ is the total volume contained between the curved surface extended upto free surface.

Hence $\rho g \int h dA \cos \theta$ is the total weight supported by the curved surface. Thus

$$F_y = \rho g \int h dA \cos \theta$$

= weight of liquid supported by the curved surface upto free surface of liquid. ...(3.18)

In Fig. 3.28, the curved surface *AB* is not supporting any fluid. In such cases, F_y is equal to the weight of the imaginary liquid supported by *AB* upto free surface of liquid. The direction of F_y will be taken in upward direction.

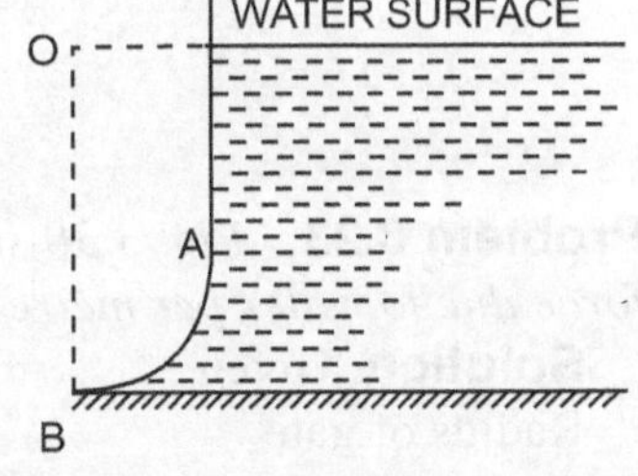

Fig. 3.28

Problem 3.22 *Compute the horizontal and vertical components of the total force acting on a curved surface AB, which is in the form of a quadrant of a circle of radius 2 m as shown in Fig. 3.29. Take the width of the gate as unity.*

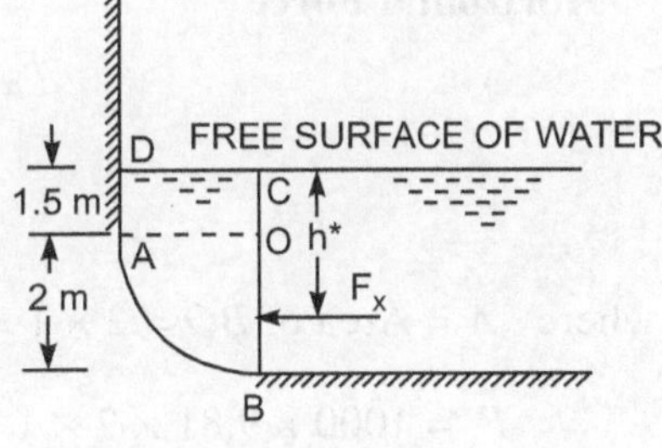

Fig. 3.29

Solution. Given :

Width of gate $= 1.0$ m

Radius of the gate $= 2.0$ m

∴ Distance $AO = OB = 2$ m

Horizontal force, F_x exerted by water on gate is given by equation (3.17) as

F_x = Total pressure force on the projected area of curved surface AB on vertical plane

= Total pressure force on *OB*

{projected area of curved surface on vertical plane = $OB \times 1$}

$$= \rho g A \bar{h}$$

$$= 1000 \times 9.81 \times 2 \times 1 \times \left(1.5 + \frac{2}{2}\right) \quad \{\because \text{ Area of } OB = A = BO \times 1 = 2 \times 1 = 2,$$

$\bar{h}$ = Depth of C.G. of *OB* from free surface = $1.5 + \frac{2}{2}$ }

$F_x = 9.81 \times 2000 \times 2.5 =$ **49050 N. Ans.**

The point of application of F_x is given by $h^* = \dfrac{I_G}{A\bar{h}} + \bar{h}$

where I_G = M.O.I. of *OB* about its C.G. $= \dfrac{bd^3}{12} = \dfrac{1 \times 2^3}{12} = \dfrac{2}{3}$ m^4

$$\therefore \qquad h^* = \frac{\frac{2}{3}}{2 \times 2.5} + 2.5 = \frac{1}{7.5} + 2.5 \text{ m}$$

$= 0.1333 + 2.5 = 2.633$ m from free surface.

Vertical force, F_y, exerted by water is given by equation (3.18)

F_y = Weight of water supported by *AB* upto free surface

= Weight of portion *DABOC*

= Weight of *DAOC* + Weight of water *AOB*

= ρg [Volume of *DAOC* + Volume of *AOB*]

$$= 1000 \times 9.81 \left[AD \times AO \times 1 + \frac{\pi}{4}(AO)^2 \times 1 \right]$$

$$= 1000 \times 9.81 \left[1.5 \times 2.0 \times 1 + \frac{\pi}{4} \times 2^2 \times 1\right]$$

$$= 1000 \times 9.81\ [3.0 + \pi]\text{N} = \mathbf{60249.1\ N.\ Ans.}$$

Problem 3.23 *Fig. 3.30 shows a gate having a quadrant shape of radius 2 m. Find the resultant force due to water per metre length of the gate. Find also the angle at which the total force will act.*

Solution. Given :

Radius of gate $= 2$ m

Width of gate $= 1$ m

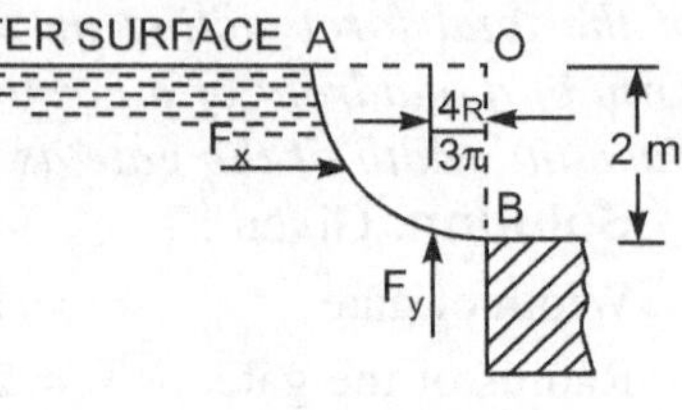

Fig. 3.30

Horizontal Force

$$F_x = \text{Force on the projected area of the curved surface on vertical plane}$$

$$= \text{Force on } BO = \rho g A\bar{h}$$

where A = Area of $BO = 2 \times 1 = 2\ \text{m}^2$, $\bar{h} = \frac{1}{2} \times 2 = 1$ m ;

$$F_x = 1000 \times 9.81 \times 2 \times 1 = 19620\ \text{N}$$

This will act at a depth of $\frac{2}{3} \times 2 = \frac{4}{3}$ m from free surface of liquid,

Vertical Force, F_y

$$F_y = \text{Weight of water (imagined) supported by } AB$$

$$= \rho g \times \text{Area of } AOB \times 1.0$$

$$= 1000 \times 9.81 \times \frac{\pi}{4}\ (2)^4 \times 1.0 = 30819\ \text{N}$$

This will act at a distance of $\frac{4R}{3\pi} = \frac{4 \times 2.0}{3\pi} = 0.848$ m from OB.

$\therefore$ Resultant force, F is given by

$$F = \sqrt{F_x^2 + F_y^2}$$

$$= \sqrt{19620 + 30819} = \sqrt{384944400 + 949810761}$$

$$= \mathbf{36534.4\ N.\ Ans.}$$

The angle made by the resultant with horizontal is given by

$$\tan\theta = \frac{F_y}{F_x} = \frac{30819}{19620} = 1.5708$$

$$\therefore \qquad \theta = \tan^{-1} 1.5708 = \mathbf{57^\circ\ 31'.\ Ans.}$$

Problem 3.24 *Find the magnitude and direction of the resultant force due to water acting on a roller gate of cylindrical form of 4.0 m diameter, when the gate is placed on the dam in such a way that water is just going to spill. Take the length of the gate as 8 m.*

Solution. Given :

Dia. of gate $= 4$ m

$\therefore$ Radius, $R = 2$ m

Length of gate, $l = 8$ m

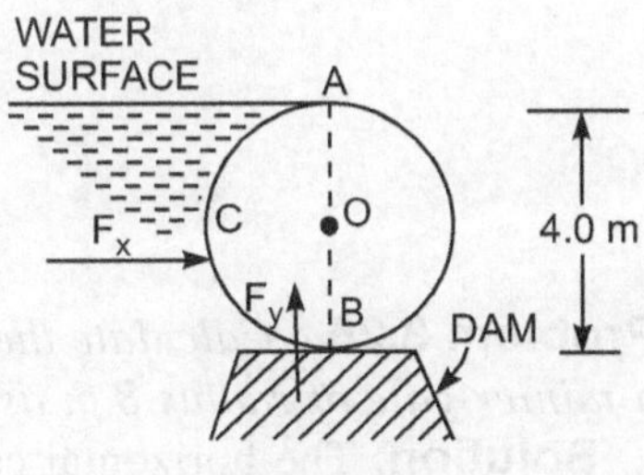

Fig. 3.31

Horizontal force, F_x acting on the gate is

$F_x = \rho g A\bar{h}$ = Force on projected area of curved surface ACB on vertical plane

= Force on vertical area AOB

where A = Area of $AOB = 4.0 \times 8.0 = 32.0$ m^2

$\bar{h}$ = Depth of C.G. of $AOB = 4/2 = 2.0$ m

$$\therefore \quad F_x = 1000 \times 9.81 \times 32.0 \times 2.0 = 627840 \text{ N}.$$

Vertical force, F_y is given by

F_y = Weight of water enclosed or supported (actually or imaginary) by the curved surface ACB

= ρg × Volume of portion ACB

= ρg × Area of $ACB \times l$

$$= 1000 \times 9.81 \times \frac{\pi}{2}(R)^2 \times 8.0 = 9810 \times \frac{\pi}{2}(2)^2 \times 8.0 = 493104 \text{ N}$$

It will be acting in the upward direction.

$$\therefore \quad \text{Resultant force,} \quad F = \sqrt{F_x^2 + F_y^2} = \sqrt{627840 + 493104} = \mathbf{798328\ N.\ Ans.}$$

Direction of resultant force is given by $\tan\theta = \frac{F_y}{F_x} = \frac{493104}{627840} = 0.7853$

$$\therefore \quad \theta = \mathbf{31° 8'.\ Ans.}$$

Problem 3.25 *Find the horizontal and vertical component of water pressure acting on the face of a tainter gate of 90° sector of radius 4 m as shown in Fig. 3.32. Take width of gate unity.*

Solution. Given :

Radius of gate, $R = 4$ m

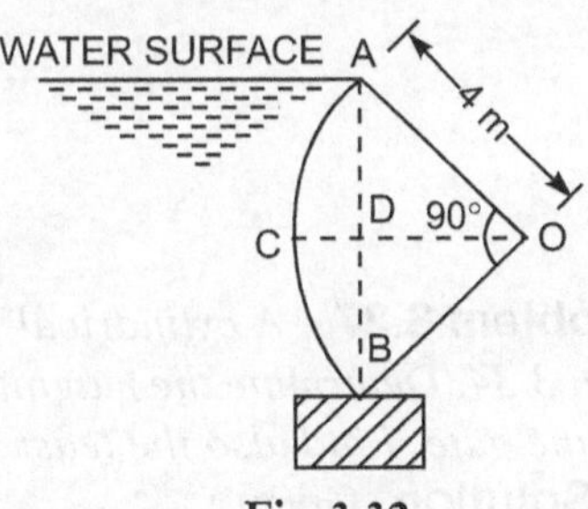

Fig. 3.32

Horizontal component, F_x of force acting on the gate is

F_x = Force on area of gate projected on vertical plane

= Force on area ADB

= $\rho g A\bar{h}$

where $A = AB$ × Width of gate

$= 2 \times AD \times 1$ $\quad (\because AB = 2AD)$

$= 2 \times 4 \times \sin 45° = 8 \times .707 = 5.656$ m^2 $\{\because AD = 4 \sin 45°\}$

$$\bar{h} = \frac{AB}{2} = \frac{5.656}{2} = 2.828 \text{ m}$$

$$\therefore \quad F_x = 1000 \times 9.81 \times 5.656 \times 2.828 \text{ N} = \mathbf{156911\ N.\ Ans.}$$

Vertical component

F_y = Weight of water supported or enclosed by the curved surface

= Weight of water in portion $ACBDA$

= ρg × Area of $ACBDA$ × Width of gate

= $1000 \times 9.81 \times$ [Area of sector $ACBOA$ – Area of ΔABO] × 1

$$= 9810 \times \left[\frac{\pi}{4} R^2 - \frac{AO \times BO}{2}\right] \qquad [\because \Delta AOB \text{ is a right angled}]$$

$$= 9810 \times \left[\frac{\pi}{4} 4^2 - \frac{4 \times 4}{2}\right] = \mathbf{44796\ N.\ Ans.}$$

Problem 3.26 *Calculate the horizontal and vertical components of the water pressure exerted on a tainter gate of radius 8 m as shown in Fig. 3.33. Take width of gate unity.*

Solution. The horizontal component of water pressure is given by

$F_x = \rho g A \bar{h}$ = Force on the area projected on vertical plane

= Force on the vertical area of BD

where $A = BD \times$ Width of gate $= 4.0 \times 1 = 4.0$ m

$$\bar{h} = \frac{1}{2} \times 4 = 2 \text{ m}$$

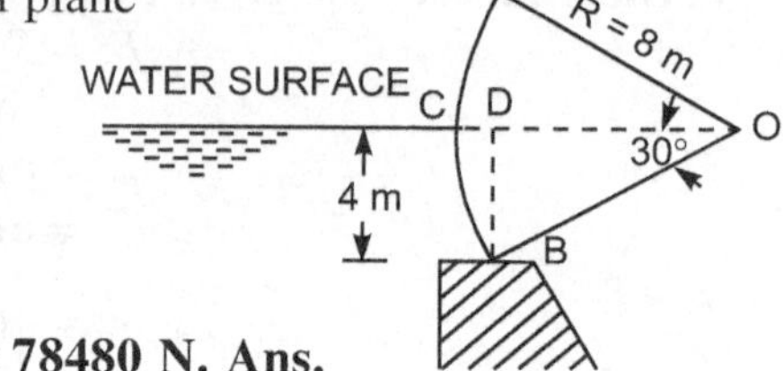

Fig. 3.33

$$F_x = 1000 \times 9.81 \times 4.0 \times 2.0 = \mathbf{78480\ N.\ Ans.}$$

Vertical component of the water pressure is given by

F_y = Weight of water supported or enclosed (imaginary) by curved surface CB

= Weight of water in the portion $CBDC$

= $\rho g \times$ [Area of portion $CBDC$] $\times$ Width of gate

= $\rho g \times$ [Area of sector CBO – Area of the triangle BOD] $\times$ 1

$$= 1000 \times 9.81 \times \left[\frac{30}{360} \times \pi R^2 - \frac{BD \times DO}{2}\right]$$

$$= 9810 \times \left[\frac{1}{12} \pi \times 8^2 - \frac{4.0 \times 8.8 \cos 30°}{2}\right]$$

$$\{\because DO = BO \cos 30° = 8 \times \cos 30°\}$$

$$= 9810 \times [16.755 - 13.856] = \mathbf{28439\ N.\ Ans.}$$

Problem 3.27 *A cylindrical gate of 4 m diameter 2 m long has water on its both sides as shown in Fig. 3.34. Determine the magnitude, location and direction of the resultant force exerted by the water on the gate. Find also the least weight of the cylinder so that it may not be lifted away from the floor.*

Solution. Given :

Dia. of gate = 4 m

Radius = 2 m

(*i*) The forces acting on the left side of the cylinder are :

The horizontal component, F_{x_1}

where F_{x_1} = Force of water on area projected on vertical plane

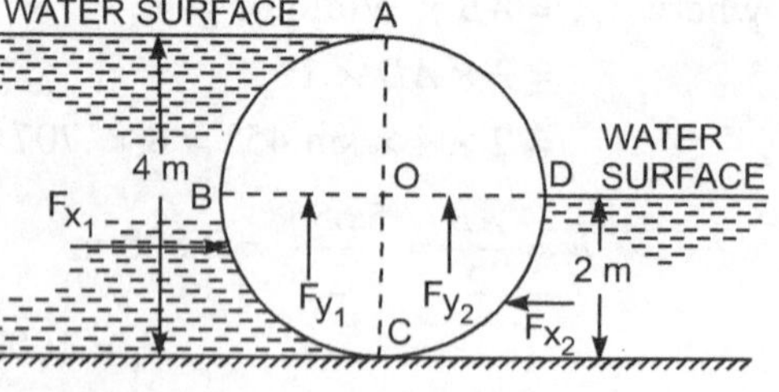

Fig. 3.34

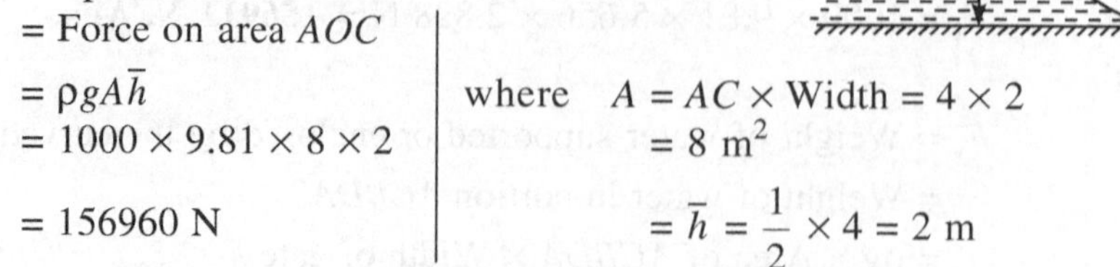

= Force on area AOC

$= \rho g A \bar{h}$ where $A = AC \times$ Width $= 4 \times 2 = 8$ m^2

$= 1000 \times 9.81 \times 8 \times 2$

$= 156960$ N $\qquad \bar{h} = \frac{1}{2} \times 4 = 2$ m

$$F_{y_1} = \text{weight of water enclosed by } ABCOA$$

$$= 1000 \times 9.81 \times \left[\frac{\pi}{2} R^2\right] \times 2.0 = 9810 \times \frac{\pi}{2} \times 2^2 \times 2.0 = \textbf{123276 N. Ans.}$$

Right Side of the Cylinder

$$F_{x_2} = \rho g A_2 \bar{h}_2 = \text{Force on vertical area } CO$$

$$= 1000 \times 9.81 \times 2 \times 2 \times \frac{2}{2} \left\{\because A_2 = CO \times 1 = 2 \times 1 = 2 \text{ m}^2, \bar{h}_2 = \frac{2}{2} = 1.0\right\}$$

$$= 39240 \text{ N}$$

$$F_{y_2} = \text{Weight of water enclosed by } DOCD$$

$$= \rho g \times \left[\frac{\pi}{4} R^2\right] \times \text{Width of gate}$$

$$= 1000 \times 9.81 \times \frac{\pi}{4} \times 2^2 \times 2 = 61638 \text{ N}$$

$\therefore$ Resultant force in the direction of x,

$$F_x = F_{x_1} - F_{x_2} = 156960 - 39240 = 117720 \text{ N}$$

Resultant force in the direction of y,

$$F_y = F_{y_1} + F_{y_2} = 123276 + 61638 = 184914 \text{ N}$$

(*i*) **Resultant force,** F is given as

$$F = \sqrt{F_x^2 + F_y^2} = \sqrt{(117720)^2 + (184914)^2} = \textbf{219206 N. Ans.}$$

(*ii*) **Direction of resultant force** is given by

$$\tan \theta = \frac{F_y}{F_x} = \frac{184914}{117720} = 1.5707$$

$$\therefore \quad \theta = \mathbf{57° \ 31'}. \textbf{ Ans.}$$

(*iii*) **Location of the resultant force**

Force, F_{x_1} acts at a distance of $\frac{2 \times 4}{3} = 2.67$ m from the top surface of water on left side, while F_{x_2} acts at a distance of $\frac{2}{3} \times 2 = 1.33$ m from free surface on the right side of the cylinder. The resultant force F_x in the direction of x will act at a distance of y from the bottom as

$$F_x \times y = F_{x_1} [4 - 2.67] - F_{x_2} [2 - 1.33]$$

or

$$117720 \times y = 156960 \times 1.33 - 39240 \times .67 = 208756.8 - 26290.8 = 182466$$

$$\therefore \quad y = \frac{182466}{117720} = 1.55 \text{ m from the bottom.}$$

Force F_{y_1} acts at a distance $\frac{4R}{3\pi}$ from AOC or at a distance $\frac{4 \times 2.0}{3\pi} = 0.8488$ m from AOC towards left of AOC.

Also F_{y_2} acts at a distance $\frac{4R}{3\pi} = 0.8488$ m from AOC towards the right of AOC. The resultant force F_y will act at a distance x from AOC which is given by

$$F_y \times x = F_{y_1} \times .8488 - F_{y_2} \times .8488$$

or $$184914 \times x = 123276 \times .8488 - 61638 \times .8488 = .8488\ [123276 - 61638] = 52318.4$$

$\therefore$ $$x = \frac{52318.4}{184914} = 0.2829 \text{ m from AOC.}$$

(*iv*) **Least Weight of Cylinder.** The resultant force in the upward direction is

$$F_y = 184914 \text{ N}$$

Thus the weight of cylinder should not be less than the upward force F_y. Hence least weight of cylinder should be at least.

$$= \mathbf{184914\ N.\ Ans.}$$

Problem 3.28 *Fig. 3.35 shows the cross-section of a tank full of water under pressure. The length of the tank is 2 m. An empty cylinder lies along the length of the tank on one of its corner as shown. Find the horizontal and vertical components of the force acting on the curved surface ABC of the cylinder.* (A.M.I.E., May 1974)

Solution. Radius, $R = 1$ m

Length of tank, $l = 2$ m

Pressure, $p = 0.2 \text{ kgf/cm}^2 = 0.2 \times 9.81 \text{ N/cm}^2$
$= 1.962 \text{ N/cm}^2 = 1.962 \times 10^4 \text{ N/m}^2$

$\therefore$ Pressure head, $$h = \frac{p}{\rho g} = \frac{1.962 \times 10^4}{1000 \times 9.81} = 2 \text{ m}$$

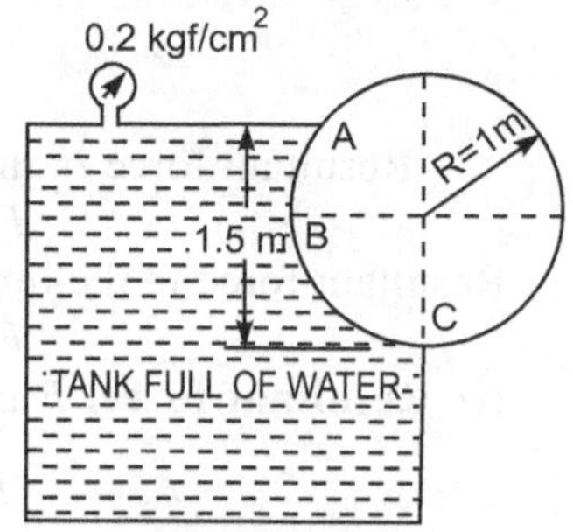

Fig. 3.35

$\therefore$ Free surface of water will be at a height of 2 m from the top of the tank.

$\therefore$ Fig. 3.36 shows the equivalent free surface of water.

(*i*) **Horizontal Component of Force**

$$F_x = \rho g A \bar{h}$$

where A = Area projected on vertical plane
$= 1.5 \times 2.0 = 3.0 \text{ m}^2$

$$\bar{h} = 2 + \frac{1.5}{2} = 2.75$$

$\therefore$ $$F_x = 1000 \times 9.81 \times 3.0 \times 2.75$$
$$= \mathbf{80932.5\ N.\ Ans.}$$

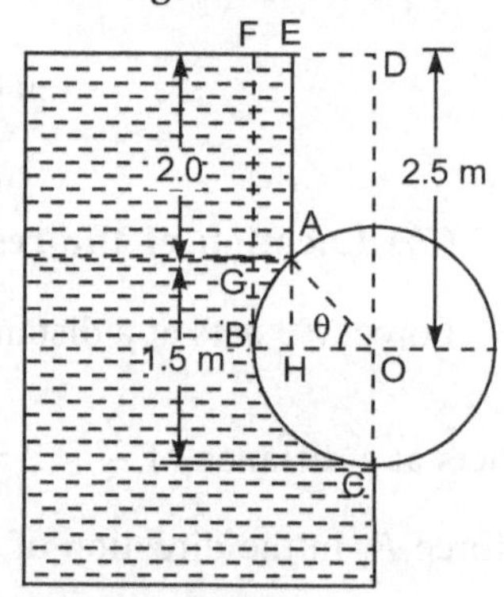

Fig. 3.36

(*ii*) **Vertical Component of Force**

F_y = Weight of water enclosed or supported actually or imaginary by curved surface *ABC*
= Weight of water in the portion *CODE ABC*
= Weight of water in *CODFBC* – Weight of water in *AEFB*

But weight of water in *CODFBC*
= Weight of water in [*COB* + *ODFBO*]

$$= \rho g \left[\frac{\pi R^2}{4} + BO \times OD\right] \times 2 = 1000 \times 9.81 \left[\frac{\pi}{4} \times 1^2 + 1.0 \times 2.5\right] \times 2$$

$$= 64458.5 \text{ N}$$

Weight of water in *AEFB* = ρg [Area of *AEFB*] × 2.0

$= 1000 \times 9.81$ [Area of $(AEFG + AGBH - AHB)$] $\times 2.0$

In ΔAHO, $\sin\theta = \dfrac{AH}{AO} = \dfrac{0.5}{1.0} = 0.5 \quad \therefore \quad \theta = 30°$

$BH = BO - HO = 1.0 - AO\cos\theta = 1.0 - 1 \times \cos 30° = 0.134$

Area, ABH = Area ABO – Area AHO

$$= \pi R^2 \times \frac{30}{360} - \frac{AH \times HO}{2.0} = \frac{\pi R^2}{12} - \frac{0.5 \times .866}{2} = 0.0453$$

$\therefore$ Weight of water in $AEFB$

$= 9810 \times [AE \times AG + AG \times AH - 0.0453] \times 2.0$

$= 9810 \times [2.0 \times .134 + .134 \times .5 - .0453] \times 2.0$

$= 9810 \times [.268 + .067 - .0453] \times 2.0 = 5684$ N

$\therefore$ $F_y = 64458.5 - 5684 =$ **58774.5 N. Ans.**

Problem 3.29 *Find the magnitude and direction of the resultant water pressure acting on a curved face of a dam which is shaped according to the relation $y = \dfrac{x^2}{9}$ as shown in Fig. 3.37. The height of the water retained by the dam is 10 m. Consider the width of the dam as unity.*

Solution. Equation of curve AB is

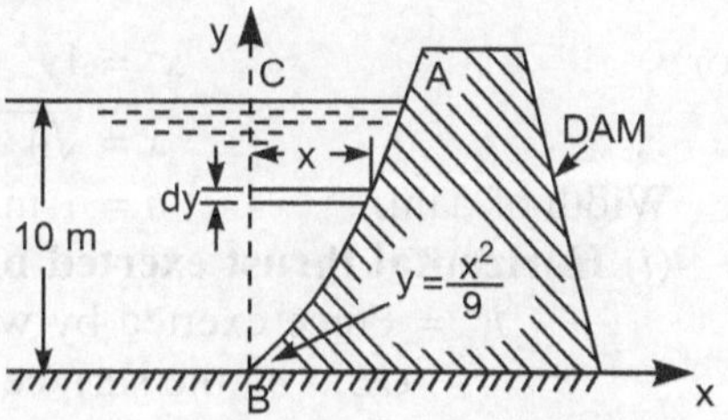

$$y = \frac{x^2}{9} \quad \text{or} \quad x^2 = 9y$$

$\therefore$ $x = \sqrt{9y} = 3\sqrt{y}$

Height of water, $h = 10$ m

Width, $b = 1$ m

Fig. 3.37

The horizontal component, F_x is given by

F_x = Pressure due to water on the curved area projected on vertical plane

= Pressure on area BC

$= \rho g A\bar{h}$

where $A = BC \times 1 = 10 \times 1\ \text{m}^2$, $\bar{h} = \frac{1}{2} \times 10 = 5$ m

$F_x = 1000 \times 9.81 \times 10 \times 5 = 490500$ N

Vertical component, F_y is given by

F_y = Weight of water supported by the curve AB

= Weight of water in the portion ABC

$= \rho g$[Area of ABC] $\times$ Width of dam

$$= \rho g \left[\int_0^{10} x \times dy\right] \times 1.0 \qquad \left\{\text{Area of strip} = xdy \quad \therefore \text{Area } ABC = \int_0^{10} xdy\right\}$$

$$= 1000 \times 9.81 \times \int_0^{10} 3\sqrt{y}\, dy \qquad \{\because x = 3\sqrt{y}\}$$

$$= 29430 \left[\frac{y^{3/2}}{3/2}\right]_0^{10} = 29430 \times \frac{2}{3}\left[y^{3/2}\right]_0^{10} = 19620\ [10^{3/2}]$$

$= 19620 \times 31.622 = 620439$ N

∴ Resultant water pressure on dam

$$F = \sqrt{F_x^2 + F_y^2} = \sqrt{(490500)^2 + (620439)^2}$$

$$= 790907 \text{ N} = \mathbf{790.907 \text{ kN. Ans.}}$$

Direction of the resultant is given by

$$\tan\theta = \frac{F_y}{F_x} = \frac{620439}{490500} = 1.265$$

∴ $\theta = \mathbf{51° \; 40'}$. **Ans.**

Problem 3.30 *A dam has a parabolic shape $y = y_0\left(\frac{x}{x_0}\right)^2$ as shown in Fig. 3.38 below having $x_0 = 6$ m and $y_0 = 9$ m. The fluid is water with density = 1000 kg/m³. Compute the horizontal, vertical and the resultant thrust exerted by water per metre length of the dam.* (A.M.I.E., Summer, 1985)

Solution. Given :

Equation of the curve *OA* is

$$y = y_0\left(\frac{x}{x_0}\right)^2 = 9\left(\frac{x}{6}\right)^2 = 9 \times \frac{x^2}{36} = \frac{x^2}{4}$$

or $x^2 = 4y$

∴ $x = \sqrt{4x} = 2y^{1/2}$

Width of dam, $b = 1$ m.

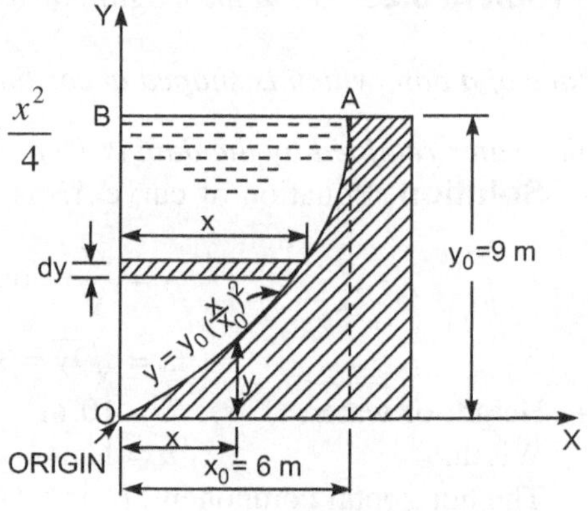

Fig. 3.38

(*i*) **Horizontal thrust exerted by water**

F_x = Force exerted by water on vertical surface *OB*, *i.e.*, the surface obtained by projecting the curved surface on vertical plane

$$= \rho g A \bar{h}$$

$$= 1000 \times 9.81 \times (9 \times 1) \times \frac{9}{2} = \mathbf{397305 \text{ N. Ans.}}$$

(*ii*) **Vertical thrust exerted by water**

F_y = Weight of water supported by curved surface *OA* upto free surface of water

= Weight of water in the portion *ABO*

= ρg × Area of *OAB* × Width of dam

$$= 1000 \times 9.81 \times \left[\int_0^9 x \times dy\right] \times 1.0$$

$$= 1000 \times 9.81 \times \left[\int_0^9 2y^{1/2} \times dy\right] \times 1.0 \qquad (\because x = 2y^{1/2})$$

$$= 19620 \times \left[\frac{y^{3/2}}{(3/2)}\right]_0^9 = 19620 \times \frac{2}{3}\,[9^{3/2}]$$

$$= 19620 \times \frac{2}{3} \times 27 = \mathbf{353160 \text{ N. Ans.}}$$

(*iii*) **Resultant thrust exerted by water**

$$F = \sqrt{F_x^2 + F_y^2} = \sqrt{397305 + 353160} = \mathbf{531574\ N.\ Ans.}$$

Direction of resultant is given by

$$\tan\theta = \frac{F_y}{F_x} = \frac{353160}{397305} = 0.888$$

$$\theta = \tan^{-1} 0.888 = \mathbf{41.63°.\ Ans.}$$

Problem 3.31 *A cylinder 3 m in diameter and 4 m long retains water on one side. The cylinder is supported as shown in Fig. 3.39. Determine the horizontal reaction at A and the vertical reaction at B. The cylinder weighs 196.2 kN. Ignore friction.*

Solution. Given :

Dia. of cylinder = 3 m

Length of cylinder = 4 m

Weight of cylinder, $W = 196.2$ kN = 196200 N

Horizontal force exerted by water

$$F_x = \text{Force on vertical area } BOC$$
$$= \rho g A \bar{h}$$

Fig. 3.39

where $A = BOC \times l = 3 \times 4 = 12\ \text{m}^2$, $\bar{h} = \frac{1}{2} \times 3 = 1.5$ m

$$F_x = 1000 \times 9.81 \times 12 \times 1.5 = 176580\ \text{N}$$

The vertical force exerted by water

$$F_y = \text{Weight of water enclosed in } BDCOB$$
$$= \rho g \times \left(\frac{\pi}{2} R^2\right) \times l = 1000 \times 9.81 \times \frac{\pi}{2} \times (1.5)^2 \times 4 = 138684\ \text{N}$$

Force F_y is acting in the upward direction.

For the equilibrium of cylinder

Horizontal reaction at $A = F_x = 176580$ N

Vertical reaction at B = Weight of cylinder $- F_y$

$= 196200 - 138684 =$ **57516 N. Ans.**

HIGHLIGHTS

1. When the fluid is at rest, the shear stress is zero.
2. The force exerted by a static fluid on a vertical, horizontal or an inclined plane immersed surface,

$$F = \rho g A \bar{h}$$

where ρ = Density of the liquid,

A = Area of the immersed surface, and

$\bar{h}$ = Depth of the centre of gravity of the immersed surface from free surface of the liquid.

3. Centre of pressure is defined as the point of application of the resultant pressure.
4. The depth of centre of pressure of an immersed surface from free surface of the liquid,

$$h^* = \frac{I_G}{A\bar{h}} + \bar{h}$$ for vertically immersed surface.

$$= \frac{I_G \sin^2 \theta}{A\bar{h}} + \bar{h}$$ for inclined immersed surface.

5. The centre of pressure for a plane vertical surface lies at a depth of two-third the height of the immersed surface.

6. The total force on a curved surface is given by $F = \sqrt{F_x^2 + F_y^2}$

   where F_x = Horizontal force on curved surface and is equal to total pressure force on the projected area of the curved surface on the vertical plane,
   $= \rho g A \bar{h}$

   and F_y = Vertical force on submerged curved surface and is equal to the weight of liquid actually or imaginary supported by the curved surface.

7. The inclination of the resultant force on curved surface with horizontal, $\tan \theta = \frac{F_y}{F_x}$.

EXERCISE 3

(A) THEORETICAL PROBLEMS

1. What do you understand by 'Total Pressure' and 'Centre of Pressure' ?
2. Derive an expression for the force exerted on a submerged vertical plane surface by the static liquid and locate the position of centre of pressure.
3. Prove that the centre of pressure of a completely submerged plane surface is always below the centre of gravity of the submerged surface or at most coincide with the centre of gravity when the plane surface is horizontal.
4. Prove that the total pressure exerted by a static liquid on an inclined plane submerged surface is the same as the force exerted on a vertical plane surface as long as the depth of the centre of gravity of the surface is unaltered.
5. Derive an expression for the depth of centre of pressure from free surface of liquid of an inclined plane surface submerged in the liquid.
6. (*a*) How would you determine the horizontal and vertical components of the resultant pressure on a submerged curved surface ?
   (*b*) Explain the procedure of finding hydrostatic forces on curved surfaces.
   (Delhi University, Dec. 2002)
7. Explain how you would find the resultant pressure on a curved surface immersed in a liquid.
   (A.M.I.E., Summer 1981)
8. Why the resultant pressure on a curved submerged surface is determined by first finding horizontal and vertical forces on the curved surface ? Why is the same method not adopted for a plane inclined surface submerged in a liquid ?
9. Describe briefly with sketches the various methods used for measured pressure exerted by fluids.
   (A.M.I.E., Summer 1980)
10. Prove that the vertical component of the resultant pressure on a submerged curved surface is equal to the weight of the liquid supported by the curved surface.
11. When will centre of pressure and centre of gravity of an immersed plane surface coincide ?
    (A.M.I.E., Summer 1990)
12. Find an expression for the force exerted and centre of pressure for a completely submerged inclined plane surface. Can the same method be applied for finding the resultant force on a curved surface immersed in the liquid ? If not, why ?
    (Delhi University, 1992)

13. What do you understand by the hydrostatic equation ? With the help of this equation derive the expressions for the total thrust on a submerged plane area and the buoyant force acting on a submerged body. *(A.M.I.E., Summer 1990)*

(B) NUMERICAL PROBLEMS

1. Determine the total pressure and depth of centre of pressure on a plane rectangular surface of 1 m wide and 3 m deep when its upper edge is horizontal and (*a*) coincides with water surface (*b*) 2 m below the free water surface. [**Ans.** (*a*) 44145 N, 2.0 m, (*b*) 103005 N, 3.714 m]

2. Determine the total pressure on a circular plate of diameter 1.5 m which is placed vertically in water in such a way that centre of plate is 2 m below the free surface of water. Find the position of centre of pressure also. [**Ans.** 34668.54 N, 2.07 m]

3. A rectangular sluice gate is situated on the vertical wall of a lock. The vertical side of the sluice is 6 m in length and depth of centroid of area is 8 m below the water surface. Prove that the depth of centre of pressure is given by 8.475 m.

4. A circular opening, 3 m diameter, in a vertical side of a tank is closed by a disc of 3 m diameter which can rotate about a horizontal diameter. Calculate : (*i*) the force on the disc, and (*ii*) the torque required to maintain the disc in equilibrium in the vertical position when the head of water above the horizontal diameter is 6 m. [**Ans.** (*i*) 416.05 kN (*ii*) 39005 Nm]

5. The pressure at the centre of a pipe of diameter 3 m is 29.43 N/cm^2. The pipe contains oil of sp. gr. 0.87 and is filled with a gate valve. Find the force exerted by the oil on the gate and position of centre of pressure. [**Ans.** 2.08 MN, .016 m below centre of pipe]

6. Determine the total pressure and centre of pressure on an isosceles triangular plate of base 5 m and altitude 5 m when the plate is immersed vertically is an oil of sp. gr. 0.8. The base of the plate is 1 m below the free surface of water. [**Ans.** 261927 N, 3.19 m]

7. The opening in a dam is 3 m wide and 2 m high. A vertical sluice gate is used to cover the opening. On the upstream of the gate, the liquid of sp. gr. 1.5, lies upto a height of 2.0 m above the top of the gate, whereas on the downstream side, the water is available upto a height of the top of the gate. Find the resultant force acting on the gate and position of centre of pressure. Assume that the gate is higher at the bottom. [**Ans.** 206010 N, 0.964 m above the hinge]

8. A caisson for closing the entrance to a dry dock is of trapezoidal form 16 m wide at the top and 12 m wide at the bottom and 8 m deep. Find the total pressure and centre of pressure on the caisson if the water on the outside is 1 m below the top level of the caisson and dock is empty.
(A.M.I.E., Winter 1980) [**Ans.** 3.164 MN, 4.56 m below water surface]

9. A sliding gate 2 m wide and 1.5 m high lies in a vertical plane and has a co-efficient of friction of 0.2 between itself and guides. If the gate weighs one tonne, find the vertical force required to raise the gate if its upper edge is at a depth of 4 m from free surface of water. [**Ans.** 37768.5 N]

10. A tank contains water upto a height of 1 m above the base. An immiscible liquid of sp. gr. 0.8 is filled on the top of water upto 1.5 m height. Calculate : (*i*) total pressure on one side of the tank, (*ii*) the position of centre of pressure for one side of the tank, which is 3 m wide.
[**Ans.** 76518 N, 1.686 m from top]

11. A rectangular tank 4 m long, 1.5 m wide contains water upto a height of 2 m. Calculate the force due to water pressure on the base of the tank. Find also the depth of centre of pressure from free surface.
[**Ans.** 117720 N, 2 m from free surface]

12. A rectangular plane surface 1 m wide and 3 m deep lies in water in such a way that its plane makes an angle of 30° with the free surface of water. Determine the total pressure and position of centre of pressure when the upper edge of the plate is 2 m below the free water surface.
[**Ans.** 80932.5 N, 2.318 m]

13. A circular plate 3.0 m diameter is immersed in water in such a way that the plane of the plate makes an angle of 60° with the free surface of water. Determine the total pressure and position of centre of pressure when the upper edge of the plate is 2 m below the free water surface.

[**Ans.** 228.69 kN, 3.427 m from free surface]

14. A rectangular gate 6 m × 2 m is hinged at its base and inclined at 60° to the horizontal as shown in Fig. 3.40. To keep the gate in a stable position, a counter weight of 29430 N is attached at the upper end of the gate. Find the depth of water at which the gate begins to fall. Neglect the weight of the gate and also friction at the hinge and pulley. [**Ans.** 3.43 m]

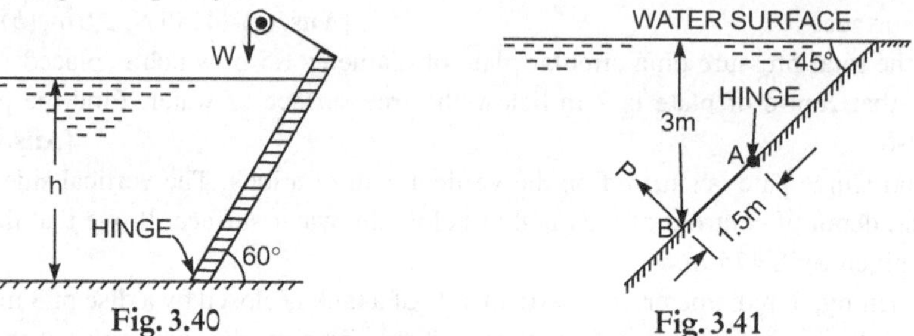

Fig. 3.40 Fig. 3.41

15. An inclined rectangular gate of width 5 m and depth 1.5 m is installed to control the discharge of water as shown in Fig. 3.41. The end *A* is hinged. Determine the force normal to the gate applied at *B* to open it.

[**Ans.** 97435.8 N]

16. A gate supporting water is shown in Fig. 3.42. Find the height '*h*' of the water so that the gate begins to tip about the hinge. Take the width of the gate as unity. (*Delhi University, 1986*) [**Ans.** $3 \times \sqrt{3}$ m]

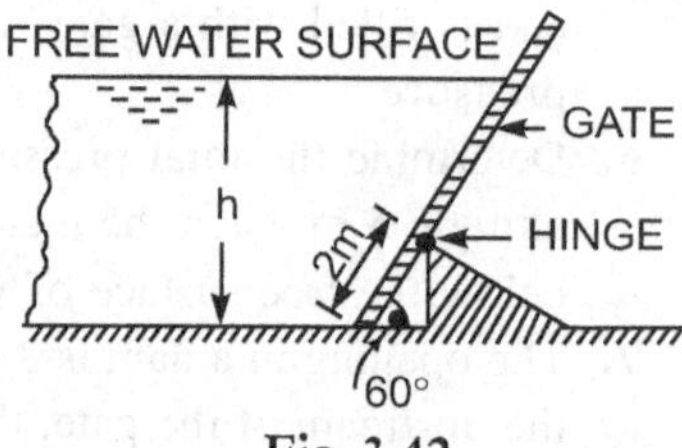

Fig. 3.42

17. Find the total pressure and depth of centre of pressure on a triangular plate of base 3 m and height 3 m which is immersed in water in such a way that plane of the plate makes an angle of 60° with the free surface. The base of the plate is parallel to water surface and at a depth of 2 m from water surface. [**Ans.** 126.52 kN, 2.996 m]

18. Find the horizontal and vertical components of the total force acting on a curved surface *AB*, which is in the form of a quadrant of a circle of radius 2 m as shown in Fig. 3.43. Take the width of the gate 2 m.

[**Ans.** $F_x = 117.72$ kN, $F_y = 140.114$ kN]

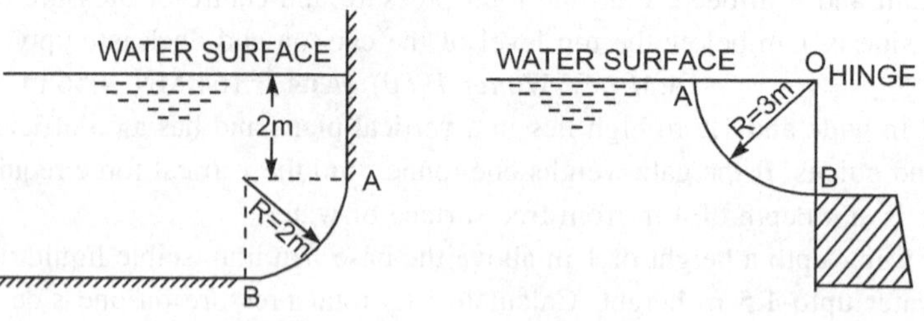

Fig. 3.43 Fig. 3.44

19. Fig. 3.44 shows a gate having a quadrant shape of radius of 3 m. Find the resultant force due to water per metre length of the gate. Find also the angle at which the total force will act. [**Ans.** 82.201 kN, θ = 57° 31′]

20. A roller gate is shown in Fig. 3.45. It is cylindrical form of 6.0 m diameter. It is placed on the dam. Find the magnitude and direction of the resultant force due to water acting on the gate when the water is just going to spill. The length of the gate is given 10 m. [**Ans.** 2.245 MN, θ = 38° 8′]

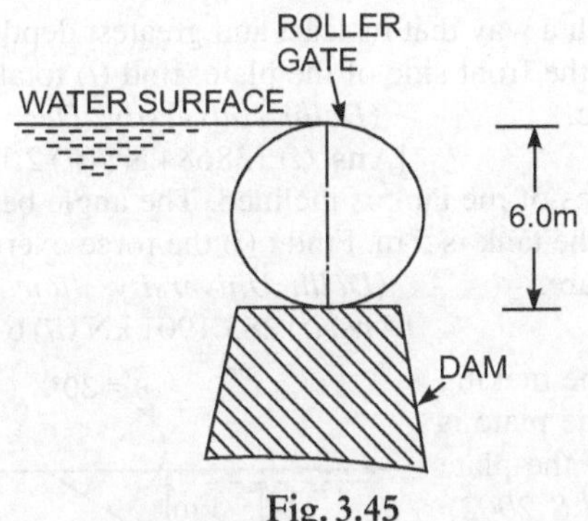

Fig. 3.45

Fig. 3.46

21. Find the horizontal and vertical components of the water pressure exerted on a tainter gate of radius 4 m as shown in Fig. 3.46. Consider width of the gate unity. [**Ans.** $F_x = 19.62$ kN, $F_y = 7102.44$ N]

22. Find the magnitude and direction of the resultant water pressure acting on a curved face of a dam which is shaped according to the relation $y = \dfrac{x^2}{6}$ as shown in Fig. 3.47. The height of water retained by the dam is 12 m. Take the width of dam as unity.

[**Ans.** 970.74 kN, $\theta = 43° \, 19'$]

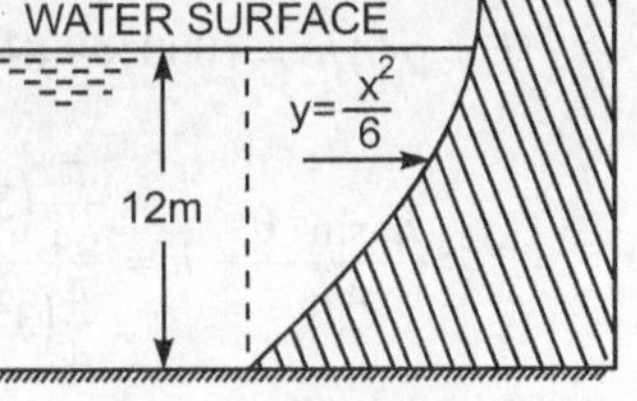

Fig. 3.47

23. A hollow circular plate of 2 m external and 1 m internal diameter is immersed vertically in water such that the centre of plate is 4 m deep from water surface. Find the total pressure and depth of centre of pressure.

(*Punjab, 1972*) [**Ans.** 92.508 kN, 4.078 m]

24. A rectangular opening 2 m wide and 1 m deep in the vertical side of a tank is closed by a sluice gate of the same size. The gate can turn about the horizontal centroidal axis. Determine : (*i*) the total pressure on the sluice gate and (*ii*) the torque on the sluice gate. The head of water above the upper edge of the gate is 1.5 m.

[**Ans.** (*i*) 39.24 kN and (*ii*) 1635 Nm]

25. Determine the total force and location of centre of pressure on one face of the plate shown in Fig. 3.48 immersed in a liquid of specific gravity 0.9. [**Ans.** 62.4 kN, 3.04 m]

26. A circular opening, 3 m diameter, in the vertical side of water tank is closed by a disc of 3 m diameter which can rotate about a horizontal diameter ? Calculate : (*i*) the force on the disc, and (*ii*) the torque required to maintain the disc in equilibrium in the vertical position when the head of water above the horizontal diameter is 4 m.

[**Ans.** (*i*) 270 kN, and (*ii*) 38 kN m]

Fig. 3.48

27. A penstock made up by a pipe of 2 m diameter contains a circular disc of same diameter to act as a valve which controls the discharge passing through it. It can rotate about a horizontal diameter. If the head of water above its centre is 20 m, find the total force acting on the disc and the torque required to maintain it in the vertical position.

(*A.M.I.E., Summer 1990*) [**Ans.** 616380 N, 7704.8 Nm]

28. A circular drum 1.8 m diameter and 1.2 m height is submerged with its axis vertical and its upper end at a depth of 1.8 m below water level. Determine :

(*i*) total pressure on top, bottom and curved surfaces of the drum,
(*ii*) resultant pressure on the whole surface, and
(*iii*) depth of centre of pressure on curved surface.

(*A.M.I.E., Winter 1991*) [**Ans.** (*i*) 44934 N, 74890 N, 0, (*ii*) 29956 N, (*iii*) 2,4 m]

29. A circular plate of diameter 3 m is immersed in water in such a way that its least and greatest depth from the free surface of water are 1 m and 3 m respectively. For the front side of the plate, find (*i*) total force exerted by water and (*ii*) the position of centre of pressure. (*Delhi University, Dec. 1996*)
[**Ans.** (*i*) 138684 N ; (*ii*) 2.125 m]

30. A tank contains water upto a height of 10 m. One of the sides of the tank is inclined. The angle between free surface of water and inclined side is 60°. The width of the tank is 5 m. Find : (*i*) the force exerted by water on inclined side and (*ii*) position of centre of pressure. (*Delhi University, June 1996*)
[**Ans.** (*i*) 283.1901 kN (*ii*) 6.67 m]

31. A circular plate of 3 m diameter is under water with its plane making an angle of 30° with the water surface. If the top edge of the plate is 1 m below the water surface, find the force on one side of the plate and its location. (*J.N.T.U., Hyderabad S 2002*)

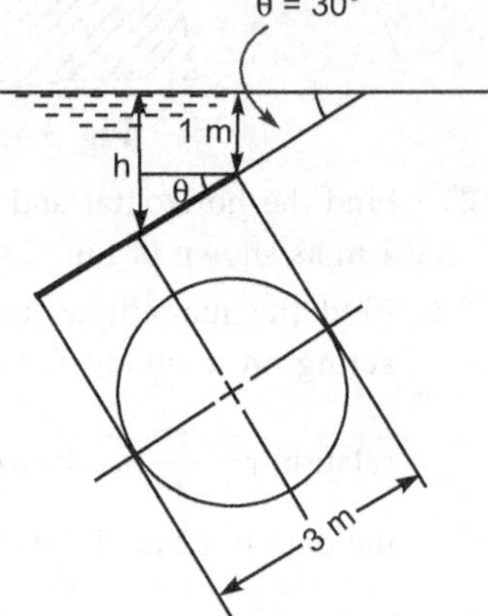

Fig. 3.49

[**Hint.** $d = 3$ m, $\theta = 30°$, height of top edge = 1 m, $\bar{h} = 1 + 1.5 \times \sin 30° = 1.75$

$$F = \rho g A\bar{h} = 1000 \times 9.81 \times \left(\frac{\pi}{4} \times 3^2\right) \times 1.75 = \mathbf{121.35\ kN.\ Ans.}$$

$$h^* = \frac{I_G \sin^2\theta}{A\bar{h}} + \bar{h} = \frac{\frac{\pi}{64}(3^4)^4 \times \frac{1}{4}}{\frac{\pi}{4}(3^2) \times 1.75} + 1.75 = 0.08 + 1.75 = \mathbf{1.83\ m.\ Ans.}]$$

Buoyancy and Floatation

▶ 4.1 INTRODUCTION

In this chapter, the equilibrium of the floating and sub-merged bodies will be considered. Thus the chapter will include : 1. Buoyancy, 2. Centre of buoyancy, 3. Metacentre, 4. Metacentric height, 5. Analytical method for determining metacentric height, 6. Conditions of equilibrium of a floating and sub-merged body, and 7. Experimental method for metacentric height.

▶ 4.2 BUOYANCY

When a body is immersed in a fluid, an upward force is exerted by the fluid on the body. This upward force is equal to the weight of the fluid displaced by the body and is called the force of buoyancy or simply buoyancy or buoyant force.

▶ 4.3 CENTRE OF BUOYANCY

It is defined as the point, through which the force of buoyancy is supposed to act. As the force of buoyancy is a vertical force and is equal to the weight of the fluid displaced by the body, the centre of buoyancy will be the centre of gravity of the fluid displaced.

Problem 4.1 *Find the volume of the water displaced and position of centre of buoyancy for a wooden block of width 2.5 m and of depth 1.5 m, when it floats horizontally in water. The density of wooden block is 650 kg/m³ and its length 6.0 m.*

Solution. Given :

Width $= 2.5$ m

Depth $= 1.5$ m

Length $= 6.0$ m

Volume of the block $= 2.5 \times 1.5 \times 6.0 = 22.50 \text{ m}^3$

Density of wood, $\rho = 650 \text{ kg/m}^3$

$\therefore$ Weight of block $= \rho \times g \times \text{Volume}$

$= 650 \times 9.81 \times 22.50 \text{ N} = 143471 \text{ N}$

Fig. 4.1

For equilibrium the weight of water displaced = Weight of wooden block

$= 143471$ N

$\therefore$ Volume of water displaced

$$= \frac{\text{Weight of water displaced}}{\text{Weight density of water}} = \frac{143471}{1000 \times 9.81} = \mathbf{14.625\ m^3.\ Ans.}$$

($\because$ Weight density of water $= 1000 \times 9.81$ N/m^3)

Position of Centre of Buoyancy. Volume of wooden block in water

= Volume of water displaced

or $2.5 \times h \times 6.0 = 14.625$ m^3, where h is depth of wooden block in water

$$\therefore \quad h = \frac{14.625}{2.5 \times 6.0} = 0.975 \text{ m}$$

$$\therefore \quad \text{Centre of Buoyancy} = \frac{0.975}{2} = \mathbf{0.4875\ m\ from\ base.\ Ans.}$$

Problem 4.2 *A wooden log of 0.6 m diameter and 5 m length is floating in river water. Find the depth of the wooden log in water when the sp. gravity of the log is 0.7.*

Solution. Given :

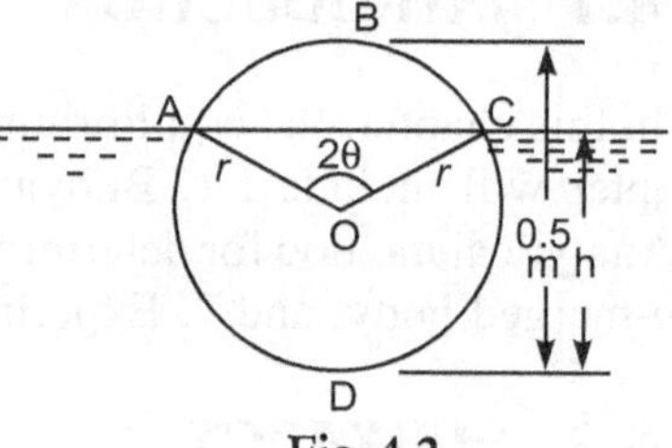

Fig. 4.2

Dia. of log $= 0.6$ m

Length, $L = 5$ m

Sp. gr., $S = 0.7$

$\therefore$ Density of log $= 0.7 \times 1000 = 700$ kg/m^3

$\therefore$ Weight density of log, $w = \rho \times g$

$= 700 \times 9.81$ N/m^3

Find depth of immersion or h

Weight of wooden log = Weight density $\times$ Volume of log

$$= 700 \times 9.81 \times \frac{\pi}{4} (D)^2 \times L$$

$$= 700 \times 9.81 \times \frac{\pi}{4} (.6)^2 \times 5 \text{ N} = 989.6 \times 9.81 \text{ N}$$

For equilibrium,

Weight of wooden log = Weight of water displaced

= Weight density of water $\times$ Volume of water displaced

$$\therefore \quad \text{Volume of water displaced} = \frac{989.6 \times 9.81}{1000 \times 9.81} = 0.9896 \text{ m}^3$$

($\because$ Weight density of water $= 1000 \times 9.81$ N/m^3)

Let h is the depth of immersion

$\therefore$ Volume of log inside water = Area of $ADCA \times$ Length

= Area of $ADCA \times 5.0$

But volume of log inside water = Volume of water displaced $= 0.9896$ m^3

$\therefore$ 0.9896 = Area of $ADCA \times 5.0$

$\therefore$ Area of $ADCA$ $= \dfrac{0.9896}{5.0} = 0.1979 \text{ m}^2$

But area of $ADCA$ = Area of curved surface $ADCOA$ + Area of ΔAOC

$$= \pi r^2 \left[\frac{360 - 2\theta}{360}\right] + \frac{1}{2} r \cos\theta \times 2r \sin\theta$$

$$= \pi r^2 \left[1 - \frac{\theta}{180}\right] + r^2 \cos\theta \sin\theta$$

$$\therefore \quad 0.1979 = \pi\,(.3)^2 \left[1 - \frac{\theta}{180}\right] + (.3)^2 \cos\theta \sin\theta$$

$$0.1979 = .2827 - .00157\,\theta + 0.09 \cos\theta \sin\theta$$

or $\quad .00157\,\theta - .09 \cos\theta \sin\theta = .2827 - .1979 = 0.0848$

$$\theta - \frac{.09}{.00157} \cos\theta \sin\theta = \frac{.0848}{.00157}$$

or $\quad \theta - 57.32 \cos\theta \sin\theta = 54.01.$

or $\quad \theta - 57.32 \cos\theta \sin\theta - 54.01 = 0$

For $\theta = 60$, $\quad 60 - 57.32 \times 0.5 \times .866 - 54.01 = 60 - 24.81 - 54.01 = -18.82$

For $\theta = 70$, $\quad 70 - 57.32 \times .342 \times 0.9396 - 54.01 = 70 - 18.4 - 54.01 = -2.41$

For $\theta = 72$, $\quad 72 - 57.32 \times .309 \times .951 - 54.01 = 72 - 16.84 - 54.01 = +1.14$

For $\theta = 71$, $\quad 71 - 57.32 \times .325 \times .9455 - 54.01 = 71 - 17.61 - 54.01 = -0.376$

$\therefore \quad \theta = 71.5°$, $\quad 71.5 - 57.32 \times .3173 \times .948 - 54.01 = 71.5 - 17.24 - 54.01 = +.248$

Then $\quad h = r + r \cos 71.5°$

$= 0.3 + 0.3 \times 0.3173 =$ **0.395 m. Ans.**

Problem 4.3 *A stone weighs 392.4 N in air and 196.2 N in water. Compute the volume of stone and its specific gravity.*

Solution. Given :

Weight of stone in air = 392.4 N

Weight of stone in water = 196.2 N

For equilibrium,

Weight in air – Weight of stone in water = Weight of water displaced

or $\quad 392.4 - 196.2 = 196.2 = 1000 \times 9.81 \times$ Volume of water displaced

$\therefore$ Volume of water displaced

$$= \frac{196.2}{1000 \times 9.81} = \frac{1}{50} \text{ m}^3 = \frac{1}{50} \times 10^6 \text{ cm}^3 = \mathbf{2 \times 10^4\ cm^3.\ Ans.}$$

= Volume of stone

$\therefore$ Volume of stone $=$ **2×10^4 cm³. Ans.**

Specific Gravity of Stone

Mass of stone $= \dfrac{\text{Weight in air}}{g} = \dfrac{392.4}{9.81} = 40 \text{ kg}$

Density of stone $= \dfrac{\text{Mass in air}}{\text{Volume}} = \dfrac{40.0 \text{ kg}}{\frac{1}{50} \text{ m}^3} = 40 \times 50 = 2000 \ \dfrac{\text{kg}}{\text{m}^3}$

$\therefore$ Sp. gr. of stone $= \dfrac{\text{Density of stone}}{\text{Density of water}} = \dfrac{2000}{1000} = $ **2.0. Ans.**

Problem 4.4 *A body of dimensions 1.5 m × 1.0 m × 2 m, weighs 1962 N in water. Find its weight in air. What will be its specific gravity ?*

Solution. Given :

Volume of body $= 1.50 \times 1.0 \times 2.0 = 3.0 \text{ m}^3$

Weight of body in water $= 1962$ N

Volume of the water displaced = Volume of the body $= 3.0 \text{ m}^3$

$\therefore$ Weight of water displaced $= 1000 \times 9.81 \times 3.0 = 29430$ N

For the equilibrium of the body

Weight of body in air – Weight of water displaced = Weight in water

$$\therefore \quad W_{air} - 29430 = 1962$$

$$W_{air} = 29430 + 1962 = 31392 \text{ N}$$

Mass of body $= \dfrac{\text{Weight in air}}{g} = \dfrac{31392}{9.81} = 3200$ kg

Density of the body $= \dfrac{\text{Mass}}{\text{Volume}} = \dfrac{3200}{3.0} = 1066.67$

$\therefore$ Sp. gravity of the body $= \dfrac{1066.67}{1000} = $ **1.067. Ans.**

Problem 4.5 *Find the density of a metallic body which floats at the interface of mercury of sp. gr. 13.6 and water such that 40% of its volume is sub-merged in mercury and 60% in water.*

Solution. Let the volume of the body $= V \text{ m}^3$

Then volume of body submerged in mercury

$$= \frac{40}{100} V = 0.4 \ V \text{ m}^3$$

Volume of body submerged in water

$$= \frac{60}{100} \times V = 0.6 \ V \text{ m}^3$$

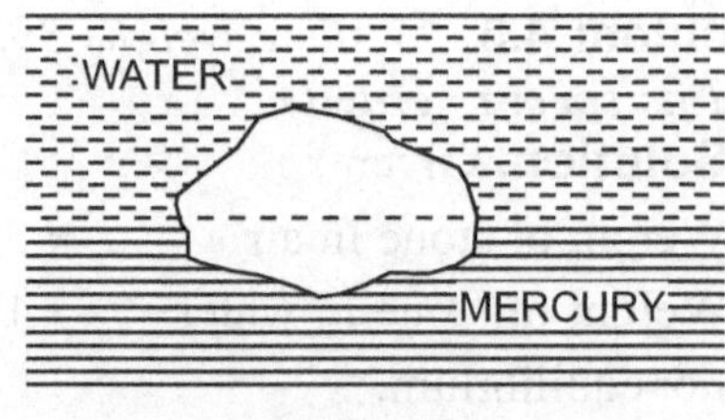

Fig. 4.3

For the equilibrium of the body

Total buoyant force (upward force) = Weight of the body

But total buoyant force = Force of buoyancy due to water + Force of buoyancy due to mercury

Force of buoyancy due to water = Weight of water displaced by body

= Density of water × g × Volume of water displaced

= 1000 × g × Volume of body in water

$= 1000 \times g \times 0.6 \times V$ N

and Force of buoyancy due to mercury = Weight of mercury displaced by body

= g × Density of mercury × Volume of mercury displaced

= g × 13.6 × 1000 × Volume of body in mercury

$= g \times 13.6 \times 1000 \times 0.4 \ V$ N

Weight of the body $= \text{Density} \times g \times \text{Volume of body} = \rho \times g \times V$

where ρ is the density of the body

$\therefore$ For equilibrium, we have

Total buoyant force $=$ Weight of the body

$1000 \times g \times 0.6 \times V + 13.6 \times 1000 \times g \times .4\ V = \rho \times g \times V$

or $\rho = 600 + 13600 \times .4 = 600 + 54400 = 6040.00 \text{ kg/m}^3$

$\therefore$ Density of the body $= \mathbf{6040.00\ kg/m^3}$. **Ans.**

Problem 4.6 *A float valve regulates the flow of oil of sp. gr. 0.8 into a cistern. The spherical float is 15 cm in diameter. AOB is a weightless link carrying the float at one end, and a valve at the other end which closes the pipe through which oil flows into the cistern. The link is mounted in a frictionless hinge at O and the angle AOB is 135°. The length of OA is 20 cm, and the distance between the centre of the float and the hinge is 50 cm. When the flow is stopped AO will be vertical. The valve is to be pressed on to the seat with a force of 9.81 N to completely stop the flow of oil into the cistern. It was observed that the flow of oil is stopped when the free surface of oil in the cistern is 35 cm below the hinge. Determine the weight of the float.* (U.P.S.C., Engg. Services, 1975)

Solution. Given :

Sp. gr. of oil $= 0.8$

$\therefore$ Density of oil, $\rho_0 = 0.8 \times 1000 = 800 \text{ kg/m}^3$

Dia. of float, $D = 15$ cm

$\angle AOB = 135°$

$OA = 20$ cm

Force, $P = 9.81$ N

$OB = 50$ cm

Fig. 4.4

Find the weight of the float. Let it is equal to W.

When the flow of oil stopped, the centre of float is shown in Fig. 4.4

The level of oil is also shown. The centre of float is below the level of oil, by a depth 'h'.

From ΔBOD, $\sin 45° = \dfrac{OD}{OB} = \dfrac{OC + CD}{OB} = \dfrac{35 + h}{50}$

$\therefore$ $50 \times \sin 45° = 35 + h$

or $h = 50 \times \dfrac{1}{\sqrt{2}} - 35 = 35.355 - 35 = 0.355 \text{ cm} = .00355 \text{ m}.$

The weight of float is acting through B, but the upward buoyant force is acting through the centre of weight of oil displaced.

Volume of oil displaced $= \dfrac{2}{3}\pi r^3 + h \times \pi r^2$ $\left\{ r = \dfrac{D}{2} = \dfrac{15}{2} = 7.5 \text{ cm} \right\}$

$= \frac{2}{3} \times \pi \times (.075)^3 + .00355 \times \pi \times (.075)^2 = 0.000945 \text{ m}^3$

$\therefore$ Buoyant force $=$ Weight of oil displaced

$= \rho_0 \times g \times \text{Volume of oil}$

$= 800 \times 9.81 \times .000945 = 7.416 \text{ N}$

The buoyant force and weight of the float passes through the same vertical line, passing through B. Let the weight of float is W. Then net vertical force on float

$$= \text{Buoyant force} - \text{Weight of float} = (7.416 - W)$$

Taking moments about the hinge O, we get

$$P \times 20 = (7.416 - W) \times BD = (7.416 - W) \times 50 \times \cos 45°$$

or

$$9.81 \times 20 = (7.416 - W) \times 35.355$$

$$\therefore \quad W = 7.416 - \frac{20 \times 9.81}{35.355} = 7.416 - 5.55 = \mathbf{1.866\ N.\ Ans.}$$

▶ 4.4 META-CENTRE

It is defined as the point about which a body starts oscillating when the body is tilted by a small angle. The meta-centre may also be defined as the point at which the line of action of the force of buoyancy will meet the normal axis of the body when the body is given a small angular displacement.

Consider a body floating in a liquid as shown in Fig. 4.5 (*a*). Let the body is in equilibrium and G is the centre of gravity and B the centre of buoyancy. For equilibrium, both the points lie on the normal axis, which is vertical.

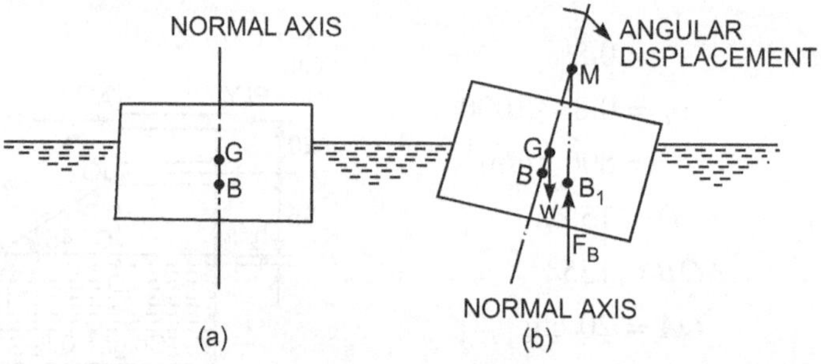

Fig. 4.5 *Meta-centre*

Let the body is given a small angular displacement in the clockwise direction as shown in Fig. 4.5 (*b*). The centre of buoyancy, which is the centre of gravity of the displaced liquid or centre of gravity of the portion of the body sub-merged in liquid, will now be shifted towards right from the normal axis. Let it is at B_1 as shown in Fig. 4.5 (*b*). The line of action of the force of buoyancy in this new position, will intersect the normal axis of the body at some point say M. This point M is called **Meta-centre.**

▶ 4.5 META-CENTRIC HEIGHT

The distance MG, *i.e.*, the distance between the meta-centre of a floating body and the centre of gravity of the body is called meta-centric height.

▶ 4.6 ANALYTICAL METHOD FOR META-CENTRE HEIGHT

Fig. 4.6 (*a*) shows the position of a floating body in equilibrium. The location of centre of gravity and centre of buoyancy in this position is at G and B. The floating body is given a small angular displacement in the clockwise direction. This is shown in Fig. 4.6 (*b*). The new centre of buoyancy is at B_1. The vertical line through B_1 cuts the normal axis at M. Hence M is the meta-centre and GM is meta-centric height.

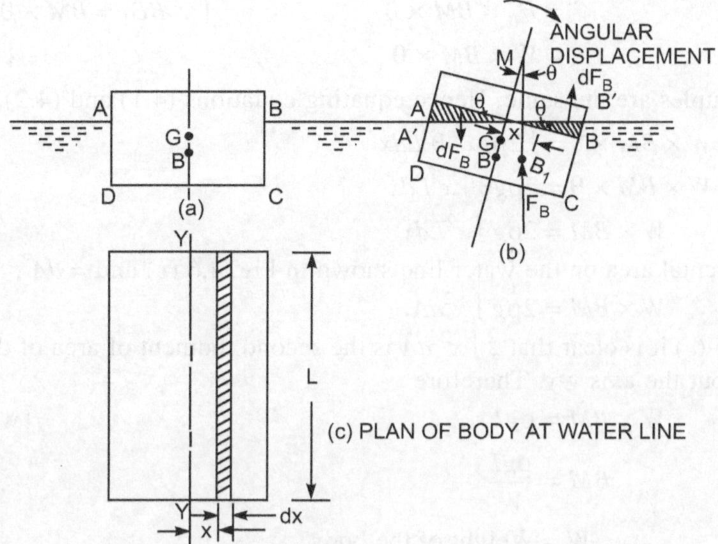

Fig. 4.6 *Meta-centre height of floating body.*

The angular displacement of the body in the clockwise direction causes the wedge-shaped prism BOB' on the right of the axis to go inside the water while the identical wedge-shaped prism represented by AOA' emerges out of the water on the left of the axis. These wedges represent a gain in buoyant force on the right side and a corresponding loss of buoyant force on the left side. The gain is represented by a vertical force dF_B acting through the $C.G.$ of the prism BOB' while the loss is represented by an equal and opposite force dF_B acting vertically downward through the centroid of AOA'. The couple due to these buoyant forces dF_B tends to rotate the ship in the counter clockwise direction. Also the moment caused by the displacement of the centre of buoyancy from B to B_1 is also in the counter clockwise direction. Thus these two couples must be equal.

Couple Due to Wedges. Consider towards the right of the axis a small strip of thickness dx at a distance x from O as shown in Fig. 4.5 (*b*). The height of strip $x \times \angle BOB' = x \times \theta$.

$$\{\because \angle BOB' = \angle AOA' = BMB_1' = \theta\}$$

$\therefore$ Area of strip $= \text{Height} \times \text{Thickness} = x \times \theta \times dx$

If L is the length of the floating body, then

Volume of strip $= \text{Area} \times L$

$$= x \times \theta \times L \times dx$$

$\therefore$ Weight of strip $= \rho g \times \text{Volume} = \rho gx\ \theta L\ dx$

Similarly, if a small strip of thickness dx at a distance x from O towards the left of the axis is considered, the weight of strip will be $\rho gx\theta\ L\ dx$. The two weights are acting in the opposite direction and hence constitute a couple.

Moment of this couple $= \text{Weight of each strip} \times \text{Distance between these two weights}$

$$= \rho gx\ \theta L\ dx\ [x + x]$$

$$= \rho gx\ \theta L\ dx \times 2x = 2\rho gx^2\ \theta L\ dx$$

$\therefore$ Moment of the couple for the whole wedge

$$= \int 2\rho gx^2\ \theta L\ dx \qquad \text{...(4.1)}$$

Moment of couple due to shifting of centre of buoyancy from B to B_1

$$= F_B \times BB_1$$

$$= F_B \times BM \times \theta \qquad \{\because BB_1 = BM \times \theta \text{ if } \theta \text{ is very small}\}$$

$$= W \times BM \times \theta \qquad \{\because F_B = W\} \quad ...(4.2)$$

But these two couples are the same. Hence equating equations (4.1) and (4.2), we get

$$W \times BM \times \theta = \int 2\rho g x^2 \, \theta \, L dx$$

$$W \times BM \times \theta = 2\rho g \theta \int x^2 L dx$$

$$W \times BM = 2\rho g \int x^2 L dx$$

Now Ldx = Elemental area on the water line shown in Fig. 4.6 (*c*) and = dA

$$\therefore \quad W \times BM = 2\rho g \int x^2 dA.$$

But from Fig. 4.5 (*c*) it is clear that $2\int x^2 \, dA$ is the second moment of area of the plan of the body at water surface about the axis *y-y*. Therefore

$$W \times BM = \rho g I \qquad \{\text{where } I = 2\int x^2 \, dA\}$$

$$\therefore \quad BM = \frac{\rho g I}{W}$$

But W = Weight of the body

= Weight of the fluid displaced by the body

= ρg × Volume of the fluid displaced by the body

= ρg × Volume of the body sub-merged in water

= $\rho g \times \forall$

$$\therefore \quad BM = \frac{\rho g \times I}{\rho g \times \forall} = \frac{I}{\forall} \quad ...(4.3)$$

$$GM = BM - BG = \frac{I}{\forall} - BG$$

$$\therefore \quad \text{Meta-centric height} = GM = \frac{I}{\forall} - BG. \quad ...(4.4)$$

Problem 4.7 *A rectangular pontoon is 5 m long, 3 m wide and 1.20 m high. The depth of immersion of the pontoon is 0.80 m in sea water. If the centre of gravity is 0.6 m above the bottom of the pontoon, determine the meta-centric height. The density for sea water = 1025 kg/m³.*

(Delhi University, 1992)

Solution. Given :

Dimension of pontoon = 5 m × 3 m × 1.20 m

Depth of immersion = 0.8 m

Distance AG = 0.6 m

Distance $AB = \frac{1}{2}$ × Depth of immersion

$= \frac{1}{2} \times .8 = 0.4$ m

Density for sea water = 1025 kg/m^3

Meta-centre height *GM*, given by Equation (4.4) is

$$GM = \frac{I}{\forall} - BG$$

where I = M.O. Inertia of the plan of the pontoon about *y-y* axis

$$= \frac{1}{12} \times 5 \times 3^3 \text{ m}^4 = \frac{45}{4} \text{ m}^4$$

$\forall$ = Volume of the body submerged in water

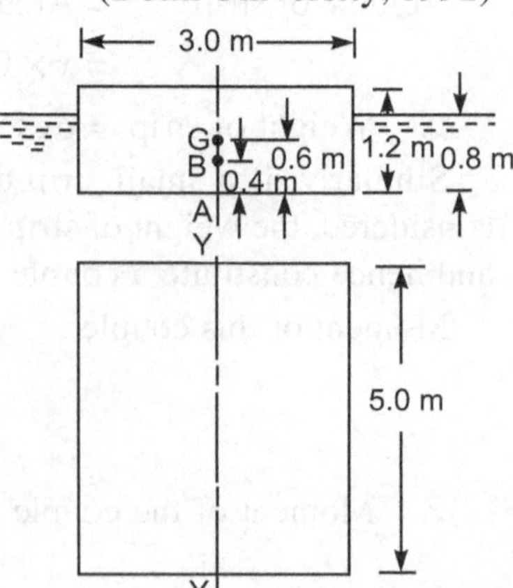

Fig. 4.7

$= 3 \times 0.8 \times 5.0 = 12.0 \text{ m}^3$

$BG = AG - AB = 0.6 - 0.4 = 0.2$ m

$$\therefore \qquad GM = \frac{45}{4} \times \frac{1}{12.0} - 0.2 = \frac{45}{48} - 0.2 = 0.9375 - 0.2 = \mathbf{0.7375\ m.}\quad \textbf{Ans.}$$

Problem 4.8 *A uniform body of size 3 m long × 2 m wide × 1 m deep floats in water. What is the weight of the body if depth of immersion is 0.8 m ? Determine the meta-centric height also.*

Solution. Given :

Dimension of body $= 3 \times 2 \times 1$

Depth of immersion $= 0.8$ m

Find (*i*) Weight of body, W

(*ii*) Meta-centric height, GM

(*i*) **Weight of Body, W**

= Weight of water displaced

$= \rho g \times$ Volume of water displaced

$= 1000 \times 9.81 \times$ Volume of body in water

$= 1000 \times 9.81 \times .3 \times 2 \times 0.8$ N

= **47088 N. Ans.**

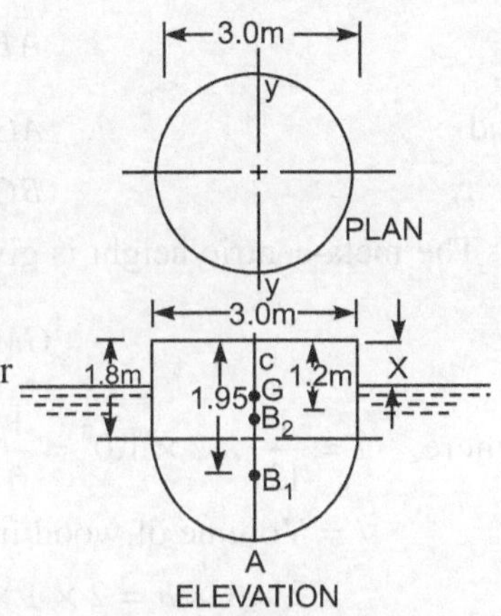

Fig. 4.8

(*ii*) **Meta-centric Height, GM**

Using equation (4.4), we get

$$GM = \frac{I}{\forall} - BG$$

where I = M.O.I about y-y axis of the plan of the body

$$= \frac{1}{12} \times 3 \times 2^3 = \frac{3 \times 2^3}{12} = 2.0 \text{ m}^4$$

$\forall$ = Volume of body in water

$= 3 \times 2 \times 0.8 = 4.8 \text{ m}^3$

$$BG = AG - AB = \frac{1.0}{2} - \frac{0.8}{2} = 0.5 - 0.4 = 0.1$$

$$\therefore \qquad GM = \frac{2.0}{4.8} - 0.1 = 0.4167 - 0.1 = \mathbf{0.3167\ m.}\quad \textbf{Ans.}$$

Problem 4.9 *A block of wood of specific gravity 0.7 floats in water. Determine the meta-centric height of the block if its size is 2 m × 1 m × 0.8 m.*

Solution. Given :

Dimension of block $= 2 \times 1 \times 0.8$

Let depth of immersion $= h$ m

Sp. gr. of wood $= 0.7$

Weight of wooden piece = Weight density of wood* × Volume

$= 0.7 \times 1000 \times 9.81 \times 2 \times 1 \times 0.8$ N

Weight of water displaced = Weight density of water

× Volume of the wood submerged in water

$= 1000 \times 9.81 \times 2 \times 1 \times h$ N

Fig. 4.9

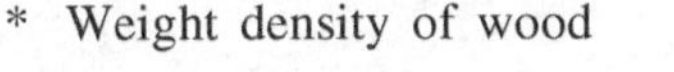

* Weight density of wood $= \rho \times g$, where ρ = density of wood

$= 0.7 \times 1000 = 700 \text{ kg/m}^3$. Hence w for wood $= 700 \times 9.81 \text{ N/m}^3$.

For equilibrium,

Weight of wooden piece = Weight of water displaced

$\therefore \quad 700 \times 9.81 \times 2 \times 1 \times 0.8 = 1000 \times 9.81 \times 2 \times 1 \times h$

$$\therefore \quad h = \frac{700 \times 9.81 \times 2 \times 1 \times 0.8}{1000 \times 9.81 \times 2 \times 1} = 0.7 \times 0.8 = 0.56 \text{ m}$$

$\therefore$ Distance of centre of Buoyancy from bottom, *i.e.*,

$$AB = \frac{h}{2} = \frac{0.56}{2} = 0.28 \text{ m}$$

and $$AG = 0.8/2.0 = 0.4 \text{ m}$$

$$\therefore \quad BG = AG - AB = 0.4 - 0.28 = 0.12 \text{ m}$$

The meta-centric height is given by equation (4.4) or

$$GM = \frac{I}{\forall} - BG$$

where $I = \frac{1}{12} \times 2 \times 1.0^3 = \frac{1}{6} \text{ m}^4$

$\forall$ = Volume of wood in water

$= 2 \times 1 \times h = 2 \times 1 \times .56 = 1.12 \text{ m}^3$

$$\therefore \quad GM = \frac{1}{6} \times \frac{1}{1.12} - 0.12 = 0.1488 - 0.12 = \mathbf{0.0288\ m.}\ \textbf{Ans.}$$

Problem 4.10 *A solid cylinder of diameter 4.0 m has a height of 3 metres. Find the meta-centric height of the cylinder when it is floating in water with its axis vertical. The sp. gr. of the cylinder = 0.6.*

Solution. Given :

Dia. of cylinder, $D = 4.0$ m

Height of cylinder, $h = 3.0$ m

Sp. gr. of cylinder = 0.6

Depth of immersion of cylinder

$= 0.6 \times 3.0 = 1.8$ m

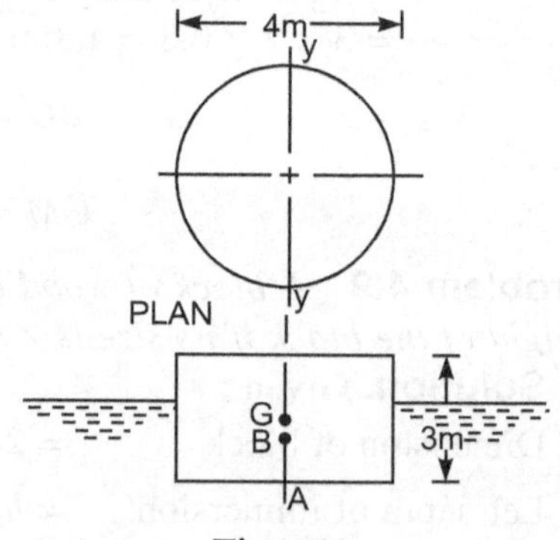

Fig. 4.10

$$\therefore \quad AB = \frac{1.8}{2} = 0.9 \text{ m}$$

and $$AG = \frac{3}{2} = 1.5 \text{ m}$$

$$\therefore \quad BG = AG - AB$$

$$= 1.5 - 0.9 = 0.6 \text{ m}$$

Now the meta-centric height GM is given by equation (4.4)

$$GM = \frac{I}{\forall} - BG$$

But $$I = \text{M.O.I. about } y\text{-}y \text{ axis of the plan of the body}$$

$$= \frac{\pi}{64} D^4 = \frac{\pi}{64} \times (4.0)^4$$

and $$\forall = \text{Volume of cylinder in water}$$

$$= \frac{\pi}{4} D^2 \times \text{Depth of immersion}$$

$$= \frac{\pi}{4} (4)^2 \times 1.8 \text{ m}^3$$

$$\therefore \quad GM = \frac{\frac{\pi}{64} \times (4.0)^4}{\frac{\pi}{4} \times (4.0)^2 \times 1.8} - 0.6$$

$$= \frac{1}{16} \times \frac{4.0^2}{1.8} - 0.6 = \frac{1}{1.8} - 0.6 = 0.55 - 0.6 = \mathbf{-0.05 \text{ m. Ans.}}$$

– ve sign means that meta-centre, (M) is below the centre of gravity (G).

Problem 4.11 *A body has the cylindrical upper portion of 3 m diameter and 1.8 m deep. The lower portion is a curved one, which displaces a volume of 0.6 m³ of water. The centre of buoyancy of the curved portion is at a distance of 1.95 m below the top of the cylinder. The centre of gravity of the whole body is 1.20 m below the top of the cylinder. The total displacement of water is 3.9 tonnes. Find the meta-centric height of the body.*

Solution. Given :

Dia. of body $= 3.0$ m

Depth of body $= 1.8$ m

Volume displaced by curved portion

$= 0.6 \text{ m}^3$ of water.

Let B_1 is the centre of buoyancy of the curved surface and G is the centre of gravity of the whole body.

Then $$CB_1 = 1.95 \text{ m}$$

$$CG = 1.20 \text{ m}$$

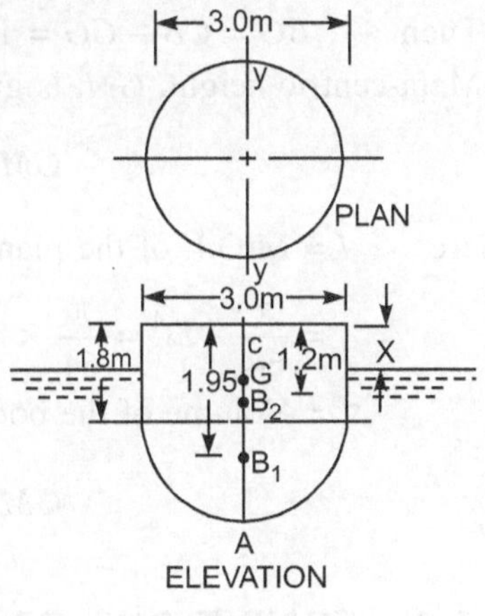

Fig. 4.11

Total weight of water displaced by body = 3.9 tonnes

$= 3.9 \times 1000 = 3900$ kgf

$= 3900 \times 9.81 \text{ N} = 38259 \text{ N}$

Find **meta-centric** height of the body.

Let the height of the body above the water surface x m. Total weight of water displaced by body

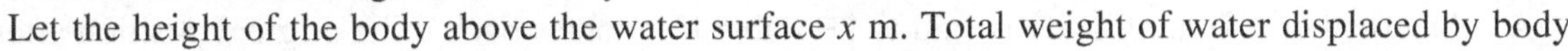

= Weight density of water × [Volume of water displaced]

= 1000 × 9.81 × [Volume of the body in water]

= 9810 [Volume of cylindrical part in water + Volume of curved portion]

$$= 9810 \left[\frac{\pi}{4} \times D^2 \times \text{Depth of cylindrical part in water} + \text{Volume displaced by curved portion} \right]$$

or $$38259 = 9810\left[\frac{\pi}{4}(3)^2 \times (1.8 - x) + 0.6\right]$$

$\therefore$ $$\frac{\pi}{4}(3)^2 \times (1.8 - x) + 0.6 = \frac{38259}{9810} = 3.9$$

$\therefore$ $$\frac{\pi}{4} \times 3^2 \times (1.8 - x) = 3.9 - 0.6 = 3.3$$

or $$1.8 - x = \frac{3.3 \times 4}{\pi \times 3 \times 3} = 0.4668$$

$\therefore$ $$x = 1.8 - .4668 = 1.33 \text{ m}$$

Let B_2 is the centre of buoyancy of cylindrical part and B is the centre of buoyancy of the whole body.

Then depth of cylindrical part in water $= 1.8 - x = 0.467$ m

$\therefore$ $$CB_2 = x + \frac{.467}{2} = 1.33 + .2335 = 1.5635 \text{ m.}$$

The distance of the centre of buoyancy of the whole body from the top of the cylindrical part is given as

$$CB = (\text{Volume of curved portion} \times CB_1 + \text{Volume of cylindrical part in water} \times CB_2)$$
$$\div (\text{Total volume of water displaced})$$
$$= \frac{0.6 \times 1.95 + 3.3 \times 1.5635}{(0.6 + 3.3)} = \frac{1.17 + 5.159}{3.9} = 1.623 \text{ m.}$$

Then $BG = CB - CG = 1.623 - 1.20 = .423$ m.

Meta-centric height, GM, is given by

$$GM = \frac{I}{\forall} - BG$$

where I = M.O.I. of the plan of the body at water surface about y-y

$$= \frac{\pi}{64} \times D^4 = \frac{\pi}{64} \times 3^4 \text{ m}^4$$

$\forall$ = Volume of the body in water = 3.9 m^3

$\therefore$ $$GM = \frac{\pi}{64} \times \frac{3^4}{3.9} - .423 = 1.019 - .423 = \mathbf{0.596 \text{ m.}} \textbf{ Ans.}$$

▶ 4.7 CONDITIONS OF EQUILIBRIUM OF A FLOATING AND SUBMERGED BODIES

A submerged or a floating body is said to be stable if it comes back to its original position after a slight disturbance. The relative position of the centre of gravity (G) and centre of buoyancy (B_1) of a body determines the stability of a submerged body.

4.7.1 Stability of a Submerged Body. The position of centre of gravity and centre of buoyancy in case of a completely submerged body are fixed. Consider a balloon, which is completely sub-merged in air. Let the lower portion of the balloon contains heavier material, so that its centre of gravity is lower than its centre of buoyancy as shown in Fig. 4.12 (a). Let the weight of the balloon is W. The weight W is acting through G, vertically in the downward direction, while the buoyant force

F_B is acting vertically up, through B. For the equilibrium of the balloon $W = F_B$. If the balloon is given an angular displacement in the clockwise direction as shown in Fig. 4.12 (*a*), then W and F_B constitute a couple acting in the anti-clockwise direction and brings the balloon in the original position. Thus the balloon in the position, shown by Fig. 4.12 (*a*) is in stable equilibrium.

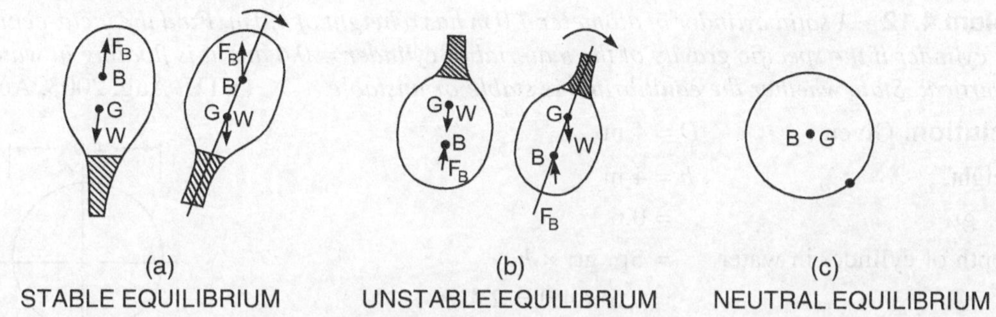

Fig. 4.12 *Stabilities of submerged bodies.*

(*a*) **Stable Equilibrium.** When $W = F_B$ and point B is above G, the body is said to be in stable equilibrium.

(*b*) **Unstable Equilibrium.** If $W = F_B$, but the centre of buoyancy (B) is below centre of gravity (G), the body is in unstable equilibrium as shown in Fig. 4.12 (*b*). A slight displacement to the body, in the clockwise direction, gives the couple due to W and F_B also in the clockwise direction. Thus the body does not return to its original position and hence the body is in unstable equilibrium.

(*c*) **Neutral Equilibrium.** If $F_B = W$ and B and G are at the same point, as shown in Fig. 4.12 (*c*), the body is said to be in Neutral Equilibrium.

4.7.2 Stability of Floating Body.

The stability of a floating body is determined from the position of Meta-centre (M). In case of floating body, the weight of the body is equal to the weight of liquid displaced.

(*a*) **Stable Equilibrium.** If the point M is above G, the floating body will be in stable equilibrium as shown in Fig. 4.13 (*a*). If a slight angular displacement is given to the floating body in the clockwise direction, the centre of buoyancy shifts from B to B_1 such that the vertical line through B_1 cuts at M. Then the buoyant force F_B through B_1 and weight W through G constitute a couple acting in the anti-clockwise direction and thus bringing the floating body in the original position.

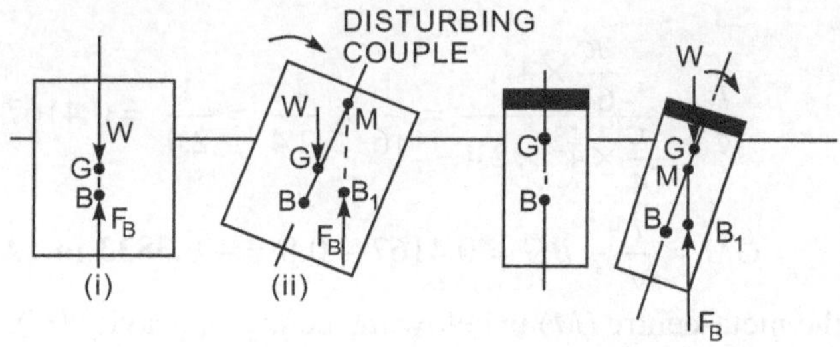

(a) Stable equilibrium *M* is above *G* (b) Unstable equilibrium *M* is below *G*.

Fig. 4.13 *Stability of floating bodies.*

(*b*) **Unstable Equilibrium.** If the point M is below G, the floating body will be in unstable equilibrium as shown in Fig. 4.13 (*b*). The disturbing couple is acting in the clockwise direction. The couple due to buoyant force F_B and W is also acting in the clockwise direction and thus overturning the floating body.

(*c*) **Neutral Equilibrium.** If the point M is at the centre of gravity of the body, the floating body will be in neutral equilibrium.

Problem 4.12 *A solid cylinder of diameter 4.0 m has a height of 4.0 m. Find the meta-centric height of the cylinder if the specific gravity of the material of cylinder = 0.6 and it is floating in water with its axis vertical. State whether the equilibrium is stable or unstable.* (VTU, Aug. 2003, Auto Engg.)

Solution. Given : $D = 4$ m

Height, $h = 4$ m

Sp. gr. $= 0.6$

Depth of cylinder in water $=$ Sp. gr. $\times h$

$= 0.6 \times 4.0 = 2.4$ m

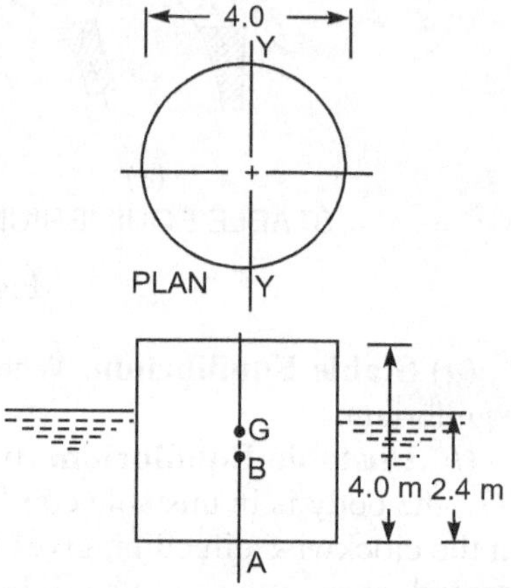

$\therefore$ Distance of centre of buoyancy (B) from A

or $$AB = \frac{2.4}{2} = 1.2 \text{ m}$$

Distance of centre of gravity (G) from A

or $$AG = \frac{h}{2} = \frac{4.0}{2} = 2.0 \text{ m}$$

$$\therefore \quad BG = AG - AB = 2.0 - 1.2 = 0.8 \text{ m}$$

Now the meta-centric height GM is given by

$$GM = \frac{I}{\forall} - BG$$

where I = M.O.I. of the plan of the body about y-y.

$$= \frac{\pi}{64} D^4 = \frac{\pi}{64} \times (4.0)^4$$

$\forall$ = Volume of cylinder in water

$$= \frac{\pi}{4.0} \times D^2 \times \text{Depth of cylinder in water} = \frac{\pi}{4} \times 4^2 \times 2.4 \text{ m}^3$$

Fig. 4.14

$$\therefore \quad \frac{I}{\forall} = \frac{\frac{\pi}{64} \times 4^4}{\frac{\pi}{4} \times 4^2 \times 2.4} = \frac{1}{16} \times \frac{4^2}{2.4} = \frac{1}{2.4} = 0.4167 \text{ m}$$

$$\therefore \quad GM = \frac{I}{\forall} - BG = 0.4167 - 0.8 = \mathbf{-\,0.3833 \text{ m. Ans.}}$$

– ve sign means that the meta-centre (M) is below the centre of gravity (G). Thus the cylinder is in unstable equilibrium. **Ans.**

Problem 4.13 *A solid cylinder of 10 cm diameter and 40 cm long, consists of two parts made of different materials. The first part at the base is 1.0 cm long and of specific gravity = 6.0. The other part of the cylinder is made of the material having specific gravity 0.6. State, if it can float vertically in water.*

Solution. Given : $D = 10$ cm

Length, $L = 40$ cm

Length of 1st part, $l_1 = 1.0$ cm

Sp. gr., $S_1 = 6.0$

Density of 1st part, $\rho_1 = 6 \times 1000 = 6000$ kg/m^3

Length of 2nd part, $l_2 = 40 - 1.0 = 39.0$ cm

Sp. gr., $S_2 = 0.6$

Density of 2nd part, $\rho_2 = 0.6 \times 1000 = 600$ kg/m^3

The cylinder will float vertically in water if its meta-centric height GM is positive. To find meta-centric height, find the location of centre of gravity (G) and centre of buoyancy (B) of the combined solid cylinder. The distance of the centre of gravity of the solid cylinder from A is given as

Fig. 4.15

AG = [(Weight of 1st part × Distance of C.G. of 1st part from A)
+ (Weight of 2nd part of cylinder
× Distance of C.G. of 2nd part from A)]
+ [Weight of 1st part + weight of 2nd part]

$$= \frac{\left(\frac{\pi}{4} D^2 \times 1.0 \times 6.0 \times 0.5\right) + \left(\frac{\pi}{4} D^2 \times 39.0 \times 0.6 \times (1.0 \times 39/2)\right)}{\left(\frac{\pi}{4} D^2 \times 1.0 \times 6.0 + \frac{\pi}{4} D^2 \times 39 \times 0.6\right)}$$

$$= \frac{1.0 \times 6.0 \times 0.5 + 39.0 \times .6 \times (20.5)}{1.0 \times 6.0 + 39.0 \times 0.6}$$

Cancel $\frac{\pi}{4} D^2$ in the Numerator and Denominator $= \frac{3.0 + 479.7}{6.0 \times 23.4} = \frac{482.7}{29.4} = 16.42.$

To find the centre of buoyancy of the combined two parts or of the cylinder, determine the depth of immersion of the cylinder. Let the depth of immersion of the cylinder is h. Then

Weight of the cylinder = Weight of water displaced

$$\frac{\pi}{4} \times (.1)^2 \times \frac{39.0}{100} \times 600 \times 9.81 + \frac{\pi}{4} (.1)^2 \times \frac{1.0}{100} \times 6000 \times 9.81 = \frac{\pi}{4} (.1)^2 \times \frac{h}{100} \times 1000 \times 9.81$$

[∵ h is in cm]

or cancelling $\frac{\pi}{4} (.1)^2 \times \frac{1000 \times 9.81}{100}$ throughout, we get

$$39.0 \times 0.6 + 1.0 \times 6.0 = h \quad \text{or} \quad h = 23.4 + 6.0 = 29.4$$

∴ The distance of the centre of the buoyancy B, of the cylinder from A is

$$AB = h/2 = \frac{29.4}{2} = 14.7$$

∴ $$BG = AG - AB = 16.42 - 14.70 = 1.72 \text{ cm.}$$

Meta-centric height GM is given by

$$GM = \frac{I}{\forall} - BG$$

where I = M.O.I. of plan of the body about y-y

$$= \frac{\pi}{64} D^4 = \frac{\pi}{.64} (10)^4 \text{ cm}^4$$

$\forall$ = Volume of cylinder in water

$$= \frac{\pi}{4} D^2 \times h = \frac{\pi}{4} (10)^2 \times 29.4 \text{ m}^3$$

$$\therefore \quad \frac{I}{\forall} = \frac{\pi}{64} (10)^4 \Big/ \frac{\pi}{4} (10)^2 \times 29.4 = \frac{1}{16} \times \frac{10^2}{29.4} = \frac{100}{19 \times 29.4} = 0.212$$

$$\therefore \quad GM = 0.212 - 1.72 = -1.508 \text{ cm.}$$

As GM – ve. It means that the Meta-centre M is below the centre of gravity (G). Thus the cylinder is in unstable equilibrium and so it cannot float vertically in water. **Ans.**

Problem 4.14 *A rectangular pontoon 10.0 m long, 7 m broad and 2.5 m deep weighs 686.7 kN. It carries on its upper deek an empty boiler of 5.0 m diameter weighing 588.6 kN. The centre of gravity of the boiler and the pontoon are at their respective centres along a vertical line. Find the meta-centric height. Weight density of sea water is 10.104 kN/m³.*

Solution. Given : Dimension of pontoon = 10 × 7 × 2.5

Weight of pontoon, $W_1 = 686.7$ kN

Dia. of boiler, $D = 5.0$ m

Weight of boiler, $W_2 = 588.6$ kN

w for sea water $= 10.104 \text{ kN/m}^3$

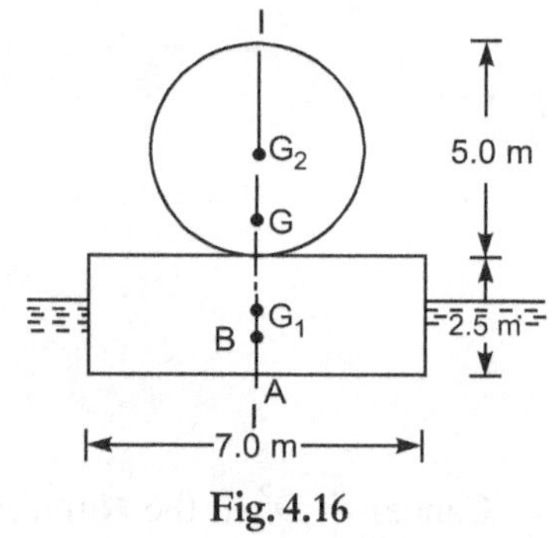

Fig. 4.16

To find the meta-centric height, first determine the common centre of gravity G and common centre of buoyancy B of the boiler and pontoon. Let G_1 and G_2 are the centre of gravities of pontoon and boiler respectively. Then

$$AG_1 = \frac{2.5}{2} = 1.25 \text{ m}$$

$$AG_2 = 2.5 + \frac{5.0}{2} = 2.5 + 2.5 = 5.0 \text{ m}$$

The distance of common centre of gravity G from A is given as

$$AG = \frac{W_1 \times AG_1 + W_2 \times AG_2}{W_1 + W_2}$$

$$= \frac{686.7 \times 1.25 + 588.6 \times 5.0}{(686.7 + 588.6)} = 2.98 \text{ m.}$$

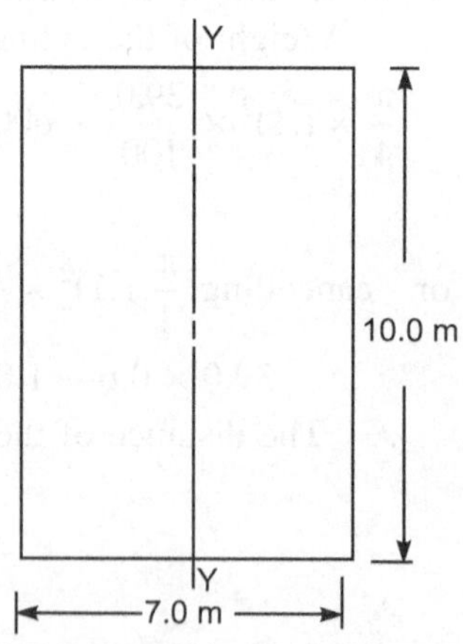

Fig. 4.17 *Plan of the body at water-line*

Let h is the depth of immersion. Then

Total weight of pontoon and boiler = Weight of sea water displaced

or $(686.7 + 588.6) = w \times$ Volume of the pontoon in water

$= 10.104 \times L \times b \times$ Depth of immersion

$$\therefore \quad 1275.3 = 10.104 \times 10 \times 7 \times h$$

$$h = \frac{1275.3}{10 \times 7 \times 10.104} = 1.803 \text{ m}$$

∴ The distance of the common centre of buoyancy B from A is

$$AB = \frac{h}{2} = \frac{1.803}{2} = .9015 \text{ m}$$

$\therefore$ $$BG = AG - AB = 2.98 - .9015 = 2.0785 \text{ m} \approx 2.078 \text{ m}$$

Meta-centric height is given by $GM = \frac{I}{\forall} - BG$

where I = M.O.I. of the plan of the body at the water level along $y = y$

$$= \frac{1}{12} \times 10.0 \times 7^3 = \frac{10 \times 49 \times 7}{12} \text{ m}^4$$

$\forall$ = Volume of the body in water

$= L \times b \times h = 10.0 \times 7 \times 1.857$

$\therefore$ $$\frac{I}{\forall} = \frac{10 \times 49 \times 7}{12 \times 10 \times 7 \times 1.857} = \frac{49}{12 \times 1.857} = 2.198 \text{ m}$$

$\therefore$ $$GM = \frac{I}{\forall} - BG = 2.198 - 2.078 = 0.12 \text{ m.}$$

$\therefore$ Meta-centric height of both the pontoon and boiler = **0.12 m. Ans.**

Problem 4.15 *A wooden cylinder of sp. gr. = 0.6 and circular in cross-section is required to float in oil (sp. gr. = 0.90). Find the L/D ratio for the cylinder to float with its longitudinal axis vertical in oil, where L is the height of cylinder and D is its diameter.* (VTU, Jan. 2004)

Solution. Given :

Dia. of cylinder $= D$

Height of cylinder $= L$

Sp. gr. of cylinder, $S_1 = 0.6$

Sp. gr. of oil $S_2 = 0.9$

Let the depth of cylinder immersed in oil $= h$

For the principle of buoyancy

Weight of cylinder = wt. of oil displaced

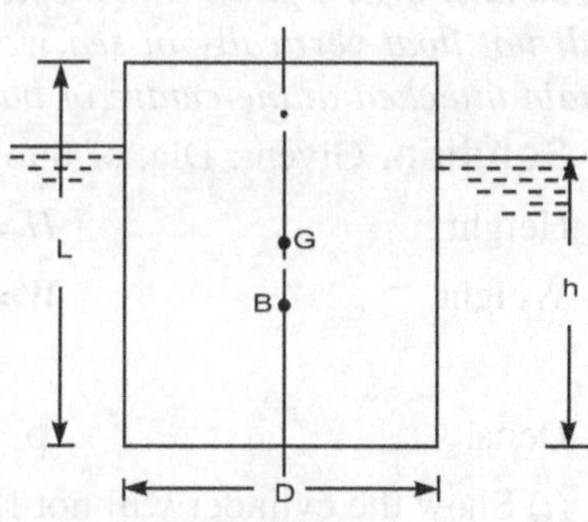

Fig. 4.18

$$\frac{\pi}{4} D^2 \times L \times 0.6 \times 1000 \times 9.81 = \frac{\pi}{4} D^2 \times h \times 0.9 \times 1000 \times 9.81$$

or $$L \times 0.6 = h \times 0.9$$

$\therefore$ $$h = \frac{0.6 \times L}{0.9} = \frac{2}{3} L.$$

The distance of centre of gravity C from A, $AG = \frac{L}{2}$

The distance of centre of buoyancy B from A,

$$AB = \frac{h}{2} = \frac{1}{2}\left[\frac{2}{3}L\right] = \frac{L}{3}$$

$\therefore$ $$BG = AG - AB = \frac{L}{2} - \frac{L}{3} = \frac{3L - 2L}{6} = \frac{L}{6}$$

The meta-centric height GM is given by

$$GM = \frac{I}{\forall} - BG$$

where $I = \frac{\pi}{64} D^4$ and $\forall$ = volume of cylinder in oil $= \frac{\pi}{4} D^2 \times h$

$$\therefore \quad \frac{I}{\forall} = \left(\frac{\pi}{64} D^4 \Big/ \frac{\pi}{4} D^2 h\right) = \frac{1}{16} \frac{D^2}{h} = \frac{D^2}{16 \times \frac{2}{3} L} = \frac{3D^2}{32L} \qquad \left\{\because \; h = \frac{2}{3} L\right\}$$

$$\therefore \quad GM = \frac{3D^3}{32L} - \frac{L}{6}.$$

For stable equilibrium, GM should be +ve or

$$GM > 0 \quad \text{or} \quad \frac{3D^2}{32L} - \frac{L}{6} > 0$$

$$\text{or} \quad \frac{3D^2}{32L} > \frac{L}{6} \quad \text{or} \quad \frac{3 \times 6}{32} > \frac{L^2}{D^2}$$

$$\text{or} \quad \frac{L^2}{D^2} < \frac{18}{32} \quad \text{or} \quad \frac{9}{16}$$

$$\therefore \quad \frac{L}{D} < \sqrt{\frac{9}{16}} = \frac{3}{4}$$

$\therefore$ **L/D < 3/4. Ans.**

Problem 4.16 *Show that a cylindrical buoy of 1 m diameter and 2.0 m height weighing 7.848 kN will not float vertically in sea water of density 1030 kg/m³. Find the force necessary in a vertical chain attached at the centre of base of the buoy that will keep it vertical.*

Solution. Given : Dia. of buoy, $D = 1$ m

Height, $H = 2.0$ m

Weight, $W = 7.848$ kN

$= 7.848 \times 1000 = 7848$ N

Density, $\rho = 1030$ kg/m³

(*i*) Show the cylinder will not float vertically.

(*ii*) Find the force in the chain.

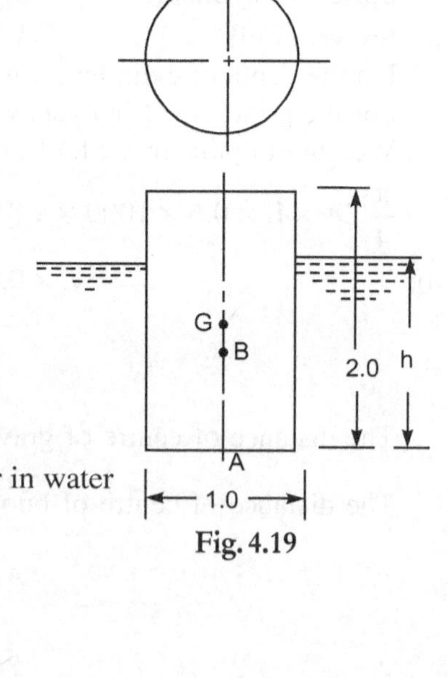

Fig. 4.19

Part I. The cylinder will not float if meta-centric height is – ve.

Let the depth of immersion be h

Then for equilibrium, Weight of cylinder

= Weight of water displaced

= Density × g × Volume of cylinder in water

$$\therefore \quad 7848 = 1030 \times 9.81 \times \frac{\pi}{4} D^2 \times h$$

$$= 10104.3 \times \frac{\pi}{4} (1)^2 \times h$$

$$\therefore \quad h = \frac{4 \times 7848}{10104.3 \times \pi} = 0.989 \text{ m.}$$

$\therefore$ The distance of centre of buoyancy B from A,

$$AB = \frac{h}{2} = \frac{0.989}{2} = 0.494 \text{ m.}$$

And the distance of centre of gravity G, from A is $AG = \dfrac{2.0}{2} = 1.0$ m

$$\therefore \qquad BG = AG - AB = 1.0 - .494 = .506 \text{ m.}$$

Now meta-centric height GM is given by $\quad GM = \dfrac{I}{\forall} - BG$

where $I = \dfrac{\pi}{64} D^4 = \dfrac{\pi}{64} \times (1)^4 \text{ m}^4$

and $\quad \forall$ = Volume of cylinder in water $= \dfrac{\pi}{4} D^2 \times h = \dfrac{\pi}{4} 1^2 \times .989$

$$\therefore \qquad \frac{I}{\forall} = \frac{\frac{\pi}{64} \times 1^4}{\frac{\pi}{4} D^2 \times h} = \frac{\frac{\pi}{64} \times 1^4}{\frac{\pi}{4} \times 1^2 \times .989}$$

$$= \frac{1}{16} \times 1^2 \times \frac{1}{.989} = \frac{1}{16 \times .989} = 0.063 \text{ m}$$

$$\therefore \qquad GM = .063 - .506 = \mathbf{-\ 0.443 \text{ m. Ans.}}$$

As the meta-centric height is – ve, the point M lies below G and hence the cylinder will be in unstable equilibrium and hence cylinder will not float vertically.

Part II. Let the force applied in a vertical chain attached at the centre of the base of the buoy is T to keep the buoy vertical.

Now find the combined position of centre of gravity (G') and centre of buoyancy (B'). For the combined centre of buoyancy, let h' = depth of immersion when the force T is applied. Then

Total downward force = Weight of water displaced

or $\quad (7848 + T)$ = Density of water × g × Volume of cylinder in water

$$= 1030 \times 9.81 \times \frac{\pi}{4} D' \times h' \text{ [where } h' = \text{depth of immersion]}$$

$$\therefore \qquad h' = \frac{7848 + T}{10104.3 \times \frac{\pi}{4} \times D^2} = \frac{7848 + T}{10104.3 \times \frac{\pi}{4} \times 1^2} = \frac{10104.3 + T}{7935.9} \text{ m}$$

Fig. 4.20

$$\therefore \qquad AB' = \frac{h'}{2} = \frac{1}{2}\left[\frac{7848 + T}{7935.9}\right] = \frac{7848 + T}{15871.8} \text{ m.}$$

The combined centre of gravity (G') due to weight of cylinder and due to tension T in the chain from A is

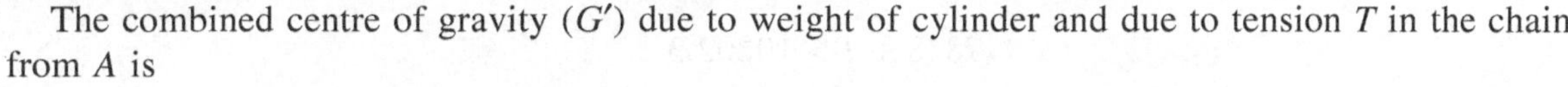

AG' = [Wt. of cylinder × Distance of C.G. of cylinder from A + T × Distance of C.G. of T from A] ÷ [Weight of cylinder + T]

$$= \left(7848 \times \frac{2}{2} + T \times 0\right) \div [7848 + T] = \frac{7848}{7848 + T} \text{ m}$$

$$\therefore \qquad B'G' = AG' - AB' = \frac{7848}{(7848 + T)} - \frac{(7848 + T)}{15871.8}$$

The meta-centric height GM is given by $GM = \dfrac{I}{\forall} - B'G'$

where $= \dfrac{\pi}{64} \times D^4 = \dfrac{\pi}{64} \times 1^4 = \dfrac{\pi}{64}\ \text{m}^4$

and $\forall = \dfrac{\pi}{4} D^2 \times h' = \dfrac{\pi}{4} \times 1^2 \times \dfrac{(7848 + T)}{7935.9} = \dfrac{\pi}{4} \times \dfrac{7848 + T}{7935.9}$

$$\therefore \quad \frac{I}{\forall} = \frac{\frac{\pi}{64}}{\frac{\pi}{4}\frac{(7848+T)}{7935.9}} = \frac{1}{16} \times \frac{7935.9}{(7848+T)}$$

$$\therefore \quad GM = \frac{7935.9}{16\,(7848+T)} - \left[\frac{7848}{(7848+T)} - \frac{(7848+T)}{15871.8}\right]$$

For stable equilibrium GM should be positive

or $$GM > 0$$

or $$\frac{7935.9}{16\,(7848+T)} - \left[\frac{7848}{(7848+T)} - \frac{(7848+T)}{15871.8}\right] > 0$$

or $$\frac{7935.9}{16\,(7848+T)} - \frac{7848}{(7848+T)} + \frac{7848+T}{15871.8} > 0$$

or $$\frac{7935.9 - 16 \times 7848}{16\,(7848+T)} + \frac{(7848+T)}{15871.8} > 0$$

or $$\frac{-117632}{16\,(7848+T)} + \frac{(7848+T)}{15871.8} > 0$$

or $$\frac{(7848+T)}{15871.8} > \frac{117632}{16\,(7848+T)}$$

or $$(7848 + T)^2 > \frac{117632}{16.0} \times 15871.8$$

$$> 116689473.5$$

$$> (10802.3)^2$$

$$\therefore \quad 7848 + T > 10802.3$$

$$\therefore \quad T > 10802.3 - 7848$$

$$> 2954.3 \text{ N. } \textbf{Ans.}$$

$\therefore$ The force in the chain must be at least 2954.3 N so that the cylindrical buoy can be kept in vertical position. **Ans.**

Problem 4.16A *A cylindrical buoy is 2 m in diameter 2.5 m long and weighs 2.2 metric tonnes. The density of sea-water is 1025 kg/m^3. Show that the body cannot float with its axis vertical.*

(VTU, Aug. 2001)

Solution. Given :

Dia., $D = 2$ m ; Height, $H = 2.5$ m ; Weight $W = 2.2$ metric tonnes

$= 2.2 \times 1000 \times 9.81$ N [$\because$ one metric tonne = 1000 kgf = 1000 × 9.81 N]

Density, $\rho = 1025$ kg/m^3.

Let h = Depth of immersion of the cylinder

Then for equilibrium,

Weight of cylinder = Weight of water displaced

$$2.2 \times 1000 \times 9.81 = \text{Density of sea water} \times g \times \text{Volume of cylinder in water}$$

$$= 1025 \times 9.81 \times \text{Volume of cylinder in water}$$

$$= 1025 \times 9.81 \times \left[\frac{\pi}{4} D^2 \times h\right] = 1025 \times 9.81 \times \frac{\pi}{4} \times 2^2 \times h$$

$$= 1025 \times 9.81 \times \pi \times h$$

$$\therefore \quad h = \frac{2.2 \times 1000 \times 9.81}{1025 \times 9.81 \times \pi} = 0.685 \text{ m}$$

$\therefore$ Distance ofcentre of buoyancy (B) from bottom,

$$AB = \frac{h}{2} = \frac{0.685}{2} = 0.3425 \text{ m}$$

Distance of centre of gravity (G) from the bottom,

$$AG = \frac{H}{2} = \frac{2.5}{2} = 1.25 \text{ m}$$

$$\therefore \quad BG = AG - AB = 1.25 - 0.3425 = 0.9075 \text{ m}$$

Now meta-centric height (GM) is given by,

$$GM = \frac{I}{\forall} - BG$$

where $I = \frac{\pi}{64} D^4 = \frac{\pi}{64} \times 2^4 = \frac{\pi}{4}$ m^4

$\forall$ = Volume of cylinder in water

$$= \frac{\pi}{4} D^2 \times h = \frac{\pi}{4} \times 2^2 \times 0.685$$

$$= 0.685\,\pi$$

$$= \frac{\left(\frac{\pi}{4}\right)}{0.685\,\pi} - 0.9075$$

$$= -\,\mathbf{0.5415 \text{ m. Ans.}}$$

As the meta-centric height is –ve, the point M lies below G and hence the cylinder will be in unstable equilibrium and hence cylinder will not float vertically.

Problem 4.17 *A solid cone floats in water with its apex downwards. Determine the least apex angle of cone for stable equilibrium. The specific gravity of the material of the cone is given 0.8.*

Solution. Given :

Sp. gr. of cone $= 0.8$

Density of cone, $\rho = 0.8 \times 1000 = 800$ kg/m^3

Let D = Dia. of the cone

d = Dia. of cone at water level

2θ = Apex angle of cone

H = Height of cone

h = Depth of cone in water

G = Centre of gravity of the cone

B = Centre of buoyancy of the cone

For the cone, the distance of centre of gravity from the apex A is

$$AC = \tfrac{3}{4} \text{ height of cone} = \tfrac{3}{4}H$$

also
$$AB = \tfrac{3}{4} \text{ depth of cone in water} = \tfrac{3}{4}h$$

Volume of water displaced $= \frac{1}{3}\pi r^2 \times h$

Volume of cone $= \frac{1}{3} \times \pi R^2 \times h$

$\therefore$ Weight of cone $= 800 \times g \times \frac{1}{3} \times \pi R^2 \times H$

Fig. 4.21

Now from ΔAEF, $\tan\theta = \dfrac{EF}{EA} = \dfrac{R}{H}$

$\therefore$ $R = H\tan\theta$

Similarly, $r = h\tan\theta$

$$\therefore \text{ Weight of cone} = 800 \times g \times \tfrac{1}{3} \times \pi \times (H\tan\theta)^2 \times H = \frac{800 \times g \times \pi \times H^3 \tan^2\theta}{3}$$

$$\therefore \text{ Weight of water displaced} = 1000 \times g \times \tfrac{1}{3} \times \pi r^2 \times h$$

$$= 1000 \times g \times \tfrac{1}{3} \times \pi (h\tan\theta)^2 \times h = \frac{1000 \times g \times \pi \times h^3 \tan^2\theta}{3.0}$$

For equilibrium

Weight of cone = Weight of water displaced

or
$$\frac{800 \times g \times \pi \times H^3 \tan^2\theta}{3.0} = \frac{1000 \times 9.81 \times \pi \times h^3 \times \tan^2\theta}{3.0}$$

or
$$800 \times H^3 = 1000 \times h^3$$

$$\therefore \quad H^3 = \frac{1000}{800} \times h^3 \text{ or } \frac{H}{h} = \left(\frac{1000}{800}\right)^{1/3}$$

For stable equilibrium, Meta-centric height GM should be positive. But GM is given by

$$GM = \frac{I}{\forall} - BG$$

where I = M.O.I. of cone at water-line $= \dfrac{\pi}{64} d^4$

$\forall$ = Volume of cone in water $= \dfrac{1}{3}\dfrac{\pi}{4} d^2 \times h$

$$\therefore \quad \frac{I}{\forall} = \frac{\pi}{64} d^4 \Big/ \frac{1}{3} \times \frac{\pi}{4} d^2 \times h$$

$$= \frac{1 \times 3}{16} \times \frac{d^2}{h} = \frac{3d^2}{16h} = \frac{3}{16h} \times (2r)^2 = \frac{3}{4}\frac{r^2}{h}$$

$$= \frac{3}{4}\frac{(h\tan\theta)^2}{h} \qquad \{\because \quad r = h\tan\theta\}$$

$$= \tfrac{3}{4} h \tan^2\theta$$

and $$BG = AG - AB = \tfrac{3}{4}H - \tfrac{3}{4}h = \tfrac{3}{4}(H - h)$$

$\therefore$ $$GM = \tfrac{3}{4} h\tan^2\theta - \tfrac{3}{4}(H - h)$$

For stable equilibrium *GM* should be positive or

$$\tfrac{3}{4} h \tan^2 - \tfrac{3}{4}(H - h) > 0 \quad \text{or} \quad h\tan^2\theta - (H - h) > 0$$

or $$h\tan^2\theta > (H - h) \quad \text{or} \quad h\tan^2\theta + h > H$$

or $$h[\tan^2\theta + 1] > H \quad \text{or} \quad 1 + \tan^2\theta > H/h \quad \text{or} \quad \sec^2\theta > \frac{H}{h}$$

But $$\frac{H}{h} = \left(\frac{1000}{800}\right)^{1/3} = 1.077$$

$\therefore$ $$\sec^2\theta > 1.077 \text{ or } \cos^2\theta > \frac{1}{1.077} = 0.9285$$

$\therefore$ $$\cos\theta > 0.9635$$

$\therefore$ $$\theta > 15°\ 30' \quad \text{or} \quad 2\theta > 31°$$

$\therefore$ Apex angle (2θ) should be at least **31°. Ans.**

Problem 4.18 *A cone of specific gravity S, is floating in water with its apex downwards. It has a diameter D and vertical height H. Show that for stable equilibrium of the cone* $H < \frac{1}{2}\left[\frac{D^2 . S^{1/3}}{2 - S^{1/3}}\right]^{1/2}$.

Solution. Given :

Dia. of cone $= D$

Height of cone $= H$

Sp. gr. of cone $= S$

Let G = Centre of gravity of cone

B = Centre of buoyancy

2θ = Apex angle and

A = Apex of the cone

h = Depth of immersion

d = Dia. of cone at water surface

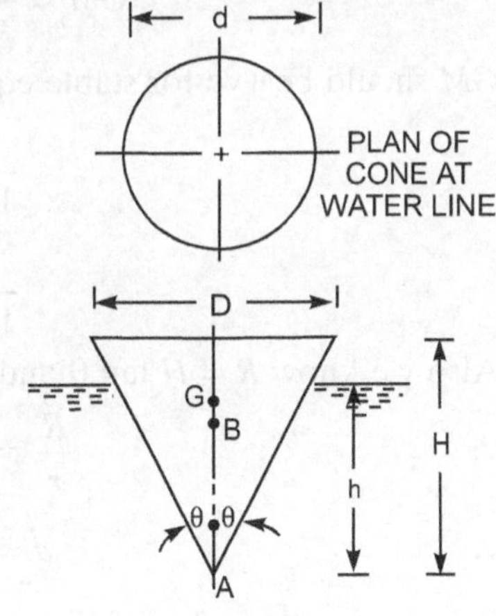

Fig. 4.22

Then $$AG = \frac{3}{4}H$$

$$AB = \frac{3}{4}h$$

Also weight of cone =Weight of water displaced.

$$1000\, S \times g \times \tfrac{1}{3}\pi R^2 \times H = 1000 \times g \times \frac{1}{3}\pi r^2 \times h \quad \text{or} \quad SR^2H = r^2h$$

$\therefore$ $$h = \frac{SR^2H}{r^2}$$

But $$\tan\theta = \frac{R}{H} = \frac{r}{h}$$

$\therefore \quad R = H \tan \theta, \; r = h \tan \theta$

$$\therefore \quad h = \frac{S \times (H \tan \theta)^2 \times H}{(h \tan \theta)^2}$$

$$h = \frac{S \times H^2 \times \tan^2 \theta \times H}{h^2 \tan^2 \theta} = \frac{SH^3}{h^2} \quad \text{or} \quad h^3 = SH^3$$

or $$h = (SH^3)^{1/3} = S^{1/3} H \qquad \text{...(1)}$$

Distance, $$BG = AG - AB$$

$$= \frac{3}{4} H - \frac{3}{4} h = \frac{3}{4} (H - h) = \frac{3}{4} (H - S^{1/3} H) \qquad \{\because h = S^{1/3} H\}$$

$$= \frac{3}{4} H[1 - S^{1/3}] \qquad \text{...(2)}$$

Also I = M.O. Inertia of the plan of body at water surface

$$= \frac{\pi}{64} d^4$$

$$\forall = \text{Volume of cone in water} = \frac{1}{3} \times \frac{\pi}{4} \times d^2 \times h = \frac{1}{3} \frac{\pi}{4} d^2 \; [H.S^{1/3}]$$

$$\therefore \quad \frac{I}{\forall} = \frac{\frac{\pi}{64} d^4}{\frac{1}{3} \times \frac{\pi}{4} d^2 \; H.S^{1/3}} = \frac{3d^2}{16.H.S^{1/3}}$$

Now Meta-centric height GM is given as

$$GM = \frac{I}{\forall} - BG = \frac{3d^2}{16.H.S^{1/3}} - \frac{3H}{4} [1 - S^{1/3}]$$

GM should be +ve for stable equilibrium or $GM > 0$

or $$\frac{3d^2}{16.H.S^{1/3}} - \frac{3H}{4} (1 - S^{1/3}) > 0$$

or $$\frac{3d^2}{16.H.S^{1/3}} > \frac{3H}{4} (1 - S^{1/3}) \qquad \text{...(3)}$$

Also we know $R = H \tan \theta$ and $r = h \tan \theta$

$$\therefore \quad \frac{R}{r} = \frac{H}{h} = \frac{D}{d}$$

$$\therefore \quad d = \frac{Dh}{H} = \frac{D}{H} \times HS^{1/3} = DS^{1/3}$$

Substituting the value of d in equation (3), we get

$$\frac{3\left(DS^{1/3}\right)^2}{16.H.S^{1/3}} > \frac{3H}{4} (1 - S^{1/3}) \quad \text{or} \quad \frac{D^2.S^{1/3}}{4.H} > H (1 - S^{1/3})$$

or $$\frac{D^2.S^{1/3}}{4\left(1 - S^{1/2}\right)} > H^2 \quad \text{or} \quad H^2 < \frac{D^2.S^{1/3}}{4\left(1 - S^{1/3}\right)}$$

or $$H < \frac{1}{2} \left[\frac{D^2.S^{1/3}}{1 - S^{1/3}} \right]^{1/2}$$ **. Proved.**

▶ 4.8 EXPERIMENTAL METHOD OF DETERMINATION OF META-CENTRIC HEIGHT

The meta-centric height of a floating vessel can be determined, provided we know the centre of gravity of the floating vessel. Let w_1 is a known weight placed over the centre of the vessel as shown in Fig. 4.23 (*a*) and the vessel is floating.

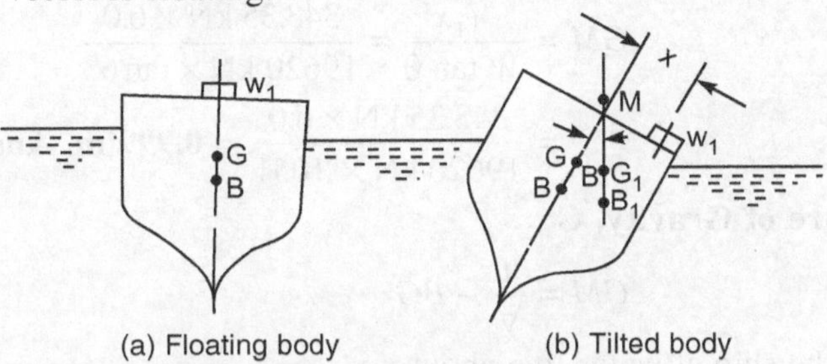

Fig. 4.23 *Meta-centric height.*

Let W = Weight of vessel including w_1

G = Centre of gravity of the vessel

B = Centre of buoyancy of the vessel

The weight w_1 is moved across the vessel towards right through a distance x as shown in Fig. 4.23 (*b*). The vessel will be tilted. The angle of heel θ is measured by means of a plumbline and a protractor attached on the vessel. The new centre of gravity of the vessel will shift to G_1 as the weight w_1 has been moved towards the right. Also the centre of buoyancy will change to B_1 as the vessel has tilted. Under equilibrium, the moment caused by the movement of the load w_1 through a distance x must be equal to the moment caused by the shift of the centre of gravity from G to G_1. Thus

The moment due to change of $G = GG_1 \times W = W \times GM \tan\theta$

The moment due to movement of $w_1 = w_1 \times x$

$$\therefore \quad w_1 x = WGM \tan\theta$$

Hence
$$GM = \frac{w_1 x}{W \tan\theta} \quad \text{...(4.5)}$$

Problem 4.19 *A ship 70 m long and 10 m broad has a displacement of 19620 kN. A weight of 343.35 kN is moved across the deck through a distance of 6 m. The ship is tilted through 6°. The moment of inertia of the ship at water-line about its fore and aft axis is 75% of M.O.I. of the circumscribing rectangle. The centre of buoyancy is 2.25 m below water-line. Find the meta-centric height and position of centre of gravity of ship. Specific weight of sea water is 10104 N/m³.*

(Anna University, May, 1986)

Solution. Given :

Length of ship, $L = 70$ m

Breadth of ship, $b = 10$ m

Displacement, $W = 19620$ kN

Angle of heel, $\theta = 6°$

M.O.I. of ship at water-line = 75% of M.O.I. of circumscribing rectangle

w for sea-water $= 10104 \text{ N/m}^3 = 10.104 \text{ kN/m}^3$

Movable weight, $w_1 = 343.35$ kN

Distance moved by w_1, $x = 6$ m

Centre of buoyancy = 2.25 m below water surface

Find (*i*) Meta-centric height, GM

(*ii*) Position of centre of gravity, G.

(*i*) **Meta-centric height, GM** is given by equation (4.5)

$$\therefore \quad GM = \frac{w_1 x}{W \tan \theta} = \frac{343.35 \text{ kN} \times 6.0}{19620 \text{ kN} \times \tan 6°}$$

$$= \frac{343.35 \text{ kN} \times 6.0}{19620 \text{ kN} \times .1051} = \mathbf{0.999 \text{ m. Ans.}}$$

(*ii*) **Position of Centre of Gravity, G**

$$GM = \frac{I}{\forall} - BG$$

where I = M.O.I. of the ship at water-line about y-y

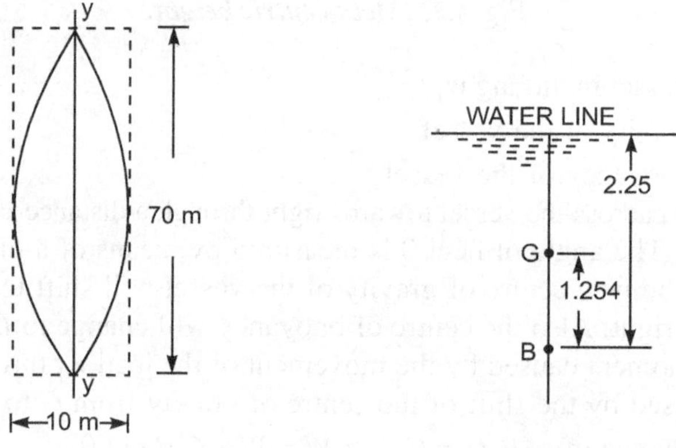

Fig. 4.24

Fig. 4.25

$$= 75\% \text{ of } \frac{1}{12} \times 70 \times 10^3 = .75 \times \frac{1}{12} \times 70 \times 10^3 = 4375 \text{ m}^4$$

and $\quad \forall$ = Volume of ship in water = $\dfrac{\text{Weight of ship}}{\text{Weight density of water}} = \dfrac{19620}{10.104} = 1941.74 \text{ m}^3$

$$\therefore \quad \frac{I}{\forall} = \frac{4375}{1941.74} = 2.253 \text{ m}$$

$$\therefore \quad GM = 2.253 - BG \text{ or } .999 = 2.253 - BG$$

$$\therefore \quad BG = 2.253 - .999 = 1.254 \text{ m.}$$

From Fig. 4.25, it is clear that the distance of G from free surface of the water = distance of B from water surface – BG

$$= 2.25 - 1.254 = \mathbf{0.996 \text{ m. Ans.}}$$

Problem 4.20 *A pontoon of 15696 kN displacement is floating in water. A weight of 245.25 kN is moved through a distance of 8 m across the deck of pontoon, which tilts the pontoon through an angle 4°. Find meta-centric height of the pontoon.*

Solution. Given :

Weight of pontoon = Displacement

or $\quad W = 15696$ kN

Movable weight, $w_1 = 245.25$ kN

Distance moved by weight w_1, $x = 8$ m

Angle of heel, $\theta = 4°$

The meta-centric height, GM is given by equation (4.5)

or
$$GM = \frac{w_1 x}{W \tan\theta} = \frac{245.25 \text{ kN} \times 8}{15696 \text{ kN} \times \tan 4°}$$
$$= \frac{1962}{15696 \times 0.0699} = \textbf{1.788 m. Ans.}$$

▶ 4.9 OSCILLATION (ROLLING) OF A FLOATING BODY

Consider a floating body, which is tilted through an angle by an overturning couple as shown in Fig. 4.26. Let the over-turning couple is suddenly removed. The body will start oscillating. Thus, the body will be in a state of oscillation as if suspended at the meta-centre M. This is similar to the case of a pendulum. The only force acting on the body is due to the restoring couple due to the weight W of the body force of buoyancy F_B.

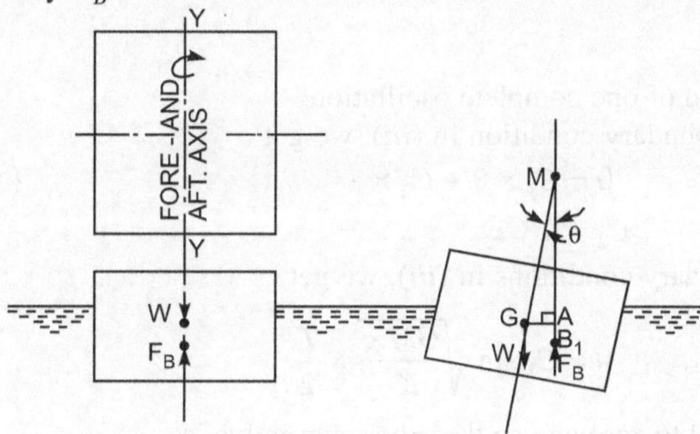

Fig. 4.26

∴ Restoring couple $= W \times$ Distance GA

$= W \times GM \sin\theta$ ...(*i*)

This couple tries to decrease the angle

Angular acceleration of the body, $\alpha = -\dfrac{d^2\theta}{dt^2}$.

– ve sign has been introduced as the restoring couple tries to decrease the angle θ.

Torque due to inertia = Moment of Inertia about Y-Y × Angular acceleration

$$= I_{Y\text{-}Y} \times \left(-\frac{d^2\theta}{dt^2}\right)$$

But
$$I_{Y\text{-}Y} = \frac{W}{g} K^2$$

where W = Weight of body, K = Radius of gyration about Y-Y

∴ Inertia torque
$$= \frac{W}{g} K^2 \left(-\frac{d^2\theta}{dt^2}\right) = -\frac{W}{g} K^2 \frac{d^2\theta}{dt^2} \quad ...(ii)$$

Equating (*i*) and (*ii*), we get

$$W \times GM \sin\theta = -\frac{W}{g} K^2 \frac{d^2\theta}{dt^2} \quad \text{or} \quad GM \sin\theta = -\frac{K^2}{g}\frac{d^2\theta}{dt^2}$$

For small angle θ, $\sin\theta \simeq \theta$

$$\therefore \quad GM \times \theta = -\frac{K^2}{g}\frac{d^2\theta}{dt^2} \quad \text{or} \quad \frac{K^2}{g}\frac{d^2\theta}{dt^2} + GM \times \theta = 0$$

Dividing by $\frac{K^2}{g}$, we get $\frac{d^2\theta}{dt^2} + \frac{GM \times g \times \theta}{K^2} = 0$

The above equation is a differential equation of degree second. The solution is

$$\theta = C_1 \sin\sqrt{\frac{GM.g}{K^2}} \times t + C_2 \cos\sqrt{\frac{GM.g \times t}{K^2}} \qquad \text{...(iii)}$$

where C_1 and C_2 are constants of integration.

The values of C_1 and C_2 are obtained from boundary conditions which are

(*i*) at $t = 0$, $\theta = 0$

(*ii*) at $t = \frac{T}{2}$, $\theta = 0$

where T is the time period of one complete oscillation.

Substituting the 1st boundary condition in (*iii*), we get

$$0 = C_1 \times 0 + C_2 \times 1 \qquad \{\because \sin\theta = 0, \cos\theta = 1\}$$

$$\therefore \quad C_2 = 0$$

Substituting 2nd boundary conditions in (*iii*), we get

$$0 = C_1 \sin\sqrt{\frac{GM.g}{K^2}} \times \frac{T}{2}$$

But C_1 cannot be equal to zero and so the other alternative is

$$\sin\sqrt{\frac{GM.g}{K^2}} \times \frac{T}{2} = 0 = \sin\pi \qquad \{\because \sin\pi = 0\}$$

$$\therefore \quad \sqrt{\frac{GM.g}{K^2}} \times \frac{T}{2} = \pi \quad \text{or} \quad T = 2\pi\sqrt{\frac{K^2}{GM.g}} \qquad \text{...(4.6)}$$

∴ Time period of oscillation is given by equation (4.6).

Problem 4.21 *The least radius of gyration of a ship is 8 m and meta-centric height 70 cm. Calculate the time period of oscillation of the ship.*

Solution. Given :

Least radius of gyration, $K = 8$ m

Meta-centric height, $GM = 70$ cm $= 0.70$ m

The time period of oscillation is given by equation (4.6).

$$T = 2\pi\sqrt{\frac{K^2}{GM.g}} = 2\pi\sqrt{\frac{8 \times 8}{0.7 \times 9.81}} = \textbf{19.18 sec. Ans.}$$

Problem 4.22 *The time period of rolling of a ship of weight 29430 kN in sea water is 10 seconds. The centre of buoyancy of the ship is 1.5 m below the centre of gravity. Find the radius of gyration of the ship if the moment of inertia of the ship at the water line about fore and aft. axis is 1000 m^4. Take specific weight of sea water as = 10100 N/m^3.*

Solution. Given :

Time period, $T = 10$ sec

Distance between centre of buoyancy and centre of gravity, $BG = 1.5$ m

Moment of Inertia, $I = 10000 \text{ m}^4$

Weight, $W = 29430 \text{ kN} = 29430 \times 1000 \text{ N}$

Let the radius of gyration = K

First calculate the meta-centric height GM, which is given as

$$GM = BM - BG = \frac{I}{\forall} - BG$$

where I = M.O. Inertia

and $\forall$ = Volume of water displaced

$$= \frac{\text{Weight of ship}}{\text{Sp. weight of sea water}} = \frac{29430 \times 1000}{10104} = 2912.6 \text{ m}^3$$

$$\therefore \quad GM = \frac{10000}{2912.6} - 1.5 = 3.433 - 1.5 = 1.933 \text{ m.}$$

Using equation (4.6), we get $T = 2\pi \sqrt{\dfrac{K^2}{GM \times g}}$

or $$10 = 2\pi \sqrt{\frac{K^2}{1.933 \times 9.81}} = \frac{2\pi K}{\sqrt{1.933 \times 9.81}}$$

or $$K = \frac{10 \times \sqrt{1.933 \times 9.81}}{2\pi} = \mathbf{6.93 \text{ m. Ans.}}$$

HIGHLIGHTS

1. The upward force exerted by a liquid on a body when the body is immersed in the liquid is known buoyancy or force of buoyancy.
2. The point through which force of buoyancy is supposed to act is called centre of buoyancy.
3. The point about which a body starts oscillating when the body is tilted is known meta-centre.
4. The distance between the meta-centre and centre of gravity is known as meta-centric height.
5. The meta-centric height (GM) is given by $GM = \dfrac{I}{\forall} - BG$

   where I = Moment of Inertia of the floating body (in plan) at water surface about the axis Y-Y

   $\forall$ = Volume of the body submerged in water

   BG = Distance between centre of gravity and centre of buoyancy.

6. Conditions of equilibrium of a floating and submerged body are :

Equilibrium	*Floating Body*	*Submerged Body*
(*i*) Stable Equilibrium	M is above G	B is above G
(*ii*) Unstable Equilibrium	M is below G	B is below G
(*iii*) Neutral Equilibrium	M and G coincide	B and G coincide

7. The value of meta-centric height GM, experimentally is given as $GM = \frac{w_1 x}{W \tan \theta}$

where w_1 = Movable weight

x = Distance through which w_1 is moved

W = Weight of the ship or floating body including w_1

θ = Angle through the ship or floating body is tilted due to the movement of w_1.

8. The time period of oscillation or rolling of a floating body is given by $T = 2\pi \sqrt{\frac{K^2}{GM \times g}}$

where K = Radius of gyration, GM = Meta-centric height

T = Time of one complete oscillation.

EXERCISE 4

(A) THEORETICAL PROBLEMS

1. Define the terms 'buoyancy' and 'centre of buoyancy'.
2. Explain the terms 'meta-centre' and 'meta-centric height'.
3. Derive an expression for the meta-centric height of a floating body.
4. Show that the distance between the meta-centre and centre of buoyancy is given by $BM = \frac{I}{\forall}$

   where I = Moment of inertia of the plan of the floating body at water surface about longitudinal axis.

   $\forall$ = Volume of the body submerged in liquid.
5. What are the conditions of equilibrium of a floating body and a submerged body ?
   (A.S.M.E., June 1992 ; Delhi University, 1982)
6. How will you determine the meta-centric height of a floating body experimentally ? Explain with neat sketch. *(VTU, Aug. 2001)*
7. Select the correct statement :

   (*a*) The buoyant force for a floating body passes through the

   (*i*) centre of gravity of the body (*ii*) centroid of volume of the body

   (*iii*) meta-centre of the body (*iv*) centre of gravity of the submerged part of the body

   (*v*) centroid of the displaced volume.

   (*b*) A body submerged in liquid is in equilibrium when :

   (*i*) its meta-centre is above the centre of gravity

   (*ii*) its meta-centre is above the centre of buoyancy

   (*iii*) its centre of gravity is above the centre of buoyancy

   (*iv*) its centre of buoyancy is above the centre of gravity

   (*v*) none of these. **[Ans.** 7 (*a*) (*v*), (*b*) (*iv*)]
8. Derive an expression for the time period of the oscillation of a floating body in terms of radius of gyration and meta-centric height of the floating body.

9. Define the terms : meta-centre, centre of buoyancy, meta-centric height, gauge pressure and absolute pressure. *(A.S.M.E., June 1992)*

10. What do you understand by the hydrostatic equation ? With the help of this equation, derive the expression for the buoyant force acting on a sub-merged body. *(A.M.I.E. S 1990)*

11. With neat sketches, explain the conditions of equilibrium for floating and sub-merged bodies. *(Delhi University, June 1996)*

12. Differentiate between :
(*i*) Dynamic viscosity and kinematic viscoscity, (*ii*) Absolute and gauge pressure (*iii*) Simple and differential manometers (*iv*) Centre of gravity and centre of buoyancy. *(Delhi University, Dec. 2002)*

13. Explain the following terms :
(*i*) Force of buoyancy and centre of buoyancy
(*ii*) Meta-centre and metacentric height
(*iii*) Stability of a floating body. *(VTU, Aug. 2003)*

(B) NUMERICAL PROBLEMS

1. A wooden block of width 2 m, depth 1.5 m and length 4 m floats horizontally in water. Find the volume of water displaced and position of centre buoyancy. The specific gravity of the wooden block is 0.7. [**Ans.** 8.4 m^3, 0.525 m from the base]

2. A wooden log of 0.8 m diameter and 6 m length is floating in river water. Find the depth of wooden log in water when the sp. gr. of the wooden log is 0.7. [**Ans.** 0.54 m]

3. A stone weighs 490.5 N in air and 196.2 N in water. Determine the volume of stone and its specific gravity. [**Ans.** 0.03 m^3 or 3×10^4 cm^3, 1.67]

4. A body of dimensions 2.0 m × 1.0 m × 3.0 m weighs 3924 N in water. Find its weight in air. What will be its specific gravity ? [**Ans.** 62784 N, 1.0667]

5. A metallic body floats at the interface of mercury of sp. gr. 13.6 and water in such a way that 30% of its volume is submerged in mercury and 70% in water. Find the density of the metallic body. [**Ans.** 4780 kg/m^3]

6. A body of dimensions 0.5 m × 0.5 m × 1.0 m and of sp. gr. 3.0 is immersed in water. Determine the least force required to lift the body. [**Ans.** 4905 N]

7. A rectangular pontoon is 4 m long, 3 m wide and 1.40 m high. The depth of immersion of the pontoon is 1.0 m in sea-water. If the centre of gravity is 0.70 m above the bottom of the pontoon, determine the meta-centric height. Take the density of sea-water as 1030 kg/m^3. [**Ans.** 0.45 m]

8. A uniform body of size 4 m long × 2 m wide × 1 m deep floats in water. What is the weight of the body if depth of immersion is 0.6 m ? Determine the meta-centric height also. [**Ans.** 47088 N, 0.355 m]

9. A block of wood of specific gravity 0.8 floats in water. Determine the meta-centric height of the block if its size is 3 m × 2 m × 1 m. [**Ans.** 0.316 m]

10. A solid cylinder of diameter 3.0 m has a height of 2 m. Find the meta-centric height of the cylinder when it is floating in water with its axis vertical. The sp. gr. of the cylinder is 0.7. [**Ans.** 0.1017 m]

11. A body has the cylindrical upper portion of 4 m diameter and 2 m deep. The lower portion is a curved one, which displaces a volume of 0.9 m^3 of water. The centre of buoyancy of the curved portion is at a distance of 2.10 m below the top of the cylinder. The centre of gravity of the whole body is 1.50 m below the top of the cylinder. The total displacement of water is 4.5 tonnes. Find the meta-centric height of the body. [**Ans.** 2.387 m]

12. A solid cylinder of diameter 5.0 m has a height of 5.0 m. Find the meta-centric height of the cylinder if the specific gravity of the material of cylinder is 0.7 and it is floating in water with its axis vertical. State whether the equilibrium is stable or unstable. [**Ans.** – 0.304 m, Unstable Equilibrium]

13. A solid cylinder of 15 cm diameter and 60 cm long, consists of two parts made of different materials. The first part at the base is 1.20 cm long and of specific gravity = 5.0. The other parts of the cylinder is made of the material having specific gravity 0.6. State, if it can float vertically in water.
[**Ans.** $GM = -5.26$, Unstable, Equilibrium]

14. A rectangular pontoon 8.0 m long, 7 m broad and 3.0 m deep weighs 588.6 kN. It carries on its upper deck an empty boiler of 4.0 m diameter weighing 392.4 kN. The centre of gravity of the boiler and the pontoon are at their respective centres along a vertical line. Find the meta-centric height. Weight density of sea-water is 10104 N/m^3.
[**Ans.** 0.325 m]

15. A wooden cylinder of sp. gr. 0.6 and circular in cross-section is required to float in oil (sp. gr. 0.8). Find the L/D ratio for the cylinder to float with its longitudinal axis vertical in oil where L is the height of cylinder and D is its diameter.
[**Ans.** $(L/D) < 0.8164$]

16. Show that a cylindrical buoy of 1.5 m diameter and 3 m long weighing 2.5 tonnes will not float vertically in sea-water of density 1030 kg/m^3. Find the force necessary in a vertical chain attached at the centre of the base of the buoy that will keep it vertical.
[**Ans.** 10609.5 N]

17. A solid cone floats in water its apex downwards. Determine the least apex angle of cone for stable equilibrium. The specific gravity of the material of the cone is given 0.7.
[**Ans.** 39° 7′]

18. A ship 60 m long and 12 m broad has a displacement of 19620 kN. A weight of 294.3 kN is moved across the deck through a distance of 6.5 m. The ship is tilted through 5°. The moment of inertia of the ship at water line about its force and aft. axis is 75% of moment of inertia the circumscribing rectangle. The centre of buoyancy is 2.75 m below water line. Find the meta-centric height and position of centre of gravity of ship. Take specific weight of sea water = 10104 N/m^3.
[**Ans.** 1.1145 m, 0.53 m below water surface]

19. A pontoon of 1500 tonnes displacement is floating in water. A weight of 20 tonnes is moved through a distance of 6 m across the deck of pontoon, which tilts the pontoon through an angle of 5°. Find meta-centric height of the pontoon.
[**Ans.** 0.9145 m]

20. Find the time period of rolling of a solid circular cylinder of radius 2.5 m and 5.0 m long. The specific gravity of the cylinder is 0.9 and is floating in water with its axis vertical.
[**Ans.** 0.35 sec]

21. A solid cylinder 2 m in diameter and 2 m high is floating in water with its axis vertical. If the sp. gr. of the cylinder is 0.65, find its meta-centric height. State whether the equilibrium is stable or not.
(*VTU., March 2001*)

[**Hint.** $D = 2$ m, $H = 2$ m, Sp. gr. = 0.65. Depth of cylinder in water, h = Sp. gr. $\times H = 0.65 \times 2 = 1.3$ m.

Distance of centre of buoyancy from bottom, $AB = \frac{1.3}{2} = 0.65$ m. Distance of centre of gravity (G) from bottom,

$$AG = \frac{H}{2} = \frac{2}{2} = 1 \text{ m} \quad \therefore \quad BG = AG - AB = 1 - 0.65 = 0.35 \text{ m}$$

Now metacentric height, $GM = \frac{I}{\forall} - BG$ where $I = \frac{\pi}{64} \times D^4 = \frac{\pi}{64} \times 2^4 = \frac{\pi}{4} \text{ m}^4$

$$\forall = \frac{\pi}{4} D^2 \times h = \frac{\pi}{4} \times 2^2 \times 1.3 = 1.3\,\pi \text{ and } BG = 0.35$$

$$\therefore \quad GM = \frac{\left(\frac{\pi}{4}\right)}{1.3\,\pi} - 0.35 = -0.158 \text{ m. Equilibrium is unstable. **Ans.**]}$$

Fluid Kinematics

▶ 5.1 INTRODUCTION

Kinematics is defined as that branch of science which deals with motion of particles without considering the forces causing the motion. The velocity at any point in a flow field at any time is studied in this branch of fluid mechanics. Once the velocity is known, then the pressure distribution and hence forces acting on the fluid can be determined. In this chapter, the methods of determining velocity and acceleration are discussed.

▶ 5.2 METHODS OF DESCRIBING FLUID MOTION

The fluid motion is described by two methods. They are —(*i*) Lagrangian Method, and (*ii*) Eulerian Method. In the Lagrangian method, a **single fluid particle** as shown in Fig.5.1 (*a*) is followed during its motion and its velocity, acceleration, density, etc. are described. In case of Eulerian method, the velocity, acceleration, pressure, density etc. are described **at a point or at a section** as shown in Fig. 5.1 (*b*) in flow field. The Eulerian method is commonly used in fluid mechanics.

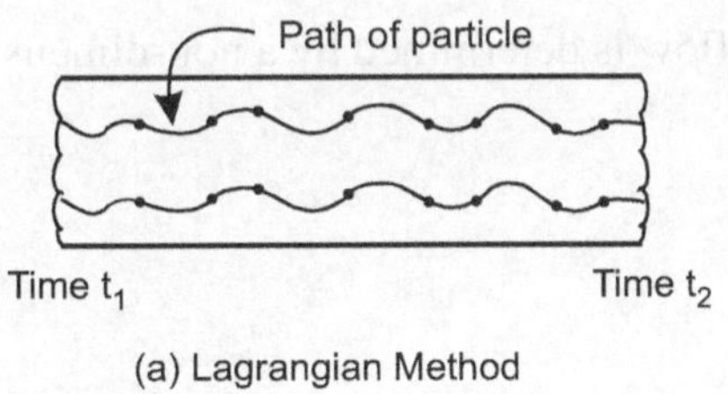

(a) Lagrangian Method

(b) Eulerian Method

Fig. 5.1

▶ 5.3 TYPES OF FLUID FLOW

The fluid flow is classified as :

(*i*) Steady and unsteady flows ;
(*ii*) Uniform and non-uniform flows ;
(*iii*) Laminar and turbulent flows ;
(*iv*) Compressible and incompressible flows ;
(*v*) Rotational and irrotational flows ; and
(*vi*) One, two and three-dimensional flows.

5.3.1 Steady and Unsteady Flows. Steady flow is defined as that type of flow in which the fluid characteristics like velocity, pressure, density, etc. at a point do not change with time. Thus for steady flow, mathematically, we have

$$\left(\frac{\partial V}{\partial t}\right)_{x_0, y_0, z_0} = 0, \left(\frac{\partial p}{\partial t}\right)_{x_0, y_0, z_0} = 0, \left(\frac{\partial \rho}{\partial t}\right)_{x_0, y_0, z_0} = 0$$

where (x_0, y_0, z_0) is a fixed point in fluid field.

Unsteady flow is that type of flow, in which the velocity, pressure of density at a point changes with respect to time. Thus, mathematically, for unsteady flow

$$\left(\frac{\partial V}{\partial t}\right)_{x_0, y_0, z_0} \neq 0, \left(\frac{\partial p}{\partial t}\right)_{x_0, y_0, z_0} \neq 0 \text{ etc.}$$

5.3.2 Uniform and Non-uniform Flows. Uniform flow is defined as that type of flow in which the velocity at any given time does not change with respect to space (*i.e.*, length of direction of the flow). Mathematically, for uniform flow

$$\left(\frac{\partial V}{\partial s}\right)_{t = \text{constant}} = 0$$

where ∂V = Change of velocity

∂s = Length of flow in the direction S.

Non-uniform flow is that type of flow in which the velocity at any given time changes with respect to space. Thus, mathematically, for non-uniform flow

$$\left(\frac{\partial V}{\partial s}\right)_{t = \text{constant}} \neq 0.$$

5.3.3 Laminar and Turbulent Flows. Laminar flow is defined as that type of flow in which the fluid particles move along well-defined paths or stream line and all the stream-lines are straight and parallel. Thus the particles move in laminas or layers gliding smoothly over the adjacent layer. This type of flow is also called stream-line flow or viscous flow.

Turbulent flow is that type of flow in which the fluid particles move in a *zig-zag* way. Due to the movement of fluid particles in a *zig-zag* way, the eddies formation takes place which are responsible for high energy loss. For a pipe flow, the type of flow is determined by a non-dimensional number $\frac{VD}{\nu}$ called the Reynold number.

where D = Diameter of pipe

V = Mean velocity of flow in pipe

and ν = Kinematic viscosity of fluid.

If the Reynold number is less than 2000, the flow is called laminar. If the Reynold number is more than 4000, it is called turbulent flow. If the Reynold number lies between 2000 and 4000, the flow may be laminar or turbulent.

5.3.4 Compressible and Incompressible Flows. Compressible flow is that type of flow in which the density of the fluid changes from point to point or in other words the density (ρ) is not constant for the fluid. Thus, mathematically, for compressible flow

$$\rho \neq \text{Constant}$$

Incompressible flow is that type of flow in which the density is constant for the fluid flow. Liquids are generally incompressible while gases are compressible. Mathematically, for incompressible flow

$$\rho = \text{Constant}.$$

5.3.5 Rotational and Irrotational Flows. Rotational flow is that type of flow in which the fluid particles while flowing along stream-lines, also rotate about their own axis. And if the fluid particles while flowing along stream-lines, do not rotate about their own axis that type of flow is called irrotational flow.

5.3.6 One, Two and Three-Dimensional Flows. One-dimensional flow is that type of flow in which the flow parameter such as velocity is a function of time and one space co-ordinate only, say x. For a steady one-dimensional flow, the velocity is a function of one-space-co-ordinate only. The variation of velocities in other two mutually perpendicular directions is assumed negligible. Hence mathematically, for one-dimensional flow

$$u = f(x),\ v = 0 \text{ and } w = 0$$

where u, v and w are velocity components in x, y and z directions respectively.

Two-dimensional flow is that type of flow in which the velocity is a function of time and two rectangular space co-ordinates say x and y. For a steady two-dimensional flow the velocity is a function of two space co-ordinates only. The variation of velocity in the third direction is negligible. Thus, mathematically for two-dimensional flow

$$u = f_1(x, y)\ v = f_2(x, y) \text{ and } w = 0.$$

Three-dimensional flow is that type of flow in which the velocity is a function of time and three mutually perpendicular directions. But for a steady three-dimensional flow the fluid parameters are functions of three space co-ordinates (x, y and z) only. Thus, mathematically, for three-dimensional flow

$$u = f_1(x, y, z),\ v = f_2(x, y, z),\ w = f_3(x, y, z).$$

▶ 5.4 RATE OF FLOW OR DISCHARGE (Q)

It is defined as the quantity of a fluid flowing per second through a section of a pipe or a channel. For an incompressible fluid (or liquid) the rate of flow or discharge is expressed as the volume of fluid flowing across the section per second. For compressible fluids, the rate of flow is usually expressed as the weight of fluid flowing across the section. Thus

(*i*) For liquids the units of Q are m^3/s or litres/s

(*ii*) For gases the units of Q is kgf/s or Newton/s

Consider a liquid flowing through a pipe in which

A = Cross-sectional area of pipe

V = Average velocity of fluid across the section

Then discharge $Q = A \times V$. ...(5.1)

▶ 5.5 CONTINUITY EQUATION

The equation based on the principle of conservation of mass is called continuity equation. Thus for a fluid flowing through the pipe at all the cross-section, the quantity of fluid per second is constant. Consider two cross-sections of a pipe as shown in Fig. 5.1(*c*).

Let V_1 = Average velocity at cross-section 1-1

ρ_1 = Density at section 1-1

A_1 = Area of pipe at section 1-1

and V_2, ρ_2, A_2 are corresponding valves at section, 2-2.

Then rate of flow at section 1-1 $= \rho_1 A_1 V_1$

Rate of flow at section 2-2 $= \rho_2 A_2 V_2$

According to law of conservation of mass

Rate of flow at section 1-1 = Rate of flow at section 2-2

or $$\rho_1 A_1 V_1 = \rho_2 A_2 V_2 \quad \text{...(5.2)}$$

Fig. 5.1(c) *Fluid flowing through a pipe.*

Equation (5.2) is applicable to the compressible as well as incompressible fluids and is called **Continuity Equation**. If the fluid is incompressible, then $\rho_1 = \rho_2$ and continuity equation (5.2) reduces to

$$A_1 V_1 = A_2 V_2 \quad \text{...(5.3)}$$

Problem 5.1 *The diameters of a pipe at the sections 1 and 2 are 10 cm and 15 cm respectively. Find the discharge through the pipe if the velocity of water flowing through the pipe at section 1 is 5 m/s. Determine also the velocity at section 2.*

Solution. Given :

At section 1, $D_1 = 10 \text{ cm} = 0.1 \text{ m}$

$$A_1 = \frac{\pi}{4}(D_1{}^2) = \frac{\pi}{4}(.1)^2 = .007854 \text{ m}^2$$

$$V_1 = 5 \text{ m/s.}$$

At section 2, $D_2 = 15 \text{ cm} = 0.15 \text{ m}$

$$A_2 = \frac{\pi}{4}(.15)^2 = 0.01767 \text{ m}^2$$

Fig. 5.2

(*i*) Discharge through pipe is given by equation (5.1)

or $$Q = A_1 \times V_1$$

$$= .007854 \times 5 = \mathbf{0.03927 \text{ m}^3\text{/s. Ans.}}$$

Using equation (5.3), we have $A_1 V_1 = A_2 V_2$

(*ii*) $\therefore$ $$V_2 = \frac{A_1 V_1}{A_1} = \frac{.007854}{.01767} \times 5.0 = \mathbf{2.22 \text{ m/s. Ans.}}$$

Problem 5.2 *A 30 cm diameter pipe, conveying water, branches into two pipes of diameters 20 cm and 15 cm respectively. If the average velocity in the 30 cm diameter pipe is 2.5 m/s, find the discharge in this pipe. Also determine the velocity in 15 cm pipe if the average velocity in 20 cm diameter pipe is 2 m/s.*

Solution. Given :

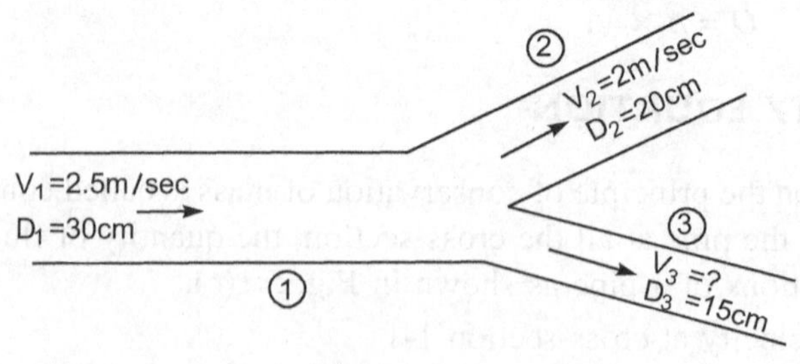

Fig. 5.3

$$D_1 = 30 \text{ cm} = 0.30 \text{ m}$$

$$\therefore \quad A_1 = \frac{\pi}{4} D_1^2 = \frac{\pi}{4} \times .3^2 = 0.07068 \text{ m}^2$$

$$V_1 = 2.5 \text{ m/s}$$

$$D_2 = 20 \text{ cm} = 0.20 \text{ m}$$

$$\therefore \quad A_2 = \frac{\pi}{4} (.2)^2 = \frac{\pi}{4} \times .4 = 0.0314 \text{ m}^2,$$

$$V_2 = 2 \text{ m/s}$$

$$D_3 = 15 \text{ cm} = 0.15 \text{ m}$$

$$\therefore \quad A_3 = \frac{\pi}{4} (.15)^2 = \frac{\pi}{4} \times 0.225 = 0.01767 \text{ m}^2$$

Find (*i*) Discharge in pipe 1 or Q_1

(*ii*) Velocity in pipe of dia. 15 cm or V_3

Let Q_1, Q_2 and Q_3 are discharges in pipe 1, 2 and 3 respectively.
Then according to continuity equation

$$Q_1 = Q_2 + Q_3 \quad \text{...(1)}$$

(*i*) **The discharge $\mathbf{Q_1}$** in pipe 1 is given by

$$Q_1 = A_1 V_1 = 0.07068 \times 2.5 \text{ m}^3\text{/s} = \mathbf{0.1767 \text{ m}^3\text{/s. Ans.}}$$

(*ii*) **Value of $\mathbf{V_3}$**

$$Q_2 = A_2 V_2 = .0314 \times 2.0 = .0628 \text{ m}^3\text{/s}$$

Substituting the values of Q_1 and Q_2 in equation (1)

$$0.1767 = 0.0628 + Q_3$$

$$\therefore \quad Q_3 = .1767 - 0.0628 = 0.1139 \text{ m}^3\text{/s}$$

But $\quad Q_3 = A_3 \times V_3 = .01767 \times V_3 \quad \text{or} \quad .1139 = .01767 \times V_3$

$$\therefore \quad V_3 = \frac{.1139}{.01767} = \mathbf{6.44 \text{ m/s. Ans.}}$$

Problem 5.3 *Water flows through a pipe AB 1.2 m diameter at 3 m/s and then passes through a pipe BC 1.5 m diameter. At C, the pipe branches. Branch CD is 0.8 m in diameter and carries one-third of the flow in AB. The flow velocity in branch CE is 2.5 m/s. Find the volume rate of flow in AB, the velocity in BC, the velocity in CD and the diameter of CE.*

(A.M.I.E., Winter 1982 ; Osmania University, 1990)

Solution. Given :

Diameter of pipe *AB*, $D_{AB} = 1.2$ m

Velocity of flow through *AB*, $V_{AB} = 3.0$ m/s

Dia. of pipe *BC*, $D_{BC} = 1.5$ m

Dia of branched pipe *CD*, $V_{CD} = 0.8$ m

Velocity of flow in pipe *CE*, $V_{CE} = 2.5$ m/s

Let the flow rate in pipe $AB = Q$ m³/s

Velocity of flow in pipe $BC = V_{BC}$ m/s

Velocity of flow in pipe $CD = V_{CD}$ m/s

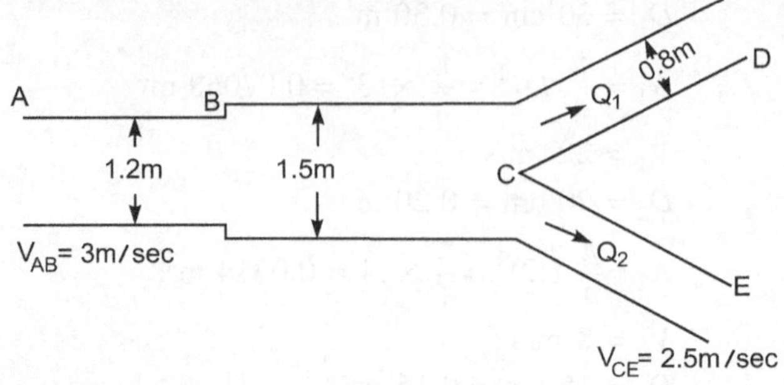

Fig. 5.4

Diameter of pipe $CE = D_{CE}$

Then flow rate through $CD = Q/3$

and flow rate through $CE = Q - Q/3 = \frac{2Q}{3}$

(*i*) Now volume flow rate through $AB = Q = V_{AB} \times$ Area of AB

$$= 3.0 \times \frac{\pi}{4} (D_{AB})^2 = 3.0 \times \frac{\pi}{4} (1.2)^2 = \mathbf{3.393\ m^3/s.\ Ans.}$$

(*ii*) Applying continuity equation to pipe AB and pipe BC,

$$V_{AB} \times \text{Area of pipe } AB = V_{BC} \times \text{Area of pipe } BC$$

or $$3.0 \times \frac{\pi}{4} (D_{AB})^2 = V_{BC} \times \frac{\pi}{4} (D_{BC})^2$$

or $$3.0 \times (1.2)^2 = V_{BC} \times (1.5)^2 \qquad \left[\text{Divide by } \frac{\pi}{4}\right]$$

or $$V_{BC} = \frac{3 \times 1.2^2}{1.5^2} = \mathbf{1.92\ m/s.\ Ans.}$$

(*iii*) The flow rate through pipe

$$C_D = Q_1 = \frac{Q}{3} = \frac{3.393}{3} = 1.131\ \text{m}^3/\text{s}$$

$\therefore$ $$Q_1 = V_{CD} \times \text{Area of pipe } C_D \times \frac{\pi}{4} (C_{CD})^2$$

or $$1.131 = V_{CD} \times \frac{\pi}{4} \times .8^2 = 0.5026\ V_{CD}$$

$\therefore$ $$V_{CD} = \frac{1.131}{0.5026} = \mathbf{2.25\ m/s.\ Ans.}$$

(*iv*) Flow rate through CE,

$$Q_2 = Q - Q_1 = 3.393 - 1.131 = 2.262\ \text{m}^3/\text{s}$$

$\therefore$ $$Q_2 = V_{CE} \times \text{Area of pipe } CE = V_{CE} \frac{\pi}{4} (D_{CE})^2$$

or $$2.263 = 2.5 \times \frac{\pi}{4} \times (D_{CE})^2$$

or $$D_{CE} = \sqrt{\frac{2.263 \times 4}{2.5 \times \pi}} = \sqrt{1.152} = 1.0735\ \text{m}$$

$\therefore$ Diameter of pipe CE = **1.0735 m. Ans.**

Problem 5.4 *A 25 cm diameter pipe carries oil of sp. gr. 0.9 at a velocity of 3 m/s. At another section the diameter is 20 cm. Find the velocity at this section and also mass rate of flow of oil.*

Solution. Given :

at section 1, $D_1 = 25 \text{ cm} = 0.25 \text{ m}$

$$A_1 = \frac{\pi}{4} D_1^2 = \frac{\pi}{4} \times .25^2 = 0.049 \text{ m}^3$$

$$V_1 = 3 \text{ m/s}$$

at section 2, $D_2 = 20 \text{ cm} = 0.2 \text{ m}$

$$A_2 = \frac{\pi}{4} (.2)^2 = 0.0314 \text{ m}^2$$

$$V_2 = ?$$

Mass rate of flow of oil = ?

Applying continuity equation at sections 1 and 2,

$$A_1 V_1 = A_2 V_2$$

or $$0.049 \times 3.0 = 0.0314 \times V_2$$

$\therefore$ $$V_2 = \frac{0.049 \times 3.0}{.0314} = \mathbf{4.68 \text{ m/s. Ans.}}$$

Mass rate of flow of oil $= \text{Mass density} \times Q = \rho \times A_1 \times V_1$

Sp. gr. of oil $= \dfrac{\text{Density of oil}}{\text{Density of water}}$

$\therefore$ Density of oil = Sp. gr. of oil × Density of water

$$= 0.9 \times 1000 \text{ kg/m}^3 = \frac{900 \text{ kg}}{\text{m}^3}$$

$\therefore$ Mass rate of flow $= 900 \times 0.049 \times 3.0 \text{ kg/s} = \mathbf{132.23 \text{ kg/s. Ans.}}$

Problem 5.5 *A jet of water from a 25 mm diameter nozzle is directed vertically upwards. Assuming that the jet remains circular and neglecting any loss of energy, that will be the diameter at a point 4.5 m above the nozzle, if the velocity with which the jet leaves the nozzle is 12 m/s.*

(A.M.I.E., Summer, 1988)

Solution. Given :

Dia. of nozzle, $D_1 = 25 \text{ mm} = 0.025 \text{ m}$

Velocity of jet at nozzle, $V_1 = 12 \text{ m/s}$

Height of point A, $h = 4.5 \text{ m}$

Let the velocity of the jet at a height 4.5 m = V_2

Consider the vertical motion of the jet from the outlet of the nozzle to the point A (neglecting any loss of energy).

Initial velocity, $u = V_1 = 12 \text{ m/s}$

Final velocity, $V = V_2$

Value of $g = -9.81 \text{ m/s}^2$ and $h = 4.5 \text{ m}$

Using, $V^2 - u^2 = 2gh$, we get

$$V_2^2 - 12^2 = 2 \times (-9.81) \times 4.5$$

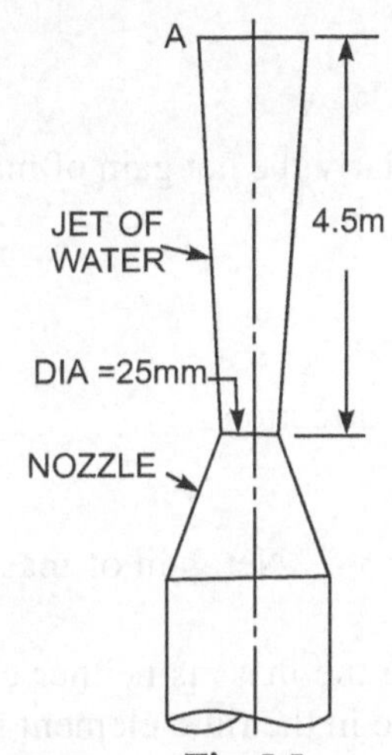

Fig. 5.5

$\therefore \qquad V_2 = \sqrt{12^2 - 2 \times 9.81 \times 4.5} = \sqrt{144 - 88.29} = 7.46 \text{ m/s}$

Now applying continuity equation to the outlet of nozzle and at point A, we get

$$A_1 V_1 = A_2 V_2$$

or
$$A_2 = \frac{A_1 V_1}{V_2} = \frac{\frac{\pi}{4} D_1^2 \times V_1}{V_2} = \frac{\pi \times (0.025)^2 \times 12}{4 \times 7.46} = 0.0007896$$

Let $\qquad D_2$ = Diameter of jet at point A.

Then $\qquad A_2 = \frac{\pi}{2} D_2^2$ or $0.0007896 = \frac{\pi}{4} \times D_2^2$

$$\therefore \qquad D_2 = \sqrt{\frac{0.0007896 \times 4}{\pi}} = 0.0317 \text{ m} = \mathbf{31.7 \text{ mm. Ans.}}$$

▶ 5.6 CONTINUITY EQUATION IN THREE-DIMENSIONS

Consider a fluid element of lengths dx, dy and dz in the direction of x, y and z. Let u, v and w are the inlet velocity components in x, y and z directions respectively. Mass of fluid entering the face $ABCD$ per second

$$= \rho \times \text{Velocity in } x\text{-direction} \times \text{Area of } ABCD$$
$$= \rho \times u \times (dy \times dz)$$

Then mass of fluid leaving the face $EFGH$ per second $= \rho u \, dydz + \frac{\partial}{\partial x} (\rho u \, dydz) \, dx$

$\therefore$ Gain of mass in x-directions

$$= \text{Mass through } ABCD - \text{Mass through } EFGH \text{ per second}$$
$$= \rho u \, dydz - \rho u \, dydz - \frac{\partial}{\partial x} (\rho u \, dydz) dx$$
$$= -\frac{\partial}{\partial x} (\rho u \, dydz) \, dx$$
$$= -\frac{\partial}{\partial x} (\rho u) \, dx \, dydz \quad \{\because dydz \text{ is constant}\}$$

Similarly the net gain of mass in y-directions

$$= -\frac{\partial}{\partial y} (\rho v) \, dxdydz$$

and in z-direction
$$= -\frac{\partial}{\partial z} (\rho w) \, dxdydz$$

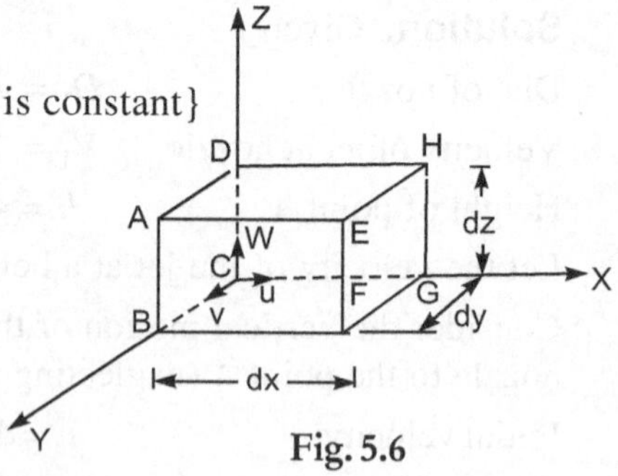

Fig. 5.6

$$\therefore \qquad \text{Net gain of masses} = -\left[\frac{\partial}{\partial x}(\rho u) + \frac{\partial}{\partial y}(\rho v) + \frac{\partial}{\partial z}(\rho w)\right] dxdydz$$

Since the mass is neither created nor destroyed in the fluid element, the net increase of mass per unit time in the fluid element must be equal to the rate of increase of mass of fluid in the element. But mass of fluid in the element is ρ. $dx.\ dy.\ dz$ and its rate of increase with time is $\frac{\partial}{\partial t}$ (ρ $dx.\ dy.\ dz$) or

$\frac{\partial \rho}{\partial t}.\ dx\ dy\ dz.$

Equating the two expressions,

$$\text{or}\qquad -\left[\frac{\partial}{\partial x}(\rho u)+\frac{\partial}{\partial y}(\rho v)+\frac{\partial}{\partial z}(\rho w)\right] dxdydz = \frac{\partial \rho}{\partial t}.\ dxdydz$$

$$\text{or}\qquad \frac{\partial \rho}{\partial t}+\frac{\partial}{\partial x}(\rho u)+\frac{\partial}{\partial y}(\rho v)+\frac{\partial}{\partial z}(\rho w) = 0 \text{ [Cancelling } dx.dy.dz \text{ to both sides]} \qquad ...(5.3A)$$

Equation (5.3*A*) is the continuity equation in cartesian co-ordinates in its most general form. This equation is applicable to :

(*i*) Steady and unsteady flow,

(*ii*) Uniform and non-uniform flow, and

(*iii*) Compressible and incompressible fluids.

For steady flow, $\frac{\partial \rho}{\partial t}$ = 0 and hence equation (5.3*A*) becomes as

$$\frac{\partial}{\partial x}(\rho u)+\frac{\partial}{\partial y}(\rho v)+\frac{\partial}{\partial z}(\rho w) = 0 \qquad ...(5.3B)$$

If the fluid is incompressible, then ρ is constant and the above equation becomes as

$$\frac{\partial u}{\partial x}+\frac{\partial v}{\partial y}+\frac{\partial w}{\partial z} = 0 \qquad ...(5.4)$$

Equation (5.4) is the continuity equation in three-dimensions. For a two-dimensional flow, the component w = 0 and hence continuity equation becomes as

$$\frac{\partial u}{\partial x}+\frac{\partial v}{\partial y} = 0. \qquad ...(5.5)$$

5.6.1 Continuity Equation in Cylindrical Polar Co-ordinate. The continuity equation in cylindrical polar co-ordinates (*i.e.*, r, θ, z co-ordinates) is derived by the procedure given below.

Consider a two-dimensional incompressible flow field. The two-dimensional polar co-ordinates are r and θ. Consider a fluid element $ABCD$ between the radii r and $r + dr$ as shown in Fig. 5.7. The angle sub-tended by the element at the centre is $d\theta$. The components of the velocity V are u_r in the radial direction and u_θ in the tangential direction. The sides of the element are having the lengths as

Side $AB = rd\theta$, $BC = dr$, $DC = (r + dr)\ d\theta$, $AD = dr$.

The thickness of the element perpendicular to the plane of the paper is assumed to be unity.

Consider the flow in radial direction

Mass of fluid entering the face AB per unit time

= ρ × Velocity in r-direction × Area

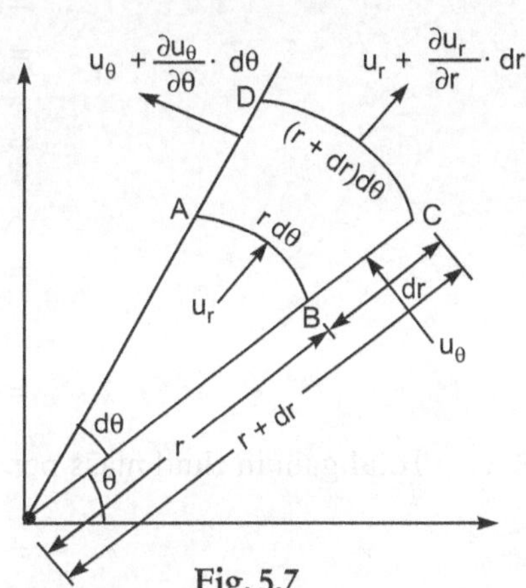

Fig. 5.7

$$= \rho \times u_r \times (AB \times 1) \qquad (\because \text{ Area} = AB \times \text{Thickness} = rd\theta \times 1)$$

$$= \rho \times u_r \times (rd\theta \times 1) = \rho \,.\, u_r .\, rd\theta$$

Mass of fluid leaving the face CD per unit time

$$= \rho \times \text{Velocity} \times \text{Area}$$

$$= \rho \times \left(u_r + \frac{\partial u_r}{\partial r}.dr\right) \times (CD \times 1) \qquad (\because \text{ Area} = CD \times 1)$$

$$= \rho \times \left(u_r + \frac{\partial u_r}{\partial r} dr\right) \times (r + dr)d\theta \qquad [\because CD = (r + dr)\, d\theta]$$

$$= \rho \times \left[u_r \times r + u_r dr + r\frac{\partial u_r}{\partial r} dr + \frac{\partial u_r}{\partial r}(dr)^2\right] d\theta$$

$$= \rho\left[u_r \times r + u_r \times dr + r\frac{\partial u_r}{\partial r}.dr\right] d\theta$$

[The term containing $(dr)^2$ is very small and has been neglected]

$\therefore$ Gain of mass in r-direction per unit time

$$= (\text{Mass through } AB - \text{Mass through } CD) \text{ per unit time}$$

$$= \rho.\, u_r .\, rd\theta - \rho\left[u_r.r + u_r.dr + r\frac{\partial u_r}{\partial r}.dr\right] d\theta$$

$$= \rho.\, u_r .\, rd\theta - \rho.\, u_r .\, r.\, d\theta - \rho\left[u_r.dr + r\frac{\partial u_r}{\partial r}.dr\right] d\theta$$

$$= -\rho\left[u_r.dr + r\frac{\partial u_r}{\partial r}.dr\right].\, d\theta$$

$$= -\rho\left[\frac{u_r}{r} + \frac{\partial u_r}{\partial r}\right] r.\, dr.\, d\theta$$

[This is written in this form because ($r.\, d\theta.\, dr.\, 1$) is equal to volume of element]

Now consider the flow in θ-direction

Gain in mass in θ-direction per unit time

$$= (\text{Mass through } BC - \text{Mass through } AD) \text{ per unit time}$$

$$= [\rho \times \text{Velocity through } BC \times \text{Area} - \rho \times \text{Velocity through } AD \times \text{Area}]$$

$$= \left[\rho.u_\theta .dr \times 1 - \rho\left(u_\theta + \frac{\partial u_\theta}{\partial \theta}.d\theta\right) \times dr \times 1\right]$$

$$= -\rho\left(\frac{\partial u_\theta}{\partial \theta}.d\theta\right) dr \times 1 \qquad (\because \text{ Area} = dr \times 1)$$

$$= -\rho\frac{\partial u_\theta}{\partial \theta}.\frac{r.\, d\theta.\, dr}{r} \qquad [\text{Multiplying and dividing by } r]$$

$\therefore$ Total gain in fluid mass per unit time

$$= -\rho\left[\frac{u_r}{r} + \frac{\partial u_r}{\partial r}\right].\, r.\, dr.\, d\theta - \rho\frac{\partial u_\theta}{\partial \theta}.\frac{rd\theta.\, dr}{r} \qquad ...(5.5A)$$

But mass of fluid element $= \rho \times$ Volume of fluid element

$$= \rho \times [rd\theta \times dr \times 1]$$
$$= \rho \times rd\theta \,.\, dr$$

Rate of increase of fluid mass in the element with time

$$= \frac{\partial}{\partial t}[\rho \,.\, rd\theta \,.\, dr] = \frac{\partial \rho}{\partial t} \,.\, rd\theta\, dr \qquad ...(5.5B)$$

($\because$ $rd\theta \,.\, dr \,.\, 1$ is the volume of element and is a constant quantity)

Since the mass is neither created nor destroyed in the fluid element, hence net gain of mass per unit time in the fluid element must be equal to the rate of increase of mass of fluid in the element.

Hence equating the two expressions given by equations (5.5 *A*) and (5.5 *B*), we get

$$-\rho\left[\frac{u_r}{r}+\frac{\partial u_r}{\partial r}\right] r\,.\,dr.\,d\theta - \rho\,\frac{\partial u_\theta}{\partial \theta}\,\frac{rd\theta\,.\,dr}{r} = \frac{\partial \rho}{\partial t}\, rd\theta\, dr$$

or
$$-\rho\left[\frac{u_r}{r}+\frac{\partial u_r}{\partial r}\right] - \rho\frac{\partial u_\theta}{\partial \theta}\,.\,\frac{1}{r} = \frac{\partial \rho}{\partial t} \qquad \text{[Cancelling } rdr\,.\,d\theta \text{ to both sides]}$$

or
$$\frac{\partial \rho}{\partial t} + \rho\left[\frac{u_r}{r}+\frac{\partial u_r}{\partial r}\right] + \rho\,\frac{\partial u_\theta}{\partial \theta}\,.\,\frac{1}{r} = 0 \qquad ...(5.5C)$$

Equation (5.5 *C*) is the continuity equation in polar co-ordinates for two-dimensional flow.

For steady flow $\frac{\partial \rho}{\partial t} = 0$ and hence equation (5.5 *C*) reduces to

$$\rho\left[\frac{u_r}{r}+\frac{\partial u_r}{\partial r}\right] + \rho\,\frac{\partial u_\theta}{\partial \theta}\,.\,\frac{1}{r} = 0$$

or
$$\frac{u_r}{r}+\frac{\partial u_r}{\partial r}+\frac{\partial u_\theta}{\partial \theta}\,.\,\frac{1}{r} = 0$$

or
$$u_r + r\frac{\partial u_r}{\partial r}+\frac{\partial u_\theta}{\partial \theta} = 0$$

or
$$\frac{\partial}{\partial r}(ru_r)+\frac{\partial}{\partial \theta}(u_\theta) = 0 \qquad \left[\because \frac{\partial}{\partial r}(r\,.\,u_r) = r\,.\,\frac{\partial u_r}{\partial r}+u_r\right] \qquad ...(5.5D)$$

Equation (5.5 *D*) represents the continuity equation in polar co-ordinates for two-dimensional steady incompressible flow.

Problem 5.5A *Examine whether the following velocity components represent a physically possible flow ?*

$$u_r = r \sin\theta,\ u_\theta = 2r\cos\theta.$$

Solution. Given : $u_r = r\sin\theta$ and $u_\theta = 2r\cos\theta$

For physically possible flow, the continuity equation,

$$\frac{\partial}{\partial r}(ru_r)+\frac{\partial}{\partial \theta}(u_\theta) = 0 \text{ should be satisfied.}$$

Now $u_r = r\sin\theta$

Multiplying the above equation by r, we get

$$ru_r = r^2 \sin\theta$$

Differentiating the above equation w.r.t. to r, we get

$$\frac{\partial}{\partial r}(ru_r) = \frac{\partial}{\partial r}(r^2 \sin\theta)$$

$$= 2r\sin\theta \qquad (\because \sin\theta \text{ is constant w.r.t. to } r)$$

Now $\qquad u_\theta = 2r\cos\theta$

Differentiating the above equation w.r.t. to θ, we get

$$\frac{\partial}{\partial\theta}(u_\theta) = \frac{\partial}{\partial\theta}(2r\cos\theta)$$

$$= 2r(-\sin\theta) \qquad (\because 2r \text{ is constant w.r.t. to } \theta)$$

$$= -2r\sin\theta$$

$$\therefore \qquad \frac{\partial}{\partial r}(ru_r) + \frac{\partial}{\partial\theta}(u_\theta) = 2r\sin\theta - 2r\sin\theta = 0$$

Hence the continuity equation is satisfied. Hence the given velocity components represent a physically possible flow.

▶ 5.7 VELOCITY AND ACCELERATION

Let V is the resultant velocity at any point in a fluid flow. Let u, v and w are its component in x, y and z directions. The velocity components are functions of space-co-ordinates and time. Mathematically, the velocity components are given as

$$u = f_1(x, y, z, t)$$
$$v = f_2(x, y, z, t)$$
$$w = f_3(x, y, z, t)$$

and Resultant velocity, $\qquad V = ui + vj + wk = \sqrt{u^2 + v^2 + w^2}$

Let a_x, a_y and a_z are the **total acceleration** in x, y and z directions respectively. Then by the chain rule of differentiation, we have

$$a_x = \frac{du}{dt} = \frac{\partial u}{\partial x}\frac{dx}{dt} + \frac{\partial u}{\partial y}\frac{dy}{dt} + \frac{\partial u}{\partial z}\cdot\frac{dz}{dt} + \frac{\partial u}{\partial t}$$

But $\qquad \frac{dx}{dt} = u, \frac{dy}{dt} = v \text{ and } \frac{dz}{dt} = w$

$$\therefore \qquad \left.\begin{aligned} a_x &= \frac{du}{dt} = u\frac{\partial u}{\partial x} + v\frac{\partial u}{\partial y} + w\frac{\partial u}{\partial z} + \frac{\partial u}{\partial t} \\ a_y &= \frac{dv}{dt} = u\frac{\partial v}{\partial x} + v\frac{\partial v}{\partial y} + w\frac{\partial v}{\partial z} + \frac{\partial v}{\partial t} \\ a_z &= \frac{dw}{dt} = u\frac{\partial w}{\partial x} + v\frac{\partial w}{\partial y} + w\frac{\partial w}{\partial z} + \frac{\partial w}{\partial t} \end{aligned}\right\} \qquad \text{...(5.6)}$$

(the second line preceded by "Similarly,")

For steady flow, $\frac{\partial V}{\partial t} = 0$, where V is resultant velocity

or
$$\frac{\partial u}{\partial t} = 0, \frac{\partial v}{\partial t} = 0 \text{ and} \frac{\partial w}{\partial t} = 0$$

Hence acceleration in x, y and z directions becomes

$$\left.\begin{aligned} a_x &= \frac{du}{dt} = u\frac{\partial u}{\partial x} + v\frac{\partial u}{\partial y} + w\frac{\partial u}{\partial z} \\ a_y &= \frac{dv}{dt} = u\frac{\partial v}{\partial x} + v\frac{\partial v}{\partial y} + w\frac{\partial v}{\partial z} \\ a_z &= \frac{dw}{dt} = u\frac{\partial w}{\partial x} + v\frac{\partial w}{\partial y} + w\frac{\partial w}{\partial z} \end{aligned}\right] \quad ...(5.7)$$

Acceleration vector
$$\left.\begin{aligned} A &= a_x i + a_y j + a_z k \\ &= \sqrt{a_x^2 + a_y^2 + a_z^2}. \end{aligned}\right] \quad ...(5.8)$$

5.7.1 Local Acceleration and Convective Acceleration. Local acceleration is defined as the rate of increase of velocity with respect to time at a given point in a flow field. In the equation given by (5.6), the expression $\frac{\partial u}{\partial t}$, $\frac{\partial v}{\partial t}$ or $\frac{\partial w}{\partial t}$ is known as local acceleration.

Convective acceleration is defined as the rate of change of velocity due to the change of position of fluid particles in a fluid flow. The expressions other than $\frac{\partial u}{\partial t}$, $\frac{\partial v}{\partial t}$ and $\frac{\partial w}{\partial t}$ in the equation (5.6) are known as convective acceleration.

Problem 5.6 *The velocity vector in a fluid flow is given*
$$V = 4x^3 i - 10x^2 yj + 2tk.$$
Find the velocity and acceleration of a fluid particle at (2, 1, 3) at time t = 1.

(Delhi University, 1987)

Solution. The velocity components u, v and w are $u = 4x^3$, $v = -10x^2 y$, $w = 2t$

For the point (2, 1, 3), we have $x = 2$, $y = 1$ and $z = 3$ at time $t = 1$.

Hence velocity components at (2, 1, 3) are

$$u = 4 \times (2)^3 = 32 \text{ units}$$
$$v = -10(2)^2(1) = -40 \text{ units}$$
$$w = 2 \times 1 = 2 \text{ units}$$

$\therefore$ Velocity vector V at (2, 1, 3) $= 32i - 40j + 2k$

or
$$\text{Resultant velocity} = \sqrt{u^2 + v^2 + w^2}$$
$$= \sqrt{32^2 + (-40)^2 + 2^2} = \sqrt{1024 + 1600 + 4} = \textbf{51.26 units. Ans.}$$

Acceleration is given by equation (5.6)

$$a_x = u\frac{\partial u}{\partial x} + v\frac{\partial u}{\partial y} + w\frac{\partial u}{\partial z} + \frac{\partial u}{\partial t}$$

$$a_y = u\frac{\partial v}{\partial x} + v\frac{\partial v}{\partial y} + w\frac{\partial v}{\partial z} + \frac{\partial v}{\partial t}$$

$$a_z = u\frac{\partial w}{\partial x} + v\frac{\partial w}{\partial y} + w\frac{\partial w}{\partial z} + \frac{\partial w}{\partial t}$$

Now from velocity components, we have

$$\frac{\partial u}{\partial x} = 12x^2, \frac{\partial u}{\partial y} = 0, \frac{\partial u}{\partial z} = 0 \text{ and } \frac{\partial u}{\partial t} = 0$$

$$\frac{\partial v}{\partial x} = -20xy, \frac{\partial v}{\partial y} = -10x^2, \frac{\partial v}{\partial z} = 0 \frac{\partial v}{\partial t} = 0$$

$$\frac{\partial w}{\partial x} = 0, \frac{\partial w}{\partial y} = 0, \frac{\partial w}{\partial z} = 0 \text{ and } \frac{\partial w}{\partial t} = 2.0$$

Substituting the values, the acceleration components at (2, 1, 3) at time $t = 1$ are

$$a_x = 4x^3(12x^2) + (-10x^2y)(0) + 2t \times (0) + 0$$

$$= 48x^5 = 48 \times (2)^5 = 48 \times 32 = 1536 \text{ units}$$

$$a_y = 4x^3(-20xy) + (-10x^2y)(-10x^2) + 2t(0) + 0$$

$$= -80x^4y + 100x^4y$$

$$= -80(2)^4(1) + 100(2)^4 \times 1 = -1280 + 1600 = 320 \text{ units.}$$

$$a_z = 4x^3(0) + (-10x^2y)(0) + (2t)(0) + 2.0 = 2.0 \text{ units}$$

∴ Acceleration is $A = a_x i + a_y j + a_z k =$ **1536i + 320j + 2k. Ans.**

or Resultant $A = \sqrt{(1536)^2 + (320)^2 + (2)^2}$ units

$= \sqrt{2359296 + 102400 + 4}$ = **1568.9 units. Ans.**

Problem 5.7 *The following cases represent the two velocity components, determine the third component of velocity such that they satisfy the continuity equation :*

(i) $u = x^2 + y^2 + z^2$; $v = xy^2 - yz^2 + xy$

(ii) $v = 2y^2$, $w = 2xyz$.

Solution. The continuity equation for incompressible fluid is given by equation (5.4) as

$$\frac{\partial u}{\partial x} + \frac{\partial v}{\partial y} + \frac{\partial w}{\partial z} = 0$$

Case I. $u = x^2 + y^2 + z^2$ ∴ $\frac{\partial u}{\partial x} = 2x$

$v = xy^2 - yz^2 + xy$ ∴ $\frac{\partial v}{\partial y} = 2xy - z^2 + x$

Substituting the values of $\frac{\partial u}{\partial x}$ and $\frac{\partial v}{\partial y}$ in continuity equation.

$$2x + 2xy - z^2 + x + \frac{\partial w}{\partial z} = 0$$

or

$$\frac{\partial w}{\partial z} = -3x - 2xy + z^2 \text{ or } \partial w = (-3x - 2xy + z^2)\,\partial z$$

Integration of both sides gives $\int dw = \int(-3x - 2xy + z^2)\,dz$

or $$w = \left(-3xz - 2xyz + \frac{z^3}{3}\right) + \text{Constant of integration,}$$

where constant of integration cannot be a function of z. But it can be a function of x and y that is $f(x, y)$.

$$\therefore \quad w = \left(\mathbf{-3xz - 2xyz + \frac{z^3}{3}}\right) + \mathbf{f(x, y)}. \textbf{ Ans.}$$

Case II. $$v = 2y^2 \quad \therefore \quad \frac{\partial v}{\partial y} = 4y$$

$$w = 2xyz \quad \therefore \quad \frac{\partial w}{\partial z} = 2xy$$

Substituting the values of $\frac{\partial v}{\partial y}$ and $\frac{\partial w}{\partial z}$ in continuity equation, we get

$$\frac{\partial u}{\partial x} + 4y + 2xy = 0$$

or $$\frac{\partial u}{\partial x} = -4y - 2xy \text{ or } du = (-4y - 2xy)\, dx$$

Integrating, we get $$u = -4xy - 2y\frac{x^2}{2} + f(y, z) = \mathbf{-4xy - x^2y^2 + f(y, z)}. \textbf{ Ans.}$$

Problem 5.8 *A fluid flow field is given by*

$$V = x^2yi + y^2zj - (2xyz + yz^2)k$$

Prove that it is a case of possible steady incompressible fluid flow. Calculate the velocity and acceleration at the point (2, 1, 3).

Solution. For the given fluid flow field $u = x^2y \quad \therefore \quad \frac{\partial u}{\partial x} = 2xy$

$$v = y^2z \quad \therefore \quad \frac{\partial v}{\partial y} = 2yz$$

$$w = -2xyz - yz^2 \quad \therefore \quad \frac{\partial w}{\partial z} = -2xy - 2yz.$$

For a case of possible steady incompressible fluid flow, the continuity equation (5.4) should be satisfied.

i.e., $$\frac{\partial u}{\partial x} + \frac{\partial v}{\partial y} + \frac{\partial w}{\partial z} = 0.$$

Substituting the values of $\frac{\partial u}{\partial x} + \frac{\partial v}{\partial y}$ and $\frac{\partial w}{\partial z}$, we get

$$\frac{\partial u}{\partial x} + \frac{\partial v}{\partial y} + \frac{\partial w}{\partial z} = 2xy + 2yz - 2xy - 2yz = 0$$

Hence the velocity field $V = x^2yi + y^2zj - (2xyz + yz^2)\, k$ is a possible case of fluid flow. **Ans.**

Velocity at (2, 1, 3)

Substituting the values $x = 2, y = 1$ and $z = 3$ in velocity field, we get

$$V = x^2yi + y^2zj - (2xyz + yz^2)\,k$$
$$= 2^2 \times 1i + 1^2 \times 3j - (2 \times 2 \times 1 \times 3 + 1 \times 3^2)\,k$$
$$= \mathbf{4i + 3j - 21k.\ Ans.}$$

and Resultant velocity $= \sqrt{4^2 + 3^2 + (-21)^2} = \sqrt{16 + 9 + 441} = \sqrt{466}$ = **21.587 units. Ans.**

Acceleration at (2, 1, 3)

The acceleration components a_x, a_y and a_z for steady flow are

$$a_x = u\frac{\partial u}{\partial x} + v\frac{\partial u}{\partial y} + w\frac{\partial u}{\partial z}$$

$$a_y = u\frac{\partial v}{\partial x} + v\frac{\partial v}{\partial y} + w\frac{\partial v}{\partial z}$$

$$a_z = u\frac{\partial w}{\partial x} + v\frac{\partial w}{\partial y} + w\frac{\partial w}{\partial z}$$

$$u = x^2y, \frac{\partial u}{\partial x} = 2xy, \frac{\partial u}{\partial y} = x^2 \text{ and } \frac{\partial u}{\partial z} = 0$$

$$v = y^2z, \frac{\partial v}{\partial x} = 0, \frac{\partial v}{\partial y} = 2yz, \frac{\partial v}{\partial z} = y^2$$

$$w = -2xyz - yz^2, \frac{\partial w}{\partial x} = -2yz, \frac{\partial w}{\partial y} = -2xz - z^2, \frac{\partial w}{\partial z} = -2xy - 2yz.$$

Substituting these values in acceleration components, we get acceleration at (2, 1, 3)

$$a_x = x^2y\,(2xy) + y^2z\,(x)^2 - (2xyz + yz^2)\,(0)$$
$$= 2x^3y^2 + x^2y^2z$$
$$= 2\,(2)^3 1^2 + 2^2 \times 1^2 \times 3 = 2 \times 8 + 12$$
$$= 16 + 12 = 28 \text{ units}$$
$$a_y = x^2y\,(0) + y^2z\,(2yz) - (2xyz + yz^2)\,(y^2)$$
$$= 2y^3z^2 - 2xy^3z - y^3z^2$$
$$= 2 \times 1^3 \times 3^2 - 2 \times 2 \times 1^3 \times 3 - 1^3 \times 3^2 = 18 - 12 - 9 = -3 \text{ units}$$
$$a_z = x^2y\,(-2yz) + y^2z\,(-2xz - z^2) - (2xyz + yz^2)\,(-2xy - 2yz)$$
$$= -2x^2y^2z - 2xy^2z^2 - y^2z^3 + [4x^2y^2z + 2xy^2z^2 + 4xy^2z^2 + 2y^2z^3]$$
$$= -2 \times 2^2 \times 1^2 \times 3 - 2 \times 2 \times 1^2 \times 3^2 - 1^2 \times 3^3 + [4 \times 2^2 \times 1^2 \times 3 + 2 \times 2 \times 1^2 \times 4^2 + 4 \times 2 \times 1^2 \times 3^2 + 2 \times 1^2 \times 3^3]$$
$$= -24 - 36 - 9 + [48 + 36 + 72 + 18]$$
$$= -24 - 36 - 9 + 48 + 36 + 72 + 18 = 105$$

∴ Acceleration $= a_xi + a_yj + a_zk =$ **28i – 3j + 105k. Ans.**

or Resultant acceleration $= \sqrt{28^2 + (-3)^2 + 105^2} = \sqrt{784 + 9 + 11025}$

$= \sqrt{11818}$ = **108.71 units. Ans.**

Problem 5.9 *Find the convective acceleration at the middle of a pipe which converges uniformly from 0.4 m diameter to 0.2 m diameter over 2 m length. The rate of flow is 20 lit/s. If the rate of flow changes uniformly from 20 l/s to 40 l/s in 30 seconds, find total acceleration at the middle of the pipe at 15th seconds.*

Solution. Given :

Dia. at section 1, $D_1 = 0.4$ m ; $D_2 = 0.2$ m, $L = 2$ m, $Q = 20$ l/s $= 0.02$ m^3/s as one litre = 0.001 m^3 = 1000 cm^3.

Find : (*i*) Convective acceleration at middle *i.e.*, at A when $Q = 20$ l/s.

(*ii*) Total acceleration at A when Q changes from 20 l/s to 40 l/s in 30 seconds.

Case I. In this case, the rate of flow is constant and equal to 0.02 m^3/s. The velocity of flow is in x-direction only. Hence this is one-dimensional flow and velocity components in y and z directions are zero or $v = 0$ and $z = 0$.

$\therefore$ Convective acceleration $= u \dfrac{\partial u}{\partial y}$ only ...(*i*)

Let us find the value of u and $\dfrac{\partial u}{\partial x}$ at a distance x from inlet.

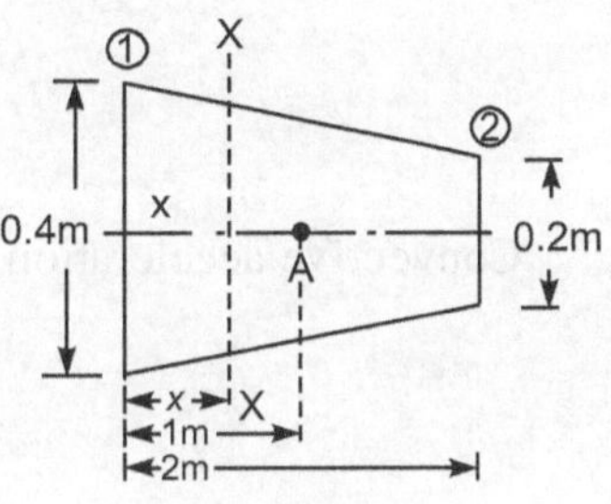

Fig. 5.8

The diameter (D_x) at a distance x from inlet or at section X-X is given by,

$$D_x = 0.4 - \frac{0.4 - 0.2}{2} \times x$$

$$= (0.4 - 0.1\,x) \text{ m}$$

The area of cross-section (A_x) at section X-X is given by

$$A_x = \frac{\pi}{4} D_x^2 = \frac{\pi}{4}(0.4 - 0.1\,x)^2$$

Velocity (u) at the section X-X in terms of Q (*i.e.*, in terms of rate of flow)

$$u = \frac{Q}{\text{Area}} = \frac{Q}{A_x} = \frac{Q}{\frac{\pi}{4} D_x^2} = \frac{4Q}{\pi (0.4 - 0.1\,x)^2}$$

$$= \frac{1.273\,Q}{(0.4 - 0.1\,x)^2} = 1.273\,Q\,(0.4 - 0.1\,x)^{-2} \text{ m/s} \quad ...(ii)$$

To find $\dfrac{\partial u}{\partial x}$, we must differentiate equation (*ii*) with respect to x.

$$\therefore \quad \frac{\partial u}{\partial x} = \frac{\partial}{\partial x}[1.273\,Q\,(0.4 - 0.1\,x)^{-2}]$$

$$= 1.273\,Q\,(-2)\,(0.4 - 0.1\,x)^{-1} \times (-0.1) \quad \text{[Here } Q \text{ is constant]}$$

$$= 0.2546\,Q\,(0.4 - 0.1\,x)^{-1} \quad ...(iii)$$

Substituting the value of u and $\dfrac{\partial u}{\partial x}$ in equation (*i*), we get

Convective acceleration $= [1.273\,Q\,(0.4 - 0.1\,x)^{-2}] \times [0.2546\,Q\,(0.4 - 0.1\,x)^{-1}]$

$= 1.273 \times 0.2546 \times Q^2 \times (0.4 - 0.1\,x)^{-3}$

$$= 1.273 \times 0.2546 \times (0.02)^2 \times (0.4 - 0.1\,x)^{-3} \quad [\because Q = 0.02 \text{ m}^3\text{/s}]$$

$\therefore$ Convective acceleration at the middle (where $x = 1$ m)

$$= 1.273 \times 0.2546 \times (0.02)^2 \times (0.4 - 0.1 \times 1)^{-3} \text{ m/s}^2$$

$$= 1.273 \times 0.2546 \times (0.02)^2 \times (0.3)^{-3} \text{ m/s}^2$$

$$= \mathbf{0.0048 \text{ m/s}^2}. \textbf{ Ans.}$$

Case II. When Q changes from 0.02 m^3/s to 0.04 m^3/s in 30 seconds, find total acceleration at $x = 1$ m and $t = 15$ seconds.

Total acceleration = Convective acceleration + Local acceleration at $t = 15$ seconds.

The rate of flow at $t = 15$ seconds is given by

$$Q = Q_1 + \frac{Q_2 - Q_1}{30} \times 15 \text{ where } Q_2 = 0.04 \text{ m}^3\text{/s and } Q_1 = 0.02 \text{ m}^3\text{/s}$$

$$= 0.02 + \frac{(0.04 - 0.02)}{30} \times 15 = 0.03 \text{ m}^3\text{/s}$$

The velocity (u) and gradient $\left(\frac{\partial u}{\partial x}\right)$ in terms of Q are given by equations (*ii*) and (*iii*) respectively.

$$\therefore \text{ Convective acceleration} = u \,.\, \frac{\partial u}{\partial x}$$

$$= [1.273\,Q\,(0.4 - 0.1\,x)^{-2}] \times [0.2546\,Q\,(0.4 - 0.1\,x)^{-1}]$$

$$= 1.273 \times 0.2546 \times Q^2 \times (0.4 - 0.1\,x)^{-3}$$

$\therefore$ Convective acceleration (when $Q = 0.03$ m^3/s and $x = 1$ m)

$$= 1.273 \times 0.2546 \times (0.03)^2 \times (0.4 - 0.1 \times 1)^{-3}$$

$$= 1.273 \times 0.2546 \times (0.03)^2 \times (0.3)^{-3} \text{ m/s}^2$$

$$= 0.0108 \text{ m/s}^2 \qquad \ldots(iv)$$

$$\text{Local acceleration} = \frac{\partial u}{\partial t} = \frac{\partial}{\partial t}[1.273\,Q\,(0.4 - 0.1\,x)^{-2}]$$

$[\because u$ from equation (*ii*) is $u = 1.273\,Q\,(0.4 - 0.1\,x)^{-2}]$

$$= 1.273 \times (0.4 - 0.1\,x)^{-2} \times \frac{\partial Q}{\partial t}$$

[$\because$ Local acceleration is at a point where x is constant but Q is changing]

Local acceleration (at $x = 1$ m)

$$= 1.273 \times (0.4 - 0.1 \times 1)^{-2} \times \frac{\partial Q}{\partial t}$$

$$= 1.273 \times (0.3)^{-2} \times \frac{0.02}{30} \left[\because \frac{\partial Q}{\partial t} = \frac{Q_2 - Q_1}{t} = \frac{0.04 - 0.02}{30} = \frac{0.02}{30}\right]$$

$$= 0.00943 \text{ m/s}^2 \qquad \ldots(v)$$

Hence adding equations (*iv*) and (*v*), we get total acceleration.

$\therefore$ Total acceleration = Convective acceleration + Local acceleration

$$= 0.0108 + 0.00943 = \mathbf{0.02023 \text{ m/s}^2}. \textbf{ Ans.}$$

▸ 5.8 VELOCITY POTENTIAL FUNCTION AND STREAM FUNCTION

5.8.1 Velocity Potential Function. It is defined as a scalar function of space and time such that its negative derivative with respect to any direction gives the fluid velocity in that direction. It is defined by ϕ (Phi). Mathematically, the velocity, potential is defined as $\phi = f(x, y, z)$ for steady flow such that

$$\left.\begin{aligned} u &= -\frac{\partial \phi}{\partial x} \\ v &= -\frac{\partial \phi}{\partial y} \\ w &= -\frac{\partial \phi}{\partial z} \end{aligned}\right\} \qquad \text{...(5.9)}$$

where u, v and w are the components of velocity in x, y and z directions respectively.

The velocity components in cylindrical polar co-ordinates in terms of velocity potential function are given by

$$\left.\begin{aligned} u_r &= \frac{\partial \phi}{\partial r} \\ u_\theta &= \frac{1}{r}\frac{\partial \phi}{\partial \theta} \end{aligned}\right\} \qquad \text{...(5.9A)}$$

where u_r = velocity component in radial direction (*i.e.* in r direction)

and u_θ = velocity component in tangential direction (*i.e.*, in θ direction)

The continuity equation for an incompressible steady flow is $\frac{\partial u}{\partial x} + \frac{\partial v}{\partial y} + \frac{\partial w}{\partial z} = 0$.

Substituting the values of u, v and w from equation (5.9), we get

$$\frac{\partial}{\partial x}\left(-\frac{\partial \phi}{\partial x}\right) + \frac{\partial}{\partial y}\left(-\frac{\partial \phi}{\partial y}\right) + \frac{\partial}{\partial z}\left(-\frac{\partial \phi}{\partial z}\right) = 0$$

or

$$\frac{\partial^2 \phi}{\partial x^2} + \frac{\partial^2 \phi}{\partial y^2} + \frac{\partial^2 \phi}{\partial z^2} = 0. \qquad \text{...(5.10)}$$

Equation (5.10) is a Laplace equation.

For two-dimension case the equation (5.10) reduces to $\frac{\partial^2 \phi}{\partial x^2} + \frac{\partial^2 \phi}{\partial y^2} = 0.$ ...(5.11)

If any value of ϕ that satisfies the Laplace equation, will correspond to some case of fluid flow.

Properties of the Potential Function. The rotational components* are given by

$$\omega_z = \frac{1}{2}\left(\frac{\partial v}{\partial x} - \frac{\partial u}{\partial y}\right)$$

---

* Please, refer to equation (5.17) on page 175.

$$\omega_y = \frac{1}{2}\left(\frac{\partial u}{\partial z} - \frac{\partial w}{\partial v}\right)$$

$$\omega_x = \frac{1}{2}\left(\frac{\partial w}{\partial y} - \frac{\partial v}{\partial z}\right)$$

Substituting the values, of u, v and w from equation (5.9) in the above rotational components, we get

$$\omega_z = \frac{1}{2}\left[\frac{\partial}{\partial x}\left(-\frac{\partial \phi}{\partial y}\right) - \frac{\partial}{\partial y}\left(-\frac{\partial \phi}{\partial x}\right)\right] = \frac{1}{2}\left[-\frac{\partial^2 \phi}{\partial x \partial y} + \frac{\partial^2 \phi}{\partial y \partial x}\right]$$

$$\omega_y = \frac{1}{2}\left[\frac{\partial}{\partial z}\left(-\frac{\partial \phi}{\partial x}\right) - \frac{\partial}{\partial x}\left(-\frac{\partial \phi}{\partial z}\right)\right] = \frac{1}{2}\left[-\frac{\partial^2 \phi}{\partial z \partial x} + \frac{\partial^2 \phi}{\partial x \partial z}\right]$$

and
$$\omega_x = \frac{1}{2}\left[\frac{\partial}{\partial y}\left(-\frac{\partial \phi}{\partial z}\right) - \frac{\partial}{\partial z}\left(-\frac{\partial \phi}{\partial y}\right)\right] = \frac{1}{2}\left[-\frac{\partial^2 \phi}{\partial y \partial z} + \frac{\partial^2 \phi}{\partial z \partial y}\right]$$

If ϕ is a continuous function, then $\frac{\partial^2 \phi}{\partial x \partial y} = \frac{\partial^2 \phi}{\partial y \partial x}$; $\frac{\partial^2 \phi}{\partial z \partial x} = \frac{\partial^2 \phi}{\partial x \partial z}$; etc.

$\therefore$
$$\omega_z = \omega_y = \omega_x = 0.$$

When rotational components are zero, the flow is called irrotational. Hence the properties of the potential function are :

1. If velocity potential (ϕ) exists, the flow should be irrotational.
2. If velocity potential (ϕ) satisfies the Laplace equation, it represents the possible steady incompressible irrotational flow.

5.8.2 Stream Function. It is defined as the scalar function of space and time, such that its partial derivative with respect to any direction gives the velocity component at right angles to that direction. It is denoted by ψ (*Psi*) and defined only for two-dimensional flow. Mathematically, for steady flow it is defined as $\psi = f(x, y)$ such that

$$\left.\begin{aligned} \frac{\partial \psi}{\partial x} &= v \\ \text{and} \quad \frac{\partial \psi}{\partial y} &= -u \end{aligned}\right\} \qquad \text{...(5.12)}$$

The velocity components in cylindrical polar co-ordinates in terms of stream function are given as

$$u_r = \frac{1}{r}\frac{\partial \psi}{\partial \theta} \text{ and } u_\theta = -\frac{\partial \psi}{\partial r} \qquad \text{...(5.12A)}$$

where u_r = radial velocity and u_θ = tangential velocity

The continuity equation for two-dimensional flow is $\frac{\partial u}{\partial x} + \frac{\partial v}{\partial y} = 0.$

Substituting the values of u and v from equation (5.12), we get

$$\frac{\partial}{\partial x}\left(-\frac{\partial \psi}{\partial y}\right)+\frac{\partial}{\partial y}\left(\frac{\partial \psi}{\partial x}\right)=0 \text{ or } -\frac{\partial^2 \psi}{\partial x \partial y}+\frac{\partial^2 \psi}{\partial x \partial y}=0.$$

Hence existence of ψ means a possible case of fluid flow. The flow may be rotational or irrotational.

The rotational component ω_z is given by $\omega_z=\frac{1}{2}\left(\frac{\partial v}{\partial x}-\frac{\partial u}{\partial y}\right)$.

Substituting the values of u and v from equation (5.12) in the above rotational component, we get

$$\omega_z=\frac{1}{2}\left[\frac{\partial}{\partial x}\left(\frac{\partial \psi}{\partial x}\right)-\frac{\partial}{\partial y}\left(-\frac{\partial \psi}{\partial y}\right)\right]=\frac{1}{2}\left[\frac{\partial^2 \psi}{\partial x^2}+\frac{\partial^2 \psi}{\partial y^2}\right]$$

For irrotational flow, $\omega_z = 0$. Hence above equation becomes as $\frac{\partial^2 \psi}{\partial x^2}+\frac{\partial \psi}{\partial y^2}=0$

which is Laplace equation for ψ.

The properties of stream function (ψ) are :

1. If stream function (ψ) exists, it is a possible case of fluid flow which may be rotational or irrotational.

2. If stream function (ψ) satisfies the Laplace equation, it is a possible case of an irrotational flow.

5.8.3 Equipotential Line.

A line along which the velocity potential ϕ is constant, is called equipotential line.

For equipotential line $\quad \phi = \text{Constant}$

$\therefore \quad d\psi = 0$

But $\quad \phi = f(x, y)$ for steady flow

$$\therefore \quad d\phi=\frac{\partial \phi}{\partial x}dx+\frac{\partial \phi}{\partial y}dy$$

$$= -udx - vdy \qquad \left\{\because \frac{\partial \phi}{\partial x}=-u, \frac{\partial \phi}{\partial x}=-v\right\}$$

$$= -(udx + vdy).$$

For equipotential line, $\quad d\phi = 0$

or $\quad -(udx + vdy) = 0$ or $udx + vdy = 0$

$$\therefore \quad \frac{dy}{dx}=-\frac{u}{v} \qquad \text{...(5.13)}$$

But $\quad \frac{dy}{dx}$ = Slope of equipotential line.

5.8.4 Line of Constant Stream Function

$\psi = \text{Constant}$

$\therefore \quad d\psi = 0$

$$\text{But} \quad d\psi=\frac{\partial \psi}{\partial x}dx+\frac{\partial \psi}{\partial y}dy=+vdx-udy \qquad \left\{\because \frac{\partial \psi}{\partial x}=v\,;\frac{\partial \psi}{\partial y}=-u\right\}$$

For a line of constant stream function

$$= d\psi = 0 \text{ or } vdx - udy = 0$$

or $$\frac{dy}{dx} = \frac{v}{u} \quad \text{...(5.14)}$$

But $\frac{dy}{dx}$ is slope of stream line.

From equations (5.13) and (5.14) it is clear that the product of the slope of the equipotential line and the slope of the stream line at the point of intersection is equal to – 1. Thus the equipotential lines are othogonal to the stream lines at all points of intersection.

5.8.5 Flow Net. A grid obtained by drawing a series of equipotential lines and stream lines is called a flow net. The flow net is an important tool in analysing two-dimensional irrotational flow problems.

5.8.6 Relation between Stream Function and Velocity Potential Function, From equation (5.9)

We have $$u = -\frac{\partial \phi}{\partial x} \text{ and } v = -\frac{\partial \phi}{\partial y}$$

From equation (5.12), we have $u = -\frac{\partial \psi}{\partial y}$ and $v = \frac{\partial \psi}{\partial x}$

Thus, we have $$u = -\frac{\partial \phi}{\partial x} = -\frac{\partial \psi}{\partial y} \text{ and } v = -\frac{\partial \phi}{\partial y} = \frac{\partial \psi}{\partial x}$$

Hence $$\frac{\partial \phi}{\partial x} = \frac{\partial \psi}{\partial y}$$

and $$\frac{\partial \phi}{\partial y} = -\frac{\partial \psi}{\partial x} \quad \text{...(5.15)}$$

Problem 5.10 *The velocity potential function (ϕ) is given by an expression*

$$\phi = -\frac{xy^3}{3} - x^2 + \frac{x^3 y}{3} + y^2$$

(*i*) *Find the velocity components in x and y direction.*
(*ii*) *Show that ϕ represents a possible case of flow.*

Solution. Given : $$\phi = -\frac{xy^3}{3} - x^2 + \frac{x^3 y}{3} + y^2$$

The partial derivatives of ϕ w.r.t. to x and y are

$$\frac{\partial \phi}{\partial x} = -\frac{y^3}{3} - 2x + \frac{3x^2 y}{3} \quad \text{...(1)}$$

and $$\frac{\partial \phi}{\partial y} = -\frac{3xy^2}{3} + \frac{x^3}{3} + 2y \quad \text{...(2)}$$

(*i*) The velocity components u and v are given by equation (5.9)

$$u = -\frac{\partial \phi}{\partial x} = -\left[\frac{y^3}{3} - 2x + \frac{3x^2 y}{3}\right] = \frac{y^3}{3} + 2x - x^2 y$$

$$\therefore \qquad u = \frac{\mathbf{y^2}}{\mathbf{3}} + \mathbf{2x - x^2y}. \textbf{ Ans.}$$

$$\therefore \qquad v = -\frac{\partial \phi}{\partial y} = -\left[-\frac{3xy^2}{3} + \frac{x^3}{3} + 2y\right] = \frac{3xy^2}{3} - \frac{x^3}{3} - 2y = \mathbf{xy^2} - \frac{\mathbf{x^3}}{\mathbf{3}} - \mathbf{2y}.$$

Ans.

(*ii*) The given value of ϕ, will represent a possible case of flow if it satisfies the Laplace equation, *i.e.*,

$$\frac{\partial^2 \phi}{\partial x^2} + \frac{\partial^2 \phi}{\partial y^2} = 0$$

From equations (1) and (2), we have

$$\text{Now} \qquad \frac{\partial \phi}{\partial x} = -y^3/3 - 2x + x^2y$$

$$\therefore \qquad \frac{\partial^2 \phi}{\partial x^2} = -2 + 2xy$$

$$\text{and} \qquad \frac{\partial \phi}{\partial y} = -xy^2 + \frac{x^3}{3} + 2y$$

$$\therefore \qquad \frac{\partial^2 \phi}{\partial y^2} = -2xy + 2$$

$$\therefore \qquad \frac{\partial^2 \phi}{\partial x^2} + \frac{\partial^2 \phi}{\partial y^2} = (-2 + 2xy) + (-2xy + 2) = 0$$

$\therefore$ Laplace equation is satisfied and hence ϕ represent a possible case of flow. **Ans.**

Problem 5.11 *The velocity potential function is given by $\phi = 5\ (x^2 - y^2)$. Calculate the velocity components at the point (4, 5).*

Solution. $\phi = 5\ (x^2 - y^2)$

$$\therefore \qquad \frac{\partial \phi}{\partial x} = 10x$$

$$\frac{\partial \phi}{\partial y} = -10y.$$

But velocity components u and v are given by equation (5.9) as

$$u = -\frac{\partial \phi}{\partial x} = -10x$$

$$v = -\frac{\partial \phi}{\partial y} = -(-10y) = 10y$$

The velocity components at the point (4, 5), *i.e.*, at $x = 4$, $y = 5$

$$u = -10 \times 4 = -40 \text{ units}$$

$$v = 10 \times 5 = \mathbf{50 \text{ units. Ans.}}$$

Problem 5.12 *A stream function is given by $\psi = 5x - 6y$.*
Calculate the velocity components and also magnitude and direction of the resultant velocity at any point.

Solution. $\psi = 5x - 6y$

$$\therefore \quad \frac{\partial \psi}{\partial x} = 5 \text{ and } \frac{\partial \psi}{\partial y} = -6.$$

But the velocity components u and v in terms of stream function are given by equation (5.12) as

$$u = -\frac{\partial \psi}{\partial y} = -(-6) = \mathbf{6 \text{ units/sec. Ans.}}$$

$$v = \frac{\partial \psi}{\partial x} = \mathbf{5 \text{ units/sec. Ans.}}$$

Resultant velocity $= \sqrt{u^2 + v^2} = \sqrt{6^2 + 5^2} = \sqrt{36 + 25} = \sqrt{61} = 7.81$ unit/sec

Direction is given by, $\tan \theta = \frac{v}{u} = \frac{5}{6} = 0.833$

$$\therefore \quad \theta = \tan^{-1} .833 = \mathbf{39° \ 48'. \text{ Ans.}}$$

Problem 5.13 *If for a two-dimensional potential flow, the velocity potential is given by*

$$\phi = x\,(2y - 1)$$

determine the velocity at the point P (4, 5). Determine also the value of stream function ψ at the point P.

(Similar Question in VTU, Aug. 2003; A.M.I.E., Winter, 1977)

Solution. Given : $\phi = x\,(2y - 1)$

(*i*) The velocity components in the direction of x and y are

$$u = -\frac{\partial \psi}{\partial x} = -\frac{\partial}{\partial x}[x\,(2y-1)] = -[2y - 1] = 1 - 2y$$

$$v = -\frac{\partial \phi}{\partial y} = -\frac{\partial}{\partial y}[x\,(2y-1)] = -[2x] = -2x$$

At the point P (4, 5), *i.e.*, at $x = 4$, $y = 5$

$$u = 1 - 2 \times 5 = -9 \text{ units/sec}$$

$$v = -2 \times 4 = -8 \text{ units/sec}$$

$\therefore$ Velocity at $P = -9i - 8j$

or Resultant velocity at $P = \sqrt{9^2 + 8^2} = \sqrt{81 + 64} = 12.04$ units/sec = **12.04 units/sec. Ans.**

(*ii*) **Value of Stream Function at P**

We know that $\frac{\partial \psi}{\partial y} = -u = -(1 - 2y) = 2y - 1$ ...(*i*)

and $\frac{\partial \psi}{\partial x} = v = -2x$ ...(*ii*)

Integrating equation (*i*) w.r.t. 'y', we get

$$\int d\psi = \int (2y - 1)\, dy \text{ or } \psi = \frac{2y^2}{2} - y + \text{Constant of integration.}$$

The constant of integration is not a function of y but it can be a function of x. Let the value of constant of integration is k. Then

$$\psi = y^2 - y + k. \quad \text{...(iii)}$$

Differentiating the above equation w.r.t. 'x', we get

$$\frac{\partial \psi}{\partial x} = \frac{\partial k}{\partial x}.$$

But $\frac{\partial \psi}{\partial x}$ from equation (*ii*) $= -2x$

Equating the value of $\frac{\partial \psi}{\partial x}$, we get $\frac{\partial k}{\partial x} = -2x$.

Integrating this equation, we get $k = \int -2x dx = -\frac{2x^2}{2} = -x^2$.

Substituting this value of k in equation (*iii*), we get $\psi = \mathbf{y^2 - y - x^2}$. **Ans.**

∴ Stream function ψ at P (4, 5) $= 5^2 - 5 - 4^2 = 25 - 5 - 16 =$ **4 units. Ans.**

Problem 5.14 *The stream function for a two-dimensional flow is given by $\psi = 2xy$, calculate the velocity at the point P (2, 3). Find the velocity potential function ϕ.*

(Similar Question VTU, Aug. 2002 and Feb. 2003)

Solution. Given : $\psi = 2xy$

The velocity components u and v in terms of ψ are

$$u = -\frac{\partial \psi}{\partial y} = -\frac{\partial}{\partial y}(2xy) = -2x$$

$$v = \frac{\partial \psi}{\partial x} = \frac{\partial}{\partial x}(2xy) = 2y.$$

At the point P (2, 3), we get $u = -2 \times 2 = -4$ units/sec

$v = 2 \times 3 = 6$ units/sec

∴ Resultant velocity at $P = \sqrt{u^2 + v^2} = \sqrt{4^2 + 6^2} = \sqrt{16 + 36} = \sqrt{52} = 7.21$ units/sec.

Velocity Potential Function ϕ

We know

$$\frac{\partial \phi}{\partial x} = -u = -(-2x) = 2x \quad \text{...(i)}$$

$$\frac{\partial \phi}{\partial y} = -v = -2y \quad \text{...(ii)}$$

Integrating equation (*i*), we get

$$\int d\phi = \int 2x dx$$

or

$$\phi = \frac{2x^2}{2} + C = x^2 + C \quad \text{...(iii)}$$

where C is a constant which is independent of x but can be a function of y.

Differentiating equation (*iii*) w.r.t. 'y', we get $\frac{\partial \phi}{\partial y} = \frac{\partial C}{\partial y}$

But from (*ii*), $$\frac{\partial \phi}{\partial y} = -2y$$

$$\therefore \qquad \frac{\partial C}{\partial y} = -2y$$

Integrating this equation, we get $C = \int -2y\, dy = -\frac{2y^2}{2} = -y^2$

Substituting this value of C in equation (*iii*), we get $\phi = \mathbf{x^2 - y^2}$. **Ans.**

Problem 5.15 *Sketch the stream lines represented by $y = x^2 + y^2$. Also find out the velocity and its direction at point (1, 2).* (A.M.I.E., Winter, 1980)

Solution. Given : $\psi = x^2 + y^2$

The velocity components u and v are

$$u = -\frac{\partial \psi}{\partial y} = -\frac{\partial}{\partial y}(x^2 + y^2) = -2y$$

$$v = \frac{\partial \psi}{\partial x} = \frac{\partial}{\partial x}(x^2 + y^2) = 2x$$

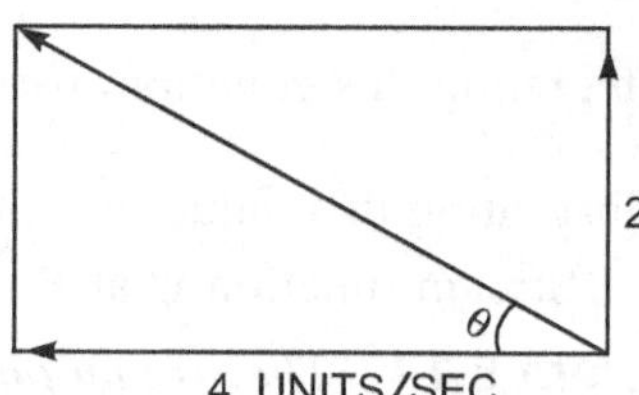

Fig. 5.9

At the point (1, 2), the velocity components are

$$u = -2 \times 2 = -4 \text{ units/sec}$$

$$v = 2 \times 1 = 2 \text{ units/sec}$$

Resultant velocity $= \sqrt{u^2 + v^2} = \sqrt{(-4)^2 + 2^2}$

$= \sqrt{20} = 4.47$ units/sec

and $$\tan \theta = \frac{v}{u} = \frac{2}{4} = \frac{1}{2}$$

$$\therefore \qquad \theta = \tan^{-1} .5 = 26° \, 34'$$

$\therefore$ Resultant velocity makes an angle of 26° 34′ with x-axis.

Sketch of Stream Lines.

$$\psi = x^2 + y^2$$

Let $\psi = 1, 2, 3$ and so on.

Then we have
$$1 = x^2 + y^2$$
$$2 = x^2 + y^2$$
$$3 = x^2 + y^2$$

and so on.

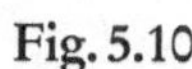

Fig. 5.10

Each equation is a equation of a circle. Thus we shall get concentric circles of different diameters as shown in Fig. 5.10.

Problem 5.16 *The velocity components in a two-dimensional flow field for an incompressible fluid are as follows :*

$$u = \frac{y^3}{3} + 2x - x^2 y \text{ and } v = xy^2 - 2y - x^3/3$$

obtain an expression for the stream function ψ. (VTU, Jan. 2004)

Solution. Given : $u = y^3/3 + 2x - x^2y$

$v = xy^2 - 2y - x^3/3.$

The velocity components in terms of stream function are

$$\frac{\partial \psi}{\partial x} = v = xy^2 - 2y - x^3/3 \quad ...(i)$$

$$\frac{\partial \psi}{\partial y} = -u = -y^3/3 - 2x + x^2y \quad ...(ii)$$

Integrating (*i*) w.r.t. x, we get $\psi = \int (xy^2 - 2y - x^3/3)\, dx$

or
$$\psi = \frac{x^2y^2}{2} - 2xy - \frac{x^4}{4 \times 3} + k, \quad ...(iii)$$

where k is a constant of integration which is independent of x but can be a function of y.

Differentiating equation (*iii*) w.r.t. to y, we get

$$\frac{\partial \psi}{\partial y} = \frac{2x^2y}{2} - 2x + \frac{\partial k}{\partial y} = x^2y - 2x + \frac{\partial k}{\partial y}$$

But from (*ii*),
$$\frac{\partial \psi}{\partial y} = -y^3/3 - 2x + x^2y$$

Comparing the value of $\frac{\partial \psi}{\partial y}$, we get $x^2y - 2x + \frac{\partial k}{\partial y} = -y^3/3 - 2x + x^2y$

$\therefore$
$$\frac{\partial k}{\partial y} = -y^3/3$$

Integrating, we get
$$k = \int (-y^3/3)\, dy = \frac{-y^4}{4 \times 3} = \frac{-y^4}{12}$$

Substituting this value in (*iii*), we get

$$\psi = \frac{\mathbf{x^2y^2}}{\mathbf{2}} - \mathbf{2xy} - \frac{\mathbf{x^4}}{\mathbf{12}} - \frac{\mathbf{y^4}}{\mathbf{12}}. \text{ **Ans.**}$$

Problem 5.17 *In a two-dimensional incompressible flow, the fluid velocity components are given by*

$u = x - 4y$ and $v = -y - 4x$.

Show that velocity potential exists and determine its form. Find also the stream function.

(Delhi University, 1996 ; A.M.I.E., Winter, 1982)

Solution. Given : $u = x - 4y$ and $v = -y - 4x$

$\therefore$
$$\frac{\partial u}{\partial x} = 1 \quad \text{and} \quad \frac{\partial v}{\partial y} = -1$$

$\therefore$
$$\frac{\partial u}{\partial x} + \frac{\partial v}{\partial y} = 1 - 1 = 0$$

Hence flow is continuous and velocity potential exists.

Let ϕ = Velocity potential.

Let velocity components in terms of velocity potential is given by

$$\frac{\partial \phi}{\partial x} = -u = -(x - 4y) = -x + 4y \qquad ...(i)$$

and

$$\frac{\partial \phi}{\partial y} = -v = -(-y - 4x) = y + 4x \qquad ...(ii)$$

Integrating equation (*i*), we get $\phi = -\frac{x^2}{2} + 4xy + C$ ...(*iii*)

where C is a constant of integration, which is independent of x.

This constant can be a function of y.

Differentiating the above equation, *i.e.*, equation (*iii*) with respect to 'y', we get

$$\frac{\partial \phi}{\partial y} = 0 + 4x + \frac{\partial C}{\partial y}$$

But from equation (*iii*), we have $\frac{\partial \phi}{\partial y} = y + 4x$

Equating the two values of $\frac{\partial \phi}{\partial y}$, we get

$$4x + \frac{\partial C}{\partial y} = y + 4x \quad \text{or} \quad \frac{\partial C}{\partial y} = y$$

Integrating the above equation, we get

$$C = \frac{y^2}{2} + C_1$$

where C_1 is a constant of integration, which is independent of x and y.

Taking it equal to zero, we get $C = \frac{y^2}{2}$.

Substituting the value of C in equation (*iii*), we get

$$\boldsymbol{\phi = -\frac{x^2}{2} + 4xy + \frac{y^2}{2}}. \textbf{ Ans.}$$

Value of Stream functions

Let ψ = Stream function

The velocity components in terms of stream function are

$$\frac{\partial \psi}{\partial x} = v = -y - 4x \qquad ...(iv)$$

and

$$\frac{\partial \psi}{\partial y} = -u = -(x - 4y) = -x + 4y \qquad ...(v)$$

Integrating equation (*iv*) w.r.t. x, we get

$$\psi = -yx - \frac{4x^2}{2} + k \qquad ...(vi)$$

where k is a constant of integration which is independent of x but can be a function of y.

Differentiating equation (*vi*) w.r.t. *y*, we get $\frac{\partial \psi}{\partial y} = -x - 0 + \frac{\partial k}{\partial y}$

But from equation (*v*), we have $\frac{\partial \psi}{\partial y} = -x + 4y$

Equating the two values of $\frac{\partial \psi}{\partial y}$, we get $-x + \frac{\partial k}{\partial y} = -x + 4y$ or $\frac{\partial k}{\partial y} = 4y$

Integrating the above equation, we get $k = \frac{4y^2}{2} = 2y^2$

Substituting the value of *k* in equation (*vi*), we get

$$\psi = -yx - 2x^2 + 2y^2. \textbf{ Ans.}$$

▶ 5.9 TYPES OF MOTION

A fluid particle while moving may undergo anyone or combination of following four types of displacements :

(*i*) Linear Translation or Pure Translation,
(*ii*) Linear Deformation,
(*iii*) Angular Deformation, and
(*iv*) Rotation.

5.9.1 Linear Translation. It is defined as the movement of a fluid element in such a way that it moves bodily from one position to another position and the two axes *ab* and *cd* represented in new positions by *a′b′* and *c′d′* are parallel as shown in Fig. 5.11 (*a*).

5.9.2 Linear Deformation. It is defined as the deformation of a fluid element in linear direction when the element moves. The axes of the element in the deformed position and un-deformed position are parallel, but their lengths change as shown in Fig. 5.11 (*b*).

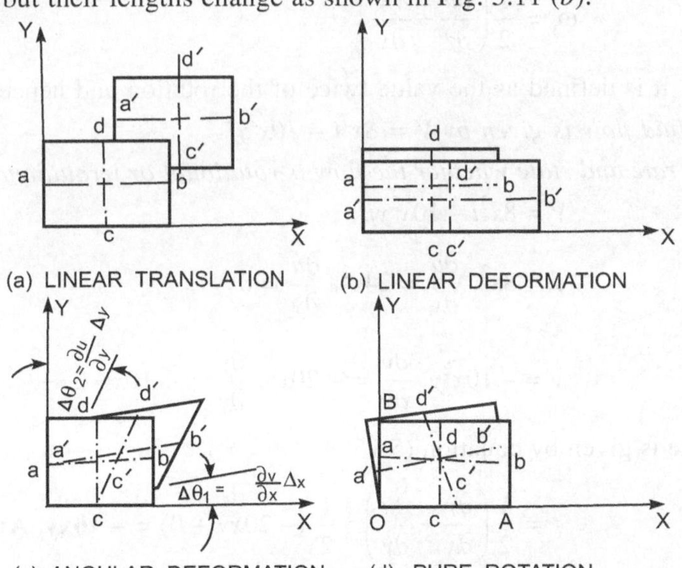

Fig. 5.11. *Displacement of a fluid element.*

5.9.3 Angular Deformation or Shear Deformation. It is defined as the average change in the angle contained by two adjacent sides. Let $\Delta\theta_1$ and $\Delta\theta_2$ is the change in angle between two adjacent side of a fluid element as shown in Fig. 5.11 (*c*), then angular deformation or shear strain rate

$$= \frac{1}{2} [\Delta\theta_1 + \Delta\theta_2]$$

Now $$\Delta\theta_1 = \frac{\partial v}{\partial x} \times \frac{\Delta x}{\Delta x} = \frac{\partial v}{\partial x} \text{ and } \Delta\theta_2 = \frac{\partial u}{\partial y} \cdot \frac{\Delta y}{\Delta y} = \frac{\partial u}{\partial y}.$$

$\therefore$ $$\textbf{Angular deformation} = \frac{1}{2} [\Delta\theta_1 + \Delta\theta_2]$$

or $$\textbf{Shear strain rate} = \frac{1}{2}\left[\frac{\partial v}{\partial x} + \frac{\partial u}{\partial y}\right] \qquad \text{...(5.16)}$$

5.9.4 Rotation. It is defined as the movement of a fluid element in such a way that both of its axes (horizontal as well as vertical) rotate in the same direction as shown in Fig. 5.11 (*d*). It is equal to $\frac{1}{2}\left(\frac{\partial v}{\partial x} - \frac{\partial u}{\partial y}\right)$ for a two-dimensional element in *x*-*y* plane. The rotational components are

$$\left.\begin{aligned} \omega_z &= \frac{1}{2}\left(\frac{\partial v}{\partial x} - \frac{\partial u}{\partial y}\right) \\ \omega_x &= \frac{1}{2}\left(\frac{\partial w}{\partial y} - \frac{\partial v}{\partial z}\right) \\ \omega_y &= \frac{1}{2}\left(\frac{\partial u}{\partial z} - \frac{\partial w}{\partial x}\right) \end{aligned}\right\} \qquad \text{...(5.17)}$$

5.9.5 Vorticity. It is defined as the value twice of the rotation and hence it is given as = 2ω.

Problem 5.18 *A fluid flow is given by* $V = 8x^3 i - 10x^2 y j$.

Find the shear strain rate and state whether the flow is rotational or irrotational.

Solution. Given : $V = 8x^3 i - 10x^2 y j$

$\therefore$ $$u = 8x^3, \quad \frac{\partial u}{\partial x} = 24x^2, \quad \frac{\partial u}{\partial y} = 0$$

and $$v = -10x^2 y, \quad \frac{\partial v}{\partial x} = -20xy, \quad \frac{\partial v}{\partial y} = -10x^3$$

(*i*) Shear strain rate is given by equation (5.16) as

$$= \frac{1}{2}\left(\frac{\partial v}{\partial x} + \frac{\partial u}{\partial y}\right) = \frac{1}{2}(-20xy + 0) = \mathbf{-10xy.\ Ans.}$$

(*ii*) Rotation in x - y plane is given by equation (5.17) or

$$\omega_z = \frac{1}{2}\left(\frac{\partial v}{\partial x} - \frac{\partial u}{\partial y}\right) = \frac{1}{2}\,(-20xy - 0) = -10xy$$

As rotation $\omega_z \neq 0$. Hence flow is rotational. **Ans.**

Problem 5.19 *The velocity components in a two-dimensional flow are*

$$u = y^3/3 + 2x - x^2 y \text{ and } v = xy^2 - 2y - x^3/3.$$

Show that these components represent a possible case of an irrotational flow.

Solution. Given : $u = y^3/3 + 2x - x^2 y$

$$\therefore \quad \frac{\partial u}{\partial x} = 2 - 2xy$$

$$\frac{\partial u}{\partial y} = \frac{3y^2}{3} - x^2 = y^2 - x^2$$

Also $v = xy^2 - 2y - x^3/3$

$$\therefore \quad \frac{\partial v}{\partial y} = 2xy - 2$$

$$\frac{\partial v}{\partial x} = y^2 - \frac{3x^2}{3} = y^2 - x^2.$$

(*i*) For a two-dimensional flow, continuity equation is $\frac{\partial u}{\partial x} + \frac{\partial v}{\partial y} = 0$

Substituting the value of $\frac{\partial u}{\partial x}$ and $\frac{\partial v}{\partial y}$, we get

$$\frac{\partial u}{\partial x} + \frac{\partial u}{\partial y} = 2 - 2xy + 2xy - 2 = 0$$

$\therefore$ It is a possible case of fluid flow.

(*ii*) Rotation, ω_z is given by $\omega_z = \frac{1}{2}\left(\frac{\partial v}{\partial x} - \frac{\partial u}{\partial y}\right) = \frac{1}{2}\,[(y^2 - x^2) - (y^2 - x^2)] = 0$

$\therefore$ Rotation is zero, which means it is case of irrotational flow. **Ans.**

HIGHLIGHTS

1. If the fluid characteristics like velocity, pressure, density etc. do not change at a point with respect to time, the fluid flow is called steady flow. If they change w.r.t. time, the fluid flow is called unsteady flow.

   Or $\left(\frac{\partial v}{\partial t}\right) = 0$ For steady flow and $\left(\frac{\partial v}{\partial t}\right) \neq 0$ For unsteady flow.

2. If the velocity in a fluid flow does not change with respect to space (length of direction of flow), the flow is said, uniform otherwise non-uniform. Thus

   $\left(\frac{\partial v}{\partial s}\right) = 0$ For uniform flow and $\left(\frac{\partial v}{\partial s}\right) \neq 0$ For non-uniform flow.

3. If the Reynolds number in a pipe is less than 2000, the flow is said to be laminar and if Reynold number is more than 4000, the flow is said to be turbulent.
4. For compressible flow, $\rho \neq$ constant
   For incompressible flow, $\rho =$ constant.
5. Rate of discharge for incompressible fluid (liquid), $Q = A \times v$.
6. Continuity equation is written as $A_1 v_1 = A_2 v_2 = A_3 v_3$.
7. Continuity equation in differential form,

$$\frac{\partial u}{\partial x} + \frac{\partial v}{\partial y} + \frac{\partial w}{\partial z} = 0 \text{ for three-dimensional flow}$$

$$\frac{\partial u}{\partial x} + \frac{\partial v}{\partial y} = 0 \text{ for two-dimensional flow.}$$

8. The components of acceleration in x, y and z direction are

$$a_x = u \frac{\partial u}{\partial x} + v \frac{\partial u}{\partial y} + w \frac{\partial u}{\partial z} + \frac{\partial u}{\partial t}$$

$$a_y = u \frac{\partial v}{\partial x} + v \frac{\partial v}{\partial y} + w \frac{\partial v}{\partial z} + \frac{\partial v}{\partial t}$$

$$a_z = u \frac{\partial w}{\partial x} + v \frac{\partial w}{\partial y} + w \frac{\partial w}{\partial z} + \frac{\partial w}{\partial t}.$$

9. The components of velocity in x, y and z direction in terms of velocity potential (ϕ) are

$$u = -\frac{\partial \phi}{\partial x}, \; v = -\frac{\partial \phi}{\partial y} \text{ and } w = -\frac{\partial \phi}{\partial z}.$$

10. The stream function (ψ) is defined only for two-dimensional flow. The velocity components in x and y directions in terms of stream function are $u = -\frac{\partial \psi}{\partial y}$ and $v = \frac{\partial \psi}{\partial x}$.
11. Angular deformation or shear strain rate is given as

$$\text{Shear strain rate} = \frac{1}{2}\left[\frac{\partial v}{\partial x} + \frac{\partial u}{\partial y}\right]$$

12. Rotational components of a fluid particle are

$$\omega_z = \frac{1}{2}\left[\frac{\partial v}{\partial x} - \frac{\partial u}{\partial y}\right]; \; \omega_x = \frac{1}{2}\left[\frac{\partial w}{\partial y} - \frac{\partial v}{\partial z}\right]; \; \omega_y = \frac{1}{2}\left[\frac{\partial u}{\partial z} - \frac{\partial w}{\partial x}\right]$$

13. Vorticity is two times the value of rotation.

EXERCISE 5

(A) THEORETICAL PROBLEMS

1. What are the methods of describing fluid flow ?
2. Explain the terms :
   (*i*) Path line, (*ii*) Streak line, (*A.M.I.E., Winter 1990, Summer 1991*)
   (*iii*) Stream line, and (*iv*) Stream tube. (*A.M.I.E., Summer 1988*)

3. Distinguish between :
   (*i*) Steady flow and un-steady flow, (*ii*) Uniform and non-uniform flow, (*VTU, Aug. 2002*)
   (*iii*) Compressible and incompressible flow, (*A.M.I.E., Winter 1988*)
   (*iv*) Rotational and irrotational flow, (*v*) Laminar and turbulent flow. (*A.S.M.E., June 1992*)
4. Define the following and give one practical example for each :
   (*i*) Laminar flow, (*ii*) Turbulent flow,
   (*iii*) Steady flow, and (*iv*) Uniform flow. (*A.M.I.E., Winter 1987*)
5. Define the equation of continuity. Obtain an expression for continuity equation for a three-dimensional steady incompressible flow. (*VTU, March 2000, Aug. 2001; R.G.P.V, S 2002*)
6. What do you understand by the terms : (*i*) Total acceleration, (*ii*) Convective acceleration, and (*iii*) Local acceleration ? (*Delhi University, Dec. 2002*)
7. (*a*) Define the terms :
   (*i*) Velocity potential function, and (*ii*) Stream function
   (*b*) What are the conditions for flow to be irrotational. (*A.M.I.E., Summer 1991*)
8. What do you mean by equipotential line and a line of constant stream function ?
9. (*a*) Describe the use and limitations of the flow nets. (*A.M.I.E., Summer 1987*)
   (*b*) Under what conditions can one draw flow net ? (*A.M.I.E., Winter 1991*)
10. Derive, from first principles, the condition for irrotational flow. Prove that, for potential flow, both the stream function and velocity potential function satisfy the Laplace equation. (*A.M.I.E., Summer 1983*)
11. Define velocity potential function and stream function. Show that the equipotential lines and lines of constant stream function are orthogonal. (*VTU, March 2001; Delhi University, 1987*)
12. Under what conditions can one treat real fluid flow as irrotational (as an approximation). (*A.M.I.E., Winter 1991*)
13. Define the following :
   (*i*) Steady flow, (*ii*) Non-uniform flow,
   (*iii*) Laminar flow, and (*iv*) Two-dimensional flow. (*A.M.I.E., Winter 1989*)
14. (*a*) Distinguish between rotational flow and irrotational flow. Give one example each
   (*b*) Cite two examples of unsteady, non-uniform flow. How can the unsteady flow be transformed to steady flow ? (*J.N.T., University, S 2002*)
15. Differentiate between :
   (*i*) Stream function and velocity potential function
   (*ii*) Stream line and streak line and
   (*iii*) Rotational and irrotational flows (*R.G.P.V., Bhopal S 2000*)
   (*iv*) Continuity equation and momentum equation (*VTU, Feb. 2003*)
   (*v*) Laminar and turbulent flow. (*VTU, Aug. 2002, Jan. 2004*)

(B) NUMERICAL PROBLEMS

1. The diameters of a pipe at the sections 1 and 2 are 15 cm and 20 cm respectively. Find the discharge through the pipe if velocity of water at section 1 is 4 m/s. Determine also the velocity at section 2. [**Ans.** 0.07068 m^3/s, 2.25 m/s]
2. A 40 cm diameter pipe, conveying water, branches into two pipes of diameters 30 cm and 20 cm respectively. If the average velocity in the 40 cm diameter pipe is 3 m/s. Find the discharge in this pipe. Also determine the velocity in 20 cm pipe if the average velocity in 30 cm diameter pipe is 2 m/s. [**Ans.** 0.3769 m^3/s, 7.5 m/s]
3. A 30 cm diameter pipe carries oil of sp. gr. 0.8 at a velocity of 2 m/s. At another section the diameter is 20 cm. Find the velocity at this section and also mass rate of flow of oil. [**Ans.** 4.5 m/s, 113 kg/s]
4. The velocity vector in a fluid flow is given by $V = 2x^3\mathbf{i} - 5x^2y\mathbf{j} + 4t\mathbf{k}$.
   Find the velocity and acceleration of a fluid particle at (1, 2, 3) at time, $t = 1$.
   (*Delhi University, June 1996*) [**Ans.** 10.95 units, 16.12 units]

5. The following cases represent the two velocity components, determine the third component of velocity such that they satisfy the continuity equation :
(*i*) $u = 4x^2$, $v = 4xyz$ (*ii*) $u = 4x^2 + 3xy$, $w = z^3 - 4xy - 2yz$.

$\left[\textbf{Ans.}\ (i)\ w = -8xz - 2xz^2 + f(x, y) \quad (ii)\ v = -8xy - \frac{y^2}{2} + 3yz^2 + f(x, z)\right]$

Calculate the unknown velocity components so that they satisfy the following equations :
(*i*) $u = 2x^2$, $v = 2xyz$, $w = ?$ (*ii*) $u = 2x^2 + 2xy$, $w = z^3 - 4xz + 2yz$, $v = ?$
(*A.M.I.E., Winter 1981*) [**Ans.** (*i*) $w = -1xz - x^2z$]

6. A fluid flow is given by : $V = xy^2i - 2yz^2j - \left(zy^2 - \frac{2z^3}{3}\right)k.$

Prove that it is a case of possible steady incompressible fluid flow.
Calculate the velocity and acceleration at the point [1, 2, 3]. [**Ans.** 36.7 units, 874.50 units]

7. Find the convective acceleration at the middle of a pipe which converges uniformly from 0.6 m diameter to 0.3 m diameter over 3 m length. The rate of flow is 40 lit/s. If the rate of flow changes uniformly from 40 lit/s to 80 lit/s in 40 seconds, find the total acceleration at the middle of the pipe at 20th seconds.
[**Ans.** .0499 m/s^2 ; .11874 m/s^2]

8. The velocity potential function, ϕ, is given by $\phi = x^2 - y^2$. Find the velocity components in x and y direction. Also show that ϕ represents a possible case of fluid flow. [**Ans.** $u = 2x$ and $v = -2y$]

9. For the velocity potential function, $\phi = x^2 - y^2$, find the velocity components at the point (4, 5).
[**Ans.** $u = 8$, $v = -10$ units]

10. A stream function is given by : $\psi = 2x - 5y$. Calculate the velocity components and also magnitude and direction of the resultant velocity at any point. [**Ans.** $u = 5$, $v = 2$, Resultant = 5.384 and $\theta = 21° 48'$]

11. If for a two-dimensional potential flow, the velocity potential is given by : $\phi = 4x(3y - 4)$, determine the velocity at the point (2, 3). Determine also the value of stream function ψ at the point (2, 3).
(*Delhi University, June 1996 ; Nagpur University, 1987*)

$\left[\textbf{Ans.}\ 40 \text{ units}, \psi = 6x^2 - 4\left(\frac{3}{2}y^2 - 4y\right), -18\right]$

12. The stream function for a two-dimensional flow is given by $\psi = 8xy$, calculate the velocity at the point $p(4, 5)$. Find the velocity potential function ϕ.
(*VTU, Feb. 2003*) [**Ans.** $u = -32$ units, $v = 40$ units, $\phi = 4y^2 - 4x^2$]

13. Sketch the stream lines represented by $\psi = xy$. Also find out the velocity and its direction at point (2, 3).
[**Ans.** 3.60 units and $\theta = 56° 18.6'$ or $123° 42'$]

14. For the velocity components given as : $u = ay \sin xy$, $v = ax \sin xy$.
Obtain an expression for the velocity potential function. [**Ans.** $\phi = a \cos xy$]

15. A fluid flow is given by : $V = 10x^3i - 8x^3yj$.
Find the shear strain rate and state whether the flow is rotational or irrotational. [**Ans.** $-8xy$, rotational]

16. The velocity components in a two-dimensional flow are :

$$u = 8x^2y - \frac{8}{3}y^3 \text{ and } v = -8xy^3 + \frac{8}{3}x^3.$$

Show that these velocity components represent a possible case of an irrotational flow.

$\left[\textbf{Ans.}\ \frac{\partial u}{\partial x} + \frac{\partial v}{\partial y} = 0,\ \omega_z = 0\right]$

17. Define rotational and irrotational flow. The stream function and velocity potential for a flow are given by :

$$\psi = 2xy, \phi = x^2 - y^2.$$

Show that the conditions of continuity and irrotational flow are satisfied. (*Osmania, University, 1990*)

18. For the steady incompressible flow, are the following valves of u and v possible ?
(*i*) $u = 4xy + y^2$, $v = 6xy + 3x$ and (*ii*) $u = 2x^2 + y^2$, $v = -4xy$. (*A.M.I.E., Winter 1988*)
[**Ans.** (*i*) No, (*ii*) Yes]

19. Define two-dimensional stream function and velocity potential. Show that following stream function :

$$\psi = 6x - 4y + 7xy + 9$$

represents an irrotational flow. Find its velocity potential. (*A.M.I.E., Summer 1990*)
[**Ans.** $\phi = 4x + 6y - 3.5x^2 + 3.5y^2 + C$]

20. Check if $\phi = x^2 - y^2 + y$ represents the velocity potential for 2-dimensional irrotational flow. If it does, then determine the stream function ψ. (*A.M.I.E., Summer 1990*) [**Ans.** Yes, $\psi = -2xy + x$]

21. If stream function for steady flow is given by $\psi = (y^2 - x^2)$, determine whether the flow is rotational or irrotational. Then determine the velocity potential ϕ.
(*A.M.I.E., Winter 1991*) [**Ans.** Irrotational, $\phi = -2xy + C$]

22. A pipe (1) 450 mm in diameter branches into two pipes (2) and (3) of diameters 300 mm and 200 mm respectively as shown in Fig. 5.12. If the average velocity in 450 mm diameter pipe is 3 m/s, find : (*i*) discharge through 450 mm dia. pipe and (*ii*) velocity in 200 mm diameter pipe if the average velocity in 300 mm pipe is 2.5 m/s. (*J.N.T.U., Hyderabad, S 2002*)

[**Hint.** Given : $d_1 = 450$ mm $= 0.45$ m, $d_2 = 300$ mm $= 0.3$ m
$d_3 = 200$ mm $= 0.2$ m, $V_1 = 3$ m/s, $V_2 = 2.5$ m/s

(*i*) $Q_1 = A_1V_1 = \frac{\pi}{4}(0.45^2) \times 3 = \mathbf{0.477\ m^3/s.\ Ans.}$

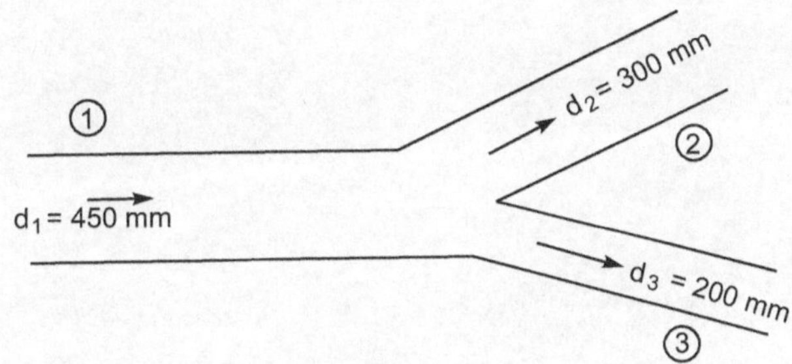

Fig. 5.12

(*ii*) $Q_2 = A_2V_2 = \frac{\pi}{4}(.3^2) \times 2.5 = 0.176\ m^3/s$

But $Q_1 = Q_2 + Q_3 \quad \therefore \quad Q_3 = Q_1 - Q_2 = 0.477 - 0.176 = 0.301$

Also $Q_3 = A_3 \times V_3 = \frac{\pi}{4}(0.2^2) \times V_3$

$\therefore$ $V_3 = \dfrac{Q_3}{\frac{\pi}{4}(.2^2)} = \dfrac{0.301}{0.0314} = \mathbf{9.6\ m/s.\ Ans.}$]

23. A stream function for a 2-D flow is given by $\psi = 8xy$. Calculate the velocity at a point P (4, 5). Find also the velocity potential function ϕ. (*VTU. Aug. 2002, Feb. 2003*)

[**Hint.** $\psi = 8xy$, $u = -\frac{\partial \psi}{\partial y} = -\frac{\partial}{\partial y}(8xy) = -8x$; $v = \frac{\partial \psi}{\partial x} = \frac{\partial}{\partial x}(8xy) = 8y$

At P (4, 5), $u = -8 \times 4 = -32$ and $v = 8 \times 5 = 40$

$\therefore$ $V = \sqrt{u^2 + v^2} = \sqrt{(-32)^2 + (40)^2}$

$= \mathbf{51.22\ units/s\ Ans.}$

Value of ϕ $\quad \frac{\partial \phi}{\partial x} = -u = -(-8x) = 8x$ and $\frac{\partial \phi}{\partial y} = -v = -8y$

Integrating $\quad \phi = \int 8x\, dx = \frac{8x^2}{2} + C = 4x^2 + C \qquad \text{... }(i)$

Now $\quad \frac{\partial \phi}{\partial y} = \frac{\partial}{\partial y}[4x^2 + C] = \frac{\partial C}{\partial y} \qquad (\because\ 4x^2 \text{ is constant w.r.t. to } y \ \therefore\ \frac{\partial}{\partial y}(4x^2) = 0)$

But $\quad \frac{\partial \phi}{\partial y} = -8y$

$\therefore \quad \frac{\partial C}{\partial y} = -8y$ Integrating we get $C = \frac{8y^2}{2} = 4y^2$

Substituting the value of C in equation (i), we get

$$\phi = \mathbf{4x^2 - 4y^2}. \text{ **Ans.**]}$$

Dimensional Analysis

▶ 6.1 INTRODUCTION

Dimensional analysis is a method of dimensions. It is a mathematical technique used in research work for design and for conducting model tests. It deals with the dimensions of the physical quantities involved in the phenomenon. All physical quantities are measured by comparison, which is made with respect to an arbitrarily fixed value. Length L, mass M and time T are three fixed diemensions which are of importance in Fluid Mechanics. If in any problem of fluid mechanics, heat is involved then temperature is also taken as fixed dimension. These fixed dimensions are called fundamental dimensions or fundamental quantity.

▶ 6.2 SECONDARY OR DERIVED QUANTITIES

Secondary or derived quantities are those quantities which possess more than one fundamental dimension. For example, velocity is denoted by distance per unit time (L/T), density by mass per unit volume $\left(\frac{M}{L^3}\right)$ and acceleration by distance per second square (L/T^2). Then velocity, density and acceleration become as secondary or derived quantities. The expressions (L/T), $\left(\frac{M}{L^3}\right)$ and $\left(\frac{L}{T^2}\right)$ are called the dimensions of velocity, density and acceleration respectively.

▶ 6.3 DIMENSIONS OF PHYSICAL QUANTITIES

The dimensions of mostly used physical quantities in Fluid Mechanics are given in Table 6.1.

Table 6.1

S. No.	*Physical Quantity*	*Symbol*	*Dimensions*
	(*a*) **Fundamental**		
1.	Length	L	L
2.	Mass	M	M
3.	Time	T	T

S.No.	Physical Quantity	Symbol	Dimensions
	(b) **Geometric**		
4.	Area	A	L^2
5.	Volume	$\forall$	L^3
	(c) **Kinematic Quantities**		
6.	Velocity	v	LT^{-1}
7.	Angular Velocity	ω	T^{-1}
8.	Acceleration	a	LT^{-2}
9.	Angular Acceleration	α	T^{-2}
10.	Discharge	Q	L^3T^{-1}
11.	Acceleration due to Gravity	g	LT^{-2}
12.	Kinematic Viscosity	ν	L^2T^{-1}
	(d) **Dynamic Quantities**		
13.	Force	F	MLT^{-2}
14.	Weight	W	MLT^{-2}
15.	Density	ρ	ML^{-3}
16.	Specific Weight	w	$ML^{-2}T^{-2}$
17.	Dynamic Viscosity	μ	$ML^{-1}T^{-1}$
18.	Pressure Intensity	p	$ML^{-1}T^{-2}$
19.	Modulus of Elasticity	$\begin{cases} K \\ E \end{cases}$	$ML^{-1}T^{-2}$
20.	Surface tension	σ	MT^{-2}
21.	Shear Stress	τ	$ML^{-1}T^{-2}$
22.	Work, Energy	W or E	$ML^{+2}T^{-2}$
23.	Power	P	MLT^{-3}
24.	Torque	T	ML^2T^{-2}
25.	Momentum	M	MLT^{-1}

Problem 6.1 *Determine the dimensions of the quantities given below : (i) Angular velocity, (ii) Angular acceleration, (iii) Discharge, (iv) Kinematic viscosity, (v) Force, (vi) Specific weight, and (vii) Dynamic viscosity.*

Solution. (*i*) Angular velocity $= \dfrac{\text{Angle covered in radians}}{\text{Times}} = \dfrac{1}{T} = T^{-1}$.

(*ii*) Angular acceleration $= \text{rad/sec}^2 = \dfrac{\text{rad}}{T^2} = \dfrac{1}{T^2} = T^{-2}$.

(*iii*) Discharge $= \text{Area} \times \text{Velocity} = L^2 \times \dfrac{L}{T} = \dfrac{L^3}{T} = L^3T^{-1}$.

(*iv*) Kinematic viscosity (ν) $= \dfrac{\mu}{\rho}$, where μ is given by $\tau = \mu \dfrac{\partial u}{\partial y}$

$$\therefore \qquad \mu = \frac{\tau}{\dfrac{\partial u}{\partial y}} = \frac{\text{Shear Stress}}{\dfrac{L}{T} \times \dfrac{1}{L}} = \frac{\dfrac{\text{Force}}{\text{Area}}}{\dfrac{1}{T}}$$

$$= \frac{\text{Mass} \times \text{Acceleration}}{\text{Area} \times \text{Time}} = \frac{M \times \frac{L}{T^2}}{L^2 \times \frac{1}{T}} = \frac{ML}{L^2T^2 \times \frac{1}{T}} = \frac{M}{LT} = ML^{-1}T^{-1}$$

and $$\rho = \frac{\text{Mass}}{\text{Volume}} = \frac{M}{L^3} = ML^{-3}$$

$\therefore$ Kinematic viscosity $$(\nu) = \frac{\mu}{\rho} = \frac{ML^{-1}T^{-1}}{ML^{-3}} = L^2T^{-1}.$$

(*v*) Force $$= \text{Mass} \times \text{Acceleration} = M \times \frac{\text{Length}}{(\text{Time})^2} = \frac{ML}{T^2} = MLT^{-2}.$$

(*vi*) Specific weight $$= \frac{\text{Weight}}{\text{Volume}} = \frac{\text{Force}}{\text{Volume}} = \frac{MLT^{-2}}{L^3} = ML^{-2}T^{2}.$$

(*vii*) Dynamic viscosity, μ is derived in (*iv*) as $\mu = ML^{-1}T^{-1}$.

▶ 6.4 DIMENSIONAL HOMOGENEITY

Dimensional homogeneity means the dimensions of each terms in an equation on both sides equal. Thus if the dimensions of each term on both sides of an equation are the same the equation is known as dimensionally homogeneous equation. The powers of fundamental dimensions (*i.e.*, *L*, *M*, *T*) on both sides of the equation will be identical for a dimensionally homogeneous equation. Such equations are independent of the system of units.

Let us consider the equation, $V = \sqrt{2gH}$

Dimension of L.H.S. $$= V = \frac{L}{T} = LT^{-1}$$

Dimension of R.H.S. $$= \sqrt{2gH} = \sqrt{\frac{L}{T^2} \times L} = \sqrt{\frac{L^2}{T^2}} = \frac{L}{T} = LT^{-1}$$

Dimension of L.H.S. = Dimension of R.H.S. = LT^{-1}

$\therefore$ Equation $V = \sqrt{2gH}$ is dimensionally homogeneous. So it can be used in any system of units.

▶ 6.5 METHODS OF DIMENSIONAL ANALYSIS

If the number of variable involved in a physical phenomenon are known, then the relation among the variables can be determined by the following two methods :

1. Rayleigh's method, and
2. Buckingham's π-theorem.

6.5.1 Rayleigh's Method. This method is used for determining the expression for a variable which depends upon maximum three or four variables only. If the number of independent variables becomes more than four, then it is very difficult to find the expression for the dependent variable.

Let X is a variable, which depends on X_1, X_2 and X_3 variables. Then according to Rayleigh's method, X is function of X_1, X_2 and X_3 and mathematically it is written as $X = f[X_1, X_2, X_3]$.

This can also be written as $X = KX_1^{\,a} \,.\, X_2^{\,b} \,.\, X_3^{\,c}$

where K is constant and a, b and c are arbitrarily powers.

The values of a, b and c are obtained by comparing the powers of the fundamental dimension on both sides. Thus the expression is obtained for dependent variable.

Problem 6.2 *The time period (t) of a pendulum depends upon the length (L) of the pendulum and acceleration due to gravity (g). Derive an expression for the time period.*

Solution. Time period t is a function of (*i*) L and (*ii*) g

$$\therefore \qquad t = KL^a \,.\, g^b, \text{ where } K \text{ is a constant} \qquad \ldots(i)$$

Substituting the dimensions on both sides $T^1 = KL^a \,.\, (LT^{-2})^b$

Equating the powers of M, L and T on both sides, we have

Power of T, $\qquad 1 = -2b \quad \therefore \quad b = -\dfrac{1}{2}$

Power of L, $\qquad 0 = a + b \quad \therefore \quad a = -b = -\left(-\dfrac{1}{2}\right) = \dfrac{1}{2}$

Substituting the values of a and b in equation (*i*),

$$t = KL^{1/2} \,.\, g^{-1/2} = K\sqrt{\frac{L}{g}}$$

The value of K is determined from experiments which is given as

$$K = 2\pi$$

$$\therefore \qquad \mathbf{t = 2\pi\sqrt{\frac{L}{g}}}\text{. } \textbf{Ans.}$$

Problem 6.3 *Find an expression for the drag force on smooth sphere of diameter D, moving with a uniform velocity V in a fluid of density ρ and dynamic viscosity μ.*

Solution. Drag force F is a function of

(*i*) Diameter, D (*ii*) Velocity, V (*iii*) Density, ρ

(*iv*) Viscocity, μ

$$\therefore \qquad F = KD^a \,.\, V^b \,.\, \rho^c \,.\, \mu^d \qquad \ldots(i)$$

where K is non-dimensional factor.

Substituting the dimensions on both sides,

$$MLT^{-2} = KL^a \,.\, (LT^{-1})^b \,.\, (ML^{-3})^c \,.\, (ML^{-1}T^{-1})^d$$

Equating the powers of M, L and T on both sides,

Power of M, $\qquad 1 = c + d$

Power of L, $\qquad 1 = a + b - 3c - d$

Power of T, $\qquad -2 = -b - d.$

There are four unknowns (a, b, c, d) but equations are three. Hence it is not possible to find the values of a, b, c and d. But three of them can be expressed in terms of fourth variable which is most important. Here viscosity is having a vital role and hence a, b, c are expressed in terms of d which is the power to viscosity.

$$\therefore \qquad c = 1 - d$$

$$b = 2 - d$$
$$a = 1 - b + 3c + d = 1 - 2 + d + 3(1 - d) + d$$
$$= 1 - 2 + d + 3 - 3d + d = 2 - d$$

Substituting these values of a, b and c in (*i*), we get

$$F = KD^{2-d} \cdot V^{2-d} \cdot \rho^{1-d} \cdot \mu^d$$

$$= KD^2V^2\rho\,(D^{-d} \cdot V^{-d} \cdot \rho^{-d} \cdot \mu^d) = K\rho D^2V^2\left(\frac{\mu}{\rho VD}\right)^d$$

$$= \mathbf{K\rho D^2V^2\phi\left(\frac{\mu}{\rho VD}\right)}. \text{ Ans.}$$

Problem 6.4 *Find the expression for the power P, developed by a pump when P depends upon the head H, the discharge Q and specific weight w of the fluid.*

Solution. Power P is a function of

(*i*) Head, H (*ii*) Discharge, Q

(*iii*) Specific weight, w

$$\therefore \quad P = KH^a \cdot Q^b \cdot w^c \quad ...(i)$$

where K = Non-dimensional constant.

Substituting the dimensions on both sides of equation (*i*)

$$ML^2T^{-3} = KL^a \cdot (L^3T^{-1})^b \cdot (ML^{-2}T^{-2})^c$$

Equating the powers of M, L and T on both sides,

Power of M, $\quad 1 = c, \quad \therefore \quad c = 1$

Power of K, $\quad 2 = a + 3b - 2c, \quad a = 2 - 3b + 2c = 2 - 3 + 2 = 1$

Power of T, $\quad -3 = -b - 2c \quad \therefore \quad b = 3 - 2c = 3 - 2 = 1$

Substituting the values of a, b and c in (*i*)

$$P = KH^1 \cdot Q^1 \cdot w^1 = \mathbf{KHQw}. \text{ Ans.}$$

Problem 6.5 *The efficiency η of a fan depends on the density ρ, the dynamic viscosity μ of the fluid, the angular velocity ω, diameter D of the rotor and the discharge Q. Express η in terms of dimensionless parameters.* (A.M.I.E., Winter, 1976)

Solution. The efficiency η depends on

(*i*) density, ρ (*ii*) viscosity, μ

(*iii*) Angular velocity, ω (*iv*) diameter D

(*v*) discharge, Q

$$\therefore \quad \eta = K\rho^a \cdot \mu^b \cdot \omega^c \cdot D^d \cdot Q^e \quad ...(i)$$

where K = Non-dimensional constant.

Substituting the dimensions on both sides of equation (*i*)

$$M^0L^0T^0 = (ML^{-3})^a \cdot (ML^{-1}T^{-1})^b \cdot (T^{-1})^c \cdot (L)^d \cdot (L^3T^{-1})^e$$

Equating powers of M, L, T on both sides,

Power of M, $\quad 0 = a + b$

Power of L, $\quad 0 = -3a - b + d + 3e$

Power of T, $\quad 0 = -b - c - e.$

There are five unknowns but equations are three. Express the three unknowns in terms of the other two unknowns which are more important. Viscosity and discharge are more important in this problem. Hence expressing a, c and d in terms of b and e, we get

$$a = -b$$

$$c = -(b + e)$$
$$d = +3a + b - 3e = -3b + b - 3e = -2b - 3e.$$

Substituting these values in equation (*i*), we get

$$\eta = K\rho^{-b}, \mu^{b}, \omega^{-(b+e)} . D^{-2b-3e} . Q^{e}$$
$$= K\rho^{-b} . \mu^{b} . \omega^{-e} . \omega^{-e} . D^{-2b} . D^{-3e} . Q^{e}$$
$$= \mathbf{K}\left(\frac{\mu}{\rho\omega \mathbf{D}^2}\right)^{\mathbf{b}} . \left(\frac{\mathbf{Q}}{\omega \mathbf{D}^3}\right)^{\mathbf{e}} = \phi\left[\left(\frac{\mu}{\rho\omega \mathbf{D}^2}\right) . \left(\frac{\mathbf{Q}}{\omega \mathbf{D}^3}\right)\right]. \textbf{ Ans.}$$

Problem 6.6 *The resisting force R of a supersonic plane during flight can be considered as dependent upon the length of the aircraft l, velocity V, air viscosity μ, air density ρ and bulk modulus of air K. Express the functional relationship between these variables and the resisting force.*

(A.M.I.E., Summer, 1977, 1990)

Solution. The resisting force R depends upon

(*i*) density, l, (*ii*) velocity, V,
(*iii*) viscosity, μ, (*iv*) density ρ,
(*v*) Bulk modulus, K.

$$\therefore \quad R = Al^{a} . V^{b} . \mu^{c} . \rho^{d} . K^{e} \quad \text{...}(i)$$

where A is the non-dimensional constant.

Substituting the dimensions on both sides of the equation (*i*),

$$MLT^{-2} = AL^{a} . (LT^{-1})^{b} . (ML^{-1}T^{-1})^{c} . (ML^{-3})^{d} . (ML^{-1}T^{-2})^{e}$$

Equating the powers of M, L, T on both sides,

Power of M, $\quad 1 = c + d + e$

Power of L, $\quad 1 = a + b - c - 3d - e$

Power of T, $\quad -2 = -b - c - 2e.$

There are five unknowns but equations are only three. Expressing the three unknowns in terms of two unknowns (μ and K).

∴ Express the values of a, b and d in terms of c and e.

Solving,

$$d = 1 - c - e$$
$$b = 2 - c - 2e$$
$$a = 1 - b + c + 3d + e = 1 - (2 - c - 2e) + c + 3(1 - c - e) + e$$
$$= 1 - 2 + c + 2e + c + 3 - 3c - 3e + e = 2 - c.$$

Substituting these values in (*i*), we get

$$R = A\, l^{2-c} . V^{2-c-2e} . \mu^{c} . \rho^{1-c-e} . K^{e}$$
$$= A\, l^{2} . V^{2} . \rho(l^{-c} V^{-c} \mu^{c} \rho^{-c}) . (V^{-2e} . \rho^{-e} . K^{e})$$
$$= A\, l^{2}V^{2}\rho \left(\frac{\mu}{\rho VL}\right)^{e} . \left(\frac{K}{\rho V^{2}}\right)^{e}$$
$$= \mathbf{A\rho l^{2}V^{2}}\, \phi\left[\left(\frac{\mu}{\rho \mathbf{VL}}\right)^{\mathbf{e}} . \left(\frac{\mathbf{K}}{\rho \mathbf{V}^{2}}\right)^{\mathbf{e}}\right]. \textbf{ Ans.}$$

Problem 6.7 *A partially submerged body is towed in water. The resistance R to its motion depends on the density ρ, the viscosity μ of water, length l of the body, velocity v of the body and the acceleration due to gravity g. Show that the resistance to the motion can be expressed in the form*

$$R = \rho L^{2}V^{2} \phi\left[\left(\frac{\mu}{\rho VL}\right), \left(\frac{lg}{V^{2}}\right)\right].$$

(A.M.I.E., Summer, 1978)

Solution. The resistance R depends on

(*i*) density ρ, (*ii*) viscosity μ,

(*iii*) length l, (*iv*) velocity V,

(*v*) acceleration g.

$$\therefore \qquad R = K\rho^a \cdot \mu^b \cdot l^c \cdot V^d \cdot g^e \qquad \text{...}(i)$$

where K = Non-dimensional constant.

Substituting the dimensions on both sides of the equation (*i*),

$$MLT^{-2} = K(ML^{-3})^a \cdot (ML^{-1}T^{-1})^b \cdot L^c \cdot (LT^{-1})^d \cdot (LT^{-2})^e$$

Equating the powers of M, L, T on both sides

Power of M, $\quad 1 = a + b$

Power of L, $\quad 1 = -3a - b + c + d + e$

Power of T, $\quad 2 = -b - d - 2e.$

There are five unknowns and equations are only three. Expressing the three unknowns in terms of two unknowns (μ and g). Hence express a, c and d in terms of b and e. Solving, we get

$$a = 1 - b$$

$$d = 2 - b - 2e$$

$$c = 1 + 3a + b - d - e = 1 + 3(1 - b) + b - (2 - b - 2e) - e$$

$$= 1 + 3 - 3b + b - 2 + b + 2e - e = 2 - b + e.$$

Substituting these values in equation (*i*), we get

$$R = K\rho^{1-b} \cdot \mu^b \cdot l^{2-b+c} \cdot V^{2-b-2e} \cdot g^e$$

$$= K\rho l^2 \cdot V^2 \cdot (\rho^{-b}\mu^b l^{-b} V^{-b}) \cdot (l^e \cdot V^{-2e} \cdot g^e)$$

$$= K\rho l^2 V^2 \cdot \left(\frac{\mu}{\rho V l}\right)^b \cdot \left(\frac{lg}{V^2}\right)^e = \rho l^2 V^2 \phi\left[\left(\frac{\mu}{\rho V l}\right), \left(\frac{lg}{V^2}\right)\right]. \quad \textbf{Ans.}$$

6.5.2 Buckingham's π-Theorem. The Rayleigh's method of dimensional analysis becomes more laborious if the variables are more than the number of fundamental dimensions (M, L, T). This difficulty is overcame by using Buckingham's π-theorem, which states, "If there are n variables (independent and dependent variables) in a physical phenomenon and if these variables contain m fundamental dimensions (M, L, T), then the variables are arranged into $(n - m)$ dimensionless terms. Each term is called π-term".

Let $X_1, X_2, X_3, ..., X_n$ are the variables involved in a physical problem. Let X_1 be the dependent variable and $X_2, X_3, ..., X_n$ are the independent variables on which X_1 depends. Then X_1 is a function of $X_2, X_3, ..., X_n$ and mathematically it is expressed as

$$X_1 = f(X_2, X_3, ..., X_n) \qquad \text{...(6.1)}$$

Equation (6.1) can also be written as

$$f_1(X_1, X_2, X_3, ..., X_n) = 0. \qquad \text{...(6.2)}$$

Equation (6.2) is a dimensionally homogeneous equation. It contains n variables. If there are m fundamental dimensions then according to Buckingham's π-theorem, equation (6.2) can be written in terms of number of dimensionless groups or π-terms in which number of π-terms is equal to $(n - m)$. Hence equation (6.2) becomes as

$$f(\pi_1, \pi_2, ..., \pi_{n-m}) = 0. \qquad \text{...(6.3)}$$

Each of π-terms is dimensionless and is independent of the system. Division or multiplication by a constant does not change the character of the π-term. Each π-term contains $m + 1$ variables, where m is the number of fundamental dimensions and is also called repeating variables. Let in the above case X_2, X_3 and X_4 are repeating variables if the fundamental dimension m (M, L, T) = 3. Then each π-term is written as

$$\left.\begin{aligned} \pi_1 &= X_2^{a_1} . X_3^{b_1} . X_4^{c_1} . X_1 \\ \pi_2 &= X_2^{a_2} . X_3^{b_2} . X_4^{c_2} . X_5 \\ &\vdots \\ \pi_{n-m} &= X_2^{a_{n-m}} . X_3^{b_{n-m}} . X_4^{c_m} . X_n \end{aligned}\right\} \quad ...(6.4)$$

Each equation is solved by the principle of dimensional homogeneity and values of a_1, b_1, c_1 etc., are obtained. These values are substituted in equation (6.4) and values of $\pi_1, \pi_2, ..., \pi_{n-m}$ are obtained. These values of π's are substituted in equation (6.3). The final equation for the phenomenon is obtained by expressing any one of the π-terms as a function of others as

$$\pi_1 = \phi\,[\pi_2, \pi_3, ..., \pi_{n-m}]$$

or

$$\pi_2 = \phi_1\,[\pi_1, \pi_3, ..., \pi_{n-m}] \quad ...(6.5)$$

6.5.3 Method of Selecting Repeating Variables. The number of repeating variables are equal to the number of fundamental dimensions of the problem. The choice of repeating variables if governed by the following considerations :

1. As far as possible, the dependent variable should not be selected as repeating variable.
2. The repeating variables should be choosen in such a way that one variable contains geometric property, other variable contains flow property and third variable contains fluid property.

Variables with Geometric Property are

(*i*) Length, l (*ii*) d (*iii*) Height, H etc.

Variables with flow property are

(*i*) Velocity, V (*ii*) Acceleration etc.

Variables with fluid property : (*i*) μ, (*ii*) ρ, (*iii*) ω etc.

3. The repeating variables selected should not form a dimensionless group.
4. The repeating variables together must have the same number of fundamental dimensions.
5. No two repeating variables should have the same dimensions.

In most of fluid mechanics problems, the choice of repeating variables may be (*i*) d, v, ρ (*ii*) l, v, ρ or (*iii*) l, v, μ or (*iv*) d, v, μ.

6.5.4 Procedure for Solving Problems by Buckingham's π-theorem. The procedure for solving problems by Buckingham's π-theorem is explained by considering the problem 6.6 which is also solved by the Rayleigh's method. The problem is :

The resisting force R of a supersonic plane during flight can be considered as dependent upon the length of the aircraft l, velocity V, air viscosity μ, air density ρ and bulk modulus of air K. Express the functional relationship between these variables and the resisting force. (*A.M.I.E., Summer, 1977*)

Solution. Step 1. The resisting force R depends upon (*i*) l, (*ii*) V, (*iii*) μ, (*iv*) ρ and (*v*) K. Hence R is a function of l, V, μ, ρ and K. Mathematically,

$$R = f(l, V, \mu, \rho, K) \quad ...(i)$$

or it can be written as $f_1\,(R, l, V, \rho, K) = 0$ ...(*ii*)

∴ Total number of variables, $n = 6$.

Number of fundamental dimensions, $m = 3$.

[m is obtained by writing dimensions of each variables as $R = MLT^{-2}$, $V = LT^{-1}$, $\mu = MT^{-1}\,T^{-1}$, $\rho = ML^{-3}$, $K = MLT^{-2}$. Thus as fundamental dimensions in the problem are M, L, T and hence $m = 3$.]

Number of dimensionless π-terms $= n - m = 6 - 3 = 3$.

Thus three π-terms say π_1, π_2 and π_3 are formed. Hence equation (*ii*) is written as

$$f_1\,(\pi_1, \pi_2, \pi_3) = 0. \quad ...(iii)$$

Step 2. Each π term $= m + 1$ variables, where m is equal to 3 and also called repeating variables. Out of six variables R, l, V, μ, ρ and K, three variables are to be selected as repeating variable. R is a dependent variable and should not be selected as a repeating variable. Out of the five remaining variables, one variable should have geometric property, the second variable should have flow property and third one fluid property. These requirements are fulfilled by selecting l, V and ρ as repeating variables. The repeating variables themselves should not form a dimensionless term and should have themselves fundamental dimensions equal to m, *i.e.*, 3 here. Dimensions of l, V and ρ are L, LT^{-1}, ML^{-3} and hence the three fundamental dimensions exist in l, V and ρ and they themselves do not form dimensionless group.

Step 3. Each π-term is written as according to equation (6.4)

$$\left.\begin{aligned} \pi_1 &= l^{a_1} . V^{b_1} . \rho^{c_1} . R \\ \pi_2 &= l^{a_2} . V^{b_2} . \rho^{c_2} . \mu \\ \pi_3 &= l^{a_3} . V^{b_3} . \rho^{c_3} . K \end{aligned}\right\} \quad ...(iv)$$

Step 4. Each π-term is solved by the principle of dimensional homogeneity. For the first π-term, we have

$$\pi_1 = M^0L^0T^0 = L^{a_1} . (LT^{-1})^{b_1} . (ML^{-3})^{c_1} . MLT\,T^{-2}.$$

Equating the powers of M, L, T on both sides, we get

Power of M, $\quad 0 = c_1 + 1 \quad \therefore \quad c_1 = -1$

Power of L, $\quad 0 = a_1 + b_1 - 3c_1 + 1,$

$\therefore \quad a_i = -b_1 + 3c_1 - 1 = 2 - 3 - 1 = -2$

Power of T, $\quad 0 = -b_1 - 2 \quad \therefore \quad b_1 = -2$

Substituting the values of a_1, b_1 and c_1 in equation (iv),

$$\pi_1 = l^{-2} . V^{-2} . \rho^{-2} . R$$

or $$\pi_1 = \frac{R}{l^2V^2\rho} = \frac{R}{\rho l^2 V^2} \quad ...(v)$$

Similarly for the 2nd π-term, we get $\pi_2 = M^0L^0T^0 = L^{a_2} . (LT^{-1})^{b_2} . (ML^{-3})^{b_2} . ML^{-1}T^{-1}$.

Equating the powers of M, L, T on both sides

Power of M, $\quad 0 = c_2 + 1, \quad \therefore \quad c_2 = -1$

Power of L, $\quad 0 = a_2 + b_2 - 3c_2 - 1,$

$\quad a_2 = -b_2 + 3c_2 + 1 = 1 - 3 + 1 = -1$

Power of T, $\quad 0 = -b_2 - 1, \quad \therefore \quad b_2 = -1$

Substituting the values of a_2, b_2 and c_2 in π_2 of (iv)

$$\pi_2 = l^{-1} . V^{-1} . \rho^{-1} . \mu = \frac{\mu}{lV\rho}.$$

3rd π-term

$$\pi_3 = l^{a_3} . V^{b_3} . \rho^{c_3} . K$$

or $$M^0L^0T^0 = L^{a_3} . (LT^{-3})^{b_3} . (ML^{-3})^{a_3} . ML^{-1}T^{-2}$$

Equating the powers of M, L, T on both sides, we have

Power of M, $\quad 0 = c_3 + 1, \quad \therefore \quad c_3 = -1$

Power of L, $\quad 0 = a_3 + b_3 - 3c_3 - 1, \quad \therefore \quad a_3 = -b_3 + 3c_3 + 1 = 2 - 3 + 1 = 0$

Power of T, $\quad 0 = -b_3 - 2, \quad \therefore \quad b_3 = -2$

Substituting the values of a_3, b_3 and c_3 in π_3 terms

$$\pi_3 = l^0 . V^{-2} . \rho^{-1} . K = \frac{K}{V^2\rho}.$$

Step 5. Substituting the values of π_1, π_2 and π_3 in equation (*ii*), we get

$$f_1\left(\frac{R}{\rho l^2V^2}, \frac{\mu}{lV\rho}, \frac{K}{V^2\rho}\right) = 0 \quad \text{or} \quad \frac{R}{\rho l^2V^2} = \phi\left[\frac{\mu}{lV\rho}, \frac{K}{V^2\rho}\right]$$

or $$R = \rho \mathbf{l}^2\mathbf{V}^2\phi\left[\frac{\mu}{\mathbf{l}\mathbf{V}\rho}, \frac{\mathbf{K}}{\mathbf{V}^2\rho}\right]. \textbf{ Ans.}$$

Problem 6.8 *(a) State Buckingham's π-theorem.*

(b) The efficiency η of a fan depends on density ρ, dynamic viscosity μ of the fluid, angular velocity ω, diameter D of the rotor and the discharge Q. Express η in terms of dimensionless parameters.

(A.M.I.E., Winter, 1976)

Solution. (*a*) Statement of Buckingham's π-theorem is given in Article 6.5.2.

(*b*) Given : η is a function of ρ, μ, ω, *D* and *Q*

$\therefore$ $$\eta = f(\rho, \mu, \omega, D, Q) \quad \text{or} \quad f_1(\eta, \rho, \mu, \omega, D, Q) = 0 \qquad \text{...(i)}$$

Hence total number of variables, η = 6.

The value of *m*, *i.e.*, number of fundamental dimensions for the problem is obtained by writing dimensions of each variable. Dimensions of each variable are

$$\eta = \text{Dimensionless}, \rho = ML^{-3}, \mu = ML^{-1}T^{-1}, \omega = T^{-1}, D = L \text{ and } Q = L^3T^{-1}$$

$\therefore$ $m = 3$

Number of π-terms $= n - m = 6 - 3 = 3$

Equation (*i*) is written as $f_1(\pi_1, \pi_2, \pi_3) = 0$ ...(*ii*)

Each π-term contains $m + 1$ variables, where *m* is equal to three and is also repeating variable.

Choosing *D*, ω and ρ as repeating variables, we have

$$\pi_1 = D^{a_1} . \omega^{b_1} . \rho^{c_1} . \eta$$
$$\pi_2 = D^{a_2} . \omega^{b_2} . \rho^{c_2} . \mu$$
$$\pi_3 = D^{a_3} . \omega^{b_3} . \rho^{c_3} . Q$$

First π-term (π_1). $\pi_1 = D^{a_1} . \omega^{b_1} . \rho^{c_1} . \eta$

Substituting dimensions on both sides of π_1,

$$M^0L^0T^0 = L^{a_1} . (T^{-1})^{b_1} . (ML^{-3})^{c_1} . M^0L^0T^0$$

Equating the powers of *M*, *L*, *T* on both sides

Power of *M*, $0 = c_1 + 0$, $\therefore$ $c_1 = 0$

Power of *L*, $0 = a_1 + 0$, $\therefore$ $a_1 = 0$

Power of *T*, $0 = -b_1 + 0$, $\therefore$ $b_1 = 0$

Substituting the values of a_1, b_1 and c_1 in π_1, we get

$$\pi_1 = D^0\omega^0\rho^0 . \eta = \eta$$

[If a variable is dimensionless, it itself is a π-term. Here the variable η is a dimensionless and hence η is a π-term. As it exists in first π-term and hence $\pi_1 = \eta$. Then there is no need of equating the powers. Directly the value can be obtained.]

Second π-term. $\pi_2 = D^{a_2} . \omega^{b_2} . \rho^{c_2} . \mu$

Substituting the dimensions on both sides

$$M^0L^0T^0 = L^{a_2} . (T^{-1})^{b_2} . (ML^{-3})^{c_2} . ML^{-1}T^{-1}$$

Equating the powers of *M*, *L*, *T* on both sides

Power of M, $0 = c_2 + 1$, $\therefore$ $c_2 = -1$

Power of L, $0 = a_2 - 3c_2 - 1$, $\therefore$ $a_2 = 3c_2 + 1 = -3 + 1 = -2$

Power of T, $0 = -b_2 - 1$, $\therefore$ $b_2 = -1$

Substituting the values of a_2, b_2 and c_2 in π_2,

$$\pi_2 = D^{-2} \cdot \omega^{-1} \cdot \rho^{-1} \cdot \mu = \frac{\mu}{D^2 \omega \rho}$$

Third π-term. $\pi_3 = D^{a_3} \cdot \omega^{b_3} \cdot \rho^{c_3} \cdot Q$

Substituting the dimensions on both sides

$$M^0 L^0 T^0 = L^{a_3} \cdot (T^{-1})^{b_3} \cdot (ML^{-3})^{c_3} \cdot L^3 T^{-1}$$

Equating the powers of M, L and T on both sides

Power of M, $0 = c_3$, $\therefore$ $c_3 = 0$

Power of L, $0 = a_3 - 3c_3 + 3$, $\therefore$ $a_3 = 3c_3 - 3 = -3$

Power of T, $0 = -b_{3} - 1$, $\therefore$ $b_3 = -1$

Substituting the values of a_3, b_3 and c_3 in π_3,

$$\pi_3 = D^{-3} \cdot \omega^{-1} \cdot \rho^0 \cdot Q = \frac{Q}{D^2 \omega}$$

Substituting the values of π_1, π_2 and π_3 in equation (*ii*)

$$f_1\left(\eta, \frac{\mu}{D^2\omega\rho}, \frac{Q}{D^2\omega}\right) = 0 \text{ or } \eta = \phi\left[\frac{\mu}{\mathbf{D^2\omega\rho}}, \frac{\mathbf{Q}}{\mathbf{D^2\omega}}\right]. \textbf{ Ans.}$$

Problem 6.9 *Using Buckingham's π-theorem, show that the velocity through a circular orifice is given by* $V = \sqrt{2gH}\ \phi\left[\frac{D}{H}, \frac{\mu}{\rho VH}\right]$, *where H is the head causing flow, D is the diameter of the orifice, μ is co-efficient of viscosity, ρ is the mass density and g is the acceleration due to gravity.*

(A.M.I.E., Winter, 1977)

Solution. Given :

V is a function of H, D, μ, ρ and g

$\therefore$ $V = f(H, D, \mu, \rho, g)$ or $f_1(V, H, D, \mu, \rho, g) = 0$

$\therefore$ Total number of variable, $n = 6$ ...(*i*)

Writing dimension of each variable, we have

$$V = LT^{-1},\ H = L,\ D = L,\ \mu = ML^{-1}T^{-1},\ \rho = ML^{-3},\ g = LT^{-2}.$$

Thus number of fundamental dimensions, $m = 3$

$\therefore$ Number of π-terms $= n - m = 6 - 3 = 3$.

Equation (*i*) can be written as $f_1(\pi_1, \pi_2, \pi_3) = 0$ ...(*ii*)

Each π-term contains $m + 1$ variables, where $m = 3$ and is also equal to repeating variables. Here V is a dependent variable and hence should not be selected as repeating variable. Choosing H, g, ρ as repeating variable, we get three π-terms as

$$\pi_1 = H^{a_1} \cdot g^{b_1} \cdot \rho^{c_1} \cdot V$$

$$\pi_2 = H^{a_2} \cdot g^{b_2} \cdot \rho^{c_2} \cdot D$$

$$\pi_3 = H^{a_3} \cdot g^{b_3} \cdot \rho^{c_3} \cdot \mu$$

First π-term. $\pi_1 = H^{a_1} \cdot g^{b_1} \cdot \rho^{c_1} \cdot V$

Substituting dimensions on both sides

$$M^0L^0T^0 = L^{a_1} . (LT^{-2})^{b_1} . (MT^{-3})^{c_1} . (LT^{-1})$$

Equating the powers of M, L, T on both sides,

Power of M, $0 = c$, $\therefore$ $c_1 = 0$

Power of L, $0 = a_1 + b_1 - 3c_1 + 1$, $\therefore$ $a_1 = -b_1 + 3c_1 - 1 = \frac{1}{2} - 1 = -\frac{1}{2}$

Power of T, $0 = -2b_1 - 1$, $\therefore$ $b_1 = -\frac{1}{2}$

Substituting the values of a_1, b_1 and c_1 in π_1,

$$\pi_1 = H^{-\frac{1}{2}} . g^{-\frac{1}{2}} . \rho^0 . V = \frac{V}{\sqrt{gH}}.$$

Second π-term. $\pi_2 = H^{a_2} . g^{b_2} . \rho^{c_2} . D$

Substituting the dimensions on both sides,

$$M^0L^0T^0 = L^{a_2} . (LT^{-2})^{b_2} . (ML^{-3})^{c_2} . L$$

Equating the powers of M, L, T,

Power of M, $0 = c_2$ $\therefore$ $c_2 = 0$

Power of L, $0 = a_2 + b_2 - 3c_2 + 1$, $a_2 = -b_2 + 3c_2 - 1 = -1$

Power of T, $0 = -2b_2$, $\therefore$ $b_2 = 0$

Substituting the values of a_2, b_2, c_2 in π_2,

$$\pi_2 = H^{-1} . g^6\rho^0 . D = \frac{D}{H}.$$

Third π-term. $\pi_3 = H^{a_3} . g^{b_3} . \rho^{c_3} . \mu$

Substituting the dimensions on both sides

$$M^0L^0T^0 = L^{a_3} . (LT^{-2})^{b_3} . (ML^{-3})^{c_3} . ML^{-1}T^{-1}$$

Equating the powers of M, L, T on both sides

Power of M, $0 = c_3 + 1$, $\therefore$ $c_3 = -1$

Power of L, $0 = a_3 + b_3 - 3c_3 - 1$, $\therefore$ $a_3 = -b_3 + 3c_3 + 1 = \frac{1}{2} - 3 + 1 = -\frac{3}{2}$

Power of T, $0 = -2b_3 - 1$, $\therefore$ $b_3 = -\frac{1}{2}$

Substituting the values of a_3, b_3 and c_3 in π_3,

$$\pi_3 = H^{-3/2} . g^{-1/2} . \rho^{-1} . \mu = \frac{\mu}{H^{3/2}\sqrt{g\rho}}$$

$$= \frac{\mu}{H\rho\sqrt{gH}} = \frac{\mu V}{H\rho V\sqrt{gH}} \quad \text{[Multiply and Divide by } V\text{]}$$

$$= \frac{\mu}{H\rho V} . \pi_1 \qquad \left\{\because \frac{V}{\sqrt{gH}} = \pi_1\right\}$$

Substituting the values of π_1, π_2 and π_3 in equation (*ii*),

$$f_1\left(\frac{V}{\sqrt{gH}}, \frac{D}{H}, \pi_1 \frac{\mu}{H\rho V}\right) = 0 \text{ or } \frac{V}{\sqrt{gH}} = \phi\left[\frac{D}{H}, \pi_1 \frac{\mu}{H\rho V}\right]$$

or
$$V = \sqrt{2gH}\ \phi\left[\frac{D}{H}, \frac{\mu}{\rho VH}\right].\ \textbf{Ans.}$$

Multiplying by a constant does not change the character of π-terms.

Problem 6.10 *The pressure difference Δp in a pipe of diameter D and length l due to turbulent flow depends on the velocity V, viscosity μ, density ρ and roughness k. Using Buckingham's π-theorem, obtain an expression for Δp.* (Delhi University, November, 1982)

Solution. Given :

Δp is a function of D, l, V, μ, ρ, k

$\therefore$
$$\Delta p = f(D, l, V, \mu, \rho, k) \text{ or } f_1(\Delta p, D, l, V, \mu, \rho, k) = 0 \quad ...(i)$$

$\therefore$ Total number of variables, $n = 7$.

Writing dimensions of each variable,

Dimension of Δp = Dimension of pressure = $ML^{-1}T^{-2}$

$D = L,\ l = L,\ V = LT^{-1},\ \mu = ML^{-1}T^{-1},\ \rho = ML^{-3},\ k = L$

$\therefore$ Number of fundamental dimensions, $m = 3$

Number of π-terms $= n - m = 7 - 3 = 4$.

Now equation (i) can be grouped in 4 π-terms as

$$f_1(\pi_1, \pi_2, \pi_3, \pi_4) = 0 \quad ...(ii)$$

Each π-term contains $m + 1$ or $3 + 1 = 4$ variables. Out of four variables, three are repeating variables. Choosing D, V, ρ as the repeating variables, we have the four π-terms as

$$\pi_1 = D^{a_1} . V^{b_1} . \rho^{c_1} . \Delta p$$
$$\pi_2 = D^{a_2} . V^{b_2} . \rho^{c_2} . l$$
$$\pi_3 = D^{a_3} . V^{b_3} . \rho^{c_3} . \mu$$
$$\pi_4 = D^{a_4} . V^{b_4} . \rho^{c_4} . k$$

First π-term $\quad \pi_1 = D^{a_1} . V^{b_1} . \rho^{c_1} . \Delta p$

Substituting dimensions on both sides,

$$M^0L^0T^0 = L^{a_1} . (LT^{-1})^{b_1} . (ML^{-3})^{c_1} . ML^{-1}T^{-2}$$

Equating the powers of M, L, T on both sides,

Power of M, $\quad 0 = c_1 + 1 \quad \therefore\ c_1 = -1$

Power of L, $\quad 0 = a_1 + b_1 - 3c_1 - 1, \quad \therefore\ a_1 = -b_1 + 3c_1 + 1 = 2 - 3 + 1 = 0$

Power of T, $\quad 0 = -b_1 - 2, \quad \therefore\ b_1 = -2$

Substituting the values of a_1, b_1 and c_1 in π_1,

$$\pi_1 = D^0 . V^{-2} . \rho^{-1} . \Delta p = \frac{\Delta p}{\rho V^2}$$

Second π-term $\quad \pi_2 = D^{a_2} . V^{b_2} . \rho^{c_2} . l$

Substituting dimensions on both sides,

$$M^0L^0T^0 = L^{a_2} . (LT^{-1})^{b_2} . (ML^{-3})^{c_2} . L$$

Equating powers of M, L, T on both sides,

Power of M, $\quad 0 = c_2, \quad \therefore\ c_2 = 0$

Power of L, $\quad 0 = a_2 - b_2 - 3c_2 + 1, \quad \therefore\ a_2 = b_2 + 3c_2 - 1 = -1$

Power of T, $\quad 0 = -b_2, \quad \therefore\ b_2 = 0$

Substituting the values of a_2, b_2, c_2 in π_2,

$$\pi_2 = D^{-1} \,.\, V^0 \,.\, \rho^0 \,.\, l = \frac{l}{D}$$

Third π-term $\quad \pi_3 = D^{a_3} .\, V^{b_3} \,.\, \rho^{c_3} \,.\, \mu$

Substituting dimensions on both sides,

$$M^0L^0T^0 = L^{a_3} \,.\, (LT^{-1})^{b_3} \,.\, (ML^{-3})^{c_3} \,.\, ML^{-1}T^{-1}$$

Equating the powers of M, L, T on both sides,

Power of M, $\quad 0 = c_3 + 1, \quad \therefore \; c_3 = -1$

Power of L, $\quad 0 = a_3 + b_3 - 3c_3 - 1, \quad \therefore \; a_3 = -b_3 + 3c_3 + 1 = 1 - 3 + 1 = -1$

Power of T, $\quad 0 = -b_3 - 1, \quad \therefore \; b_3 = -1$

Substituting the values of a_3, b_3 and c_3 in π_3,

$$\pi_3 = D^{-1} \,.\, V^{-1} \,.\, \rho^{-1} \,.\, \pi = \frac{\mu}{DV\rho}.$$

Fourth π-term $\quad \pi_4 = D^{a_4} \,.\, V^{b_4} \,.\, \rho^{c_4} \,.\, k$

or $\quad M^0L^0T^0 = L^{a_4} \,.\, (LT^{-1})^{b_4} \,.\, (ML^{-3})^{c_4} \,.\, L \qquad$ {Dimension of $k = L$}

Equating the power of M, L, T on both sides,

Power of M, $\quad 0 = c_4, \quad \therefore \; c_4 = 0$

Power of L, $\quad 0 = a_4 - b_4 - 3c_4 + 1, \quad \therefore \; a_4 = b_4 + 3c_4 - 1 = -1$

Power of T, $\quad 0 = -b_4, \quad \therefore \; b_4 = 0$

Substituting the values of a_4, b_4, c_4 in π_4,

$$\pi_4 = D^{-1} \,.\, V^0 \,.\, \rho^0 \,.\, k = \frac{k}{D}$$

Substituting the values of π_1, π_2, π_3 and π_4 in (*ii*), we get

$$f_1\left(\frac{\Delta p}{\rho V^2}, \frac{1}{D}, \frac{\mu}{DV\rho}, \frac{k}{D}\right) = 0 \quad \text{or} \quad \frac{\Delta p}{\rho V^2} = \phi\left[\frac{\mathbf{1}}{\mathbf{D}}, \frac{\mu}{\mathbf{DV\rho}}, \frac{\mathbf{k}}{\mathbf{D}}\right]. \textbf{ Ans.}$$

Expression for h_f (Difference of pressure-head). From experiments, it was observed that pressure difference, Δp is a linear function of $\frac{l}{D}$ and hence it is taken out of function

$$\therefore \quad \frac{\Delta p}{\rho V^2} = \frac{l}{D}\phi\left[\frac{\mu}{DV\rho}, \frac{k}{D}\right]$$

$$\therefore \quad \frac{\Delta p}{\rho} = V^2 \,.\, \frac{l}{D}\phi\left[\frac{\mu}{DV\rho}, \frac{k}{D}\right]$$

Dividing by g to both sides, we have $\dfrac{\Delta p}{\rho g} = \dfrac{V^2 \,.\, l}{g \,.\, D}\phi\left[\dfrac{\mu}{DV\rho}, \dfrac{k}{D}\right].$

Now $\phi\left[\dfrac{\mu}{DV\rho}, \dfrac{k}{D}\right]$ contains two terms. First one is $\dfrac{\mu}{DV\rho}$ which is $\dfrac{1}{\text{Reynold number}}$ or $\dfrac{1}{R_e}$ and second one is $\dfrac{k}{D}$ which is called roughness factor. Now $\phi\left[\dfrac{1}{R_e}, \dfrac{k}{D}\right]$ is put equal to f, where f is the co-efficient of friction which is a function of Reynold number and roughness factor.

$$\therefore \qquad \frac{\Delta p}{\rho g} = \frac{4f}{2} \cdot \frac{V^2 l}{gD} \qquad \left\{\because \ f = \phi\left(\frac{\mu}{DV\rho}, \frac{K}{D}\right)\right\}$$

Multiplying or dividing by any constant does not change the character of π-terms.

$$\therefore \qquad \frac{\Delta p}{\rho g} = h_f = \frac{\mathbf{4f \cdot LV^2}}{\mathbf{D \times 2g}}. \textbf{ Ans.}$$

Problem 6.11 *The pressure difference Δp in a pipe of diameter D and length l due to viscous flow depends on the velocity V, viscosity μ and density ρ. Using Buckingham's π-theorem, obtain an expression for Δp.*

Solution. This problem is similar to problem 6.10. The only difference is that Δp is to be calculated for viscous flow. Then in the repeating variable instead of ρ, the fluid property μ is to be chosen.

Now Δp is a function of D, l, V, μ, ρ or $\Delta p = f(D, l, V, \mu, \rho)$

or $\qquad f_1 (\Delta p, D, l, V, \mu, \rho) = 0 \qquad$ ...(*i*)

Total number of variables, $n = 6$

Number of fundamental dimension, $m = 3$

Number of π-terms $= n - 3 = 6 - 3 = 3$

Hence equation (*i*) is written as $f_1(\pi_1, \pi_2, \pi_3) = 0 \qquad$ ...(*ii*)

Each π-term contains $m + 1$ variables, *i.e.*, $3 + 1 = 4$ variable. Out of four variables, three are repeating variables.

Choosing D, V, μ as repeating variables, we have π-terms as

$$\pi_1 = D^{a_1} \cdot V^{b_1} \cdot \mu^{c_1} \cdot \Delta p$$

$$\pi_2 = D^{a_2} \cdot V^{b_2} \cdot \mu^{c_2} \cdot l$$

$$\pi_3 = D^{a_3} \cdot V^{b_3} \cdot \mu^{c_3} \cdot \rho$$

First π-term $\qquad \pi_1 = D^{a_1} \cdot V^{b_1} \cdot \mu^{c_1} \cdot \Delta p$

Substituting the dimensions on both sides,

$$M^0L^0T^0 = L^{a_1} \cdot (LT^{-1})^{b_1} \cdot (ML^{-1}T^{-1})^{c_1} \cdot ML^{-1}T^{-2}$$

Equating the powers of M, L, T on both sides,

Power of M, $\qquad 0 = c_1 + 1, \qquad \therefore \ c_1 = -1$

Power of L, $\qquad 0 = a_1 + b_1 - c_1 - 1, \qquad \therefore \ a_1 = -b_1 + c_1 + 1 = 1 - 1 + 1 = 1$

Power of T, $\qquad 0 = -b_1 - c_1 - 2, \qquad \therefore \ b_1 = -c_1 - 2 = 1 - 2 = -1$

Substituting the values of a_1, b_1 and c_1 in π_1,

$$\pi_1 = D^1 \cdot V^{-1} \cdot \mu^{-1} \cdot \Delta p = \frac{D\Delta p}{\mu V}.$$

Second π-term $\qquad \pi_2 = D^{a_2} \cdot V^{b_2} \cdot \mu^{c_2} \cdot l$

Substituting the dimensions on both sides,

$$M^0L^0T^0 = L^{a_2} \cdot (LT^{-1})^{b_2} \cdot (ML^{-1}T^{-1})^{c_2} \cdot L.$$

Equating the powers of M, L, T on both sides

Power of M, $\qquad 0 = c_2, \qquad \therefore \ c_2 = 0$

Power of L, $\qquad 0 = a_2 + b_2 - c_2 + 1, \qquad \therefore \ a_2 = -b_2 + c_2 - 1 = -1$

Power of T, $\qquad 0 = -b_2 - c_2, \qquad \therefore \ b_2 = -c_2 = 0$

Substituting the values of a_2, b_2 and c_2 in π_2,

$$\pi_2 = D^{-1} . V^0 . \mu^0 . l = \frac{l}{D}.$$

Third π-term $\pi_3 = D^{a_3} . V^{b_3} . \mu^{c_3} . \rho$

Substituting the dimension on both sides,

$$M^0L^0T^0 = L^{a_3} . (LT^{-1})^{b_3} . (ML^{-1}T^{-1})^{c_3} . ML^{-3}.$$

Equating the powers of M, L, T on both sides

Power of M, $0 = c_3 + 1$, $\therefore$ $c_3 = -1$

Power of L, $0 = a_3 + b_3 - c_3 - 3$, $\therefore$ $a_3 = -b_3 + c_3 + 3 = -1 - 1 + 3 = 1$

Power of T, $0 = -b_3 - c_3$, $\therefore$ $b_3 = -c_3 = -(-1) = 1$

Substituting the values of a_3, b_3 and c_3 in π_3,

$$\pi_3 = D^1 . V^1 . \mu^{-1} . \rho = \frac{\rho DV}{\mu}.$$

Substituting the values of π_1, π_2 and π_3 in equation (*ii*),

$$f_1\left(\frac{D\Delta p}{\mu V}, \frac{l}{D}, \frac{\rho DV}{\mu}\right) = 0 \quad \text{or} \quad \frac{D\Delta p}{\mu V} = \phi\left[\frac{l}{D}, \frac{\rho DV}{\mu}\right] \quad \text{or} \quad \Delta p = \frac{\mu V}{D}\phi\left[\frac{l}{D}, \frac{\rho DV}{\mu}\right]$$

Experiments show that the pressure difference Δp is a linear function $\frac{l}{D}$. Hence $\frac{l}{D}$ can be taken out of the functional as

$$\Delta p = \frac{\mathbf{\mu V}}{\mathbf{D}} \times \frac{\mathbf{L}}{\mathbf{D}} \phi\left[\frac{\mathbf{\rho DV}}{\mu}\right]. \textbf{ Ans.}$$

Expression for difference of pressure head for viscous flow

$$h_f = \frac{\Delta p}{\rho g} = \frac{\mu V}{D} \times \frac{l}{D} \times \frac{1}{\rho g} \phi\,[R_e] \qquad \left\{\because \frac{\rho DV}{\mu} = R_e\right\}$$

$$= \frac{\mathbf{\mu VL}}{\mathbf{wD^2}} \phi\,[\mathbf{R_e}]. \textbf{ Ans.}$$

Problem 6.12 *Derive on the basis of dimensional analysis suitable parameters to present the thrust developed by a propeller. Assume that the thrust P depends upon the angular velocity ω, speed of advance V, diameter D, dynamic viscosity μ, mass density ρ, elasticity of the fluid medium which can be denoted by the speed of sound in the medium C.* (A.M.I.E., Winter, 1981)

Solution. Thrust P is a function of ω, V, D, μ, ρ, C

or $$P = f(\omega, V, D, \mu, \rho, C)$$

or $$f_1 = (P, \omega, V, D, \mu, \rho, C) = 0 \qquad \ldots(i)$$

$\therefore$ Total number of variables, $n = 7$

Writing dimensions of each variable, we have

$$P = MLT^{-2},\ \omega = T^{-1},\ V = LT^{-1},\ D = L,$$
$$\mu = ML^{-1}T^{-1},\ \rho = ML^{-3},\ C = LT^{-1}$$

$\therefore$ Number of fundamental dimensions, $m = 3$

$\therefore$ Number of π-terms $= n - m = 7 - 3 = 4$

Hence equation (*i*) can be written as $f_1(\pi_1, \pi_2, \pi_3, \pi_4) = 0$ ...(*ii*)

Each π-term contains $m + 1$, *i.e.*, $3 + 1 = 4$ variables. Out of four variables, three are repeating variables.

Choosing D, V, ρ as repeating variables, we get π-terms as

$$\pi_1 = D^{a_1} . V^{b_1} . \rho^{c_1} . P$$

$$\pi_2 = D^{a_2} . V^{b_2} . \rho^{c_2} . \omega$$

$$\pi_3 = D^{a_3} . V^{b_3} . \rho^{c_3} . \mu$$

$$\pi_4 = D^{a_4} . V^{b_4} . \rho^{c_4} . C$$

First π-term $\pi_1 = D^{a_1} . V^{b_1} . \rho^{c_1} . P$

Writing dimensions on both sides,

$$M^0L^0T^0 = L^{a_1} . (LT^{-1})^{b_1} . (ML^{-3})^{c_1} . MLT^{-2}.$$

Equating powers of M, L, T on both sides,

Power of M, $0 = c_1 + 1$, $\because$ $C_1 = -1$

Power of L, $0 = a_1 + b_1 - 3c_1 + 1$,

$a_1 = -b_1 + 3c_1 - 1 = 2 - 3 - 1 = -2$

Power of T, $0 = -b_1 - 2$, $\therefore$ $b_1 = -2$

Substituting the values of a_1, b_1 and c_1 in π_1,

$$\pi_1 = D^{-2} . V^{-2} . \rho^{-1} P = \frac{P}{D^2V^2\rho}.$$

Second π-term $\pi_2 = D^{a_2} . V^{b_2} . \Delta^{c_2} . \omega$

Writing dimensions on both sides,

$$M^0L^0T^0 = L^{a_2} . (LT^{-1})^{b_2} . (ML^{-3})^{c_2} . T^{-1}$$

Equating the powers of M, L, T on both sides,

Power of M, $0 = c_2$, $\therefore$ $c_2 = 0$

Power of L, $0 = a_2 + b_2 - 3c_2$, $\therefore$ $a_2 = -b_2 + 3c_2 = 1 + 0 = 1$

Power of T, $0 = -b_2 - 1$, $\therefore$ $b_2 = -1$

Substituting the values of a_2, b_2, c_2 in π_2,

$$\pi_2 = D^1 . V^{-1} . \rho^0 . \omega = \frac{D\omega}{V}.$$

Third π-term $\pi_3 = D^{a_3} . V^{b_3} . \rho^{c_3} . \mu.$

Writing dimensions on both sides,

$$M^0L^0T^0 = D^{a_3} . (LT^{-1})^{b_3} . (ML^{-3})^{c_3} . ML^{-1}T^{-1}.$$

Equating the powers of M, L, T on both sides,

Power of M, $0 = c_3 + 1$, $\therefore$ $c_3 = -1$

Power of L, $0 = a_3 + b_3 - 3c_3 - 1$, $\therefore$ $a_3 = -b_3 + 3c_3 + 1 = 1 - 3 + 1 = -1$

Power of T, $0 = -b_3 - 1$ $\therefore$ $b_3 = -1$

Substituting the values of a_3, b_3 and c_3 in π_3,

$$\pi_3 = D^{-1} . V^{-1} . \rho^{-1} . \mu = \frac{\mu}{DV\rho}.$$

Fourth π-term $\pi_4 = D^{a_4} . V^{b_4} . \rho^{c_4} C.$

Substituting dimensions on both sides,

$$M^0L^0T^0 = L^{a_4} . (LT^{-1})^{b_4} . (ML^{-3})^{c_4} . LT^{-1}$$

Equating the powers of M, L, T on both sides,

Power of M, $\quad 0 = c_4$, $\quad \therefore \quad c_4 = 0$

Power of L, $\quad 0 = a_4 + b_4 - 3c_4 + 1$, $\quad \therefore \quad a_4 = -b_4 + 3c_4 - 1 = 1 + 0 - 1 = 0$

Power of T, $\quad 0 = -a_4 - 1$, $\quad \therefore \quad b_4 = -1$

Substituting the values of a_4, b_4 and c_4 in π_4,

$$\pi_4 = D^0 . V^{-1} . \rho^0 . C = \frac{C}{V}.$$

Substituting the values of π_1, π_2, π_3 and π_4 in equation (*ii*),

$$f_1\left(\frac{P}{D^2V^2\rho}, \frac{D\omega}{V}, \frac{\mu}{DV\rho}, \frac{C}{V}\right) = 0 \quad \text{or} \quad \frac{P}{D^2V^2\rho} = f_2\left(\frac{D\omega}{V}, \frac{\mu}{DV\rho}, \frac{C}{V}\right)$$

or
$$\mathbf{P = D^2V^2\rho\, f_2\left(\frac{D\omega}{V}, \frac{\mu}{DV\rho}, \frac{C}{V}\right). \text{ Ans.}}$$

Problem 6.13 *The frictional torque T of a disc of diameter D rotating at a speed N in a fluid of viscosity μ and density ρ in a turbulent flow is given by* $T = D^5N^2\rho\phi\left[\frac{\mu}{D^2N\rho}\right]$.

Prove this by the method of dimensions. (A.M.I.E., May, 1975)

Solution. Given : $T = f(D, N, \mu, \rho)$ or $f_1(T, D, N, \mu, \rho) = 0$ ...(*i*)

$\therefore$ Total number of variable, $n = 5$

Dimensions of each variable are expressed as

$$T = ML^2T^{-2},\ D = L,\ N = T^{-1},\ \mu = ML^{-1}T^{-1},\ \rho = ML^{-3}$$

$\therefore$ Number of fundamental dimensions, $m = 3$

Number of π-terms $= n - m = 5 - 3 = 2$

Hence equation (*i*) can be written as $f_1(\pi_1, \pi_2) = 0$ ...(*ii*)

Each π-term contains $m + 1$ variable, *i.e.*, $3 + 1 = 4$ variables. Three variables are repeating variables.

Choosing D, N, ρ as repeating variables, the π-terms are

$$\pi_1 = D^{a_1}. N^{b_1}. \rho^{c_1}. T$$
$$\pi_2 = D^{a_2}. N^{b_2}. \rho^{c_2} . \mu$$

Dimensional Analysis of π_1

$$\pi_1 = D^{a_1} . N^{b_1} . \rho^{c_1} . T$$

Substituting dimensions on both sides,

$$M^0L^0T^0 = L^{a_1} . (T^{-1})^{b_1}\ (ML^{-3})^{c_1} . ML^2T^{-2}.$$

Equating the powers of M, L, T on both sides,

Power of M, $\quad 0 = c_1 + 1$, $\quad \therefore \quad c_1 = -1$

Power of L, $\quad 0 = a_1 - 3c_1 + 2$, $\quad \therefore \quad a_1 = 3c_1 - 2 = -3 - 2 = -5$

Power of T, $\quad 0 = -b_1 - 2$, $\quad \therefore \quad b_1 = -2$

Substituting the values of a_1, b_1, c_1 in π,

$$\pi_1 = D^{-5} . N^{-2} . \rho^{-1} . T = \frac{T}{D^2N^2\rho}.$$

Dimensional Analysis of π_2

$$\pi_2 = D^{a_2} . N^{b_2} . \rho^{c_2} . \mu$$

Substituting dimension on both sides,

$$M^0L^0T^0 = L^{a_2} . (T^{-1})^{b_2} . (ML^{-3})^{c_2} . ML^{-1}T^{-1}.$$

Equating the powers of M, L, T on both sides,

Power of M, $0 = c_2 + 1$, $\therefore$ $c_2 = -1$

Power of L, $0 = a_2 - 3c_2 - 1$, $\therefore$ $a_2 = 3c_2 + 1 = -3 + 1 = -2$

Power of T, $0 = -b_2 - 1$, $\therefore$ $b_2 = -1$

Substituting the values of a_2, b_2 and c_2 in π_2,

$$\pi_2 = D^{-2}N^{-1}\rho^{-1} . \mu = \frac{\mu}{D^2N\rho}.$$

Substituting the values of π_1 and π_2 in equation (*ii*),

$$f_1\left(\frac{T}{D^2N^2\rho}, \frac{\mu}{D^2N\rho}\right) = 0 \quad \text{or} \quad \frac{T}{D^5N^2\rho} = \phi\left(\frac{\mu}{D^2N\rho}\right)$$

or

$$T = \mathbf{D^5N^2\,\rho\,\phi}\left[\frac{\mu}{\mathbf{D^2N\rho}}\right]. \textbf{ Ans.}$$

Problem 6.14 *Using Buckingham's π-theorem, shown that the discharge Q consumed by an oil ring is given by*

$$Q = Nd^3\phi\left[\frac{\mu}{\rho Nd^2}, \frac{\sigma}{\rho N^2d^3}, \frac{w}{\rho N^2d}\right]$$

where d is the internal diameter of the ring, N is rotational speed, ρ is density, μ is viscosity, σ is surface tension and w is the specific weight of oil.

Solution. Given : $Q = f(d, N, \rho, \mu, \sigma, w)$ or $f_1(Q, d, N, \rho, \mu, \sigma, w) = 0$ ...(*i*)

$\therefore$ Total number of variables, $n = 7$

Dimensions of each variables are

$$Q = L^3T^{-1},\ d = L,\ N = T^{-1},\ \rho = ML^{-3},\ \mu = ML^{-1}T^{-1},\ \sigma = MT^{-2}$$

and $w = ML^{-2}T^{-2}$

$\therefore$ Total number of fundamental dimensions, $m = 3$

$\therefore$ Total number of π-terms $= n - m = 7 - 3 = 4$

$\therefore$ Equation (*i*) becomes as $f_1(\pi_1, \pi_2, \pi_3, \pi_4) = 0$ ...(*ii*)

Choosing d, N, ρ as repeating variables, the π-terms are

$$\pi_1 = d^{a_1} . N^{b_1} . \rho^{c_1} . Q$$
$$\pi_2 = d^{a_2} . N^{b_2} . \rho^{c_2} . \mu$$
$$\pi_3 = d^{a_3} . N^{b_3} . \rho^{c_3} . \sigma$$
$$\pi_4 = d^{a_4} . N^{b_4} . \rho^{c_4} . w.$$

First π-term $\pi_1 = d^{a_1}. N^{b_1}. \rho^{c_1}. Q.$

Substituting dimensions on both sides,

$$M^0L^0T^0 = L^{a_1} . (T^{-1})^{b_1} . (ML^{-3})^{c_1} . L^3T^{-1}.$$

Equating the powers of M, L, T on both sides,

Power of M, $0 = c_1$, $\therefore$ $c_1 = 0$

Power of L, $0 = a_1 - 3c_1 + 3$, $\therefore$ $a_1 = 3c_1 - 3 = 0 - 3 = -3$

Power of T, $0 = -b_1 - 1$, $\therefore$ $b_1 = -1$

Substituting a_1, b_1, c_1 in π_1, $\pi_1 = d^{-3} . N^{-1} . \rho^0 . Q = \dfrac{Q}{d^3N}$

Second π-term $\pi_2 = d^{a_2} . N^{b_2} . \rho^{c_2} . \mu.$

Substituting the dimensions on both sides,

$$M^0L^0T^0 = L^{a_2} . (T^{-1})^{b_2} . (ML^{-3})^{c_2} . ML^{-1}T^{-1}$$

Equating the powers of M, L, T on both sides,

Power of M, $0 = c_2 + 1, \quad \therefore \quad c_2 = -1$

Power of L, $0 = a_2 - 3c_2 - 1, \quad \therefore \quad c_2 = -1$

$\therefore \quad a_2 = 3c_2 + 1 = -3 + 1 = -2$

Power of T, $0 = -b_2 - 1, \quad \therefore \quad b_2 = -1$

Substituting the values of a_2, b_2, c_2 in π_2,

$$\pi_2 = d^{-2} . N^{-1} . \rho^{-1} . \mu = \frac{\mu}{d^2 N\rho} \text{ or } \frac{\mu}{\rho N d^2}.$$

Third π-term $\pi_3 = d^{a_3} . N^{b_3} . \rho^{c_3} . \sigma.$

Substituting dimensions on both sides,

$$M^0L^0T^0 = L^{a_3} . (T^{-1})^{b_3} . (ML^{-3})^{c_3} . MT^{-2}.$$

Equating the powers of M, L, T on the sides,

Power of M, $0 = c_3 + 1, \quad \therefore \quad c_3 = -1$

Power of L, $0 = a_3 - 3c_3, \quad \therefore \quad a_3 = 3c_3 = -3$

Power of T, $0 = -b_3 - 2, \quad \therefore \quad b_3 = -2$

Substituting the values of a_3, b_3, c_3 in π_3,

$$\pi_3 = d^{-3} . N^{-2} . \rho^{-1} . \sigma = \frac{\sigma}{d^3 N^2 \rho}.$$

Fourth π-term $\pi_4 = d^{a_4} . N^{b_4} . \rho^{c_4} . w$

Substituting dimensions on both sides,

$$M^0L^0T^0 = L^{a_4} . (T^{-1})^{b_4} . (ML^{-3})^{c_4} . ML^{-2}T^{-2}.$$

Equating the powers of M, L, T on both sides,

Power of M, $0 = c_4 + 1, \quad \therefore \quad c_4 = -1$

Power of L, $0 = a_4 - 3c_4 - 2, \quad \therefore \quad a_4 = 3c_4 + 2 = -3 + 2 = -1$

Power of T, $0 = -b_4 - 2, \quad \therefore \quad b_4 = -2$

Substituting the values of a_4, b_4 and c_4 in π_4,

$$\pi_4 = d^{-1} . N^{-2} . \rho^{-1} . w = \frac{w}{dN^2\rho}.$$

Now substituting the values of π_1, π_2, π_3, π_4 in (*ii*),

$$f\left(\frac{Q}{d^3N}, \frac{\mu}{\rho N d^2}, \frac{\sigma}{d^3N^2\rho}, \frac{w}{dN^2\rho}\right) = 0 \text{ or } \frac{Q}{d^3N} = f_1\left[\frac{\mu}{\rho N d^2}, \frac{\sigma}{d^3N^2\rho}, \frac{w}{dN^2\rho}\right]$$

or

$$\mathbf{Q = d^3N\phi\left[\frac{\mu}{\rho Nd^2}, \frac{\sigma}{d^3N^2d}, \frac{w}{dN^2\rho}\right]. \text{ Ans.}}$$

▶ 6.6 MODEL ANALYSIS

For predicting the performance of the hydraulic structures (such as dams, spill ways etc.) or hydraualic machines (such as turbines, pumps etc.), before actually constructing or manufacturing, models of the structures or machines are made and tests are performed on them to obtain the desired information.

The **model** is the small scale replica of the actual structure or machine. The actual structure or machine is called **Prototype**. It is not necessary that the models should be smaller than the prototypes (though in most of cases it is), they may be larger than the prototype. The study of models of actual machines is called **Model analysis**. Model analysis is actually an experimental method of finding solutions of complex flow problems. Exact analytical solutions are possible only for a limited number of flow problems. The followings are the advantages of the dimensional and model analysis :

1. The performance of the hydraulic structure or hydraulic machine can be easily predicted, in advance, from its model.
2. With the help of dimensional analysis, a relationship between the variables influencing a flow problem in terms of dimensionless parameters is obtained. This relationship helps in conducting tests on the model.
3. The merits of alternative designs can be predicted with the help of model testing. The most economical and safe design may be, finally, adopted.
4. The tests performed on the models can be utilized for obtaining, in advance, useful information about the performance of the prototypes only if a complete similarity exists between the model and the prototype.

▶ 6.7 SIMILITUDE-TYPES OF SIMILARITIES

Similitude is defined as the similarity between the model and its prototype in every respect, which means that the model and prototype have similar properties or model and prototype are completely similar. Three types of similarities must exist between the model and prototype. They are

1. Geometric Similarity, 2. Kinematic Similarity, and 3. Dynamic Similarity.

1. Geometric Similarity. The geometric similarity is said to exist between the model and the prototype. The ratio of all corresponding linear dimension in the model and prototype are equal.

Let L_m = Length of model, b_m = Breadth of model,

D_m = Diameter of model, A_m = Area of model,

$\forall_m$ = Volume of model,

and $L_P, B_P, D_P, A_P, \forall_P$ = Corresponding values of the prototype.

For geometric similarity between model and prototype, we must have the relation,

$$\frac{L_P}{L_m} = \frac{b_P}{b_m} = \frac{D_P}{D_m} = L_r \qquad \text{...(6.6)}$$

where L_r is called the scale ratio.

For area's ratio and volume's ratio the relation should be as given below :

$$\frac{A_P}{A_m} = \frac{L_P \times b_P}{L_m \times b_m} = L_r \times L_r = L_r^{\,2} \qquad \text{...(6.7)}$$

and

$$\frac{\forall_P}{\forall_m} = \left(\frac{L_P}{L_m}\right)^3 = \left(\frac{b_P}{b_m}\right)^3 = \left(\frac{D_P}{D_m}\right)^3 \qquad \text{...(6.8)}$$

2. Kinematic Similarity. Kinematic similarity means the similarity of motion between model and prototype. Thus kinematic similarity is said to exist between the model and the prototype if the ratios of the velocity and acceleration at the corresponding points in the model and at the corresponding points in the prototype are the same. Since velocity and acceleration are vector quantities, hence not only the ratio of magnitude of velocity and acceleration at the corresponding points in model and prototype should be same ; but the directions of velocity and accelerations at the corresponding points in the model and prototype also should be parallel.

Let V_{P_1} = Velocity of fluid at point 1 in prototype,

V_{P_2} = Velocity of fluid at point 2 in prototype,

a_{P_1} = Acceleration of fluid at 1 in prototype,

a_{P_2} = Acceleration of fluid at 2 in prototype, and

$V_{m_1}, V_{m_2}, a_{m_1}, a_{m_2}$ = Corresponding values at the corresponding points of fluid velocity and acceleration in the model.

For kinematic similarity, we must have

$$\frac{V_{P_1}}{V_{m_1}} = \frac{V_{P_2}}{V_{m_2}} = V_r \qquad ...(6.9)$$

where V_r is the velocity ratio.

For acceleration, we must have $\dfrac{a_{P_1}}{a_{m_1}} = \dfrac{a_{P_2}}{a_{m_2}} = a_r$ ...(6.10)

where a_r is the acceleration ratio.

Also the directions of the velocities in the model and prototype should be same.

3. Dynamic Similarity. Dynamic similarity means the similarity of forces between the model and prototype. Thus dynamic similarity is said to exist between the model and the prototype if the ratios of the corresponding forces acting at the corresponding points are equal. Also the directions of the corresponding forces at the corresponding points should be same.

Let $(F_i)_P$ = Inertia force at a point in prototype,

$(F_v)_P$ = Viscous force at the point in prototype,

$(F_g)_P$ = Gravity force at the point in prototype,

and $(F_i)_m$ $(F_v)_m$, $(F_g)_m$ = Corresponding values of forces at the corresponding point in model.

Then for dynamic similarity, we have

$$\frac{(F_i)_P}{(F_i)_m} = \frac{(F_v)_P}{(F_v)_m} = \frac{(F_g)_P}{(F_g)_m} \ldots = F_r, \text{ where } F_r \text{ is the force ratio.}$$

Also the directions of the corresponding forces at the corresponding points in the model and prototype should be same.

▶ 6.8 TYPES OF FORCES ACTING IN MOVING FLUID

For the fluid flow problems, the forces acting on a fluid mass may be any one, or a combination of the several of the following forces :

1. Inertia force, F_i.
2. Viscous force, F_v.
3. Gravity force, F_g.
4. Pressure force, F_p.
5. Surface tension force, F_s.
6. Elastic force, F_e.

1. **Inertia Force** (F_i). It is equal to the product of mass and acceleration of the flowing fluid and acts in the direction opposite to the direction of acceleration. It is always existing in the fluid flow problems.

2. **Viscous Force** (F_v). It is equal to the product of shear stress (τ) due to viscosity and surface area of the flow. It is present in fluid flow problems where viscosity is having an important role to play.

3. **Gravity Force** (F_g). It is equal to the product of mass and acceleration due to gravity of the flowing fluid. It is present in case of open surface flow.

4. **Pressure Force** (F_p). It is equal to the product of pressure intensity and cross-sectional area of the flowing fluid. It is present in case of pipe-flow.

5. **Surface Tension Force** (F_s). It is equal to the product of surface tension and length of surface of the flowing fluid.

6. **Elastic Force** (F_e). It is equal to the product of elastic stress and area of the flowing fluid.

For a flowing fluid, the above-mentioned forces may not always be present. And also the forces, which are present in a fluid flow problem, are not of equal magnitude. There are always one or two forces which dominate the other forces. These dominating forces govern the flow of fluid.

▶ 6.9 DIMENSIONLESS NUMBERS

Dimensionless numbers are those numbers which are obtained by dividing the inertia force by viscous force or gravity force or pressure force or surface tension force or elastic force. As this is a ratio of one force to the other force, it will be a dimensionless number. These dimensionless numbers are also called non-dimensional parameters. The followings are the important dimensionless numbers :

1. Reynold's number,
2. Froude's number,
3. Euler's number,
4. Weber's number,
5. Mach's number.

6.9.1 Reynold's Number (R_e).

It is defined as the ratio of inertia force of a flowing fluid and the viscous force of the fluid. The expression for Reynold's number is obtained as

Inertia force (F_i) = Mass × Acceleration of flowing fluid

$$= \rho \times \text{Volume} \times \frac{\text{Velocity}}{\text{Time}} = \rho \times \frac{\text{Volume}}{\text{Time}} \times \text{Velocity}$$

$$= \rho \times AV \times V \quad \{\because \text{ Volume per sec} = \text{Area} \times \text{Velocity} = A \times V\}$$

$$= \rho A V^2 \qquad \text{...(6.11)}$$

Viscous force (F_v) = Shear stress × Area $\quad \left\{\because \tau = \mu \dfrac{du}{dy} \quad \therefore \text{ Force} = \tau \times \text{Area}\right\}$

$$= \tau \times A$$

$$= \left(\mu \frac{du}{dy}\right) \times A = \mu \,.\, \frac{V}{L} \times A \qquad \left\{\because \frac{du}{dy} = \frac{V}{L}\right\}$$

By definition, Reynold's number,

$$R_e = \frac{F_i}{F_v} = \frac{\rho A V^2}{\mu \,.\, \dfrac{V}{L} \times A} = \frac{\rho V L}{\mu}$$

$$= \frac{V \times L}{(\mu / \rho)} = \frac{V \times L}{\nu} \qquad \left\{\because \frac{\mu}{\rho} = \nu = \text{Kinematic viscosity}\right\}$$

In case of pipe flow, the linear dimension L is taken as diameter, d. Hence Reynold's number for pipe flow,

$$R_e = \frac{V \times d}{\nu} \quad \text{or} \quad \frac{\rho V d}{\mu}. \qquad \text{...(6.12)}$$

6.9.2 Froude's Number (F_e). The Froude's number is defined as the square root of the ratio of inertia force of a flowing fluid to the gravity force. Mathematically, it is expressed as

$$F_e = \sqrt{\frac{F_i}{F_g}}$$

where F_i from equation (6.11) $= \rho A V^2$

and F_g = Force due to gravity

= Mass × Acceleration due to gravity

$= \rho \times \text{Volume} \times g = \rho \times L^3 \times g$ $\{\because \ \text{Volume} = L^3\}$

$= \rho \times L^2 \times L \times g = \rho \times A \times L \times g$ $\{\because L^2 = A = \text{Area}\}$

$$\therefore \quad F_e = \sqrt{\frac{F_i}{F_g}} = \sqrt{\frac{\rho A V^2}{\rho A L g}} = \sqrt{\frac{V^2}{Lg}} = \frac{V}{\sqrt{Lg}} \qquad \text{...(6.13)}$$

6.9.3 Euler's Number (E_u). It is defined as the square root of the ratio of the inertia force of a flowing fluid to the pressure force. Mathematically, it is expressed as

$$E_u = \sqrt{\frac{F_i}{F_P}}$$

where F_P = Intensity of pressure × Area $= p \times A$

and $F_i = \rho A V^2$

$$\therefore \quad E_u = \sqrt{\frac{\rho A V^2}{p \times A}} = \sqrt{\frac{V^2}{p/\rho}} = \frac{V}{\sqrt{p/\rho}} \qquad \text{...(6.14)}$$

6.9.4 Weber's Number (W_e). It is defined as the square root of the ratio of the inertia force of a flowing fluid to the surface tension force. Mathematically, it is expressed as

Weber's Number, $W_e = \sqrt{\frac{F_i}{F_s}}$

where F_i = Inertia force $= \rho A V^2$

and F_s = Surface tension force

= Surface tension per unit length × Length $= \sigma \times L$

$$\therefore \quad W_e = \sqrt{\frac{\rho A V^2}{\sigma \times L}} = \sqrt{\frac{\rho \times L^2 \times V^2}{\sigma \times L}} \qquad \{\because \ A = L^2\}$$

$$= \sqrt{\frac{\rho L \times V^2}{\sigma}} = \sqrt{\frac{V^2}{\sigma / \rho L}} = \frac{V}{\sqrt{\sigma / \rho L}}. \qquad \text{...(6.15)}$$

6.9.5 Mach's Number (M). Mach's number is defined as the square root of the ratio of the inertia force of a flowing fluid to the elastic force. Mathematically, it is defined as

$$M = \sqrt{\frac{\text{Inertia force}}{\text{Elastic force}}} = \sqrt{\frac{F_i}{F_e}}$$

where $F_i = \rho A V^2$

and F_e = Elastic force = Elastic stress × Area

$= K \times A = K \times L^2$ {∵ K = Elastic stress}

$$\therefore \quad M = \sqrt{\frac{\rho A V^2}{K \times L^2}} = \sqrt{\frac{\rho \times L^2 \times V^2}{K \times L^2}} = \sqrt{\frac{V^2}{K/\rho}} = \frac{V}{\sqrt{K/\rho}}$$

But $\sqrt{\dfrac{K}{\rho}} = C$ = Velocity of sound in the fluid

$$\therefore \quad M = \frac{V}{C}. \quad \text{...(6.16)}$$

▶ 6.10 MODEL LAWS OR SIMILARITY LAWS

For the dynamic similarity between the model and the prototype the ratio of the corresponding forces acting at the corresponding points in the model and prototype should be equal. The ratio of the forces are dimensionless numbers. It means for dynamic similarity between the model and prototype, the dimensionless numbers should be same for model and the prototype. But it is quite difficult to satisfy the condition that all the dimensionless numbers (*i.e.*, R_e, F_e, W_e, E_u and M) are the same for the model and prototype. Hence models are designed on the basis of ratio of the force, which is dominating in the phenomenon. The laws on which the models are designed for dynamic similarity are called model laws or laws of similarity. The followings are the model laws :

1. Reynold's model law,
2. Froude model law,
3. Euler model law,
4. Weber model law,
5. Mach model law.

6.10.1 Reynold's Model Law.

Reynold's model law is the law in which models are based on Reynold's number. Models based on Reynold's number includes :

(*i*) Pipe flow

(*ii*) Resistance experienced by sub-marines, airplanes, fully immersed bodies etc.

As defined earlier that Reynold number is the ratio of inertia force and viscous force, and hence fluid flow problems where viscous forces alone are predominent, the models are designed for dynamic similarity on Reynolds law, which states that the Reynold number for the model must be equal to the Reynold number for the prototype.

Let V_m = Velocity of fluid in model,

ρ_m = Density of fluid in model,

L_m = Length or linear dimension of the model,

μ_m = Viscosity or fluid in model,

and V_P, ρ_P, L_P and μ_P are the corresponding values of velocity, density, linear dimension and viscosity of fluid in prototype. Then according to Reynold's model law,

$$[R_e]_m = [R_e]_P \text{ or } \frac{\rho_m V_m L_m}{\mu_m} = \frac{\rho_P V_P L_P}{\mu_P} \quad \text{...(6.17)}$$

or $$\frac{\rho_P \,.\, V_P \,.\, L_P}{\rho_m \,.\, V_m \,.\, L_m} \times \frac{1}{\frac{\mu_P}{\mu_m}} = 1 \quad \text{or} \quad \frac{\rho_r \,.\, V_r \,.\, L_r}{\mu_r} = 1$$

$$\left\{ \text{where} \quad \rho_r = \frac{\rho_P}{\rho_m},\; V_r = \frac{V_P}{V_m} \text{ and } L_r = \frac{L_P}{L_m},\; \frac{\mu_P}{\mu_m} = \mu_r \right\}$$

And also ρ_r, V_r, L_r and μ_r are called the scale ratios for density, velocity, linear dimension and viscosity.

The scale ratios for time, acceleration, force and discharge for Reynold's model law are obtained as

$$t_r = \text{Time scale ratio} = \frac{L_r}{V_r} \qquad \left\{ \because \quad V = \frac{L}{t} \quad \therefore \quad t = \frac{V}{L} \right\}$$

$$a_r = \text{Acceleration scale ratio} = \frac{V_r}{t_r}$$

$$F_r = \text{Force scale ratio} = (\text{Mass} \times \text{Acceleration})_r$$

$$= m_r \times a_r = \rho_r A_r\, V_r \times a_r \qquad \{A_r = \text{Area ratio}\}$$

$$= \rho_r L_r^2 V_r \times a_r$$

$$Q_r = \text{Discharge scale ratio} = (\rho A V)_r$$

$$= \rho_r A_r\, V_r = \rho_r : L_r^2 \,.\, V_r.$$

Problem 6.15 *A pipe of diameter 1.5 m is required to transport an oil of sp. gr. 0.90 and viscosity 3×10^{-2} poise at the rate of 3000 litre/s. Tests were conducted on a 15 cm diameter pipe using water at 20°C. Find the velocity and rate of flow in the model. Viscosity of water at 20°C = 0.01 poise.*

(Delhi University, 1992)

Solution. Given :

Dia. of prototype, $D_P = 1.5$ m

Viscosity of fluid, $\mu_P = 3 \times 10^{-2}$ poise

Q for prototype, $Q_P = 3000$ lit/s $= 3.0$ m^3/s

Sp. gr. of oil, $S_P = 0.9$

∴ Density of oil, $\rho_P = S_P \times 1000 = 0.9 \times 1000 = 900$ kg/m^3

Dia. of the model, $D_m = 15$ cm $= 0.15$ m

Viscosity of water at 20°C $= .01$ poise $= 1 \times 10^{-2}$ poise or $\mu_m = 1 \times 10^{-2}$ poise

Density of water or $\rho_m = 1000$ kg/m^3.

For pipe flow, the dynamic similarity will be obtained if the Reynold's number in the model and prototype are equal

Hence using equation (6.17), $$\frac{\rho_m V_m D_m}{\mu_m} = \frac{\rho_P V_P D_P}{\mu_P} \qquad \{\text{For pipe, linear dimension is } D\}$$

$$\therefore \qquad \frac{V_m}{V_P} = \frac{\rho_P}{\rho_m} \cdot \frac{D_P}{D_m} \cdot \frac{\mu_P}{\mu_m}$$

$$= \frac{900}{1000} \times \frac{1.5}{0.15} \times \frac{1 \times 10^{-2}}{3 \times 10^{-2}} = \frac{900}{1000} \times 10 \times \frac{1}{3} = 3.0$$

But $$V_P = \frac{\text{Rate of flow in prototype}}{\text{Area of prototype}} = \frac{3.0}{\frac{\pi}{4}(D_P)^2} = \frac{3.0}{\frac{\pi}{4}(1.5)^2}$$

$$= \frac{3.0 \times 4}{\pi \times 2.25} = 1.697 \text{ m/s}$$

$$\therefore \quad V_m = 3.0 \times V_P = 3.0 \times 1.697 = \mathbf{5.091 \text{ m/s. Ans.}}$$

Rate of flow through model, $$Q_m = A_m \times V_m = \frac{\pi}{4}(D_m)^2 \times V_m = \frac{\pi}{4}(0.15)^2 \times 5.091 \text{ m}^3\text{/s}$$

$$= 0.0899 \text{ m}^3\text{/s} = 0.0899 \times 1000 \text{ lit/s} = \mathbf{89.9 \text{ lit/s. Ans.}}$$

Problem 6.16 *Water is flowing through a pipe of diameter 30 cm at a velocity of 4 m/s. Find the velocity of oil flowing in another pipe of diameter 10 cm, if the condition of dynamic similarity is satisfied between the two pipes. The viscosity of water and oil is given as 0.01 poise and .025 poise. The sp. gr. of oil = 0.8.*

Solution. Given :

Two pipes having different liquids.

Let for pipe 1, Liquid = Water

Dia. of pipe, $d_1 = 30 \text{ cm} = 0.30 \text{ m}$

Velocity of flow, $V_1 = 4 \text{ m/s}$

Viscosity, $\mu_1 = 0.01 \text{ poise} = \frac{0.01}{10}$ (S.I. Units)

Density, $\rho_1 = 1000 \text{ kg/m}^3$

For pipe 2, Liquid = Oil

Dia. of pipe, $d_2 = 10 \text{ cm} = 0.1 \text{ m}$

Velocity of flow, $V_2 = ?$

Viscosity, $\mu_2 = 0.025 \text{ poise} = \frac{0.025}{10}$ (S.I. Units)

Sp. gr. of oil $= 0.8$

∴ Density, $\rho_2 = 0.8 \times 1000 = 800 \text{ kg/m}^3$

If the pipes are dynamically similar, the Reynold's number for both the pipes should be same.

$$\therefore \quad \frac{\rho_1 V_1 d_1}{\mu_1} = \frac{\rho_2 V_2 d_2}{\mu_2} \quad \text{or} \quad V_2 = \frac{\rho_1}{\rho_2} \cdot \frac{d_1}{d_2} \cdot \frac{\mu_2}{\mu_1} V_1$$

$$= \frac{1000}{800} \times \frac{0.30}{0.10} \times \frac{\frac{.025}{10}}{\frac{.01}{10}} \times 4.0 = \frac{1000}{800} \times 3 \times \frac{.025}{.01} \times 4.0$$

$$= \mathbf{37.5 \text{ m/s. Ans.}}$$

Problem 6.17 *The ratio of lengths of a sub-marine and its model is 30 : 1. The speed of sub-marine (prototype) is 10 m/s. The model is to be tested in a wind tunnel. Find the speed of air in wind tunnel. Also determine the ratio of the drag (resistance) between the model and its prototype. Take the value of kinematic viscosities for sea water and air as .012 stokes and .016 stokes respectively. The density for sea-water and air is given as 1030 kg/m³ and 1.24 kg/m³ respectively.*

Solution. Given :

Prototype (sub-marine) and its model.

For prototype, Speed $V_P = 10$ m/s

Fluid = Sea - water

Kinematic viscosity, $\nu_P = 0.012$ stokes $= .012$ cm²/s

$= .012 \times 10^{-4}$ m²/s $\{\because$ Stoke = cm²/s$\}$

Density, $\rho_P = 1030$ kg/m³

For model Fluid = Air

Kinematic viscosity, $\nu_m = 0.016$ stokes $= 0.016$ cm²/s $= .016 \times 10^{-4}$ m²/s

Density, $\rho_m = 1.24$ kg/m³

Also $\frac{\text{Length of prototype}}{\text{Length of model}} = \frac{L_P}{L_m} = 30.0$

Let the velocity of air in model $= V_m$.

For dynamic similarity between model and sub-marine, the viscous resistance is to be overcame and hence for fully submerged sub-marine, the Reynold's number for model and prototype should be same.

$$\therefore \quad \frac{\rho_P V_P D_P}{\mu_P} = \frac{\rho_m V_m D_m}{\mu_m} \quad \text{or} \quad \frac{V_P D_P}{(\mu/\rho)_P} = \frac{V_m D_m}{(\mu/\rho)_m}\ ; \quad \frac{V_P D_P}{\nu_P} = \frac{V_m D_m}{\nu_m}$$

$$\therefore \quad V_m = \frac{\nu_m}{\nu_P} \times \frac{D_P}{D_m} \times V_P$$

$$= \frac{0.016 \times 10^{-4}}{.012 \times 10^{-4}} \times 30 \times 10 \text{ m/s} \qquad \left\{\because \frac{D_P}{D_m} = \frac{L_P}{L_m} = 30\right\}$$

$$= \frac{0.016}{.012} \times 30 \times 10 = \textbf{400 m/s. Ans.}$$

Ratio of drag force (resistance) :

Drag force = Mass × Acceleration

$$= \rho L^3 \times \frac{V}{t} = \rho \,.\, L^2 \,.\, \frac{L}{t} \times V = \rho\, L^2 V^2 \qquad \left\{\because \frac{L}{t} = V\right\}$$

Let F_P and F_m denote the drag force for the prototype and for the model respectively then,

$$\frac{F_P}{F_m} = \frac{\rho_P L_P^2 V_P^2}{\rho_m L_m^2 V_m^2} = \frac{\rho_P}{\rho_m} \times \left(\frac{L_P}{L_m}\right)^2 \times \left(\frac{V_P}{V_m}\right)^2$$

$$= \frac{1030}{1.24} \times 30^2 \times \left(\frac{10}{400}\right)^2 = \textbf{467.22. Ans.}$$

Problem 6.18 *A ship 300 m long moves in sea-water, whose density is 1030 kg/m³, A 1 : 100 model of this ship is to be tested in a wind tunnel. The velocity of air in the wind tunnel around the model is 30 m/s and the resistance of the model is 60 N. Determine the velocity of ship in sea-water and also the resistance of the ship in sea-water. The density of air is given as 1.24 kg/m³. Take the kinematic viscosity of sea-water and air as 0.012 stokes and 0.018 stokes respectively.*

(Delhi University, 1996)

Solution. Given :

For Prototype,

Length, $L_P = 300$ m

Fluid = Sea-water

Density of water = 1030 kg/m^3

Kinematic viscosity, $\nu_P = 0.018$ stokes $= 0.018 \times 10^{-4}$ m^2/s

Let velocity of ship $= V_P$

Resistance $= F_P$

For model

Length, $L_m = \dfrac{1}{100} \times 300 = 3$ m

Velocity, $V_m = 30$ m/s

Resistance, $F_m = 60$ N

Density of air, $\rho_m = 1.24$ kg/m^3

Kinematic viscosity of air, $\nu_m = 0.018$ stokes $= .018 \times 10^{-4}$ m^2/s.

For dynamic similarity between the prototype and its model, the Reynolds number for both of them should be equal.

$$\therefore \quad \frac{V_P \times L_P}{\nu_P} = \frac{V_m \times L_m}{\nu_m} \quad \text{or} \quad V_P = \frac{\nu_P}{\nu_m} \times \frac{L_P}{L_m} \times V_m$$

$$= \frac{.012 \times 10^{-4}}{.018 \times 10^{-4}} \times \frac{3}{300} \times 30 = \frac{1}{1.5} \times \frac{1}{100} \times 30 = \mathbf{0.2\ m/s.\ Ans.}$$

Resistance = Mass × Acceleration

$$= \rho L^3 \times \frac{V}{t} = \rho L^2 \times \frac{V}{1} \times \frac{L}{t} = \rho L^2 V^2$$

Then
$$\frac{F_P}{F_m} = \frac{(\rho L^2 V^2)_P}{(\rho L^2 V^2)_m} = \frac{\rho_P}{\rho_m} \times \left(\frac{L_P}{L_m}\right)^2 \times \left(\frac{V_P}{V_m}\right)^2$$

But
$$\frac{\rho_P}{\rho_m} = \frac{1030}{1.24}$$

$$\therefore \quad \frac{F_P}{F_m} = \frac{1030}{1.24} \times \left(\frac{300}{3}\right)^2 \times \left(\frac{0.2}{30}\right)^2 = 369.17$$

$$\therefore \quad F_P = 369.17 \times F_m = 369.17 \times 60 = \mathbf{22150.2\ N.\ Ans.}$$

6.10.2 Froude Model Law. Froude model law is the law in which the models are based on Froude number which means for dynamic similarity between the model and prototype, the Froude number for both of them should be equal. Froude model law is applicable when the gravity force is only predominent force which controls the flow in addition to the force of inertia. Froude model law is applied in the following fluid flow problems :

1. Free surface flows such as flow over spillways, weirs, sluices, channels etc.,
2. Flow of jet from an orifice or nozzle,
3. Where waves are likely to be formed on surface,
4. Where fluids of different densities flow over one another.

Let V_m = Velocity of fluid in model,
L_m = Linear dimension or length of model,
g_m = Acceleration due to gravity at a place where model is tested.

and V_P, L_P and g_P are the corresponding values of the velocity, length and acceleration due to gravity for the prototype. Then according to Froude model law,

$$(F_e)_{model} = (F_e)_{prototype} \text{ or } \frac{V_m}{\sqrt{g_m L_m}} = \frac{V_P}{\sqrt{g_P L_P}} \quad ...(6.18)$$

If the tests on the model are performed on the same place where prototype is to operate, then $g_m = g_P$ and equation (6.18) becomes as

$$\frac{V_m}{\sqrt{L_m}} = \frac{V_P}{\sqrt{L_P}} \quad ...(6.19)$$

or
$$\frac{V_m}{V_P} \times \frac{1}{\sqrt{\frac{L_m}{L_P}}} = 1$$

$$\frac{V_P}{V_m} = \sqrt{\frac{L_P}{L_m}} = \sqrt{L_r} \qquad \left\{\because \frac{L_P}{L_m} = L_r\right\}$$

where L_r = Scale ratio for length

$\frac{V_P}{V_m} = V_r$ = Scale ratio for velocity.

$$\therefore \quad \frac{V_P}{V_m} = V_r = \sqrt{L_r}\,. \quad ...(6.20)$$

Scale ratios for various physical quantities based on Froude model law are :

(*a*) **Scale ratio for time**

$$\text{As time} = \frac{\text{Length}}{\text{Velocity}},$$

then ratio of time for prototype and model is

$$T_r = \frac{T_P}{T_m} = \frac{\left(\frac{L}{V}\right)_P}{\left(\frac{L}{V}\right)_m} = \frac{\frac{L_P}{V_P}}{\frac{L_m}{V_m}} = \frac{L_P}{L_m} \times \frac{V_m}{V_P} = L_r \times \frac{1}{\sqrt{L_r}} \qquad \left\{\because \frac{V_P}{V_m} = \sqrt{L_r}\right\}$$

$$= \sqrt{L_r}\,. \quad ...(6.21)$$

(*b*) **Scale ratio for acceleration**

$$\text{Acceleration} = \frac{V}{T}$$

$$\therefore \quad a_r = \frac{a_P}{a_m} = \frac{\left(\frac{V}{T}\right)_P}{\left(\frac{V}{T}\right)_m} = \frac{V_P}{T_P} \times \frac{T_m}{V_m} = \frac{V_P}{V_m} \times \frac{T_m}{T_P}$$

$$= \sqrt{L_r} \times \frac{1}{\sqrt{L_r}} \qquad \left\{\because \frac{V_P}{V_m} = \sqrt{L_r},\ \frac{T_P}{T_m} = \sqrt{L_r}\right\}$$

$$= 1. \qquad \text{...(6.22)}$$

(*c*) **Scale ratio for discharge**

$$Q = A \times V = L^2 \times \frac{L}{T} = \frac{L^3}{T}$$

$$\therefore \quad Q_r = \frac{Q_P}{Q_m} = \frac{\left(\frac{L^3}{T}\right)_P}{\left(\frac{L^3}{T}\right)_m} = \left(\frac{L_P}{L_m}\right)^3 \times \left(\frac{T_m}{T_P}\right) = L_r^{\ 3} \times \frac{1}{\sqrt{L_r}} = L_r^{2.5} \qquad \text{...(6.23)}$$

(*d*) **Scale ratio for force**

$$\text{As Force} = \text{Mass} \times \text{Acceleration} = PL^3 \times \frac{V}{T} = PL^2 . \frac{L}{T} . V = \rho L^2 V^2$$

$\therefore$ Ratio for force,

$$F_r = \frac{F_P}{F_m} = \frac{\rho_P L_P^2 V_P^2}{\rho_m L_m^2 V_m^2} = \frac{\rho_P}{\rho_m} \times \left(\frac{L_P}{L_m}\right)^2 \times \left(\frac{V_P}{V_m}\right)^2.$$

If the fluid used in model and prototype is same, then

$$\frac{\rho_P}{\rho_m} = 1 \quad \text{or} \quad \rho_P = \rho_m$$

and hence

$$F_r = \left(\frac{L_P}{L_m}\right)^2 \times \left(\frac{V_P}{V_m}\right)^2 = L_r^{\ 2} \times \left(\sqrt{L_r}\right)^2 = L_r^{\ 2} . L_r = L_r^{\ 3}. \qquad \text{...(6.24)}$$

(*e*) **Scale ratio for pressure intensity**

As

$$p = \frac{\text{Force}}{\text{Area}} = \frac{\rho L^2 V^2}{L^2} = \rho L^2$$

$\therefore$ Pressure ratio,

$$p_r = \frac{p_P}{p_m} = \frac{\rho_P V_P^2}{\rho_m V_m^2}$$

If fluid is same, then $\rho_P = \rho_m$

$$\therefore \quad p_r = \frac{V_P^2}{V_m^2} = \left(\frac{V_P}{V_m}\right)^2 = L_r. \qquad \text{...(6.25)}$$

(*f*) **Scale ratio for work, energy, torque, moment etc.**

$$\text{Torque} = \text{Force} \times \text{Distance} = F \times L$$

$\therefore$ Torque ratio,

$$T_r{}^* = \frac{T_P{}^*}{T_m{}^*} = \frac{(F \times L)_P}{(F \times L)_m} = F_r \times L_r = L_r^{\ 3} \times L_r = L_r^{\ 4}. \qquad \text{...(6.26)}$$

(*g*) **Scale ratio for power**

As Power = Work per unit time

$$= \frac{F \times L}{T}$$

$\therefore$ Power ratio,
$$P_r = \frac{P_P}{P_m} = \frac{\dfrac{F_P \times L_P}{T_P}}{\dfrac{F_m \times L_m}{T_m}} = \frac{F_P}{F_m} \times \frac{L_P}{L_m} \times \frac{1}{\dfrac{T_P}{T_m}}$$

$$= F_r \,.\, L_r \,.\, \frac{1}{T_r} = L_r^3 \,.\, L_r \,.\, \frac{1}{\sqrt{L_r}} = L^{3.5}. \qquad \text{...(6.27)}$$

Problem 6.19 *In 1 in 40 model of a spillway, the velocity and discharge are 2 m/s and 2.5 m³/s. Find the corresponding velocity and discharge in the prototype.*

Solution. Given :

Scale ratio of length, $L_r = 40$

Velocity in model, $V_m = 2$ m/s

Discharge in model, $Q_m = 2.4$ m³/s

Let V_P and Q_P are the velocity and discharge in prototype.

Using equation (6.20) for velocity ratio, $\dfrac{V_P}{V_m} = \sqrt{L_r} = \sqrt{40}$

$\therefore$ $V_P = V_m \times \sqrt{40} = 2 \times \sqrt{40} =$ **12.65 m/s. Ans.**

Using equation (6.23) for discharge ratio,

$$\frac{Q_P}{Q_m} = L_r^{2.5} = (40)^{2.5}$$

$\therefore$ $Q_P = Q_m \times 40^{2.5} = 2.5 \times 40^{2.5} =$ **25298.2 m³/s. Ans.**

Problem 6.20 *A ship model of scale $\frac{1}{50}$ is towed through sea water at a speed of 1 m/s. A force of 2 N is required to tow the model. Determine the speed of ship and the propulsive force on the ship, if prototype is subjected to wave resistance only.*

Solution. Given :

Scale ratio of length, $L_r = 50$

Speed of model, $V_m = 1$ m/s

Force required for model, $F_m = 2$ N

Let the speed of ship $= V_P$

and the propulsive force for ship $= F_P$.

As prototype is subjected to wave resistance only for dynamic similarity, the Froude number should be same for model and prototype. Hence for velocity ratio, for Froude model law using equation (6.20), we have

$\therefore$
$$\frac{V_P}{V_m} = \sqrt{L_r} = \sqrt{50}$$

$\therefore$ $V_P = \sqrt{50} \times V_m = \sqrt{50} \times 1 =$ **7.071 m/s. Ans.**

Force scale ratio is given by equation (6.24),

$\therefore$
$$F_r = \frac{F_P}{F_m} = L_r^3$$

$\therefore$ $F_P = F_m \times L_r^3 = 2 \times (50)^3 =$ **250000 N. Ans.**

Problem 6.21 *In the model test of a spillway the discharge and velocity of flow over the model were 2 m³/s and 1.5 m/s respectively. Calculate the velocity and discharge over the prototype which is 36 times the model size.* (A.M.I.E., Summer, 1980)

Solution. Given :

Discharge over model, $Q_m = 2 \text{ m}^3/\text{s}$

Velocity over model, $V_m = 1.5 \text{ m/s}$

Linear scale ratio, $L_r = 36.$

For dynamic similarity, Froude model law is used. Using equation (6.20), we have

$$\frac{V_P}{V_m} = \sqrt{L_r} = \sqrt{36} = 6.0$$

$\therefore$ V_P = Velocity over prototype = $V_m \times 6.0 = 1.5 \times 6.0 =$ **9 m/s. Ans.**

For discharge, using equation (6.23), we get

$$\frac{Q_P}{Q_m} = L_r^{2.5} = (36)^{2.5}.$$

$\therefore$ $Q_P = Q_m \times (36)^{2.5} = 2 \times 36^{2.5} =$ **15552 m³/s. Ans.**

Problem 6.22 *In a geometrically similar model of spillway the discharge per metre length is $\frac{1}{6}$ m³/s. If the scale of the model is $\frac{1}{36}$, find the discharge per metre run of the prototype.* (A.M.I.E., Summer, 1979)

Solution. Given :

Discharge per metre length for model, $q_m = \frac{1}{6} \text{ m}^3/\text{s}$

Linear scale ratio, $L_r = 36$

Discharge per metre length for prototype, $q_P = ?$

The discharge ratio for spillway is given by equation (6.23), $\frac{Q_P}{Q_m} = L_r^{2.5}.$

But discharge ratio per metre length is given as

$$\frac{q_P}{q_m} = \frac{Q_P / L_P}{Q_m / L_m} = \frac{Q_P}{Q_m} \times \frac{L_m}{L_P} = L_r^{2.5} \times \frac{1}{L_r} = L_r^{1.5}$$

$$\therefore \quad q_P = q_m \times L_r^{1.5} = \frac{1}{6} \times (36)^{1.5} = \frac{1}{2} \times 6^{2 \times 1.5}$$

$= 6^{3-1} = 6^2 =$ **36 m³/s per metre sum. Ans.**

Problem 6.23 *A spillway model is to be built to a geometrically similar scale of $\frac{1}{50}$ across a flume of 600 mm width. The prototype is 15 m high and maximum head on it is expected to be 1.5 m. (i) What height of model and what head on the model should be used ? (ii) If the flow over the model at a particular head is 12 litres per second, what flow per metre length of the prototype is expected ? (iii) If the negative pressure in the model is 200 mm, what is the negative pressure in prototype ? Is it practicable ?* (A.M.I.E., Winter, 1977)

Solution. Given :

Scale ratio for length, $L_r = 50$

Width of model, $B_m = 600 \text{ mm} = 0.6 \text{ m}$

Flow over model, $Q_m = 12$ litres/s

Pressure in model, $h_m = -\ 200$ mm of water $= -\ 0.2$ m

Height of prototype, $H_P = 15$ m

Head on prototype, $H_P{}^* = 1.5$ m

(*i*) Let the height of model $= H_m$

and head on model $= H_m{}^*$

Linear scale ratio, $$L_r = \frac{H_P}{H_m} = \frac{H_{P*}}{H_{m*}} = 50$$

$\therefore$ Height of model, $$H_m = \frac{H_P}{50} = \frac{15}{50} = \textbf{0.3 m. Ans.}$$

And head on model, $$H_m{}^* = \frac{H_{P*}}{50} = \frac{1.50}{50} = \textbf{0.03 m. Ans.}$$

Width of prototype, $B_P = L_r \times B_m = 50 \times 0.6 = 30$ m.

(*ii*) Discharge ratio is given by equation (6.23) as

$$\frac{Q_P}{Q_m} = L_r^{2.5} = (50)^{2.5} = 17677.67$$

$\therefore$ $$Q_P = Q_m \times 17677.67 = 12 \times 17677.67 = 212132.04 \text{ lit/s}$$

Discharge per metre length of prototype $= \dfrac{Q_P}{\text{Length of prototype}} = \dfrac{212132.04}{\text{Width of prototype}}$

$$= \frac{212132.04}{30} = \textbf{7071.078 litres/s. Ans.}$$

(*iii*) Negative pressure head in prototype,

$$h_P = L_r \times h_m = 50 \times (-\ 0.2) = \textbf{– 10.0 m. Ans.}$$

This negative pressure is not practicable. Maximum practicable negative pressure head is – 7.50 m of water.

Problem 6.24 *In a 1 in 20 model of stilling basin, the height of the hydraulic jump in the model is observed to be 0.20 metre. What is the height of the hydraulic jump in the prototype ? If the energy dissipated in the model is $\frac{1}{10}$ kW, what is the corresponding value in prototype ?*

(A.M.I.E., Summer, 1977)

Solution. Given :

Linear scale ratio, $L_r = 20$

Height of hydraulic jump in model, $h_m = 0.20$ m

Energy dissipated in model, $P_m = \dfrac{1}{10}$ kW

(*i*) Let the height of hydraulic jump in the prototype $= h_P$

Then $$\frac{h_P}{h_m} = L_r = 20$$

$\therefore$ $$h_P = h_m \times 20 = 0.20 \times 20 = \textbf{4 m. Ans.}$$

(*ii*) Let the energy dissipated in prototype $= F_P$

Using equation (6.27) for power ratio, $$\frac{P_P}{P_m} = L_r^{3.5} = 20^{3.5} = 35777.088$$

$\therefore$ $$P_P = P_m \times 35777.088 = \frac{1}{10} \times 35777.088 = \textbf{3577.708 kW. Ans.}$$

Problem 6.25 *The characteristics of the spillway are to be studied by means of a geometrically similar model constructed to the scale ratio of 1 : 10.*

(i) If the maximum rate of flow in the prototype is 28.3 cumecs, what will be the corresponding flow in model ?

(ii) If the measured velocity in the model at a point on the spillway is 2.4 m/s, what will be the corresponding velocity in prototype ?

(iii) If the hydraulic jump at the foot of the model is 50 mm high, what will be the height of jump in prototype ?

(iv) If the energy dissipated per second in the model is 3.5 N m, what energy is dissipated per second in the prototype ? (A.M.I.E., Summer, 1988)

Solution. Given :

$$\frac{\text{Linear dimension of model}}{\text{Linear dimension of prototype}} = \frac{1}{10}$$

$\therefore$ Scale ratio, $L_r = 10.$

(*i*) Discharge in prototype, $Q_P = 28.3 \text{ m}^3\text{/s}$

Let Q_m = Discharge in model

For discharge using equation (6.23), we get

$$\frac{Q_P}{Q_m} = L_r^{2.5}$$

$$\therefore \quad Q_m = \frac{Q_P}{L_r^{2.5}} = \frac{28.3}{10^{2.5}} = \mathbf{0.0895\ m^3/s.\ Ans.}$$

(*ii*) Velocity in the model, $V_m = 2.4$ m/s

Let V_P = Velocity in the prototype

For velocity using equation (6.20), we get

$$\frac{V_P}{V_m} = \sqrt{L_r}$$

$$\therefore \quad V_P = V_m \times \sqrt{L_r} = 2.4 \times \sqrt{10} = \mathbf{7.589\ m/s.\ Ans.}$$

(*iii*) Hydraulic jump in model, $H_m = 50$ mm

Let H_P = Hydraulic jump in prototype

Now scale ratio $= \dfrac{H_P}{H_m}$

$$\therefore \quad H_P = H_m \times \text{Scale ratio} = 50 \times 10 = \mathbf{500\ mm.\ Ans.}$$

(*iv*) Energy dissipated/s in model, $E_m = 3.5$ N m/s

Let E_P = Energy dissipated/s in prototype

Now using equation (6.27), we get $\dfrac{E_P}{E_m} = L_r^{3.5}$

$$\therefore \quad E_P = E_m \times L_r^{3.5} = 3.5 \times 10^{3.5} = \mathbf{11067.9\ N\ m/s.\ Ans.}$$

Problem 6.26 *A 1 : 64 model is constructed of an open channel in concrete which has Manning's N = 0.014. Find the value of N for the model.* (A.M.I.E., Winter, 1987)

Solution. Given :

Linear scale ratio, $L_r = 64$

Value of N for prototype, $N_P = 0.014$

Let N_m = Value of N for model.

The Manning's formula is given by, $V = \frac{1}{N} m^{3/2} . i^{1/2}$

in which m = Hydraulic mean depth in m

i = Slope of the bed of the channel

Now for the model, the Manning's formula becomes as

$$V_m = \frac{1}{N_m} . (m_m)^{2/3} . (i_m)^{1/2} \quad ...(i)$$

and for the prototype, the Manning's formula is written as

$$V_P = \frac{1}{N_P} . (m_p)^{2/3} . (i_p)^{1/2} \quad ...(ii)$$

Dividing equation (*ii*) by equation (*i*), we get

$$\frac{V_P}{V_m} = \frac{\frac{1}{N_P} . (m_P)^{2/3} . (i_P)^{1/2}}{\frac{1}{N_m} . (m_m)^{2/3} . (i_m)^{1/2}} = \frac{N_m}{N_P} . \left(\frac{m_P}{m_m}\right)^{2/3} . \left(\frac{i_P}{i_m}\right)^{1/2} \quad ...(iii)$$

For dynamic similarity, Froude model law is used. Using equation (6.20), we have

$$\frac{V_P}{V_m} = \sqrt{L_r} = \sqrt{64} = 8$$

But $\frac{m_p}{m_m} = L_r$ and $\frac{i_p}{i_m} = 1$ as i_p and i_m are dimensionless.

Substituting these values in equation (*iii*), we get

$$8 = \frac{N_m}{N_P} \times (L_r)^{2/3} \times 1 = \frac{N_m}{0.014} \times (64)^{2/3} \quad (\because N_P = 0.014)$$

$$\therefore \quad N_m = \frac{8 \times 0.014}{64^{2/3}} = \frac{8 \times 0.014}{16} = \mathbf{0.007. \ Ans.}$$

Problem 6.27 *A 7.2 m height and 15 m long spillway discharges 94 m³/s discharge under a head of 2.0 m. If a 1 : 9 scale model of this spillway is to be constructed, determine model dimensions, head over spillway model and the model discharge. If model experiences a force of 7500 N (764.53 kgf), determine force on the prototype.* (A.M.I.E., Summer, 1991)

Solution. Given :

For prototype : Height $h_P = 7.2$ m

Length, $L_P = 15$ m

Discharge, $Q_P = 94 \text{ m}^3/\text{s}$

Head, $H_P = 2.0$ m

Size of model $= \frac{1}{9}$ of the size of prototype.

$\therefore$ Linear scale ratio, $L_r = 9$

Force experienced by model, $F_P = 7500$ N

Find : (*i*) Model dimensions *i.e.*, height and length of model (h_m and L_m)

(*ii*) Head over model *i.e.*, H_m

(*iii*) Discharge through model *i.e.*, Q_m

(*iv*) Force on prototype (*i.e.*, F_p)

(i) Model dimensions (h_m and L_m)

$$\frac{h_P}{h_m} = \frac{L_P}{L_m} = L_r = 9$$

$$\therefore \quad h_m = \frac{h_P}{9} = \frac{7.2}{9} = \textbf{0.8 m. Ans.}$$

$$\text{And} \quad L_m = \frac{L_P}{9} = \frac{15}{9} = \textbf{1.67 m. Ans.}$$

(ii) Head over model (H_m)

$$\frac{h_P}{H_m} = L_r = 9$$

$$\therefore \quad H_m = \frac{H_P}{9} = \frac{2}{9} = \textbf{0.222 m. Ans.}$$

(iii) Discharge through model (Q_m)

Using equation (6.23), we get $\frac{Q_P}{Q_m} = L_r^{2.5}$

$$\therefore \quad Q_m = \frac{Q_P}{L_r^{2.5}} = \frac{94}{9^{2.5}} = \frac{94}{243} = \textbf{0.387 m}^3\textbf{/s. Ans.}$$

(iv) Force on the Prototype (F_P)

Using equation (6.24), we get $F_r = \frac{F_P}{F_m} = L_r^{\ 3}$

$$\therefore \quad F_P = F_m \times L_r^{\ 3} = 7500 \times 9^3 = \textbf{5467500 N. Ans.}$$

6.10.3 Euler's Model Law. Euler's model law is the law in which the models are designed on Euler's number which means for dynamic similarity between the model and prototype, the Euler number for model and prototype should be equal. Euler's model law is applicable when the pressure forces are alone predominant in addition to the inertia force. According to this law :

$$(E_u)_{\text{model}} = (E_u)_{\text{prototype}} \quad \text{...(6.28)}$$

If V_m = Velocity of fluid in model,

p_m = Pressure of fluid in model,

ρ_m = Density of fluid in model,

and V_P, p_P, ρ_P = Corresponding values in prototype, then substituting these values in equation (6.28), we get

$$\frac{V_m}{\sqrt{p_m/\rho_m}} = \frac{V_P}{\sqrt{p_P/\rho_P}} \qquad \text{...(6.29)}$$

If fluid is same in model and prototype, then equation (6.29) becomes as

$$\frac{V_m}{\sqrt{p_m}} = \frac{V_P}{\sqrt{p_P}} \qquad \text{...(6.30)}$$

Euler's model law is applied for fluid flow problems where flow is taking place in a closed pipe in which case turbulence is fully developed so that viscous forces are negligible and gravity force and surface tension force is absent. This law is also used where the phenomenon of cavitation takes place.

6.10.4 Weber Model Law. Weber model law is the law in which models are based on Weber's number, which is the ratio of the square root of inertia force to surface tension force. Hence where surface tension effects predominate in addition to inertia force, the dynamic similarity between the model and prototype is obtained by equating the Weber number of the model and its prototype. Hence according to this law :

$$(W_e)_{\text{model}} = (W_e)_{\text{prototype}}, \qquad \text{where } W_e \text{ is Weber number and } = \frac{V}{\sqrt{\sigma/\rho L}}$$

If V_m = Velocity of fluid in model,

σ_m = Surface tensile force in model,

ρ_m = Density of fluid in model,

L_m = Length of surface in model,

and V_P, σ_P, ρ_P, L_P = Corresponding values of fluid in prototype.

Then according to Weber law, we have

$$\frac{V_m}{\sqrt{\sigma_m/\rho_m L_m}} = \frac{V}{\sqrt{\sigma_P/\rho_P L_P}} \qquad \text{...(6.31)}$$

Weber model law is applied in following cases :

1. Capillary rise in narrow passages,
2. Capillary movement of water in soil,
3. Capillary waves in channels,
4. Flow over weirs for small heads.

6.10.5 Mach Model Law. Mach model law is the law in which models are designed on Mach number, which is the ratio of the square root of inertia force to elastic force of a fluid. Hence where the forces due to elastic compression predominate in addition to inertia force, the dynamic similarity between the model and its prototype is obtained by equating the Mach number of the model and its prototype. Hence according to this law :

$$(M)_{\text{model}} = (M)_{\text{prototype}}$$

where M = Mach number = $\dfrac{V}{\sqrt{K/\rho}}$

If V_m = Velocity of fluid in model,

K_m = Elastic stress for model,

ρ_m = Density of fluid in model,

and V_P, K_P and ρ_P = Corresponding values for prototype. Then according to Mach law,

$$= \frac{V_m}{\sqrt{K_m/\rho_m}} = \frac{V}{\sqrt{K_P/\rho_P}} \qquad \text{...(6.32)}$$

Mach model law is applied in the following cases :

1. Flow of aeroplane and projectile through air at supersonic speed, *i.e.*, at a velocity more than the velocity of sound,
2. Aerodynamic testing,
3. Under water testing of torpedoes,
4. Water-hammer problems.

Problem 6.28 *The pressure drop in an aeroplane model of size $\frac{1}{10}$ of its prototype is 80 N/cm². The model is tested in water. Find the corresponding pressure drop in the prototype. Take density of air = 1.24 kg/m³. The viscosity of water is 0.01 poise while the viscosity of air is 0.00018 poise.*

Solution. Given :

Pressure drop in model, $p_m = 80 \text{ N/cm}^2 = 80 \times 10^4 \text{ N/m}^2$

Linear scale ratio, $L_r = 40$

Fluid in model = Water, while in prototype = Air

Viscosity of water, $\mu_m = 0.01$ poise

Density of water, $\rho_m = 1000 \text{ kg/m}^3$

Viscosity of air, $\mu_P = .00018$ poise

Density of air, $\rho_P = 1.24 \text{ kg/m}^3$

Let the corresponding pressure drop in prototype = p_P.

As the problem involves pressure force and viscous force and hence for dynamic similarity between the model and prototype, Euler's number and Reynold's number should be considered. Making first of all, Reynold's number equal, we get from equation (6.17)

$$\frac{\rho_m V_m L_m}{\mu_m} = \frac{\rho_P V_P L_P}{\mu_P} \quad \textit{or} \quad \frac{V_m}{V_P} = \frac{\rho_P}{\rho_m} \times \frac{L_P}{L_m} \times \frac{\mu_m}{\mu_P}$$

But $$\frac{\rho_P}{\rho_m} = \frac{1.24}{1000}$$

$$\frac{L_P}{L_m} = L_r = 40, \quad \frac{\mu_m}{\mu_P} = \frac{0.01}{.00018}$$

$$\therefore \quad \frac{V_m}{V_P} = \frac{1.24}{1000} \times 40 \times \frac{.01}{.00018} = 2.755.$$

Now making Euler's number equal, we get from equation (6.29) as

$$\frac{V_m}{\sqrt{\frac{p_m}{\rho_m}}} = \frac{V_P}{\sqrt{\frac{p_P}{\rho_P}}} \quad \text{or} \quad \frac{V_m}{V_P} = \frac{\sqrt{p_m \rho_m}}{\sqrt{p_P / \rho_P}} = \sqrt{\frac{p_m}{p_P}} \times \sqrt{\frac{\rho_P}{\rho_m}}$$

But $$\frac{V_m}{V_P} = 2.755 \text{ and } \frac{\rho_P}{\rho_m} = \frac{1.24}{1000}$$

$$\therefore \quad 2.755 = \sqrt{\frac{p_m}{p_P}} \times \sqrt{\frac{1.24}{1000}} = \sqrt{\frac{p_m}{p_P}} \times .0352$$

$$\therefore \quad \sqrt{\frac{p_m}{p_P}} = \frac{2.755}{.0352} = 78.267$$

$$\therefore \quad \frac{p_m}{p_P} = (78.267)^2 \text{ or } p_P = \frac{p_m}{(78.267)^2} = \frac{80}{(78.267)^2}$$

$$= \mathbf{0.01306 \text{ N/cm}^2. \text{ Ans.}}$$

HIGHLIGHTS

1. Dimensional analysis is the method of dimensions, in which fundamental dimensions are M, L and T.
2. Dimensional analysis is performed by two methods namely Rayleigh's Method and Buckingham's π-theorem.
3. Rayleigh's method is used for finding an expression for a variable which depends on maximum three or four variables while there is no restriction on the number of variables for Buckingham's π-theorem.
4. Model analysis is an experimental method of finding solutions of complex flow problems. A model is a small scale replica of the actual machine or structure. The actual machine or structure is called prototype.
5. Three types of similarities must exist between the model and prototype. They are : (*i*) Geometric Similarity, (*ii*) Kinematic Similarity, and (*iii*) Dynamic Similarity.
6. For geometric similarity, the ratio of all linear dimensions of the model and of the prototype should be equal.
7. Kinematic similarity means the similarity of motion between model and prototype.
8. Dynamic similarity means the similarity of forces between the model and prototype.
9. Reynold's number is defined as the ratio of inertia force and viscous force of a flowing fluid. It is given by,

$$R_e = \frac{\rho VL}{\mu} = \frac{VL}{\nu} = \frac{V \times d}{\nu} \text{ for pipe flow}$$

where V = Velocity of flow, d = Diameter of pipe and
ν = Kinematic viscosity of fluid.

10. Froude's Number is the ratio of the square root of inertia force and gravity force and is given by

$$F_e = \sqrt{\frac{F_i}{F_g}} = \frac{V}{\sqrt{Lg}}.$$

11. Euler's number is the ratio of the square root of inertia force and pressure force and is given by,

$$E_u = \sqrt{\frac{F_i}{F_P}} = \frac{V}{\sqrt{p/\rho}}.$$

12. Mach number is the ratio of the square root of inertia force and elastic force and is given by

$$M = \sqrt{\frac{F_i}{F_e}} = \frac{V}{\sqrt{K/\rho}} = \frac{V}{C}.$$

13. The laws on which the models are designed for dynamic similarity are called model laws or laws of similarity. The model laws are (*i*) Reynold's Model Law, (*ii*) Froude Model Law, (*iii*) Euler Model Law, (*iv*) Weber Model Law, (*v*) Mach Model Law.

EXERCISE 6

(A) THEORETICAL PROBLEMS

1. Define the terms dimensional analysis and model analysis.
2. What do you mean by fundamental units and derived units ? Give examples.
3. Explain the term, 'dimensionally homogeneous equation'.
4. What are the methods of dimensional analysis ? Describe the Rayleigh's method for dimensional analysis.

5. State Buckingham's p-theorem. Why this theorem is considered superior over the Rayleigh's method for dimensional analysis ? *(A.M.I.E., Summer 1990)*

6. What do you mean by repeating variables ? How are the repeating variables selected for dimensional analysis ?

7. Define the terms : model, prototype, model analysis, hydraulic similitude.

8. Explain the different types of hydraulic similarities that must exist between a prototype and its model. *(Delhi University, 1982, 87, 96)*

9. What do you mean by dimensionless numbers ? Name any four dimensionless numbers.
Define and explain Reynold's number, Froude's number's and Mach number. Derive expressions for any above two numbers. *(Delhi University, 1987)*

10. What is meant by geometric, kinematic and dynamic similarities ? Are these similarities truly attainable ? If not why ? *(A.M.I.E., Winter, 1983)*

11. Define the following non-dimensional numbers : Reynold's number, Froude's number and Mach's number. What are their significances for fluid flow problems ?

12. What are the different laws on which models are designed for dynamic similarity ? Where are they used ?

13. How will you determine the total drag of a ship or partially submerged bodies ?

14. Show that ratio of inertia force to viscous force gives the Reynold's' number. *(A.M.I.E., Winter, 1983)*

15. State Buckingham's π-theorem. What do you mean by repeating variables ? How are the repeating variables selected in dimensional analysis ? *(A.M.I.E., June, 1992)*

16. What is the significance of the non-dimensional numbers : Reynolds number, Froude number and Mach number in the theory of similarity ? What is the dimensional analysis ? How is this analysis related to the theory of similarity ? *(Delhi University, December 2001)*

17. Define and explain : (*i*) Froude's number, (*ii*) Mach number, (*iii*) Hydraulic similarities (*iv*) Distorted and undistorted models. *(Delhi University, December 2002)*

(B) NUMERICAL PROBLEMS

1. Give the dimensions of : (*i*) Force (*ii*) Viscosity (*iii*) Power and (*iv*) Kinematic viscosity.
[**Ans.** (*i*) MLT^{-2}, $ML^{-1}T^{-1}$, $ML^{+2}T^{-3}$, L^2T^{-1}]

2. The variables controlling the motion of a floating vessel through water are the drag force F, the speed V, the length L, the density ρ and dynamic viscosity μ of water and acceleration due to gravity g. Derive an expression for F by dimensional analysis. $\left[\textbf{Ans. } F = \rho L^2 V^2 \phi\left[\frac{\mu}{\rho VL}, \frac{Lg}{V^2}\right]\right]$ *(University of Jodhpur, 1983)*

3. The resistance R, to the motion of a completely sub-merged body depends upon the length of the body L, velocity of flow V, mass density of fluid ν and kinematic viscosity of fluid ν. By dimensional analysis prove that

$$R = \rho V^2 L^2 \phi\left(\frac{VL}{\nu}\right).$$

4. A pipe of diameter 1.8 m is required to transport an oil of sp. gr. 0.8 and viscosity .04 poise at the rate of 4 m^3/s. Tests were conducted on a 20 cm diameter pipe using water at 20°C. Find the velocity and rate of flow in the model. Viscosity of water at 20°C = .01 poise. [**Ans.** 2.829 m/s, 88.8 litres/s]

5. A model of a sub-marine of scale $\frac{1}{40}$ is tested in a wind tunnel. Find the speed of air in wind tunnel if the speed of sub-marine in sea-water is 15 m/s. Also find the ratio of the resistance between the model and its prototype. Take the values of kinematic viscosities for sea-water and air as .012 stokes and 0.016 stokes respectively. The density of sea-water and of air are given as 1030 kg/m^3 and 1.24 kg/m^3 respectively. $\left[\textbf{Ans.} \text{ 800 m/s}, \frac{F_m}{F_P} = 0.00214\right]$

6. A ship 250 m long moves in sea-water, whose density is 1030 kg/m^3. A 1 : 125 model of this ship is to be tested in wind tunnel. The velocity of air in the wind tunnel around the model is 20 m/s and the resistance of the model is 50 N. Determine the velocity of ship in sea-water and also the resistance of the ship in sea-water. The density of air is given as 1.24 kg/m^3. Take the kinematic viscosity of sea-water and air as 0.012 stokes and 0.018 stokes respectively. [**Ans.** 0.106 m/s, 18228.7 N]
7. In 1 : 30 model of a spillway, the velocity and discharge are 1.5 m/s and 2.0 m^3/s. Find the corresponding velocity and discharge in the prototype. [**Ans.** 8.216 m/s, 9859 m^3/s]
8. A ship-model of scale $\frac{1}{60}$ is towed through sea-water at a speed of 0.5 m/s. A force of 1.5 N is required to tow the model. Determine the speed of the ship and propulsive force on the ship, if prototype is subjected to wave resistance only. [**Ans.** 3.873 m/s, 324000 N]
9. A spillway model is, to be built to a geometrically similar scale of $\frac{1}{40}$ across a flume, of 50 cm width. The prototype is 20 m high and maximum head on it is expected to be 2 m. (*i*) What height of model and what head on the model should be used ? (*ii*) If the flow over the model at a particular head is 10 litres/s, what flow per metre length of the prototype is expected ? (*iii*) If the negative pressure in the model is 150 mm, what is the negative pressure in the prototype ? Is it practicable ?
[**Ans.** (*i*) 0.5 m, 0.05 m, (*ii*) 5059.64 litres/s (*iii*) 6 m. Yes]
10. The pressure drop in an aeroplane model of size $\frac{1}{50}$ of its prototype is 4 N/cm^2. The model is tested in water. Find the corresponding pressure drop in the prototype. Take density of air = 1.24 kg/m^3. The viscosity of water is 0.01 poise while the viscosity of air is 0.00018 poise. [**Ans.** .00042 N/cm^2]
11. A 1 : 20 model of a flying boat is towed through water. The prototype is moving in sea-water of density 1024 kg/m^3 at a velocity of 15 m/s. Find the corresponding speed of the model. Also determine the resistance due to waves on model, if the resistance due to waves of prototype is 500 N.
[**Ans.** 3.354 m/s, .061 N]
12. A 1 : 50 model of an ocean tanker is dragged through fresh water at 1.5 m/s with a total measured drag of 10 N. The frictional drag co-efficient '*f*' for model and prototype are 0.03 and 0.02 respectively in the equation,
$$R_f = f \, . \, A \, . \, V^2 \, ;$$
the wetted surface area of the model is 20 m^2. Determine the total drag on the prototype and the power required to derive the prototype. Take density of sea-water and of fresh water as 1024 kg/m^3 and 1000 kg/m^2 respectively. [**Ans.** 1118436 N, 158072 h.p.]
13. If model prototype ratio is 1 : 75, show that the ratio of discharges per unit width of spillway is given by $\left(\frac{1}{75}\right)^{3/2}$. (*A.M.I.E., Winter, 1983*)
14. A fluid of density ρ and viscosity μ, flows at an average velocity *V* through a circular pipe of diameter *D*. Show by dimensional analysis, that the shear stress at the pipe wall is given as
$$\tau_0 = \rho V^2 \phi \left[\frac{\rho V D}{\mu}\right].$$
(*Delhi University, 1981*)
15. The drag force exerted by a flowing fluid on a solid body depends upon the length of the body, *L*, velocity of flow *V*, density of fluid ρ, and viscosity μ. Find an expression for drag force using Buckingham's theorem. (*Delhi University, 1987*)
16. The efficiency η of geometrically similar fans depends upon the mass density of air ρ, its viscosity μ, speed of fan *N* (revolutions per sec), diameter of blades *D* and discharge *Q*. Perform dimensional analysis. (*A.M.I.E., Summer 1991*)
[**Hint.** Take *D*, *N*, ρ as repeating variables. Then three π-terms will be

$$\pi_1 = D^{a_1} . N^{b_1} . \rho^{c_1} . \eta = \eta \;;$$

$$\pi_2 = D^{a_2} . N^{b_2} . \rho^{c_2} . \mu = \frac{\mu}{D^2 NP} \text{ and } \pi_3 = D^{a_3} . N^{b_3} . \rho^{c_3} . Q = \frac{Q}{D^3 N}$$

$$\therefore \qquad \eta = \phi\,(\pi_2, \pi_3) = \phi\left(\frac{\mu}{D^2 NP}, \frac{Q}{D^3 N}\right) \textbf{ Ans.} \;]$$

17. The discharge through an orifice depends on the diameter D of the orifice, head H over the orifice, density ρ of liquid, viscosity μ of the liquid and acceleration g due to gravity. Using dimensional analysis, find an expression for the discharge. Hence find the dimensionless parameters on which the discharge co-efficient of an orifice meter depend.

[**Hint.** $Q = f(D, H, \rho, \mu, g)$. Hence $N = 6$, $m = 3$ and number of π-terms = 3. Take ρ, D, g as repeating variables. Then

$$\pi_1 = \rho^{a_1} D^{b_1} g^{c_1} Q,\ \pi_2 = \rho^{a_2} . D^{b_2} . g^{c_2} . \mu \text{ and } \pi_3 = \rho^{a_3} . D^{b_3} . g^{c_3} . H.$$

Find π_1, π_2 and π_3. They will be $\pi_1 = \dfrac{Q}{D^{2.5} \times g^{1/2}}$, $\pi_2 = \dfrac{\mu}{\rho . D^{3/2} . g^{1/2}}$ and $\pi_3 = \dfrac{H}{D}$.

Hence $$Q = D^{2.5} . g^{1/2} . \left(\frac{H}{D}, \frac{\mu}{\rho . D^{3/2} . g^{1/2}}\right) = D^2 . g^{1/2} . H^{1/2} \left(\frac{H}{D}, \frac{\mu}{\rho . D^{3/2} . g^{1/2}}\right).$$

Also $$Q = C_d \times A \times \sqrt{2gh}\,.$$

Comparing the two values of Q.

We get $C_d = \phi\left(\dfrac{H}{D}, \dfrac{\mu}{\rho . D^{3/2} . g^{1/2}}\right)$. Hence C_d depends upon $\dfrac{H}{D}$ and $\dfrac{\mu}{\rho . D^{3/2} . g^{1/2}}$]

18. The force exerted by a flowing fluid on a stationary body depends upon the length (L) of the body, velocity (V) of the fluid, density (ρ) of fluid, viscosity (μ) of the fluid and acceleration (g) due to gravity. Find an expression for the force using dimensional analysis. (*Delhi University, May 1998*)

$$\left[\textbf{Ans. } F = \rho L^2 V^2 \phi\left(\frac{\mu}{\rho V L}, \frac{L \times g}{V^2}\right)\right]$$

19. The pressure difference Δp in a pipe of diameter D and length L due to turbulent flow depends upon the velocity V, viscosity μ, density ρ and roughness k. Using Buckingham's π-theorem or otherwise obtain an expression for Δp. (*Delhi University, December 2002*)

7 CHAPTER

Fluid Dynamics

▶ 7.1 INTRODUCTION

In the previous chapter, we studied the velocity and acceleration at a point in a fluid flow, without taking into consideration the forces causing the flow. This chapter includes the study of forces causing fluid flow. Thus dynamics of fluid flow is the study of fluid motion with the forces causing flow. The dynamic behaviour of the fluid flow is analysed by the Newton's second law of motion, which relates the acceleration with the forces. The fluid is assumed to be incompressible and non-viscous.

▶ 7.2 EQUATIONS OF MOTION

According to Newton's second law of motion, the net force F_x acting on a fluid element in the direction of x is equal to mass m of the fluid element multiplied by the acceleration a_x in the x-direction. Thus mathematically,

$$F_x = m.a_x \qquad \text{...(7.1)}$$

In the fluid flow, the following forces are present :

(i) F_g, gravity force.
(ii) F_p, the pressure force.
(iii) F_v, force due to viscosity.
(iv) F_t, force due to turbulence.
(v) F_c, force due to compressibility.

Thus in equation (7.1), the net force

$$F_x = (F_g)_x + (F_p)_x + (F_v)_x + (F_t)_x + (F_c)_x.$$

(i) If the force due to compressibility, F_c is negligible, the resulting net force

$$F_x = (F_g)_x + (F_p)_x + (F_v)_x + (F_t)_x$$

and equation of motions are called **Reynold's equations of motion.**

(ii) For flow, where (F_t) is negligible, the resulting equations of motion are known as **Navier-Stokes Equation.**

(iii) If the flow is assumed to be ideal, viscous force (F_v) is zero and equation of motions are known as **Euler's equation of motion.**

▶ 7.3 EULER'S EQUATION OF MOTION

This is equation of motion in which the forces due to gravity and pressure are taken into consideration. This is derived by considering the motion of a fluid element along a stream-line as :

Consider a stream-line in which flow is taking place in *s*-direction as shown in Fig. 7.1. Consider a cylindrical element of cross-section *dA* and length *dS*. The forces acting on the cylindrical element are :

1. Pressure force *pdA* in the direction of flow.
2. Pressure force $\left(p + \frac{\partial p}{\partial s} ds\right) dA$ opposite to the direction of flow.
3. Weight of element $\rho g dA ds$.

Let θ is the angle between the direction of flow and the line of action of the weight of element.

The resultant force on the fluid element in the direction of *s* must be equal to the mass of fluid element × acceleration in the direction *s*.

$$\therefore \quad pdA - \left(p + \frac{\partial p}{\partial s} ds\right) dA - \rho g dA ds \cos\theta$$

$$= \rho dA ds \times a_s \qquad \text{...(7.2)}$$

where a_s is the acceleration in the direction of *s*.

Now $\qquad a_s = \frac{dv}{dt}$, where *v* is a function of *s* and *t*.

$$= \frac{\partial v}{\partial s}\frac{ds}{dt} + \frac{\partial v}{\partial t} = \frac{v \partial v}{\partial s} + \frac{\partial v}{\partial t} \quad \left\{\because \frac{dS}{dt} = v\right\}$$

If the flow is steady, $\frac{\partial v}{\partial t} = 0$

$$\therefore \quad a_s = \frac{v\partial v}{\partial s}$$

Substituting the value of a_s in equation (7.2) and simplifying the equation, we get

$$-\frac{\partial p}{\partial s} ds dA - \rho g\, dA ds \cos\theta = \rho dA ds \times \frac{v\partial v}{\partial s}$$

Fig. 7.1 *Forces on a fluid element.*

Dividing by $\rho ds dA$, $-\frac{\partial p}{\rho \partial s} - g\cos\theta = \frac{v\partial v}{\partial s}$

or $\qquad \frac{\partial p}{\rho \partial s} + g\cos\theta + v\frac{\partial v}{\partial s} = 0$

But from Fig. 7.1 (*b*), we have $\cos\theta = \frac{dz}{ds}$

$$\therefore \quad \frac{1}{\rho}\frac{\partial p}{\partial \rho} + g\frac{dz}{ds} + \frac{v\partial v}{\partial s} = 0 \quad \text{or} \quad \frac{\partial p}{\rho} + gdz + vdv = 0$$

or $$\frac{\partial p}{\rho} + gdz + vdv = 0 \qquad \text{...(7.3)}$$

Equation (7.3) is known as Euler's equation of motion.

▶ 7.4 BERNOULLI'S EQUATION FROM EULER'S EQUATION

Bernoulli's equation is obtained by integrating the Euler's equation of motion (7.3) as

$$\int \frac{dp}{\rho} + \int gdz + \int vdv = \text{constant}$$

If flow is incompressible, ρ is constant and

$$\therefore \quad \frac{p}{\rho} + gz + \frac{v^2}{2} = \text{constant}$$

or

$$\frac{p}{\rho g} + z + \frac{v^2}{2g} = \text{constant}$$

or

$$\frac{p}{\rho g} + \frac{v^2}{2g} + z = \text{constant} \qquad \text{...(7.4)}$$

Equation (7.4) is a Bernoulli's equation in which

$\dfrac{p}{\rho g}$ = pressure energy per unit weight of fluid or pressure head.

$v^2/2g$ = kinetic energy per unit weight or kinetic head.

z = potential energy per unit weight or potential head.

7.4.1. Aussmptions. The following are the assumptions made in the derivation of Bernoulli's equation :

(*i*) The fluid is ideal, *i.e.*, viscosity is zero (*ii*) The flow is steady

(*iii*) The flow is incompressible (*iv*) The flow is irrotational.

Problem 7.1 *Water is flowing through a pipe of 5 cm diameter under a pressure of 29.43 N/cm² (gauge) and with mean velocity of 2.0 m/s. Find the total head or total energy per unit weight of the water at a cross-section, which is 5 m above the datum line.*

Solution. Given :

Diameter of pipe = 5 cm = 0.5 m

Pressure, $p = 29.43 \text{ N/cm}^2 = 29.43 \times 10^4 \text{ N/m}^2$

Velocity, $v = 2.0$ m/s

Datum head, $z = 5$ m

Total head = pressure head + kinetic head + datum head

Pressure head $= \dfrac{p}{\rho g} = \dfrac{29.43 \times 10^4}{1000 \times 9.81} = 30$ m $\left\{ \rho \text{ for water} = 1000 \dfrac{\text{kg}}{\text{m}^3} \right\}$

Kinetic head $= \dfrac{v^2}{2g} = \dfrac{2 \times 2}{2 \times 9.81} = 0.204$ m

∴ Total head $= \dfrac{p}{\rho g} + \dfrac{v^2}{2g} + z = 30 + 0.204 + 5 =$ **35.204 m. Ans.**

Problem 7.2 *A pipe, through which water is flowing, is having diameters, 20 cm and 10 cm at the cross-sections 1 and 2 respectively. The velocity of water at section 1 is given 4.0 m/s. Find the velocity head at sections 1 and 2 and also rate of discharge.*

Solution. Given :

$$D_1 = 20 \text{ cm} = 0.2 \text{ m}$$

∴ Area, $$A_1 = \frac{\pi}{4} D_1^2 = \frac{\pi}{4} (.2)^2 = 0.0314 \text{ m}^2$$

$$V_1 = 4.0 \text{ m/s}$$

$$D_2 = 0.1 \text{ m}$$

∴ $$A_2 = \frac{\pi}{4} (.1)^2 = .00785 \text{ m}^2$$

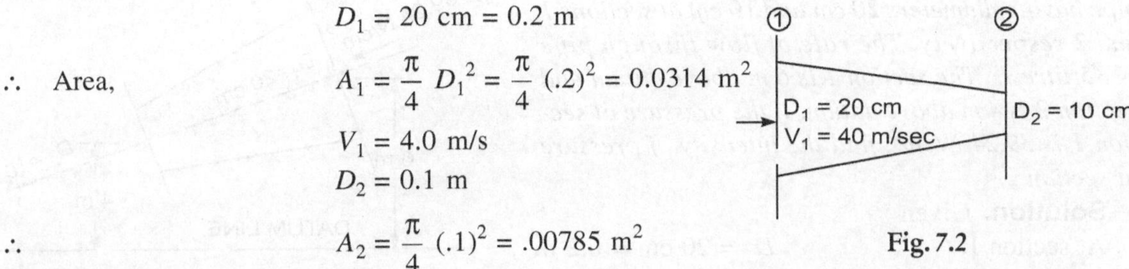

Fig. 7.2

(*i*) Velocity head at section 1

$$= \frac{V_1^2}{2g} = \frac{4.0 \times 4.0}{2 \times 9.81} = \mathbf{0.815 \text{ m. Ans.}}$$

(*ii*) Velocity head at section 2 = $V_2^2/2g$

To find V_2, apply continuity equation at 1 and 2

∴ $$A_1V_1 = A_2V_2 \quad \text{or} \quad V_2 = \frac{A_1V_1}{A_2} = \frac{.0314}{.00785} \times 4.0 = 16.0 \text{ m/s}$$

∴ Velocity head at section 2 $$= \frac{V_2^2}{2g} = \frac{16.0 \times 16.0}{2 \times 9.81} = \mathbf{83.047 \text{ m. Ans.}}$$

(*iii*) Rate of discharge $= A_1V_1$ or A_2V_2

$$= 0.0314 \times 4.0 = 0.1256 \text{ m}^3\text{/s}$$

$$= \mathbf{125.6 \text{ litres/s. Ans.}} \qquad \{\because 1 \text{ m}^3 = 1000 \text{ litres}\}$$

Problem 7.3 *State Bernoulli's theorem for steady flow of an incompressible fluid. Derive an expression for Bernoulli's equation from first principle and state the assumptions made for such a derivation.* (A.M.I.E., May 1974)

Solution. Statement of Bernoulli's Theorem. It states that in a steady, ideal flow of an incompressible fluid, the total energy at any point of the fluid is constant. The total energy consists of pressure energy, kinetic energy and potential energy or datum energy. These energies per unit weight of the fluid are :

$$\text{Pressure energy} = \frac{p}{\rho g}$$

$$\text{Kinetic energy} = \frac{v^2}{2g}$$

$$\text{Datum energy} = z$$

Thus mathematically, Bernoulli's theorem is written as

$$\frac{p}{w} + \frac{v^2}{2g} + z = \text{Constant.}$$

Derivation of Bernoulli's theorem. For derivation of Bernoulli's theorem, the Articles 7.3 and 7.4 should be written.

Assumptions are given in Article 7.4.1.

Problem 7.4 *The water is flowing through a pipe having diameters 20 cm and 10 cm at sections 1 and 2 respectively. The rate of flow through pipe is 35 litres/s. The section 1 is 6 m above datum and section 2 is 4 m above datum. If the pressure at section 1 is 39.24 N/cm², find the intensity of pressure at section 2.*

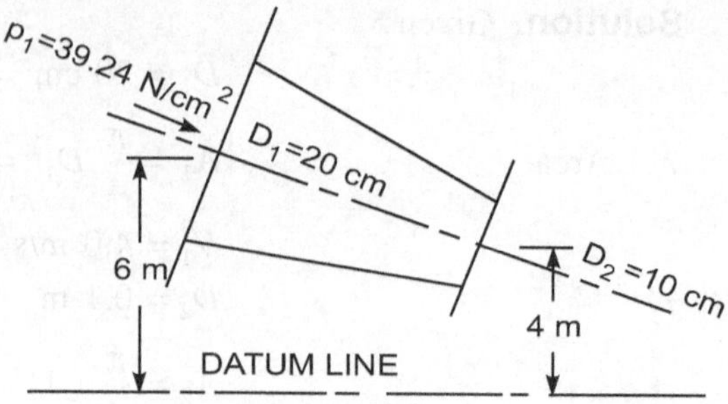

Fig. 7.3

Solution. Given :

At section 1, $D_1 = 20 \text{ cm} = 0.2 \text{ m}$

$$A_1 = \frac{\pi}{4}(.2)^2 = .0314 \text{ m}^2$$

$$p_1 = 39.24 \text{ N/cm}^2 = 39.24 \times 10^4 \text{ N/m}^2$$

$$z_1 = 6.0 \text{ m}$$

At section 2, $D_2 = 0.10 \text{ m}$

$$A_2 = \frac{\pi}{4}(0.1)^2 = .00785 \text{ m}^2$$

$$z_2 = 4 \text{ m}$$

$$p_2 = ?$$

Rate of flow, $Q = 35 \text{ lit/s} = \dfrac{35}{1000} = .035 \text{ m}^3\text{/s}$

Now $Q = A_1V_1 = A_2V_2$

$$\therefore \quad V_1 = \frac{Q}{A_1} = \frac{.035}{.0314} = 1.114 \text{ m/s}$$

and

$$V_2 = \frac{Q}{A_2} = \frac{.035}{.00785} = 4.456 \text{ m/s}$$

Applying Bernoulli's equation at sections 1 and 2, we get

$$\frac{p_1}{\rho g} + \frac{V_1^2}{2g} + z_1 = \frac{p_2}{\rho g} + \frac{V_2^2}{2g} + z_2$$

or

$$\frac{39.24 \times 10^4}{1000 \times 9.81} + \frac{(1.114)^2}{2 \times 9.81} + 6.0 = \frac{p_2}{1000 \times 9.81} + \frac{(4.456)^2}{2 \times 9.81} + 4.0$$

or

$$40 + 0.063 + 6.0 = \frac{p_2}{9810} + 1.012 + 4.0$$

or

$$46.063 = \frac{p_2}{9810} + 5.012$$

$$\therefore \quad \frac{p_2}{9810} = 46.063 - 5.012 = 41.051$$

$$\therefore \quad p_2 = 41.051 \times 9810 \text{ N/m}^2$$

$$= \frac{41.051 \times 9810}{10^4} \text{ N/cm}^2 = \mathbf{40.27 \text{ N/cm}^2}. \textbf{ Ans.}$$

Problem 7.5 *Water is flowing through a pipe having diameter 300 mm and 200 mm at the bottom and upper end respectively. The intensity of pressure at the bottom end is 24.525 N/cm² and the pressure at the upper end is 9.81 N/cm². Determine the difference in datum head if the rate of flow through pipe is 40 lit/s.*

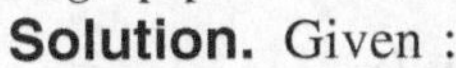

Solution. Given :

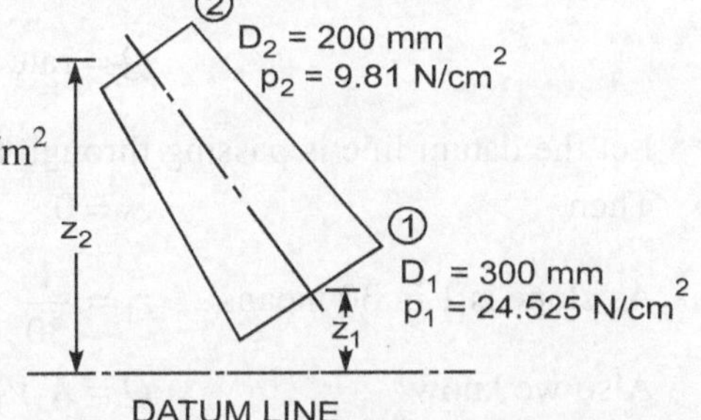

Fig. 7.4

Section 1, $D_1 = 300 \text{ mm} = 0.3 \text{ m}$

$p_1 = 24.525 \text{ N/cm}^2 = 24.525 \times 10^4 \text{ N/m}^2$

Section 2, $D_2 = 200 \text{ mm} = 0.2 \text{ m}$

$p_2 = 9.81 \text{ N/cm}^2 = 9.81 \times 10^4 \text{ N/m}^2$

Rate of flow $= 40$ lit/s

or $$Q = \frac{40}{1000} = 0.04 \text{ m}^3\text{/s}$$

Now $$A_1V_1 = A_2V_2 = \text{rate of flow} = 0.04$$

$$\therefore \quad V_1 = \frac{.04}{A_1} = \frac{.04}{\frac{\pi}{4}D_1^2} = \frac{0.04}{\frac{\pi}{4}(0.3)^2} = 0.5658 \text{ m/s}$$

$$\simeq 0.566 \text{ m/s}$$

$$V_2 = \frac{.04}{A_2} = \frac{.04}{\frac{\pi}{4}(D_2)^2} = \frac{0.04}{\frac{\pi}{4}(0.2)^2} = 1.274 \text{ m/s}$$

Applying Bernoulli's equation at (1) and (2), we get

$$\frac{p_1}{\rho g} + \frac{V_1^2}{2g} + z_1 = \frac{p_2}{\rho g} + \frac{V_2^2}{2g} + z_2$$

or $$\frac{24.525 \times 10^4}{1000 \times 9.81} + \frac{.566 \times .566}{2 \times 9.81} + z_1 = \frac{9.81 \times 10^4}{1000 \times 9.81} + \frac{(1.274)^2}{2 \times 9.81} + z_2$$

or $$25 + .32 + z_1 = 10 + 1.623 + z_2$$

or $$25.32 + z_1 = 11.623 + z_2$$

$$\therefore \quad z_2 - z_1 = 25.32 - 11.623 = 13.697 = 13.70 \text{ m}$$

$\therefore$ Difference in datum head $= z_2 - z_1 =$ **13.70 m. Ans.**

Problem 7.6 *The water is flowing through a taper pipe of length 100 m having diameters 600 mm at the upper end and 300 mm at the lower end, at the rate of 50 litres/s. The pipe has a slope of 1 in 30. Find the pressure at the lower end if the pressure at the higher level is 19.62 N/cm².*

Solution. Given :

Length of pipe, $L = 100$ m

Dia. at the upper end, $D_1 = 600 \text{ mm} = 0.6 \text{ m}$

$\therefore$ Area, $$A_1 = \frac{\pi}{4} D_1^2 = \frac{\pi}{4} \times (.6)^2$$

$$= 0.2827 \text{ m}^2$$

p_1 = pressure at upper end

$= 19.62 \text{ N/cm}^2$

$= 19.62 \times 10^4 \text{ N/m}^2$

Fig. 7.5

Dia. at lower end, $D_2 = 300 \text{ mm} = 0.3 \text{ m}$

∴ Area, $A_2 = \frac{\pi}{4} D_2^2 = \frac{\pi}{4} (.3)^2 = 0.07068 \text{ m}$

$$Q = \text{rate of flow} = 50 \text{ litres/s} = \frac{50}{1000} = 0.05 \text{ m}^3\text{/s}$$

Let the datum line is passing through the centre of the lower end.

Then $z_2 = 0$

As slope is 1 in 30 means $z_1 = \frac{1}{30} \times 100 = \frac{10}{3} \text{ m}$

Also we know $Q = A_1 V_1 = A_2 V_2$

∴ $V_1 = \frac{Q}{A} = \frac{0.05}{.2827} = 0.1768 \text{ m/sec} = 0.177 \text{ m/s}$

and $V_2 = \frac{Q}{A_2} = \frac{0.5}{.07068} = 0.7074 \text{ m/sec} = 0.707 \text{ m/s}$

Applying Bernoulli's equation at sections (1) and (2), we get

$$\frac{p_1}{\rho g} + \frac{V_1^2}{2g} + z_1 = \frac{p_2}{\rho g} + \frac{V_2^2}{2g} + z_2$$

or $$\frac{19.62 \times 10^4}{1000 \times 9.81} + \frac{.177^2}{2 \times 9.81} + \frac{10}{3} = \frac{p_2}{\rho g} + \frac{.707^2}{2 \times 9.81} + 0$$

or $$20 + 0.001596 + 3.334 = \frac{p_2}{\rho g} + 0.0254$$

or $$23.335 - 0.0254 = \frac{p_2}{1000 \times 9.81}$$

or $p_2 = 23.3 \times 9810 \text{ N/m}^2 = 228573 \text{ N/m}^2 = \mathbf{22.857 \text{ N/cm}^2}$. **Ans.**

Problem 7.7 *A non-uniform part of a pipe line 5 m long is laid at a slope of 2 in 5. Two pressure gauges each fitted at upper and lower ends read 20 N/cm² and 12.5 N/cm². If the diameters at the upper and lower ends are 15 cm and 10 cm respectively, determine the quantity of water flowing per second.* (VTU, Jan. 2004)

Solution. Given :

$$L = 5 \text{ m}, D_1 = 15 \text{ cm} = 0.15 \text{ m}, D_2 = 10 \text{ cm} = 0.1 \text{ m}, p_1 = 20 \text{ N/cm}^2$$
$$= 20 \times 10^4 \text{ N/m}^2, p_2 = 12.5 \text{ N/cm}^2 = 12.5 \times 10^4 \text{ N/m}^2$$

∴ $$A_1 = \frac{\pi}{4} D_1^2 = \frac{\pi}{4} \times (0.15)^2 = 0.01767 \text{ m}^2, A_2 = \frac{\pi}{4} D_2^2 = \frac{\pi}{4} \times (0.1)^2$$
$$= 0.00785 \text{ m}^2$$

Let the datum line is passing through the centre of lower end.

Then $z_2 = 0$

As the slope is 2 in 5, hence $z_1 = \frac{2}{5} \times 5 = 2$ m

Using continuity equation,

$$A_1 V_1 = A_2 V_2$$

$$\therefore \quad V_1 = \frac{A_2 V_2}{A_1} = \frac{0.00785}{0.01767} \times V_2$$

$$= 0.444 \ V_2$$

Applying Bernoulli's equation at section (1) and (2), we get

Fig. 7.5 (*a*)

$$\frac{p_1}{\rho g} + \frac{V_1^2}{2g} + z_1 = \frac{p_2}{\rho g} + \frac{V_2^2}{2g} + z_2$$

$$\text{or} \quad \frac{20 \times 10^4}{1000 \times 9.81} + \frac{(0.444 \ V_2)^2}{2g} + 2 = \frac{12.5 \times 10^4}{1000 \times 9.81} + \frac{V_2^2}{2g} + 0$$

$$\text{or} \quad \frac{20 \times 10^4}{1000 \times 9.81} - \frac{12.5 \times 10^4}{1000 \times 9.81} + 2 = \frac{V_2^2}{2g} - \frac{(0.444)^2 \times V_2^2}{2g}$$

$$\text{or} \quad \frac{7.5 \times 10^4}{1000 \times 9.81} + 2 = \frac{V_2^2}{2g}\left(1 - (0.444)^2\right)$$

$$\text{or} \quad 7.645 + 2 = \frac{V_2^2}{2g} \times 0.8028$$

$$\text{or} \quad 9.645 = \frac{V_2^2}{2g} \times 0.8028$$

$$\therefore \quad V_2 = \sqrt{\frac{9.645 \times 2 \times 9.81}{0.8028}} = 15.35 \text{ m/s}$$

$$\therefore \quad \text{Discharge, } Q = A_2 V_2$$

$$= (0.00785) \times 15.35 = 0.1205 \text{ m}^3\text{/s} = \textbf{120.5 l/s. Ans.}$$

Problem 7.8 *A pump has a tappering pipe running full of water. The pipe is placed vertically with the diameters at the base and top being 1.2 m and 0.6 m respectively. The pressure at the upper end is 240 mm of Hg vacuum, while the pressure at the lower end is 15 kN/m². Assume the head loss to be 20% of difference of velocity head. Calculate the discharge, the flow is vertically upwards and difference of elevation is 3.9 m.*

(VTU, Aug. 2002)

Solution. Given : $D_1 = 1.2$ m, $D_2 = 0.6$ m, $p_1 = 15$ kN/m^2 = 15×1000 N/m^2, $\frac{p_2}{\rho_m g} = 240$ mm of Hg (vacuum)

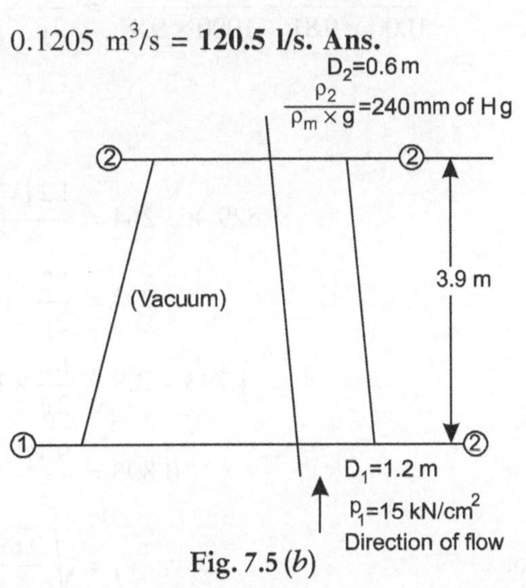

Fig. 7.5 (*b*)

$= -240$ mm of Hg

where ρ_m = Density of Hg = (13.6×1000) kg/m^3.

Head loss, h_L = 20% of difference of velocity head

$$= \frac{20}{100}\left[\frac{V_2^2}{2g} - \frac{V_1^2}{2g}\right] = 0.2\left[\frac{V_2^2 - V_1^2}{2g}\right]$$

(Here V_2 is more than V_1)

Difference of vertical height, $z_2 - z_1 = 3.9$ m

As pressure head at the upper end is 240 mm of Hg (vacuum)

$$\text{hence } \frac{p_2}{\rho_m g} = -240 \text{ mm of Hg} = -0.24 \text{ m of Hg}$$

where ρ_m = Density of mercury
= (13.6×1000) kg/m^3

$\therefore \quad p_2 = -0.24 \times \rho_m \times g$
$= -0.24 \times (13.6 \times 1000) \times 9.81$ N/m^2 = -32019.8 N/m^2

Using continuity equation,

$$A_1V_1 = A_2V_2$$

$$\therefore \quad V_2 = \frac{A_1V_1}{A_2} = \frac{\frac{\pi}{4}(D_1^2) \times V_1}{\frac{\pi}{4}(D_2^2)} = \left(\frac{D_1}{D_2}\right)^2 \times V_1 = \left(\frac{1.2}{0.6}\right)^2 \times V_1 = 4V_1$$

Now applying Bernoulli's equation at sections (1) and (2), we get

$$\frac{p_1}{\rho g} + \frac{V_1^2}{2g} + z_1 = \frac{p_2}{\rho g} + \frac{V_2^2}{2g} + z_2 + h_L$$

or
$$\frac{p_1}{\rho g} - \frac{p_2}{\rho g} = \frac{V_2^2}{2g} - \frac{V_1^2}{2g} + z_2 - z_1 + h_L$$

$$\frac{15 \times 1000}{1000 \times 9.81} - \frac{(-32019.8)}{1000 \times 9.81} = \frac{V_2^2}{2g} - \frac{V_1^2}{2g} + 3.9 + 0.2\left(\frac{V_2^2 - V_1^2}{2g}\right)$$

$$\left[\because z_2 - z_1 = 3.9 \text{ m and } h_L = 0.2\,\frac{(V_2^2 - V_1^2)}{2g}\right]$$

$$1.529 + 3.264 = \frac{1.2\left(V_2^2 - V_1^2\right)}{2g} + 3.9$$

$$4.793 = \frac{1.2}{2g}\,[(4V_1)^2 - V_1^2] + 3.9 \qquad (\because V_2 = 4V_1)$$

or
$$4.793 - 3.9 = \frac{1.2}{2g} \times 15\,V_1^2$$

or
$$0.893 = \frac{9\,V_1^2}{g}$$

or
$$V_1 = \sqrt{\frac{0.893 \times 9.81}{9}} = \sqrt{0.97337} = 0.9865 \text{ m/s}$$

$\therefore$ Discharge, $Q = A_1V_1 = \dfrac{\pi}{4} D_1^2 \times V_1 = \dfrac{\pi}{4} \times (1.2)^2 \times 0.9865 =$ **1.1157 m³/s. Ans.**

▶ 7.5 BERNOULLI'S EQUATION FOR REAL FLUID

The Bernoulli's equation was derived on the assumption that fluid is inviscid (non-viscous) and therefore frictionless. But all the real fluids are viscous and hence offer resistance to flow. Thus there are always some losses in fluid flows and hence in the application of Bernoulli's equation, these losses have to be taken into consideration. Thus the Bernoulli's equation for real fluids between points 1 and 2 is given as

$$\frac{p_1}{\rho g} + \frac{v_1^2}{2g} + z_1 = \frac{p_2}{\rho g} + \frac{v_2^2}{2g} + z_2 + h_L \qquad \text{...(7.5)}$$

where h_L is loss of energy between points 1 and 2.

Problem 7.9 *A pipe of diameter 400 mm carries water at a velocity of 25 m/s. The pressures at the points A and B are given as 29.43 N/cm² and 22.563 N/cm² respectively while the datum head at A and B are 28 m and 30 m. Find the loss of head between A and B.*

Solution. Given :

Dia. of pipe, $D = 400$ mm $= 0.4$ m

Velocity, $V = 25$ m/s

At point A, $p_A = 29.43$ N/cm² $= 29.43 \times 10^4$ N/m²

$z_A = 28$ m

$v_A = v = 25$ m/s

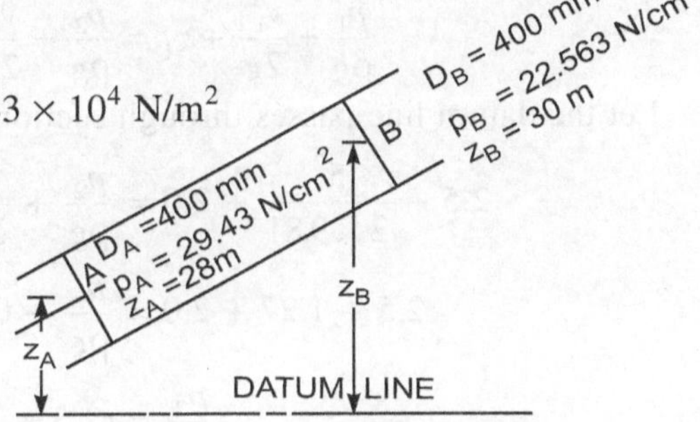

Fig. 7.6

$\therefore$ Total energy at A, $E_A = \dfrac{p_A}{\rho g} + \dfrac{v_A^2}{2g} + z_A$

$$= \frac{29.43 \times 10^4}{1000 \times 9.81} + \frac{25^2}{2 \times 9.81} + 28$$

$= 30 + 31.85 + 28 = 89.85$ m

At point B, $p_B = 22.563$ N/cm² $= 22.563 \times 10^4$ N/m²

$z_B = 30$ m

$v_B = v = v_A = 25$ m/s

$\therefore$ Total energy at B, $E_B = \dfrac{p_B}{\rho g} + \dfrac{v_B^2}{2g} + z_B$

$$= \frac{22.563 \times 10^4}{1000 \times 9.81} + \frac{25^2}{2 \times 9.81} + 30 = 23 + 31.85 + 30 = 84.85 \text{ m}$$

$\therefore$ Loss of energy $= E_A - E_B = 89.85 - 84.85 =$ **5.0 m. Ans.**

Problem 7.10 *A conical tube of length 2.0 m is fixed vertically with its smaller end upwards. The velocity of flow at the smaller end is 5 m/s while at the lower end it is 2 m/s. The pressure head at the* *smaller end is 2.5 m of liquid. The loss of head in the tube is* $\frac{0.35(v_1 - v_2)^2}{2g}$*, where* v_1 *is the velocity at the smaller end and* V_2 *at the lower end respectively. Determine the pressure head at the lower end. Flow takes place in the downward direction.*

Solution. Let the smaller end is represented by (1) and lower end by (2)

Given :

Length of tube, $L = 2.0$ m

$v_1 = 5$ m/s

$p_1/\rho g = 2.5$ m of liquid

$v_2 = 2$ m/s

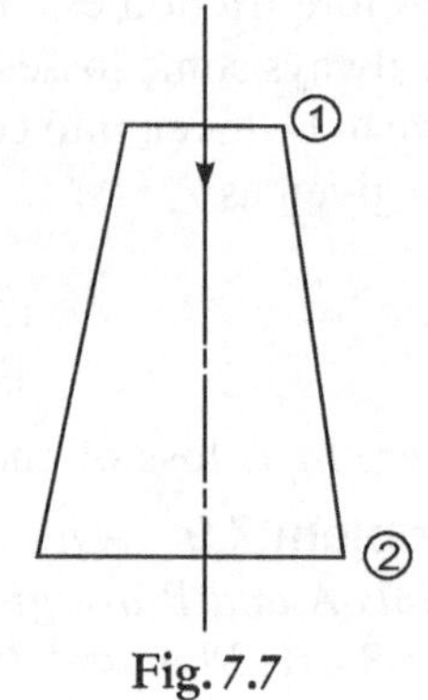

Fig. 7.7

Loss of head $= h_L = \frac{0.35(v_1 - v_2)^2}{2g}$

$= \frac{0.35\,[5-2]^2}{2g} = \frac{0.35 \times 9}{2 \times 9.81} = 0.16$ m

Pressure head, $\frac{p_2}{\rho g} = ?$

Applying Bernoulli's equation at sections (1) and (2), we get

$$\frac{p_1}{\rho g} + \frac{v_1^2}{2g} + z_1 = \frac{p_2}{\rho g} + \frac{v_2^2}{2g} + z_2 + h_L$$

Let the datum line passes through section (2). Then $z_2 = 0$, $z_1 = 2.0$

$$\therefore \quad 2.5 + \frac{5^2}{2 \times 9.81} + 2.0 = \frac{p_2}{\rho g} + \frac{2^2}{2 \times 9.81} + 0 + 0.16$$

$$2.5 + 1.27 + 2.0 = \frac{p_2}{\rho g} + 0.203 + .16$$

or $$\frac{p_2}{\rho g} = (2.5 + 1.27 + 2.0) - (.203 + .16)$$

$= 5.77 - .363 =$ **5.407 m of fluid. Ans.**

Problem 7.11 *A pipe line carrying oil of specific gravity 0.87, changes in diameter from 200 mm diameter at a position A to 500 mm diameter at a position B which is 4 metres at a higher level. If the pressures at A and B are 9.81 N/cm² and 5.886 N/cm² respectively and the discharge is 200 litres/s determine the loss of head and direction of flow.* (A.M.I.E., Summer 1976)

Solution. Discharge, $Q = 200$ lit/s $= 0.2$ m³/s

Sp. gr. of oil $= 0.87$

$\therefore$ ρ for oil $= .87 \times 1000 = 870 \frac{\text{kg}}{\text{m}^3}$

Given : **At section A,** $D_A = 200$ mm $= 0.2$ m

Area, $A_A = \frac{\pi}{4}(D_A)^2 = \frac{\pi}{4}(.2)^2$

$= 0.0314$ m²

$p_A = 9.81$ N/cm²

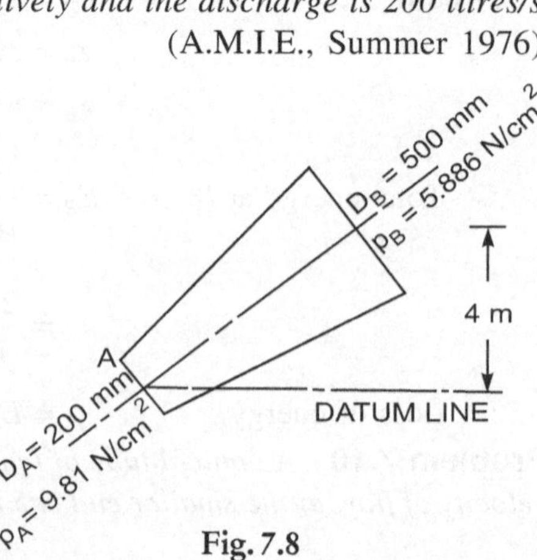

Fig. 7.8

$$= 9.81 \times 10^4 \text{ N/m}^2$$

If datum line is passing through A, then

$$Z_A = 0$$

$$V_A = \frac{Q}{A_A} = \frac{0.2}{0.0314} = 6.369 \text{ m/s}$$

At section B, $D_B = 500 \text{ mm} = 0.50 \text{ m}$

Area, $A_B = \frac{\pi}{4} D_B^{\ 2} = \frac{\pi}{4} (.5)^2 = 0.1963 \text{ m}^2$

$$p_B = 5.886 \text{ N/cm}^2 = 5.886 \times 10^4 \text{ N/m}^2$$

$$Z_B = 4.0 \text{ m}$$

$$V_B = \frac{Q}{\text{Area}} = \frac{0.2}{.1963} = 1.018 \text{ m/s}$$

Total energy at A $= E_A = \frac{p_A}{\rho g} + \frac{V_A^2}{2g} + Z_A$

$$= \frac{9.81 \times 10^4}{870 \times 9.81} + \frac{(6.369)^2}{2 \times 9.81} + 0 = 11.49 + 2.067 = 13.557 \text{ m}$$

Total energy at B $= E_B = \frac{p_B}{\rho g} + \frac{V_B^2}{2g} + Z_B$

$$= \frac{5.886 \times 10^4}{870 \times 9.81} + \frac{(1.018)^2}{2 \times 9.81} + 4.0 = 6.896 + 0.052 + 4.0 = 10.948 \text{ m}$$

(*i*) **Direction of flow.** As E_A is more than E_B and hence flow is taking place from A to B. **Ans.**

(*ii*) Loss of head $= h_L = E_A - E_B = 13.557 - 10.948 =$ **2.609 m. Ans.**

▶ 7.6 THE MOMENTUM EQUATION

It is based on the law of conservation of momentum or on the momentum principle, which states that the net force acting on a fluid mass is equal to the change in momentum of flow per unit time in that direction. The force acting on a fluid mass 'm' is given by the Newton's second law of motion,

$$F = m \times a$$

where a is the acceleration acting in the same direction as force F.

But $a = \frac{dv}{dt}$

$\therefore$ $F = m \frac{dv}{dt}$

$= \frac{d(mv)}{dt}$ {m is constant and can be taken inside the differential}

$\therefore$ $$F = \frac{d(mv)}{dt} \quad \text{...(7.6)}$$

Equation (7.6) is known as the momentum principle.

Equation (7.6) can be written as $F.dt = d(mv)$ ...(7.7)

which is known as the *impulse-momentum equation* and states that the impulse of a force F acting on a fluid of mass m in a short interval of time dt is equal to the change of momentum $d(mv)$ in the direction of force.

▶ 7.7 FORCE EXERTED BY A FLOWING FLUID ON A PIPE-BEND

The impulse-momentum equation (7.7) is used to determine the resultant force exerted by a flowing fluid on a pipe bend.

Consider two sections (1) and (2), as shown in Fig. 7.9.

Let v_1 = velocity of flow at section (1),

p_1 = pressure intensity at section (1),

A_1 = area of cross-section of pipe at section (1) and

v_2, p_2, A_2 = corresponding values of velocity, pressure and area at section (2).

Let F_x and F_y be the components of the forces exerted by the flowing fluid on the bend in x-and y-directions respectively. Then the force exerted by the bend on the fluid in the directions of x and y will be equal to F_x and F_y but in the opposite directions. Hence component of the force exerted by bend on the fluid in the x-direction $= -F_x$ and in the direction of $y = -F_y$. The other external forces acting on the fluid are p_1A_1 and p_2A_2 on the sections (1) and (2) respectively. Then momentum equation in x-direction is given by

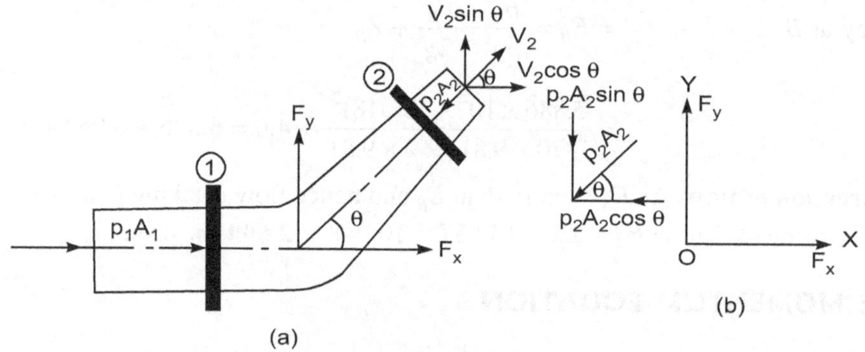

Fig. 7.9 *Forces on bend.*

Net force acting on fluid in the direction of x = Rate of change of momentum in x-direction

$$\therefore \quad p_1A_1 - p_2A_2 \cos\theta - F_x = \text{(Mass per sec) (change of velocity)}$$

$$= \rho Q \text{ (Final velocity in the direction of } x - \text{Initial velocity in the direction of } x)$$

$$= \rho Q (V_2 \cos\theta - V_1) \quad \text{...(7.8)}$$

$$\therefore \quad F_x = \rho Q (V_1 - V_2 \cos\theta) + p_1A_1 - p_2A_2 \cos\theta \quad \text{...(7.9)}$$

Similarly the momentum equation in y-direction gives

$$0 - p_2A_2 \sin\theta - F_y = \rho Q (V_2 \sin\theta - 0) \quad \text{...(7.10)}$$

$$\therefore \quad F_y = \rho Q (-V_2 \sin\theta) - p_2A_2 \sin\theta \quad \text{...(7.11)}$$

Now the resultant force (F_R) acting on the bend

$$= \sqrt{F_x^2 + F_y^2} \quad \text{...(7.12)}$$

And the angle made by the resultant force with horizontal direction is given by

$$\tan\theta = \frac{F_y}{F_x} \qquad \text{...(7.13)}$$

Problem 7.12 *A 45° reducing bend is connected in a pipe line, the diameters at the inlet and outlet of the bend being 600 mm and 300 mm respectively. Find the force exerted by water on the bend if the intensity of pressure at inlet to bend is 8.829 N/cm² and rate of flow of water is 600 litres/s.*

(Similar Question in VTU, March 2001)

Solution. Given :

Angle of bend, $\theta = 45°$

Dia. at inlet, $D_1 = 600 \text{ mm} = 0.6 \text{ m}$

$\therefore$ Area, $A_1 = \frac{\pi}{4} D_1^2 = \frac{\pi}{4} (.6)^2$

$= 0.2827 \text{ m}^2$

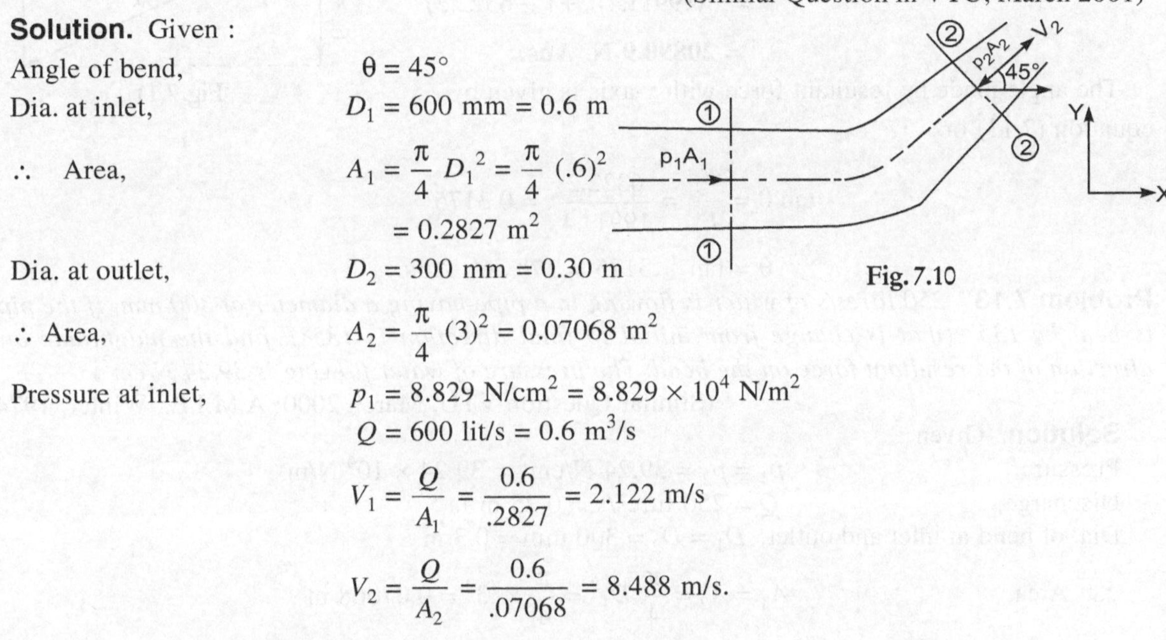

Fig. 7.10

Dia. at outlet, $D_2 = 300 \text{ mm} = 0.30 \text{ m}$

$\therefore$ Area, $A_2 = \frac{\pi}{4} (3)^2 = 0.07068 \text{ m}^2$

Pressure at inlet, $p_1 = 8.829 \text{ N/cm}^2 = 8.829 \times 10^4 \text{ N/m}^2$

$Q = 600 \text{ lit/s} = 0.6 \text{ m}^3\text{/s}$

$$V_1 = \frac{Q}{A_1} = \frac{0.6}{.2827} = 2.122 \text{ m/s}$$

$$V_2 = \frac{Q}{A_2} = \frac{0.6}{.07068} = 8.488 \text{ m/s.}$$

Applying Bernoulli's equation at section (1) and (2), we get

$$\frac{p_1}{\rho g} + \frac{V_1^2}{2g} + z_1 = \frac{p_2}{\rho g} + \frac{V_2^2}{2g} + z_2$$

But $z_1 = z_2$

$$\therefore \quad \frac{p_1}{\rho g} + \frac{V_1^2}{2g} = \frac{p_2}{\rho g} + \frac{V_2^2}{2g} \quad \text{or} \quad \frac{8.829 \times 10^4}{1000 \times 9.81} + \frac{2.122^2}{2 \times 9.81} = \frac{p_2}{\rho g} + \frac{8.488^2}{2 \times 9.81}$$

$$9 + .2295 = p_2/\rho g + 3.672$$

$$\therefore \quad \frac{p_2}{\rho g} = 9.2295 - 3.672 = 5.5575 \text{ m of water}$$

$$\therefore \quad p_2 = 5.5575 \times 1000 \times 9.81 \text{ N/m}^2 = 5.45 \times 10^4 \text{ N/m}^2$$

Forces on the bend in x- and y-directions are given by equations (7.9) and (7.11) as

$$F_x = \rho Q\,[v_1 - v_2 \cos\theta] + p_1A_1 - p_2A_2 \cos\theta$$

$$= 1000 \times 0.6\,[2.122 - 8.488 \cos 45°]$$

$$+ 8.829 \times 10^4 \times .2827 - 5.45 \times 10^4 \times .07068 \times \cos 45°$$

$$= -2327.9 + 24959.6 - 2720.3 = 24959.6 - 5048.2$$

$$= 19911.4 \text{ N}$$

and $F_y = \rho Q\,[-\,V_2 \sin \theta] - p_2 A_2 \sin \theta$

$= 1000 \times 0.6\,[-\,8.488 \sin 45°[\,-\,5.45 \times 10^4 \times .07068 \times \sin 45°$

$= -\,3601.1 - 2721.1 = -\,6322.2$ N

$-$ve sign means F_y is acting in the downward direction

$\therefore$ Resultant force, $F_R = \sqrt{F_x^2 + F_y^2}$

$= \sqrt{(19911.4)^2 + (-\,6322.2)^2}$

= **20890.9 N. Ans.**

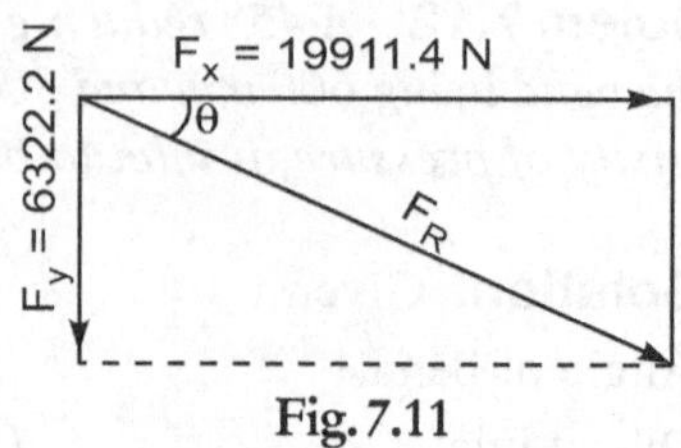

Fig. 7.11

The angle made by resultant force with x-axis is given by equation (7.13) or

$$\tan \theta = \frac{F_y}{F_x} = \frac{6322.2}{19911.4} = 0.3175$$

$\therefore$ $\theta = \tan^{-1} .3175 =$ **17° 36′. Ans.**

Problem 7.13 *250 litres/s of water is flowing in a pipe having a diameter of 300 mm. If the pipe is bent by 135° (that is change from initial to final direction is 135°), find the magnitude and direction of the resultant force on the bend. The pressure of water flowing is 39.24 N/cm².*

(Similar Question VTU, March 2000; A.M.I.E., Winter, 1974)

Solution. Given :

Pressure, $p_1 = p_2 = 39.24$ N/cm² $= 39.24 \times 10^4$ N/m²

Discharge, $Q = 250$ litres/s $= 0.25$ m³/s

Dia. of bend at inlet and outlet, $D_1 = D_2 = 300$ mm $= 0.3$ m

$\therefore$ Area, $A_1 = A_2 = \frac{\pi}{4} D_1^2 = \frac{\pi}{4} \times .3^2 = 0.07068$ m²

Velocity of water at (1) and (2), $V = V_1 = V_2 = \frac{Q}{\text{Area}} = \frac{0.25}{.07068} = 3.537$ m/s.

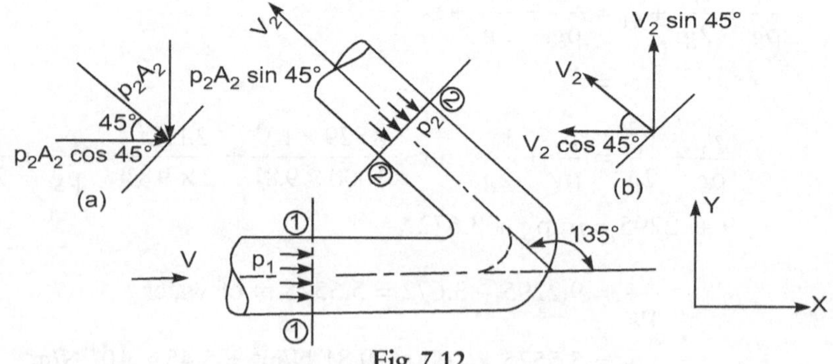

Fig. 7.12

Force along x-axis

$= F_x = \rho Q[V_{1x} - V_{2x}] + p_{1x} A_1 + p_{2x} A_2$

where V_{1x} = initial velocity in the direction of $x = 3.537$ m/s

V_{2x} = final velocity in the direction of $x = -\,V_2 \cos 45° = -\,3.537 \times .7071$

p_{1x} = pressure at (1) in x-direction

$= 39.24 \text{ N/cm}^2 = 39.24 \times 10^4 \text{ N/m}^2$

p_{2x} = pressure at (2) in x-direction

$= p_2 \cos 45° = 39.24 \times 10^4 \times .7071$

$\therefore \quad F_x = 1000 \times .25[3.537 - (-3.537 \times .7071)] + 39.24 \times 10^4 \times .07068 + 39.24 \times 10^4 \times .07068 \times .7071$

$= 1000 \times .25[3.537 + 3.537 \times .7071] + 39.24 \times 10^4 \times .07068\ [1 + .7071]$

$= 1509.4 + 47346 = 48855.4 \text{ N}$

Force along y-axis

$= F_y = \rho Q[V_{1y} - V_{2y}] + (p_1A_1)_y + (p_2A_2)_y$

where V_{1y} = initial velocity in y-direction = 0

V_{2y} = final velocity in y-direction = $- V_2 \sin 45° = 3.537 \times .7071$

$(p_1A_1)_y$ = pressure force in y-direction = 0

$(p_2A_2)_y$ = pressure force at (2) in y-direction

$= - p_2A_2 \sin 45° = - 39.24 \times 10^4 \times .07068 \times .7071$

$\therefore \quad F_y = 1000 \times .25[0 - 3.537 \times .7071] + 0 + (- 39.24 \times 10^4 \times .07068 \times .7071)$

$= - 625.2 - 19611.1 = - 20236.3 \text{ N}$

– ve sign means F_y is acting in the downward direction

$\therefore$ Resultant force, $F_R = \sqrt{F_x^2 + F_y^2}$

$= \sqrt{48855.4^2 + 20236.3^2}$

$=$ **52880.6 N. Ans.**

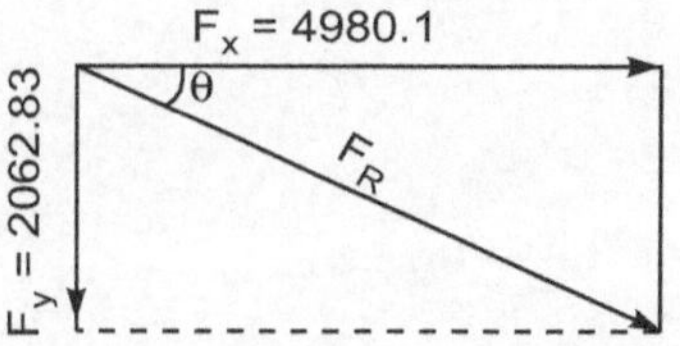

Fig. 7.13

The direction of the resultant force F_R, with the x-axis is given as

$$\tan \theta = \frac{F_y}{F_x} = \frac{20236.3}{48855.4} = 0.4142$$

$\therefore \quad \theta =$ **22° 30′. Ans.**

Problem 7.14 *A 300 mm diameter pipe carries water under a head of 20 metres with a velocity of 3.5 m/s. If the axis of the pipe turns through 45°, find the magnitude and direction of the resultant force at the bend.* (A.M.I.E., Summer, 1978)

Solution. Given :

Dia. of bend, $D = D_1 = D_2 = 300 \text{ mm} = 0.30 \text{ m}$

$\therefore$ Area, $A = A_1 = A_2 = \frac{\pi}{4} D^2 = \frac{\pi}{4} \times .3^2 = 0.07068 \text{ m}^2$

Velocity, $V = V_1 = V_2 = 3.5 \text{ m/s}$

$\theta = 45°$

Discharge, $Q = A \times V = 0.07068 \times 3.5 = 0.2475 \text{ m}^3\text{/s}$

Pressure head $= 20$ m of water or $\frac{p}{\rho g} = 20$ m of water

$\therefore \quad p = 20 \times \rho g = 20 \times 1000 \times 9.81 \text{ N/m}^2 = 196200 \text{ N/m}^2$

$\therefore$ Pressure intensity, $p = p_1 = p_2 = 196200 \text{ N/m}^2$

Now $V_{1x} = 3.5 \text{ m/s},\ V_{2x} = V_2 \cos 45° = 3.5 \times .7071$

$$V_{1y} = 0,\ V_{2y} = V_2 \sin 45° = 3.5 \times .7071$$

$$(p_1A_1)_x = p_1A_1 = 196200 \times .07068,\ (p_1A_1)_y = 0$$

$$(p_2A_2)_x = -\ p_2A_2 \cos 45°,\ (p_2A_2)_y = -\ p_2A_2 \sin 45°$$

Force along x-axis,

$$\begin{aligned} F_x &= \rho Q[V_{1x} - V_{2x}] + (p_1A_1)_x + (p_2A_2)_x \\ &= 1000 \times .2475[3.5 - 3.5 \times .7071] + 196200 \times .07068 - p_2A_2 \times \cos 45° \\ &= 253.68 + 196200 \times .07068 - 196200 \times .07068 \times 0.7071 \\ &= 253.68 + 13871.34 - 9808.04 = 4316.98 \text{ N} \end{aligned}$$

Force along y-axis,

$$\begin{aligned} F_y &= \rho Q\ [v_1y - v_2y] + (p_1A_1)_y + (p_2A_2)_y \\ &= 1000 \times .2475[0 - 3.5 \times .7071] + 0 + [-\ p_2A_2 \sin 45°] \\ &= -\ 612.44 - 196200 \times .07068 \times .7071 \\ &= -\ 612.44 - 9808 = -\ 10420.44 \text{ N} \end{aligned}$$

∴ Resultant force $F_R = \sqrt{F_x^2 + F_y^2} = \sqrt{(4316.98)^2 + (10420.44)^2}$ = **11279 N. Ans.**

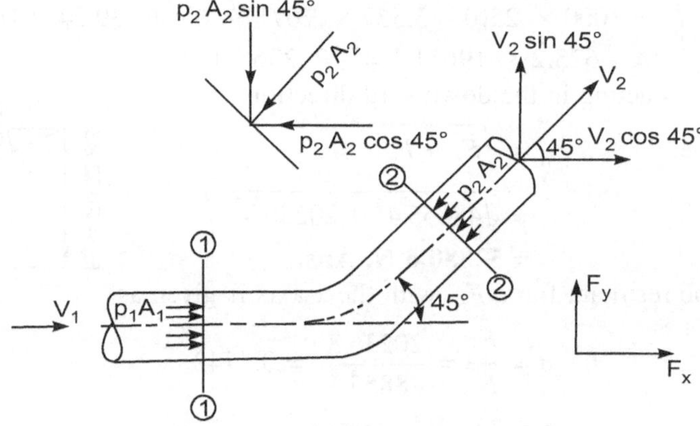

Fig. 7.14

The angle made by F_R with x-axis

$$\tan \theta = \frac{F_y}{F_x} = \frac{10420.44}{4316.98} = 2.411$$

∴ $\theta = \tan^{-1} 2.411$ = **67° 28′. Ans.**

Problem 7.15 *In a 45° bend a rectangular air duct of 1 m² cross-sectional area is gradually reduced to 0.5 m² area. Find the magnitude and direction of the force required to hold the duct in position if the velocity of flow at the 1 m² section is 10 m/s, and pressure is 2.943 N/cm². Take density of air as 1.16 kg/m³.* (A.M.I.E., Winter, 1980)

Solution. Given :

Area at section (1), $A_1 = 1 \text{ m}^2$

Area at section (2), $A_2 = 0.5 \text{ m}^2$

Velocity at section (1), $V_1 = 10 \text{ m/s}$

Pressure at section (1), $p_1 = 2.943 \text{ N/cm}^2 = 2.943 \times 10^4 \text{ N/m}^2 = 29430 \text{ N/m}^2$

Density of air, $\rho = 1.16 \text{ kg/m}^3$

Applying continuity equation at sections (1) and (2)

$$A_1V_1 = A_2V_2$$

$$\therefore \quad V_2 = \frac{A_1V_1}{A_2} = \frac{1}{0.5} \times 10 = 20 \text{ m/s}$$

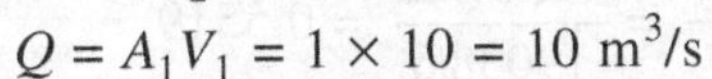

Discharge $\quad Q = A_1V_1 = 1 \times 10 = 10 \text{ m}^3/\text{s}$

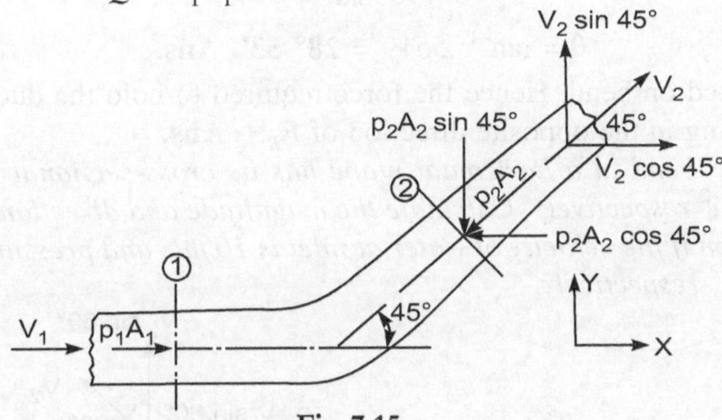

Fig. 7.15

Applying Bernoulli's equation at (1) and (2)

$$\therefore \quad \frac{p_1}{\rho g} + \frac{V_1^2}{2g} = \frac{p_2}{\rho g} + \frac{V_2^2}{2g} \qquad \{\because \; Z_1 = Z_2\}$$

or

$$\frac{2.943 \times 10^4}{1.16 \times 9.81} + \frac{10^2}{2 \times 9.81} = \frac{p_2}{\rho g} + \frac{20^2}{2 \times 9.81}$$

$$\therefore \quad \frac{p_2}{\rho g} = \frac{2.943 \times 10^4}{1.16 \times 9.81} + \frac{10^2}{2 \times 9.81} - \frac{20^2}{2 \times 9.81}$$

$$= 2586.2 + 5.0968 - 20.387 = 2570.90 \text{ m}$$

$$\therefore \quad p_2 = 2570.90 \times 1.16 \times 9.81 = 29255.8 \text{ N}$$

Force along x-axis, $\quad F_x = \rho Q\,[V_{1x} - V_{2x}] + (p_1A_1)_x + (p_2A_2)_x$

where $\quad V_{1x} = 10$ m/s, $V_{2x} = V_2 \cos 45° = 20 \times .7071$,

$(p_1A_1)_x = p_1A_1 = 29430 \times 1 = 29430$ N

and $\quad (p_2A_2)_x = -\,p_2A_2 \cos 45° = -\,p_2A_2 \cos 45° = -\,29255.8 \times 0.5 \times .7071$

$$\therefore \quad F_x = 1.16 \times 10[10 - 20 \times .7071] + 29430 \times 1 - 29255.8 \times .5 \times .7071$$

$$= -\,48.04 + 29430 - 10343.37 = 0 - 19038.59 \text{ N}$$

Similarly force along y-axis, $F_y = \rho Q[V_{1y} - V_{2y}] + (p_1A_1)_y + (p_2A_2)_y$

where $\quad V_{1y} = 0$, $V_{2y} = V_2 \sin 45° = 20 \times .7071 = 14.142$

$(p_1A_1)_y = 0$, $(p_2A_2)_y = -\,p_2A_2 \sin 45° = -\,29255.8 \times .5 \times .7071 = -\,10343.37$

$$F_y = 1.16 \times 10\;[0 - 14.142] + 0 - 10343.37$$

$$= -\,164.05 - 10343.37 = -\,10507.42 \text{ N}$$

$\therefore$ Resultant force, $F_R = \sqrt{F_x^2 + F_y^2} = \sqrt{(19038.6)^2 + (10507.42)^2}$ = **21746.6 N. Ans.**

The direction of F_R with x-axis is given as

$$\tan\theta = \frac{F_y}{F_x} = \frac{10507.42}{19038.6} = 0.5519$$

$$\therefore \quad \theta = \tan^{-1} .5519 = \mathbf{28° \; 53'. \; Ans.}$$

F_R is the force exerted on bend. Hence the force required to hold the duct in position is equal to 21746.6 N but it is acting in the opposite direction of F_R. **Ans.**

Problem 7.16 *A 60° bend in a horizontal plane has its cross-sectional area at inlet and outlet equal to 1 m² and 0.5 m² respectively. Calculate the magnitude and direction of the force required to hold the bend in position if the velocity of water at inlet is 10 m/s and pressure at inlet and outlet are 40 kN/m² and 30 kN/m² respectively.* (VTU, Aug. 2003)

Solution.

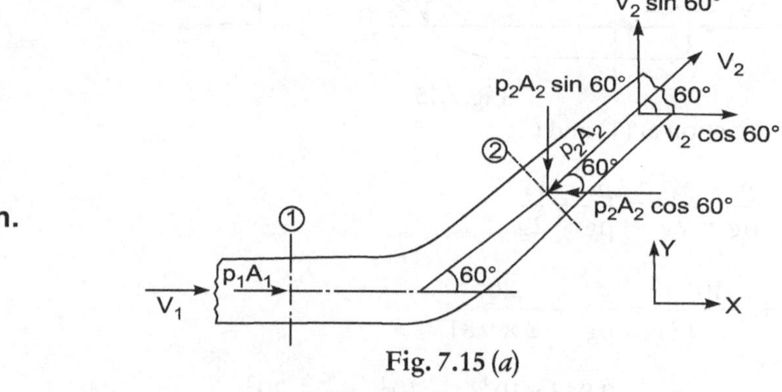

Fig. 7.15 (*a*)

Given :

$$A_1 = 1 \text{ m}^2 \; ; \; A_2 = 0.5 \text{ m}^2 \; ; \; V_1 = 10 \text{ m/s}$$

$$p_1 = 40 \text{ kN/m}^2 = 40 \times 1000 \text{ N/m}^2,$$

$$p_2 = 30 \text{ kN/m}^2 = 30 \times 1000 \text{ N/m}^2$$

$$\theta = 60°$$

Applying continuity equation of section (1) and (2),

$$A_1V_1 = A_2V_2$$

or $$V_2 = \frac{A_1V_1}{A_2} = \frac{1 \times 10}{0.5} = 20 \text{ m/s}$$

Rate of flow, $Q = A_1V_1 = 1 \times 10 = 10 \text{ m}^3/\text{s}$

Force along x-axis is given by,

$$F_x = PQ\,(V_{1x} - V_{2x}) + (p_1A_1)_x + (p_2A_2)_x$$

where $V_{1x} = V_1 = 10$ m/s ; $V_{2x} = V_2 \cos 60° = 20 \times 0.5 = 10$ m/s

$(p_1A_1)_x = p_1A_1 = (40 \times 1000) \times 1 = 40 \times 1000$ N

and $(p_2A_2)_x = -\,p_2A_2 \cos 60°$ [– ve sign is due to pressure force ($p_2A_2 \cos 60°$) acting in opposite direction of x]

$$= -\,(30 \times 1000) \times 0.5 \times 0.5$$

$$= -\,7500 \text{ N}$$

$\therefore \qquad F_x = 1000 \times 10\,(10 - 10) + 40 \times 1000 - 7500$

$[\because\ \rho = 1000 \text{ kg/m}^3 \text{ and } Q = 10 \text{ m}^3\text{/s}]$

$= 40 \times 1000 - 7500 = 32500 \text{ N}$

Similarly force along y-axis is given by,

$$F_y = PQ\,[V_{1y} - V_{2y}] + (p_1A_1)_y + (p_2A_2)_y$$
$$= 1000 \times 10\,[0 - 17.32] + 0 + (-12990)$$
$$= -173200 - 12990 = -186190 \text{ N}$$

(– ve sign means F_y is acting in the downward direction)

where $V_{1y} = 0,\ V_{2y} = V_2 \sin 60°$
$= 20 \times 0.866 = 17.32$ m/s
$(p_1A_1)_y = 0,\ (p_2A_2)_y = -p_2A_2 \sin 60°$
$= -(30 \times 1000) \times 0.5 \times 0.866$
$= -12990$ N

$\therefore$ Resultant force, $F_R = \sqrt{F_x^2 + F_y^2}$

$$= \sqrt{(32500)^2 + (-186190)^2}$$

= 189014 N. Ans.

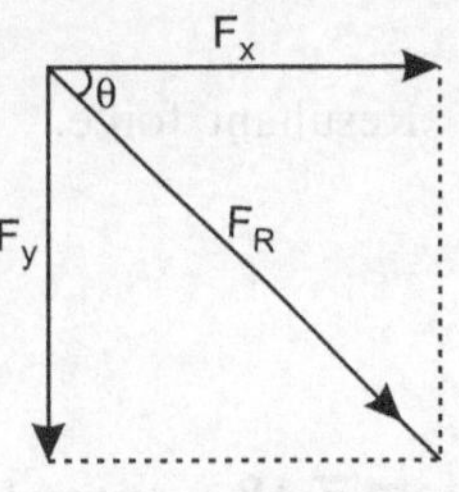

Fig. 7.15 (*b*)

The direction of resultant force with x-axis is given by

$$\tan\theta = \frac{F_y}{F_x} = \frac{186190}{32500} = 5.7289$$

$\therefore \qquad \theta = \tan^{-1} 5.7289 = $ **80.09° Ans.**

The resultant force exerted by water on the bend is $F_R = 189014$ N. Hence the force required to hold the bend in position is equal to 189014 N but it will be acting in the opposite direction of F_R.

Problem 7.17 *A pipe of 300 mm diameter conveying 0.30 m³/s of water has a right angled bend in a horizontal plane. Find the resultant force exerted on the bend if the pressure at inlet and outlet of the bend are 24.525 N/cm² and 23.544 N/cm².*

Solution. Given :

Dia. of bend, $\qquad D = 300 \text{ mm} = 0.3 \text{ m}$

$\therefore$ Area, $\qquad A = A_1 = A_2 = \frac{\pi}{4}(.3)^2 = 0.07068 \text{ m}^2$

$\therefore$ Discharge, $\qquad Q = 0.30 \text{ m}^3\text{/s}$

$\therefore$ Velocity, $\qquad V = V_1 = V_2 = \frac{Q}{A} = \frac{0.30}{.07068} = 4.244 \text{ m/s}$

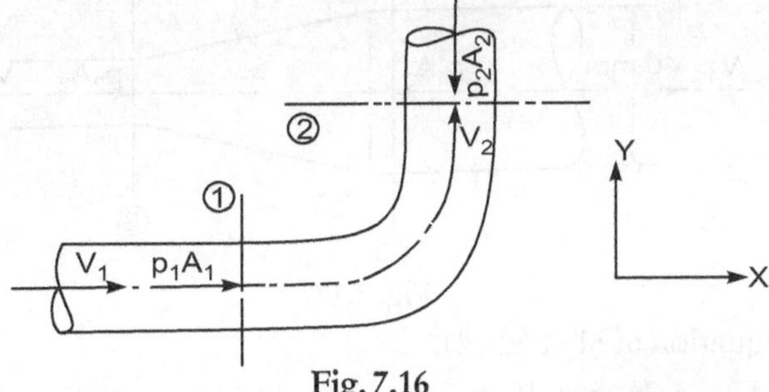

Fig. 7.16

Angle of bend, $\qquad \theta = 90°$

$p_1 = 24.525 \text{ N/cm}^2 = 24.525 \times 10^4 \text{ N/m}^2 = 245250 \text{ N/m}^2$

$p_2 = 23.544 \text{ N/cm}^2 = 23.544 \times 10^4 \text{ N/m}^2 = 235440 \text{ N/m}^2$

Force of bend along x-axis $F_x = \rho Q\,[V_{1x} - V_{2x}] + (p_1A_1)_x + (p_2A_2)_x$

where $\rho = 1000$, $V_{1x} = V_1 = 4.244$ m/s, $V_{2x} = 0$

$(p_1A_1)_x = p_1A_1 = 245250 \times .07068$

$(p_2A_2)_x = 0$

$$\therefore \quad F_x = 1000 \times 0.30\,[4.244 - 0] + 245250 \times .07068 + 0$$
$$= 1273.2 + 17334.3 = 18607.5 \text{ N}$$

Force on bend along y-axis, $F_y = \rho Q\,[V_{1y} - V_{2y}] + (p_1A_1)_y + (p_2A_2)_y$

where $V_{1y} = 0$, $V_{2y} = V_2 = 4.244$ m/s

$(p_1A_1)_y = 0$, $(p_2A_2)_y = -p_2A_2 = -235440 \times .07068 = -16640.9$

$$\therefore \quad F_y = 1000 \times 0.30[0 - 4.244] + 0 - 16640.9$$
$$= -1273.2 - 16640.9 = -17914.1 \text{ N}$$

$\therefore$ Resultant force, $$F_R = \sqrt{F_x^2 + F_y^2} = \sqrt{(18607.5)^2 + (17914.1)^2} = -25829.3 \text{ N}$$

and $$\tan\theta = \frac{F_y}{F_x} = \frac{17914.1}{18607.5} = 0.9627$$

$\therefore$ $\theta = $ **43° 54′. Ans.**

Problem 7.18 *A nozzle of diameter 20 mm is fitted to a pipe of diameter 40 mm. Find the force exerted by the nozzle on the water which is flowing through the pipe at the rate of 1.2 m³/minute.*

Solution. Given :

Dia. of pipe, $D_1 = 40 \text{ mm} = 40 \times 10^{-3} \text{ m} = .04 \text{ m}$

$\therefore$ Area, $$A_1 = \frac{\pi}{4} D_1^2 = \frac{\pi}{4}(.04)^2 = 0.001256 \text{ m}^2$$

Dia. of nozzle, $D_2 = 20 \text{ mm} = 0.02 \text{ m}$

$\therefore$ Area, $$A_2 = \frac{\pi}{4}(.02)^2 = .000314 \text{ m}^2$$

Discharge, $$Q = 1.2 \text{ m}^3/\text{minute} = \frac{1.2}{60} \text{ m}^3/\text{s} = 0.02 \text{ m}^3/\text{s}$$

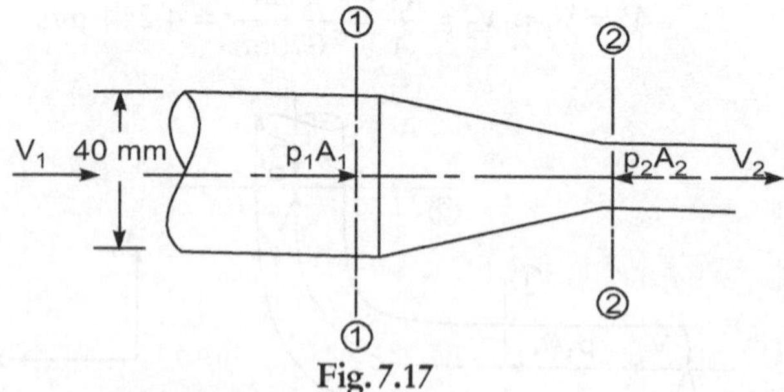

Fig. 7.17

Applying continuity equation at (1) and (2),

$$A_1V_1 = A_2V_2 = Q$$

$$\therefore \quad V_1 = \frac{Q}{A_1} = \frac{0.2}{.001256} = 15.92 \text{ m/s}$$

and $$V_2 = \frac{Q}{A_2} = \frac{0.2}{.000314} = 63.69 \text{ m/s}$$

Applying Bernoulli's equation at sections (1) and (2), we get

$$\frac{p_1}{\rho g} + \frac{V_1^2}{2g} + z_1 = \frac{p_2}{\rho g} + \frac{V_2^2}{2g} + z_2$$

Now $$z_1 = z_2, \frac{p_2}{\rho g} = \text{atmospheric pressure} = 0$$

$$\therefore \quad \frac{p_1}{\rho g} + \frac{V_1^2}{2g} = \frac{V_2^2}{2g}$$

$$\therefore \quad \frac{p_1}{\rho g} = \frac{V_2^2}{2g} - \frac{V_1^2}{2g} = \frac{(63.69^2)}{2 \times 9.81} - \frac{(15.92^2)}{2 \times 9.81} = 206.749 - 12.917$$

$$= 193.83 \text{ m of water}$$

$$\therefore \quad p_1 = 193.83 \times 1000 \times 9.81 \frac{\text{N}}{\text{m}^2} = 1901472 \frac{\text{N}}{\text{m}^2}$$

Let the force exerted by the nozzle on water = F_x

Net force in the direction of x = rate of change of momentum in the direction of x

$$\therefore \quad p_1A_1 - p_2A_2 + F_x = \rho Q(V_2 - V_1)$$

where p_2 = atmospheric pressure = 0 and ρ = 1000

$\therefore$ $1901472 \times .001256 - 0 + F_x = 1000 \times 0.02(63.69 - 15.92)$ or $2388.24 + F_x = 916.15$

$$\therefore \quad F_x = -2388.24 + 916.15 = \mathbf{-1472.09.\ Ans.}$$

– ve sign indicates that the force exerted by the nozzle on water is acting from right to left.

Problem 7.19 *The diameter of a pipe gradually reduces from 1 m to 0.7 m as shown in Fig. 7.18. The pressure intensity at the centre-line of 1 m section 7.848 kN/m² and rate of flow of water through the pipe is 600 litres/s. Find the intensity of pressure at the centre-line of 0.7 m section. Also determine the force exerted by flowing water on transition of the pipe.*

Solution. Given :

Dia. of pipe at section 1, $D_1 = 1$ m

$\therefore$ Area, $$A_1 = \frac{\pi}{4}(1)^2 = 0.7854 \text{ m}^2$$

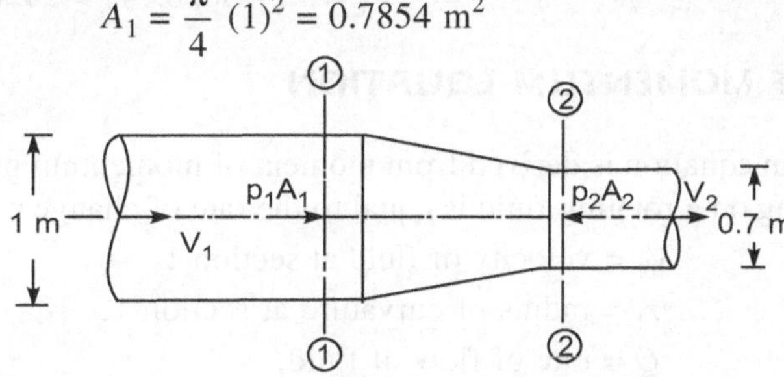

Fig. 7.18

Dia. of pipe at section 2, $D_2 = 0.7$ m

$\therefore$ Area, $\qquad A_2 = \frac{\pi}{4}(0.7)^2 = 0.3848 \text{ m}^2$

Pressure at section 1, $\qquad p_1 = 7.848 \text{ kN/m}^2 = 7848 \text{ N/m}^2$

Discharge, $\qquad Q = 600 \text{ litres/s} = \frac{600}{1000} = 0.6 \text{ m}^3\text{/s}$

Applying continuity equation,

$$A_1V_1 = A_2V_2 = Q$$

$$\therefore \quad V_1 = \frac{Q}{A_2} = \frac{0.6}{0.7854} = 0.764 \text{ m/s}$$

$$V_2 = \frac{Q}{A_2} = \frac{0.6}{.3854} = 1.55 \text{ m/s}$$

Applying Bernoulli's equation at sections (1) and (2),

$$\frac{p_1}{\rho g} + \frac{V_1^2}{2g} = \frac{p_2}{\rho g} + \frac{V_2^2}{2g} \qquad \{\because \text{ pipe is horizontal, } \therefore \; z_1 = z_2\}$$

or
$$\frac{7848}{1000 \times 9.81} + \frac{(.764)^2}{2 \times 9.81} = \frac{p_2}{\rho g} + \frac{(1.55)^2}{2 \times 9.81}$$

$$\therefore \quad \frac{p_2}{\rho g} = 0.8 + \frac{(.764)^2}{2 \times 9.81} - \frac{(1.55)^2}{2 \times 9.81}$$

$$= 0.8 + 0.0297 - 0.122 = 0.7077 \text{ m of water}$$

$$\therefore \quad p_2 = 0.7077 \times 9.81 \times 1000$$

$$= \mathbf{6942.54 \text{ N/m}^2 \text{ or } 6.942 \text{ kN/m}^2. \text{ Ans.}}$$

Let F_x = the force exerted by pipe transition on the flowing water in the direction of flow

Then net force in the direction of flow = rate of change of momentum in the direction of flow

or
$$p_1A_1 - p_2A_2 + F_x = \rho(V_2 - V_1)$$

$\therefore$ $7848 \times .7854 - 6942.54 \times .3848 + F_x = 1000 \times 0.6[1.55 - .764]$

or
$$6163.8 - 2671.5 + F_x = 471.56$$

$$\therefore \quad F_x = 471.56 - 6163.8 + 2671.5 = -\,3020.74 \text{ N}$$

$\therefore$ The force exerted by water on pipe transition

$$= -F_x = -(-\,3020.74) = \mathbf{3020.74 \text{ N. Ans.}}$$

▸ 7.8 MOMENT OF MOMENTUM EQUATION

Moment of momentum equation is derived from moment of momentum principle which states that the resulting torque acting on a rotating fluid is equal to the rate of change of moment of momentum.

Let $\quad V_1$ = velocity of fluid at section 1

r_1 = radius of curvature at section 1,

Q = rate of flow of fluid,

ρ = density of fluid,

and $\quad V_2$ and r_2 = velocity and radius of curvature at section 2

Momentum of fluid at section 1 = mass × velocity = $\rho Q \times V_1/s$

∴ Moment of momentum per second at section 1,

$$= \rho Q \times V_1 \times r_1$$

Similarly moment of momentum per second of fluid at section 2

$$= \rho Q \times V_2 \times r_2$$

∴ Rate of change of moment of momentum

$$= \rho Q V_2 r_2 - \rho Q V_1 r_1 = \rho Q[V_2 r_2 - V_1 r_1]$$

According to moment of momentum principle

Resultant torque = rate of change of moment of momentum

or $$T = \rho Q[V_2 r_2 - V_1 r_1] \quad \text{...(7.14)}$$

Equation (7.14) is known as moment of momentum equation. This equation is applied :

1. For analysis flow problems in turbines and centrifugal pumps.
2. For finding torque exerted by water on sprinkler.

HIGHLIGHTS

1. The study of fluid motion with the forces causing flow is called dynamics of fluid flow, which is analysed by the Newton's second law of motion.
2. Bernoulli's equation is obtained by integrating the Euler's equation of motion. Bernoulli's equation states "For a steady, ideal flow of an incompressible fluid, the total energy which consists of pressure energy, kinetic energy and datum energy, at any point of the fluid is constant". Mathematically,

$$\frac{p_1}{\rho g} + \frac{v_1^2}{2g} + z_1 = \frac{p_2}{\rho g} + \frac{v_2^2}{2g} + z_2$$

where $\frac{p_1}{\rho g}$ = pressure energy per unit weight = pressure head

$\frac{v_1^2}{2g}$ = kinetic energy per unit weight = kinetic head

z_1 = datum energy per unit weight = datum head.

3. Bernoulli's equation for real fluids

$$\frac{p_1}{\rho g} + \frac{v_1^2}{2g} + z_1 = \frac{p_2}{\rho g} + \frac{v_2^2}{2g} + z_2 + h_L$$

where h_L = loss of energy between sections 1 and 2.

4. The momentum equation states that the net force acting on a fluid mass is equal to the change in momentum per second in that direction. This is given as $F = \frac{d}{dt}(mV)$

The impulse-momentum equation is given by $F \cdot dt = d(mV)$.

5. The force exerted by a fluid on a pipe bend in the directions of x and y are given by

$$F_x = \frac{\text{mass}}{\text{sec}}(\text{Initial velocity in the direction of } x - \text{Final velocity in } x\text{-direction})$$

$$+ \text{Initial pressure force in } x\text{-direction} + \text{Final pressure force in } x\text{-direction}$$

$$= \rho Q[V_{1x} - V_{2x}] + (p_1 A_1)_x + (p_2 A_2)_x$$

and $$F_y = \rho Q[V_{1y} - V_{2y}] + (p_1 A_1)_y + (p_2 A_2)_y$$

Resultant force, $$F_R = \sqrt{F_x^2 + F_y^2}$$

and the direction of the resultant with horizontal is $\tan\theta = \frac{F_y}{F_x}$.

6. The force exerted by the nozzle on the water is given by $F_x = \rho Q[V_{2x} - V_{1x}]$ and force exerted by the water on the nozzle is $= -F_x = \rho Q[V_{1x} - V_{2x}]$.
7. Moment of momentum equation states that the resultant torque acting on a rotating fluid is equal to the rate of change of moment of momentum. Mathematically, it is given by $T = \rho Q[V_2 r_2 - V_1 r_1]$.

EXERCISE 7

(A) THEORETICAL PROBLEMS

1. Name the different forces present in a fluid flow. For the Euler's equation of motion, which forces are taken into consideration.
2. What is Euler's equation of motion ? How will you obtain Bernoulli's equation from it ?
3. (*a*) Derive Bernoulli's equation for the flow of an incompressible frictionless fluid from consideration of momentum. *(A.M.I.E., Summer, 1988)*
   (*b*) Derive Bernoulli's equation starting from or fundamental and state all the assumptions made. *(VTU, 2000, VTU, Aug. 2001, VTU, Aug. 2002)*
4. State Bernoulli's theorem for steady flow of an incompressible fluid. Derive an expression for Bernoulli's theorem from first principle and state the assumptions made for such a derivation. *(Delhi University, 1987)*
5. State the momentum equation. How will you apply momentum equation for determining the force exerted by a flowing liquid on a pipe bend ? *(VTU, Aug. 2002)*
6. What is the difference between momentum equation and impulse momentum equation ?
7. Define moment of momentum equation. Where this equation is used ?
8. Which of the following statement is correct in case of pipe flow :
   (*a*) flow takes place from higher pressure to lower pressure ;
   (*b*) flow takes place from higher velocity to lower velocity ;
   (*c*) flow takes place from higher elevation to lower elevation ;
   (*d*) flow takes place from higher energy to lower energy.
9. Derive Euler's equation of motion along a stream line for an ideal fluid stating clearly the assumptions. Explain how this is integrated to get Bernoulli's equation along a stream-line. *(A.M.I.E., Summer, 1984)*
10. State Bernoulli's theorem. Mention the assumptions made. How is it modified while applying in practice? List out its engineering application. *(A.M.I.E., Winter, 1987)*
11. Define continuity equation and Bernoulli's equation. *(Delhi University, 1992)*
12. What are the different forms of energy in a flowing fluid ? Represent schematically the Bernoulli's equation for flow through a tapering pipe and show the position of total energy line and the datum line. *(Osmania University, 1990)*
13. Write Euler's equation of motion long a streamline and integrate it to obtain Bernoulli's equation. State all assumptions made. *(A.M.I.E., Winter, 1990)*
14. Starting with Euler's equation of motion along a stream line, obtain Bernoulli's equation by its integration. List all the assumptions made. *(A.M.I.E., Summer, 1991)*
15. Derive Bernoulli's equation from fundamentals. *(J.N.T.U., Hyderabad, S 2002)*

(B) NUMERICAL PROBLEMS

1. Water is flowing through a pipe of 100 mm diameter under a pressure of 19.62 N/cm^2 (gauge) and with mean velocity of 3.0 m/s. Find the total head of the water at a cross-section, which is 8 m above the datum line. [**Ans.** 28.458 m]

2. A pipe, through which water is flowing is having diameters 40 cm and 20 cm at the cross-sections 1 and 2 respectively. The velocity of water at section 1 is given 5.0 m/s. Find the velocity head at the sections 1 and 2 and also rate of discharge. [**Ans.** 1.274 m ; 20.387 m ; 0.628 m^3/s]
3. The water is flowing through a pipe having diameters 20 cm and 15 cm at sections 1 and 2 respectively. The rate of flow through pipe is 40 litres/s. The section 1 is 6 m above datum line and section 2 is 3 m above the datum. If the pressure at section 1 is 29.43 N/cm^2, find the intensity of pressure at section 2. [**Ans.** 32.19 N/cm^2]
4. Water is flowing through a pipe having diameters 30 cm and 15 cm at the bottom and upper end respectively. The intensity of pressure at the bottom end is 29.43 N/cm^2 and the pressure at the upper end is 14.715 N/cm^2. Determine the difference in datum head if the rate of flow through pipe is 50 lit/s. [**Ans.** 14.618 m]
5. The water is flowing through a taper pipe of length 50 m having diameters 40 cm at the upper end and 20 cm at the lower end, at the rate of 60 litres/s. The pipe has a slope of 1 in 40. Find the pressure at the lower end if the pressure at the higher level is 24.525 N/cm^2. [**Ans.** 25.58 N/cm^2]
6. A pipe of diameter 30 cm carries water at a velocity of 20 m/sec. The pressures at the points *A* and *B* are given as 34.335 N/cm^2 and 29.43 N/cm^2 respectively, while the datum head at *A* and *B* are 25 m and 28 m. Find the loss of head between *A* and *B*. [**Ans.** 2 m]
7. A conical tube of length 3.0 m is fixed vertically with its smaller end upwards. The velocity of flow at the smaller end is 4 m/s while at the lower end it is 2 m/s. The pressure head at the smaller end is 2.0 m of liquid. The loss of head in the tube is $0.95\,(v_1 - v_2)^2/2g$, where v_1 is the velocity at the smaller end and v_2 at the lower end respectively. Determine the pressure head at the lower end. Flow takes place in downward direction. [**Ans.** 5.56 m of fluid]
8. A pipe line carrying oil of specific gravity 0.8, changes in diameter from 300 mm at a position *A* to 500 mm diameter to a position *B* which is 5 m at a higher level. If the pressures at *A* and *B* are 19.62 N/cm^2 and 14.91 N/cm^2 respectively, and the discharge is 150 litres/s, determine the loss of head and direction of flow. [**Ans.** 1.45 m, Flow takes place from *A* to *B*]
9. A 45° reducing bend is connected in a pipe line, the diameters at the inlet and outlet of the bend being 40 cm and 20 cm respectively. Find the force exerted by water on the bend if the intensity of pressure at inlet of bend is 21.58 N/cm^2. The rate of flow of water is 500 litres/s. [**Ans.** 22696.5 N ; 20° 3.5′]
10. The discharge of water through a pipe of diameter 40 cm is 400 litres/s. If the pipe is bend by 135°, find the magnitude and direction of the resultant force on the bend. The pressure of flowing water is 29.43 N/cm^2. [**Ans.** 7063.2 N, θ = 22° 29.9′ with *x*-axis clockwise]
11. A 30 cm diameter pipe carries water under a head of 15 metres with a velocity of 4 m/s. If the axis of the pipe turns through 45°, find the magnitude and direction of the resultant force at the bend. [**Ans.** 8717.5 N, θ = 67° 30′]
12. A pipe of 20 cm diameter conveying 0.20 m^3/sec of water has a right angled bend in a horizontal plane. Find the resultant force exerted on the bend if the pressure at inlet and outlet of the bend are 22.563 N/cm^2 and 21.582 N/cm^2 respectively. [**Ans.** 11604.7 N, θ = 43° 54.2′]
13. A nozzle of diameter 30 mm is fitted to a pipe of 60 mm diameter. Find the force exerted by the nozzle on the water which is flowing through the pipe at the rate of 4.0 m^3/minute. [**Ans.** 7057.7 N]

Fluid Flow Measurements

▶ 8.1 INTRODUCTION

Fluid flow measurements means the measuring the rate of flow of a fluid flowing through a pipe or through an open channel. The rate of flow of a fluid through a pipe is measured by venturimeter, orifice meter, pitot-tube etc., whereas through an open channel the rate of flow is measured by notches and weirs.

▶ 8.2 VENTURIMETER

A venturimeter is a device used for measuring the rate of a flow of a fluid flowing through a pipe. It consists of three parts :

(*i*) A short converging part, (*ii*) Throat, and (*iii*) Diverging part. It is based on the Principle of Bernoulli's equation.

8.2.1 Expression for Rate of Flow Through Venturimeter. Consider a venturimeter fitted in a horizontal pipe through which a fluid is flowing (say water), as shown in Fig. 8.1.

Let d_1 = diameter at inlet or at section (1),

p_1 = pressure at section (1)

v_1 = velocity of fluid at section (1),

a = area at section (1) = $\frac{\pi}{4} d_1^2$

and d_2, p_2, v_2, a_2 are corresponding values at section (2).

Fig. 8.1 *Venturimeter.*

Applying Bernoulli's equation at sections (1) and (2), we get

$$\frac{p_1}{\rho g} + \frac{v_1^2}{2g} + z_1 = \frac{p_2}{\rho g} + \frac{v_2^2}{2g} + z_2$$

As pipe is horizontal, hence $z_1 = z_2$

$$\therefore \qquad \frac{p_1}{\rho g} + \frac{v_1^2}{2g} = \frac{p_2}{\rho g} + \frac{v_2^2}{2g} \quad \text{or} \quad \frac{p_1 - p_2}{\rho g} = \frac{v_2^2}{2g} - \frac{v_1^2}{2g}$$

But $\frac{p_1 - p_2}{\rho g}$ is the difference of pressure heads at sections 1 and 2 and it is equal to h or $\frac{p_1 - p_2}{\rho g} = h$

Substituting this value of $\frac{p_1 - p_2}{\rho g}$ in the above equation, we get

$$h = \frac{v_2^2}{2g} - \frac{v_1^2}{2g} \quad \text{...(8.1)}$$

Now applying continuity equation at sections 1 and 2

$$a_1 v_1 = a_2 v_2 \quad \text{or} \quad v_1 = \frac{a_2 v_2}{a_1}$$

Substituting this value of v_1 in equation (8.1)

$$h = \frac{v_2^2}{2g} - \frac{\left(\frac{a_2 v_2}{a_1}\right)^2}{2g} = \frac{v_2^2}{2g}\left[1 - \frac{a_2^2}{a_1^2}\right] = \frac{v_2^2}{2g}\left[\frac{a_1^2 - a_2^2}{a_1^2}\right]$$

or

$$v_2{}^2 = 2gh \frac{a_1^2}{a_1^2 - a_2^2}$$

$\therefore$

$$v_2 = \sqrt{2gh \frac{a_1^2}{a_1^2 - a_2^2}} = \frac{a_1}{\sqrt{a_1^2 - a_2^2}} \sqrt{2gh}$$

$\therefore$ Discharge,

$$Q = a_2 v_2$$

$$= a_2 \frac{a_1}{\sqrt{a_1^2 - a_2^2}} \times \sqrt{2gh} = \frac{a_1 a_2}{\sqrt{a_1^2 - a_2^2}} \times \sqrt{2gh} \quad \text{...(8.2)}$$

Equation (8.2) gives the discharge under ideal conditions and is called, theoretical discharge. Actual discharge will be less than theoretical discharge.

$\therefore$

$$Q_{\text{act}} = C_d \times \frac{a_1 a_2}{\sqrt{a_1^2 - a_2^2}} \times \sqrt{2gh} \quad \text{...(8.3)}$$

where C_d = Co-efficient of venturimeter and its value is less than 1.

8.2.2 Value of 'h' given by differential U-tube manometer

Case I. Let the differential manometer contains a liquid which is heavier than the liquid flowing through the pipe.

Let S_h = Sp. gravity of the heavier liquid

S_o = Sp. gravity of the liquid flowing through pipe

x = Difference of the heavier liquid column in U-tube

Then

$$h = x\left[\frac{S_h}{S_o} - 1\right] \quad \text{...(8.4)}$$

Case II. If the differential manometer contains a liquid which is lighter than the liquid flowing through the pipe, the value of h is given by

$$h = x\left[1 - \frac{S_l}{S_o}\right] \qquad ...(8.5)$$

where S_1 = Sp. gr. of lighter liquid in U-tube

S_0 = Sp. gr. of fluid flowing through pipe

x = Difference of the lighter liquid columns in U-tube.

Case III. Inclined Venturimeter with Differential U-tube manometer. The above two cases are given for a horizontal venturimeter. This case is related to inclined venturimeter having differential U-tube manometer. Let the differential manometer contains heavier liquid then h is given as

$$h = \left(\frac{p_1}{\rho g} + z_1\right) - \left(\frac{p_2}{\rho g} + z_2\right) = x\left[\frac{S_h}{S_o} - 1\right] \qquad ...(8.6)$$

Case IV. Similarly, for inclined venturimeter in which differential manometer contains a liquid which is lighter than the liquid flowing through the pipe, the value of h is given as

$$h = \left(\frac{p_1}{\rho g} + z_1\right) - \left(\frac{p_2}{\rho g} + z_2\right) = x\left[1 - \frac{S_l}{S_o}\right] \qquad ...(8.7)$$

Problem 8.1 *A horizontal venturimeter with inlet and throat diameters 30 cm and 15 cm respectively is used to measure the flow of water. The reading of differential manometer connected to the inlet and the throat is 20 cm of mercury. Determine the rate of flow. Take C_d = 0.98.*

Solution. Given :

Dia. at inlet, $d_1 = 30$ cm

∴ Area at inlet, $a_1 = \frac{\pi}{4} d_1^2 = \frac{\pi}{4}(30)^2 = 706.85 \text{ cm}^2$

Dia. at throat, $d_2 = 15$ cm

∴ $a_2 = \frac{\pi}{4} \times 15^2 = 176.7 \text{ cm}^2$

$C_d = 0.98$

Reading of differential manometer = x = 20 cm of mercury.

∴ Difference of pressure head is given by (8.4)

or $$h = x\left[\frac{S_h}{S_o} - 1\right]$$

where S_h = Sp. gravity of mercury = 13.6, S_0 = Sp. gravity of water = 1

$$= 20\left[\frac{13.6}{1} - 1\right] = 20 \times 12.6 \text{ cm} = 252.0 \text{ cm of water.}$$

The discharge through venturimeter is given by eqn. (8.3)

$$Q = C_d \frac{a_1 a_2}{\sqrt{a_1^2 - a_2^2}} \times \sqrt{2gh}$$

$$= 0.98 \times \frac{706.85 \times 176.7}{\sqrt{(706.85)^2 - (176.7)^2}} \times \sqrt{2 \times 9.81 \times 252}$$

$$= \frac{86067593.36}{\sqrt{499636.9 - 31222.9}} = \frac{86067593.36}{684.4}$$

$$= 125756 \text{ cm}^3\text{/s} = \frac{125756}{1000} \text{ lit/s} = \textbf{125.756 lit/s. Ans.}$$

Problem 8.2 *An oil of sp. gr. 0.8 is flowing through a venturimeter having inlet diameter 20 cm and throat diameter 10 cm. The oil-mercury differential manometer shows a reading of 25 cm. Calculate the discharge of oil through the horizontal venturimeter. Take C_d = 0.98.*

Solution. Given :

Sp. gr. of oil, $S_o = 0.8$

Sp. gr. of mercury, $S_h = 13.6$

Reading of differential manometer, $x = 25$ cm

$\therefore$ Difference of pressure head, $h = x\left[\frac{S_h}{S_o} - 1\right]$

$$= 25\left[\frac{13.6}{0.8} - 1\right] \text{ cm of oil} = 25\ [17 - 1] = 400 \text{ cm of oil.}$$

Dia. at inlet, $d_1 = 20$ cm

$$\therefore \quad a_1 = \frac{\pi}{4}\, d_1^{\,2} = \frac{\pi}{4} \times 20^2 = 314.16 \text{ cm}^2$$

$$d_2 = 10 \text{ cm}$$

$$\therefore \quad a_2 = \frac{\pi}{4} \times 10^2 = 78.54 \text{ cm}^2$$

$$C_d = 0.98$$

$\therefore$ The discharge Q is given by equation (8.3)

or
$$Q = C_d \frac{a_1 a_2}{\sqrt{a_1^2 - 7a_2^2}} \times \sqrt{2gh}$$

$$= 0.98 \times \frac{314.16 \times 78.54}{\sqrt{(314.16)^2 - (78.54)^2}} \times \sqrt{2 \times 981 \times 400}$$

$$= \frac{21421375.68}{\sqrt{98696 - 6168}} = \frac{21421375.68}{304} \text{ cm}^3\text{/s}$$

$$= 70465 \text{ cm}^3\text{/s} = \textbf{70.465 litres/s. Ans.}$$

Problem 8.3 *A horizontal venturimeter with inlet diameter 20 cm and throat diameter 10 cm is used to measure the flow of oil of sp. gr. 0.8. The discharge of oil through venturimeter is 60 litres/s. Find the reading of the oil-mercury differential manometer. Take C_d = 0.98.*

Solution. Given : $d_1 = 20$ cm

$$\therefore \quad a_1 = \frac{\pi}{4}\, 20^2 = 314.16 \text{ cm}^2$$

$$d_2 = 10 \text{ cm}$$

$\therefore \quad a_2 = \frac{\pi}{4} \times 10^2 = 78.54 \text{ cm}^2$

$C_d = 0.98$

$Q = 60 \text{ litres/s} = 60 \times 1000 \text{ cm}^3\text{/s}$

Using the equation (8.3), $Q = C_d \frac{a_1 a_2}{\sqrt{a_1^2 - a_2^2}} \times \sqrt{2gh}$

or
$$60 \times 1000 = 9.81 \times \frac{314.16 \times 78.54}{\sqrt{(314.16)^2 - (78.54)^2}} \times \sqrt{2 \times 981 \times h}$$

$$= \frac{1071068.78\sqrt{h}}{304}$$

or
$$\sqrt{h} = \frac{304 \times 60000}{1071068.78} = 17.029$$

$\therefore \quad h = (17.029)^2 = 289.98$ cm of oil

But
$$h = x\left[\frac{S_h}{S_o} - 1\right]$$

where S_h = Sp. gr. of mercury = 13.6

S_o = Sp. gr. of oil = 0.8

x = Reading of manometer

$$\therefore \quad 289.98 = x\left[\frac{13.6}{0.8} - 1\right] = 16x$$

$$\therefore \quad x = \frac{289.98}{16} = 18.12 \text{ cm.}$$

$\therefore$ Reading of oil-mercury differential manometer = **18.12 cm. Ans.**

Problem 8.4 *A horizontal venturimeter with inlet diameter 20 cm and throat diameter 10 cm is used to measure the flow of water. The pressure at inlet is 17.658 N/cm² and the vacuum pressure at the throat is 30 cm of mercury. Find the discharge of water through venturimeter. Take C_d = 0.98.*

Solution. Given :

Dia. at inlet, $d_1 = 20$ cm

$\therefore \quad a_1 = \frac{\pi}{4} \times (20)^2 = 314.16 \text{ cm}^2$

Dia. at throat, $d_2 = 10$ cm

$\therefore \quad a_2 = \frac{\pi}{4} \times 10^2 = 78.74 \text{ cm}^2$

$p_1 = 17.658 \text{ N/cm}^2 = 17.658 \times 10^4 \text{ N/m}^2$

ρ for water $= 1000 \frac{\text{kg}}{\text{m}^3}$ and $\therefore \frac{p_1}{\rho g} = \frac{17.658 \times 10^4}{9.81 \times 1000} = 18$ m of water

$$\frac{p_2}{\rho g} = -\ 30 \text{ cm of mercury}$$

$$= -\ 0.30 \text{ m of mercury} = -\ 0.30 \times 13.6 = -\ 4.08 \text{ m of water}$$

$\therefore$ Differential head $= h = \dfrac{p_1}{\rho g} - \dfrac{p_2}{\rho g} = 18 - (-\ 4.08)$

$$= 18 + 4.08 = 22.08 \text{ m of water} = 2208 \text{ cm of water}$$

The discharge Q is given by equation (8.3)

$$Q = C_d \frac{a_1 a_2}{\sqrt{a_1^2 - a_2^2}} \times \sqrt{2gh}$$

$$= 0.98 \times \frac{314.16 \times 78.54}{\sqrt{(314.16)^2 - (78.74)^2}} \times \sqrt{2 \times 981 \times 2208}$$

$$= \frac{50328837.21}{304} \times 165555 \text{ cm}^3\text{/s} = \textbf{165.555 lit/s. Ans.}$$

Problem 8.5 *The inlet and throat diameters of a horizontal venturimeter are 30 cm and 10 cm respectively. The liquid flowing through the meter is water. The pressure intensity at inlet is 13.734 N/cm² while the vacuum pressure head at the throat is 37 cm of mercury. Find the rate of flow. Assume that 4% of the differential head is lost between the inlet and throat. Find also the value of C_d for the venturimeter.* (A.M.I.E., Winter, 1980)

Solution. Given :

Dia. at inlet, $d_1 = 30$ cm

$\therefore$ $a_1 = \dfrac{\pi}{4}(30)^2 = 706.85 \text{ cm}^2$

Dia. at throat, $d_2 = 10$ cm

$\therefore$ $a_2 = \dfrac{\pi}{4}(10)^2 = 78.54 \text{ cm}^2$

Pressure, $p_1 = 13.734 \text{ N/cm}^2 = 13.734 \times 10^4 \text{ N/m}^2$

$\therefore$ Pressure head, $\dfrac{p_1}{\rho g} = \dfrac{13.734 \times 10^4}{1000 \times 9.81} = 14$ m of water

$$\frac{p_2}{\rho g} = -\ 37 \text{ cm of mercury}$$

$$= \frac{-37 \times 13.6}{100} \text{ m of water} = -\ 5.032 \text{ m of water}$$

Differential head, $h = \dfrac{p_1}{\rho g} - \dfrac{p_2}{\rho g} = 14.0 - (-\ 5.032) = 14.0 + 5.032$

$$= 19.032 \text{ m of water} = 1903.2 \text{ cm}$$

Head lost, $h_f = 4\% \text{ of } h = \dfrac{4}{100} \times 19.032 = 0.7613$ m

$$\therefore \qquad C_d = \sqrt{\frac{h - h_f}{h}} = \sqrt{\frac{19.032 - .7613}{19.032}} = 0.98$$

$$\therefore \quad \text{Discharge} \qquad = C_d \frac{a_1 a_2 \sqrt{2gh}}{\sqrt{a_1^2 - a_2^2}}$$

$$= \frac{0.98 \times 706.85 \times 78.54 \times \sqrt{2 \times 981 \times 1903.2}}{\sqrt{(706.85)^2 - (78.54)^2}}$$

$$= \frac{105132247.8}{\sqrt{499636.9 - 6168}} = 149692.8 \text{ cm}^3\text{/s} = \mathbf{0.14969\ m^3/s.\ Ans.}$$

Problem 8.6 *A venturimeter is used for measurement of discharge of water in a horizontal pipe-line. If the ratio of upstream pipe diameter to that of throat is 2 : 1, upstream diameter is 300 mm, the difference of pressure between the throat and upstream is equal to 3 m head of water and loss of head through meter is one eighth of the throat velocity head, calculate discharge in the pipe.*

(VTU, Jan. 2004)

Solution. Given :

Ratio of inlet dia to throat dia *i.e.*, $\frac{d_1}{d_2} = \frac{2}{1}$

$d_1 = 300 \text{ mm} = 0.3 \text{ m}, \; d_2 = \frac{300}{2} = 150 \text{ mm} = 0.15 \text{ m}$

$\left(\frac{p_1}{\rho g} - \frac{p_2}{\rho g}\right) = 3$ m of water, loss of head,

$$h_f = \frac{1}{8} \text{ of throat velocity head} = \frac{1}{8} \times \frac{V_2^2}{2g}$$

Using continuity equation $a_1 \times v_1 = a_2 \times v_2$

$$\therefore \qquad v_2 = \frac{a_1 \times v_1}{a_2} = 4v_1$$

Using Bernoulli's equation at inlet and throat, we get

$$\frac{p_1}{\rho g} + \frac{v_1^2}{2g} + z_1 = \frac{p_2}{\rho g} + \frac{v_2^2}{2g} + z_2 + h_f$$

But $z_1 = z_2$ as pipe is horizontal

$$\frac{p_1}{\rho g} + \frac{v_1^2}{2g} = \frac{p_2}{\rho g} + \frac{v_2^2}{2g} + \frac{1}{8} \times \frac{v_2^2}{2g}$$

or

$$\frac{p_1}{\rho g} - \frac{p_2}{\rho g} = \frac{v_2^2}{2g} - \frac{v_1^2}{2g} + \frac{1}{8}\frac{v_2^2}{2g}$$

But

$$\frac{p_1}{\rho g} - \frac{p_2}{\rho g} = 3 \text{ m}$$

$$\therefore \quad 3 = \frac{v_2^2}{2g} - \frac{v_1^2}{2g} + \frac{1}{8}\frac{v_2^2}{2g}$$

or $$3 = \frac{(4v_1)^2}{2g} - \frac{v_1^2}{2g} + \frac{1}{8} \times \frac{(4v_1)^2}{2g} \quad [v_2 = 4v_1]$$

$$= \frac{16v_1^2}{2g} - \frac{v_1^2}{2g} + \frac{2v_1^2}{2g} = \frac{17v_1^2}{2g}$$

or $$v_1 = \sqrt{\frac{3 \times 2 \times 9.81}{17}} = 1.86 \text{ m/s}$$

$\therefore$ Discharge, $Q = a_1 \times v_1$

$$= \left(\frac{\pi}{4} d_1^2\right) \times v_1$$

$$= \left(\frac{\pi}{4} \times (0.3)^2\right) \times 1.86 = 0.1315 \text{ m}^3\text{/s}$$

$$= \textbf{131.5 l/s. Ans.}$$

PROBLEMS ON INCLINED VENTURIMETER

Problem 8.7 *A 30 cm × 15 cm venturimeter is inserted in a vertical pipe carrying water, flowing in the upward direction. A differential mercury manometer connected to the inlet and throat gives a reading of 20 cm. Find the discharge. Take C_d = 0.98.*

Solution. Given :

Dia. at inlet, $d_1 = 30$ cm

$\therefore$ $a_1 = \frac{\pi}{4} (30)^2 = 706.85 \text{ cm}^2$

Dia. at throat, $d_2 = 15$ cm

$\therefore$ $a_2 = \frac{\pi}{4} (15)^2 = 176.7 \text{ cm}^2$

$$h = x\left[\frac{S_h}{S_o} - 1\right] = 20\left[\frac{13.6}{1.0} - 1.0\right] = 20 \times 12.6 = 252.0 \text{ cm of water}$$

$C_d = 0.98$

Discharge, $$Q = C_d \frac{a_1 a_2}{\sqrt{a_1^2 - a_2^2}} \times \sqrt{2gh}$$

$$= 0.98 \times \frac{706.85 \times 176.7}{\sqrt{(706.85)^2 - (176.7)^2}} \times \sqrt{2 \times 981 \times 252}$$

$$= \frac{86067593.36}{\sqrt{499636.3 - 31222.9}} = \frac{86067593.36}{684.4}$$

$$= 125756 \text{ cm}^3\text{/s} = \textbf{125.756 lit/s. Ans.}$$

Problem 8.8 *A vertical venturimeter has an area ratio 5. It has a throat diameter of 10 cm. When oil of sp. gr. 0.8 flows through it the mercury in the differential gauge indicates a difference in height of 12 cm. Find the discharge through the venturimeter. Take $C_d = 0.98$.* (VTU, Aug. 2001)

Solution. Given :

Area ratio, $k = \dfrac{a_1}{a_2} = 5$, throat diameter, $d_2 = 10$ cm $= 0.1$ m. Hence $a_2 = \dfrac{\pi}{4} \times (0.1)^2$

sp. gr. of oil, $S_o = 0.8$, difference of Hg level, $x = 12$ cm $= 0.12$ m

Now differential head (h) is given by,

$$h = x \left[\frac{S_g}{S_o} - 1 \right]$$

$$= 0.12 \left[\frac{13.6}{0.8} - 1 \right] = 1.92 \text{ m of oil}$$

The discharge is given by,

$$Q = \frac{C_d \,.\, a_1 \, a_2}{\sqrt{a_1^2 - a_2^2}} \times \sqrt{2gh}$$

The discharge can be expressed in terms of area ratio (k) as

$$Q = \frac{C_d \, a_1 \, a_2}{\sqrt{a_1^2 - a_2^2}} \times \sqrt{2gh} = \frac{C_d \,.\, \dfrac{a_1}{a_2} \,.\, a_2}{\sqrt{\dfrac{a_1^2}{a_2^2} - \dfrac{a_2^2}{a_2^2}}} \times \sqrt{2gh}$$

$$= \frac{C_d \,.\, k \,.\, a_2}{\sqrt{k^2 - 1}} \times \sqrt{2gh} \qquad \left(\because \frac{a_1}{a_2} = k \right)$$

$$= \frac{0.98 \times 5 \times \dfrac{\pi}{4} (0.1)^2}{\sqrt{(5)^2 - 1}} \times \sqrt{2 \times 9.81 \times 1.92}$$

$= \mathbf{0.0482 \ m^3/s.}$ **Ans.**

Problem 8.9 *A 20 cm × 10 cm venturimeter is inserted in a vertical pipe carrying oil of sp. gr. 0.8, the flows of oil is in upward direction. The difference of levels between the throat and inlet section is 50 cm. The oil mercury differential manometer gives a reading of 30 cm of mercury. Find the discharge of oil. Neglect losses.*

Solution. Dia. at inlet, $d_1 = 20$ cm

$\therefore$ $a_1 = \dfrac{\pi}{4} (20)^2 = 314.16 \text{ cm}^2$

Dia. at throat, $d_2 = 10$ cm

$\therefore$ $a_2 = \dfrac{\pi}{4} (10)^2 = 78.54 \text{ cm}^2$

Sp. gr. of oil, $S_o = 0.8$

Sp. gr. of mercury, $S_g = 13.6$

Differential manometer reading, $x = 30$ cm

$$\therefore \quad h = \left(\frac{p_1}{\rho g} + z_1\right) - \left(\frac{p_2}{\rho g} + z_2\right) = x\left[\frac{S_g}{S_o} - 1\right]$$

$$= 30\left[\frac{13.6}{0.8} - 1\right] = 30\,[17 - 1] = 30 \times 16 = 480 \text{ cm of oil}$$

$$C_d = 1.0$$

The discharge,

$$Q = C_d \frac{a_1 a_2}{\sqrt{a_1^2 - a_2^2}} \times \sqrt{2gh}$$

$$= \frac{1.0 \times 314.16 \times 78.54}{\sqrt{(314.16)^2 - (78.54)^2}} \times \sqrt{2 \times 981 \times 480} \text{ cm}^3/\text{s}$$

$$= \frac{23932630.7}{304} = 78725.75 \text{ cm}^3/\text{s} = \mathbf{78.725 \text{ litres/s. Ans.}}$$

Problem 8.10 *In a vertical pipe conveying oil of specific gravity 0.8, two pressure gauges have been installed at A and B where the diametres are 16 cm and 8 cm respectively. A is 2 metres above B. The pressure gauge readings have shown that the pressure at B is greater than at A by 0.981 N/cm². Neglecting all losses, calculate the flow rate. If the gauges at A and B are replaced by tubes filled with the same liquid and connected to a U-tube containing mercury, calculate the difference of level of mercury in the two limbs of the U-tube.* (A.M.I.E., Summer 1985)

Solution. Given :

Sp. gr. of oil, $S_o = 0.8$

$\therefore$ Density, $\rho = 0.8 \times 1000 = 800 \dfrac{\text{kg}}{\text{m}^3}$

Dia. at A, $D_A = 16$ cm $= 0.16$ m

$\therefore$ Area at A, $A_1 = \dfrac{\pi}{4}(.16)^2 = 0.0201 \text{ m}^2$

Dia. at B, $D_B = 8$ cm $= 0.08$ m

$\therefore$ Area at B, $A_2 = \dfrac{\pi}{4}(.08)^2 = 0.005026 \text{ m}^2$

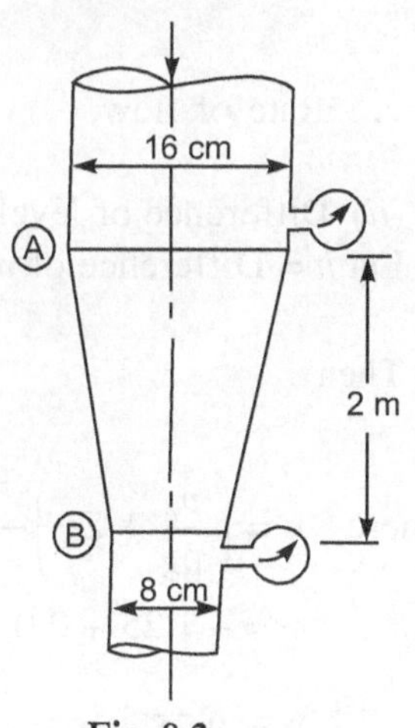

Fig. 8.2

(*i*) Difference of pressures, $p_B - p_A = 0.981 \text{ N/cm}^2$

$$= 0.981 \times 10^4 \text{ N/m}^2 = \frac{9810 \text{N}}{\text{m}^2}$$

Difference of pressure head

$$\therefore \quad \frac{p_B - p_A}{\rho g} = \frac{9810}{800 \times 9.81} = 1.25 \qquad (\because \rho = 800 \text{ kg/m}^3)$$

Applying Bernoulli's theorem at A and B and taking the reference line passing through section B, we get

$$\frac{p_A}{\rho g} + \frac{V_A^2}{2g} + Z_A = \frac{p_B}{\rho g} + \frac{V_B^2}{2g} + Z_B$$

or $$\frac{p_A}{\rho g} + \frac{p_B}{p_g} + Z_A - Z_B = \frac{V_B^2}{2g} - \frac{V_A^2}{2g}$$

or $$\left(\frac{p_A - p_B}{\rho g}\right) + 2.0 - 0.0 = \frac{V_B^2}{2g} - \frac{V_A^2}{2g}$$

or $$-1.25 + 2.0 = \frac{V_B^2}{2g} - \frac{V_A^2}{2g} \qquad \left(\because \frac{p_B - p_A}{\rho g} = 1.25\right)$$

$$0.75 = \frac{V_B^2}{2g} - \frac{V_A^2}{2g} \qquad \text{...}(i)$$

Now applying continuity equation at A and B, we get

$$V_A \times A_1 = V_B \times A_2$$

or $$V_B = \frac{V_A \times A_1}{A_2} = \frac{V_A \times \frac{\pi}{4}(.16)^2}{\frac{\pi}{4}(.08)^2} = 4V_A$$

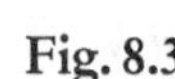

Fig. 8.3

Substituting the value of V_B in equation (*i*), we get

$$0.75 = \frac{16V_A^2}{2g} - \frac{V_A^2}{2g} = \frac{15V_A^2}{2g}$$

$$\therefore \quad V_A = \sqrt{\frac{0.75 \times 2 \times 9.81}{15}} = 0.99 \text{ m/s}$$

$\therefore$ Rate of flow, $$Q = V_A \times A_1$$
$$= 0.99 \times 0.0201 = \mathbf{0.01989\ m^3/s.\ Ans.}$$

(*ii*) Difference of level of mercury in the *U*-tube.

Let h = Difference of mercury level.

Then $$h = x\left(\frac{S_g}{S_o} - 1\right)$$

where $$h = \left(\frac{p_A}{\rho g} + Z_A\right) - \left(\frac{p_B}{\rho g} + Z_B\right) = \frac{p_A - p_B}{\rho g} + Z_A - Z_B$$

$$= -1.25 + 2.0 - 0$$

$$= 0.75 \qquad \left(\because \frac{p_B - p_A}{\rho g} = 1.25\right)$$

$$\therefore \quad 0.75 = x\left[\frac{13.6}{0.8} - 1\right] = x \times 16$$

$$\therefore \quad x = \frac{0.75}{16} = 0.04687 \text{ m} = \mathbf{4.687\ cm.\ Ans.}$$

Problem 8.11 *Find the discharge of water flowing through a pipe 30 cm diameter placed in an inclined position where a venturimeter is inserted, having a throat diameter of 15 cm. The difference of pressure between the main and throat is measured by a liquid of sp. gr. 0.6 in an inverted U-tube which gives a reading of 30 cm. The loss of head between the main and throat is 0.2 times the kinetic head of the pipe.*

Solution. Dia. at inlet $= 30$ cm

$\therefore$ $d_1 = 30$ cm

$$a_1 = \frac{\pi}{4}(30)^2 = 706.85 \text{ cm}^2$$

Dia. at throat, $d_2 = 15$ cm

$\therefore$ $$a_2 = \frac{\pi}{4}(15)^2 = 176.7 \text{ cm}^2$$

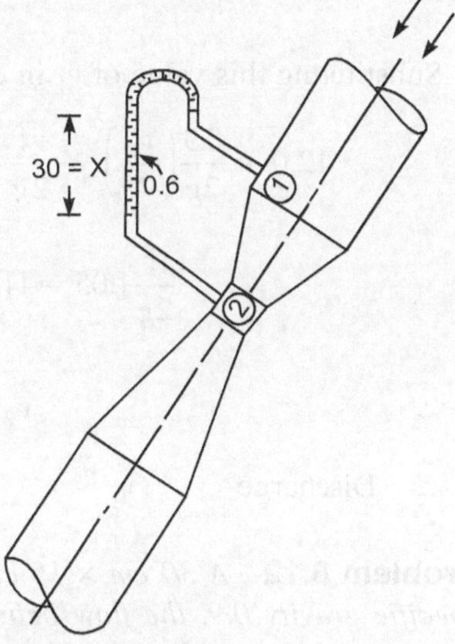

Fig. 8.4

Reading of differential manometer, $x = 30$ cm

Difference of pressure head, h is given by

$$\left(\frac{p_1}{\rho g} + z_1\right) - \left(\frac{p_2}{\rho g} + z_2\right) = h$$

Also $$h = x\left[1 - \frac{S_l}{S_o}\right]$$

where $S_l = 0.6$ and $S_o = 1.0$

$$= 30\left[1 - \frac{0.6}{1.0}\right] = 30 \times .4 = 12.0 \text{ cm of water}$$

Loss of head, $h_L = 0.2 \times$ kinetic head of pipe $= 0.2 \times \frac{v_1^2}{2g}$

Now applying Bernoulli's equation at sections (1) and (2), we get

$$\frac{p_1}{\rho g} + z_1 + \frac{v_1^2}{2g} = \frac{p_2}{\rho g} + z_2 + \frac{v_2^2}{2g} + h_L$$

or $$\left(\frac{p_1}{\rho g} + z_1\right) - \left(\frac{p_2}{\rho g} + z_2\right) + \frac{v_1^2}{2g} - \frac{v_2^2}{2g} = h_L$$

But $$\left(\frac{p_1}{\rho g} + z_1\right) - \left(\frac{p_2}{\rho g} + z_2\right) = h = 12.0 \text{ cm of water}$$

and $$h_L = 0.2 \times v_1^2/2g$$

$\therefore$ $$12.0 + \frac{v_1^2}{2g} - \frac{v_2^2}{2g} = 0.2 \times \frac{v_1^2}{2g}$$

$\therefore$ $$12.0 + 0.8\frac{v_1^2}{2g} - \frac{v_2^2}{2g} = 0 \qquad \text{...(1)}$$

Applying continuity equation at (1) and (2), we get

$$a_1 v_1 = a_2 v_2$$

$$\therefore \qquad v_1 = \frac{a_2}{a_1} v_2 = \frac{\frac{\pi}{4}(15)^2 v_2}{\frac{\pi}{4}(30)^2} = \frac{v_2}{4}$$

Substituting this value of v_1 in equation (1), we get

$$12.0 + \frac{0.9}{2g}\left(\frac{v_2}{4}\right)^2 - \frac{v_2^2}{2g} = 0 \text{ or } 12.0 + \frac{v_2^2}{2g}\left[\frac{0.8}{16} - 1\right] = 0$$

or

$$\frac{v_2^2}{2g}[.05 - 1] = -12.0 \text{ or } \frac{0.95\, v_2^2}{2g} = 12.0$$

$$\therefore \qquad v_2 = \sqrt{\frac{2 \times 981 \times 12.0}{0.95}} = 157.4 \text{ cm/s}$$

$\therefore$ Discharge $= a_2 v_2$

$= 176.7 \times 157.4 \text{ cm}^3/\text{s} = 27800 \text{ cm}^3/\text{s} =$ **27.8 litres/s. Ans.**

Problem 8.12 *A 30 cm × 15 cm venturimeter is provided in a vertical pipe line carrying oil of specific gravity 0.9, the flow being upwards. The difference in elevation of the throat section and entrance section of the venturimeter is 30 cm. The differential U-tube mercury manometer shows a gauge deflection of 25 cm. Calculate :*

(i) the discharge of oil, and

(ii) the pressure difference between the entrance section and the throat section. Take the co-efficient of meter as 0.98 and specific gravity of mercury as 13.6. (A.M.I.E., Summer, 1978)

Solution. Given :

Dia. at inlet, $d_1 = 30$ cm

$\therefore$ Area, $a_1 = \frac{\pi}{4}(30)^2 = 706.85 \text{ cm}^2$

Dia. at throat, $d_2 = 15$ cm

$\therefore$ Area, $a_2 = \frac{\pi}{4}(15)^2 = 176.7 \text{ cm}^2$

Let section (1) represents inlet and section (2) represents throat. Then $z_2 - z_1 = 30$ cm

Sp. gr. of oil, $S_o = 0.9$

Sp. gr. of mercury, $S_g = 13.6$

Reading of diff. manometer, $x = 25$ cm

The differential head, h is given by

$$h = \left(\frac{p_1}{\rho g} + z_1\right) - \left(\frac{p_2}{\rho g} + z_2\right)$$

$$= x\left[\frac{S_g}{S_o} - 1\right] = 25\left[\frac{13.6}{0.9} - 1\right] = 352.77 \text{ cm of oil}$$

(*i*) The discharge, Q of oil $= C_d \dfrac{a_1 a_2}{\sqrt{a_1^2 - a_2^2}} \times \sqrt{2gh}$

$$= \frac{0.98 \times 706.85 \times 176.7}{\sqrt{(706.85)^2 - (176.7)^2}} = \sqrt{2 \times 981 \times 352.77}$$

$$= \frac{101832219.9}{684.4} = 148790.5 \text{ cm}^3/\text{s}$$

$= \textbf{148.79 litres/s. Ans.}$

(*ii*) Pressure difference between entrance and throat section

$$h = \left(\frac{p_1}{\rho g} + z_1\right) - \left(\frac{p_2}{\rho g} + z_2\right) = 352.77$$

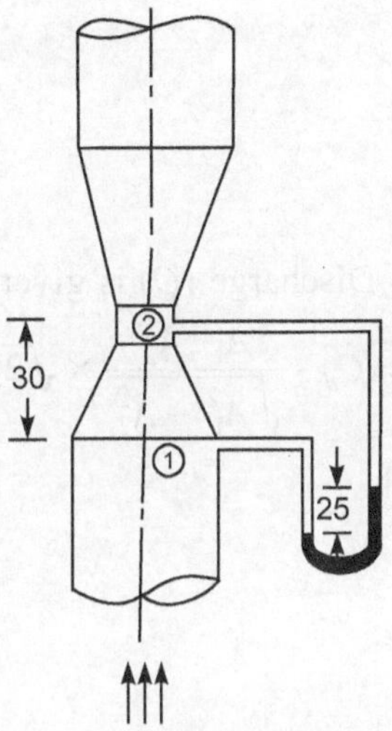

or $$\left(\frac{p_1}{\rho g} - \frac{p_2}{\rho g}\right) + z_1 - z_2 = 352.77$$

But $$z_2 - z_1 = 30 \text{ cm}$$

$\therefore$ $$\left(\frac{p_1}{\rho g} - \frac{p_2}{\rho g}\right) - 30 = 352.77$$

Fig. 8.5

$\therefore$ $$\frac{p_1}{\rho g} - \frac{p_2}{\rho g} = 352.77 + 30 = 382.77 \text{ cm of oil} = \textbf{3.8277 m of oil. Ans.}$$

or $$(p_1 - p_2) = 3.8277 \times \rho g$$

But density of oil $= \text{Sp. gr. of oil} \times 1000 \text{ kg/m}^3$

$= 0.9 \times 1000 = 900 \text{ kg/cm}^3$

$\therefore$ $$(p_1 - p_2) = 3.8277 \times 900 \times 9.81 \frac{\text{N}}{\text{m}^2}$$

$$= \frac{33795}{10^4} \text{ N/cm}^2 = \textbf{3.3795 N/cm}^2\textbf{. Ans.}$$

Problem 8.13 *A venturimeter is to be placed in a vertical line to measure the rate of flow of benzene (S = 0.899). The inlet diameter of venturimeter is 200 mm and the throat diameter is 87.5 mm. Benzene-mercury differential gauge is used to measure the difference of pressure between the inlet and the throat of the meter. When the gauge reading is 100 mm, find : (i) gauge reading in mm of benzene (ii) velocity at throat and discharge. Take* $C_d = 0.98$ *and for mercury sp. gr. = 13.55.*

(VTU, Aug. 2003)

Solution. Given :

Sp. gr. of benzene, $S_f = 0.899$, $D_1 = 200$ mm $= 0.2$ m, $D_2 = 87.5$ mm $= 0.0875$ m

Benzene-mercury differential gauge reading, $x = 100$ mm $= 0.1$ m

Sp. gr. of mercury, $S_g = 13.55$, $C_d = 0.98$

Area at inlet of venturimeter, $A_1 = \dfrac{\pi}{4} D_1^2 = \dfrac{\pi}{4}(0.2)^2 = 0.0314 \text{ m}^2$

Area at throat of venturimeter, $A_2 = \dfrac{\pi}{4} D_2^2 = \dfrac{\pi}{4}(0.0875)^2 = 0.00601 \text{ m}^2$

(i) Gauge reading in mm of benzene

The differential head (h), by a differential manometer in terms of mm of fluid flowing through venturimeter is given by,

$$h = x\left[\frac{S_g}{S_f} - 1\right] \text{ where } x = 100 \text{ mm}$$

$$= 100\left[\frac{13.55}{0.899} - 1\right] \text{ mm of benzene} \qquad [\because S_g = 13.55, S_f = 0.899]$$

$$= \mathbf{1407.24 \text{ mm of benzene. Ans.}}$$

(*ii*) Discharge (Q) is given by,

$Q = C_d \cdot \dfrac{A_1 \cdot A_2}{\sqrt{A_1^2 - A_2^2}} \times \sqrt{2gh}$ [This formula is used when value of C_d is given. Here value of C_d is given]

$$= 0.98 \times \frac{0.0314 \times 0.00601}{\sqrt{(0.0314)^2 - (0.00601)^2}} \times \sqrt{2 \times 9.81 \times 1.40724}$$

$$= \frac{0.98 \times 0.0314 \times 0.00601 \times 5.2545}{\sqrt{0.000985 - 0.00003612}} \qquad [\because h = 1407.24 \text{ mm or } 1.40724 \text{ m}]$$

$$= 0.03154 \text{ m}^3/\text{s} = \mathbf{31.54 \text{ l/s. Ans.}}$$

Alternate Method

Bernoulli's equation can also be used. But as the value of C_d is given, this means there is some loss of head. This loss of head will be calculated from the given value of C_d.

Let h = Differential head

h_f = Loss of head in venturimeter

Then $(h - h_f)$ = Actual differential head.

Now
$$Q_{th} = \frac{A_1 A_2}{\sqrt{A_1^2 - A_2^2}} \times \sqrt{2gh}$$

and
$$Q_{act} = \frac{A_1 A_2}{\sqrt{A_1^2 - A_2^2}} \times \sqrt{2g(h - h_f)}$$

$\therefore$
$$\frac{Q_{act}}{Q_{th}} = \frac{\sqrt{h - h_f}}{\sqrt{h}} = \sqrt{\frac{h - h_f}{h}} = \sqrt{\left(1 - \frac{h_f}{h}\right)}$$

But
$$\frac{Q_{act}}{Q_{th}} = C_d$$

$\therefore$
$$C_d = \sqrt{\left(1 - \frac{h_f}{h}\right)}$$

or $$0.98 = \sqrt{\left(1 - \frac{h_f}{h}\right)}$$

Squaring we get, $(0.98)^2 = 1 - \dfrac{h_f}{h}$

or $$\frac{h_f}{h} = 1 - (0.98)^2 = 0.0396$$

or $$h_f = 0.0396 \times h = 0.0396 \times 1.407 \qquad (\because h = 1.407 \text{ m})$$

$$= 0.0557 \text{ m}$$

Now using Bernoulli's equation between inlet and throat, we get

$$\frac{p_1}{\rho g} + \frac{v_1^2}{2g} + z_1 = \frac{p_2}{\rho g} + \frac{v_2^2}{2g} + z_2 + h_f$$

or $$\left(\frac{p_1}{\rho g} - \frac{p_2}{\rho g}\right) + (z_1 - z_2) = \left(\frac{v_2^2}{2g} - \frac{v_1^2}{2g}\right) + h_f$$

But $$\left(\frac{p_1}{\rho g} - \frac{p_2}{\rho g}\right) + (z_1 - z_2) = h$$

$\therefore$ $$h = \left(\frac{v_2^2}{2g} - \frac{v_1^2}{2g}\right) + h_f$$

or $$h - h_f = \frac{v_2^2}{2g} - \frac{v_1^2}{2g}$$

$$1.407 - 0.0557 = \frac{v_2^2}{2g} - \frac{v_1^2}{2g} \qquad (\because h = 1.407 \text{ m and } h_f = 0.0557 \text{ m})$$

or $$1.3513 = \frac{v_2^2}{2g} - \frac{v_1^2}{2g} \qquad ...(i)$$

From continuity equation, we have $A_1 V_1 = A_2 V_2$

or $$V_2 = \frac{A_1 V_1}{A_2} = \frac{\frac{\pi}{4} D_1^2 \times v_1}{\frac{\pi}{4} D_2^2} = \left(\frac{D_1}{D_2}\right)^2 \times v_1$$

$$= \left(\frac{0.2}{0.0875}\right)^2 \times v_1 = 5.2245\, v_1 \qquad ...(ii)$$

Substituting this value of v_2 in equation (i), we get

$$1.3513 = \frac{(5.2245\, v_1)^2}{2g} - \frac{v_1^2}{2g} = \frac{26.3\, v_1^2}{2g}$$

$$\therefore \qquad v_1 = \sqrt{\frac{1.3513 \times 2 \times 9.81}{26.3}} = 1.004 \text{ m/s}$$

Now discharge, $Q = A_1V_1 = 0.0314 \times 1.004 = 0.03152 \text{ m}^3/\text{s}$ = **31.52 l/s. Ans.**

(iii) Velocity at throat (v_2)

From equation *(ii)*, $v_2 = 5.224 \times v_1 = 5.224 \times 1.004$ = **5.245 m/s. Ans.**

Problem 8.14 *Crude oil of specific gravity 0.85 flows upwards at a volume rate of flow of 60 litre per second through a vertical venturimeter with an inlet diameter of 200 mm and a throat diameter of 100 mm. The co-efficient of discharge of the venturimeter is 0.98. The vertical distance between the pressure tappings is 300 mm.*

(i) If two pressure gauges are connected at the tappings such that they are positioned at the levels of their corresponding tapping points, determine the difference of readings in N/cm² of the two pressure gauges.

(ii) If a mercury differential manometer is connected, in place of pressure gauges, to the tappings such that the connecting tube upto mercury are filled with oil, determine the difference in the level of the mercury column. (A.M.I.E., Summer 1986)

Solution. Given :

Specific gravity of oil, $S_o = 0.85$

$\therefore$ Density, $\rho = 0.85 \times 1000 = 850 \text{ kg/m}^3$

Discharge, $Q = 60$ litre/s

$$= \frac{60}{1000} = 0.06 \text{ m}^3/\text{s}$$

Inlet dia, $d_1 = 200 \text{ mm} = 0.2 \text{ m}$

$\therefore$ Area, $a_1 = \frac{\pi}{4}(.2)^2 = 0.0314 \text{ m}^2$

Throat dia., $d_2 = 100 \text{ mm} = 0.1 \text{ m}$

$\therefore$ Area, $a_2 = \frac{\pi}{4}(0.1)^2 = 0.00785 \text{ m}^2$

Value of $C_d = 0.98$

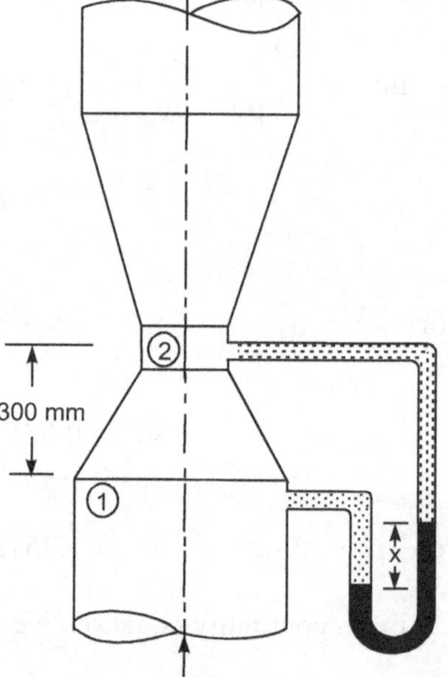

Fig. 8.6

Let section (1) represents inlet and section (2) represents throat. Then

$$z_2 - z_1 = 300 \text{ mm} = 0.3 \text{ m}$$

(i) Difference of readings in N/cm² of the two pressure gauges

The discharge Q is given by,

$$Q = C_d \frac{a_1 a_2}{\sqrt{a_1^2 - a_2^2}} \times \sqrt{2gh}$$

or

$$0.06 = \frac{0.98 \times 0.0314 \times 0.00785}{\sqrt{0.0314^2 - 0.00785^2}} \times \sqrt{2 \times 9.81 \times h}$$

$$= \frac{0.98 \times 0.00024649}{0.0304} \times 4.429\sqrt{h}$$

$$\therefore \qquad \sqrt{h} = \frac{0.06 \times 0.0304}{0.98 \times 0.00024649 \times 4.429} = 1.705$$

$\therefore \qquad h = 1.705^2 = 2.908 \text{ m}$

But for a vertical venturimeter, $h = \left(\frac{p_1}{\rho g} + z_1\right) - \left(\frac{p_2}{\rho g} + z_2\right)$

$$\therefore \qquad 2.908 = \left(\frac{p_1}{\rho g} + z_1\right) - \left(\frac{p_2}{\rho g} + z_2\right) = \left(\frac{p_1}{\rho g} - \frac{p_2}{\rho g}\right) + z_1 - z_2$$

$$\frac{p_1 - p_2}{\rho g} = 2.908 + z_2 - z_1 = 2.908 + 0.3 \qquad (\because \quad z_2 - z_1 = 0.3 \text{ m})$$

$$= 3.208 \text{ m of oil}$$

$$\therefore \qquad p_1 - p_2 = \rho g \times 3.208$$

$$= 850 \times 9.81 \times 3.208 \text{ N/m}^2 = \frac{850 \times 9.81 \times 3.208}{10^4} \text{ N/cm}^2$$

$$= \mathbf{2.675 \text{ N/cm}^2}. \textbf{ Ans.}$$

(ii) Difference in the levels of mercury columns (i.e., x)

The value of h is given by, $h = x\left[\frac{S_g}{S_o} - 1\right]$

$$\therefore \qquad 2.908 = x\left[\frac{13.6}{0.85} - 1\right] = x\,[16 - 1] = 15\,x$$

$$\therefore \qquad x = \frac{2.908}{15} = 0.1938 \text{ m} = \mathbf{19.38 \text{ cm of oil. Ans.}}$$

Problem 8.15 *In a 100 mm diameter horizontal pipe a venturimeter of 0.5 contraction ratio has been fixed. The head of water on the metre when there is no flow is 3 m (gauge). Find the rate of flow for which the throat pressure will be 2 metres of water absolute. The co-efficient of meter is 0.97. Take atmospheric pressure head = 10.3 m of water.* (A.M.I.E., Winter 1976)

Solution. Given :

Dia. of pipe, $d_1 = 100 \text{ mm} = 10 \text{ cm}$

$\therefore$ Area, $a_1 = \frac{\pi}{4} d_1^2 = \frac{\pi}{4} (10)^2 = 78.54 \text{ cm}^2$

Dia. at throat, $d_2 = 0.5 \times d_1 = 0.5 \times 10 = 5 \text{ cm}$

$\therefore$ Area, $a_2 = \frac{\pi}{4} (5)^2 = 19.635 \text{ cm}^2$

Head of water for no flow $= \frac{p_1}{\rho g} = 3 \text{ m (gauge)} = 3 + 10.3 = 13.3 \text{ m (abs.)}$

Throat pressure head $= \frac{p_2}{\rho g} = 2 \text{ m of water absolute.}$

$\therefore$ Difference of pressure head, $h = \frac{p_1}{\rho g} - \frac{p_2}{\rho g} = 13.3 - 2.0 = 11.3 \text{ m} = 1130 \text{ cm}$

∴ Rate of flow, Q is given by $Q = C_d \dfrac{a_1 a_2}{\sqrt{a_1^2 - a_2^2}} \times \sqrt{2gh}$

$$= 0.97 \times \frac{78.54 \times 19.635}{\sqrt{(78.54)^2 - (19.635)^2}} \times \sqrt{2 \times 981 \times 1130}$$

$$= \frac{2227318.17}{76} = 29306.8 \text{ cm}^3\text{/s} = \textbf{29.306 litres/s. Ans.}$$

▶ 8.3 ORIFICE METER OR ORIFICE PLATE

It is a device used for measuring the rate of flow of a fluid through a pipe. It is a cheaper device as compared to venturimeter. It also works on the same principle as that of venturimeter. It consists of a flat circular plate which has a circular sharp edged hole called orifice, which is concentric with the pipe. The orifice diameter is kept generally 0.5 times the diameter of the pipe, though it may vary from 0.4 to 0.8 times the pipe diameter.

A differential manometer is connected at section (1), which is at a distance of about 1.5 to 2.0 times the pipe diameter upstream from the orifice plate, and at section (2), which is at a distance of about half the diameter of the orifice on the down stream side from the orifice plate.

Let p_1 = pressure at section (1),
v_1 = velocity at section (1),
a_1 = area of pipe at section (1), and

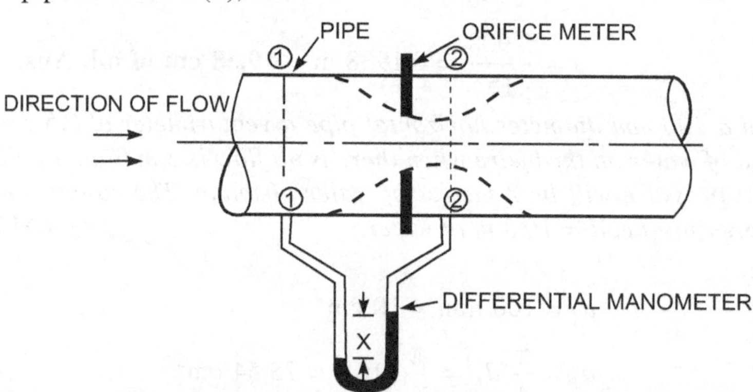

Fig. 8.7. *Orifice meter.*

p_2, v_2, a_2 are corresponding values at section (2). Applying Bernoulli's equation at sections (1) and (2), we get

$$\frac{p_1}{\rho g} + \frac{v_1^2}{2g} + z_1 = \frac{p_2}{\rho g} + \frac{v_2^2}{2g} + z_2$$

or $$\left(\frac{p_1}{\rho g} + z_1\right) - \left(\frac{p_2}{\rho g} + z_2\right) = \frac{v_2^2}{2g} - \frac{v_1^2}{2g}$$

But $$\left(\frac{p_1}{\rho g} + z_1\right) - \left(\frac{p_2}{\rho g} + z_2\right) = h = \text{Differential head}$$

$$\therefore \qquad h = \frac{v_2^2}{2g} - \frac{v_1^2}{2g} \quad \text{or} \quad 2gh = v_2^{\,2} - v_1^{\,2}$$

or
$$v_2 = \sqrt{2gh + v_1^2} \qquad \text{...(i)}$$

Now section (2) is at the vena contracta and a_2 represents the area at the vena contracta. If a_0 is the area of orifice then, we have

$$C_c = \frac{a_2}{a_0}$$

where C_c = Co-efficient of contraction

$$\therefore \qquad a_2 = a_0 \times C_c \qquad \text{...(ii)}$$

By continuity equation, we have

$$a_1 v_1 = a_2 v_2 \quad \text{or} \quad v_1 = \frac{a_2}{a_1} v_2 = \frac{a_0\, C_c}{a_1} v_2 \qquad \text{...(iii)}$$

Substituting the value of v_1 in equation (*i*), we get

$$v_2 = \sqrt{2gh + \frac{a_0^2 C_c^2 v_2^2}{a_1^2}}$$

or
$$v_2^{\,2} = 2gh + \left(\frac{a_0}{a_1}\right)^2 C_c^{\,2} v_2^{\,2} \text{ or } v_2^{\,2} = \left[1 - \left(\frac{a_0}{a_1}\right)^2 C_c^2\right] = 2hg$$

$$\therefore \qquad v_2 = \frac{\sqrt{2gh}}{\sqrt{1 - \left(\dfrac{a_0}{a_1}\right)^2 C_c^2}}$$

∴ The discharge $Q = v_2 \times a_2 = v_2 \times a_0\; C_c$ $\qquad \{\because a_2 = a_0 C_c \text{ from } (ii)\}$

$$= \frac{a_0 C_c \sqrt{2gh}}{\sqrt{1 - \left(\dfrac{a_0}{a_1}\right)^2 C_c^2}} \qquad \text{...(iv)}$$

The above expression is simplified by using

$$C_d = C_c \frac{\sqrt{1 - \left(\dfrac{a_0}{a_1}\right)^2}}{\sqrt{1 - \left(\dfrac{a_0}{a_1}\right)^2 C_c^2}}$$

$$\therefore \qquad C_c = C_d \frac{\sqrt{1 - \left(\dfrac{a_0}{a_1}\right)^2 C_c^2}}{\sqrt{1 - \left(\dfrac{a_0}{a_1}\right)^2}}$$

Substituting this value of C_c in equation (*iv*), we get

$$Q = a_0 \times C_d \frac{\sqrt{1-\left(\frac{a_0}{a_1}\right)^2 C_c^2}}{\sqrt{1-\left(\frac{a_0}{a_1}\right)^2}} \times \frac{\sqrt{2gh}}{\sqrt{1-\left(\frac{a_0}{a_1}\right)^2 C_c^2}}$$

$$= \frac{C_d a_0 \sqrt{2gh}}{\sqrt{1-\left(\frac{a_0}{a_1}\right)^2}} = \frac{C_d a_0 a_1 \sqrt{2gh}}{\sqrt{a_1^2 - a_0^2}}. \quad \text{...(8.8)}$$

where C_d = Co-efficient of discharge for orifice meter.

The co-efficient of discharge for orifice meter is much smaller than that for a venturimeter.

8.3.1. Merits and Demerits of Orifice meter

Merits :

1. It is simple in construction and hence cost of installation and replacement is cheaper as compared to venturimeter.
2. It requires less space to instal.

Demerits :

The rate of flow of a liquid measured by orificemeter is not so accurate as measured by venturimeter.

This is due to more losses in orifice plate.

Problem 8.16 *An orifice meter with orifice diameter 10 cm is inserted in a pipe of 20 cm diameter. The pressure gauges fitted upstream and downstream of the orifice meter gives readings of 19.62 N/cm² and 9.81 N/cm² respectively. Co-efficient of discharge for the meter is given as 0.6. Find the discharge of water through pipe.* (VTU, Aug. 2003 Auto Engg.)

Solution. Given :

Dia. of orifice, $d_0 = 10$ cm

∴ Area, $a_0 = \frac{\pi}{4}(10)^2 = 78.54 \text{ cm}^2$

Dia. of pipe, $d_1 = 20$ cm

∴ Area, $a_1 = \frac{\pi}{4}(20)^2 = 314.16 \text{ cm}^2$

$$p_1 = 19.62 \text{ N/cm}^2 = 19.62 \times 10^4 \text{ N/m}^2$$

∴ $$\frac{p_1}{\rho g} = \frac{19.62 \times 10^4}{1000 \times 9.81} = 20 \text{ m of water}$$

Similarly $$\frac{p_2}{\rho g} = \frac{9.81 \times 10^4}{1000 \times 9.81} = 10 \text{ m of water}$$

∴ $$h = \frac{p_1}{\rho g} - \frac{p_2}{\rho g} = 20.0 - 10.0 = 10 \text{ m of water} = 1000 \text{ cm of water}$$

$C_d = 0.6$

The discharge, Q is given by equation (8.8)

$$Q = C_d \frac{a_0 a_1}{\sqrt{a_1^2 - a_0^2}} \times \sqrt{2gh}$$

$$= 0.6 \times \frac{78.54 \times 314.16}{\sqrt{(314.16)^2 - (78.54)^2}} \times \sqrt{2 \times 981 \times 1000}$$

$$= \frac{20736838.09}{304} = 68213.28 \text{ cm}^3/\text{s} = \textbf{68.21 litres/s. Ans.}$$

Problem 8.17 *An orifice meter with orifice diameter 15 cm is inserted in a pipe of 30 cm diameter. The pressure difference measured by a mercury oil differential manometer on the two sides of the orifice meter gives a reading of 50 cm of mercury. Find the rate of flow of oil of sp. gr. 0.9 when the co-efficient of discharge of the meter = 0.64.*

Solution. Given :

Dia. of orifice, $d_0 = 15$ cm

∴ Area, $a_0 = \frac{\pi}{4}(15)^2 = 176.7 \text{ cm}^2$

Dia. of pipe, $d_1 = 30$ cm

∴ Area, $a_1 = \frac{\pi}{4}(30)^2 = 706.85 \text{ cm}^2$

Sp. gr. of oil, $S_o = 0.9$

Reading of diff. manometer, $x = 50$ cm of mercury

∴ Differential head, $h = x\left[\frac{S_g}{S_o} - 1\right] = 50\left[\frac{13.6}{0.9} - 1\right]$ cm of oil

$= 50 \times 14.11 = 705.5$ cm of oil

$C_d = 0.64$

∴ The rate of the flow, Q is given equation (8.8)

$$Q = C_d \cdot \frac{a_0 a_1}{\sqrt{a_1^2 - a_0^2}} \times \sqrt{2gh}$$

$$= 0.64 \times \frac{176.7 \times 706.85}{\sqrt{(706.85)^2 - (176.7)^2}} \times \sqrt{2 \times 981 \times 705.5}$$

$$= \frac{94046317.78}{684.4} = 137414.25 \text{ cm}^3/\text{s} = \textbf{137.414 litres/s. Ans.}$$

▶ 8.4 PITOT-TUBE

It is a device used for measuring the velocity of flow at any point in a pipe or a channel. It is based on the principle that if the velocity of flow at a point becomes zero, the pressure there is increased due to the conversion of the kinetic energy into pressure energy. In its simplest form, the pitot-tube consists of a glass tube, bent at right angles as shown in Fig. 8.8.

The lower end, which is bent through 90° is directed in the upstream direction as shown in Fig. 8.8. The liquid rises up in the tube due to the conversion of kinetic energy into pressure energy. The velocity is determined by measuring the rise of liquid in the tube.

Consider two points (1) and (2) at the same level in such a way that point (2) is just as the inlet of the pitot-tube and point (1) is far away from the tube.

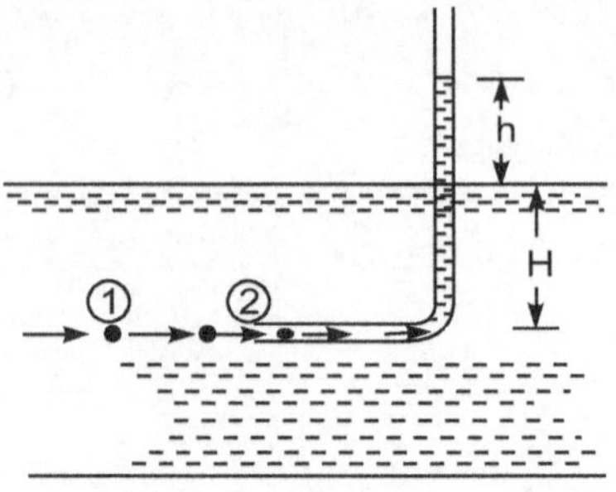

Fig. 8.8 *Pitot-tube.*

Let p_1 = intensity of pressure at point (1)

v_1 = velocity of flow at (1)

p_2 = pressure at point (2)

v_2 = velocity at point (2), which is zero

H = depth of tube in the liquid

h = rise of liquid in the tube above the free surface.

Applying Bernoulli's equations at points (1) and (2), we get

$$\frac{p_1}{\rho g} + \frac{v_1^2}{2g} + z_1 = \frac{p_2}{\rho g} + \frac{v_2^2}{2g} + z_2$$

But $z_1 = z_2$ as points (1) and (2) are on the same line and $v_2 = 0$.

$$\frac{p_1}{\rho g} = \text{pressure head at (1)} = H$$

$$\frac{p_2}{\rho g} = \text{pressure head at (2)} = (h + H)$$

Substituting these values, we get

$$\therefore \qquad H + \frac{v_1^2}{2g} = (h + H) \quad \therefore \quad h = \frac{v_1^2}{2g} \quad \text{or} \quad v_1 = \sqrt{2gh}$$

This is theoretical velocity. Actual velocity is given by

$$(v_1)_{act} = C_v \sqrt{2gh}$$

where C_v = Co-efficient of pitot-tube

$\therefore$ Velocity at any point $\qquad v = C_v \sqrt{2gh}$ ...(8.9)

Velocity of flow in a pipe by pitot-tube. For finding the velocity at any point in a pipe by pitot-tube, the following arrangements are adopted :

1. Pitot-tube along with a vertical piezometer tube as shown in Fig. 8.9.
2. Pitot-tube connected with piezometer tube as shown in Fig. 8.10.
3. Pitot-tube and vertical piezometer tube connected with a differential U-tube manometer as shown in Fig. 8.11.

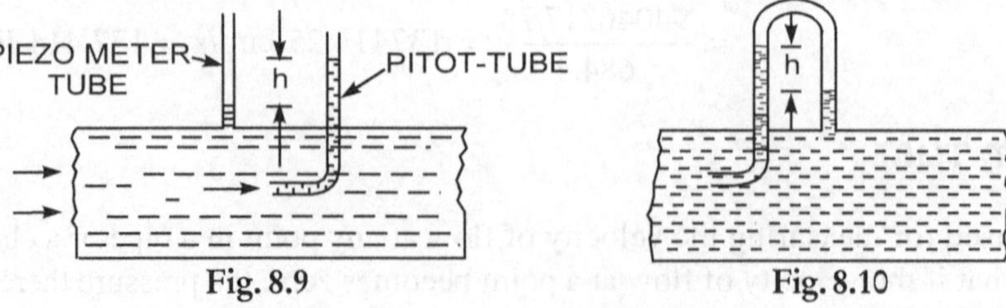

Fig. 8.9 Fig. 8.10

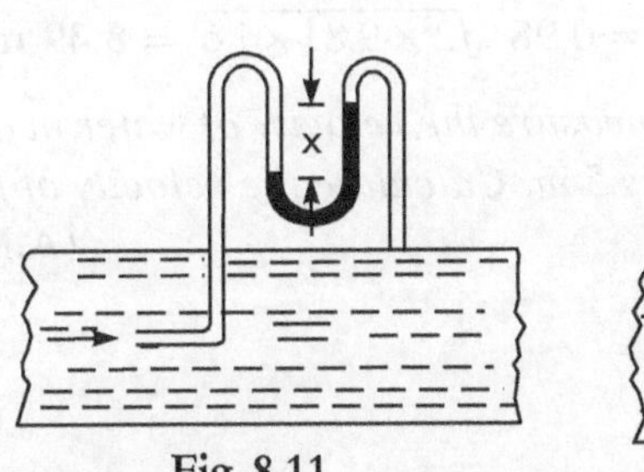

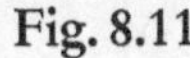
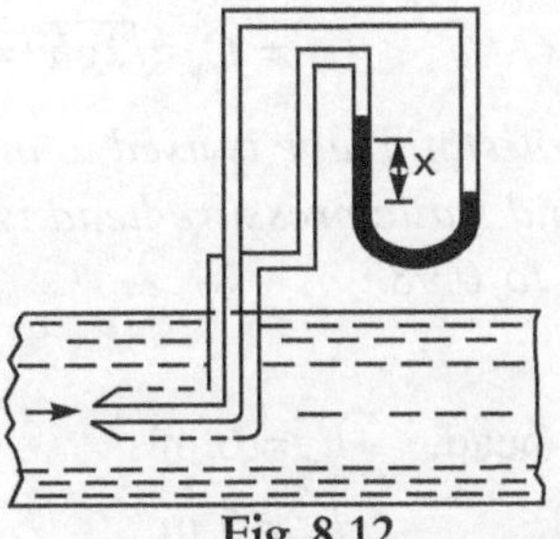

Fig. 8.11 Fig. 8.12

4. Pitot-static tube, which consists of two circular concentric tubes one inside the other with some annular space in between as shown in Fig. 8.12. The outlet of these two tubes are connected to the differential manometer where the difference of pressure head 'h' is measured by knowing the difference of the levels of the manometer liquid say x. Then $h = x\left[\frac{S_g}{S_o} - 1\right]$.

Problem 8.18 *A pitot-static tube placed in the centre of a 300 mm pipe line has one orifice pointing upstream and other perpendicular to it. The mean velocity in the pipe is 0.80 of the central velocity. Find the discharge through the pipe if the pressure difference between the two orifices is 60 mm of water. Take the co-efficient of pitot tube as $C_v = 0.98$.*

Solution. Given :

Dia. of pipe, $d = 300 \text{ mm} = 0.30 \text{ m}$

Diff. of pressure head, $h = 60 \text{ mm of water} = .06 \text{ m of water}$

$C_v = 0.98$

Mean velocity, $\bar{V} = 0.80 \times \text{Central velocity}$

Central velocity is given by equation (8.9)

$$= C_v \sqrt{2gh} = 0.98 \times \sqrt{2 \times 9.81 \times .06} = 1.063 \text{ m/s}$$

$\therefore$ $\bar{V} = 0.80 \times 1.063 = 0.8504 \text{ m/s}$

Discharge, $Q = \text{Area of pipe} \times \bar{V}$

$$= \frac{\pi}{4} d^2 \times \bar{V} = \frac{\pi}{4} (.30)^2 \times 0.8504 = \mathbf{0.06 \ m^3/s.\ Ans.}$$

Problem 8.19 *Find the velocity of the flow of an oil through a pipe, when the difference of mercury level in a differential U-tube manometer connected to the two tappings of the pitot-tube is 100 mm. Take co-efficient of pitot-tube 0.98 and sp. gr. of oil = 0.8.*

Solution. Given :

Diff. of mercury level, $x = 100 \text{ mm} = 0.1 \text{ m}$

Sp. gr. of oil, $S_o = 0.8$

Sp. gr. of mercury, $S_g = 13.6$

$C_v = 0.98$

Diff. of pressure head, $h = x\left[\frac{S_g}{S_o} - 1\right] = .1\left[\frac{13.6}{0.8} - 1\right] = 1.6 \text{ m of oil}$

∴ Velocity of flow $= C_v \sqrt{2gh} = 0.98 \sqrt{2 \times 9.81 \times 1.6}$ = **5.49 m/s. Ans.**

Problem 8.20 *A pitot-static tube is used to measure the velocity of water in a pipe. The stagnation pressure head is 6 m and static pressure head is 5 m. Calculate the velocity of flow assuming the co-efficient of tube equal to 0.98.* (A.M.I.E., Winter, 1979)

Solution. Given :

Stagnation pressure head, $h_s = 6$ m

Static pressure head, $h_t = 5$ m

∴ $h = 6 - 5 = 1$ m

Velocity of flow, $V = C_v \sqrt{2gh} = 0.98 \sqrt{2 \times 9.81 \times 1}$ = **4.34 m/s. Ans.**

Problem 8.21 *A sub-marine moves horizontally in sea and has its axis 15 m below the surface of water. A pitot-tube properly placed just in front of the sub-marine and along its axis is connected to the two limbs of a U-tube containing mercury. The difference of mercury level is found to be 170 mm. Find the speed of the sub-marine knowing that the sp. gr. of mercury is 13.6 and that of sea-water is 1.026 with respect of fresh water.* (A.M.I.E., Winter, 1975)

Solution. Given :

Diff. of mercury level, $x = 170$ mm $= 0.17$ m

Sp. gr. of mercury, $S_g = 13.6$

Sp. gr. of sea-water, $S_o = 1.026$

$$\therefore \quad h = x\left[\frac{S_g}{S_o} - 1\right] = 0.17\left[\frac{13.6}{1.026} - 1\right] = 2.0834 \text{ m}$$

$$\therefore \quad V = \sqrt{2gh} = \sqrt{2 \times 9.81 \times 2.0834} = 6.393 \text{ m/s}$$

$$= \frac{6.393 \times 60 \times 60}{1000} \text{ km/hr} = \mathbf{23.01\ km/hr.\ Ans.}$$

Problem 8.22 *A pitot-tube is inserted in a pipe of 300 mm diameter. The static pressure in pipe is 100 mm of mercury (vacuum). The stagnation pressure at the centre of the pipe, recorded by the pitot-tube is 0.981 N/cm². Calculate the rate of flow of water through pipe, if the mean velocity of flow is 0.85 times the central velocity. Take $C_v = 0.98$.* (Converted to S.I. Units, A.M.I.E., Summer, 1987)

Solution. Given :

Dia. of pipe, $d = 300$ mm $= 0.30$ m

∴ Area, $a = \frac{\pi}{4} d^2 = \frac{\pi}{4} (.3)^2 = 0.07068 \text{ m}^2$

Static pressure head $= 100$ mm of mercury (vacuum)

$$= -\frac{100}{1000} \times 13.6 = -1.36 \text{ m of water}$$

Stagnation pressure $= .981 \text{ N/cm}^2 = .981 \times 10^4 \text{ N/m}^2$

$\therefore$ Stagnation pressure head $= \dfrac{.981\times10^4}{\rho g} = \dfrac{.981\times10^4}{1000\times9.81} = 1$ m

$\therefore$ h = Stagnation pressure head – Static pressure head

$= 1.0 - (-1.36) = 1.0 + 1.36 = 2.36$ m of water

$\therefore$ Velocity at centre $= C_v\sqrt{2gh}$

$= 0.98 \times \sqrt{2\times9.81\times2.36} = 6.668$ m/s

Mean velocity, $\overline{V} = 0.85 \times 6.668 = 5.6678$ m/s

$\therefore$ Rate of flow of water $= \overline{V} \times$ area of pipe

$= 5.6678 \times 0.07068$ m^3/s $=$ **0.4006 m^3/s. Ans.**

8.5 NOTCHES AND WEIRS

A notch is a device used for measuring the rate of flow of a liquid through a small channel or a tank. The notch be defined as an opening in the side of a tank or a small channel in such a way that the liquid surface in the tank or channel is below the top edge of the opening.

A weir is a concrete or masonary structure, placed in an open channel over which the flow occurs. It is generally in the form of vertical wall, with a sharp edge at the top, running all the way across the open channel. The notch is of small size while the weir is of a bigger size. The notch is generally made of metallic plate while weir is made of concrete or masonary structure.

1. Nappe or Vein. The sheet of water flowing through a notch or over a weir is called Nappe or Vein.

2. Crest or Sill. The bottom edge of a notch or a top of a weir over which the water flows, is known as the sill or crest.

8.5.1 Classification of Notches and Weirs. The notches are classified according to the shape of the opening as :

(*a*) Rectangular notch,

(*b*) Triangular notch,

(*c*) Trapezoidal notch, and

(*d*) Stepped notch.

Weirs are classified according to the shape of the opening as :

(*i*) Rectangular weir,

(*ii*) Triangular weir, and

(*iii*) Trapezoidal weir (Cippoletti weir)

8.5.2 Discharge over a Rectangular Notch or Weir. The expression for discharge over a rectangular notch or weir is the same.

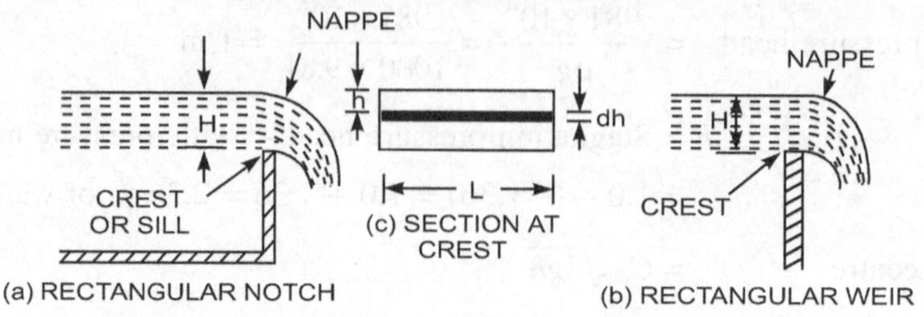

Fig. 8.13 *Rectangular notch and weir.*

Consider a rectangular notch or weir provided in a channel carrying water as shown in Fig. 8.13.

Let H = Head of water over the crest

L = Length of the notch or weir

For finding the discharge of water flowing over the weir or notch, consider an elementary horizontal strip of water of thickness dh and length L at a depth h form the free surface of water as shown in Fig. 8.13(*c*).

The area of strip $= L \times dh$

and theoretical velocity of water flowing through strip $= \sqrt{2gh}$

The discharge dQ, through strip is

$$dQ = C_d \times \text{Area of strip} \times \text{Theoretical velocity}$$

$$= C_d \times L \times dh \times \sqrt{2gh} \quad \text{...(i)}$$

where C_d = Co-efficient of discharge.

The total discharge, Q , for the whole notch or weir is determined by integrating equation (*i*) between the limits 0 and H.

$$\therefore \quad Q = \int_0^H C_d \cdot L \cdot \sqrt{2gh}\ .\ dh = C_d \times L \times \sqrt{2g} \int_0^H h^{1/2}\, dh$$

$$= C_d \times L \times \sqrt{2g} \left[\frac{h^{1/2+1}}{\frac{1}{2}+1} \right]_0^H = C_d \times L \times \sqrt{2g} \left[\frac{h^{3/2}}{3/2} \right]_0^H$$

$$= \frac{2}{3} C_d \times L \times \sqrt{2g}\ [H]^{3/2}. \quad \text{...(8.10)}$$

Problem 8.23 *Find the discharge of water flowing over a rectangular notch of 2 m length when the constant head over the notch is 300 mm. Take C_d = 0.60.*

Solution. Given :

Length of the notch, $L = 2.0$ m

Head over notch, $H = 300$ m $= 0.30$ m

$C_d = 0.60$

Discharge, $Q = \frac{2}{3} C_d \times L \times \sqrt{2g}\ [H^{3/2}]$

$$= \frac{2}{3} \times 0.6 \times 2.0 \times \sqrt{2 \times 9.81} \times [.30]^{1.5}\ \text{m}^3/\text{s}$$

$$= 3.5435 \times 0.1643 = \mathbf{0.582\ m^3/s.\ Ans.}$$

Problem 8.24 *Determine the height of a rectangular weir of length 6 m to be built across a rectangular channel. The maximum depth of water on the upstream side of the weir is 1.8 m and discharge is 2000 litres/s. Take C_d = 0.6 and neglect end contractions.*

Solution. Given :

Length of weir, $L = 6$ m

Depth of water, $H_1 = 1.8$ m

Discharge, $Q = 2000$ lit/s $= 2$ m^3/s

$C_d = 0.6$

Let H is height of water above the crest of weir, and H_2 = height of weir (Fig. 8.14)

The discharge over the weir is given by the equation (8.10) as

$$Q = \frac{2}{3} C_d \times L \times \sqrt{2g}\ H^{3/2}$$

or
$$2.0 = \frac{2}{3} \times 0.6 \times 6.0 \times \sqrt{2 \times 9.81} \times H^{3/2}$$

$$= 10.623\ H^{3/2}$$

$\therefore$
$$H^{3/2} = \frac{2.0}{10.623}$$

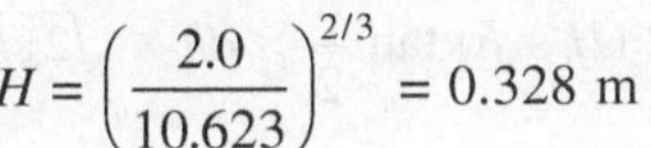

$\therefore$
$$H = \left(\frac{2.0}{10.623}\right)^{2/3} = 0.328 \text{ m}$$

Fig. 8.14

$\therefore$ Height of weir,
$$H_2 = H_1 - H$$

$$= \text{Depth of water on upstream side} - H$$

$$= 1.8 - .328 = \mathbf{1.472\ m.\ Ans.}$$

Problem 8.25 *The head of water over a rectangular notch is 900 mm. The discharge is 300 litres/s. Find the length of the notch, when C_d = 0.62.*

Solution. Given :

Head over notch, $H = 90$ cm $= 0.9$ m

Discharge, $Q = 300$ lit/s $= 0.3$ m^3/s

$C_d = 0.62$

Let length of notch $= L$

Using equation (8.10), we have

$$Q = \frac{2}{3} \times C_d \times L \times \sqrt{2g} \times H^{3/2}$$

or
$$0.3 = \frac{2}{3} \times 0.62 \times L \times \sqrt{2 \times 9.81} \times (0.9)^{3/2} = 1.83 \times L \times 0.8538$$

$\therefore$
$$L = \frac{0.3}{1.83 \times .8538} = .192 \text{ m} = \mathbf{192\ mm.\ Ans.}$$

8.5.3 Discharge Over a Triangular Notch or Weir. The expression for the discharge over a triangular notch or weir is the same. It is derived as :

Let H = head of water above the V-notch

θ = angle of notch

Consider a horizontal strip of water of thickness 'dh' at a depth of h from the free surface of water as shown in Fig. 8.15.

From Fig. 8.15 (*b*), we have

$$\tan\frac{\theta}{2} = \frac{AC}{OC} = \frac{AC}{(H-h)}$$

$$\therefore \quad AC = (H-h)\tan\frac{\theta}{2}$$

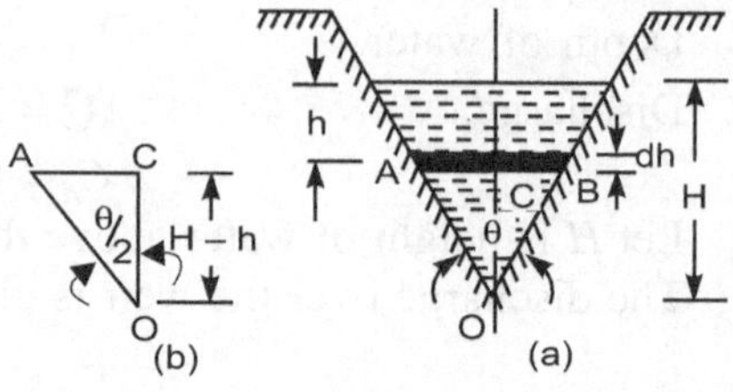

Width of strip $= AB = 2AC = 2\,(H-h)\tan\frac{\theta}{2}$

Fig. 8.15 *The triangular notch.*

$\therefore$ Area of strip $= 2\,(H-h)\tan\frac{\theta}{2}\times dh$

The theoretical velocity of water through strip $= \sqrt{2gh}$

$\therefore$ Discharge, dQ, through the strip is

$$dQ = C_d \times \text{Area of strip} \times \text{Velocity (theoretical)}$$

$$= C_d \times 2\,(H-h)\tan\frac{\theta}{2}\times dh \times \sqrt{2gh}$$

$$= 2C_d\,(H-h)\tan\frac{\theta}{2}\times \sqrt{2gh}\times gh$$

$\therefore$ Total discharge, Q is

$$Q = \int_0^H 2C_d\,(H-h)\tan\frac{\theta}{2}\times\sqrt{2gh}\times dh$$

$$= 2C_d \times \tan\frac{\theta}{2}\times\sqrt{2g}\int_0^H (H-h)h^{1/2}\,dh$$

$$= 2\times C_d\times\tan\frac{\theta}{2}\times\sqrt{2g}\int_0^H (Hh^{1/2}-h^{3/2})\,dh$$

$$= 2\times C_d\times\tan\frac{\theta}{2}\times\sqrt{2g}\left[\frac{Hh^{3/2}}{3/2}-\frac{h^{5/2}}{5/2}\right]_0^H$$

$$= 2\times C_d\times\tan\frac{\theta}{2}\times\sqrt{2g}\left[\frac{2}{3}H.H^{3/2}-\frac{2}{5}H^{5/2}\right]$$

$$= 2\times C_d\times\tan\frac{\theta}{2}\times\sqrt{2g}\left[\frac{2}{3}H^{5/2}-\frac{2}{5}H^{5/2}\right]$$

$$= 2\times C_d\times\tan\frac{\theta}{2}\times\sqrt{2g}\left[\frac{4}{15}H^{5/2}\right]$$

$$= \frac{8}{15}C_d\times\tan\frac{\theta}{2}\times\sqrt{2g}\times H^{5/2} \qquad \text{...(8.11)}$$

For a right-angled V-notch, if $C_d = 0.6$

$$\theta = 90°, \quad \therefore \quad \tan\frac{\theta}{2} = 1$$

Discharge
$$Q = \frac{8}{15} \times 0.6 \times 1 \times \sqrt{2 \times 9.81} \times H^{5/2} \quad \text{...(8.12)}$$
$$= 1.417\ H^{5/2}.$$

Problem 8.26 *Find the discharge over a triangular notch of angle 60° when the head over the V-notch is 0.3 m. Assume $C_d = 0.6$.*

Solution. Given :

Angle of V-notch, $\theta = 60°$

Head over notch, $H = 0.3$ m

$C_d = 0.6$

Discharge, Q over a V-notch is given by equation (8.11)

$$Q = \frac{8}{15} \times C_d \times \tan\frac{\theta}{2} \times \sqrt{2g} \times H^{5/2}$$

$$= \frac{8}{15} \times 0.6 \tan\frac{60}{2} \times \sqrt{2 \times 9.81} \times (0.3)^{5/2}$$

$$= 0.8182 \times 0.0493 = \mathbf{0.040\ m^3/s.\ Ans.}$$

Problem 8.27 *Water flows over a rectangular weir 1 m wide at a depth of 150 mm and afterwards passes through a triangular right-angled weir. Taking C_d for the rectangular and triangular weir as 0.62 and 0.59 respectively, find the depth over the triangular weir.*

(VTU, March 2000 ; Osmania University, 1990 ; A.M.I.E., Winter, 1975)

Solution. Given :

For rectangular weir, length, $L = 1$ m

Depth of water, $H = 150$ mm $= 0.15$ m

$C_d = 0.62$

For triangular weir, $\theta = 90°$

$C_d = 0.59$

Let depth over triangular weir $= H_1$

The discharge over the rectangular weir is given by equation (8.10) as

$$Q = \frac{2}{3} \times C_d \times L \times \sqrt{2g} \times H^{3/2}$$

$$= \frac{2}{3} \times 0.62 \times 1.0 \times \sqrt{2 \times 9.81} \times (.15)^{3/2}\ \text{m}^3\text{/s} = 0.10635\ \text{m}^3\text{/s}$$

The same discharge passes through the triangular right-angled weir. But discharge, Q, is given by equation (8.11) for a triangular weir as

$$Q = \frac{8}{15} \times C_d \times \tan\frac{\theta}{2} \times \sqrt{2g} \times H^{5/2}$$

$$\therefore \quad 0.10635 = \frac{8}{15} \times .59 \times \tan\frac{90}{2} \times \sqrt{2g} \times {H_1}^{5/2} \quad \{\because \theta = 90° \text{ and } H = H_1\}$$

$$= \frac{8}{15} \times .59 \times 1 \times 4.429 \times H_1^{5/2} = 1.3936\ H_1^{5/2}$$

$$\therefore \quad H_1^{5/2} = \frac{0.10635}{1.3936} = 0.07631$$

$$\therefore \quad H_1 = (.07631)^{0.4} = \mathbf{0.3572\ m.\ Ans.}$$

Problem 8.28 *Water flows through a triangular right-angled weir first and then over a rectangular weir of 1 m width. The discharge co-efficients of the triangular and rectangular weirs are 0.6 and 0.7 respectively. If the depth of water over the triangular weir is 360 mm, find the depth of water over the rectangular weir.* (A.M.I.E., Summer, 1990)

Solution. Given :

For triangular weir : $\theta = 90°$, $C_d = 0.6$, $H = 360$ mm $= 0.36$ m

For rectangular weir : $L = 1$ m, $C_d = 0.7$, $H = ?$

The discharge for a triangular weir is given by equation (8.11) as

$$Q = \frac{8}{15} \times C_d \times \tan\frac{\theta}{2} \times \sqrt{2g} \times H^{5/2}$$

$$= \frac{8}{15} \times 0.6 \times \tan\left(\frac{90}{2}\right) \times \sqrt{2 \times 9.81} \times (0.36)^{5/2} = 0.1102\ \text{m}^3/\text{s}$$

The same discharge is passing through the rectangular weir. But discharge for a rectangular weir is given by equation (8.10) as

$$Q = \frac{2}{3} \times C_d \times L \times \sqrt{2g} \times H^{3/2}$$

or
$$0.1102 = \frac{2}{3} \times 0.7 \times 1 \times \sqrt{2 \times 9.81} \times H^{3/2} = 2.067\ H^{3/2}$$

or
$$H^{3/2} = \frac{0.1102}{2.067} = 0.0533$$

$$\therefore \quad H = (0.0533)^{2/3} = 0.1415\ \text{m} = \mathbf{141.5\ mm.\ Ans.}$$

Problem 8.29 *A rectangular channel 2.0 m wide has a discharge of 250 litres per second, which is measured by a right-angled V-notch weir. Find the position of the apex of the notch from the bed of the channel if maximum depth of water is not to exceed 1.3 m. Take $C_d = 0.62$.*

Solution. Given :

Width of rectangular channel, $L = 2.0$ m

Discharge, $Q = 250$ lit/s $= 0.25$ m^3/s

Depth of water in channel $= 1.3$ m

Let the height of water over *V*-notch $= H$

The rate of flow through *V*-notch is given by equation (8.11) as

$$Q = \frac{8}{15} \times C_d \times \sqrt{2g} \times \tan\frac{\theta}{2} \times H^{5/2}$$

where $C_d = 0.62$, $\theta = 90°$

$$\therefore \quad Q = \frac{8}{15} \times .62 \times \sqrt{2 \times 9.81} \times \tan\frac{90}{2} \times H^{5/2}$$

or $$0.25 = \frac{8}{15} \times .62 \times 4.429 \times 1 \times H^{5/2}$$

or $$H^{5/2} = \frac{.25 \times 15}{8 \times .62 \times 4.429} = 0.1707$$

$\therefore$ $$H = (.1707)^{2/5} = (.1707)^{0.4} = 0.493 \text{ m}$$

Position of apex of the notch from the bed of channel

= depth of water in channel – height of water over V-notch

= 1.3 – .493 = **0.807 m. Ans.**

8.5.4 Advantages of Triangular Notch or Weir Over Rectangular Notch or Weir. A triangular notch or weir is preferred to a rectangular weir or notch due to following reasons :

1. The expression for discharge for a right-angled V-notch or weir is very simple.
2. For measuring low discharge, a triangular notch gives more accurate results than a rectangular notch.
3. In case of triangular notch, only one reading, *i.e.*, (H) is required for the computation of discharge.
4. Ventilation of a triangular notch is not necessary.

8.5.5 Discharge Over a Trapezoidal Notch or Weir. As shown in Fig. 8.16, a trapezoidal notch or weir is a combination of a rectangular and triangular notch or weir. Thus the total discharge will be equal to the sum of discharge through a rectangular weir or notch and discharge through a triangular notch or weir.

Let H = Height of water over the notch

L = Length of the crest of the notch

C_{d_1} = Co-efficient or discharge for rectangular portion $ABCD$ of Fig. 8.16.

C_{d_2} = Co-efficient of discharge for triangular portion [FAD and BCE]

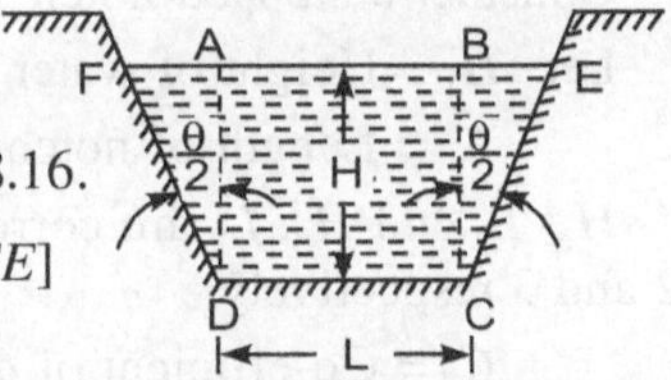

Fig. 8.16 *The trapezoidal notch.*

The discharge through rectangular portion $ABCD$ is given by (8.10)

or $$Q_1 = \frac{2}{3} \times C_{d_1} \times L \times \sqrt{2g} \times H^{3/2}$$

The discharge through two triangular notches FDA and BCE is equal to the discharge through a single triangular notch of angle θ and it is given by equation (8.11) as

$$Q_2 = \frac{8}{15} \times C_{d_2} \times \tan \frac{\theta}{2} \times \sqrt{2g} \times H^{5/2}$$

$\therefore$ Discharge through trapezoidal notch or weir $FDCEF = Q_1 + Q_2$

$$= \frac{2}{3} C_{d_1} L \sqrt{2g} \times H^{3/2} + \frac{8}{15} C_{d_2} \times \tan \theta/2 \times \sqrt{2g} \times H^{5/2}. \quad ...(8.13)$$

Problem 8.30 *Find the discharge through a trapezoidal notch which is 1 m wide at the top and 0.40 m at the bottom and is 30 cm in height. The head of water on the notch is 20 cm. Assume C_d for rectangular portion as = 0.62 while for triangular portion = 0.60.*

Solution. Given :

Top width, $AE = 1$ m

Base width, $CD = L = 0.4$ m

Head of water, $H = 0.20$ m

For rectangular portion, $C_{d_1} = 0.62$

For triangular portion, $C_{d_2} = 0.60$

From ΔABC, we have

Fig. 8.17

$$\tan\frac{\theta}{2} = \frac{AB}{BC} = \frac{(AE - CD)/2}{H}$$

$$= \frac{(1.0 - 0.4)/2}{0.3} = \frac{0.6/2}{0.3} = \frac{0.3}{0.3} = 1$$

Discharge through trapezoidal notch is given by equation (8.13)

$$Q = \frac{2}{3}\, C_{d_1} \times L \times \sqrt{2g} \times H^{3/2} + \frac{8}{15}\, C_{d_2} \times \tan\frac{\theta}{2} \times \sqrt{2g} \times H^{5/2}$$

$$= \frac{2}{3} \times 0.62 \times 0.4 \times \sqrt{2 \times 9.81} \times (0.2)^{3/2} + \frac{8}{15} \times .60 \times 1 \times \sqrt{2 \times 9.81} \times (0.2)^{5/2}$$

$= 0.06549 + 0.02535 = 0.09084$ m^3/s $=$ **90.84 litres/s. Ans.**

8.5.6 Discharge Over a Stepped Notch. A stepped notch is a combination of rectangular notches. The discharge through stepped notch is equal to the sum of the discharges through the different rectangular notches.

Consider a stepped notch as shown in Fig. 8.18.

Let H_1 = Height of water above the crest of notch (1),

L_1 = Length of notch 1,

H_2, L_2 and H_3, L_3 are corresponding values for notches 2 and 3 respectively.

C_d = Co-efficient of discharge for all notches

$\therefore$ Total discharge $Q = Q_1 + Q_2 + Q_3$

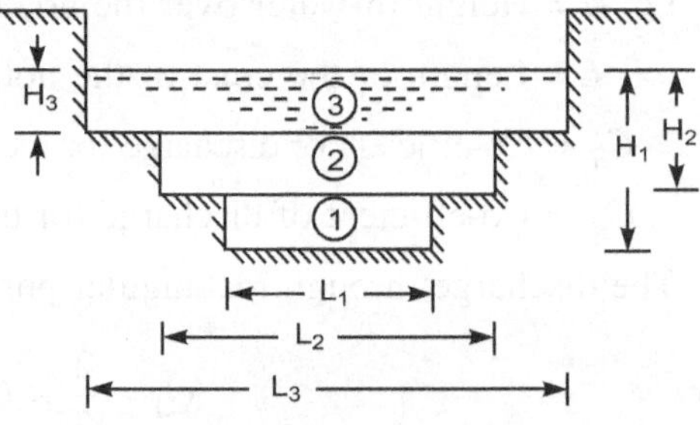

Fig. 8.18 *The stepped notch.*

or

$$Q = \frac{2}{3} \times C_d \times L_1 \times \sqrt{2g}\,[H_1^{3/2} - H_2^{3/2}]$$

$$+ \frac{2}{3}\, C_d \times L_2 \times \sqrt{2g}\,[H_2^{3/2} - H_3^{3/2}] + \frac{2}{3}\, C_d \times L_3 \times \sqrt{2g} \times H_3^{3/2}. \qquad \text{...(8.14)}$$

Problem 8.31 *Fig. 8.19 shows a stepped notch. Find the discharge through the notch if C_d for all section = 0.62.*

Solution. Given :

$L_1 = 40$ cm, $L_2 = 80$ cm,

$L_3 = 120$ cm

$H_1 = 50 + 30 + 15 = 95$ cm,

$H_2 = 80$ cm, $H_3 = 50$ cm,

$C_d = 0.62$

Total discharge, $Q = Q_1 + Q_2 + Q_3$

where $$Q_1 = \frac{2}{3} \times C_d \times L_1 \times \sqrt{2g}\ [H_1^{3/2} - H_2^{3/2}]$$

$$= \frac{2}{3} \times 0.62 \times 40 \times \sqrt{2 \times 981} \times [95^{3/2} - 80^{3/2}]$$

$$= 732.26[925.94 - 715.54]$$

$$= 154067 \text{ cm}^3/\text{s} = 154.067 \text{ lit/s}$$

Fig. 8.19

$$Q_2 = \frac{2}{3} \times C_d \times L_2 \times \sqrt{2g} \times [H_2^{3/2} - H_3^{3/2}]$$

$$= \frac{2}{3} \times 0.62 \times 80 \times \sqrt{2 \times 981} \times [80^{3/2} - 50^{3/2}]$$

$$= 1464.52[715.54 - 353.55] \text{ cm}^3/\text{s} = 530141 \text{ cm}^3/\text{s} = 530.144 \text{ lit/s}$$

and $$Q_3 = \frac{2}{3} \times C_d \times L_3 \times \sqrt{2g} \times H_3^{3/2}$$

$$= \frac{2}{3} \times 0.62 \times 120 \times \sqrt{2 \times 981} \times 50^{3/2} = 776771 \text{ cm}^3/\text{s} = 776.771 \text{ lit/s}$$

$\therefore$ $$Q = Q_1 + Q_2 + Q_3 = 154.067 + 530.144 + 776.771$$

$$= \mathbf{1460.98 \text{ lit/s. Ans.}}$$

8.5.7 Effect on Discharge Over a Notch Due to Error in the Measurement of Head. For an accurate value of the discharge over a weir or notch, an accurate measurement of head over the weir or notch is very essential as the discharge over a triangular notch is proportional to $H^{5/2}$ and in case of rectangular notch it is proportional to $H^{3/2}$. A small error in the measurement of head, will affect the discharge considerably. The following cases of error in the measurement of head will be considered :

(*i*) For Rectangular Weir or Notch.

(*ii*) For Triangular Weir or Notch.

1. For Rectangular Weir. The discharge for a rectangular weir or notch is given by equation (8.10) as

$$Q = \frac{2}{3} \times C_d \times L \times \sqrt{2g} \times H^{3/2}$$

$$= KH^{3/2} \quad \ldots(i)$$

where $K = \frac{2}{3} C_d \times L \times \sqrt{2g}$

Differentiating the above equation, we get

$$dQ = K \times \frac{3}{2} H^{1/2}\, dH \quad \ldots(ii)$$

Dividing (*ii*) by (*i*), $$\frac{dQ}{Q} = \frac{K \times \frac{3}{2} \times H^{1/2} dH}{KH^{3/2}} = \frac{3}{2}\frac{dH}{H} \quad \text{...(8.15)}$$

Equation (8.15) shows that an error of 1% in measuring H will produce 1.5% error in discharge over a rectangular weir or notch.

2. For Triangular Weir. The discharge over a triangular weir or notch is given by equation (8.11) as

$$Q = \frac{8}{15} C_d . \tan\frac{\theta}{2} \sqrt{2g} \times H^{5/2}$$

$$= KH^{5/2} \quad \text{...(iii)}$$

where $K = \frac{8}{15} C_d . \tan\frac{\theta}{2}\sqrt{2g}$

Differentiating equation (*iii*), we get

$$dQ = K\frac{5}{2} H^{3/2} \times dH \quad \text{...(iv)}$$

Dividing (*iv*) by (*iii*), we get $$\frac{dQ}{Q} = \frac{K\frac{5}{2} H^{3/2}\ dH}{KH^{5/2}} = \frac{5}{2}\frac{dH}{H} \quad \text{...(8.16)}$$

Equation (8.16) shows that an error of 1% in measuring H will produce 2.5% error in discharge over a triangular weir or notch.

Problem 8.32 *A rectangular notch 40 cm long is used for measuring a discharge of 30 litres per second. An error of 1.5 mm wars made, while measuring the head over the notch. Calculate the percentage error in the discharge. Take $C_d = 0.60$.*

Solution. Given :

Length of notch, $L = 40$ cm

Discharge, $Q = 30$ lit/s $= 30{,}000$ cm^3/s

Error in head, $dH = 1.5$ mm $= 0.15$ cm

$C_d = 0.60$

Let the height of water over rectangular notch $= H$

The discharge through a rectangular notch is given by (8.10)

or $$Q = \frac{2}{3} \times C_d \times L \times \sqrt{2g} \times H^{3/2}$$

or $$30{,}000 = \frac{2}{3} \times 0.60 \times 40 \times \sqrt{2 \times 981} \times H^{3/2}$$

or $$H^{3/2} = \frac{3 \times 30000}{2 \times .60 \times 40 \times \sqrt{2 \times 981}} = 42.33$$

$\therefore$ $$H = (42.33)^{2/3} = 12.16 \text{ cm}$$

Using equation (8.15), we get

$$\frac{dQ}{Q} = \frac{3}{2}\frac{dH}{H} = \frac{3}{2} \times \frac{0.15}{12.16} = 0.0185 = \mathbf{1.85\%.\ Ans.}$$

Problem 8.33 *A right-angled V-notch is used for measuring a discharge of 30 litres/s. An error of 1.5 mm was made while measuring the head over the notch. Calculate the percentage error in the discharge. Take $C_d = 0.62$.*

Solution. Given :

Angle of *V*-notch, $\theta = 90°$

Discharge, $Q = 30 \text{ lit/s} = 30000 \text{ cm}^3/\text{s}$

Error in head, $dH = 1.5 \text{ mm} = 0.15 \text{ cm}$

$C_d = 0.62$

Let the head over the *V*-notch $= H$

The discharge Q through a triangular notch is given by equation (8.11)

$$Q = \frac{8}{15} C_d \cdot \tan\frac{\theta}{2} \times \sqrt{2g} \times H^{5/2}$$

or

$$30000 = \frac{8}{15} \times 0.62 \times \tan\left(\frac{90°}{2}\right) \times \sqrt{2 \times 981} \times H^{5/2}$$

$$= \frac{8}{15} \times .62 \times 1 \times 44.29 \times H^{5/2}$$

$$\therefore \quad H^{5/2} = \frac{30000 \times 15}{8 \times .62 \times 44.29} = 2048.44$$

$$\therefore \quad H = (2048.44) = 21.11 \text{ cm}$$

Using equation (8.16), we get

$$\frac{dQ}{Q} = \frac{5}{2}\frac{dH}{H} = 2.5 \times \frac{0.15}{21.11} = 0.01776 = \mathbf{1.77\%.} \quad \textbf{Ans.}$$

Problem 8.34 *The head of water over a triangular notch of angle 60° is 50 cm and co-efficient of discharge is 0.62. The flow measured by it is to be within an accuracy of 1.5% up or down. Find the limiting values of the head.*

Solution. Given :

Angle of *V*-notch, $\theta = 60°$

Head of water, $H = 50 \text{ cm}$

$C_d = 0.62$

$$\frac{dQ}{Q} = \pm 1.5\% = \pm 0.015$$

The discharge Q over a triangular notch is

$$Q = \frac{8}{15} C_d \sqrt{2g} \tan\frac{\theta}{2} H^{5/2}$$

$$= \frac{8}{15} \times 0.62 \times \sqrt{2 \times 981} \times \tan \frac{60°}{2} \times (50)^{5/2}$$

$$= 14.64 \times 0.5773 \times 17677.67 = 149405.86 \text{ cm}^3/\text{s}$$

Now applying equation (8.16), we get

$$\frac{dQ}{Q} = \frac{5}{2}\frac{dH}{H} \quad \text{or } \pm .015 = 2.5 \frac{dH}{H} \quad \text{or} \quad \frac{dH}{H} = \pm \frac{.015}{2.5}$$

$$\therefore \qquad dH = \pm \frac{.015}{2.5} \times H = \pm \frac{.015}{2.5} \times 50 = \pm 0.3$$

∴ The limiting values of the head

$$= H \pm dH = 50 \pm 0.3 = 50.3 \text{ cm, } 49.7 \text{ cm}$$

= **50.3 cm and 49.7 cm. Ans.**

HIGHLIGHTS

1. The discharge, Q, through a venturimeter or an orifice meter is given by

$$Q = C_d \frac{a_1 a_2}{\sqrt{a_1^2 - a_2^2}} \times \sqrt{2gh}$$

where a_1 = area at the inlet of venturimeter,
a_2 = area at the throat of venturimeter,
C_d = co-efficient of venturimeter,
h = difference of pressure head in terms of fluid head flowing through venturimeter.

2. The value of h is given by differential U-tube manometer $h = x\left[\frac{S_h}{S_o} - 1\right]$

when differential manometer contains heavier liquid $= x\left[1 - \frac{S_l}{S_o}\right]$

when differential manometer contains lighter liquid $= \left(\frac{p_1}{\rho g} + z_1\right) - \left(\frac{p_2}{\rho g} + z_2\right) = x\left[\frac{S_h}{S_o} - 1\right]$

for inclined venturimeter and differential manometer contains heavier liquid

$$= \left(\frac{p_1}{\rho g} + z_1\right) - \left(\frac{p_2}{\rho g} + z_2\right) = x\left[1 - \frac{S_l}{S_o}\right]$$

for inclined venturimeter and manometer contains lighter liquid.
where x = difference in the readings of differential manometer,
S_h = sp. gr. of heavier liquid
S_o = sp. gr. of fluid flowing through venturimeter
S_l = sp. gr. of lighter liquid.

3. Pitot-tube is used to find the velocity of a flowing fluid at any point in a pipe or a channel. The velocity is given by the relation

$$V = C_v \sqrt{2gh}$$

where C_v = co-efficient of Pitot-tube
h = rise of liquid in the tube above free surface of liquid – for channels
$= x\left[\frac{S_g}{S_o} - 1\right]$ – for pipes.

4. A notch is a device used for measuring the rate of flow of a liquid through a small channel. A weir is a concrete or masonary structure placed in the open channel over which the flow occurs.
5. The discharge through a rectangular notch or weir is given by

$$Q = \frac{2}{3} C_d \times L \times H^{3/2}$$

where C_d = co-efficient of discharge,
L = length of notch or weir,
H = head of water over the notch or weir.

6. The discharge over a triangular notch or weir is given by

$$Q = \frac{8}{15} C_d \tan \frac{\theta}{2} \times \sqrt{2g} \times H^{3/2}$$

where θ = total angle of triangular notch.

7. The discharge through a trapezoidal notch or weir is equal to the sum of discharge through a rectangular notch and the discharge through a triangular notch. It is given as

$$Q = \frac{2}{3} C_{d_1} \times L \times H^{3/2} + \frac{8}{15} C_{d_2} \times \tan \frac{\theta}{2} \times \sqrt{2g} \times H^{5/2}$$

where C_{d_1} = co-efficient of discharge for rectangular notch,
C_{d_2} = co-efficient of discharge for triangular notch,
$\theta/2$ = slope of the side of trapezoidal notch.

8. The error in discharge due to the error in the measurement of head over a rectangular and triangular notch or weir is given by

$$\frac{dQ}{Q} = \frac{3}{2} \frac{dH}{H} \text{ ...... For a rectangular weir or notch}$$

$$= \frac{5}{2} \frac{dH}{H} \text{ ...... For a triangular weir or notch}$$

where Q = discharge through rectangular or triangular notch or weir
H = head over the notch or weir.

EXERCISE 8

(A) THEORETICAL PROBLEMS

1. What is a venturimeter ? Derive an expression for the discharge through a venturimeter. *(VTU, March 2000, Feb. 2003; A.S.M.E., June 1992 ; A.M.I.E., Winter, 1980)*
2. Explain the principle of venturimeter with a neat sketch. Derive the expression for the rate of flow of fluid through it.
3. Discuss the relative merits and demerits of venturimeter with respect to orifice-meter. *(Delhi University, Dec. 2002)*
4. Define an orifice-meter. Prove that the discharge through an orifice-meter is given by the relation

$$Q = C_d \frac{a_0 a_1}{\sqrt{a_1^2 - a_0^2}} \times \sqrt{2gh}$$

where a_1 = area of pipe in which orifice-meter is fitted
a_0 = area of orifice *(Technical University of M.P., S 2002)*

5. What is a pitot-tube ? How will you determine the velocity at any point with the help of pitot-tube ? *(VTU, March 2000, March 2001; Delhi University, Dec. 2002)*
6. What is the difference between pitot-tube and pitot-static tube ?

7. State the momentum equation. How will you apply momentum equation for determining the force exerted by a flowing liquid on a pipe bend ?
8. What is the difference between momentum equation and impulse momentum equation ?
9. Define moment of momentum equation. Where this equation is used ?
10. Define the terms : notch, weir, nappe and crest.
11. How are the weirs and notches classified ?
12. Find an expression for the discharge over a rectangular weir in terms of head of water over the crest of the weir.
13. Prove that the discharge through a triangular notch or weir is given by

$$Q = \frac{8}{15} C_d \times \tan \frac{\theta}{2} \times \sqrt{2g} \; H^{3/2}$$

where H = head of water over the notch or weir

θ = angle of notch or weir.

14. What are the advantages of triangular notch or weir over rectangular notch ?
15. Prove that the error in discharge due to the error in the measurement of head over a rectangular notch is given by

$$\frac{dQ}{Q} = \frac{3}{2} \frac{dH}{H}$$

where Q = discharge through rectangular notch

and H = head over the rectangular notch.

16. State the different devices that can be used to measure the discharge through a pipe also through an open channel. Describe one of such devices with a neat sketch and explain how one can obtain the actual discharge with its help. *(A.M.I.E., Summer, 1992)*
17. What is the difference between a notch and a weir ?
18. Define velocity of approach. How does the velocity of approach affect the discharge over a weir ?

(B) NUMERICAL PROBLEMS

VENTURIMETER ORIFICEMETER AND PITOT-TUBE

1. A horizontal venturimeter with inlet and throat diameters 30 cm and 15 cm respectively is used to measure the flow of water. The reading of differential manometer connected to inlet and throat is 10 cm of mercury. Determine the rate of flow. Take $C_d = 0.98$. [**Ans.** 88.92 litres/s]
2. An oil of sp. gr. 0.9 is flowing through a venturimeter having inlet diameter 20 cm and throat diameter 10 cm. The oil-mercury differential manometer shows a reading of 20 cm. Calculate the discharge of oil through the horizontal venturimeter. Take $C_d = 0.98$. *(Delhi University, 1987)* [**Ans.** 59.15 litres/s]
3. A horizontal venturimeter with inlet diameter 30 cm and throat diameter 15 cm is used to measure the flow of oil of sp. gr. 0.8. The discharge of oil through venturimeter is 50 litres/s, find the reading of the oil-mercury differential manometer. Take $C_d = 0.98$. [**Ans.** 2.489 cm]
4. A horizontal venturimeter with inlet diameter 20 cm and throat diameter 10 cm is used to measure the flow of water. The pressure at inlet is 14.715 N/cm^2 and vacuum pressure at the throat is 40 cm of mercury. Find the discharge of water through venturimeter. [**Ans.** 162.539 lit/s]
5. A 30 cm × 15 cm venturimeter is inserted in a vertical pipe carrying water, flowing in the upward direction. A differential mercury-manometer connected to the inlet and throat gives a reading of 30 cm. Find the discharge. Take $C_d = 0.98$. [**Ans.** 154.02 lit/s]
6. If in the problem 5, instead of water, oil of sp. gr. 0.8 is flowing through the venturimeter, determine the rate of flow of oil in litres/s. [**Ans.** 173.56 lit/s]

7. The water is flowing through a pipe of diameter 30 cm. The pipe is inclined and a venturimeter is inserted in the pipe. The diameter of venturimeter at throat is 15 cm. The difference of pressure between the inlet and throat of the venturimeter is measured by a liquid of sp. gr. 0.8 in an inverted *U*-tube which gives a reading of 40 cm. The loss of head between the inlet and throat is 0.3 times the kinetic head of the pipe. Find the discharge. [**Ans.** 22.64 lit./s]
8. A 20 × 10 cm venturimeter is provided in a vertical pipe line carrying oil of sp. gr. 0.8, the flow being upwards. The difference in elevation of the throat section and entrance section of the venturimeter is 50 cm. The differential *U*-tube mercury manometer shows a gauge deflection of 40 cm. Calculate : (*i*) the discharge of oil, and (*ii*) the pressure difference between the entrance section and the throat section. Take $C_d = 0.98$ and sp. gr. of mercury as 13.6. [**Ans.** (*i*) 89.132 lit/s, (*ii*) 5.415 N/cm^2]
9. In a 200 mm diameter horizontal pipe a venturimeter of 0.5 contraction ratio has been fixed. The head of water on the venturimeter when there is no flow is 4 m (gauge) Find the rate of flow for which the throat pressure will be 4 metres of water absolute. Take $C_d = 0.97$ and atmospheric pressure head = 10.3 m of water. [**Ans.** 111.92 lit/s]
10. An orifice-meter with orifice diameter 15 cm is inserted in a pipe of 30 cm diameter. The pressure gauges fitted upstream and downstream of the orifice meter give readings of 14.715 N/cm^2 and 9.81 N/cm^2 respectively. Find the rate of flow of water through the pipe in litres/s. Take $C_d = 0.6$. [**Ans.** 108.434 lit/s]
11. If in problem 18, instead of water, oil of sp. gr. 0.8 is flowing through the orifice meter in which the pressure difference is measured by a mercury oil differential manometer on the two sides of the orifice meter, find the rate of flow of oil when the reading of manometer is 40 cm. [**Ans.** 122.68 lit/s]
12. The pressure difference measured by the two tappings of a pitotstatic tube, one tapping pointing upstream and other perpendicular to the flow, placed in the centre of a pipe line of diameter 40 cm is 10 cm of water. The mean velocity in the pipe is 0.75 times the central velocity. Find the discharge through the pipe. Take co-efficient of pitot-tube as 0.98. [**Ans.** 0.1293 m^3/s]
13. Find the velocity of flow of an oil through a pipe, when the difference of mercury level in a differential *U*-tube manometer connected to the two tappings of the pitot-tube is 15 cm. Take sp. gr. of oil = 0.8 and co-efficient of pitot-tube as 0.98. [**Ans.** 6.72 m/s]
14. A sub-marine moves horizontally in sea and has its axis 20 m below the surface of water. A pitot-static tube placed in front of sub-marine and along its axis, is connected to the two limbs of a *U*-tube containing mercury. The difference of mercury level is found to be 20 cm. Find the speed of sub-marine. Take sp. gr. of mercury 13.6 and of sea-water 1.026. [**Ans.** 24.958 km/hr.]
15. Explain in brief the working of a pitot-tube. Calculate the velocity of flow of water in a pipe of diameter 300 mm at a point, where the stagnation pressure head is 5 m and static pressure head is 4 m. Given the co-efficient of pitot-tube = 0.97. (*Delhi University, Nov. 1982*) [**Ans.** 4.3 m/sec]
16. Find the rate of flow of water through a venturimeter fitted in a pipeline of diameter 30 cm. The ratio of diameter of throat and inlet of the venturimeter is *. The pressure at the inlet of the venturimeter is 13.734 N/cm^2 (gauge) and vacuum in the throat is 37.5 cm of mercury. The co-efficient of venturimeter is given as 0.98. (*Delhi University, April, 1982*) [**Ans.** 0.15 m^3/s]
17. A 30 cm × 15 cm venturimeter is inserted in a vertical pipe carrying an oil of sp. gr. 0.8, flowing in the upward direction. A differential mercury manometer connected to the inlet and throat gives a reading of 30 cm. The difference in the elevation of the throat section and inlet section is 50 cm. Find the rate of flow of oil. (*Delhi University, 1988*)
18. A venturimeter is used for measurement of discharge of water in horizontal pipe line. If the ratio of upstream pipe diameter to that of throat is 2 : 1, upstream diameter is 300 mm, the difference in pressure between the throat and upstream is equal to 3 m head of water and loss of head through meter is one-eighth of the throat velocity head, calculate the discharge in the pipe. (*A.M.I.E., Winter, 1987*) [**Ans.** 0.107 m^3/s]
19. A liquid of specific gravity 0.8 is flowing upwards at the rate of 0.08 m^3/s, through a vertical venturimeter with an inlet diameter of 200 mm and throat diameter of 100 mm. The $C_d = 0.98$ and the vertical distance between pressure tappings is 300 mm. Find :

(*i*) the difference in readings of the two pressure gauges, which are connected to the two pressure tappings, and

(*ii*) the difference in the level of the mercury columns of the differential manometer which is connected to the tappings, in place of pressure gauges. (*Delhi University, 1992*)

[**Ans.** (*i*) 42.928 kN/m^2, (*ii*) 32.3 cm]

[**Hint.** $Q = 0.08$ m^3/s, $d_1 = 200$ mm $= 0.2$ m, $d_2 = 100$ mm $= 0.1$ m,

$C_d = 0.98$, $z_2 - z_1 = 300$ mm $= 0.3$ m, $a_1 = \frac{\pi}{4}(.2^2) = 0.0314$ m^2

$a_2 = \frac{\pi}{4}(.1^2) = 0.007854$ m^2. Using $Q = C_d \frac{a_1 \times a_2}{\sqrt{a_1^2 - a_2^2}} \times \sqrt{2gh}$

Find 'h'. This value of $h = 5.17$ m.

Now use $h = \left(\frac{p_1}{\rho g} - \frac{p_2}{\rho g}\right) + (z_1 - z_2)$, where $\rho = 800$ kg/m^3. Find $(p_1 - p_2)$.

Now use the formula $h = x\left[\frac{S_g}{S_f} - 1\right]$,

where $h = 5.17$ m, $S_g = 13.6$ and $S_f = 0.8$. Find the value of x which will be 32.3 cm.]

20. A venturimeter is installed in a 300 mm diameter horizontal pipe line. The throat pipe rates is 1/3. Water flows through the installation. The pressure in the pipe line is 13.783 N/cm^2 (gauge) and vacuum in the throat is 37.5 cm of mercury. Neglecting head loss in the venturimeter, determine the rate of flow in the pipe line. (*Osmania University, 1990*) [**Ans.** 0.153 m^3/sec)

[**Hint.** $d_1 = 300$ mm $= 0.3$ m, $d_2 = \frac{1}{3} \times 300 = 100$ mm $= 0.1$ m, $p_1 = 13.783$ N/cm^2 $= 13.783 \times 10^4$ N/m^2.

Hence $p_1/\rho \times g = 13.783 \times 10^4/1000 \times 9.81$

$= 14.05$ m, $p_2/\rho g = -37.5$ cm of Hg $= -0.375 \times 13.6$ m of water

$= -5.1$ m of water. Hence h $= 14.05 - (-5.1) = 19.15$ m of water.

Value of $C_d = 1.0$. Now use the formula $Q = C_d \frac{a_1 a_2}{\sqrt{a_1^2 - a_2^2}} \times \sqrt{2gh}$]

21. The maximum flow through a 300 mm diameter horizontal main pipe line is 18200 litre/minute. A venturimeter is introduced at a point of the pipe line where the pressure head is 4.6 m of water. Find the smallest dia. of throat so that the pressure at the throat is never negative. Assume co-efficient of meter as unity. (*A.M.I.E., Winter, 1989*), [**Ans.** $d_2 = 192.4$ mm]

[**Hint.** $d_1 = 300$ mm $= 0.3$ m, $Q = 18200$ litres/minute $= 18200/60 = 303.33$ litres/s $= 0.3033$ m^3/s, $p_1/\rho g$ $= 4.6$ m, $p_2/\rho g = 0$. Hence $h = 4.6$ m, $C_d = 1$. d_2 = dia. at throat. Use formula $Q = C_d \frac{a_1 \times a_2}{\sqrt{a_1^2 - a_2^2}} \times \sqrt{2gh}$ and find the value of a_2. Then $a_2 = \frac{\pi}{4} d_2^2$ and find d_2.]

22. The following are the data given of a change in diameter effected in laying a water supply pipe. The change in diameter is gradual from 20 cm at A to 50 cm at B. Pressures at A and B are 7.848 N/cm^2 and 5.886 N/cm^2 respectively with the end B being 3 m higher than A. If the flow in the pipe line is 200 litre/s, find :

(*i*) direction of flow, (*ii*) the head lost in friction between A and B.

(*Osmania University, 1990*) [**Ans.** (*i*) From A to B, (*ii*) 1.015 m]

[**Hint.** $D_A = 20$ cm $= 0.2$ m, $D_B = 50$ cm $= 0.5$ m, $p_A = 7.848$ N/cm^2 $= 7.848 \times 10^4$ N/m^2

$p_B = 5.886$ N/cm^2 $= 5.886 \times 10^4$ N/m^2, $Z_A = 0$, $Z_B = 3$ m, $Q = 0.2$ m^3/s]

23. A venturimeter of inlet diameter 300 mm and throat diameter 150 mm is fixed in a vertical pipe line. A liquid of sp. gr. 0.8 is flowing upward through the pipe line. A differential manometer containing mercury gives a reading of 100 mm when connected at inlet and throat. The vertical difference between inlet and throat is 500 mm. If $C_d = 0.98$, then find : (*i*) rate of flow of liquid in litre per second and (*ii*) difference of pressure between inlet and throat in N/m^2. *(Delhi University, 1988)*

[**Ans.** (*i*) 100 litre/s, (*ii*) 15980 N/m^2]

24. A venturimeter with a throat diameter of 7.5 cm is installed in a 15 cm diameter pipe. The pressure at the entrance to the meter is 70 kPa (gauge) and it is desired that the pressure at any point should not fall below 2.5 m of water absolute. Determine the maximum flow rate of water through the meter. Take $C_d = 0.97$ and atmospheric pressure as 100 kPa. *(J.N.T.U., Hyderabad S 2002)*

[**Hint.** The pressure at the throat will be minimum. Hence $\frac{p_2}{\rho g} = 2.5$ m (abs.)

Given : $d_1 = 15 \text{ cm} \therefore A_1 = \frac{\pi}{4}(15^2) = 176.7 \text{ cm}^2$

$d_2 = 7.5 \text{ cm} \therefore A_2 = \frac{\pi}{4}(7.5^2) = 44.175 \text{ cm}^2$

$p_1 = 70 \text{ kPa} = 70 \times 10^3 \text{ N/m}^2$ (gauge), $p_{atm} = 100 \text{ kPa} = 100 \times 10^3 \text{ N/m}^2$

$\therefore \quad p_1 \text{ (abs.)} = 70 \times 10^3 + 100 \times 10^3 = 170 \times 10^3 \text{ N/m}^2$ (abs.)

$$\therefore \quad \frac{p_1}{\rho g} = \frac{170 \times 10^3}{1000 \times 9.81} = 17.33 \text{ m of water (abs.)}$$

$$\therefore \quad h = \frac{p_1}{\rho g} - \frac{p_2}{\rho g} = 17.33 - 2.5 = 14.83 \text{ m of water} = 1483 \text{ cm of water}$$

Now
$$Q = \frac{C_d A_1 A_2}{\sqrt{A_1^2 - A_2^2}} \times \sqrt{2gh} = \frac{0.97 \times 176.7 \times 44.175 \times \sqrt{2 \times 981 \times 1483}}{\sqrt{176.7^2 - 44 \cdot 175^2}} = 75488 \text{ cm}^3/\text{s}$$

$= \mathbf{75.488}$ **litre/s. Ans.**]

25. Find the discharge of water flowing through a pipe 20 cm diameter placed in an inclined position, where a venturimeter is inserted, having a throat diameter of 10 cm. The difference of pressure between the main and throat is measured by a liquid of specific gravity 0.4 in an inverted *U*-tube, which gives a reading of 30 cm. The loss of head between the main and throat is 0.2 times the kinetic head of pipe. *(Delhi University, Dec. 2002)*

[**Hint.** Given : $d_1 = 20 \text{ cm} \therefore a_1 = \frac{\pi}{4}(20^2) = 100\,\pi \text{ cm}^2$; $d_2 = 10 \text{ cm} \therefore a_2 = \frac{\pi}{4}(10^2) = 25\,\pi \text{ cm}^2$.

$$x = 30 \text{ cm}, h = x\left(1 - \frac{S_l}{S_o}\right) = 30\left(1 - \frac{0.4}{1.0}\right) = 18 \text{ cm} = 0.18 \text{ m}$$

But h is also
$$= \left(\frac{p_1}{\rho g} + z_1\right) - \left(\frac{p_2}{\rho g} + z_2\right) \therefore \left(\frac{p_1}{\rho g} + z_1\right) - \left(\frac{p_2}{\rho g} + z_2\right) = 18 \text{ cm} = 0.18 \text{ m}$$

$$h_L = 0.2 \times \frac{V_1^2}{2g}$$

From Bernoulli's equation, $\frac{p_1}{\rho g} + \frac{V_1^2}{2g} + z_1 = \frac{p_2}{\rho g} + \frac{V_2^2}{2g} + z_2 + h_L$

or
$$\left(\frac{p_1}{\rho g} + z_1\right) - \left(\frac{p_2}{\rho g} + z_2\right) + \frac{V_1^2}{2g} - \frac{V_2^2}{2g} = h_L$$

or $$0.18+\frac{V_1^2}{2g}-\frac{V_2^2}{2g}=\frac{0.2\,V_1^2}{2g}\qquad\left(\because\left(\frac{p_1}{\rho g}+z_1\right)-\left(\frac{p_2}{\rho g}+z_2\right)=0.18\text{ m and } h_L=\frac{0.2\,V_1^2}{2g}\right)$$

or $$0.18+\frac{V_1^2}{2g}-\frac{V_2^2}{2g}-\frac{0.2\,V_1^2}{2g}=0 \text{ or } 0.18+\frac{0.8\,V_1^2}{2g}-\frac{V_2^2}{2g}=0$$

From continuity equation, $a_1V_1=a_2V_2$ or $V_2=\dfrac{a_1V_1}{a_2}=\dfrac{\frac{\pi}{4}(20^2)\,V_1}{\frac{\pi}{4}(10^2)}=4V_1$

Now $$0.18+\frac{0.8\,V_1^2}{2g}-\frac{V_2^2}{2g}=0 \text{ or } 0.18+\frac{0.8\,V_1^2}{2g}-\frac{(4V_1)^2}{2g}=0$$

or $$0.18+\frac{0.8\,V_1^2}{2g}-\frac{16V_1^2}{2g}=0 \text{ or } 0.18=\frac{16V_1^2}{2g}-\frac{0.8\,V_1^2}{2g}=\frac{15.2V_1^2}{2g}$$

$\therefore$ $$V_1=\sqrt{\frac{0.18\times2\times9.81}{15.2}}=0.48\text{ m/s}=48\text{ cm/s}$$

$\therefore$ $$Q=A_1V_1=\frac{\pi}{4}(20^2)\times48=15140\text{ cm}^3\text{/s}=\textbf{15.14 litre/s. Ans.]}$$

NOTCH

26. Find the discharge of water flowing over rectangular notch of 3 m length when the constant head of water over the notch is 40 cm. Take $C_d = 0.6$. [**Ans.** 1.344 m^3/s]

27. Determine the height of a rectangular weir of length 5 m to be built across a rectangular channel. The maximum depth of water on the upstream side of the weir is 1.5 m and discharge is 1.5 m^3 per second. Take $C_d = 0.6$ and neglect end contractions. [**Ans.** 1.194 m]

28. Find the discharge over a triangular notch of angle 60° when the head over the triangular notch is 0.20 m. Take $C_d = 0.6$. [**Ans.** 0.0164 m^3/s]

29. A rectangular channel 1.5 m wide has a discharge of 200 litres per second, which is measured by a right-angled *V*-notch weir. Find the position of the apex of the notch from the bed of the channel if maximum depth of water is not be exceed 1 m. Take $C_d = 0.62$. [**Ans.** .549 m]

30. Find the discharge through a trapezoidal notch which is 1.2 m wide at the top and 0.50 m at the bottom and is 40 cm in height. The head of water on the notch is 30 cm. Assume C_d for rectangular portion as 0.62 while for triangular portion = 0.60. [**Ans.** 0.22 m^3/s]

31. A rectangular notch 50 cm long is used for measuring a discharge of 40 litres per second. An error of 2 mm was made in measuring the head over the notch. Calculate the percentage error in the discharge. Take $C_d = 0.6$. [**Ans.** 2.37%]

32. A right-angled *V*-notch is used for measuring a discharge of 30 litres/s. An error of 2 mm was made in measuring the head over the notch. Calculate the percentage error in the discharge. Take $C_d = 0.62$. [**Ans.** 2.37%]

Flow Through Pipes

▶ 9.1 INTRODUCTION

In the previous chapters, laminar flow and turbulent flow have been discussed. We have seen that when the Reynold number is less than 2000 for pipe flow, the flow is known as laminar flow whereas when the Reynold number is more than 4000, the flow is known as turbulent flow. In this chapter, the turbulent flow of fluids through pipes running full will be considered. If the pipes are partially full as in the case of sewer lines, the pressure inside the pipe is same and equal to atmospheric pressure. Then the flow of fluid in the pipe is not under pressure. This case will be taken in the chapter of flow of water through open channels. Here we will consider flow of fluids through pipes under pressure only.

▶ 9.2 LOSS OF ENERGY IN PIPES

When a fluid is flowing through a pipe, the fluid experiences some resistance due to which some of the energy of fluid is lost. This loss of energy is classified as :

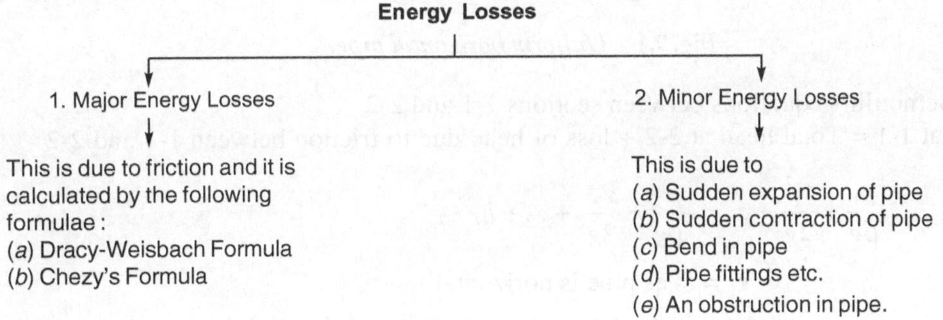

▶ 9.3 FRICTIONAL LOSS IN PIPE FLOW

When a liquid is flowing through a pipe, the velocity of the liquid layer adjacent to the pipe wall is zero. The velocity of liquid goes on increasing from the wall and thus velocity gradient and hence shear stresses are produced in the whole liquid due to viscosity. This viscous action causes loss of energy which is usually known as frictional loss.

On the basis of his experiments, William Froude gave the following laws of fluid fraction for turbulent flow.

The friction resistance for turbulent flow is :

(*i*) proportional to V^n, where n varies from 1.5 to 2.0,
(*ii*) proportional to the density of fluid,
(*iii*) proportional to the area of surface in contact,
(*iv*) independent of pressure,
(*v*) dependent on the nature of the surface in contact.

▶ 9.4 DARCY-EQUATION FOR LOSS OF HEAD DUE TO FRICTION IN PIPES

Consider a uniform horizontal pipe, having steady flow as shown in Fig. 9.1. Let 1-1 and 2-2 are two sections of pipe.

Let p_1 = pressure intensity at section 1-1,
V_1 = velocity of flow at section 1-1,
L = length of the pipe between sections 1-1 and 2-2,
d = diameter of pipe,
f' = frictional resistance per unit wetted area per unit velocity,
h_f = loss of head due to friction,
and p_2, V_2 = are values of pressure intensity and velocity at section 2-2.

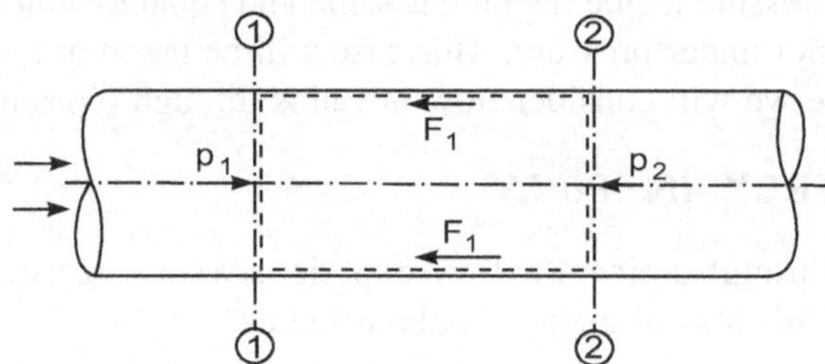

Fig. 9.1 *Uniform horizontal pipe.*

Applying Bernoulli's equations between sections 1-1 and 2-2,

Total head at 1-1 = Total head at 2-2 + loss of head due to friction between 1-1 and 2-2

or $$\frac{p_1}{\rho g} + \frac{V_1^2}{2g} + z_1 = \frac{p_2}{\rho g} + \frac{V_2^2}{2g} + z_2 + h_f$$

But $z_1 = z_2$ as pipe is horizontal

$V_1 = V_2$ as dia. of pipe is same at 1-1 and 2-2

∴ $$\frac{p_1}{\rho g} = \frac{p_2}{\rho g} + h_f \text{ or } h_f = \frac{p_1}{\rho g} - \frac{p_2}{\rho g} \quad ...(i)$$

But h_f is the head lost due to friction and hence intensity of pressure will be reduced in the direction of flow by frictional resistance.

Now frictional resistance = frictional resistance per unit wetted area per unit velocity × wetted area × velocity^2

or $$F_1 = f' \times \pi dL \times V^2 \quad [\because \text{ wetted area} = \pi d \times L \text{ velocity} = V = V_1 = V_2]$$
$$= f' \times P \times L \times V^2 \quad [\because \pi d = \text{Perimeter} = P] \quad ...(ii)$$

The forces acting on the fluid between sections 1-1 and 2-2 are :

1. pressure force at section 1-1 = $p_1 \times A$

where A = Area of pipe

2. pressure force at section 2-2 = $p_2 \times A$
3. frictional force F_1 as shown in Fig. 9.1.

Resolving all forces in the horizontal direction, we have

$$p_1 A - p_2 A - F_1 = 0$$

or $$(p_1 - p_2)A = F_1 = f' \times P \times L \times V^2 \quad [\because \text{From } (ii),\ F_1 = f'PLV^2]$$

or $$p_1 - p_2 = \frac{f' \times P \times L \times V^2}{A}$$

But from equation (*i*), $p_1 - p_2 = \rho g h_f$

Equating the value of $(p_1 - p_2)$, we get

$$\rho g h_f = \frac{f' \times P \times L \times V^2}{A}$$

or $$h_f = \frac{f'}{\rho g} \times \frac{P}{A} \times L \times V^2 \quad ...(iii)$$

In equation (*iii*), $\dfrac{P}{A} = \dfrac{\text{Wetted perimeter}}{\text{Area}} = \dfrac{\pi d}{\frac{\pi}{4} d^2} = \dfrac{4}{d}$

$\therefore$ $$h_f = \frac{f'}{\rho g} \times \frac{4}{d} \times L \times V^2 = \frac{f'}{\rho g} \times \frac{4LV^2}{d} \quad ...(iv)$$

Putting $\dfrac{f'}{\rho g} = \dfrac{f}{2}$, where f is known as **co-efficient of friction.**

Equation (*iv*), becomes as $$h_f = \frac{4 \cdot f}{2g} \cdot \frac{LV^2}{d} = \frac{4f \cdot L \cdot V^2}{d \times 2g} \quad ...(9.1)$$

Equation (9.1) is known as **Darcy-Weisbach equation.** This equation is commonly used for finding loss of head due to friction in pipes.

Sometimes equation (9.1) is written as

$$h_f = \frac{f^* \cdot L \cdot V^2}{d \times 2g} \quad ...(9.2)$$

Then f^* is known as *friction factor.*

Co-efficient of friction (f) which is a function of Reynold number is given by

$$f = \frac{16}{R_e} \text{ for } R_e < 2000 \text{ (viscous flow)}$$

$$= \frac{0.079}{R_e^{1/4}} \text{ for } R_e \text{ varying from 4000 to } 10^6$$

L = length of pipe,
V = mean velocity of flow,
d = diameter of pipe.

▶ 9.5 CHEZY'S EQUATION FOR LOSS OF HEAD DUE TO FRICTION IN PIPES

Refer to article 9.4 in which expression for loss of head due to friction in pipes is derived. The equation (*iii*) of article 9.4, is

$$h_f = \frac{f'}{\rho g} \times \frac{P}{A} \times L \times V^2 \qquad \text{...(9.2A)}$$

where h_f = loss of head due to friction, P = wetted perimeter of pipe,
A = area of cross-section of pipe, L = length of pipe,
and V = mean velocity of flow.

Now the ratio of $\frac{A}{P}\left(= \frac{\text{Area of flow}}{\text{Perimeter (wetted)}}\right)$ is called **hydraulic mean depth** or **hydraulic radius** and is denoted by m.

$$\therefore \text{ Hydraulic mean depth, } m = \frac{A}{P} = \frac{\frac{\pi}{4}d^2}{\pi d} = \frac{d}{4}$$

Substituting $\frac{A}{P} = m$ or $\frac{P}{A} = \frac{1}{m}$ in equation (9.2A), we get

$$h_f = \frac{f'}{\rho g} \times L \times V^2 \text{ or } V^2 = h_f \times \frac{\rho g}{f'} \times m \times \frac{1}{L} = \frac{\rho g}{f'} \times m \times \frac{h_f}{L}$$

$$\therefore \qquad V = \sqrt{\frac{\rho g}{f'} \times m \times \frac{h_f}{L}} = \sqrt{\frac{\rho g}{f'}}\sqrt{m\frac{h_f}{L}} \qquad \text{...(9.3)}$$

Let $\sqrt{\frac{\rho g}{f'}} = C$, where C is a constant known as Chezy's constant and $\frac{h_f}{L} = i$, where i is loss of head per unit length of pipe.

Substituting the values of $\sqrt{\frac{\rho g}{f'}}$ and $\sqrt{\frac{h_f}{L}}$ in equation (9.3), we get

$$V = C\sqrt{mi} \qquad \text{...(9.4)}$$

Equation (9.4) is known as Chezy's formula. Thus the loss of head due to friction in pipe from Chezy's formula can be obtained if the velocity of flow through pipe and also the value of C is known. The value of m for pipe is always equal to $d/4$.

Problem 9.1 *Find the head lost due to friction in a pipe of diameter 300 mm and length 50 m, through which water is flowing at a velocity of 3 m/s using (i) Darcy formula, (ii) Chezy's formula for which C = 60.*

Take v for water = 0.01 stoke.

Solution. Given :

Dia. of pipe, $d = 300$ mm $= 0.30$ m

Length of pipe, $L = 50$ m

Velocity of flow, $V = 3$ m/s

Chezy's constant, $C = 60$

Kinematic viscosity, $\nu = 0.01$ stoke $= 0.01$ cm^2/s
$= 0.01 \times 10^{-4}$ m^2/s.

(*i*) **Darcy Formula** is given by equation (9.1) as

$$h_f = \frac{4 . f . L . V^2}{d \times 2g}$$

where 'f' = co-efficient of friction is a function of Reynold number, R_e

But R_e is given by $$R_e = \frac{V \times d}{\nu} = \frac{3.0 \times 0.30}{.01 \times 10^{-4}} = 9 \times 10^5$$

$\therefore$ Value of $$f = \frac{0.079}{R_e^{1/4}} = \frac{0.079}{\left(9 \times 10^5\right)^{1/4}} = .00256$$

$\therefore$ Head lost, $$h_f = \frac{4 \times .00256 \times 50 \times 3^2}{0.3 \times 2.0 \times 9.81} = \mathbf{.7828 \text{ m. Ans.}}$$

(*ii*) **Chezy's Formula.** Using equation (9.4)

$$V = C\sqrt{mi}$$

where $C = 60$, $m = \frac{d}{4} = \frac{0.30}{4} = 0.075$ m

$$\therefore \quad 3 = 60\sqrt{.075 \times i} \text{ or } i = \left(\frac{3}{60}\right)^2 \times \frac{1}{.075} = 0.0333$$

But $$i = \frac{h_f}{L} = \frac{h_f}{50}$$

Equating the two values of i, we have $\frac{h_f}{50} = .0333$

$$\therefore \quad h_f = 50 \times .0333 = \mathbf{1.665 \text{ m. Ans.}}$$

Problem 9.2 *Find the diameter of a pipe of length 2000 m when the rate of flow of water through the pipe is 200 litres/s and the head lost due to friction is 4 m. Take the value of C = 50 in Chezy's formulae.*

Solution. Given :

Length of pipe, $L = 2000$ m

Discharge, $Q = 200$ litre/s $= 0.2$ m^3/s

Head lost due to friction, $h_f = 4$ m

Value of Chezy's constant, $C = 50$

Let the diameter of pipe = d

Velocity of flow, $$V = \frac{\text{Discharge}}{\text{Area}} = \frac{Q}{\frac{\pi}{4}d^2} = \frac{0.2}{\frac{\pi}{4}d^2} = \frac{0.2 \times 4}{\pi d^2}$$

Hydraulic mean depth, $$m = \frac{d}{4}$$

Loss of head per unit length, $i = \frac{h_f}{L} = \frac{4}{2000} = .002$

Chezy's formula is given by equation (9.4) as $V = C\sqrt{mi}$

Substituting the values of V, m, i and C, we get

$$\frac{0.2 \times 4}{\pi d^2} = 50\sqrt{\frac{d}{4} \times .002} \text{ or } \sqrt{\frac{d}{4} \times .002} = \frac{0.2 \times 4}{\pi d^2 \times 50} = \frac{.00509}{d^2}$$

Squaring both sides, $\frac{d}{4} \times .002 = \frac{.00509^2}{d^4} = \frac{.0000259}{d^4}$ or $d^5 = \frac{4 \times .0000259}{.002} = 0.0518$

$$\therefore \quad d = \sqrt[5]{0.0518} = (.0518)^{1/5} = 0.553 \text{ m} = \mathbf{553\ mm.\ Ans.}$$

Problem 9.3 *A crude oil of kinematic viscosity 0.4 stoke is flowing through a pipe of diameter 300 mm at the rate of 300 litres per sec. Find the head lost due to friction for a length of 50 m of the pipe.*

Solution. Given :

Kinematic viscosity, $\nu = 0.4$ stoke $= 0.4$ cm²/s $= .4 \times 10^{-4}$ m²/s

Dia. of pipe, $d = 300$ mm $= 0.30$ m

Discharge, $Q = 300$ litres/s $= 0.3$ m³/s

Length of pipe, $L = 50$ m

Velocity of flow, $$V = \frac{Q}{\text{Area}} = \frac{0.3}{\frac{\pi}{4}(0.3)^2} = 4.24 \text{ m/s}$$

$\therefore$ Reynold number, $$R_e = \frac{V \times d}{\nu} = \frac{4.24 \times 0.30}{0.4 \times 10^{-4}} = 3.18 \times 10^4$$

As R_e lies between 4000 and 100,000, the value of f is given by

$$f = \frac{.079}{(R_e)^{1/4}} = \frac{.079}{(3.18 \times 10^4)^{1/4}} = .00591$$

$\therefore$ Head lost due to friction, $$h_f = \frac{4 . f . L . V^2}{d \times 2g} = \frac{4 \times .00591 \times 50 \times 4.24^2}{0.3 \times 2 \times 9.81} = \mathbf{3.61\ m.\ Ans.}$$

Problem 9.4 *An oil of sp. gr. 0.7 is flowing through a pipe of diameter 300 mm at the rate of 500 litres/s. Find the head lost due to friction and power required to maintain the flow for a length of 1000 m. Take* $\nu = .29$ *stokes.*

Solution. Given :

Sp. gr. of oil, $S = 0.7$

Dia. of pipe, $d = 300 \text{ mm} = 0.3 \text{ m}$

Discharge, $Q = 500 \text{ litres/s} = 0.5 \text{ m}^3\text{/s}$

Length of pipe, $L = 1000 \text{ m}$

Velocity, $$V = \frac{Q}{\text{Area}} = \frac{0.5}{\frac{\pi}{4}d^2} = \frac{0.5 \times 4}{\pi \times 0.3^2} = 7.073 \text{ m/s}$$

$\therefore$ Reynold number, $$R_e = \frac{V \times d}{\nu} = \frac{7.073 \times 0.3}{0.29 \times 10^{-4}} = 7.316 \times (10)^4$$

$\therefore$ Co-efficient of friction, $$f = \frac{.079}{R_e^{1/4}} = \frac{0.79}{\left(7.316 \times 10^4\right)^{1/4}} = .0048$$

$\therefore$ Head lost due to friction, $$h_f = \frac{4 \times f \times L \times V^2}{d \times 2g} = \frac{4 \times .0048 \times 1000 \times 7.073^2}{0.3 \times 2 \times 9.81} = 163.18 \text{ m}$$

Power required $$= \frac{\rho g \,.\, Q \,.\, h_f}{1000} \text{ kW}$$

where ρ = density of oil = $0.7 \times 1000 = 700 \text{ kg/m}^3$

$\therefore$ Power required $$= \frac{700 \times 9.81 \times 0.5 \times 163.18}{1000} = \mathbf{560.28 \text{ kW. Ans.}}$$

Problem 9.5 *Calculate the discharge through a pipe of diameter 200 mm when the difference of pressure head between the two ends of a pipe 500 m apart is 4 m of water. Take the value of 'f' = 0.009 in the formula* $h_f = \dfrac{4 \,.\, f \,.\, L \,.\, V^2}{d \times 2g}$.

Solution. Given :

Dia. of pipe, $d = 200 \text{ mm} = 0.20 \text{ m}$

Length of pipe, $L = 500 \text{ m}$

Difference of pressure head, $h_f = 4 \text{ m of water}$

$f = .009$

Using equation (9.1), we have $$h_f = \frac{4 \times f \times L \times V^2}{d \times 2g}$$

or $$4.0 = \frac{4 \times .009 \times 500 \times V^2}{0.2 \times 2 \times 9.81} \text{ or } V^2 = \frac{4.0 \times 0.2 \times 2 \times 9.81}{4.0 \times .009 \times 500} = 0.872$$

$\therefore$ $$V = \sqrt{0.872} = 0.9338 \simeq 0.934 \text{ m/s}$$

$\therefore$ Discharge, $$Q = \text{velocity} \times \text{area}$$

$$= 0.934 \times \frac{\pi}{4} d^2 = 0.934 \times \frac{\pi}{4} (0.2)^2$$

$$= 0.0293 \text{ m}^3\text{/s} = \mathbf{29.3 \text{ litres/s. Ans.}}$$

Problem 9.6 *Water is flowing through a pipe of diameter 200 mm with a velocity of 3 m/s. Find the head lost due to friction for a length of 5 m if the co-efficient of friction is given by* $f = .002 + \frac{.09}{R_e^{0.3}}$, *where* R_e *is Reynold number. The kinematic viscosity of water = .01 stoke.*

Solution. Given :

Dia. of pipe, $d = 200 \text{ mm} = 0.20 \text{ m}$

Velocity, $V = 3 \text{ m/s}$

Length, $L = 5 \text{ m}$

Kinematic viscosity, $\nu = 0.01 \text{ stoke} = .01 \times 10^{-4} \text{ m}^2\text{/s}$

$\therefore$ Reynold number, $R_e = \frac{V \times d}{\nu} = \frac{3 \times 0.20}{.01 \times 10^{-4}} = 6 \times 10^5$

Value of $f = .02 + \frac{0.9}{R_e^{0.3}} = .02 + \frac{.09}{(6 \times 10^5)^{.3}} = .02 + \frac{0.9}{54.13}$

$= .02 + .00166 = 0.02166$

$\therefore$ Head lost due to friction, $h_f = \frac{4 \times f \times L \times V^2}{d \times 2g} = \frac{4.0 \times .02166 \times 5.0 \times 3^2}{0.20 \times 2.0 \times 9.81}$

$=$ **0.993 m of water. Ans.**

Problem 9.7 *An oil of sp. gr. 0.9 and viscosity 0.06 poise is flowing through a pipe of diameter 200 mm at the rate of 60 litres/s. Find the head lost due to friction for a 500 m length of pipe. Find the power required to maintain this flow.*

Solution. Given :

Sp. gr. of oil $= 0.9$

Viscosity, $\mu = 0.06 \text{ poise} = \frac{0.06}{10} \text{ Ns/m}^2$

Dia. of pipe, $d = 200 \text{ mm} = 0.2 \text{ m}$

Discharge, $Q = 60 \text{ litres/s} = 0.06 \text{ m}^3\text{/s}$

Length, $L = 500 \text{ m}$

Density $\rho = 0.9 \times 1000 = 900 \text{ kg/m}^3$

$\therefore$ Reynold number, $R_e = \frac{\rho V d}{\mu} = 900 \times \frac{V \times 0.2}{\frac{0.06}{10}}$

where $V = \frac{Q}{\text{Area}} = \frac{0.06}{\frac{\pi}{4} d^2} = \frac{0.06}{\frac{\pi}{4} (.2)^2} = 1.909 \text{ m/s} \simeq 1.91 \text{ m/s}$

$\therefore$ $R_e = 900 \times \frac{1.91 \times 0.2 \times 10}{0.06} = 57300$

As R_e lies between 4000 and 10^5, the value of co-efficient of friction, f is given by

$$f = \frac{0.079}{R_e^{0.25}} = \frac{0.079}{(57300)^{0.25}} = .0051$$

$\therefore$ Head lost due to friction, $h_f = \dfrac{4 \times f \times L \times V^2}{d \times 2g} = \dfrac{4 \times .0051 \times 500 \times 1.91^2}{0.2 \times 2 \times 9.81}$

$=$ **9.48 m of water. Ans.**

$\therefore$ Power required $= \dfrac{\rho g \cdot Q \cdot h_f}{1000} = \dfrac{900 \times 9.81 \times 0.06 \times 9.48}{1000} =$ **5.02 kW. Ans.**

▶ 9.6 MINOR ENERGY (HEAD) LOSSES

The loss of head or energy due to friction in a pipe is known as major loss while the loss of energy due to change of velocity of the following fluid in magnitude or direction is called minor loss of energy. The minor loss of energy (or head) includes the following cases :

1. Loss of head due to sudden enlargement,
2. Loss of head due to sudden contraction,
3. Loss of head at the entrance to a pipe,
4. Loss of head at the exit of a pipe,
5. Loss of head due to an obstruction in a pipe,
6. Loss of head due to bend in the pipe,
7. Loss of head in various pipe fittings.

In case of long pipe the above losses are small as compared with the loss of head due to friction and hence they are called minor losses and even may be neglected without serious error. But in case of a short pipe, these losses are comparable with the loss of head due to friction.

9.6.1 Loss of Head Due to Sudden Enlargement. Consider a liquid flowing through a pipe which has sudden enlargement as shown in Fig. 9.2. Consider two sections (1)-(1) and (2)-(2) before and after the enlargement.

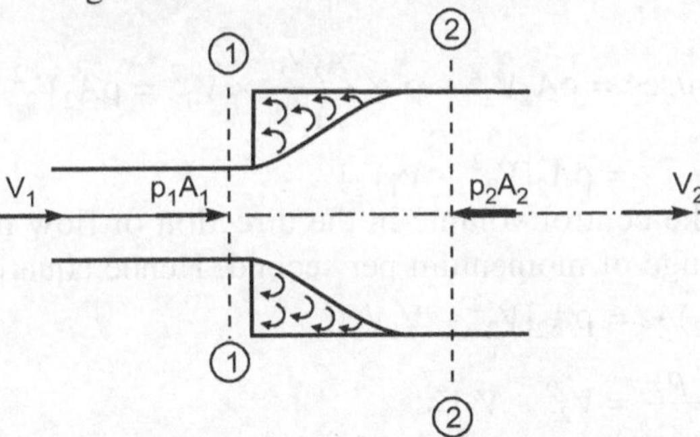

Fig. 9.2 *Sudden enlargement.*

Let p_1 = pressure intensity at section 1-1,
V_1 = velocity of flow at section 1-1,
A_1 = area of pipe at section 1-1,
p_2, V_2 and A_2 = corresponding values at section 2-2.

Due to sudden change of diameter of the pipe from D_1 to D_2, the liquid flowing from the smaller pipe is not able to follow the abrupt change of the boundary. Thus the flow separates from the boundary and turbulent eddies are formed as shown in Fig. 9.2. The loss of head (or energy) takes place due to the formation of these eddies.

Let p' = pressure intensity of the liquid eddies on the area $(A_2 - A_1)$
h_e = loss of head due to sudden enlargement

Applying Bernoulli's equation to sections 1-1 and 2-2,

$$\frac{p_1}{\rho g} + \frac{V_1^2}{2g} + z_1 = \frac{p_2}{\rho g} + \frac{V_2^2}{2g} + z_2 + \text{loss of head due to sudden enlargement}$$

But $z_1 = z_2$ as pipe is horizontal

$$\therefore \quad \frac{p_1}{\rho g} + \frac{V_1^2}{2g} = \frac{p_2}{\rho g} + \frac{V_2^2}{2g} + h_e$$

or
$$h_e = \left(\frac{p_1}{\rho g} - \frac{p_2}{\rho g}\right) + \left(\frac{V_1^2}{2g} - \frac{V_2^2}{2g}\right) \quad ...(i)$$

Consider the control volume of liquid between sections 1-1 and 2-2. Then the force acting on the liquid in the control volume in the direction of flow is given by

$$F_x = p_1A_1 + p'(A_2 - A_1) - p_2A_2$$

But experimentally it is found that $p' = p_1$

$$\therefore \quad F_x = p_1A_1 + p_1(A_2 - A_1) - p_2A_2 = p_1A_2 - p_2A_2$$
$$= (p_1 - p_2)A_2 \quad ...(ii)$$

Momentum of liquid/sec at section 1-1 = mass × velocity

$$= \rho A_1 V_1 \times V_1 = \rho A_1 V_1^2$$

Momentum of liquid/sec at section 2-2 = $\rho A_2 V_2 \times V_2 = \rho A_2 V_2^2$

$\therefore$ Change of momentum/sec = $\rho A_2 V_2^2 - \rho A_1 V_1^2$

But from continuity equation, we have

$$A_1V_1 = A_2V_2 \text{ or } A_1 = \frac{A_2V_2}{V_1}$$

$$\therefore \quad \text{Change of momentum/sec} = \rho A_2 V_2^2 - \rho \times \frac{A_2V_2}{V_1} \times V_1^2 = \rho A_2 V_2^2 - \rho A_2 V_1 V_2$$

$$= \rho A_2[V_2^2 - V_1V_2] \quad ...(iii)$$

Now net force acting on the control volume in the direction of flow must be equal to the rate of change of momentum or change of momentum per second. Hence equating (*ii*) and (*iii*)

$$(p_1 - p_2)A_2 = \rho A_2[V_2^2 - V_1V_2]$$

or
$$\frac{p_1 - p_2}{\rho} = V_2^2 - V_1V_2$$

Dividing by g to both sides, we have $\frac{p_1 - p_2}{\rho g} = \frac{V_2^2 - V_1V_2}{g}$ or $\frac{p_1}{\rho g} - \frac{p_2}{\rho g} = \frac{V_2^2 - V_1V_2}{g}$

Substituting the value of $\left(\frac{p_1}{\rho g} - \frac{p_2}{\rho g}\right)$ in equation (*i*), we get

$$h_e = \frac{V_2^2 - V_1V_2}{g} + \frac{V_1^2}{2g} - \frac{V_2^2}{2g} = \frac{2V_2^2 - 2V_1V_2 + V_1^2 - V_2^2}{2g}$$

$$= \frac{V_2^2 + V_1^2 - 2V_1V_2}{2g} = \left(\frac{V_1 - V_2}{2g}\right)^3$$

$$\therefore \qquad h_e = \frac{(V_1 - V_2)^2}{2g}. \qquad \text{...(9.5)}$$

9.6.2 Loss of Head due to Sudden Contraction. Consider a liquid flowing in a pipe which has a sudden contraction in area as shown in Fig. 9.2 (*a*). Consider two sections 1-1 and 2-2 before and after, contraction. As the liquid flows from large pipe to smaller pipe, the area of flow goes on decreasing and becomes minimum at a section *C-C* as shown in Fig. 9.2 (*a*). This section *C-C* is called Vena-contracta. After section *C-C* a sudden enlargement of the area takes place. The loss of head due to sudden contraction is actually due to sudden enlargement from Vena-contracta to smaller pipe.

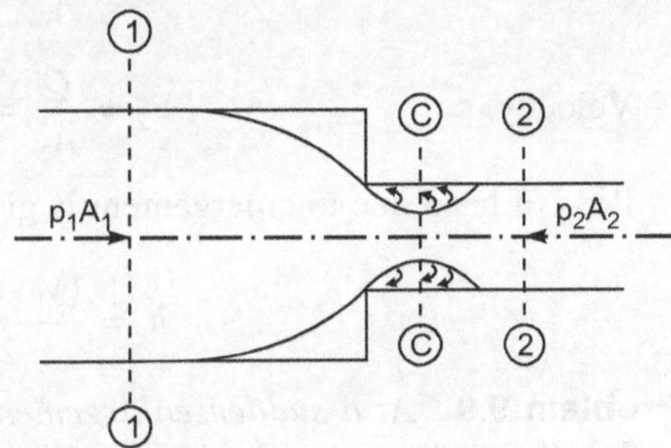

Fig. 9.2 (*a*) ***Sudden contraction.***

Let A_c = Area of flow at section *C-C*

V_c = Velocity of flow at section *C-C*

A_2 = Area of flow at section 2-2

V_2 = Velocity of flow at section 2-2

h_c = Loss of head due to sudden contraction.

Now h_c = actually loss of head due to enlargement from section *C-C* to section 2-2 and is given by equation (9.5) as

$$= \frac{(V_c - V_2)^2}{2g} = \frac{V_2^2}{2g}\left[\frac{V_c}{V_2} - 1\right]^2 \qquad \text{...}(i)$$

From continuity equation, we have

$$A_cV_c = A_2V_2 \quad \text{or} \quad \frac{V_c}{V_2} = \frac{A_2}{A_c} = \frac{1}{(A_c/A_2)} = \frac{1}{C_c} \qquad \left[\because V_c = \frac{A_c}{A_2}\right]$$

Substituting the value of $\frac{V_c}{V_2}$ in (*i*), we get

$$h_c = \frac{V_2^2}{2g}\left[\frac{1}{C_c} - 1\right]^2 \qquad \text{...(9.6)}$$

$$= \frac{kV_2^2}{2g}, \text{ where } k = \left[\frac{1}{C_c} - 1\right]^2$$

If the value of C_c is assumed to be equal to 0.62, then

$$k = \left[\frac{1}{0.62} - 1\right]^2 = 0.375$$

Then h_c becomes as $\qquad h_c = \frac{kV_2^2}{2g} = 0.375\,\frac{V_2^2}{2g}$

If the value of C_c is not given then the head loss due to contraction is taken as

$$= 0.5\,\frac{V_2^2}{2g} \text{ or } h_c = 0.5\,\frac{V_2^2}{2g}. \qquad \text{...(9.7)}$$

Problem 9.8 *Find the loss of head when a pipe of diameter 200 mm is suddenly enlarged to a diameter of 400 mm. The rate of flow of water through the pipe is 250 litres/s.*

Solution. Given :

Dia. of smaller pipe, $D_1 = 200$ mm $= 0.20$ m

$\therefore$ Area, $$A_1 = \frac{\pi}{4} D_1^{\,2} = \frac{\pi}{4} (.2)^2 = 0.03141 \text{ m}^2$$

Dia. of large pipe, $D_2 = 400$ mm $= 0.4$ m

$\therefore$ Area, $$A_2 = \frac{\pi}{4} \times (0.4)^2 = 0.12564 \text{ m}^2$$

Discharge, $Q = 250$ litres/s $= 0.25$ m^3/s

Velocity, $$V_1 = \frac{Q}{A_1} = \frac{0.25}{.03141} = 7.96 \text{ m/s}$$

Velocity, $$V_2 = \frac{Q}{A_2} = \frac{0.25}{.12564} = 1.99 \text{ m/s}$$

Loss of head due to enlargement is given by equation (9.5) as

$$h_e = \frac{(V_1 - V_2)^2}{2g} = \frac{(7.96 - 1.99)^2}{2g} = \textbf{1.816 m of water. Ans.}$$

Problem 9.9 *At a sudden enlargement of a water main from 240 mm to 480 mm diameter, the hydraulic gradient rises by 10 mm. Estimate the rate of flow.*

(J.N.T.U., S 2002 ; A.M.I.E., Summer, 1977)

Solution. Given :

Dia. of smaller pipe, $D_1 = 240$ mm $= 0.24$ m

$\therefore$ Area, $$A_1 = \frac{\pi}{4} D_1^{\,2} = \frac{\pi}{4} (.24)^2$$

Dia. of large pipe, $D_2 = 480$ mm $= 0.48$ m

$\therefore$ Area, $$A_2 = \frac{\pi}{4} (0.48)^2$$

Rise of hydraulic gradient*, *i.e.*, $\left(z_2 + \frac{p_2}{\rho g}\right) - \left(\frac{p_1}{\rho g} + z_1\right) = 10 \text{ mm} = \frac{10}{1000} = \frac{1}{100}$ m

Let the rate of flow $= Q$

Applying Bernoulli's equation to both sections, *i.e.*, smaller pipe section, and large pipe section.

$$\frac{p_1}{\rho g} + \frac{V_1^2}{2g} + z_1 = \frac{p_2}{\rho g} + \frac{V_2^2}{2g} + z_2 + \text{Head loss due to enlargement} \quad ...(i)$$

But head loss due to enlargement,

$$h_e = \frac{(V_1 - V_2)^2}{2g} \quad ...(ii)$$

From continuity equation, we have $A_1 V_1 = A_2 V_2$

*Please refer Art. 9.7.1

$$\therefore \quad V_1 = \frac{A_2 V_2}{A_1} = \frac{\frac{\pi}{4} D_2^2 \times V_2}{\frac{\pi}{4} D_1^2} = \left(\frac{D_2}{D_1}\right)^2 \times V_2 = \left(\frac{.48}{.24}\right)^2 \times V_2 = 2V_2^{\,2} = 4V_2$$

Substituting this value in (*ii*), we get

$$h_e = \frac{(4V_2 - V_2)^2}{2g} = \frac{(3V_2)^2}{2g} = \frac{9V_2^2}{2g}$$

Now substituting the value of h_e and V_1 in equation (*i*),

$$\frac{p_1}{\rho g} + \frac{(4V_2)^2}{2g} + z_1 = \frac{p_2}{\rho g} + \frac{V_2^2}{2g} + z_2 + \frac{9V_2^2}{2g}$$

or
$$\frac{16V_2^2}{2g} - \frac{V_2^2}{2g} - \frac{9V_2^2}{2g} = \left(\frac{p_2}{\rho g} + z_2\right) - \left(\frac{p_1}{\rho g} + z_1\right)$$

But hydraulic gradient rise
$$= \left(\frac{p_2}{\rho g} + z_2\right) - \left(\frac{p_1}{\rho g} + z_1\right) = \frac{1}{100}$$

$$\therefore \quad \frac{16V_2^2}{2g} - \frac{V_2^2}{2g} - \frac{9V_2^2}{2g} = \frac{1}{100} \quad \text{or} \quad \frac{6V_2^2}{2g} = \frac{1}{100}$$

$$\therefore \quad V_2 = \sqrt{\frac{2 \times 9.81}{6 \times 100}} = 0.1808 \simeq 0.181 \text{ m/s}$$

$\therefore$ Discharge,
$$Q = A_2 \times V_2$$

$$= \frac{\pi}{4} D_2^{\,2} \times V_1 = \frac{\pi}{4} (.48)^2 \times .181 = 0.03275 \text{ m}^3\text{/s}$$

$$= \mathbf{32.75 \text{ litres/s. Ans.}}$$

Problem 9.10 *The rate of flow of water through a horizontal pipe is 0.25 m³/s. The diameter of the pipe which is 200 mm is suddenly enlarged to 400 mm. The pressure intensity in the smaller pipe is 11.772 N/cm². Determine :*

(i) loss of head due to sudden enlargement, (ii) pressure intensity in the large pipe,
(iii) power lost due to enlargement. (VTU, Jan. 2004)

Solution. Given :

Discharge, $Q = 0.25 \text{ m}^3\text{/s}$

Dia. of smaller pipe, $D_1 = 200 \text{ mm} = 0.20 \text{ m}$

$\therefore$ Area, $A_1 = \frac{\pi}{4} (.2)^2 = 0.03141 \text{ m}^2$

Dia. of large pipe, $D_2 = 400 \text{ mm} = 0.40 \text{ m}$

$\therefore$ Area, $A_2 = \frac{\pi}{4} (0.4)^2 = 0.12566 \text{ m}^2$

Pressure in smaller pipe, $p_1 = 11.772 \text{ N/cm}^2 = 11.772 \times 10^4 \text{ N/m}^2$

Now velocity, $$V_1 = \frac{Q}{A_1} = \frac{0.25}{.03141} = 7.96 \text{ m/s}$$

Velocity, $$V_2 = \frac{Q}{A_2} = \frac{0.25}{.12564} = 1.99 \text{ m/s}$$

(*i*) Loss of head due to sudden enlargement,

$$h_e = \frac{(V_1 - V_2)^2}{2g} = \frac{(7.96 - 1.99)^2}{2 \times 9.81} = \mathbf{1.816 \text{ m. Ans.}}$$

(*ii*) Let the pressure intensity in large pipe = p_2.

Then applying Bernoulli's equation before and after the sudden enlargement,

$$\frac{p_1}{\rho g} + \frac{V_1^2}{2g} + z_1 = \frac{p_2}{\rho g} + \frac{V_2^2}{2g} + z_2 + h_e$$

But $$z_1 = z_2 \qquad \text{(Given horizontal pipe)}$$

$$\therefore \quad \frac{p_1}{\rho g} + \frac{V_1^2}{2g} = \frac{p_2}{\rho g} + \frac{V_2^2}{2g} + h_e \text{ or } \frac{p_2}{\rho g} = \frac{p_1}{\rho g} + \frac{V_1^2}{2g} - \frac{V_2^2}{2g} - h_e$$

$$= \frac{11.772 \times 10^4}{1000 \times 9.81} + \frac{7.96^2}{2 \times 9.81} - \frac{1.99^2}{2 \times 9.81} - 1.816$$

$$= 12.0 + 3.229 - 0.2018 - 1.8160$$

$$= 15.229 - 2.0178 = 13.21 \text{ m of water}$$

$$\therefore \quad p_2 = 13.21 \times \rho g = 13.21 \times 1000 \times 9.81 \text{ N/m}^2$$

$$= 13.21 \times 1000 \times 9.81 \times 10^{-4} \text{ N/cm}^2 = \mathbf{12.96 \text{ N/cm}^2\text{. Ans.}}$$

(*iii*) Power lost due to sudden enlargement,

$$P = \frac{\rho g \cdot Q \cdot h_e}{1000} = \frac{1000 \times 9.81 \times 0.25 \times 1.816}{1000} = \mathbf{4.453 \text{ kW. Ans.}}$$

Problem 9.11 *A horizontal pipe of diameter 500 mm is suddenly contracted to a diameter of 250 mm. The pressure intensities in the large and smaller pipe is given as 13.734 N/cm² and 11.772 N/cm² respectively. Find the loss of head due to contraction if C_e = 0.62. Also determine the rate of flow of water.*

Solution. Given :

Dia. of large pipe, $D_1 = 500 \text{ mm} = 0.5 \text{ m}$

Area, $A_1 = \frac{\pi}{4}(0.5)^2 = 0.1963 \text{ m}^2$

Dia. of smaller pipe, $D_2 = 250 \text{ mm} = 0.25 \text{ m}$

∴ Area, $A_2 = \frac{\pi}{4}(.25)^2 = 0.04908 \text{ m}^2$

Pressure in large pipe, $p_1 = 13.734 \text{ N/cm}^2 = 13.734 \times 10^4 \text{ N/m}^2$

Pressure in smaller pipe, $p_2 = 11.772 \text{ N/cm}^2 = 11.772 \times 10^4 \text{ N/m}^2$

$C_c = 0.62$

Head lost due to contraction $= \frac{V_2^2}{2g}\left[\frac{1}{C_c} - 1.0\right]^2 = \frac{V_2^2}{2g}\left[\frac{1}{0.62} - 1.0\right]^2 = 0.375 \frac{V_2^2}{2g}$

From continuity equation, we have $A_1 V_1 = A_2 V_2$

or
$$V_1 = \frac{A_2 V_2}{A_1} = \frac{\frac{\pi}{4} D_2^2 \times V_2}{\frac{\pi}{4} D_1^2} = \left(\frac{D_2}{D_1}\right)^2 \times V^2 = \left(\frac{0.25}{0.50}\right)^2 V_2 = \frac{V_2}{4}$$

Applying Bernoulli's equation before and after contraction,

$$\frac{p_1}{\rho g} + \frac{V_1^2}{2g} + z = \frac{p_2}{\rho g} + \frac{V_2^2}{2g} + z_2 + h_c$$

But $z_1 = z_2$ (pipe is horizontal)

$$\therefore \quad \frac{p_1}{\rho g} + \frac{V_1^2}{2g} = \frac{p_2}{\rho g} + \frac{V_2^2}{2g} + h_c$$

But $h_c = 0.375 \frac{V_2^2}{2g}$ and $V_1 = \frac{V_2}{4}$

Substituting these values in the above equation, we get

$$\frac{13.734 \times 10^4}{9.81 \times 1000} + \frac{(V_2 / 4)^2}{2g} = \frac{11.772 \times 10^4}{1000 \times 9.81} + \frac{V_2^2}{2g} + 0.375 \frac{V_2^2}{2g}$$

or
$$14.0 + \frac{V_2^2}{16 \times 2g} = 12.0 + 1.375 \frac{V_2^2}{2g}$$

or
$$14 - 12 = 1.375 \frac{V_2^2}{2g} - \frac{1}{16} \frac{V_2^2}{2g} = 1.3125 \frac{V_2^2}{2g}$$

or
$$2.0 = 1.3125 \times \frac{V_2^2}{2g} \text{ or } V_2 = \sqrt{\frac{2.0 \times 2 \times 9.81}{1.3125}} = 5.467 \text{ m/s.}$$

(*i*) Loss of head due to contraction, $h_c = 0.375 \frac{V_2^2}{2g} = \frac{0.375 \times (5.467)^2}{2 \times 9.81}$ = **0.571 m. Ans.**

(*ii*) Rate of flow of water, $Q = A_2 V_2 = 0.04908 \times 5.467 = 0.2683 \text{ m}^3/\text{s}$ = **268.3 lit/s. Ans.**

Problem 9.12 *If in the problem 9.11, the rate of flow of water is 300 litres/s, other data remaining the same, find the value of co-efficient of contraction, C_c.*

Solution. Given :

$$D_1 = 0.5 \text{ m}, D_2 = 0.25 \text{ m}, p_1 = 13.734 \times 10^4 \text{ N/m}^2,$$
$$p_2 = 11.772 \times 10^4 \text{ N/m}^2, Q = 300 \text{ lit/s} = 0.3 \text{ m}^3/\text{s}$$

Also from problem, 9.11, $V_1 = \frac{V_2}{4}$, where $V_1 = \frac{Q}{A_1} = \frac{0.30}{\frac{\pi}{4}(0.5)^2} = 1.528$ m/s

$\therefore \qquad V_2 = 4 \times V_1 = 4 \times 1.528 = 6.112$ m/s

From Bernoulli's equation, we have

$$\frac{p_1}{\rho g} + \frac{V_1^2}{2g} = \frac{p_2}{\rho g} + \frac{V_2^2}{2g} + h_e. \qquad [\because z_1 = z_2]$$

or
$$\frac{13.734 \times 10^4}{9.81 \times 1000} + \frac{(1.528)^2}{2 \times 9.81} = \frac{11.772 \times (10)^4}{9.81 \times 1000} + \frac{(6.112)^2}{2 \times 9.81} + h_c$$

or
$$14.0 + 0.119 = 12.0 + 1.904 + h_c$$

$$14.119 = 13.904 + h_c$$

$$\therefore \qquad h_c = 14.119 - 13.904 = 0.215$$

But from equation (9.6), $\quad h_c = \frac{V_2^2}{2g}\left[\frac{1}{C_c} - 1\right]^2$

Hence equating the two values of h_c, we get

$$\frac{V_2^2}{2g}\left[\frac{1}{C_c} - 1\right]^2 = 0.215$$

$$V_2 = 6.112, \therefore \frac{6.112^2}{2 \times 9.81}\left[\frac{1}{C_c} - 1\right]^2 = 0.215$$

or
$$\left[\frac{1}{C_c} - 1\right]^2 = \frac{0.215 \times 2.0 \times 9.81}{6.112 \times 6.112} = 0.1129$$

or
$$\frac{1.0}{C_c} - 1.0 = \sqrt{0.1129} = 0.336 \text{ or } \frac{1.0}{C_c} = 1.0 + 0.336 + 1.336$$

$$\therefore \qquad C_c = \frac{1.0}{1.336} = \mathbf{0.748.\ Ans.}$$

Problem 9.13 *A 150 mm diameter pipe reduces in diameter abruptly to 100 mm diameter. If the pipe carries water at 30 litres per second, calculate the pressure loss across the contraction. Take the co-efficient of contraction as 0.6.* (A.M.I.E., Summer, 1983)

Solution. Given :

Dia. of large pipe, $\quad D_1 = 150$ mm $= 0.15$ m

Area of large pipe, $\quad A_1 = \frac{\pi}{4}(.15)^2 = 0.01767 \text{ m}^2$

Dia. of smaller pipe, $\quad D_2 = 100$ mm $= 1.10$ m

Area of smaller pipe, $\quad A_2 = \frac{\pi}{4}(.10)^2 = 0.007854 \text{ m}^2$

Discharge, $Q = 30 \text{ litres/s} = .03 \text{ m}^3\text{/s}$

Co-efficient of contraction, $C_c = 0.6$

From continuity equation, we have $A_1V_1 = A_2V_2 = Q$

$$\therefore \quad V_1 = \frac{Q}{A_1} = \frac{0.03}{.01767} = 1.697 \text{ m/s}$$

and
$$V_2 = \frac{Q}{A_2} = \frac{.03}{.007854} = 3.82 \text{ m/s}$$

Applying Bernoulli's equation before and after contraction,

$$\frac{p_1}{\rho g} + \frac{V_1^2}{2g} + Z_1 = \frac{p_2}{\rho g} + \frac{V_2^2}{2g} + Z_2 + h_c \qquad \ldots(i)$$

But $Z_1 = Z_2$

and h_c, the head loss due to contraction is given by equation (9.6) as

$$h_1 = \frac{V_2^2}{2g}\left[\frac{1}{C_c} - 1\right]^2 = \frac{3.82^2}{2 \times 9.81}\left[\frac{1}{0.6} - 1\right]^2 = 0.33$$

Substituting these values in equation (*i*), we get

$$\frac{p_1}{\rho g} + \frac{1.697^2}{2 \times 9.81} = \frac{p_2}{\rho g} + \frac{3.82^2}{2 \times 9.81} + 0.33$$

or
$$\frac{p_1}{\rho g} + 0.1467 = \frac{p_2}{\rho g} + .7438 + .33$$

$$\therefore \quad \frac{p_1}{\rho g} - \frac{p_2}{\rho g} = .7438 + .33 - .1467 = 0.9271 \text{ m of water}$$

$$\therefore \quad (p_1 - p_2) = \rho g \times 0.9271 = 1000 \times 9.81 \times 0.9271 \text{ N/m}^2$$

$$= 0.909 \times 10^4 \text{ N/m}^2 = 0.909 \text{ N/cm}^2$$

$\therefore$ Pressure loss across contraction

$= p_1 - p_2 =$ **0.909 N/cm². Ans.**

Problem 9.14 *In Fig. 9.3 below, when a sudden contraction is introduced in a horizontal pipe line from 50 cm to 25 cm, the pressure changes from 10,500 kg/m² (103005 N/m²) to 6900 kg/m² (67689 N/m²). Calculate the rate of flow. Assume co-efficient of contraction of jet to be 0.65.*

Following this if there is a sudden enlargement from 25 cm to 50 cm and if the pressure at the 25 cm section is 6900 kg/m² (67689 N/m²) what is the pressure at the 50 cm enlarged section ?

(A.M.I.E., Winter, 1983)

Solution. Given :

Dia. of large pipe, $D_1 = 50 \text{ cm} = 0.5 \text{ m}$

Area, $A_1 = \frac{\pi}{4}(.5)^2 = 0.1963 \text{ m}^2$

Dia. of smaller pipe, $D_2 = 25 \text{ cm} = 0.25 \text{ m}$

$\therefore$ Area, $A_2 = \frac{\pi}{4}(.25)^2 = 0.04908 \text{ m}^2$

Pressure in large pipe, $p_1 = 10500 \text{ kg/m}^2$ or (103005 N/m^2)

Pressure in smaller pipe, $p_2 = 6900 \text{ kg/m}^2$ or 67689 N/m^2

Co-efficient of contraction, $C_c = 0.65$

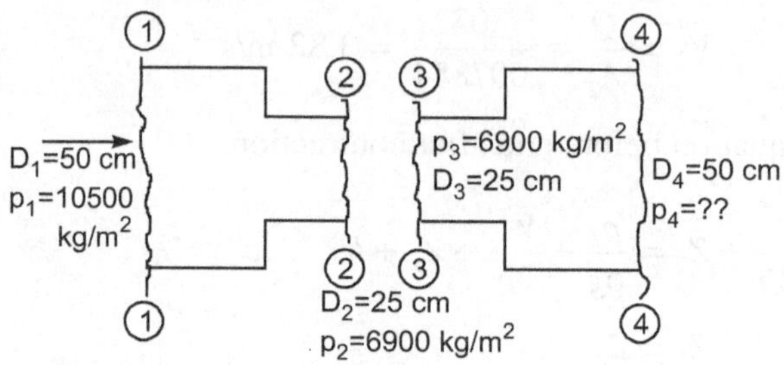

Fig. 9.3

Head lost due to contraction is given by equation (9.6),

$$h_c = \frac{V_2^2}{2g}\left(\frac{1}{C_c} - 1.0\right)^2 = \frac{V_2^2}{2g}\left(\frac{1}{0.65} - 1\right)^2 = 0.2899\frac{V_2^2}{2g} \quad \text{...(i)}$$

From continuity equation, we have

$$A_1V_1 = A_2V_2 \text{ or } V_1 = \frac{A_2V_2}{A_1} = \frac{\frac{\pi}{4}D_2^2 \times V_2}{\frac{\pi}{4}D_1^2}$$

$$= \left(\frac{D_2}{D_1}\right)^2 \times V_2 = \left(\frac{0.50}{0.25}\right)^2 \times V_2 = \frac{V_2}{4} \quad \text{...(ii)}$$

Applying Bernoulli's equation to sections 1-1 and 2-2,

$$\frac{p_1}{\rho g} + \frac{V_1^2}{2g} + Z_1 = \frac{p_2}{\rho g} + \frac{V_2^2}{2g} + Z_2 + h_c$$

But $Z_1 = Z_2$ (as pipe is horizontal)

$$\therefore \quad \frac{p_1}{\rho g} + \frac{V_1^2}{2g} = \frac{p_2}{\rho g} + \frac{V_2^2}{2g} + h_c$$

Substituting the values of p_1, p_2, h_c and V_1, we get

$$\frac{103005}{1000 \times 9.81} + \frac{(V_2/4)^2}{2g} = \frac{67689}{1000 \times 9.81} + \frac{V_2^2}{2g} + .2899\frac{V_2^2}{2g}$$

or

$$10.5 + \frac{V_2^2}{16 \times 2g} = 6.9 + 1.2899\frac{V_2^2}{2g}$$

or $$10.5 - 6.9 = 1.2899 \frac{V_2^2}{2g} - \frac{1}{16} \times \frac{V_2^2}{2g} = 1.2274 \frac{V_2^2}{2g}$$

or $$3.6 = 1.2274 \times \frac{V_2^2}{2g}$$

$$\therefore \quad V_2 = \sqrt{\frac{3.6 \times 2 \times 9.81}{1.2274}} = 7.586 \text{ m/s}$$

(*i*) Rate of flow of water, $Q = A_2 V_2 = 0.04908 \times 7.586$

$= \mathbf{0.3723\ m^3/s\ or\ 372.3\ lit/s.\ Ans.}$

(*ii*) Applying Bernoulli's equation to sections 3-3 and 4-4,

$$\frac{p_3}{\rho g} + \frac{V_3^2}{2g} + Z_3 = \frac{p_4}{\rho g} + \frac{V_4^2}{2g} + Z_4 + \text{head loss due to sudden enlargement } (h_e)$$

But $p_3 = 6900 \text{ kg/m}^2$, or 67689 N/m^2

$V_3 = V_2 = 7.586$ m/s

$V_4 = V_1 = \frac{V_2}{4} = \frac{7.586}{4} = 1.8965$

$Z_3 = Z_4$

And head loss due to sudden enlargement is given by equation (9.5) as

$$h_e = \frac{(V_3 - V_4)^2}{2g} = \frac{(7.586 - 1.8965)^2}{2 \times 9.81} = 1.65 \text{ m}$$

Substituting these values in Bernoulli's equation, we get

$$\frac{67689}{1000 \times 9.81} + \frac{7.586^2}{2 \times 9.81} = \frac{p_4}{1000 \times 9.81} + \frac{1.8965^2}{2 \times 9.81} + 1.65$$

or $$6.9 + 2.933 = \frac{p_4}{1000 \times 9.81} + 0.183 + 1.65$$

$$\therefore \quad \frac{p_4}{1000 \times 9.81} = 6.9 + 2.933 - 0.183 - 1.65 = 9.833 - 1.833 = 8.00$$

$$\therefore \quad p_4 = 8 \times 1000 \times 9.81 = \mathbf{78480\ N/m^2.\ Ans.}$$

9.6.3 Loss of Head at the Entrance of a Pipe. This is the loss of energy which occurs when a liquid enters a pipe which is connected to a large tank or reservoir. This loss is similar to the loss of head due to sudden contraction. This loss depends on the form of entrance. For a sharp edge entrance, this loss is slightly more than a rounded or bell mouthed entrance. In practice the value of loss of head at the entrance (or inlet) of a pipe with sharp cornered entrance is taken = $0.5 \frac{V^2}{2g}$, where V = velocity of liquid in pipe.

This loss is denoted by h_i

$$\therefore \qquad h_i = 0.5 \frac{V^2}{2g} \qquad \text{...(9.8)}$$

9.6.4 Loss of Head at the Exit of Pipe. This is the loss of head (or energy) due to the velocity of liquid at outlet of the pipe which is dissipated either in the form of a free jet (if outlet of the pipe is free) or it is lost in the tank or reservoir (if the outlet of the pipe is connected to the tank or reservoir). This loss is equal to $\frac{V^2}{2g}$, where V is the velocity of liquid at the outlet of pipe. This loss is denoted h_0.

$$\therefore \qquad h_0 = \frac{V^2}{2g} \qquad \text{...(9.9)}$$

where V = velocity at outlet of pipe.

9.6.5 Loss of Head Due to an Obstruction in a Pipe. Whenever there is an obstruction in a pipe, the loss of energy takes place due to reduction of the area of the cross-section of the pipe at the place where obstruction is present. There is a sudden enlargement of the area of flow beyond the obstruction due to which loss of head takes place as shown in Fig. 9.3 (*a*)

Consider a pipe of area of cross-section A having an obstruction as shown in Fig. 9.3 (*a*)

Let a = Maximum area of obstruction

A = Area of pipe

V = Velocity of liquid in pipe

Then $(A - a)$ = Area of flow of liquid at section 1-1.

As the liquid flows and passes through section 1-1, a vena-contracta is formed beyond section 1-1, after which the stream of liquid widens again and velocity of flow at section 2-2 becomes uniform and equal to velocity, V in the pipe. This situation is similar to the flow of liquid through sudden enlargement.

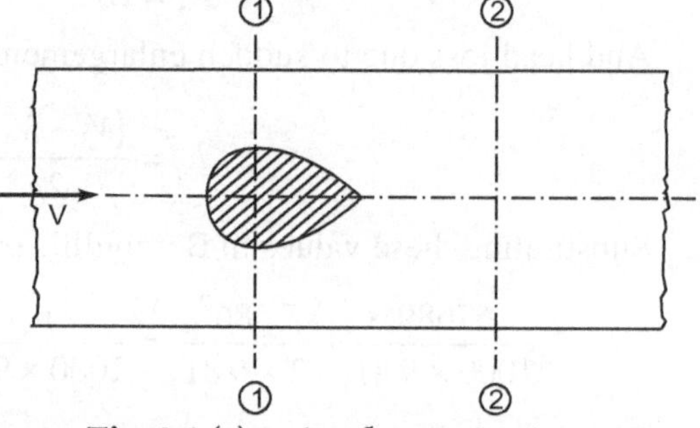

Fig. 9.3 (*a*) *An obstruction in a pipe.*

Let V_c = Velocity of liquid at vena-contracta.

Then loss of head due to obstruction = loss of head due to enlargement from vena-contracta to section 2-2.

$$= \frac{(V_c - V)^2}{2g} \qquad \text{...}(i)$$

From continuity, we have $\quad a_c \times V_c = A \times V \qquad \text{...}(ii)$

where a_c = area of cross-section at vena-contracta

If C_c = co-efficient of contraction

Then
$$C_c = \frac{\text{area at vena-contracta}}{(A - a)} = \frac{a_c}{(A - a)}$$

$$\therefore \qquad a_c = C_c \times (A - a)$$

Substituting this value in (*ii*), we get

$$C_c \times (A - a) \times V_c = A \times V \quad \therefore \quad V_c = \frac{A \times V}{C_c\,(A - a)}$$

Substituting this value of V_c in equation (*iii*), we get

$$\text{Head loss due to obstruction} = \frac{\left(V_c - V^2\right)}{2g} = \frac{\left(\dfrac{A \times V}{C_c\,(A-a)} - V\right)^2}{2g} = \frac{V^2}{2g}\left(\frac{A}{C_c\,(A-a)} - 1\right)^2 \quad ...(9.10)$$

9.6.6 Loss of Head due to Bend in Pipe. When there is any bend in a pipe, the velocity of flow changes, due to which the separation of the flow from the boundary and also formation of eddies takes place. Thus the energy is lost. Loss of head in pipe due to bend is expressed as

$$h_b = \frac{kV^2}{2g}$$

where h_b = loss of head due to bend, V = velocity of flow, k = co-efficient of bend

The value of k depends on

(*i*) Angle of bend, (*ii*) Radius of curvature of bend, (*iii*) Diameter of pipe.

9.6.7 Loss of Head in Various Pipe Fittings. The loss of head in the various pipe fittings such as valves, couplings etc., is expressed as

$$= \frac{kV^2}{2g} \quad ...(9.11)$$

where V = velocity of flow, k = co-efficient of pipe fitting.

Problem 9.15 *Water is flowing through a horizontal pipe of diameter 200 mm at a velocity of 3 m/s. A circular solid plate of diameter 150 mm is placed in the pipe to obstruct the flow. Find the loss of head due to obstruction in the pipe if C_c = 0.62.*

Solution. Given :

Dia. of pipe, $D = 200 \text{ mm} = 0.20 \text{ m}$

Velocity, $V = 3.0 \text{ m/s}$

Area of pipe, $A = \frac{\pi}{4} D^2 = \frac{\pi}{4} (0.2)^2 = 0.03141 \text{ m}^2$

Dia. of obstruction, $d = 150 \text{ mm} = 0.15 \text{ m}$

∴ Area of obstruction, $a = \frac{\pi}{4} (.15)^2 = 0.01767 \text{ m}^2$

$C_c = 0.62$

The head lost due to obstruction is given by equation (9.10) as

$$= \frac{V^2}{2g}\left(\frac{A}{C_c\,(A-a)} - 1.0\right)^2$$

$$= \frac{3 \times 3}{2 \times 9.81}\left[\frac{.03141}{0.62\,[.03141 - .01767]} - 1.0\right]^2$$

$$= \frac{9}{2 \times 9.81}\,[3.687 - 1.0]^2 = \mathbf{3.311 \text{ m. Ans.}}$$

Problem 9.16 *Determine the rate of flow of water through a pipe of diameter 20 cm and length 50 m when one end of the pipe is connected to a tank and other end of the pipe is open to the atmosphere. The pipe is horizontal and the height of water in the tank is 4 m above the centre of the pipe. Consider all minor losses and take f = .009 in the formula* $h_f = \dfrac{4 . f . L . V^2}{d \times 2g}$.

Solution. Dia. of pipe, $d = 20 \text{ cm} = 0.20 \text{ m}$

Length of pipe, $L = 50 \text{ m}$

Height of water, $H = 4 \text{ m}$

Co-efficient of friction, $f = .009$

Let the velocity of water in pipe $= V$ m/s.

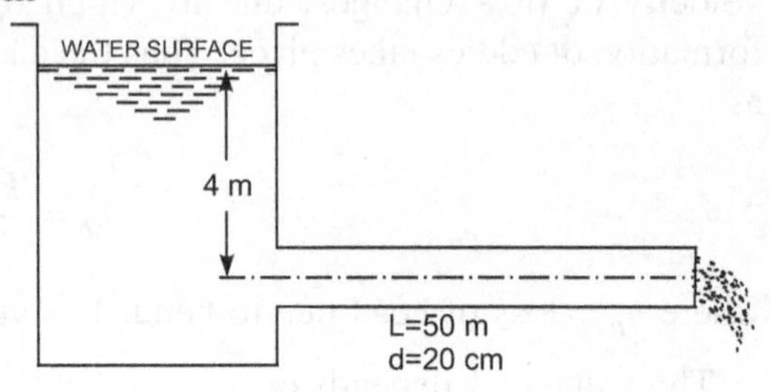

Fig. 9.4

Applying Bernoulli's equation at the top of the water surface in the tank and at the outlet of pipe, we have [Taking point 1 on the top and point 2 at the outlet of pipe].

$$\frac{p_1}{\rho g} + \frac{V_1^2}{2g} + z_1 = \frac{p_2}{\rho g} + \frac{V_2^2}{2g} + z_2 + \text{all losses}$$

Considering datum line passing through the centre of pipe

$$0 + 0 + 4.0 = 0 + \frac{V_2^2}{2g} + 0 + (h_i + h_f)$$

or

$$4.0 = \frac{V_2^2}{2g} + h_i + h_f$$

But the velocity in pipe $= V$, $\therefore V = V_2$

$$\therefore \quad 4.0 = \frac{V^2}{2g} + h_i + h_f \qquad \ldots(i)$$

From equation (9.8), $h_i = 0.5 \dfrac{V^2}{2g}$ and h_f from equation (9.1) is given as

$$h_f = \frac{4 . f . L . V^2}{d \times 2g}$$

Substituting these values, we have

$$4.0 = \frac{V^2}{g} + \frac{0.5\, V^2}{2g} + \frac{4 . f . L . V^2}{d \times 2g}$$

$$= \frac{V^2}{2g}\left[1.0 + 0.5 + \frac{4 \times .009 \times 50}{0.2}\right] = \frac{V^2}{2g}\,[1.0 + 0.5 + 9.0]$$

$$= 10.5 \times \frac{V^2}{2g}$$

$\therefore \qquad V = \sqrt{\dfrac{4 \times 2 \times 9.81}{10.5}} = 2.734$ m/sec

$\therefore$ Rate of flow, $\qquad Q = A \times V = \dfrac{\pi}{4} \times (0.2)^2 \times 2.734 = 0.08589 \text{ m}^3\text{/s}$

$= $ **85.89 litres/s. Ans.**

Problem 9.17 *A horizontal pipe line 40 m long is connected to a water tank at one end and discharges freely into the atmosphere at the other end. For the first 25 m of its length from the tank, the pipe is 150 mm diameter and its diameter is suddenly enlarged to 300 mm. The height of water level in the tank is 8 m above the centre of the pipe. Considering all losses of head which occur, determine the rate of flow. Take f = .01 for both sections of the pipe.*

(Osmania University, 1992 ; A.M.I.E., Summer, 1978)

Solution. Given :

Total length of pipe, $\quad L = 40$ m

Length of 1st pipe, $\quad L_1 = 25$ m

Dia. of 1st pipe, $\quad d_1 = 150$ mm $= 0.15$ m

Length of 2nd pipe, $\quad L_2 = 40 - 25 = 15$ m

Dia. of 2nd pipe, $\quad d_2 = 300$ mm $= 0.3$ m

Height of water, $\quad H = 8$ m

Co-efficient of friction, $\quad f = 0.01$

Applying Bernoulli's theorem to the free surface of water in the tank and outlet of pipe as shown in Fig. 9.5 and taking reference line passing through the centre of pipe.

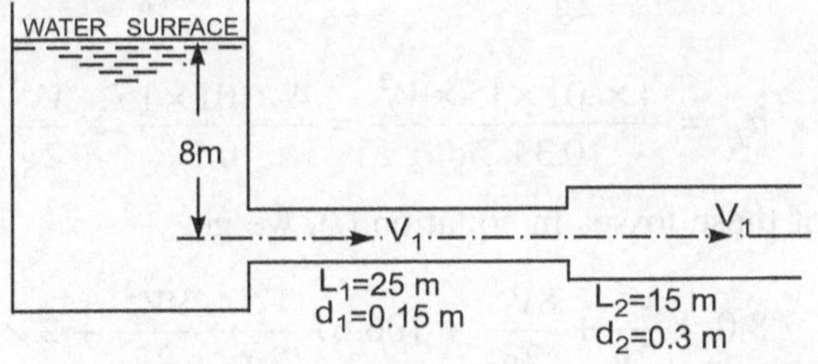

Fig. 9.5

$$0 + 0 + 8 = \frac{p_2}{\rho g} + \frac{V_2^2}{2g} + 0 + \text{all losses}$$

or
$$8.0 = 0 + \frac{V_2^2}{2g} + h_i + h_{f_1} + h_e + h_{f_2} \qquad \text{...}(i)$$

where $\quad h_i = $ loss of head at entrance $= 0.5 \dfrac{V_1^2}{2g}$

$h_{f_1} = $ head lost due to friction in pipe 1 $= \dfrac{4 \times f \times L_1 \times V_1^2}{d_1 \times 2g}$

$h_e = $ loss head due to sudden enlargement $= \dfrac{(V_1 - V_2)^2}{2g}$

h_{f_2} = Head lost due to friction in pipe 2 = $\dfrac{4 \times f \times l_2 \times V_2^2}{d_2 \times 2g}$

But from continuity equation, we have

$$A_1 V_1 = A_2 V_2$$

$$\therefore \quad V_1 = \frac{A_2 V_2}{A_1} = \frac{\frac{\pi}{4} d_2^2 \times V_2}{\frac{\pi}{4} d_1^2} = \left(\frac{d_2}{d_1}\right)^2 \times V_2 = \left(\frac{0.3}{.15}\right)^2 \times V_2 = 4V_2$$

Substituting the value of V_1 in different head losses, we have

$$h_i = \frac{0.5\, V_1^2}{2g} = \frac{0.5 \times (4V_2)^2}{2g} = \frac{8V_2^2}{2g}$$

$$h_{f_1} = \frac{4 \times 0.01 \times 25 \times (4V_2^2)}{0.15 \times 2 \times g}$$

$$= \frac{4 \times .01 \times 25 \times 16}{0.15} \times \frac{V_2^2}{2g} = 106.67 \frac{V_2^2}{2g}$$

$$h_e = \frac{(V_1 - V_2)^2}{2g} = \frac{(4V_2 - V_2)^2}{2g} = \frac{9V_2^2}{2g}$$

$$h_{f_2} = \frac{4 \times .01 \times 15 \times V_2^2}{0.3 \times 2g} = \frac{4 \times .01 \times 15}{0.3} \times \frac{V_2^2}{2g} = 2.0 \times \frac{V_2^2}{2g}$$

Substituting the values of these losses in equation (*i*), we get

$$8.0 = \frac{V_2^2}{2g} + \frac{8V_2^2}{2g} + 106.67 \frac{V_2^2}{2g} + \frac{9V_2^2}{2g} + 2 \times \frac{V_2^2}{2g}$$

$$= \frac{V_2^2}{2g} [1 + 8 + 106.67 + 9 + 2] = 126.67 \frac{V_2^2}{2}$$

$$\therefore \quad V_2 = \sqrt{\frac{8.0 \times 2 \times g}{126.67}} = \sqrt{\frac{8.0 \times 2 \times 9.81}{126.67}} = \sqrt{1.2391} = 1.113 \text{ m/s}$$

$\therefore$ Rate of flow, $Q = A_2 \times V_2 = \dfrac{\pi}{4} (0.3)^2 \times 1.113 = 0.07867 \text{ m}^3/\text{s}$ = **78.67 litres/s. Ans.**

Problem 9.18 *Determine the difference in the elevations between the water surfaces in the two tanks which are connected by a horizontal pipe of diameter 300 mm and length 400 m. The rate of flow of water through the pipe is 300 litres/s. Consider all losses and take the value of f = .008.*

Solution. Given :

Dia. of pipe, $d = 300 \text{ mm} = 0.30 \text{ m}$

Length, $L = 400 \text{ m}$

Discharge, $Q = 300$ lit/s $= 0.3\ m^3/s$

Co-efficient of friction, $f = 0.008$

Velocity, $$V = \frac{Q}{\text{Area}} = \frac{0.3}{\frac{\pi}{4} \times (.3)^2} = 4.244 \text{ m/s}$$

Let the two tanks are connected by a pipe as shown in Fig. 9.6.

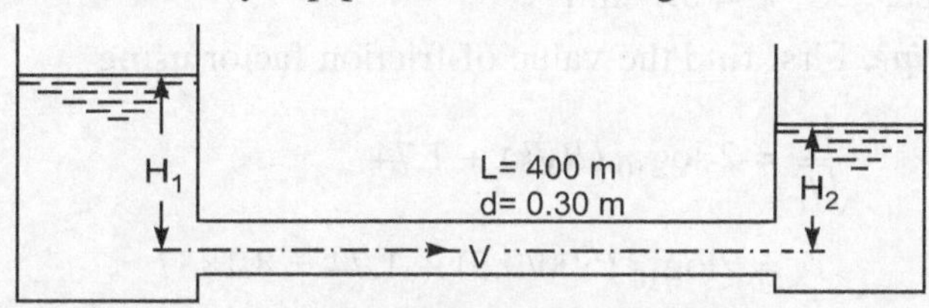

Fig. 9.6

Let H_1 = height of water in 1st tank above the centre of pipe

H_2 = height of water in 2nd tank above the centre of pipe

Then difference in elevations between water surfaces $= H_1 - H_2$

Applying Bernoulli's equation to the free surface of water in the two tanks, we have

$$H_1 = H_2 + \text{losses}$$

$$= H_2 + h_i + H_{f_1} + h_0 \qquad \ldots(i)$$

where h_i = Loss of head at entrance $= 0.5 \dfrac{V^2}{2g} = \dfrac{0.5 \times 4.244^2}{2 \times 9.81} = 0.459$ m

h_{f_1} = Loss of head due to friction $= \dfrac{4 \times f \times L \times V^2}{d \times 2g} = \dfrac{4 \times .008 \times 400 \times 4.244^2}{0.3 \times 2 \times 9.81} = 39.16$ m

h_0 = Loss of head at outlet $= \dfrac{V^2}{2g} = \dfrac{4.244^2}{2 \times 9.81} = 0.918$ m

Substituting these values in (*i*), we get

$$H_1 = H_2 + 0.459 + 39.16 + 0.918 = H_2 + 40.537$$

$\therefore$ $H_1 - H_2$ = Difference in elevations

$= \mathbf{40.537}$ **m. Ans.**

Problem 9.19 *The friction factor for turbulent flow through rough pipes can be determined by Karman-Prandtl equation,* $\dfrac{1}{\sqrt{f}} = 2 \log_{10} (R_0/k) + 1.74$

where f = friction factor, R_0 = pipe radius, k = average roughness.

Two reservoirs with a surface level difference of 20 metres are to be connected by 1 metre diameter pipe 6 km long. What will be the discharge when a cast iron pipe of roughness k = 0.3 mm is used ? What will be the percentage increase in the discharge if the cast iron pipe is replaced by a steel pipe of roughness k = 0.1 mm ? Neglect all local losses. (A.M.I.E., Winter, 1976)

Solution. Given :

Difference in levels, $h = 20$ m

Dia. of pipe, $d = 1.0$ m ∴ Radius, $R_0 = 0.5$ m = 500 mm

Length of pipe, $L = 6$ km = 6 × 1000 = 6000 m

Roughness of cast iron pipe, $k = 0.3$ mm

Roughness of steel pipe, $k = 0.1$ mm

1st Case. ***Cast Iron Pipe.*** First find the value of friction factor using

$$\frac{1}{\sqrt{f}} = 2 \log_{10} (R_0/k) + 1.74 \quad \ldots(i)$$

$$= 2\log_{10} (500/0.3) + 1.74 = 8.1837$$

$$\therefore \quad f = \left(\frac{1}{8.1837}\right)^2 = 0.0149$$

Local losses are to be neglected. This means only head loss due to friction is to be considered. Head loss due to friction is

$$20 = \frac{f \times L \times V^2}{d \times 2g}$$

[Here f is the friction factor and not co-efficient of friction
∵ Friction factor = 4 × co-efficient of friction]

$$20 = \frac{.0149 \times 6000 \times V^2}{1.0 \times 2 \times 9.81} = 4.556\ V^2$$

$$\therefore \quad V = \sqrt{\frac{20}{4.556}} = 2.095 \text{ m/s}$$

$$\therefore \quad \text{Discharge, } Q = V \times \text{Area} = 2.095 \times \frac{\pi}{4} \times d^2 = 2.095 \times \frac{\pi}{4} \times 1^2 = 1.645 \text{ m}^3/\text{s}$$

2nd Case. ***Steel Pipe.*** $k = 0.1$ mm, $R_0 = 500$ mm

Substituting these values in equation (i), we get

$$\frac{1}{\sqrt{f}} = 2 \log_{10} (500/0.1) + 1.74 = 9.1379$$

$$\therefore \quad f = \left(\frac{1}{9.1379}\right)^2 = 0.0119$$

$$\text{Head loss due to friction, } 20 = \frac{f \times L \times V^2}{d \times 2g} \text{ or } 20 = \frac{.0119 \times 6000 \times V^2}{1.0 \times 2 \times 9.81} = 3.639\ V^2$$

$$\therefore \quad V = \sqrt{\frac{60}{3.639}} = 2.344 \text{ m/s}$$

∴ Discharge, $Q_2 = V \times \text{Area} = 2.344 \times \frac{\pi}{4} \times 1^2 = 1.841 \text{ m}^3/\text{s}$

percentage increase in the discharge $= \frac{Q_2 - Q_1}{Q_1} \times 100 = \frac{(1.841 - 1.645)}{1.645} \times 100 =$ **11.91%. Ans.**

Problem 9.20 *Design the diameter of a steel pipe to carry water having kinematic viscosity $\nu = 10^{-6}$ m²/s with a mean velocity of 1 m/s. The head loss is to be limited to 5 m per 100 m length of pipe. Consider the equivalent sand roughness height of pipe $k_s = 45 \times 10^{-4}$ cm. Assume that the Darcy Weisbach friction factor over the whole range of turbulent flow can be expressed as*

$$f = 0.0055\left[1+\left(20\times10^3\,\frac{k_s}{D}+\frac{10^6}{R_e}\right)^{1/3}\right]$$

where D = Diameter of pipe and R_e = Reynolds number. (A.M.I.E., Summer, 1985)

Solution. Given :

Kinematic viscosity, $\nu = 10^{-6}$ m²/s

Mean velocity, $V = 1$ m/s

Head loss, $h_f = 5$ m in a length $L = 100$ m

Value of $k_s = 45 \times 10^{-4}$ cm $= 45 \times 10^{-6}$ m

Value of
$$f = 0.0055\left[1+\left(20\times10^3\,\frac{k_s}{D}+\frac{10^6}{R_e}\right)^{1/3}\right] \quad ...(i)$$

Using Darcy Weisbach equation, $h_f = \frac{4\times f\times L\times V^2}{D\times 2g}$

or
$$f = \frac{h_f\times D\times 2g}{4\times L\times V^2} = \frac{5\times D\times 2\times 9.81}{4\times 100\times 1^2} = 0.2452\,D$$

Now the Reynolds number is given by,

$$R_e = \frac{\rho VD}{\mu} = \frac{V\times D}{\nu} \qquad \left(\because \nu = \frac{\mu}{\rho}\right)$$

$$= \frac{1\times D}{10^{-6}} = 10^6\,D$$

Substituting the values of f, R_e and k_s in equation (*i*), we get

$$0.2452\,D = 0.0055\left[1+\left(20\times10^3\times\frac{45\times10^{-6}}{D}+\frac{10^6}{10^6\,D}\right)^{1/3}\right]$$

or
$$\frac{0.2452}{0.0055}\,D = \left[1+\left(\frac{0.9}{D}+\frac{1}{D}\right)^{1/3}\right]$$

or $$44.58\,D = \left[1+\left(\frac{1.9}{D}\right)^{1/3}\right] \quad \text{or} \quad 44.58\,D - 1 = \left(\frac{1.9}{D}\right)^{1/3}$$

or $$(44.58\,D - 1)^3 = \frac{1.9}{D} \quad \text{or} \quad D\,(44.58\,D - 1)^3 = 1.9 \qquad ...(ii)$$

The equation (*ii*) is solved by hit and trial method.

(*i*) Let $D = 0.1$ m, then L.H.S. of equation (*ii*) becomes as

L.H.S. $= 0.1\,(44.58 \times 0.1 - 1)^3 = 0.1 \times 3.458^3 = 4.135$

This is more than the R.H.S.

(*ii*) Let $D = 0.08$ m, then L.H.S. of equation (*ii*) becomes as

L.H.S. $= 0.08\,(44.58 \times 0.08 - 1)^3 = 0.08\,(2.5664)^3 = 1.352$

This is less than the R.H.S.

Hence value of D lies between 0.1 and 0.08

(*iii*) Let $D = 0.085$, then L.H.S. of equation (*ii*) becomes as

L.H.S. $= 0.085\,(44.58 \times 0.085 - 1)^3 = 1.844$

This value is slightly less than R.H.S. Hence increase the value of D slightly.

(*iv*) Let $D = 0.0854$ m, then L.H.S. of equation (*ii*) becomes as

L.H.S. $= 0.0854\,(44.58 \times 0.0854 - 1)^3 = 1.889$

This value is nearly equal to R.H.S.

$\therefore$ Correct value of D = **0.0854 m. Ans.**

Problem 9.21 *A pipe line AB of diameter 300 mm and of length 400 m carries water at the rate of 50 litres/s. The flow takes place from A to B where point B is 30 metres above A. Find the pressure at A if the pressure at B is 19.62 N/cm². Take f = .008.*

Solution. Given :

Dia. of pipe, $d = 300 \text{ mm} = 0.30 \text{ m}$

Length of pipe, $L = 400 \text{ m}$

Discharge, $Q = 50 \text{ litres/s} = 0.05 \text{ m}^3/\text{s}$

$\therefore$ Velocity, $$V = \frac{Q}{\text{Area}} = \frac{0.05}{\frac{\pi}{4}d^2} = \frac{0.05}{\frac{\pi}{4}\times(.3)^2} = 0.7074 \text{ m/s}$$

Pressure at B, $P_B = 19.62 \text{ N/cm}^2 = 19.62 \times 10^4 \text{ N/m}^2$

$f = .008$

Applying Bernoulli's equations at points A and B and taking datum line passing through A, we have

$$\frac{p_A}{\rho g} + \frac{V_A^2}{2g} + z_A = \frac{p_B}{\rho g} + \frac{V_B^2}{2g} + z_B + h_f$$

But $V_A = V_B$ [$\because$ Dia. is same]

$z_A = 0,\; z_B = 30$

and $$h_f = \frac{4 \times f \times L \times V^2}{d \times 2g}$$

$$\therefore \quad \frac{p_A}{\rho g} + 0 = \frac{19.62 \times 10^4}{1000 \times 9.81} + 30 + \frac{4 \times .008 \times 400 \times .7074^2}{0.3 \times 2 \times 9.81}$$

$$= 20 + 30 + 1.088 = 51.088$$

$$\therefore \quad p_A = 51.088 \times 1000 \times 9.81 \text{ N/m}^2$$

$$= \frac{51.088 \times 1000 \times 9.81}{10^4}$$

$$= \mathbf{50.12\ N/cm^2.\ Ans.}$$

Fig. 9.7

▶ 9.7 HYDRAULIC GRADIENT AND TOTAL ENERGY LINE

The concept of hydraulic gradient line and total energy line is very useful in the study of flow of fluids through pipes. They are defined as :

9.7.1 Hydraulic Gradient Line. It is defined as the line which gives the sum of pressure head $\left(\frac{p}{w}\right)$ and datum head (z) of a flowing fluid in a pipe with respect to some reference line or it is the line which is obtained by joining the top of all vertical ordinates, showing the pressure head (p/w) of a flowing fluid in a pipe from the centre of the pipe. It is briefly written as H.G.L. (Hydraulic Gradient Line).

9.7.2 Total Energy Line. It is defined as the line which gives the sum of pressure head, datum head and kinetic head of a flowing fluid in a pipe with respect to some reference line. It is also defined as the line which is obtained by joining the tops of all vertical ordinates showing the sum of pressure head and kinetic head from the centre of the pipe. It is briefly written as T.E.L. (Total Energy Line).

Problem 9.22 *For the problem 9.16, draw the Hydraulic Gradient Line (H.G.L.) and Total Energy Line (T.E.L.).*

Solution. Given :

$$L = 50 \text{ m}, \ d = 200 \text{ mm} = 0.2 \text{ m}$$

$$H = 4 \text{ m}, \ f = .009$$

Velocity, V through pipe is calculated in problem 9.16 and its value is $V = 2.734$ m/s

Now, $\quad h_i$ = Head lost at entrance of pipe

$$= 0.5 \frac{V^2}{2g} + \frac{0.5 \times 2.734^2}{2 \times 9.81} = 0.19 \text{ m}$$

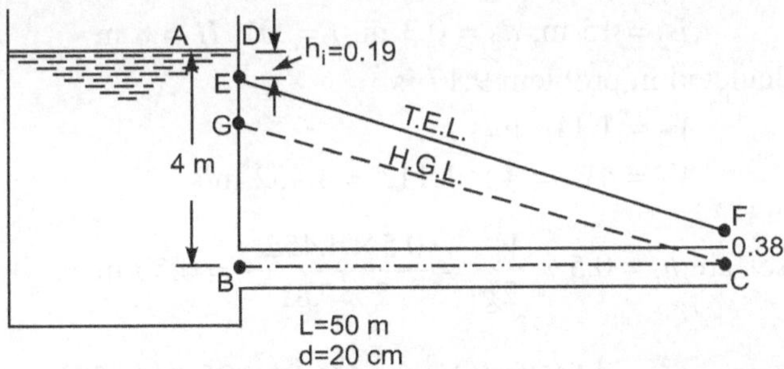

Fig. 9.8

and h_f = Head loss due to friction

$$= \frac{4 \times f \times L \times V^2}{d \times 2g} = \frac{4 \times 0.009 \times 50 \times (2.734)^2}{0.2 \times 2 \times 9.81} = 3.428 \text{ m.}$$

(*a*) **Total Energy Line (T.E.L.).** Consider three points, *A*, *B* and *C* on the free surface of water in the tank, at the inlet of the pipe and at the outlet of the pipe respectively as shown in Fig. 9.8. Let us find total energy at these points, taking the centre of pipe as reference line.

1. Total energy at $A = \frac{p}{\rho g} + \frac{V^2}{2g} + z = 0 + 0 + 4.0 = 4$ m
2. Total energy at B = Total energy at $A - h_i = 4.0 - 0.19 = 3.81$ m
3. Total energy at $C = \frac{p_c}{\rho g} + \frac{V_c^2}{2g} + z_c = 0 + \frac{V^2}{2g} + 0 = \frac{2.734^2}{2 \times 9.81} = 0.38$ m.

Hence total energy line will coincide with free surface of water in the tank. At the inlet of the pipe, it will decrease by h_i (= 0.19 m) from free surface and at outlet of pipe total energy is 0.38 m. Hence in the Fig. 9.8,

(*i*) Point *D* represents total energy at *A*

(*ii*) Point *E*, where $DE = h_i$, represents total energy at inlet of the pipe

(*iii*) Point *F*, where $CF = 0.38$ represents total energy at outlet of pipe.

Join *D* to *E* and *E* to *F*. Then *DEF* represents the total energy line.

(*b*) **Hydraulic Gradient Line (H.G.L.).** H.G.L. gives the sum of $(p/w + z)$ with reference to the datum-line. Hence hydraulic gradient line is obtained by subtracting $\frac{V^2}{2g}$ from total energy line. At outlet of the pipe, total energy $= \frac{V^2}{2g}$. By subtracting $\frac{V^2}{2g}$ from total energy at this point, we shall get point *C*, which lies on the centre line of pipe. From *C*, draw a line *CG* parallel to *EF*. Then *CG* represents the hydraulic gradient line.

Problem 9.23 *For the problem 9.17, draw the hydraulic gradient and total energy line.*

(A.M.I.E., Summer, 1978)

Solution. Refer to problem 9.17.

Given : $L_1 = 25$ m, $d_1 = 0.15$ m

$L_2 = 15$ m, $d_2 = 0.3$ m, $f = .01$, $H = 8$ m

The velocity V_2 as calculated in problem 9.17 is

$$V_2 = 1.113 \text{ m/s}$$

$$V_1 = 4V_2 = 4 \times 1.113 = 4.452 \text{ m/s}$$

The various head losses are $h_i = 0.5 \times \frac{V_1^2}{2g} = \frac{0.5 \times 4.452^2}{2 \times 9.81} = 0.50$ m

$$h_{f_1} = \frac{4f \times L_1 \times V_1^2}{d_1 \times 2g} = \frac{4 \times .01 \times 25 \times (4.452)^2}{0.15 \times 2 \times 9.81} = 6.73 \text{ m}$$

$$h_e = \frac{(V_1 - V_2)^2}{2g} = \frac{(4.452 - 1.11)^2}{2 \times 9.81} = 0.568 \text{ m}$$

$$h_{f_2} = \frac{4 \times f \times L_2 \times V_2^2}{d_2 \times 2g} = \frac{4 \times .01 \times 15 \times (1.113)^2}{0.3 \times 2 \times 9.81} = 0.126 \text{ m}$$

$$h_o = \frac{V_2^2}{2g} = \frac{1.113^2}{2 \times 9.81} = 0.063 \text{ m}$$

Also $\qquad V_1^2/2g = \dfrac{4.452^2}{2 \times 9.81} = 1.0$ m.

Total Energy Line

(*i*) Point A lies on free surface of water.

(*ii*) Take $AB = h_i = 0.5$ m.

(*iii*) From B, draw a horizontal line. Take BL equal to the length of pipe, *i.e.*, L_1. From L draw a vertical line downward.

(*iv*) Cut the line $LC = h_{f_1} = 6.73$ m.

(*v*) Join the point B to C. From C, take a line CD vertically downward equal to $h_e = 0.568$ m.

(*vi*) From D, draw DM horizontal and from point F which is lying on the centre of the pipe, draw a vertical line in the upward direction, meeting at M. From M, take a distance $ME = h_{f_2} = 0.126$. Join DE.

Then line $ABCDE$ represents the total energy line.

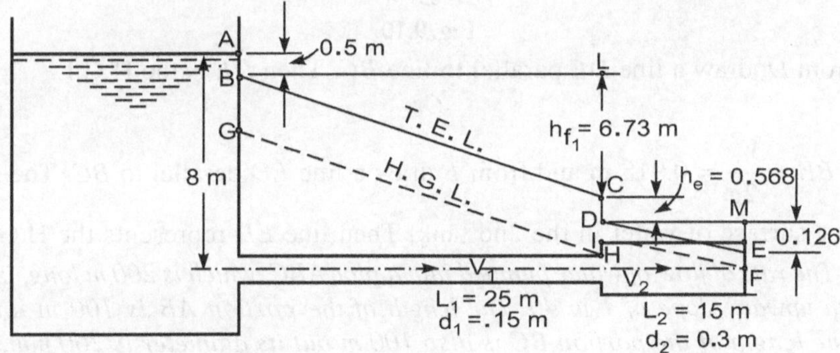

Fig. 9.9

Hydraulic Gradient Line (H.G.L.)

(*i*) From B, take $BG = \dfrac{V_1^2}{2g} = 1.0$ m.

(*ii*) Draw the line GH parallel to the line BC.

(*iii*) From F, draw a line FI parallel to the line ED.

(*iv*) Join the point H and I.

Then the line $GHIF$ represents the hydraulic gradient line (H.G.L.).

Problem 9.24 *For the problem 9.18, draw the hydraulic gradient and total energy line.*

Solution. Refer to problem, 9.18,

Given : $d = 300$ mm $= 0.3$ m

$L = 400$ m, $Q = 300$ litres/s $= 0.3$ m^3/s

$f = .008$

Let $H_1 = 50$ m. But $H_1 - H_2 = 40.537$ m (Calculated in problem 9.18)

$\therefore$ $H_2 = 50 - 40.537 = 9.463$ m.

The calculated losses are :

(*i*) $h_i = 0.459$ m (*ii*) $h_{f_1} = 39.16$ m

(*iii*) $h_e = 0.918$ m

(*a*) **T.E.L.**

(*i*) Point A is on the free surface of water in 1st tank. From A, take $AB = h_i = 0.459$ m.

(*ii*) Draw a horizontal line BF. Take BF equal to the length of pipe. From F, draw a vertical line in the downward direction. Cut $FC = h_{f_1} = 39.16$ m.

(*iii*) Join BC. From C take $CD = h_0 = 0.918$ m. The point D should coincide with free surface of water in 2nd tank. Then line $ABCD$ is the total energy line.

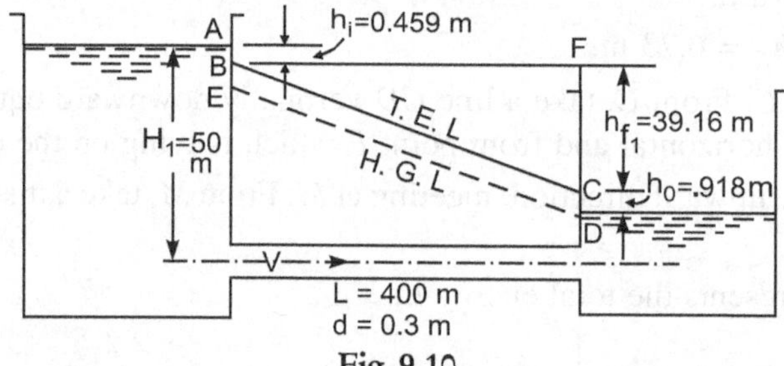

Fig. 9.10

(*b*) **H.G.L.** From D, draw a line DE parallel to line BC. Then DE is the H.G.L.

Or

From B, take $BE = \dfrac{V^2}{2g} = 0.918$ m and from E draw a line ED parallel to BC. The point D should coincide with free surface of water in the 2nd tank. Then line ED represents the H.G.L.

Problem 9.25 *The rate of flow of water pumped into a pipe ABC, which is 200 m long, is 20 litres/s. The pipe is laid on an upward slope of 1 in 40. The length of the portion AB is 100 m and its diameter 100 mm, while the length of the portion BC is also 100 m but its diameter is 200 mm. The change of diameter at B is sudden. The flow is taking place from A to C, where the pressure at A is 19.62 N/cm*^{2} *and end C is connected to a tank. Find the pressure at C and draw the hydraulic gradient and total energy line. Take f = .008.*

Solution. Given :

Length of pipe, $ABC = 200$ m

Discharge, $Q = 20$ litres/s $= 0.02$ m^3/s

Slope of pipe, $i = 1$ in $40 = \dfrac{1}{40}$

Length of pipe, $AB = 100$ m, Dia. of pipe $AB = 100$ mm $= 0.1$ m

Length of pipe, $BC = 100$ m, Dia. of pipe $BC = 200$ mm $= 0.2$ m

Pressure at A, $p_A = 19.62 \text{ N/cm}^2 = 19.62 \times 10^4 \text{ N/m}^2$

Co-efficient of friction, $f = .008$

Velocity of water in pipe AB, $V_1 = \dfrac{\text{Discharge}}{\text{Area of } AB} = \dfrac{0.02}{\frac{\pi}{4}(.1)^2} = 2.54 \text{ m/s}$

Velocity of water in pipe BC, $V_2 = \dfrac{Q}{\text{Area of } BC} = \dfrac{0.02}{\frac{\pi}{4}(.2)^2} = 0.63 \text{ m/s}$

Applying Bernoulli's equation to points A and C,

$$\frac{p_A}{\rho g} + \frac{V_A^2}{2g} + z_c = \frac{p_c}{\rho g} + \frac{V_c^2}{2g} + z_c + \text{total loss from } A \text{ to } C \qquad ...(i)$$

Total loss from A to C = Loss due to friction in pipe AB + loss of head due to enlargement at B + loss of head due to friction in pipe BC. ...(*ii*)

Now loss of head due to friction in pipe AB,

$$h_{f_1} = \frac{4fLV^2}{d \times 2g} = \frac{4 \times .008 \times 100 \times (2.54)^2}{0.1 \times 2 \times 9.81} = 10.52 \text{ m}$$

Loss of head due to friction in pipe BC,

$$h_{f_2} = \frac{4 \times .008 \times 100 \times (0.63)^2}{0.2 \times 2 \times 9.81} = 0.323 \text{ m}$$

Loss of head due to enlargement at B,

$$h_e = \frac{(V_1 - V_2)^2}{2g} = \frac{(2.54 - .63)^2}{2 \times 9.81} = 0.186 \text{ m}$$

$\therefore$ Total loss from A to C $= h_{f_1} + h_e + h_{f_2} = 10.52 + .186 + .323 = 11.029 \simeq 11.03 \text{ m}$

Substituting this value in (*i*), we get

$$\frac{p_A}{\rho g} + \frac{V_A^2}{2g} + z_A = \frac{p_c}{\rho g} + \frac{V_c^2}{2g} + z_c + 11.03 \qquad ...(iii)$$

Taking datum line passing through A, we have

$$z_A = 0$$

$$z_c = \frac{1}{40} \times \text{total length of pipe} = \frac{1}{40} \times 200 = 5 \text{ m}$$

Also $p_A = 19.62 \times 10^4 \text{ N/m}^2$

$V_A = V_1 = 2.54 \text{ m/s}, \; V_c = V_2 = 0.63 \text{ m/s}$

Substituting these values in (*iii*), we get

$$\frac{19.62 \times (10)^4}{1000 \times 9.81} + \frac{(2.54)^2}{2 \times 9.81} + 0 = \frac{p_c}{\rho g} + \frac{(0.63)^2}{2 \times 9.81} + 5.0 + 11.03$$

or $$20 + 0.328 = \frac{p_c}{\rho g} + 0.02 + 5.0 + 11.03$$

$\therefore$ $$20.328 = \frac{p_c}{\rho g} + 16.05$$

$\therefore$ $$\frac{p_c}{\rho g} = 20.328 - 16.05 = 4.278 \text{ m}$$

or $$p_c = 4.278 \times 1000 \times 9.81 \text{ N/m}^2$$

$$= \frac{4.278 \times 1000 \times 9.81}{10^4} \text{ N/cm}^2 = \mathbf{4.196 \text{ N/cm}^2. \text{ Ans.}}$$

Hydraulic Gradient and Total Energy Line

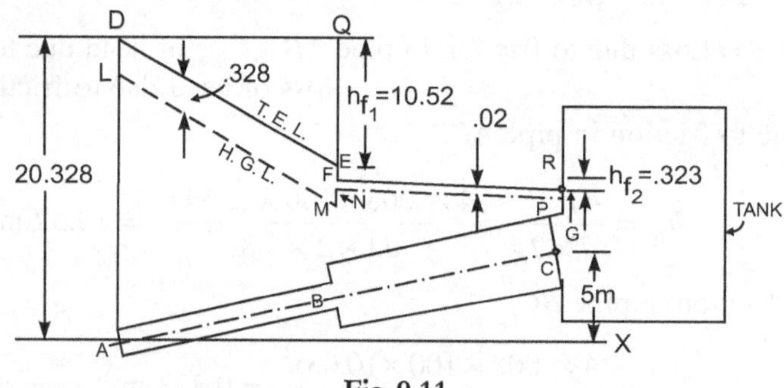

Fig. 9.11

Pipe AB. Assuming the datum line passing through *A*, then total energy at *A*

$$= \frac{p_A}{\rho g} + \frac{V_A^2}{2g} + z_A = \frac{19.62 \times 10^4}{1000 \times 9.81} + \frac{(2.54)^2}{2 \times 9.81} + 0 = 20 + 0.328$$

$$= 20.328 \text{ m}$$

Total energy at *B* $= \text{Total energy at } A - h_{f_1} = 20.328 - 10.52 = 9.808 \text{ m}$

Also $$V_c^2/2g = \frac{(0.63)^2}{2 \times 9.81} = 0.02.$$

Total energy line. Draw a horizontal line *AX* as shown in Fig. 9.11. The centre-line of the pipe is drawn in such a way that slope of pipe is 1 in 40. Thus the point *C* will be at a height of $\frac{1}{40} \times 200 = 5$ m from the line *AX*. Now draw a vertical line *AD* equal to total energy at *A*, *i.e.*, *AD* = 20.328 m. From point *D*, draw a horizontal line and from point *B*, a vertical line, meeting at *Q*. From *Q*, take vertical distance $QE = h_{f_1} = 10.52$ m. Join *DE*. From *E*, take $EF = h_e = 0.186$ m. From *F*, draw a horizontal line and from *C*, a vertical line meeting at *R*. From *R* take $RG = h_{f_2} = 0.323$ m. Join *F* to *G*. Then *DEFG* represents the total energy line.

Hydraulic Gradient Line. Draw the line *LM* parallel to the line *DE* at a distance in the downward directions equal to 0.328 m. Also draw the line *PN* parallel to the line *GF* at a distance of $\frac{V_c^2}{2g} = 0.02$. Join point *M* to *N*. Then line *LMNP* represents the hydraulic gradient line.

Problem 9.26 *A pipe line, 300 mm in diameter and 3200 m long is used to pump up 50 kg per second of an oil whose density is 950 kg/m^3 and whose kinematic viscosity is 2.1 stokes. The centre of the pipe line at the upper end is 40 m above than that at the lower end. The discharge at the upper end is atmospheric. Find the pressure at the lower end and draw the hydraulic gradient and the total energy line.*

Solution. Given :

Dia. of pipe, $d = 300 \text{ mm} = 0.3 \text{ m}$

Length of pipe, $L = 3200 \text{ m}$

Mass, $M = 50 \text{ kg/s} = \rho \,.\, Q$

$\therefore$ Discharge, $Q = \dfrac{50}{\rho} = \dfrac{50}{950} = 0.0526 \text{ m}^3/\text{s}$

$\because$ Density, $\rho = 950 \text{ kg/m}^3$

Kinematic viscosity, $\nu = 2.1 \text{ stokes} = 2.1 \text{ cm}^2/\text{s}$
$= 2.1 \times 10^{-4} \text{ m}^2/\text{s}$

Height of upper end $= 40$ m

Pressure at upper end $=$ atmospheric $= 0$

Reynold number, $R_e = \dfrac{V \times d}{\nu}$, where $V = \dfrac{\text{Discharge}}{\text{Area}} = \dfrac{0.0526}{\frac{\pi}{4}(0.3)^2} = 0.744 \text{ m/s}$

$$\therefore \quad R_e = \frac{0.744 \times 0.30}{2.1 \times 10^{-4}} = 1062.8$$

$\therefore$ Co-efficient of friction, $f = \dfrac{16}{R_e} = \dfrac{16}{1062.8} = 0.015$

Head lost due to friction, $h_f = \dfrac{4 \times f \times L \times V^2}{d \times 2g}$

$$= \frac{4 \times 0.015 \times 3200 \times (0.744)^2}{0.3 \times 2 \times 9.81} = 18.05 \text{ m of oil}$$

Applying the Bernoulli's equation at the lower and upper end of the pipe and taking datum line passing through the lower end, we have

$$\frac{p_1}{\rho g} + \frac{V_1^2}{2g} + z_1 = \frac{p_2}{\rho g} + \frac{V_2^2}{2g} + z_2 + h_f$$

But $z_1 = 0$, $z_2 = 40$ m, $V_1 = V_2$ as diameter is same

$p_2 = 0$, $h_f = 18.05$ m

$\therefore$ Substituting these values, we have

$$\frac{p_1}{\rho g} = 40 + 18.05 = 58.05 \text{ m of oil}$$

$$\therefore \quad p_1 = 58.05 \times \rho g = 58.05 \times 950 \times 9.81 \qquad [\because \rho \text{ for oil} = 950]$$

$$= 540997 \text{ N/m}^2 = \frac{540997}{10^{-4}} \text{ N/cm}^2 = \mathbf{54.099 \text{ N/cm}^2. \text{ Ans.}}$$

H.G.L. and T.E.L.

$$\frac{V^2}{2g} = \frac{(.744)^2}{2 \times 9.81} = 0.0282 \text{ m}$$

$$\frac{p_1}{\rho g} = 58.05 \text{ m of oil}$$

$$\frac{p_2}{\rho g} = 0$$

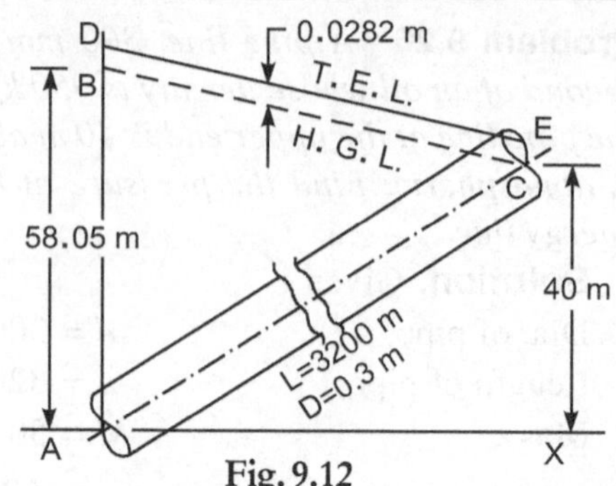

Fig. 9.12

Draw a horizontal line *AX* as shown in Fig. 9.12. From *A*, draw the centre line of the pipe in such a way that point *C* is a distance of 40 m above the horizontal line. Draw a vertical line *AB* through *A* such that *AB* = 58.05 m. Join *B* with *C*. Then *BC* is the hydraulic gradient line.

Draw a line *DE* parallel to *BC* at a height of 0.0282 m above the hydraulic gradient line. Then *DE* is the total energy line.

▶ 9.8 FLOW THROUGH SYPHON

Syphon is a long bent pipe which is used to transfer liquid from a reservoir at a higher elevation to another reservoir at a lower level when the two reservoirs are separated by a hill or high level ground as shown in Fig. 9.13.

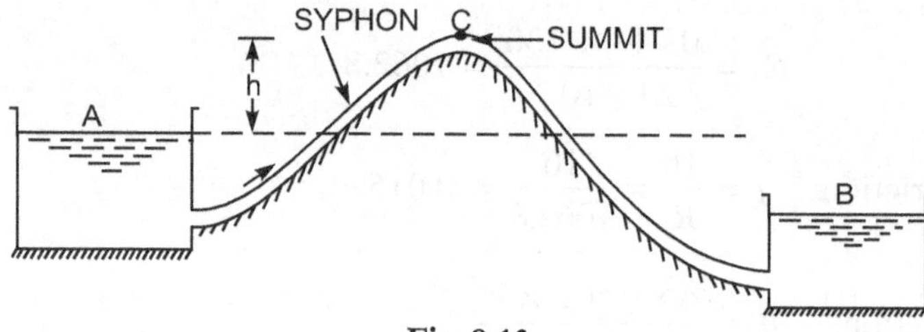

Fig. 9.13

The point *C* which is at the highest of the syphon is called the summit. As the point *C* is above the free surface of the water in the tank *A*, the pressure at *C* will be less than atmospheric pressure. Theoretically, the pressure at *C* may be reduced to – 10.3 m of water but in actual practice this pressure is only – 7.6 m of water or 10.3 – 7.6 = 2.7 m of water absolute. If the pressure at *C* becomes less than 2.7 m of water absolute, the dissolved air and other gases would come out from water and collect at the summit. The flow of water will be obstructed. Syphon is used in the following cases :

1. To carry water from one reservoir to another reservoir separated by a hill or ridge.
2. To take out the liquid from a tank which is not having any outlet.
3. To empty a channel not provided with any outlet sluice.

Problem 9.27 *A syphon of diameter 200 mm connects two reservoirs having a difference in elevation of 20 m. The length of the syphon is 500 m and the summit is 3.0 m above the water level in the upper reservoir. The length of the pipe from upper reservoir to the summit is 100 m. Determine the discharge through the syphon and also pressure at the summit. Neglect minor losses. The co-efficient of friction, f = .005.*

Solution. Given :

Dia. of syphon, $d = 200 \text{ mm} = 0.20 \text{ m}$

Difference in level of two reservoirs, $H = 20 \text{ m}$

Length of syphon, $L = 500$ m
Height of summit from upper reservoir, $h = 3.0$ m
Length of syphon up to summit, $L_1 = 100$ m
Co-efficient of friction, $f = .005$

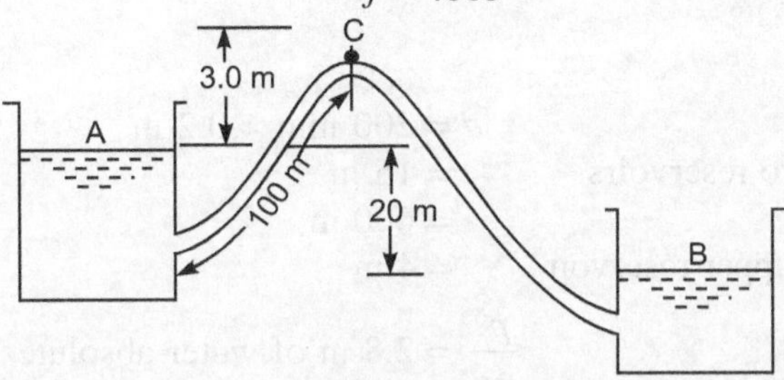

Fig. 9.14

If minor losses are neglected then the loss of head takes place only due to friction.
Applying Bernoulli's equation to point A and B,

$$\frac{p_A}{\rho g} + \frac{V_A^2}{2g} + z_A = \frac{p_B}{\rho g} + \frac{V_B^2}{2g} + z_B + \text{Loss of head due to friction from } A \text{ to } B$$

or $$0 + 0 + z_A = 0 + 0 + z_B + h_f \quad [\because p_A = p_B = \text{atmospheric pressure, } V_A = V_B = 0]$$

$\therefore$ $$z_A - z_B = h_f = \frac{4 \times f \times L \times V^2}{d \times 2g}$$

But $$z_A - z_B = 20 \text{ m}$$

$\therefore$ $$20 = \frac{4 \times .005 \times 500 \times V^2}{0.20 \times 2 \times 9.81} = 2.548\ V^2$$

$\therefore$ $$V = \sqrt{\frac{20}{2.548}} = 2.80 \text{ m/s}$$

$\therefore$ Discharge, $$Q = \text{Velocity} \times \text{Area}$$

$$= 2.80 \times \frac{\pi}{4} (.2)^2 = 0.0879 \text{ m}^3\text{/s} = \textbf{87.9 litres/s. Ans.}$$

Pressure at Summit. Applying Bernoulli's equation to point A and C,

$$\frac{p_A}{\rho g} + \frac{V_A^2}{2g} + z_A = \frac{p_c}{\rho g} + \frac{V_c^2}{2g} + z_c + \text{Loss of head due to friction between } A \text{ and } C$$

or $$0 + 0 + 0 = \frac{p_c}{\rho g} + \frac{V^2}{2g} + 3.0 + h_{f_1} \quad [\text{Taking datum line passing through } A]$$

$\therefore$ $$0 = \frac{p_c}{\rho g} + \frac{2.8^2}{2 \times 9.81} + 3.0 + \frac{4 \times .005 \times 100 \times (2.8)^2}{0.2 \times 2 \times 9.81} \quad [V_c = V = 2.80]$$

$$= \frac{p_c}{\rho g} + 0.399 + 3.0 + 4.00 = \frac{p_c}{\rho g} + 7.399$$

$\therefore$ $$\frac{p_c}{\rho g} = \textbf{– 7.399 m of water. Ans.}$$

Problem 9.28 *A syphon of diameter 200 mm connects two reservoirs having a difference in elevation of 15 m. The total length of the syphon is 600 m and the summit is 4 m above the water level in the upper reservoir. If the separation takes place at 2.8 m of water absolute, find the maximum length of syphon from upper reservoir to the summit. Take f = .004 and atmospheric pressure = 10.3 m of water.*

Solution. Given :

Dia. of syphon, $d = 200 \text{ mm} = 0.2 \text{ m}$

Difference of level in two reservoirs $= 15 \text{ m}$

Total length of pipe $= 600 \text{ m}$

Height of summit from upper reservoir $= 4 \text{ m}$

Pressure head at summit, $\frac{p_c}{\rho g} = 2.8$ m of water absolute

Atmospheric pressure head, $\frac{p_c}{\rho g} = 10.3$ m of water absolute

Co-efficient of friction, $f = .004$

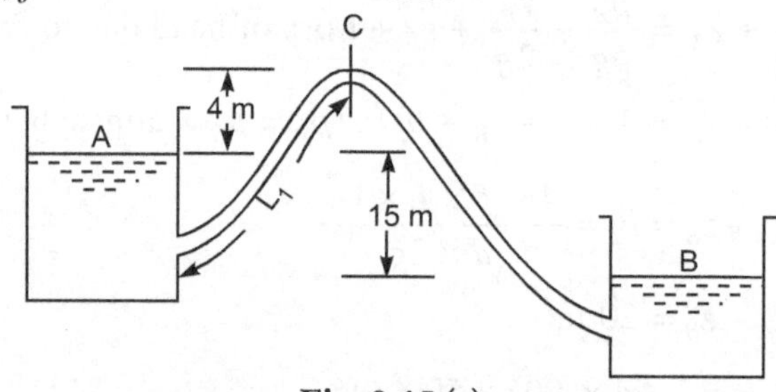

Fig. 9.15 (*a*)

Applying Bernoulli's equation to points *A* and *C* and taking the datum line passing through, *A*,

$$\frac{p_A}{\rho g} + \frac{V_A^2}{2g} + z_A = \frac{p_C}{\rho g} + \frac{V_c^2}{2g} + z_c + \text{Loss of head due to friction between } A \text{ and } C$$

Substituting the values of pressures in terms of absolute, we have

$$10.3 + 0 + 0 = 2.8 + \frac{V^2}{2g} + 4.0 + h_{f_1} \qquad [\because V_c = \text{velocity in pipe} = V]$$

$$\therefore \qquad h_{f_1} = 10.3 - 2.8 - 4.0 - \frac{V^2}{2g} = 3.5 - \frac{V^2}{2g} \qquad \text{...(i)}$$

Applying Bernoulli's equation to points *A* and *B* and taking datum line passing through *B*,

$$\frac{p_A}{\rho g} + \frac{V_A^2}{2g} + z_A = \frac{p_B}{\rho g} + \frac{V_B^2}{2g} + z_B + \text{Loss of head due to friction from } A \text{ to } B$$

But $\qquad \frac{p_A}{\rho g} = \frac{p_B}{\rho g}$ = atmospheric pressure

$$V_A = 0,\ V_B = 0,\ z_A = 15,\ z_B = 0$$

$$\therefore \qquad 0 + 0 + 15 = 0 + 0 + 0 + h_f$$

$$\therefore \qquad h_f = 15 \text{ or } \frac{4 \times f \times L \times V^2}{d \times 2g} = 15$$

or $$\frac{4 \times .004 \times 600 \times V^2}{0.2 \times 2 \times 9.81} = 15 \text{ or } V = \sqrt{\frac{15 \times 0.2 \times 2 \times 9.81}{4 \times .004 \times 600}} = 2.47 \text{ m/s}$$

Substituting this value of V in equation (*i*), we get

$$h_{f_1} = 3.35 - \frac{2.47^2}{2 \times 9.81} = 3.5 - 0.311 = 3.189 \text{ m} \quad \text{...(ii)}$$

But $$h_{f_2} = \frac{4 \times f \times L_1 \times V^2}{d \times 2g} \quad \text{...(iii)}$$

where L_1 = inlet leg of syphon or length of syphon from upper reservoir to the summit.

$$h_{f_1} = \frac{4 \times .004 \times L_1 \times (2.47)^2}{0.2 \times 2 \times 9.81} = 0.0248 \times L_1$$

Substituting this value in equation (*ii*),

$$0.0248\ L_1 = 3.189$$

$$\therefore \quad L_1 = \frac{3.189}{.0248} = \textbf{128.58 m. Ans.}$$

Problem 9.29 *A syphon of diameter 200 mm connects two reservoirs whose water surface level differ by 40 m. The total length of the pipe is 8000 m. The pipe crosses a ridge. The summit of ridge is 8 m above the level of water in the upper reservoir. Determine the minimum depth of the pipe below the summit of the ridge, if the absolute pressure head at the summit of syphon is not to fall below 3.0 m of water. Take f = 0.006 and atmospheric pressure head = 10.3 m of water. The length of syphon from the upper reservoir to the summit is 500 m. Find the discharge also.*

Solution. Given :

Dia. of syphon, d = 200 mm = 0.20 m

Difference in levels of two reservoirs, H = 40 m

Total length of pipe, L = 8000 m

Height of ridge summit from water level in upper reservoir = 8 m

Let the depth of the pipe below the summit of ridge = x m

$\therefore$ Height of syphon from water surface in the upper reservoir = $(8 - x)$ m

Pressure head at C, $\frac{p_c}{\rho g}$ = 3.0 m of water absolute

Atmospheric pressure head, $\frac{p_a}{\rho g}$ = 10.3 m of water

Co-efficient of friction f = .006

Length of syphon from upper reservoir to the summit, L_1 = 500 m

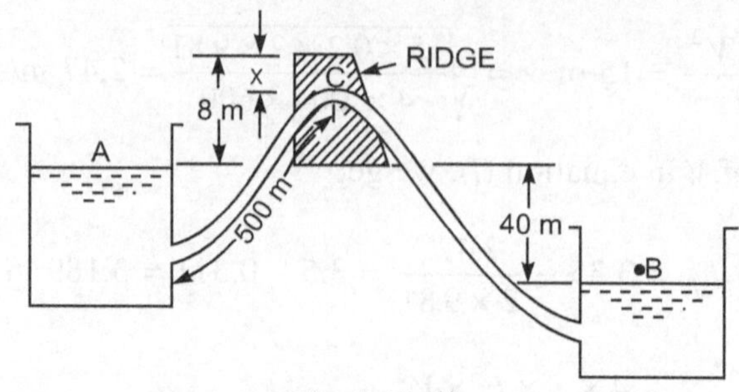

Fig. 9.15 (*b*)

Applying Bernoulli's equation to points *A* and B and taking datum line passing through *B*, we have

$$\frac{p_A}{\rho g} + \frac{V_A^2}{2g} + z_A = \frac{p_B}{\rho g} + \frac{V_B^2}{2g} + z_B + \text{head loss due to friction } A \text{ to } B$$

or $$0 + 0 + 40 = 0 + 0 + 0 + \frac{4 \times f \times L \times V^2}{d \times 2g}$$

$\therefore$ $$40 = \frac{4 \times 0.006 \times 8000 \times V^2}{0.2 \times 2 \times 9.81}$$

$\therefore$ $$V = \sqrt{\frac{40 \times 0.2 \times 2 \times 9.81}{4 \times .006 \times 8000}} = 0.904 \text{ m/s}$$

Now applying Bernoulli's equation to points *A* and *C* and assuming datum line passing through *A*, we have

$$\frac{p_A}{\rho g} + \frac{V_A^2}{2g} + z_A = \frac{p_c}{\rho g} + \frac{V_c^2}{2g} + z_e + \text{head loss due to friction from } A \text{ to } C$$

Substituting $\frac{p_A}{\rho g}$ and $\frac{p_c}{\rho g}$ in terms of absolute pressure

$$10.3 + 0 + 0 = 3.0 + \frac{V^2}{2g} + (8 - x) + \frac{4 \times f \times L_1 \times V^2}{d \times 2g}$$

or $$10.3 = 3.0 + \frac{(0.904)^2}{2 \times 9.81} + (8 - x) + \frac{4 \times .006 \times 500 \times (0.904)^2}{0.2 \times 2 \times 9.81}$$

$$= 3.0 + 0.041 + (8 - x) + 2.499 = 13.54 - x$$

$\therefore$ $$x = 13.54 - 10.3 = \mathbf{3.24 \text{ m. Ans.}}$$

Discharge, $$Q = \text{Area} \times \text{Velocity} = \frac{\pi}{4} \times (.2)^2 \times 0.904 = \mathbf{0.0283 \text{ m}^3\text{/s. Ans.}}$$

▶ 9.9 FLOW THROUGH PIPES IN SERIES OR FLOW THROUGH COMPOUND PIPES

Pipes in series or compound pipes is defined as the pipes of different lengths and different diameters connected end to end (in series) to form a pipe line as shown in Fig. 9.16.

Let, L_1, L_2, L_3 = length of pipes 1, 2 and 3 respectively

d_1, d_2, d_3 = diameter of pipes 1, 2, 3 respectively

V_1, V_2, V_3 = velocity of flow through pipes 1, 2, 3

f_1, f_2, f_3 = co-efficient of frictions for pipes 1, 2, 3

H = difference of water level in the two tanks.

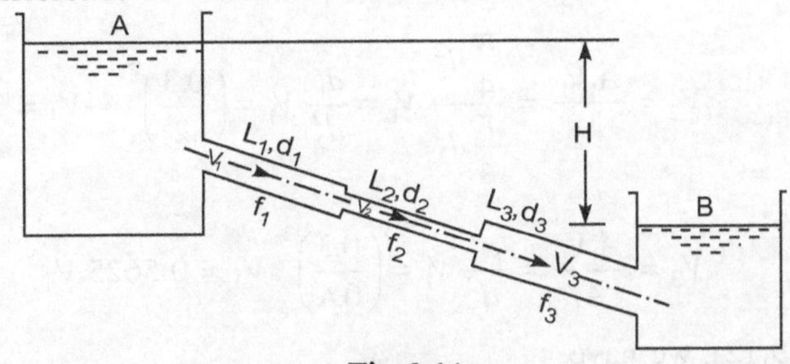

Fig. 9.16

The discharge passing through each pipe is same.

$$\therefore \qquad Q = A_1V_1 = A_2V_2 = A_3V_3$$

The difference in liquid surface levels is equal to the sum of the total head loss in the pipes.

$$\therefore \qquad I = \frac{0.5\,V_1^2}{2g} + \frac{4f_1L_1V_1^2}{d_1 \times 2g} + \frac{0.5\,V_2^2}{2g} + \frac{4f_2L_2V_2^2}{d_2 \times 2g}$$

$$+ \frac{(V_2 - V_3)^2}{2g} + \frac{4f_3L_3V_3^2}{d_3 \times 2g} + \frac{V_2^2}{2g} \qquad ...(9.12)$$

If minor losses are neglected, then above equation becomes as

$$H = \frac{4f_1L_1V_1^2}{d_1 \times 2g} + \frac{4f_2L_2V_2^2}{d_2 \times 2g} + \frac{4f_3L_3V_3^2}{d_3 \times 2g} \qquad ...(9.13)$$

If the co-efficient of friction is same for all pipes

i.e., $f_1 = f_2 = f_3 = f$, then equation (9.13) becomes as

$$H = \frac{4fL_1V_1^2}{d_1 \times 2g} + \frac{4fL_2V_2^2}{d_2 \times 2g} + \frac{4fL_3V_3^2}{d_3 \times 2g}$$

$$= \frac{4f}{2g}\left[\frac{L_1V_1^2}{d_1} + \frac{L_2V_2^2}{d_2} + \frac{L_3V_3^2}{d_3}\right] \qquad ...(9.14)$$

Problem 9.30 *The difference in water surface levels in two tanks, which are connected by three pipes in series of lengths 300 m, 170 m and 210 m and of diameters 300 mm, 200 mm and 400 mm respectively, is 12 m. Determine the rate of flow of water if co-efficient of friction are .005, .0052 and .0048 respectively, considering : (i) minor losses also (ii) neglecting minor losses.*

(Similar Question, VTU March 2001; Delhi University, 1987)

Solution. Given :

Difference of water level, $H = 12$ m

Length of pipe 1, $L_1 = 300$ m and dia., $d_1 = 300$ mm $= 0.3$ m

Length of pipe 2, $L_2 = 170$ m and dia., $d_2 = 200$ mm $= 0.2$ m

Length of pipe 3, $L_3 = 210$ m and dia., $d_3 = 400$ mm $= 0.4$ m

Also, $f_1 = .005$, $f_2 = .0052$ and $f_3 = .0048$

(*i*) **Considering Minor Losses.** Let V_1, V_2 and V_3 are the velocities in the 1st, 2nd and 3rd pipe respectively.

From continuity, we have $A_1V_1 = A_2V_2 = A_3V_3$

$$\therefore \quad V_2 = \frac{A_1V_1}{A_2} = \frac{\frac{\pi}{4}d_1^2}{\frac{\pi}{4}d_2^2}V_1 = \frac{d_1^2}{d_2^2}V_1 = \left(\frac{0.3}{.2}\right)^2 \times V_1 = 2.25\ V_1$$

and
$$V_3 = \frac{A_1V_1}{A_3} = \frac{d_1^2}{d_3^2}V_1 = \left(\frac{0.3}{0.4}\right)^2 V_1 = 0.5625\ V_1$$

Now using equation (9.12), we have

$$H = \frac{0.5\,V_1^2}{2g} + \frac{4f_1L_1V_1^2}{d_1 \times 2g} + \frac{0.5\,V_2^2}{2g} + \frac{4f_2L_2V_2^2}{d_2 \times 2g} + \frac{(V_2 - V_3)^2}{2g} + \frac{4f_3L_3V_3^2}{d_3 \times 2g} + \frac{V_3^2}{2g}$$

Substituting V_2 and V_3,
$$12.0 = \frac{0.5\,V_1^2}{2g} + \frac{4 \times .005 \times 300 \times V_1^2}{0.3 \times 2g} + \frac{0.5 \times (2.25\,V_1^2)^2}{2g}$$

$$+ 4 \times 0.0052 \times 170 \times \frac{(2.25\,V_1)^2}{0.2 \times 2g} + \frac{(2.25\,V_1 - .562\,V_1)^2}{2g} + \frac{4 \times .0048 \times 210 \times (.5625\,V_1)^2}{0.4 \times 2g} + \frac{(.5625\,V_1)^2}{2g}$$

or
$$12.0 = \frac{V_1^2}{2g}\,[0.5 + 20.0 + 2.53 + 89.505 + 2.847 + 3.189 + 0.316]$$

$$= \frac{V_1^2}{2g}\,[118.887]$$

$$\therefore \quad V_1 = \sqrt{\frac{12 \times 2 \times 9.81}{118.887}} = 1.407 \text{ m/s}$$

$\therefore$ Rate of flow, Q = Area × Velocity = $A_1 \times V_1$

$$= \frac{\pi}{4}(d_1)^2 \times V_1 = \frac{\pi}{4}(.3)^2 \times 1.407 = 0.09945 \text{ m}^3\text{/s}$$

$$= \mathbf{99.45\ litres/s.\ Ans.}$$

(*ii*) **Neglecting Minor Losses.** Using equation (9.13), we have

$$H = \frac{4f_1L_1V_1^2}{d_1 \times 2g} + \frac{4f_2L_2V_2^2}{d_2 \times 2g} + \frac{4f_3L_3V_3^2}{d_3 \times 2g}$$

or
$$12.0 = \frac{V_1^2}{2g}\left[\frac{4 \times .005 \times 300}{0.3} + \frac{4 \times .0052 \times 170 \times (2.25)^2}{0.2} + \frac{4 \times .0048 \times 210 \times (.5625)^2}{0.4}\right]$$

$$= \frac{V_1^2}{2g}\,[20.0 + 89.505 + 3.189] = \frac{V_1^2}{2g} \times 112.694$$

$$\therefore \qquad V_1 = \sqrt{\frac{2 \times 9.81 \times 12.0}{112.694}} = 1.445 \text{ m/s}$$

$$\therefore \qquad \text{Discharge, } Q = V_1 \times A_1 = 1.445 \times \frac{\pi}{4} (.3)^2 = 0.1021 \text{ m}^3\text{/s} = \mathbf{102.1 \text{ litres/s. Ans.}}$$

Problem 9.30A *Three pipes of 400 mm, 200 mm and 300 mm diameters have lengths of 400 m, 200 m, and 300 m respectively. They are connected in series to make a compound pipe. The ends of this compound pipe are connected with two tanks whose difference of water levels is 16 m. If co-efficient of friction for these pipes is same and equal to 0.005, determine the discharge through the compound pipe neglecting first the minor losses and then including them.* (A.M.I.E., Summer, 1990)

Solution. Given :

Difference of water levels, $H = 16$ m

Length and dia. of pipe 1, $L_1 = 400$ m and $d_1 = 400$ mm $= 0.4$ m

Length and dia. of pipe 2, $L_2 = 200$ m and $d_2 = 200$ mm $= 0.2$ m

Length and dia. of pipe 3, $L_3 = 300$ m and $d_3 = 300$ mm $= 0.3$ m

Also $f_1 = f_2 = f_3 = 0.005$

(*i*) **Discharge through the compound pipe first neglecting minor losses.**

Let V_1, V_2 and V_3 are the velocities in the 1st, 2nd and 3rd pipe respectively.

From continuity, we have $A_1V_1 = A_2V_2 = A_3V_3$

$$\therefore \qquad V_2 = \frac{A_1V_1}{A_2} = \frac{\frac{\pi}{4}d_1^2}{\frac{\pi}{4}d_2^2} \times V_1 = \frac{d_1^2}{d_2^2} V_1 = \left(\frac{0.4}{0.2}\right)^2 V_1 = 4V_1$$

and

$$V_3 = \frac{A_1V_1}{A_3^2} = \frac{\frac{\pi}{4}d_1^2}{\frac{\pi}{4}d_3^2} \times V_1 = \frac{d_1^2}{d_3^2} V_1 = \left(\frac{0.4}{0.2}\right)^2 V_1 = 1.77V_1$$

Now using equation (9.13), we have

$$H = \frac{4f_1L_1V_1^2}{d_1 \times 2g} + \frac{4f_2L_2V_2^2}{d_2 \times 2g} + \frac{4f_3L_3V_3^2}{d_3 \times 2g}$$

or

$$16 = \frac{4 \times 0.005 \times 400 \times V_1^2}{0.4 \times 2 \times 9.81} + \frac{4 \times 0.005 \times 200 \times (4V_1)^2}{0.2 \times 2 \times 9.81} + \frac{4 \times 0.005 \times 300}{0.3 \times 2 \times 9.81} \times (1.77V_1)^2$$

$$= \frac{V_1^2}{2 \times 9.81}\left(\frac{4 \times 0.005 \times 400}{0.4} + \frac{4 \times 0.005 \times 200 \times 16}{0.2} + \frac{4 \times 0.005 \times 300 \times 3.157}{0.3}\right)$$

$$16 = \frac{V_1^2}{2 \times 9.81}(20 + 320 + 63.14) = \frac{V_1^2}{2 \times 9.81} \times 403.14$$

$$\therefore \qquad V_1 = \sqrt{\frac{16 \times 2 \times 9.81}{403.14}} = 0.882 \text{ m/s}$$

$$\therefore \quad \text{Discharge,} \qquad Q = A_1 \times V_1 = \frac{\pi}{4}(0.4)^2 \times 0.882 = \mathbf{0.1108 \text{ m}^3\text{/s. Ans.}}$$

(*ii*) **Discharge through the compound pipe considering minor losses also.**

Minor losses are :

(*a*) At inlet, $h_i = \dfrac{0.5V_1^2}{2g}$

(*b*) Between 1st pipe and 2nd pipe, due to contraction,

$$h_c = \frac{0.5V_2^2}{2g} = \frac{0.5(4V_1^2)}{2g} \qquad (\because\ V_2 = 4V_1)$$

$$= \frac{0.5 \times 16 \times V_1^2}{2g} = 8 \times \frac{V_1^2}{2g}$$

(*c*) Between 2nd pipe and 3rd pipe, due to sudden enlargement,

$$h_e = \frac{(V_2 - V_3)^2}{2g} = \frac{(4V_1 - 1.77V_1)^2}{2g} \qquad (\because\ V_3 = 1.77V_1)$$

$$= (2.23)^2 \times \frac{V_1^2}{2g} = 4.973\ \frac{V_1^2}{2g}$$

(*d*) At the outlet of 3rd pipe, $h_o = \dfrac{V_3^2}{2g} = \dfrac{(1.77V_1)^2}{2g} = 1.77^2 \times \dfrac{V_1^2}{2g} = 3.1329\ \dfrac{V_1^2}{2g}$

The major losses are
$$= \frac{4f_1 \times L_1 \times V_1^2}{d_1 \times 2g} + \frac{4f_2 \times L_2 \times V_2^2}{d_2 \times 2g} + \frac{4f_3 \times L_3 \times V_3^2}{d_3 \times 2g}$$

$$= \frac{4 \times 0.005 \times 400 \times V_1^2}{0.4 \times 2 \times 9.81} + \frac{4 \times 0.005 \times 200 \times (4V_1)^2}{0.2 \times 2 \times 9.81} + \frac{4 \times 0.005 \times 300 \times (1.77V_1)^2}{0.3 \times 2 \times 9.81}$$

$$= 403.14 \times \frac{V_1^2}{2 \times 9.81}$$

$\therefore$ Sum of minor losses and major losses

$$= \left[\frac{0.5\,V_1^2}{2g} + 8 \times \frac{V_1^2}{2g} + 4.973\frac{V_1^2}{2g} + 3.1329\frac{V_1^2}{2g}\right] + 403.14\ \frac{V_1^2}{2g}$$

$$= 419.746\ \frac{V_1^2}{2g}$$

But total loss must be equal to H (or 16 m)

$$\therefore \qquad 419.746 \times \frac{V_1^2}{2g} = 16 \quad \therefore \quad V_1 = \sqrt{\frac{16 \times 2 \times 9.81}{419.746}} = 0.864 \text{ m/s}$$

$\therefore$ Discharge, $Q = A_1V_1 = \dfrac{\pi}{4}\ (0.4)^2 \times 0.864 =$ **0.1085 m³/s. Ans.**

▶ 9.10 EQUIVALENT PIPE

This is defined as the pipe of uniform diameter having loss of head and discharge equal to the loss of head and discharge of a compound pipe consisting of several pipes of different lengths and diameters. The uniform diameter of the equivalent pipe is called equivalent size of the pipe. The length of equivalent pipe is equal to sum of lengths of the compound pipe consisting of different pipes.

Let L_1 = length of pipe 1 and d_1 = dia.
L_2 = length of pipe 2 and d_2 = dia.
L_3 = length of pipe 3 and d_3 = dia.
H = total head loss
L = length of equivalent pipe
d = diameter of the equivalent pipe

Then $L = L_1 + L_2 + L_3$

Total head loss in the compound pipe, neglecting minor losses

$$H = \frac{4f_1L_1V_1^2}{d_1 \times 2g} + \frac{4f_2L_2V_2^2}{d_2 \times 2g} + \frac{4f_3L_3V_3^2}{d_3 \times 2g} \quad \text{...(9.14}A\text{)}$$

Assuming $\quad f_1 = f_2 = f_3 = f$

Discharge, $\quad Q = A_1V_1 = A_2V_2 = A_3V_3 = \frac{\pi}{4}{d_1}^2V_1 = \frac{\pi}{4}{d_2}^2V_2 = \frac{\pi}{4}{d_3}^2V_3$

$$\therefore \quad V_1 = \frac{4Q}{\pi d_1^2},\; V_2 = \frac{4Q}{\pi d_2^2} \text{ and } V_3 = \frac{4Q}{\pi d_3^2}$$

Substituting these values in equation (9.14*A*), we have

$$H = \frac{4fL_1 \times \left(\frac{4Q}{\pi d_1^2}\right)^2}{d_1 \times 2g} + \frac{4fL_2\left(\frac{4Q}{\pi d_2^2}\right)^2}{d_2 \times 2g} + \frac{4fL_3\left(\frac{4Q}{\pi d_3^2}\right)^2}{d_3 \times 2g}$$

$$= \frac{4 \times 16fQ^2}{\pi^2 \times 2g}\left[\frac{L_1}{d_1^5} + \frac{L_2}{d_2^5} + \frac{L_3}{d_3^5}\right] \quad \text{...(9.15)}$$

Head loss in the equivalent pipe, $H = \frac{4f\,.\,L\,.\,V^2}{d \times 2g}$ [Taking same value of f as in compound pipe]

where $V = \frac{Q}{A} = \frac{Q}{\frac{\pi}{4}d^2} = \frac{4Q}{\pi d^2}$

$$\therefore \quad H = \frac{4\,f\,.\,L\,.\left(\frac{4Q}{\pi d^2}\right)^2}{d \times 2g} = \frac{4 \times 16Q^2 f}{\pi^2 \times 2g}\left[\frac{L}{d^5}\right] \quad \text{...(9.16)}$$

Head loss in compound pipe and in equivalent pipe is same hence equating equations (9.15) and (9.16), we have

$$\frac{4\times 16fQ^2}{\pi^2\times 2g}\left[\frac{L_1}{d_1^5}+\frac{L_2}{d_2^5}+\frac{L_3}{d_3^5}\right]=\frac{4\times 16Q^2 f}{\pi^2\times 2g}\left[\frac{L}{d^5}\right]$$

or $$\frac{L_1}{d_1^5}+\frac{L_2}{d_2^5}+\frac{L_3}{d_3^5}=\frac{L}{d^5} \quad \text{or} \quad \frac{L}{d^5}=\frac{L_1}{d_1^5}+\frac{L_2}{d_2^5}+\frac{L_3}{d_3^5} \qquad ...(9.17)$$

Equation (9.17) is known as Dupuit's equation. In this equation $L = L_1 + L_2 + L_3$ and d_1, d_2 and d_3 are known. Hence the equivalent size of the pipe, *i.e.*, value of d can be obtained.

Problem 9.31 *Three pipes of lengths 800 m, 500 m and 400 m and of diameters 500 mm, 400 mm and 300 mm respectively are connected in series. These pipes are to be replaced by a single pipe of length 1700 m. Find the diameter of the single pipe.*

Solution. Given :

Length of pipe 1, $L_1 = 800$ m and dia., $d_1 = 500$ mm $= 0.5$ m

Length of pipe 2, $L_2 = 500$ m and dia., $d_2 = 400$ mm $= 0.4$ m

Length of pipe 3, $L_3 = 400$ m and dia., $d_3 = 300$ mm $= 0.3$ m

Length of single pipe, $L = 1700$ m

Let the diameter of equivalent single pipe $= d$

Applying equation (9.17), $$\frac{L}{d^5}=\frac{L_1}{d_1^5}+\frac{L_2}{d_2^5}+\frac{L_3}{d_3^5}$$

or $$\frac{1700}{d^5}=\frac{800}{.5^5}+\frac{500}{.4^5}+\frac{400}{0.3^5}=25600+48828.125+164609=239037$$

$$\therefore \quad d^5=\frac{1700}{239037}=.007118$$

$$\therefore \quad d=(.007188)^{0.2}=0.3718=\mathbf{371.8\ mm.\ Ans.}$$

▶ 9.11 FLOW THROUGH PARALLEL PIPES

Consider a main pipe which divides into two or more branches as shown in Fig. 9.17 and again join together downstream to form a single pipe, then the branch pipes are said to be connected in parallel. The discharge through the main is increased by connecting pipes in parallel.

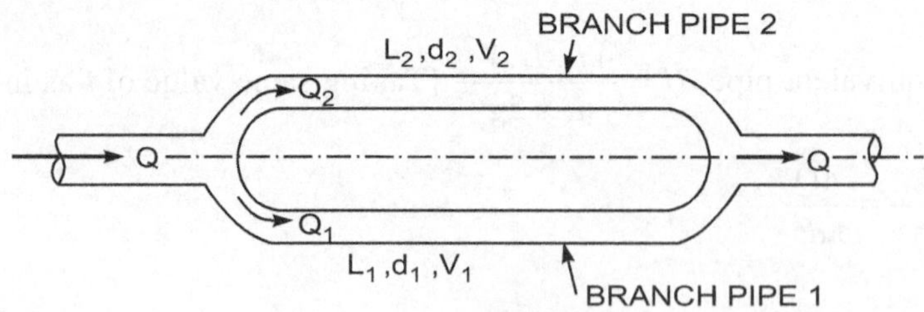

Fig. 9.17

The rate of flow in the main pipe is equal to the sum of rate of flow through branch pipes. Hence from Fig. 9.17, we have

$$Q = Q_1 + Q_2 \qquad ...(9.18)$$

In this, arrangement, the loss of head for each branch pipe is same.

$\therefore$ Loss of head for branch pipe 1 = Loss of head for branch pipe 2

or
$$\frac{4 f_1 L_1 V_1^2}{d_1 \times 2g} = \frac{4 f_2 L_2 V_2^2}{d_2 \times 2g} \qquad ...(9.19)$$

If
$$f_1 = f_2, \text{ then } \frac{L_1 V_1^2}{d_1 \times 2g} = \frac{L_2 V_2^2}{d_2 \times 2g} \qquad ...(9.20)$$

Problem 9.32 *A main pipe divides into two parallel pipes which again forms one pipe as shown in Fig. 9.17. The length and diameter for the first parallel pipe are 2000 m and 1.0 m respectively, while the length and diameter of 2nd parallel pipe are 2000 m and 0.8 m. Find the rate of flow in each parallel pipe, if total flow in the main is 3.0 m*^{3}*/s. The co-efficient of friction for each parallel pipe is same and equal to .005.*

Solution. Given :

Length of pipe 1, $L_1 = 2000$ m

Dia. of pipe 1, $d_1 = 1.0$ m

Length of pipe 2, $L_2 = 2000$ m

Dia. of pipe 2, $d_2 = 0.8$ m

Total flow, $Q = 3.0$ m^3/s

$f_1 = f_2 = f = .005$

Let Q_1 = discharge in pipe 1

Q_2 = discharge in pipe 2

From equation (9.18), $Q = Q_1 + Q_2 = 3.0 \qquad ...(i)$

Using equation (9.19), we have

$$\frac{4 f_1 L_1 V_1^2}{d_1 \times 2g} = \frac{4 f_2 L_2 V_2^2}{d_2 \times 2g}$$

$$\frac{4 \times .005 \times 2000 \times V_1}{1.0 \times 2 \times 9.81} = \frac{4 \times .005 \times 2000 \times V_2^2}{0.8 \times 2 \times 9.81}$$

or
$$\frac{V_1^2}{1.0} = \frac{V_2^2}{0.8} \text{ or } V_1^{\,2} = \frac{V_2^2}{0.8}$$

$\therefore$
$$V_1 = \frac{V_2}{\sqrt{0.8}} = \frac{V_2}{.894} \qquad ...(ii)$$

Now
$$Q_1 = \frac{\pi}{4} d_1^{\,2} \times V_1 = \frac{\pi}{4} (1)^2 \times \frac{V_2}{.894} \qquad \left[\because V_1 = \frac{V_2}{.894}\right]$$

and
$$Q_2 = \frac{\pi}{4} d_2^{\,2} \times V_2 = \frac{\pi}{4} (.8)^2 \times V_2 = \frac{\pi}{4} \times .64 \times V_2$$

Substituting the value of Q_1 and Q_2 in equation (*i*), we get

$$\frac{\pi}{4} \times \frac{V_2}{0.894} + \frac{\pi}{4} \times .64\ V_2 = 3.0 \quad \text{or} \quad 0.8785\ V_2 + 0.5026\ V_2 = 3.0$$

or $$V_2[.8785 + .5026] = 3.0 \quad \text{or} \quad V = \frac{3.0}{1.3811} = 2.17 \text{ m/s.}$$

Substituting this value in equation (*ii*),

$$V_1 = \frac{V_2}{.894} = \frac{2.17}{0.894} = 2.427 \text{ m/s}$$

Hence $$Q_1 = \frac{\pi}{4} d_1^2 \times V_1 = \frac{\pi}{4} \times 1^2 \times 2.427 = \mathbf{1.906\ m^3/s.\ Ans.}$$

$\therefore$ $$Q_2 = Q - Q_1 = 3.0 - 1.906 = \mathbf{1.094\ m^3/s.\ Ans.}$$

Problem 9.33 *A pipe line of 0.6 m diameter is 1.5 km long. To increase the discharge, another line of the same diameter is introduced parallel to the first in the second half of the length. Neglecting minor losses, find the increase in discharge if 4f = 0.04. The head at inlet is 300 mm.*

(A.M.I.E., December, 1975)

Solution. Given :

Dia. of pipe line, $D = 0.6$ m

Length of pipe line, $L = 1.5$ km $= 1.5 \times 1000 = 1500$ m

$4f = 0.04$ or $f = .01$

Head at inlet, $h = 300$ mm $= 0.3$ m

Head at outlet $=$ atmospheric head $= 0$

$\therefore$ Head lost, $h_f = 0.3$ m

Length of another parallel pipe, $L_1 = \dfrac{1500}{2} = 750$ m

Dia. of another parallel pipe, $d_1 = 0.6$ m

Fig. 9.18 shows the arrangement of pipe system.

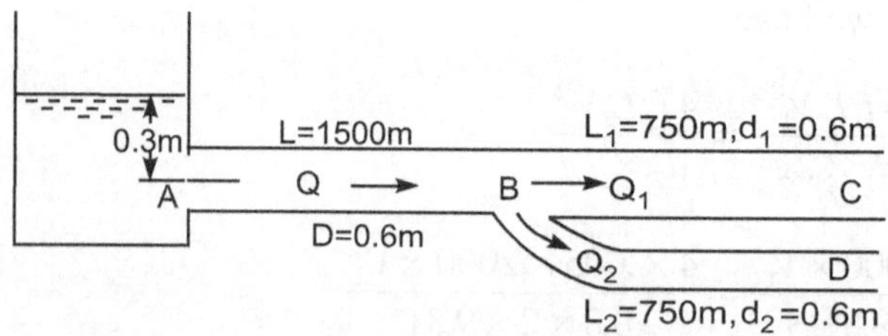

Fig. 9.18

1st Case. Discharge for a single pipe of length 1500 m and dia. = 0.6 m.

This head lost due to friction in single pipe is $h_f = \dfrac{4fLV^{*2}}{d \times 2g}$

where V^* = velocity of flow for single pipe

or $$0.3 = \frac{4 \times .01 \times 1500 \times V^{*2}}{0.6 \times 2g}$$

$\therefore$ $$V^* = \sqrt{\frac{0.3 \times 0.6 \times 2 \times 9.81}{4 \times .01 \times 1500}} = 0.2426 \text{ m/s}$$

$\therefore$ Discharge, $Q^* = V^* \times \text{Area} = 0.2426 \times \frac{\pi}{4}(.6)^2 = 0.0685 \text{ m}^3/\text{s}$ ...(*i*)

2nd Case. When an addition pipe of length 750 m and diameter 0.6 m is connected in parallel with the last half length of the pipe.

Let Q_1 = discharge in 1st parallel pipe

Q_2 = discharge in 2nd parallel pipe

$\therefore$ $Q = Q_1 + Q_2$

where Q = discharge in main pipe when pipes are parallel.

But as the length and diameters of each parallel pipe is same

$\therefore$ $Q_1 = Q_2 = Q/2$

Consider the flow through pipe *ABC* or *ABD*

Head loss through *ABC* = Head lost through *AB* + head lost through *BC* ...(*ii*)

But head lost due to friction through *ABC* = 0.3 m given

Head lost due to friction through $AB = \frac{4 \times f \times 750 \times V^2}{0.6 \times 2 \times 9.81}$, where V = velocity of flow through *AB*

$$= \frac{Q}{\text{Area}} = \frac{Q}{\frac{\pi}{4}(0.6)^2} = \frac{40}{\pi \times .36}$$

$\therefore$ Head lost due to friction through *AB*

$$= \frac{4 \times .01 \times 750}{0.6 \times 2 \times 9.81} \times \left(\frac{4Q}{\pi \times .36}\right)^2 = 31.87\ Q^2$$

Head lost due to friction through *BC*

$$= \frac{4 \times f \times L_1 \times V_1^2}{d \times 2g}$$

$$= \frac{4 \times .01 \times 750}{0.6 \times 2 \times 9.81} \times \left[\frac{Q}{2 \times \frac{\pi}{4}(.6)^2}\right] \left[\because V_1 = \frac{\text{Distance}}{\frac{\pi}{4}(.6)^2} = \frac{Q}{2 \times \frac{\pi}{4} \times (.6)^2}\right]$$

$$= \frac{4 \times .01 \times 750}{0.6 \times 2 \times 9.81} \times \frac{16}{4 \times \pi^2 \times .36^2} Q^2 = 7.969\ Q^2$$

Substituting these values in equation (*ii*), we get

$$0.3 = 31.87\ Q^2 + 7.969\ Q^2 = 39.839\ Q^2$$

$$\therefore \quad Q = \sqrt{\frac{0.3}{39.839}} = 0.0867 \text{ m}^3/\text{s}$$

$\therefore$ Increase in discharge = $Q - Q^* = 0.0867 - 0.0685 =$ **0.0182 m³/s. Ans.**

Problem 9.34 *A pumping plant forces water through a 600 mm diameter main, the friction head being 27 m. In order to reduce the power consumption, it is proposed to lay another main of appropriate diameter along the side of the existing one, so that two pipes may work in parallel for the entire length and reduce the friction head to 9.6 m only. Find the diameter of the new main if, with the exception of diameter, it is similar to the existing one in every respect.* (A.M.I.E., Winter, 1974)

Solution. Given :

Dia. of single main pipe, $d = 600$ mm $= 0.6$ m

Friction head, $h_f = 27$ m

Friction head for two parallel pipes $= 9.6$ m

1st Case.

For a single main [Fig. 9.19 (*a*)]

$$h_f = \frac{4 . f . L . V^2}{d \times 2g} \text{ or } 27.0 = \frac{4 \times f \times L \times V^2}{0.6 \times 2 \times 9.81}$$

$$\therefore \quad fLV^2 = \frac{27.0 \times 0.6 \times 2 \times 9.81}{4} = \frac{317.844}{5} = 79.461, \text{ where } V = \frac{Q}{A}$$

$$\therefore \quad f.L. \frac{Q^2}{A^2} = 79.461 \qquad ...(i)$$

2nd Case. Two pipes are in parallel [Fig. 9.19 (*b*)]

Loss of head in any one pipe = 9.6 m

$$\therefore \text{ For 1st pipe, } \quad h_{f_1} = \frac{4 . f . L . V_1^2}{d_1 \times 2g} = 9.6$$

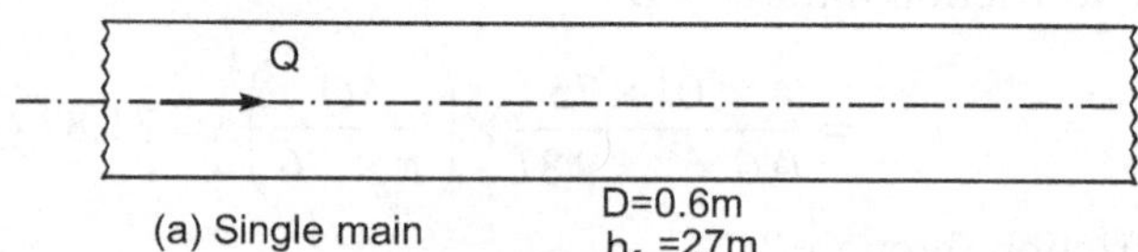

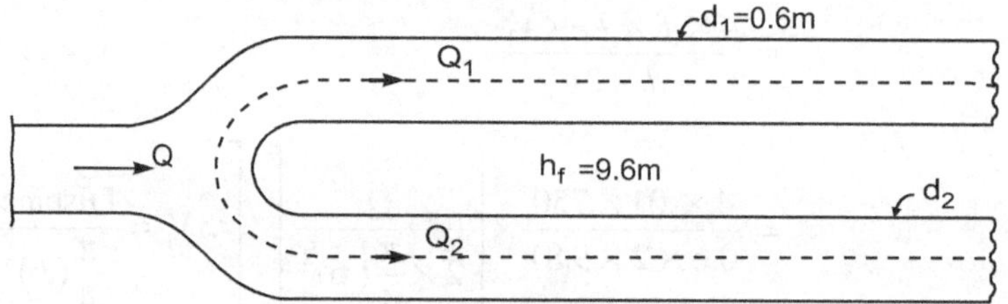

Fig. 9.19

But

$$L_1 = L, V_1 = \frac{Q_1}{A_1} = \frac{Q_1}{A} \qquad \left[\because A_1 = A = \frac{\pi}{4}(.6)^2\right]$$

$$d_1 = d = 0.6$$

$$\therefore \quad \frac{4 . f . L}{0.6 \times 2 \times 9.81} \frac{Q_1^2}{A^2} = 0.6$$

or

$$f \times L \times \frac{Q_1^2}{A^2} = \frac{9.6 \times 0.6 \times 2 \times 9.81}{4} = 28.2528 \qquad ...(ii)$$

For the 2nd pipe,

$$h_{f_2} = \frac{4f \times L_2 \times V_2^2}{d_2 \times 2g} = 9.6, \quad \text{where } L_2 = L, V_2 = \frac{Q_2}{A_2}$$

$$\therefore \quad \frac{4f \times L \times Q_2^2}{d_2 \times 2g \times A_2^2} = 9.6$$

or
$$\frac{f \times L \times Q_2^2}{d_2 \times A_2^2} = \frac{9.6 \times 2 \times 9.81}{4} = 47.088 \quad \text{...(iii)}$$

Dividing equation (*i*) by equation (*ii*), we get

$$\frac{Q^2}{Q_1^2} = \frac{79.416}{28.2528} = 2.8125$$

$$\therefore \quad \frac{Q}{Q_1} = \sqrt{2.8125} = 1.667$$

$$\therefore \quad Q_1 = \frac{Q}{1.667} = .596\ Q$$

But $\quad Q_1 + Q_2 = Q$

$$\therefore \quad Q_2 = Q - Q_1 = Q - .596\ Q = 0.404\ Q$$

Dividing equation (*ii*) by equation (*iii*),

$$\frac{Q_1^2 \times d_2 \times A_2^2}{A^2 \times Q_2^2} = \frac{28.2528}{47.088} = 0.6$$

But $\quad A_2 = \frac{\pi}{4} d_2^2 \text{ and } A = \frac{\pi}{4} d^2 = \frac{\pi}{4} (.6)^2 = \frac{\pi}{4} \times .36$

$$\therefore \quad \frac{Q_1^2}{Q_2^2} \times \frac{d_2 \times \left(\frac{\pi}{4}\right)^2 \times d_2^4}{\left(\frac{\pi}{4}\right)^2 \times (.36)^2} = 0.6 \quad \text{or} \quad \left(\frac{.596\ Q}{.404\ Q}\right)^2 \times \frac{d_2^5}{.36^2} = 0.6$$

or
$$d_2^5 = 0.6 \times .36^2 \times \left(\frac{.404}{.596}\right)^2 = 0.03537$$

$$\therefore \quad d_2 = (.03537)^{1/5} = 0.5125 \text{ m} = \mathbf{512.5\ mm.\ Ans.}$$

Problem 9.35 *A pipe of diameter 20 cm and length 2000 m connects two reservoirs, having difference of water levels as 20 m. Determine the discharge through the pipe.*

If an additional pipe of diameter 20 cm and length 1200 m is attached to the last 1200 m length of the existing pipe, find the increase in the discharge. Take f = .015 and neglect minor losses.

Solution. Given :

Dia. of pipe, $\quad d = 20 \text{ cm} = 0.20 \text{ m}$

Length of pipe, $\quad L = 2000 \text{ m}$

Difference of water levels, $H = 20 \text{ m}$

Co-efficient of friction, $\quad f = 0.015$

1st Case. When a single pipe connects the two reservoirs

$$H = \frac{4 . f . L . V^2}{d \times 2g} = \frac{4f . L}{d \times 2g} \left(\frac{Q}{\frac{\pi}{4} d^2} \right)^2 \qquad \left[\because V = \frac{Q}{\frac{\pi}{4} d^2} \right]$$

$$= \frac{32 f . L . Q^2}{\pi^2 \times g \times d^5}$$

or

$$20 = \frac{32 \times .015 \times 2000 \times Q^2}{\pi^2 \times 9.81 \times (0.2)^5} = 30985.07\ Q^2$$

$\therefore$

$$Q = \sqrt{\frac{20}{30985.07}} = \mathbf{0.0254\ m^3/s.\ Ans.}$$

2nd Case.

Let Q_1 = discharge through pipe *CD*,

Q_2 = discharge through pipe *DE*,

Q_3 = discharge through pipe *DF*

Length of pipe *CD*, $L_1 = 800$ m and its dia., $d_1 = 0.20$ m

Length of pipe *DE*, $L_2 = 1200$ m and its dia., $d_2 = 0.20$ m

Length of pipe *DF*, $L_3 = 1200$ m and its dia., $d_3 = 0.20$ m.

Since the diameters and lengths of the pipes *DE* and *DF* are equal. Hence Q_2 will be equal to Q_3. Also for parallel pipes, we have

$$Q_1 = Q_2 + Q_3 = Q_2 + Q_2 = 2Q_2 \qquad [\because Q_2 = Q_3]$$

$\therefore$

$$Q_2 = \frac{Q_1}{2}$$

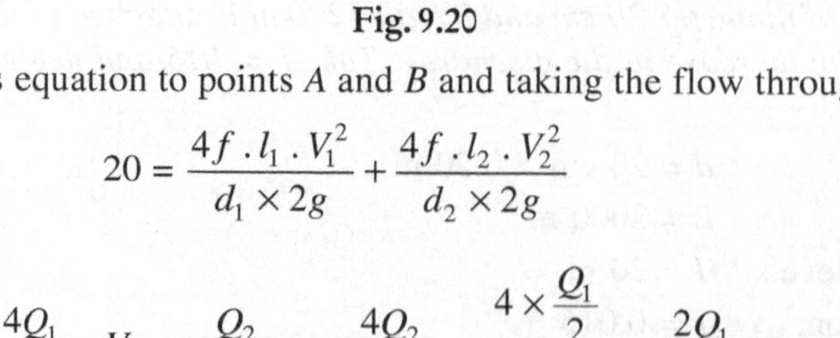

Fig. 9.20

Applying Bernoulli's equation to points *A* and *B* and taking the flow through *CDE*, we have

$$20 = \frac{4f . l_1 . V_1^2}{d_1 \times 2g} + \frac{4f . l_2 . V_2^2}{d_2 \times 2g}$$

where $V_1 = \dfrac{Q_1}{\frac{\pi}{4}(.2)^2} = \dfrac{4Q_1}{\pi \times .04}, V_2 = \dfrac{Q_2}{\frac{\pi}{4}(.2)^2} = \dfrac{4Q_2}{\pi \times .04} = \dfrac{4 \times \frac{Q_1}{2}}{\pi \times .04} = \dfrac{2Q_1}{\pi \times .04}$

$$= \frac{4 \times .015 \times 800}{0.2 \times 2 \times 9.81} \times \left(\frac{4Q_1}{\pi \times .04}\right)^2 + \frac{4 \times .015 \times 1200}{0.2 \times 2 \times 9.81} \times \left(\frac{2Q_1}{\pi \times .04}\right)^2$$

$$= 12394\ Q_1{}^2 + 4647\ Q_1{}^2 = 17041\ Q_1{}^2$$

$$\therefore \qquad Q_1 = \sqrt{\frac{20}{17041}} = 0.0342 \text{ m}^3\text{/s}$$

Increase in discharge = $Q_1 - Q = 0.0342 - 0.0254 =$ **.0088 m³/s. Ans.**

Problem 9.36 *Two pipes have a length l each. One of them has a diameter D, and the other a diameter d. If the pipes are arranged in parallel the loss of head when a total quantity of water Q flows through them is h, but, if the pipes are arranged in series and the same quantity Q flows through them, the loss of head is H. If* $d = \frac{D}{2}$, *find the ratio of H to h, neglecting secondary losses and assuming the pipe co-efficient f has a constant value.* (A.M.I.E., Summer, 1981)

Solution. Given :

Length of pipe 1, $L_1 = L$ and its dia. $d_1 = D$

Length of pipe 2, $L_2 = L$ and its dia., $d_2 = d$

Total discharge $= Q$

Head loss when pipes are arranged parallel = h

Head loss when pipes are arranged in series = H

$$d = \frac{D}{2} \text{ and } f \text{ is constant}$$

1st Case. When pipes are connected to parallel

$$Q = Q_1 + Q_2 \qquad \text{...(i)}$$

Loss of head in each pipe = h

For pipe AB, $$\frac{4fL_1V_1^2}{d_1 \times 2g} = h, \text{ where } V_1 = \frac{Q_1}{A_1} = \frac{Q_1}{\frac{\pi}{4}D^2} = \frac{4Q_1}{\pi D^2}$$

$$d_1 = D$$

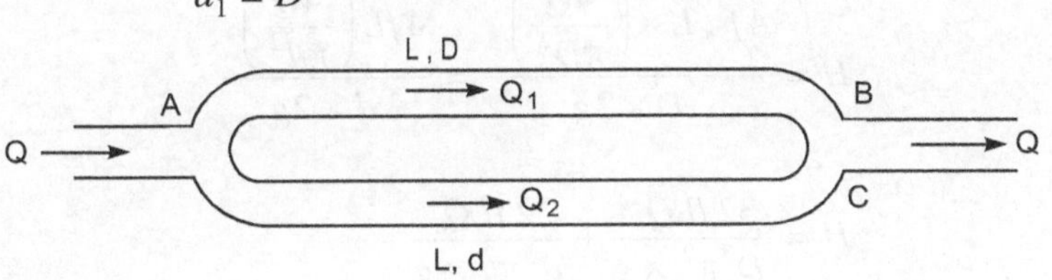

Fig. 9.21

$$\therefore \qquad \frac{4fL \times \left(\frac{4Q_1}{\pi D^2}\right)^2}{D \times 2g} = h \text{ or } \frac{32fLQ_1^2}{\pi^2 D^5 \times g} = h \qquad \text{...(ii)}$$

For pipe AC, $$\frac{32fLQ_2^2}{\pi^2 d^5 \times g} = h \qquad \text{...(iii)}$$

$$\therefore \quad \frac{32 fLQ_1^2}{\pi^2 D^5 g} = \frac{32 fLQ_2^2}{\pi^2 d^5 g} \text{ or } \frac{Q_1^2}{D^5} = \frac{Q_2^2}{d^5}$$

or
$$\left(\frac{Q_1}{Q_2}\right)^2 = \frac{D^5}{d^5} = \frac{(2d)^5}{d^5} \qquad [\because D = 2d]$$

$$= 2^5 = 32$$

$$\therefore \quad \frac{Q_1}{Q_2} = \sqrt{32} = 5.657 \text{ or } Q_1 = 5.657\ Q_2$$

Substituting the values of Q_1 in equation (*i*), we get

$$Q = 5.657\ Q_2 + Q_2 = 6.657\ Q_2$$

$$\therefore \quad Q_2 = \frac{Q}{6.657} = 0.15\ Q \qquad \text{...(iv)}$$

From (*i*) ∴
$$Q_1 = Q - Q_2 = Q - 0.15\ Q = 0.85\ Q \qquad \text{...(v)}$$

2nd Case. When the pipes are connected in series.

Total loss = Sum of head losses in the two pipes

$$\therefore \quad H = \frac{5f \cdot L \cdot V_1^2}{d_1 \times 2g} + \frac{4f \cdot L \cdot V_2^2}{d_2 \times 2g}$$

where $V_1 = \frac{Q}{\frac{\pi}{4} D^2} = \frac{4Q}{\pi D^2}$, $V_2 = \frac{Q}{\frac{\pi}{4} d^2} = \frac{4Q}{\pi d^2}$

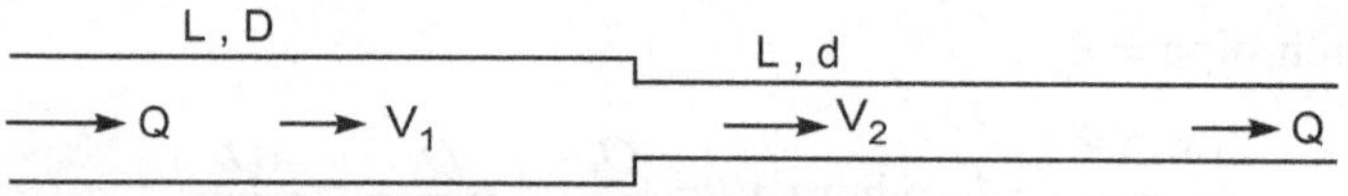

Fig. 9.22

$$H = \frac{4f \cdot L \times \left(\frac{4Q}{\pi D^2}\right)^2}{D \times 2g} = \frac{4fL\left(\frac{4Q}{\pi d^2}\right)^2}{d \times 2g}$$

or
$$H = \frac{32 fLQ^2}{D^5 \pi^2 \times g} + \frac{32 fLQ^2}{d^5 \pi^2 \times g} \qquad \text{...(vi)}$$

From equation (*ii*), $\frac{32 fL}{\pi^2 D^5 \times g} = \frac{h}{Q_1^2}$

and from equation (*iii*), $\frac{32 fL}{\pi^2 d^5 \times g} = \frac{h}{Q_2^2}$

Substituting these values in equation (*iv*), we have

$$H = Q^2 \times \frac{h}{Q_1^2} + Q^2 \times \frac{h}{Q_2^2} = \frac{Q^2}{Q_1^2}h + \frac{Q^2}{Q_2^2}h = h\left[\frac{Q^2}{Q_1^2} + \frac{Q^2}{Q_2^2}\right]$$

$$\therefore \quad \frac{H}{h} = \frac{Q^2}{Q_1^2} + \frac{Q^2}{Q_2^2}$$

But from equations (*iv*) and (*v*), $Q_1 = .85\ Q$ and $Q_2 = 0.15\ Q$

$$\therefore \quad \frac{H}{h} = \frac{Q^2}{.85^2 Q^2} + \frac{Q^2}{.15^2 Q^2} = \frac{1}{.85^2} + \frac{1}{15^2} = 1.384 + 44.444 = \mathbf{45.828.\ Ans.}$$

Problem 9.36A *Three pipes of the same length L, diameter D, and friction factor f are connected in parallel. Determine the diameter of the pipe of length L and friction factor f which will carry the same discharge for the same head loss. Use the formula* $h_f = f \times L \times V^2/2g\ D$.

(A.M.I.E., Winter, 1990)

Solution. Given :

Length of each pipe $= L$

Diameter of each pipe $= D$

Friction factor of each pipe $= f$

Head loss, $h_f = f \times L \times V^2/2gD$

When the three pipes are connected in parallel, then head loss in each pipe will be same. And total head loss will be equal to the head loss in each pipe.

Let h_f = Total head loss,

h_{f_1} = Head loss in 1st pipe,

h_{f_2} = Head loss in 2nd pipe, and h_{f_3} = Head loss in 3rd pipe.

Then $$h_f = h_{f_1} = h_{f_2} = h_{f_3} \text{ or } h_f = \frac{f \times L \times V^2}{2gD} \quad ...(i)$$

Let Q_1 = Discharge through 1st pipe, Q_2 = Discharge through 2nd pipe,

Q_3 = Discharge through 3rd pipe, and Q = Total discharge.

When the three pipes are connected in parallel, then

$$Q = Q_1 + Q_2 + Q_3 = 3 \times Q_1 \qquad (\because Q_1 = Q_2 = Q_3)$$

$$= 3 \times A_1 \times V_1$$

$$= 3 \times \frac{\pi}{4} D^2 \times V \left(\text{where } A_1 = \frac{\pi}{4} D^2 \text{ and } V_1 = V\right) \quad ...(ii)$$

For a single pipe (or length L ; friction factor f) which will carry same discharge as the three pipes in parallel

Let d = dia. of the single pipe

v = velocity through single pipe

Then discharge, $$Q = \text{Area} \times \text{Velocity} = \left(\frac{\pi}{4} d^2\right) \times v \quad ...(iii)$$

Equating the two valves of discharge, given by equations (*ii*) and (*iii*), we get

$$3 \times \frac{\pi}{4} D^2 \times V = \frac{\pi}{4} d^2 \times v \text{ or } 3 \times \frac{D^2}{d^2} = \frac{v}{V} \quad ...(iv)$$

The head loss for the single pipe is also equal to the total head loss for three pipes when they are in parallel.

But head loss for the single pipe of length L, dia. d, friction factor f and velocity v is given by

$$h_f = \frac{f \times L \times v^2}{d \times 2g} \quad \text{...}(v)$$

Equating the two values of h_f given by equations (i) and (v), we get

$$\frac{f \times L \times V^2}{D \times 2g} = \frac{f \times L \times v^2}{d \times 2g} \text{ or } \frac{V^2}{D} = \frac{v^2}{d}$$

or $$\frac{d}{D} = \frac{v^2}{V^2} \text{ or } \left(\frac{d}{D}\right)^{1/2} = \frac{v}{V}$$

Substituting the value of v/V in equation (iv), we get

$$3 \times \frac{D^2}{d^2} = \left(\frac{d}{D}\right)^{1/2} \text{ or } 3 = \left(\frac{d}{D}\right)^{1/2} \times \left(\frac{d}{D}\right)^2 = \left(\frac{d}{D}\right)^{5/2}$$

or $$\frac{d}{D} = 3^{2/5} = 3^{0.4} = 1.55$$

$\therefore$ $$d = \mathbf{1.55\ D.\ Ans.}$$

Hence dia. of single pipe should be 1.55 times the dia. of the three pipes connected in parallel.

Problem 9.37 *For a town water supply, a main pipe line of diameter 0.4 m is required. As pipes more than 0.35 m diameter are not readily available, two parallel pipes of the same diameter were used for water supply. If the total discharge in the parallel pipes is same as in the single main pipe, find the diameter of the parallel pipe. Assume the co-efficient of friction same for all pipes.*

Solution. Given :

Dia. of single main pipe line, $d = 0.4$ m

Let the length of single pipe line $= L$

Co-efficient of friction $= f$

Loss of head due to friction in single pipe $= \frac{4fLV^2}{d \times 2g} = \frac{4fLV^2}{0.4 \times 2 \times g}$ ...(i)

where V = Velocity of flow in the single pipe.

In case of parallel pipe, as the diameters and lengths of the two pipes are same. Hence discharge in each pipe will be half the discharge of single main pipe. As discharge in each parallel pipe is same, hence velocity will also be same.

Let V_* = Velocity in each parallel pipe

d_* = Dia. of each parallel pipe

Then loss of head due to friction in parallel pipes $= \frac{4f \times L \times V_*^2}{d_* \times 2g}$ ...(ii)

Equating the two losses given by equations (i) and (ii), we have

$$\frac{4f\,.\,L\,.\,V^2}{0.4 \times 2g} = \frac{4f \times L \times V_*^2}{d_* \times 2g}$$

Cancelling $\frac{4fL}{2g}$, $$\frac{V^2}{0.4} = \frac{V_*^2}{d_*^2} \text{ or } \frac{V^2}{V_*^2} = \frac{0.4}{d_*} \qquad ...(iii)$$

From continuity

Total flow in single main = sum of flow in two parallel pipes

or Velocity of main × Area = 2 × Velocity in each parallel pipe × Area

$$V \times \frac{\pi}{4}(0.4)^2 = 2 \times V_* \times \frac{\pi}{4} d^2{}_* \text{ or } \frac{V}{V_*} = \frac{2 \times \frac{\pi}{4} d_*^2}{\frac{\pi}{4}(0.4)^2} = \frac{2d_*^2}{0.16}$$

Squaring both sides, $$\frac{V^2}{V_*^2} = \frac{4d_*^4}{0.0256} \qquad ...(iv)$$

Comparing equations (*iii*), and (*iv*), we get

$$\frac{0.4}{d_*} = \frac{4d_*^4}{.0256} \text{ or } d_*^5 = \frac{0.4 \times .0256}{4} = .00256$$

$\therefore$ $$d_* = (.00256)^{1/5} = 0.303 \text{ m} = \mathbf{30.3 \text{ cm. Ans.}}$$

$\therefore$ Use two pipes of 30.3 cm diameter.

Problem 9.38 *An old water supply distribution pipe of 250 mm diameter of a city is to be replaced by two parallel pipes of smaller equal diameter having equal lengths and identical friction factor values. Find out the new diameter required.* (A.M.I.E., Summer, 1985)

Solution. Given :

Dia. of old pipe, D = 250 mm = 0.25 m

Let d = Dia. of each of parallel pipes

Q = Discharge in old pipe

Q_1 = Discharge in first parallel pipe

Q_2 = Discharge in second parallel pipe

f = Friction factor.

When a single pipe is replaced by two parallel pipes, the head loss will be same in the single pipe and in each of the parallel pipes. Also the discharge in single pipe will be equal to the total discharge in two parallel pipes *i.e.*,

$$h_f = h_{f_1} = h_2 \qquad ...(i)$$

and $$Q = Q_1 + Q_2 \qquad ...(ii)$$

As the dia. of each parallel pipe is same and also length of each parallel pipe is equal, hence

$$Q_1 = Q_2 \text{ or } Q_1 = Q_2 = Q/2$$

Now h_f = Head loss in single pipe

$$= \frac{f \times L \times V^2}{D \times 2g}, \qquad \text{where } f = \text{Friction factor}$$

$$= \frac{f \times L \times \left(\frac{Q}{\frac{\pi}{4} \times 0.25^2}\right)^2}{0.25 \times 2 \times 9.81} \qquad \left(\because V = \frac{Q}{\text{Area}} = \frac{Q}{\frac{\pi}{4}D^2}\right)$$

$$= \frac{f \times L \times (4Q)^2}{0.25 \times 2 \times 9.81 \times \left(\pi \times 0.25^2\right)^2} \quad \text{...(iii)}$$

$$h_{f_1} = \text{Head loss in 1st parallel pipe}$$

$$= \frac{f \times L \times (4Q)^2}{d \times 2g}$$ (∵ Dia. of parallel pipe = d and V_1 is the velocity in 1st parallel pipe)

$$= \frac{f \times L \times \left(\dfrac{Q}{2 \times \dfrac{\pi}{4} d^2}\right)^2}{d \times 2g} \qquad \left(\because \; V_1 = \frac{Q_1}{A_1} = \frac{\dfrac{Q}{2}}{\dfrac{\pi}{4} d^2} \text{ as } Q_1 = \frac{Q}{2} \right)$$

$$= \frac{f \times L \times (4Q)^2}{d \times 2 \times 9.81 \times \left(2 \times \pi \times d^2\right)^2} \quad \text{...(iv)}$$

But $h_f = h_{f_1}$

or $$\frac{f \times L \times (4Q)^2}{0.25 \times 2 \times 9.81 \times (\pi \times 0.25)^2} = \frac{f \times L \times (4Q)^2}{d \times 2 \times 9.81 \times \left(2 \times \pi \times d^2\right)^2}$$

or $$d \times (2\pi d^2)^2 = 0.25 \times (\pi \times 0.25^2)^2$$

or $$d \times 4 \times d^4 = 0.25 \times 0.25^4$$

or $$d^5 = \frac{0.25^5}{4} \text{ or } d = \frac{0.25}{(4)^{1/5}} = \frac{0.25}{1.3195} = \mathbf{0.1894 \text{ m} \simeq 0.19 \text{ m. Ans.}}$$

Problem 9.39 *A pipe of diameter 0.4 m and of length 2000 m is connected to a reservoir at one end. The other end of the pipe is connected to a junction from which two pipes of lengths 1000 m and diameter 300 mm run in parallel. These parallel pipes are connected to another reservoir, which is having level of water 10 m below the water level of the above reservoir. Determine the total discharge if f = 0.015. Neglect minor losses.*

Solution. Given :

Dia. of pipe, $d = 0.4$ m

Length of pipe, $L = 2000$ m

Dia. of parallel pipes, $d_1 = d_2 = 300$ mm $= 0.30$ m

Length of parallel pipes, $L_1 = L_2 = 1000$ m

Difference of water level in two reservoir, $H = 10$ m, $f = .015$

Applying Bernoulli's equation to points E and F. Taking flow through ABC.

$$10 = \frac{4fLV^2}{d \times 2g} + \frac{4f \times L_1 \times V_1^2}{d_2 \times 2g}$$

$$= \frac{4 \times .015 \times 2000 \times V^2}{0.4 \times 2 \times 9.81} + \frac{4 \times .015 \times 1000 \times V_1^2}{0.3 \times 2 \times 9.81}$$

$$= 15.29 \, V^2 + 10.19 \, V_1^2 \quad \text{...(i)}$$

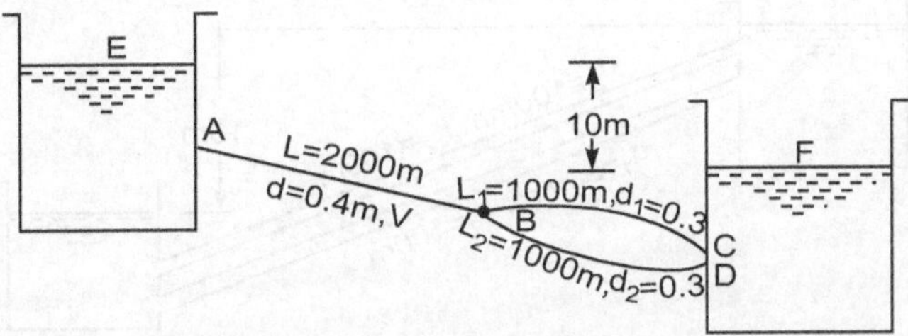

Fig. 9.23

From continuity equation

Discharge through AB = discharge through BC + discharge through BD

or
$$\frac{\pi}{4}d^2 \times V = \frac{\pi}{4}d_1^{\,2} \times V_1^{\,2} + \frac{\pi}{4}d_1^{\,2}V_2$$

But $d_1 = d_2$ and also the lengths of pipes BC and BD are equal and hence discharge through BC and BD will be same. This means $V_1 = V_2$ also

$$\therefore \quad \frac{\pi}{4}d^2V = \frac{\pi}{4}d_1^{\,2} \times V_1 + \frac{\pi}{4}d_1^{\,2} \times V_1 \qquad [\because d_1 = d_2,\ V_1 = V_2]$$

$$= 2 \times \frac{\pi}{4}d_1^{\,2} \times V_2 \text{ or } d^2V = 2d_1^{\,2}V_1$$

or
$$(0.4)^2 \times V = 2 \times (0.3)^2 V_1 \text{ or } .16V = 0.18V_1$$

$$\therefore \quad V_1 = \frac{0.16}{0.18}V = 0.888V$$

Substituting this value of V_1 in equation (*i*), we get

$$10 = 15.29V^2 + (10.19)(.888)^2V^2 = 15.29V^2 + 8.035V^2 = 23.325V^2$$

$$\therefore \quad V = \sqrt{\frac{10}{23.325}} = 0.654 \text{ m/s}$$

$\therefore$ Discharge $= V \times \text{Area}$

$$= 0.654 \times \frac{\pi}{4}d^2 = 0.654 \times \frac{\pi}{4}(0.4)^2 = \mathbf{.0822\ m^3/s.\ Ans.}$$

Problem 9.40 *Two sharp ended pipes of diameters 50 mm and 100 mm respectively, each of length 100 m are connected in parallel between two reservoirs which have a difference of level of 10 m. If the co-efficient of friction for each pipe is (4f) 0.32, calculate the rate of flow for each pipe and also the diameter of a single pipe 100 m long which would give the same discharge, if it were substituted for the original two pipes.* (A.M.I.E., Winter 1982, Summer 1987)

Solution. Given :

Dia. of 1st pipe, $d_1 = 50$ mm $= 0.05$ m

Length of 1st pipe, $L_1 = 100$ m

Dia. of 2nd pipe, $d_2 = 100$ mm $= 0.10$ m

Length of 2nd pipe, $L_2 = 100$ m

Difference in level in reservoirs, $H = 10$ m

Co-efficient of friction $4f = 0.32$

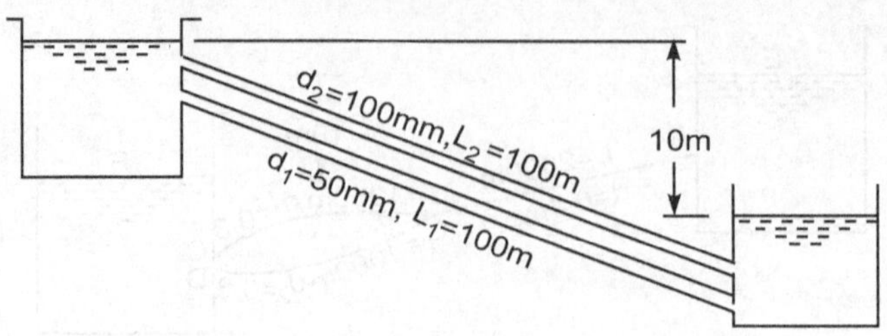

Fig. 9.24

Let V_1 = velocity of flow in pipe 1, and
V_2 = velocity of flow in pipe 2.

When the pipes are connected in parallel, the loss of head will be same in both the pipes.

For the first pipe, loss of head is given as

$$H = \frac{4f \times L_1 \times V_1^2}{d_1 \times 2g} = \frac{0.32 \times 100 \times V_1^2}{0.05 \times 2 \times 9.81} \qquad (\because 4f = .32)$$

or
$$10 = 32.619\ V_1^2$$

$$\therefore \qquad V_1 = \sqrt{\frac{10}{32.619}} = 0.5535 \text{ m/s}$$

$\therefore$ Rate of flow in 1st pipe, $Q_1 = V_1 \times A_1 = 0.5536 \times \frac{\pi}{4}(d_1)^2$

$$= .5536 \times \frac{\pi}{4}(0.05)^2 = .001087 \text{ m}^3\text{/s} = \textbf{1.087 litres/s. Ans.}$$

For the 2nd pipe, loss of head is given by,

$$10 = H = \frac{4f \times L_2 \times V_2^2}{d_2 \times 2g} = \frac{0.32 \times 100 \times V_2^2}{0.10 \times 2 \times 9.81}$$

$$\therefore \qquad V_2 = \sqrt{\frac{10 \times .10 \times 2 \times 9.81}{.32 \times 100}} = 0.783 \text{ m/s}$$

$\therefore$ Rate of flow in 2nd pipe, $Q_2 = A_2 \times V_2 = \frac{\pi}{4} d_2^2 \times V_2$

$$= \frac{\pi}{4}(.1)^2 \times .783 = 0.00615 \text{ m}^3\text{/s} = \textbf{6.15 litres/s. Ans.}$$

Let D = diameter of a single pipe which is substituted for the two original pipes
L = length of single pipe = 100 m
V = velocity through pipe

The discharge through single pipe,

$$Q = Q_1 + Q_2 = 1.087 + 6.15 = 7.237 \text{ litres/s} = .007237 \text{ m}^3\text{/s}$$

$$\therefore \qquad V = \frac{Q}{\text{Area}} = \frac{.007237}{\frac{\pi}{4}D^2} = \frac{4 \times .007237}{\pi D^2} = \frac{.009214}{D^2} \text{ m/s}$$

Loss of head through single pipe is

$$H = \frac{4f \times L \times V^2}{D \times 2g} = \frac{0.32 \times 100 \times \left(\frac{.009214}{D^2}\right)^2}{D \times 2 \times 9.81}$$

or $$10.0 = \frac{.32 \times 100 \times .009214^2}{2 \times 9.81 \times D^5} = \frac{.0001384}{D^5}$$

or $$D^5 = \frac{.0001384}{10} = .00001384$$

$\therefore$ $$D = (.00001384)^{1/5} = 0.1067 \text{ m} = \mathbf{106.7 \text{ mm. Ans.}}$$

Problem 9.41 *Two reservoir are connected by a pipe line of diameter 600 mm and length 4000 m. The difference of water level in the reservoirs is 20 m. At a distance of 1000 m from the upper reservoir, a small pipe is connected to the pipe line. The water can be taken from the small pipe. Find the discharge to the lower reservoir, if*

(i) No water is taken from the small pipe, and

(ii) 100 litres/s of water is taken from small pipe.

Take f = .005 and neglect minor losses.

Solution. Given :

Dia. of pipe, $d = 600 \text{ mm} = 0.60 \text{ m}$

Length of pipe, $L = 400 \text{ m}$

Difference of water level, $H = 20 \text{ m}, f = .005$

(*i*) **No water is taken from small pipe**

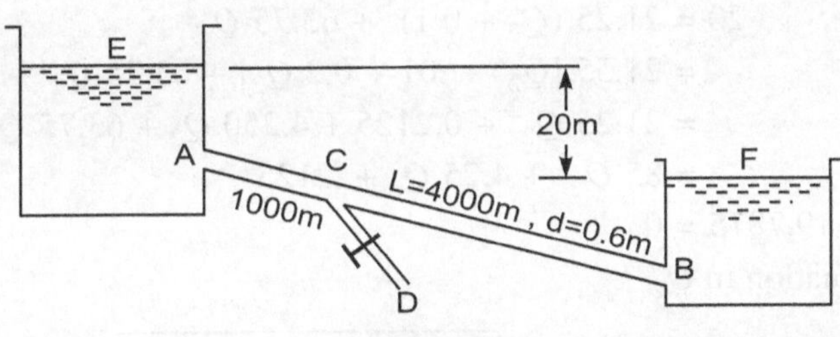

Fig. 9.25

The head loss due to friction in pipe $AB = \frac{4f \times L \times V^2}{d \times 2g}$ or $20 = \frac{4 \times .005 \times 4000 \times V^2}{0.6 \times 2 \times 9.81}$

$\therefore$ $$V = \sqrt{\frac{20 \times 0.6 \times 2 \times 9.81}{4 \times .005 \times 4000}} = \sqrt{2.94} = 1.715 \text{ m/s}$$

$\therefore$ Discharge, $$Q = \text{Area} \times V = \frac{\pi}{4}(0.6)^2 \times 1.715 = \mathbf{0.485 \text{ m}^3/\text{s. Ans.}}$$

(*ii*) **100 litres of water is taken from small pipe**

Let Q_1 = discharge through pipe AC

Q_2 = discharge through pipe CB

Then for parallel pipes $Q_1 = Q_2 + 100 \text{ litres/s} = Q_2 + 0.1 \text{ m}^3/\text{s}$

$\therefore$ $Q_2 = (Q_1 - 0.1) \text{ m}^3/\text{s}$ ...(*i*)

Length of pipe AC, $L_1 = 1000$ m

Length of pipe CB, $L_2 = 4000 - 1000 = 3000$ m

Applying Bernoulli's equation to points E and F and taking flow through ABC, we have

$$20 = \frac{4fL_1V_1^2}{d_1 \times 2g} + \frac{4fL_2V_2^2}{d_2 \times 2g} \quad \text{...(ii)}$$

where V_1 = velocity through pipe $AC = \dfrac{Q_1}{\frac{\pi}{4}(0.6)^2} = \dfrac{4Q_1}{\pi \times .36}$

d_1 = dia. of pipe $AC = 0.6$

V_2 = velocity through pipe $CB = \dfrac{Q_2}{\frac{\pi}{4}(0.6)^2} = \dfrac{4Q_2}{\pi \times .36}$

d_2 = dia. of pipe $CB = 0.6$

Substituting these values in equation (*ii*), we get

$$20 = \frac{4 \times .005 \times 1000}{0.6 \times 2 \times 9.81} \times \left(\frac{4Q_1}{\pi \times .36}\right)^2 + \frac{4 \times .005 \times 3000}{0.6 \times 2 \times 9.81} \times \left(\frac{4Q_2}{\pi \times .36}\right)^2$$

$$20 = 21.25\ Q_1^2 + 63.75\ Q_2^2 \quad \text{...(iii)}$$

But from (*i*), $Q_2 = Q_1 - 0.1$ or $Q_1 = Q_2 + 0.1$

Substituting the value of Q_1 in equation (*iii*), we get

$$20 = 21.25\ (Q_2 + 0.1)^2 + 63.75\ Q_2^2$$
$$= 21.55\ [Q_2^2 + .01 + 0.2\ Q_2] + 63.75\ Q_2^2$$
$$= 21.25\ Q_2^2 + 0.2125 + 4.250\ Q_2 + 63.75\ Q_2^2$$
$$= 85\ Q_2^2 + 4.25\ Q_2 + .2125$$

or $85\ Q_2^2 + 4.25\ Q_2 - 19.7875 = 0$

This is a quadratic equation in Q_2

$$\therefore \quad Q_2 = \frac{-4.25 \pm \sqrt{4.25^2 + 4 \times 85 \times 19.7875}}{2 \times 85}$$

$$= \frac{-4.25 \pm \sqrt{18.0625 + 6727.75}}{170} = \frac{-44.25 \pm 82.13}{170} = \frac{82.13 - 4.25}{170}$$

$$= 0.458\ \text{m}^3/\text{s} \qquad \text{(Neglecting negative root)}$$

$\therefore$ Discharge to lower reservoir = Q_2 = **0.458 m³/s. Ans.**

▶ 9.12 FLOW THROUGH BRANCHED PIPES

When three or more reservoirs are connected by means of pipes, having one or more junctions, the system is called a branching pipe system. Fig. 9.26 shows three reservoirs at different levels connected to a single junction, by means of pipes which are called branched pipes. The lengths, diameters and co-efficient of friction of each pipes is given. It is required to find the discharge and direction of flow in each pipe. The basic equations used for solving such problems are :

1. **Continuity equation** which means the inflow of fluid at the junction should be equal to the outflow of fluid.
2. **Bernoulli's equation, and**
3. **Darcy-Weisbach equation**

Also it is assumed that reservoirs are very larged and the water surface levels in the reservoirs are constant so that steady conditions exist in the pipes. Also minor losses are assumed very small. The flow from reservoir A takes place to junction D. The flow from junction D is towards reservoirs C. Now the flow from junction D towards reservoir B will take place only when piezometric head at D $\left(\text{which is equal to } \frac{p_D}{\rho g} + Z_D\right)$ is more than the piezometric head at B (*i.e.*, Z_B). Let us consider that flow is from D to reservoir B.

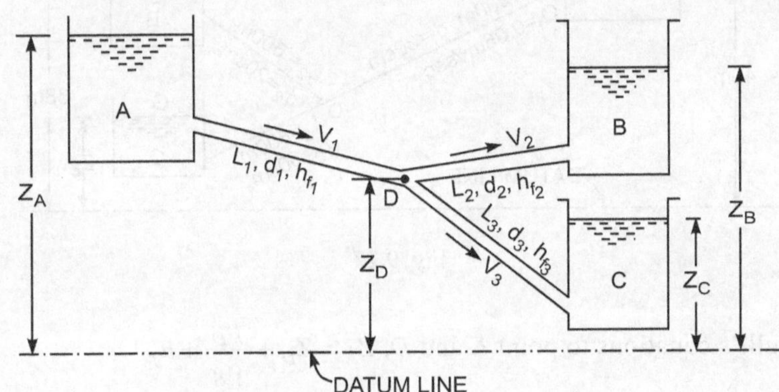

Fig. 9.26

For flow from A to D from Bernoulli's equation

$$Z_A = Z_D + \frac{p_D}{\rho g} + h_{f_1} \qquad \text{...}(i)$$

For flow from D to B from Bernoulli's equation

$$Z_D + \frac{p_D}{\rho g} = Z_B + h_{f_2} \qquad \text{...}(ii)$$

For flow from D to C from Bernoulli's equation

$$Z_D + \frac{p_D}{\rho g} = Z_C + h_{f_3} \qquad \text{...}(iii)$$

From continuity equation,
Discharge through AD = Discharge through DB + Discharge through DC

$$\therefore \qquad \frac{\pi}{4} d_1^2 V_1 = \frac{\pi}{4} d_2^2 \times V_2 + \frac{\pi}{4} d_3^2 V_3$$

or

$$d_1^2 V_1 = d_2^2 V_2 + d_3^2 V_3 \qquad \text{...}(iv)$$

There are four unknown *i.e.*, V_1, V_2, V_3 and $\frac{p_D}{\rho g}$ and there are four equations (*i*), (*ii*), (*iii*) and (*iv*). Hence unknown can be calculated.

Problem 9.42 *Three reservoirs A, B and C are connected by a pipe system shown in Fig. 9.27. Find the discharge into or from the reservoirs B and C if the rate of flow from reservoirs A is 60 litres/s. Find the height of water level in the reservoir C. Take f = .006 for all pipes.*

Solution. Given :

Length of pipe *AD*, $L_1 = 1200$ m

Dia. of pipe *AD*, $d_1 = 30$ cm $= 0.30$ m

Discharge through *AD*, $Q_1 = 60$ litres/s $= 0.06$ m^3/s

Height of water level in *A* from reference line, $Z_A = 40$ m

For pipe *DB*, length $L_2 = 600$ m, dia., $d_2 = 20$ cm $= 0.20$ m, $Z_B = 38.0$

For pipe *DC*, length $L_3 = 800$ m, dia., $d_3 = 30$ cm $= 0.30$ m

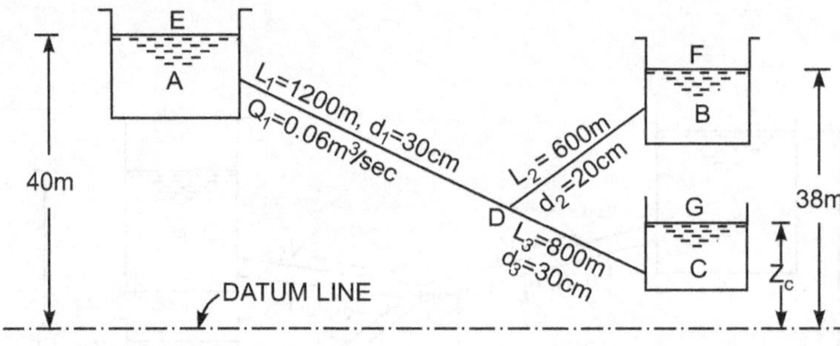

Fig. 9.27

Applying Bernoulli's equations to point *E* and *D*, $Z_A = Z_D + \dfrac{p_D}{\rho g} + h_{f_1}$

where $h_{f_1} = \dfrac{4 \,.\, f \,.\, L_1 \,.\, V_1^2}{d_1 \times 2g}$, where $V_1 = \dfrac{Q_1}{\text{Area}} = \dfrac{0.06}{\frac{\pi}{4}(.3)^2} = 0.848$ m/sec

$$h_f = \frac{4 \times .006 \times 1200 \times .848^2}{0.3 \times 2 \times 9.81} = 3.518 \text{ m}$$

$\therefore$ $$Z_A = Z_D + \frac{p_D}{\rho g} + 3.518 \text{ or } 40.0 = Z_D + \frac{p_D}{\rho g} + 3.518$$

$\therefore$ $$\left(Z_D + \frac{p_1}{\rho g}\right) = 40.0 - 3.518 = 36.482 \text{ m}$$

Hence piezometric head at *D* = 36.482. But $Z_B = 38$ m. Hence water flows from *B* to *D*.

Applying Bernoulli's equation to point *B* and *D*

$$Z_B = \left(Z_D + \frac{p_D}{\rho g}\right) + h_{f_2} \text{ or } 38 = 36.482 + h_{f_2}$$

$\therefore$ $$h_{f_2} = 38 - 36.482 = 1.518 \text{ m}$$

But $$h_{f_2} = \frac{4 \,.\, f \,.\, L_2 \,.\, V_2^2}{d_2 \times 2g} = \frac{4 \times .006 \times 600 \times V_2^2}{0.2 \times 2 \times 9.81}$$

$$\therefore \quad 1.518 = \frac{4 \times .006 \times 600 \times V_2^2}{0.2 \times 2 \times 9.81}$$

$$\therefore \quad V_2 = \sqrt{\frac{1.518 \times 0.2 \times 2 \times 9.81}{4 \times .006 \times 600}} = 0.643 \text{ m/s.}$$

$\therefore$ Discharge,
$$Q_2 = V_2 \times \frac{\pi}{4}(d_2)^2 = 0.643 \times \frac{\pi}{2} \times (.2)^2$$
$$= 0.0202 \text{ m}^3\text{/s} = \textbf{20.2 litres/s. Ans.}$$

Applying Bernoulli's equation to D and C

$$Z_D + \frac{p_D}{\rho g} = Z_C + h_{f_3}$$

or
$$36.482 = Z_C + \frac{4f \cdot L_3 \cdot V_3^2}{d_3 \times 2g}, \text{ where } V_3 = \frac{Q_3}{\frac{\pi}{4} d_3^2}$$

But from continuity $Q_1 + Q_2 = Q_3$

$$\therefore \quad Q_3 = Q_1 + Q_2 = 0.06 + 0.0202 = 0.0802 \text{ m}^3\text{/s}$$

$$\therefore \quad V_3 = \frac{Q_3}{\frac{\pi}{4}(.3)^2} = \frac{0.0802}{\frac{\pi}{4}(.09)} = 1.134 \text{ m/s}$$

$$\therefore \quad 36.482 = Z_C + \frac{4 \times .006 \times 800 \times 1.134^2}{0.3 \times 2 \times 9.81} = Z_C + 4.194$$

$$\therefore \quad Z_C = 36.482 - 4.194 = \textbf{32.288 m. Ans.}$$

Problem 9.43 *Three reservoirs, A, B and C are connected by a pipe system shown in Fig. 9.28. The lengths and diameters of pipes 1, 2 and 3 are 800 m, 1000 m, 800 m, and 300 mm, 200 mm and 150 mm respectively. Determine the piezometric head at junction D. Take f = .005.*

Solution. Given :

The length of pipe 1, $L_1 = 800$ m and its dia., $d_1 = 300$ mm $= 0.3$ m

The length of pipe 2, $L_2 = 1000$ m and its dia., $d_2 = 200$ mm $= 0.2$ m

The length of pipe 3, $L_3 = 800$ m and its dia., $d_3 = 150$ mm $= 0.15$ m

Height of reservoir, A from datum line, $Z_A = 60$ m

Similarly $Z_B = 40$ m and $Z_C = 30$ m.

The direction of flow in pipes are shown (given) in Fig. 9.28. Applying Bernoulli's equation to A and D

$$Z_A = \left(Z_D + \frac{p_D}{\rho g}\right) + h_{f_1}$$

or
$$\left[Z_A - \left(Z_D + \frac{p_D}{\rho g}\right)\right] = h_{f_1} = \frac{4 \times f \times L_1 \times V_1^2}{d_1 \times 2g} = \frac{4 \times .005 \times 800 \times V_1^2}{0.3 \times 2 \times 9.81}$$

or
$$60 - \left(Z_D + \frac{p_D}{\rho g}\right) = 2.718 \, V_1^2 \quad \text{...(i)}$$

Applying Bernoulli's equation to D and B

$$\left(Z_D + \frac{p_D}{\rho g}\right) = Z_B + h_{f_2} = 40 + \frac{4f \times L_2 \times V_2^2}{d_2 \times 2g}$$

$$= 40 + \frac{4 \times .005 \times 1000 \times V_2^2}{0.2 \times 2 \times 9.81} = 40.0 + 5.09\ V_2^2$$

or $$\left(Z_D + \frac{p_D}{\rho g}\right) - 40.0 = 5.09\ V_2^2 \qquad \ldots(ii)$$

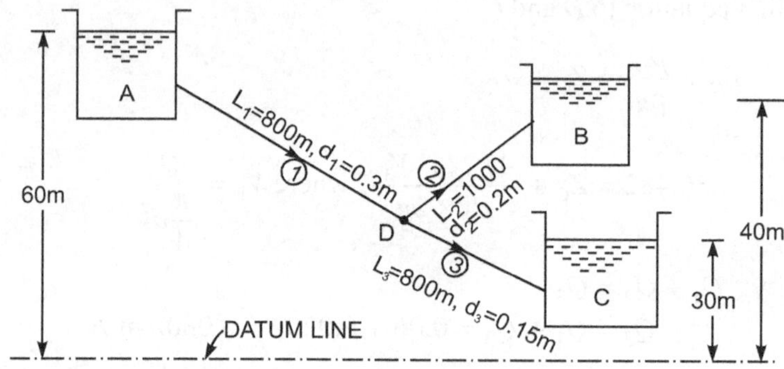

Fig. 9.28

Applying Bernoulli's equation to D and C

$$\left(Z_D + \frac{p_D}{\rho g}\right) = Z_C + h_{f_3} = 30 + \frac{4f \times L_3 \times V_3^2}{d_3 \times 2g} = 30 + \frac{4 \times .005 \times 800 \times V_3^2}{0.15 \times 2 \times 9.81}$$

or $$\left(Z_D + \frac{p_D}{\rho g}\right) = 30.0 - 5.436\ V_3^2 \qquad \ldots(iii)$$

Adding (i) and (ii), we have $60 - 40 = 2.718\ V_1^2 + 5.09\ V_2^2$

or $$20 = 2.718\ V_1^2 + 5.09\ V_2^2 \qquad \ldots(iv)$$

Adding (i) and (ii), we have $60 = 2.718\ V_1^2 + 30.0 + 5.436\ V_3^2$

or $$60 - 30 = 30 = 2.718\ V_1^2 + 5.436\ V_3^2 \qquad \ldots(v)$$

Also from continuity equation, we have

$$Q_1 = Q_2 + Q_3$$

or $$\frac{\pi}{4}\, d_1^2 \times V_1 = \frac{\pi}{4}\, d_2^2 V_2 + \frac{\pi}{4}\, d_3^2 V_3 \quad \text{or} \quad d_1^2 V_1 = d_2^2 V_2 + d_3^2 V_3$$

or $$0.3^2 V_1 = 0.2^2 V_2 + 0.15^2 \times V_3 \quad \text{or} \quad .09 V_1 = .04\ V_2 + .0225\ V_3 \qquad \ldots(vi)$$

Now from (iv), $$V_2 = \sqrt{\frac{20 - 2.718\ V_1^2}{5.09}} \qquad \ldots(vii)$$

And from (v), $$V_3 = \sqrt{\frac{30 - 2.718\ V_1^2}{5.436}} \qquad \ldots(viii)$$

Substituting the value of V_2 and V_3 in (vi), we get

$$0.09\ V_1 = .04\sqrt{\frac{20 - 2.718\ V_1^2}{5.09}} + .0225\sqrt{\frac{30 - 2.718\ V_1^2}{5.09}}$$

Squaring both sides, we get

$$(0.09\ V_1)^2 = (.04)^2 \times \left(\frac{20 - 2.718\ V_1^2}{5.09}\right) + (0.0225)^2 \times \frac{30 - 2.718\ V_1^2}{5.436} + 2 \times .04$$

$$\times .0225 \times \sqrt{\frac{20 - 2.718\ V_1^2}{5.09}} \times \sqrt{\frac{30 - 2.718\ V_1^2}{5.436}}$$

or $\quad .0081\ V_1^{\ 2} = .00628 - .000854\ V_1^{\ 2} + .00279 - .000253\ V_1^{\ 2} + .0018$

or $\quad .0081\ V_1^{\ 2} + .000854\ V_1^{\ 2} + .000253\ V_1^{\ 2} = .00628 + .00279 + .0018 = .01087$

or $\quad .009207\ V_1^{\ 2} = .01087$

$$\therefore \quad V_1 = \sqrt{\frac{.01087}{.009207}} = 1.086 \text{ m/s}$$

Substituting this value of V_1 in (*vii*) and (*viii*)

$$V_2 = \sqrt{\frac{20 - 2.718 \times V_1^2}{5.09}} = \sqrt{\frac{20 - 2.718 \times 1.086^2}{5.09}} = 1.816 \text{ m/s}$$

$$\therefore \quad V_3 = \sqrt{\frac{30 - 2.718 \times 1.086^2}{5.436}} = 2.22 \text{ m/s}$$

Piezometric head at $\quad D = Z_D + \dfrac{p_D}{\rho g} = 30.0 + 5.436 \times V_3^{\ 2}$

$$= 3.00 + 5.426 \times (2.22)^2 = \mathbf{56.79\ m.\ Ans.}$$

Problem 9.44 *A pipe line 60 cm diameter bifurcates at a Y-junction into two branches 40 cm and 30 cm in diameter. If the rate of flow in the main pipe is 1.5 m^3/s and mean velocity of flow is 30 cm diameter pipe is 7.5 m/s, determine the rate of flow in the 40 cm diameter pipe.*

(A.M.I.E., Winter, 1987)

Solution. Given :

Dia. of main pipe, $\quad D = 60$ cm = 0.6 m

Dia. of branch pipe 1, $\quad D_1 = 40$ cm = 0.4 m

Dia. of branch pipe 2, $\quad D_2 = 30$ cm = 0.3 m

Velocity in branch pipe 2, $\quad V_3 = 7.5$ m/s

Rate of flow in main pipe, $\quad Q = 1.5$ m^3/s

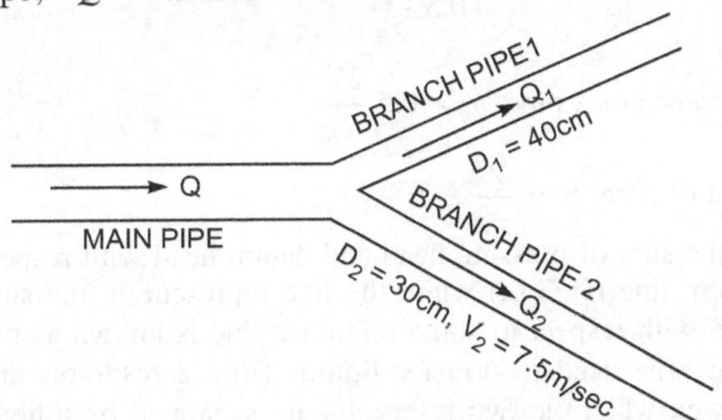

Fig. 9.29

Let Q_1 = Rate of flow in branch pipe 1,
Q_2 = Rate of flow in branch pipe 2,
Q = Rate of flow in main pipe,

Now rate of flow in main pipe is equal to the sum of rate of flow in branch pipes.

$$\therefore \quad Q = Q_1 + Q_2 \quad \ldots(i)$$

But Q_2 = Area of branch pipe 2 × Velocity in branch pipe 2

$$= A_2 \times V_2 = \frac{\pi}{4} D_2^2 \times V_2 = \frac{\pi}{4} (0.3)^2 \times 7.5 = 0.53 \text{ m}^3/\text{s}$$

Substituting the values of Q and Q_2 in equation (*i*), we get

$$1.5 = Q_1 + 0.53$$

$$\therefore \quad Q_1 = 1.5 - 0.53 = \mathbf{0.97\ m^3/s.\ Ans.}$$

HIGHLIGHTS

1. The loss energy in pipe is classifical as major energy loss and minor energy losses. Major energy loss is due to friction while minor energy losses are due to sudden expansion of pipe, sudden contraction of pipe, bend in pipe and an obstruction in pipe.
2. Energy loss due to friction is given by Darcy Formula, $h_f = \dfrac{4fLV^2}{d \times 2g}$.
3. The head loss due to friction in pipe can also be calculated by Chezy's formula.

   $V = C\sqrt{mi}$ Chezy's formula

   where C = Chezy's constant

   m = Hydraulic mean depth = $\dfrac{d}{4}$ (for pipe running full)

   V = Velocity of flow

   i = Loss of head per unit length = $\dfrac{h_f}{L}$

   $\therefore$ $h_f = L \times i$, where i is obtained from Chezy's formula.
4. Loss of head due to sudden expansion of pipe, $h_c = \dfrac{(V_1 - V_2)^2}{2g}$

   where V_1 = Velocity in small pipe, V_2 = Velocity in large pipe.
5. Loss of head due to sudden contraction of pipe, $h_c = \left(\dfrac{1}{C_c} - 1\right)\dfrac{V_2^2}{2g}$

   where C_c = co-efficient of contraction $= 0.375 \dfrac{V_2^2}{2g}$ ...(For $C_c = 0.62$)

   $= 0.5 \dfrac{V_2^2}{2g}$ ...(if value of C_c is not given)
6. Loss of head at the entrance of a pipe, $h_e = 0.5 \dfrac{V^2}{2g}$.
7. Loss of head at the exit of pipe, $h_c = \dfrac{V^2}{2g}$.
8. The line representing the sum of pressure head and datum head with respect to some reference line is called hydraulic gradient line (H.G.L.) while the line representing the sum of pressure head, datum head and velocity head with respect to some reference line is known as total energy line (T.E.L.).
9. Syphon is a long bent pipe used to transfer liquids from a reservoir at a higher level to another reservoir at a lower level, when the two reservoirs are separated by a high level grounds.

10. The maximum vacuum created at the summit of syphon is only 7.4 m of water.

11. When pipes of different lengths and different diameters are connected end to end, pipes are called in series or compound pipe. The rate of flow through each pipe connected is series is same.

12. A single pipe of uniform diameter, having same discharge and same loss of head as compound pipe consisting of several pipes of different lengths and diameters, is known as equivalent pipe. The diameter of equivalent pipe is called equivalent size of the pipe.

13. The equivalent size of the pipe is obtained from

$$\frac{L}{d^5} = \frac{L_1}{d_1^5} + \frac{L_2}{d_2^5} + \frac{L_3}{d_3^5}$$

where L = equivalent length of pipe = $L_1 + L_2 + L_3$

d_1, d_2, d_3 = are diameters of pipes connected in series

d = equivalent size of the pipes.

14. When the pipes are connected in parallel, the loss of head in each pipe is same. The rate of flow in main pipe is equal to sum of the rate of flow in each pipe, connected in parallel.

EXERCISE 9

(A) THEORETICAL PROBLEMS

1. How will you determine the loss of head due to friction in pipes by using (*i*) Darcy Formula and (*ii*) Chezy's formula ?
2. (*a*) What do you understand by the terms : Major energy loss and minor energy losses in pipes ? (*VTU, Aug. 2002, Jan. 2004*)

   (*b*) What do you understand by total energy line, hydraulic gradient line, pipes in series, pipes in parallel and equivalent pipe ? (*Delhi University, 1996*)
3. (*a*) Derive an expression for the loss of head due to : (*i*) Sudden enlargement and (*ii*) Sudden contraction of a pipe. (*A.M.I.E., Winter, 1983*)

   (*b*) Obtain expression for head loss in a sudden expansion in the pipe. List all the assumptions made in the derivation. (*A.M.I.E., Winter, 1990*)
4. Define and explain the terms : (*i*) Hydraulic gradient line and (*ii*) Total energy line. (*Delhi University, 1982, 88*)

   (*a*) Explain how you draw these lines for a pipe connecting two veservoirs having an elevation difference of *H* metres with a suitable sketch. (*VTU, March 2001, Jan. 2004*)

   (*b*) Drive Darcy's formula to calculate the frictional head loss in a pipe. (*VTU, Feb. 20003*)
5. Show that the loss of head due to sudden expansion in pipe line is a function of velocity head. (*A.M.I.E., Summer, 1988*)
6. What is a syphon ? On what principle it works ?
7. What is a compound pipe ? What will be loss of head when pipes are connected in series ?
8. Explain the terms : (*i*) Pipes in parallel (*ii*) Equivalent pipe and (*iii*) Equivalent size of the pipe.
9. Three pipes of different diameters and different lengths are connected in series of make a compound pipe. The ends of this compound pipe are connected with two tanks whose difference of water level is *H*. If co-efficient of friction for these pipes is same, then derive the formula for the total head loss, neglecting first the minor losses and then including them. (*A.M.I.E., Summer, 1990*)
10. For the two cases of flow in a sudden contraction in a pipeline and flow in a sudden expansion in a pipe line, draw the flow pattern, piezometric grade line and total energy line. (*A.M.I.E., Summer, 1991*)
11. What do you mean by "equivalent pipe" and "flow through parallel pipes" ? (*Delhi University, May 1998*)
12. (*a*) Define and explain the terms : (*i*) Hydraulic gradient line and (*ii*) total energy line.

    (*b*) What do you mean by equivalent pipe ? Obtain an expression for equivalent pipe. (*Delhi University, December 2002*)

(B) NUMERICAL PROBLEMS

1. Find the head lost due to friction in a pipe of diameter 250 mm and length 60 m, through which water is flowing at a velocity of 3.0 m/s using (*i*) Darcy formula and (*ii*) Chezy's formula for which $C = 55$. Take ν for water = .01 stoke. [**Ans.** (*i*) 1.182, (*ii*) 2.856]
2. Find the diameter of a pipe of length 2500 m when the rate of flow of water through the pipe is 0.25 m^3/s and head lost due to friction is 5 m. Take $C = 50$ in Chezy's formula. [**Ans.** 605 mm]
3. An oil of Kinematic Viscosity 0.5 stoke is flowing through a pipe of diameter 300 mm at the rate of 320 litres per sec. Find the head lost due to friction for a length of 60 m of the pipe. [**Ans.** 5.14 m]
4. Calculate the rate of flow of water through a pipe of diameter 300 mm, when the difference of pressure head between the two ends of a pipe 400 m apart is 5 m of water. Take the value of $f = .009$ in the formula
$$h_f = \frac{4fLV^2}{d \times 2g}$$
[**Ans.** 0.101 m^3/s]
5. The discharge through a pipe is 200 litres/s. Find the loss of head when the pipe is suddenly enlarged from 150 mm to 300 mm diameter. [**Ans.** 3.672 m]
6. The rate of flow of water through a horizontal pipe is 0.3 m^3/s. The diameter of the pipe is suddenly enlarged from 250 mm to 500 mm. The pressure intensity in the smaller pipe is 13.734 N/cm^2. Determine : (*i*) loss of head due to sudden enlargement, (*ii*) pressure intensity in the large pipe and (*iii*) power lost due to enlargement. [**Ans.** (*i*) 1.07 m, (*ii*) 14.43 N/cm^2, (*iii*) 3.15 kW]
7. A horizontal pipe of diameter 400 mm is suddenly contracted to a diameter of 200 mm. The pressure intensities in the large and smaller pipe is given as 14.715 N/cm^2 and 12.753 N/cm^2 respectively. If $C_c = 0.62$, find the loss of head due to contraction. Also determine the rate of flow of water. [**Ans.** (*i*) 0.571 m, (*ii*) 171. 7 litres/s]
8. Water is flowing through a horizontal pipe of diameter 300 mm at a velocity of 4 m/s. A circular solid plate of diameter 200 mm is placed in the pipe to obstruct the flow. If $C_c = 0.62$, find the loss of head due to obstruction in the pipe. [**Ans.** 2.953 m]
9. Determine the rate of flow of water through a pipe of diameter 10 cm and length 60 cm when one end of the pipe is connected to a tank and other end of the pipe is open to the atmosphere. The height of water in the tank from the centre of the pipe is 5 cm. Pipe is given as horizontal and value of $f = .01$. Consider minor losses. [**Ans.** 15.4 litres/s]
10. A horizontal pipe-line 50 m long is connected to a water tank at one end and discharges freely into the atmosphere at the other end. For the first 30 m of its length from the tank, the pipe is 200 mm diameter and its diameter is suddenly enlarged to 400 mm. The height of water level in the tank is 10 m above the centre of the pipe. Considering all minor losses, determine the rate of flow. Take $f = .01$ for both sections of the pipe. [**Ans.** 164.13 litres/s]
11. Determine the difference in the elevations between the water surfaces in the two tanks which are connected by a horizontal pipe of diameter 400 mm and length 500 m. The rate of flow of water through the pipe is 200 litres/s. Consider all losses and take the value of $f = .009$. [**Ans.** 11.79 m]
12. For the problems 9, 10 and 11 draw the hydraulic gradient lines (H.G.L.) and total energy lines (T.E.L.)
13. A syphon of diameter 150 mm connects two reservoirs having a difference in elevation of 15 m. The length of the syphon is 400 m and summit is 4.0 m above the water level in the upper reservoir. The length of the pipe from upper reservoir to the summit is 80 m. Determine the discharge through the syphon and also pressure at the summit. Neglect minor losses. The co-efficient of friction, $f = .005$. [**Ans.** 41.52 litres/s, – 7.281 m of water]
14. A syphon of diameter 200 mm connects two reservoirs having a difference in elevation as 20 m. The total length of the syphon is 800 m and the summit is 5 m above the water level in the upper reservoir. If the separation takes place at 2.8 m of water absolute find the maximum length of syphon from upper reservoir to the summit. Take $f = .004$ and atmospheric pressure = 10.3 m of water. [**Ans.** 87.52 m]
15. Three pipes of lengths 800 m, 600 m and 300 m and of diameters 400 mm, 300 mm and 200 mm respectively are connected in series. The ends of the compound pipe is connected to two tanks, whose water

surface levels are maintained at a difference of 15 m. Determine the rate of flow of water through the pipes if $f = .005$. What will be diameter of a single pipes of length 1700 m and $f = .005$, which replaces the three pipes. [**Ans.** 0.0848 m^3/s, 266.5 mm]

16. Two pipes of lengths 2500 m each and diameters 80 cm and 60 cm respectively, are connected in parallel. The co-efficient of friction for each pipe is 0.006. The total flow is equal to 250 litres/s. Find the rate of flow in each pipe. [**Ans.** 0.1683 m^3/s, 0.0817 m^3/s]

17. A pipe of diameter 300 mm and length 1000 m connects two reservoirs, having difference of water levels as 15 m. Determine the discharge through the pipe. If an additional pipe of diameter 300 mm and length 600 m is attached to the last 600 m length of the existing pipe, find the increase in the discharge. Take $f = .02$ and neglect minor losses. [**Ans.** 0.0742 m^3/s, 0.0258 m^3/s]

18. Two sharp ended pipes of diameters 60 mm and 100 mm respectively, each of length 150 m are connected in parallel between two reservoir which have a difference of level of 15 m. If co-efficient of friction for each pipe is 0.08, calculate the rate of flow for each pipe and also the diameter of a single pipe 150 m long which would give the same discharge if it were substituted for the original two pipes. [**Ans.** 0.0017, .00615, 110 mm]

19. The difference in water surface levels in two tanks, which are connected by two pipes in series of lengths 600 m and 400 m and of diameters 30 cm and 20 cm respectively, is 15 m. Determine the rate of flow of water if the co-efficient of friction is 0.005 for both the pipes. Neglect minor losses. *(Delhi University, 1988)*

20. Water is flowing vertically downwards through a 10 cm diameter pipe at the rate of 50 l.p.s. At a particular location the pipe suddenly enlarges to 20 cm diameter. A point P is located 50 cm above the section of enlargement and another point Q is located 50 cm below it in the enlarged portion. A pressure gauge connected at P gives a reading of 19.62 N/cm^2. Calculate the pressure at location Q neglecting friction loss between P and Q but considering the loss due to sudden enlargement. What will be the pressure at Q if the same discharge flows upwards assuming that the pressure P remains the same ? Consider the loss due to contraction with $C_c = 0.60$ but neglect friction loss between P and Q. [**Ans.** 21.36 N/cm^2, 23.4 N/cm^2] *(A.M.I.E., Summer, 1985)*

21. Two tanks are connected with the help of two pipes in series. The lengths of the pipes are 1000 m and 800 m whereas the diameters are 400 mm and 200 mm respectively. The co-efficient of friction for both the pipes is 0.008. The difference of water level in the two tanks is 15 m. Find the rate of flow of water through the pipes, considering all losses. Also draw the total energy line and hydraulic gradient lines for the system. *(Delhi University, May 1998)* [**Ans.** 0.0464 m^3/s]

[**Hint.** $L_1 = 1000$ m ; $L_2 = 800$ m ; $d_1 = 400$ mm $= 0.4$ m ; $d_2 = 200$ mm $= 0.2$ m, $f = 0.008$; $H = 15$ m.

Now $$H = h_i + hf_1 + h_c + hf_2 + h_0,$$

$$H = \frac{0.5\,V_1^2}{2g} + \frac{4f \times L_1 \times V_1^2}{d \times 2g} + \frac{0.5\,V_2^2}{2g} + \frac{4f \times L_2 \times V_2^2}{d_2 \times 2g} + \frac{V_2^2}{2g}$$

or $$15 = \frac{0.5\,V_1^2}{2 \times 9.81} + \frac{4 \times 0.008 \times 1000 \times V_1^2}{0.4 \times 2 \times 9.81} + \frac{0.5\,V_2^2}{2 \times 9.81} + \frac{4 \times 0.008 \times 800 \times V_2^2}{0.2 \times 2 \times 9.81} + \frac{V_2^2}{2g}$$

Also $$A_1V_1 = A_2V_2 \text{ or } V_2 = 4V_1$$

$\therefore$ $$15 = 0.02548\,V_1^2 + 4.0775\,V_1^2 + 0.02548\,V_2^2 + 6.524\,V_2^2 + 0.05097\,V_2^2$$
$$= 4.103\,V_1^2 + 6.6\,V_2^2 = 4.103\,V_1^2 + 6.6 \times (4V_1)^2$$
$$= 4.103\,V_1^2 + 105.607\,V_1^2 = 109.71\,V_1^2$$

$\therefore$ $$V_1 = \sqrt{\frac{15}{109.71}} = 0.3697 \text{ m/s} \therefore Q = A_1V_1 = \frac{\pi}{4}(.4)^2 \times 0.3697 = 0.0464 \text{ m}^3\text{/s}$$]

22. A pipe of diameter 25 cm and length 2000 m connects two reservoirs, having difference of water level 25 m. Determine the discharge through the pipe. If an additional pipe of diameter 25 cm and length 1000 m is attached to the last 1000 m length of the existing pipe, find the increase in discharge. Take $f = 0.015$. Neglect minor losses. *(Delhi University, December 2002)* [**Ans.** (*i*) 49.62 l/s, (*ii*) 13.14 l/s]

Viscous Flow

▶ 10.1 INTRODUCTION

This chapter deals with the flow of fluids which are viscous and flowing at very low velocity. At low velocity the fluid moves in layers. Each layer of fluid slides over the adjacent layer. Due to relative velocity between two layers the velocity gradient $\frac{du}{dy}$ exists and hence a shear stress $\tau = \mu \frac{du}{dy}$ acts on the layers.

The following cases will be considered in this chapter :

1. Flow of viscous fluid through circular pipe.
2. Flow of viscous fluid between two parallel plates.
3. Kinetic energy correction and momentum correction factors.
4. Power absorbed in viscous flow through
   (*a*) Journal bearings, (*b*) Foot-step bearings, and (*c*) Collar bearings.

▶ 10.2 TYPES OF FLOW

The flow of a fluid is mainly classified as :

(*i*) Laminar flow and
(*ii*) Turbulent flow.

In laminar flow the fluid particles move along straight parallel path in layers or laminae, such that the paths of individual fluid particles do not cross those of neighbouring particles. Laminar flow is possible only at low velocities and when the fluid is highly viscous. But when the velocity is increased or fluid is less viscous, the fluid particles do not move in straight paths. The fluid particles move in random manner resulting in general mixing of the particles. This type of flow is called turbulent flow.

A laminar flow changes to turbulent flow when (*i*) velocity is increased or (*ii*) diameter of a pipe is increased or (*iii*) the viscosity of fluid is decreased O. Reynold was first to demonstrate that the transition from laminar to turbulent depends not only on the mean velocity but on the quantity $\frac{\rho VD}{\mu}$. This quantity $\frac{\rho VD}{\mu}$ is a dimensionless quantity and is called Reynolds number (R_e). In case of circular pipe if $R_e < 2000$ the flow is said to be laminar and if $R_e > 4000$, the flow is said to be turbulent. If R_e lies between 2000 to 4000, the flow changes from laminar to turbulent.

▶ 10.3 REYNOLDS EXPERIMENT

The type of flow is determined from the Reynold number *i.e.*, $\frac{\rho V \times d}{\mu}$. This was demonstrated by O. Reynold in 1883. His apparatus is shown in Fig. 10.1.

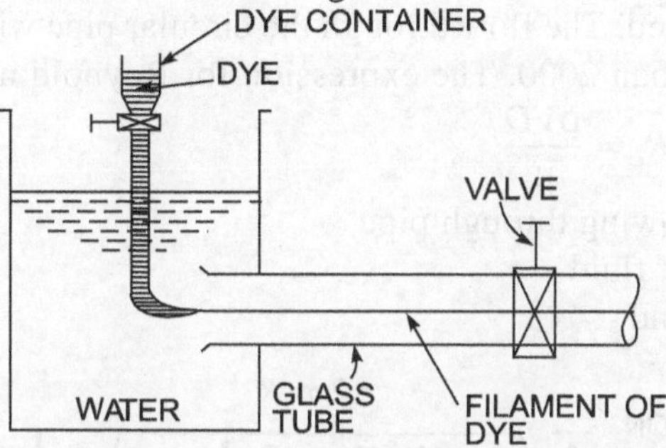

Fig. 10.1 *Reynold apparatus.*

The apparatus consists of :

(*i*) A tank containing water at constant head,

(*ii*) A small tank containing some dye,

(*iii*) A glass tube having a bell-mouthed entrance at one end and a regulating value at other ends.

The water from the tank was allowed to flow through the glass tube. The velocity of flow was varied by the regulating valve. A liquid dye having same specific weight as water was introduced into the glass tube as shown in Fig. 10.1.

The following observations were made by Reynold :

(*i*) When the velocity of flow was low, the dye filament in the glass tube was in the form of a straight line. This straight line of dye filament was parallel to the glass tube, which was the case of laminar flow as shown in Fig. 10.2 (*a*).

(*ii*) With the increase of velocity of flow, the dye-filament was no longer a straight-line but it became a wavy one as shown in Fig. 10.2 (*b*). This shows that flow is no longer laminar.

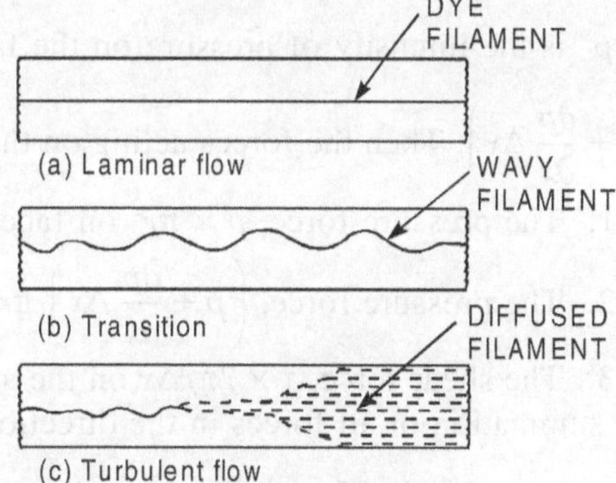

Fig. 10.2 *Different stages of filament.*

(*iii*) With further increase of velocity of flow, the wavy dye-filament broke-up and finally diffused in water as shown in Fig. 10.2 (*c*). This means that the fluid particles of the dye at this higher velocity are moving in random fashion, which shows the case of turbulent flow. Thus in case of turbulent flow the mixing of dye-filament and water is intense and flow is irregular, random and disorderly.

In case of laminar flow, the loss of pressure head was found to be proportional to the velocity but in case of turbulent flow, Reynold observed that loss of head is approximately proportional to the square of velocity. More exactly the loss of head, $h_f \propto V^n$, where n varies from 1.75 to 2.0

▶ 10.4 LAMINAR FLOW THROUGH CIRCULAR PIPE (HAGEN-POISEUILLE EQUATION)

For the flow of viscous fluid through circular pipe, the velocity distribution across a section, the ratio of maximum velocity to average velocity, the shear stress distribution and drop of pressure for a given length is to be determined. The flow through the circular pipe will be viscous or laminar, if the Reynolds number (R_e) is less than 2000. The expression for Reynold number is given by

$$R_e = \frac{\rho V D}{\mu}$$

where ρ = Density of fluid flowing through pipe
V = Average velocity of fluid
D = Diameter of pipe and
μ = Viscosity of fluid.

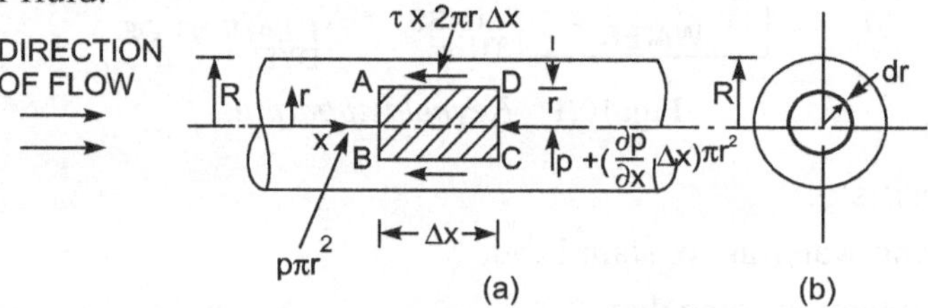

Fig. 10.3 *Viscous flow through a pipe.*

10.4.1 Shear stress Distribution.

Consider a horizontal pipe of radius R. The viscous fluid is flowing from left to right in the pipe as shown in Fig. 10.3 (a). Consider a fluid element of radius r, sliding in a cylindrical fluid element of radius ($r + dr$). Let the length of fluid element be Δx. If 'p' is the intensity of pressure on the face AB, then the intensity of pressure on face CD will be $\left(p + \frac{\partial p}{\partial x}\Delta x\right)$. Then the forces acting on the fluid element are :

1. The pressure force, $p \times \pi r^2$ on face AB.
2. The pressure force, $\left(p + \frac{\partial p}{\partial x}\Delta x\right)\pi r^2$ on face CD.
3. The shear force, $\tau \times 2\pi r \Delta x$ on the surface of fluid element. As there is no acceleration, hence the summation of all forces in the direction of flow must be zero *i.e.*,

$$p\pi r^2 - \left(p + \frac{\partial p}{\partial x}\Delta x\right)\pi r^2 - \tau \times 2\pi r \times \Delta x = 0$$

or
$$-\frac{\partial p}{\partial x}\Delta x \pi r^2 - \tau \times 2\pi r \times \Delta x = 0$$

or
$$-\frac{\partial p}{\partial x} . r - 2\tau = 0$$

∴
$$\tau = -\frac{\partial p}{\partial x}\frac{r}{2} \qquad \text{...(10.1)}$$

The shear stress τ across a section varies with 'r' as $\frac{\partial p}{\partial x}$ across a section is constant. Hence shear stress distribution across a section is linear as shown in Fig. 10.4 (a).

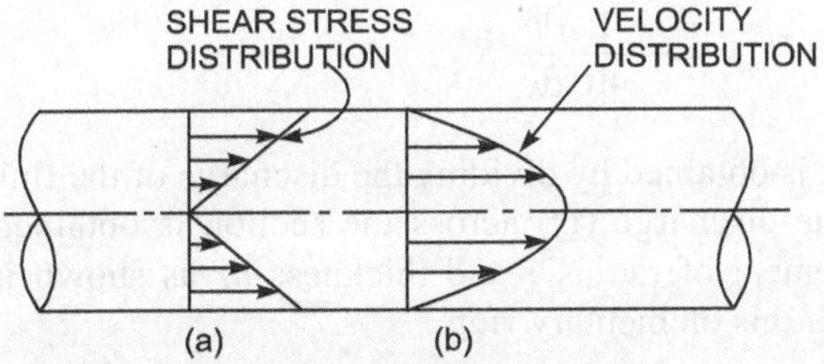

SHEAR STRESS DISTRIBUTION VELOCITY DISTRIBUTION

(a) (b)

Fig. 10.4 *Shear stress and velocity distribution across a section.*

10.4.2 Velocity Distribution. To obtain the velocity distribution across a section, the value of shear stress $\tau = \mu \dfrac{du}{dy}$ is substituted in equation (10.1).

But in the relation $\tau = \mu \dfrac{du}{dy}$, y is measured from the pipe wall. Hence

$$y = R - r \quad \text{and} \quad dy = -dr$$

$$\therefore \qquad \tau = \mu \frac{du}{-dr} = -\mu \frac{du}{dr}$$

Substituting this value in (10.1), we get

$$-\mu \frac{du}{dr} = -\frac{\partial p}{\partial x}\frac{r}{2} \quad \text{or} \quad \frac{du}{dr} = \frac{1}{2\mu}\frac{\partial p}{\partial x} r$$

Integrating this above equation w.r.t., 'r', we get

$$u = \frac{1}{4\mu}\frac{\partial p}{\partial x} r^2 + C \qquad \text{...(10.2)}$$

where C is the constant of integration and its value is obtained from the boundary condition that at $r = R$, $u = O$.

$$\therefore \qquad O = \frac{1}{4\mu}\frac{\partial p}{\partial x} R^2 + C$$

$$\therefore \qquad C = \frac{1}{4\mu}\frac{\partial p}{\partial x} R^2$$

Substituting this value of C in equation (10.2), we get

$$u = \frac{1}{4\mu}\frac{\partial p}{\partial x} r^2 - \frac{1}{4\mu}\frac{\partial p}{\partial x} R^2$$

$$= -\frac{1}{4\mu}\frac{\partial p}{\partial x}[R^2 - r^2] \qquad \text{...(10.3)}$$

In equation (10.3), values of μ, $\dfrac{\partial p}{\partial x}$ and R are constant, which means the velocity, u varies with the square of r. Thus equation (10.3) is a equation of parabola. This shows that the velocity distribution across the section of a pipe is parabolic. This velocity distribution is shown in Fig. 10.4 (*b*).

10.4.3. Ratio of Maximum Velocity to Average Velocity. The velocity is maximum, when $r = 0$ in equation (10.3). Thus maximum velocity, U_{max} is obtained as

$$U_{max} = -\frac{1}{4\mu}\frac{\partial p}{\partial x} R^2 \quad \text{...(10.4)}$$

The average velocity, $\bar{u}$, is obtained by dividing the discharge of the fluid across the section by the area of the pipe (πR^2). The discharge (Q) across the section is obtained by considering the flow through a circular ring element of radius r and thickness dr as shown in Fig. 10.3 (b). The fluid flowing per second through this elementary ring

$$dQ = \text{velocity at a radius } r \times \text{area of ring element}$$

$$= u \times 2\pi r\, dr$$

$$= -\frac{1}{4\mu}\frac{\partial p}{\partial x}[R^2 - r^2] \times 2\pi r\, dr$$

$$\therefore \quad Q = \int_0^R dQ = \int_0^R -\frac{1}{4\mu}\frac{\partial p}{\partial x}(R^2 - r^2) \times 2\pi r\, dr$$

$$= \frac{1}{4\mu}\left(\frac{-\partial p}{\partial x}\right) \times 2\pi \int_0^R (R^2 - r^2)\, r dr$$

$$= \frac{1}{4\mu}\left(\frac{-\partial p}{\partial x}\right) \times 2\pi \int_0^R (R^2 r - r^3)\, dr$$

$$= \frac{1}{4\mu}\left(\frac{-\partial p}{\partial x}\right) \times 2\pi \left[\frac{R^2 r^2}{2} - \frac{r^4}{4}\right]_0^R = \frac{1}{4\mu}\left(\frac{-\partial p}{\partial x}\right) \times 2\pi \left[\frac{R^4}{2} - \frac{R^4}{4}\right]$$

$$= \frac{1}{4\mu}\left(\frac{-\partial p}{\partial x}\right) \times 2\pi \times \frac{R^4}{4} = \frac{\pi}{8\mu}\left(\frac{-\partial p}{\partial x}\right) R^4$$

$\therefore$ Average velocity,
$$\bar{u} = \frac{Q}{\text{Area}} = \frac{\frac{\pi}{8\mu}\left(\frac{-\partial p}{\partial x}\right) R^4}{\pi R^2}$$

or
$$\bar{u} = \frac{1}{8\mu}\left(\frac{-\partial p}{\partial x}\right) R^2 \quad \text{...(10.5)}$$

Dividing equation (10.4) by equation (10.5),

$$\frac{U_{max}}{\bar{u}} = \frac{-\frac{1}{4\mu}\frac{\partial p}{\partial x} R^2}{\frac{1}{8\mu}\left(-\frac{\partial p}{\partial x}\right) R^2} = 2.0$$

$\therefore$ Ratio of maximum velocity to average velocity = 2.0.

10.4.4 Drop of Pressure for a given Length (L) of a Pipe

From equation (10.5), we have

$$\bar{u} = \frac{1}{8\mu}\left(\frac{-\partial p}{\partial x}\right)R^2 \quad \text{or} \quad \left(\frac{-\partial p}{\partial x}\right) = \frac{8\mu\bar{u}}{R^2}$$

Integrating the above equation w.r.t. x, we get

$$-\int_2^1 dp = \int_2^1 \frac{8\mu\bar{u}}{R^2}\,dx$$

$$\therefore \quad -[p_1 - p_2] = \frac{8\mu\bar{u}}{R^2}[x_1 - x_2] \text{ or } (p_1 - p_2) = \frac{8\mu\bar{u}}{R^2}[x_2 - x_1]$$

$$= \frac{8\mu\bar{u}}{R^2}L \qquad \{\because x_2 - x_1 = L \text{ from Fig. 10.5 } (a)\}$$

$$= \frac{8\mu\bar{u}L}{(D/2)^2} \qquad \left\{\because R = \frac{D}{2}\right\}$$

or $$(p_1 - p_2) = \frac{32\mu\bar{u}L}{D^2}, \quad \text{where } p_1 - p_2 \text{ is the drop of pressure.}$$

Fig. 10.5 (*a*)

$$\therefore \quad \text{Loss of pressure head} = \frac{p_1 - p_2}{\rho g}$$

$$\therefore \quad \frac{p_1 - p_2}{\rho g} = h_f = \frac{32\mu\bar{u}L}{\rho g D^2} \qquad \text{...(10.6)}$$

Equation (10.6) is called *Hagen Poiseuille Formula.*

Problem 10.1 *A crude oil of viscosity 0.97 poise and relative density 0.9 is flowing through a horizontal circular pipe of diameter 100 mm and of length 10 m. Calculate the difference of pressure at the two ends of the pipe, if 100 kg of the oil is collected in a tank in 30 seconds. Assume laminar flow.* (VTU, Aug. 2001, Aug. 2003)

Solution. Given : $\mu = 0.97 \text{ poise} = \frac{0.97}{10} = 0.097 \text{ Ns/m}^2$

Relative density $= 0.9$

$\therefore$ ρ_0, or density, $= 0.9 \times 1000 = 900 \text{ kg/m}^3$

Dia. of pipe, $D = 100 \text{ mm} = 0.1 \text{ m}$

$L = 10 \text{ m}$

Mass of oil collected, $M = 100 \text{ kg}$

in time, $t = 30 \text{ seconds}$

Calculate difference of pressure or $(p_1 - p_2)$.

The difference of pressure $(p_1 - p_2)$ for viscous or laminar flow is given by

$$p_1 - p_2 = \frac{32\mu\bar{u}L}{D^2}, \text{ where } \bar{u} = \text{average velocity} = \frac{Q}{\text{Area}}$$

Now, mass of oil/sec $= \frac{100}{30}$ kg/s

$$= \rho_0 \times Q = 900 \times Q \qquad (\because \quad \rho_0 = 900)$$

$$\therefore \quad \frac{100}{30} = 900 \times Q$$

$$\therefore \quad Q = \frac{100}{30} \times \frac{1}{900} = 0.0037 \text{ m}^3/\text{s}$$

$$\therefore \quad \bar{u} = \frac{Q}{\text{Area}} = \frac{.0037}{\frac{\pi}{4} D^2} = \frac{.0037}{\frac{\pi}{4}(.1)^2} = 0.471 \text{ m/s.}$$

For laminar or viscous flow, the Reynolds number (R_e) is less than 2000. Let us calculate the Reynolds number for this problem.

Reynolds number, $$R_e = \frac{\rho V D}{\mu}$$

where $\rho = \rho_0 = 900$, $V = \bar{u} = 0.471$, $D = 0.1$ m, $\mu = 0.097$

$$\therefore \quad R_e = 900 \times \frac{.471 \times 0.1}{0.097} = 436.91$$

As Reynolds number is less than 2000, the flow is laminar.

$$\therefore \quad p_1 - p_2 = \frac{32 \mu \bar{u} L}{D^2} = \frac{32 \times 0.097 \times .471 \times 10}{(.1)^2} \text{ N/m}^2$$

$$= 1462.28 \text{ N/m}^2 = 1462.28 \times 10^{-4} \text{ N/cm}^2 = \mathbf{0.1462 \text{ N/cm}^2. \text{ Ans.}}$$

Problem 10.2 *An oil of viscosity 0.1 Ns/m^2 and relative density 0.9 is flowing through a circular pipe of diameter 50 mm and of length 300 m. The rate of flow of fluid through the pipe is 3.5 litres/s. Find the pressure drop in a length of 300 m and also the shear stress at the pipe wall.*

(Delhi University, 1987)

Solution. Given : Viscosity, $\mu = 0.1$ Ns/m^2

Relative density $= 0.9$

$\therefore$ ρ_0 or density of oil $= 0.9 \times 1000 = 900$ kg/m^3 ($\because$ Density of water $= 1000$ kg/m^3)

$D = 50$ mm $= .05$ m

$L = 300$ m

$$Q = 3.5 \text{ litres/s} = \frac{3.5}{1000} = .0035 \text{ m}^3/\text{s}$$

Find (*i*) Pressure drop, $p_1 - p_2$

(*ii*) Shear stress at pipe wall, τ_0

(*i*) **Pressure drop ($p_1 - p_2$)** $= \dfrac{32 \mu \bar{u} L}{D^2}$, where $\bar{u} = \dfrac{Q}{\text{Area}} = \dfrac{.0035}{\frac{\pi}{4} D^2} = \dfrac{.0035}{\frac{\pi}{4}(.05)^2} = 1.782$ m/s

The Reynolds number (R_e) is given by, $R_e = \dfrac{\rho V D}{\mu}$

where $\rho = 900$ kg/m^3, V = average velocity $= \bar{u} = 1.782$ m/s

$$\therefore \quad R_e = 900 \times \frac{1.782 \times .05}{0.1} = 801.9$$

As Reynold number is less than 2000, the flow is viscous or laminar

$$\therefore \qquad p_1 - p_2 = \frac{32 \times 0.1 \times 1.782 \times 3000}{(.05)^2}$$

$$= 684288 \text{ N/m}^2 = 68428 \times 10^{-4} \text{ N/cm}^2 = \mathbf{68.43 \text{ N/cm}^2. \text{ Ans.}}$$

(*ii*) **Shear Stress at the pipe wall (τ_0)**

The shear stress at any radius r is given by the equation (10.1)

i.e.,
$$\tau = -\frac{\partial p}{\partial x}\frac{r}{2}$$

$\therefore$ Shear stress at pipe wall, where $r = R$ is given by

$$\tau_0 = \frac{-\partial p}{\partial x}\frac{R}{2}$$

Now
$$\frac{-\partial p}{\partial x} = \frac{-(p_2 - p_1)}{x_2 - x_1} = \frac{p_1 - p_2}{x_2 - x_1} = \frac{p_1 - p_2}{L}$$

$$= \frac{684288}{300}\frac{\text{N/m}^2}{\text{m}} = 2280.96 \text{ N/m}^3$$

and
$$R = \frac{D}{2} = \frac{.05}{2} = .025 \text{ m}$$

$$\tau_0 = 2280.96 \times \frac{.025}{2}\frac{\text{N}}{\text{m}^2} = \mathbf{28.512 \text{ N/m}^2. \text{ Ans.}}$$

Problem 10.3 *A laminar flow is taking place in a pipe of diameter of 200 mm. The maximum velocity is 1.5 m/s. Find the mean velocity and the radius at which this occurs. Also calculate the velocity at 4 cm from the wall of the pipe.* (Delhi University, April 1986)

Solution. Given : Dia. of pipe, $D = 200$ mm $= 0.20$ m

$U_{max} = 1.5$ m/s

Find (*i*) Mean velocity, $\bar{u}$

(*ii*) Radius at which $\bar{u}$ occurs

(*iii*) Velocity at 4 cm from the wall.

(*i*) **Mean velocity, $\bar{u}$**

Ratio of
$$\frac{U_{max}}{\bar{u}} = 2.0 \quad \text{or} \quad \frac{1.5}{\bar{u}} = 2.0 \quad \therefore \quad \bar{u} = \frac{1.5}{2.0} = \mathbf{0.75 \text{ m/s. Ans.}}$$

(*ii*) **Radius at which $\bar{u}$** occurs

The velocity, u, at any radius 'r' is given by (10.3)

$$u = -\frac{1}{4\mu}\frac{\partial p}{\partial x}[R^2 - r^2] = -\frac{1}{4\mu}\frac{\partial p}{\partial x}R^2\left[1 - \frac{r^2}{R^2}\right]$$

But from equation (10.4) U_{max} is given by

$$U_{max} = -\frac{1}{4\mu}\frac{\partial p}{\partial x}R^2 \quad \therefore \quad u = U_{max}\left[1 - \left(\frac{r}{R}\right)^2\right] \qquad ...(1)$$

Now the radius r at which $u = \bar{u} = 0.75$ m/s

$$\therefore \quad 0.75 = 1.5\left[1-\left(\frac{r}{D/2}\right)^2\right]$$

$$= 1.5\left[1-\left(\frac{r}{0.2/2}\right)^2\right] = 1.5\left[1-\left(\frac{r}{0.1}\right)^2\right]$$

$$\therefore \quad \frac{0.75}{0.50} = 1-\left(\frac{r}{0.1}\right)^2$$

$$\therefore \quad \left(\frac{r}{0.1}\right)^2 = 1-\frac{.75}{1.50} = 1-\frac{1}{2} = \frac{1}{2}$$

$$\therefore \quad \frac{r}{0.1} = \sqrt{\frac{1}{2}} = \sqrt{0.5}$$

$$\therefore \quad r = 0.1 \times \sqrt{.5} = 0.1 \times .707 = .0707 \text{ m}$$

$$= \mathbf{70.7\ mm.\ Ans.}$$

(*iii*) **Velocity at 4 cm from the wall**

$$r = R - 4.0 = 10 - 4.0 = 6.0 \text{ cm} = 0.06 \text{ m}$$

∴ The velocity at a radius = 0.06 m

or 4 cm from pipe wall is given by equation (1)

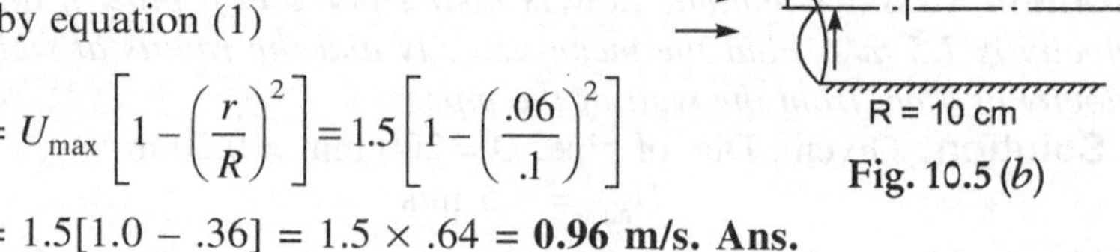

Fig. 10.5 (*b*)

$$= U_{max}\left[1-\left(\frac{r}{R}\right)^2\right] = 1.5\left[1-\left(\frac{.06}{.1}\right)^2\right]$$

$$= 1.5[1.0 - .36] = 1.5 \times .64 = \mathbf{0.96\ m/s.\ Ans.}$$

Problem 10.4 *Crude oil of μ = 1.5 poise and relative density 0.9 flows through a 20 mm diameter vertical pipe. The pressure gauges fixed 20 m apart read 58.86 N/cm² and 19.62 N/cm² as shown in Fig. 10.5 (c). Find the direction and rate of flow through the pipe.* (A.M.I.E., Winter, 1976)

Solution. Given : $\mu = 1.5 \text{ poise} = \frac{1.5}{10} = 0.15 \text{ Ns/m}^2$

Relative density $= 0.9$

∴ Density of oil $= 0.9 \times 1000 = 900 \text{ kg/m}^3$

Dia. of pipe, $D = 20 \text{ mm} = 0.02 \text{ m}$

$L = 20 \text{ m}$

$p_A = 58.86 \text{ N/cm}^2 = 58.86 \times 10^4 \text{ N/m}^2$

$p_B = 19.62 \text{ N/cm}^2 = 19.62 \times 10^4 \text{ N/m}^2.$

Find (*i*) Direction of flow

(*ii*) Rate of flow.

(*i*) **Direction of flow.** To find the direction of flow, the total energy $\left(\frac{p}{\rho g} + \frac{v^2}{2g} + Z\right)$ at the lower end A and at the upper end B is to be calculated. The direction of flow will be given from the higher energy to the lower energy. As here the diameter of the pipe is same and hence kinetic energy at A and B will be same. Hence to find the direction of flow, calculate $\left(\frac{p}{\rho g}+Z\right)$ at A and B.

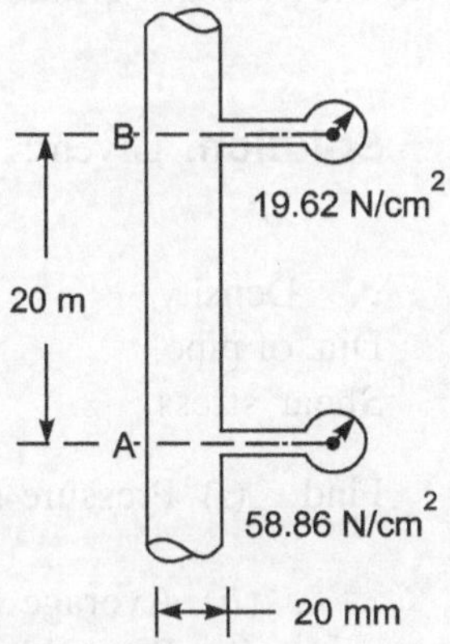

Taking the level at A as datum. The value of $\left(\frac{p_A}{\rho g}+Z\right)$ at

$$A = \frac{p_A}{\rho g} + Z_A$$

$$= \frac{6 \times 10^4 \times 9.81}{900 \times 9.81} + 0 \quad \{\because \ r = 900 \text{ kg/cm}^2\}$$

$$= 66.67 \text{ m}$$

The value of $\left(\frac{p}{\rho g}+Z\right)$ at $B = \frac{p_B}{\rho g} + Z_B$

$$= \frac{2 \times 10^4 \times 9.81}{900 \times 9.81} + 20 = 22.22 + 20 = 42.22 \text{ m}$$

Fig. 10.5 (*c*)

As the value of $\left(\frac{p}{\rho g}+Z\right)$ is higher at A and hence flow takes place from A to B. **Ans.**

(*ii*) **Rate of flow.** The loss of pressure head for viscous flow through circular pipe is given by

$$h_f = \frac{32\mu \bar{u} L}{\rho g D^2}$$

For a vertical pipe h_f = Loss of peizometric head

$$= \left(\frac{p_A}{\rho g} + Z_A\right) - \left(\frac{p_B}{\rho g} + Z_B\right) = 66.67 - 42.22 = 24.45 \text{ m}$$

$$\therefore \quad 24.45 = \frac{32 \times 0.15 \times \bar{u} \times 20.0}{900 \times 9.81 \times (.02)^2}$$

or
$$\bar{u} = \frac{24.45 \times 900 \times 9.81 \times .0004}{32 \times 0.15 \times 20.0} = 0.889 \simeq 0.9 \text{ m/s.}$$

The Reynolds number should be calculated. If Reynolds number is less than 2000, the flow will be laminar and the above expression for loss of pressure head for laminar flow can be used.

Now Reynolds number $= \frac{\rho V D}{\mu}$

where $\rho = 900$ kg/m^3 and $V = \bar{u}$

$$\therefore \quad \text{Reynolds number} = 900 \times \frac{0.9 \times .02}{0.15} = 108$$

As Reynolds number is less than 2000, the flow is laminar.

$\therefore$ Rate of flow = average velocity × area

$$= \bar{u} \times \frac{\pi}{4} D^2 = 0.9 \times \frac{\pi}{4} \times (.02)^2 \text{ m}^3\text{/s} = 2.827 \times 10^{-4} \text{ m}^3\text{/s}$$

= **0.2827 litres/s. Ans.** $(\because \ 10^{-3} \text{ m}^3 = 1 \text{ litre})$

Problem 10.5 *A fluid of viscosity 0.7 Ns/m² and specific gravity 1.3 is flowing through a circular pipe of diameter 100 mm. The maximum shear stress at the pipe wall is given as 196.2 N/m², find (a) the pressure gradient (b) the average velocity and (c) Reynold number of the flow.*

(Similar Question VTU, Jan. 2004)

Solution. Given : $\mu = 0.7 \frac{\text{Ns}}{\text{m}^2}$

Sp. gr. = 1.3

∴ Density = 1.3 × 1000 = 1300 kg/m³

Dia. of pipe, $D = 100$ mm = 0.1 m

Shear stress, $\tau_0 = 196.2$ N/m²

Find (*i*) Pressure gradient, $\frac{dp}{dx}$

(*ii*) Average velocity, $\bar{u}$

(*iii*) Reynold number, R_e

(*i*) **Pressure gradient,** $\frac{dp}{dx}$

The maximum shear stress (τ_0) is given by

$$\tau_0 = -\frac{\partial p}{\partial x}\frac{R}{2} \text{ or } 196.2 = -\frac{\partial p}{\partial x} \times \frac{D}{4} = -\frac{\partial p}{\partial x} \times \frac{0.1}{4}$$

$$\therefore \quad \frac{\partial p}{\partial x} = -\frac{196.2 \times 4}{0.1} = -7848 \text{ N/m}^2 \text{ per m}$$

∴ Pressure Gradient = **– 7848 N/m² per m. Ans.**

(*ii*) **Average velocity,** $\bar{u}$

$$\bar{u} = \frac{1}{2} U_{max} = \frac{1}{2}\left[-\frac{1}{4\mu}\frac{\partial p}{\partial x}R^2\right] \qquad \left\{\because \; U_{max} = -\frac{1}{8\mu}\frac{\partial p}{\partial x}R^2\right\}$$

$$= \frac{1}{8\mu}\left(-\frac{\partial p}{\partial x}\right)R^2$$

$$= \frac{1}{8 \times 0.7} \times (7848) \times (.05)^2 \qquad \left\{\because \; R = \frac{D}{2} = \frac{1}{2} = .05\right\}$$

$$= 3.50 \text{ m/s}$$

(*iii*) **Reynold number,** R_e

$$R_e = \frac{\bar{u} \times D}{\nu} = \frac{\bar{u} \times D}{\mu/\rho} = \frac{\rho \times \bar{u} \times D}{\mu}$$

$$= 1300 \times \frac{3.50 \times 0.1}{0.7} = \mathbf{650.00.\ Ans.}$$

Problem 10.6 *What power is required per kilometre of a line to overcome the viscous resistance to the flow of glycerine through a horizontal pipe of diameter 100 mm at the rate of 10 litres/s ? Take μ = 8 poise and kinematic viscosity (ν) = 6.0 stokes.* (Delhi University, 1982)

Solution. Given :

Length of pipe, $L = 1$ km = 1000 m

Dia. of pipe, $D = 100 \text{ mm} = 0.1 \text{ m}$

Discharge, $Q = 10 \text{ litres/s} = \frac{10}{1000} \text{ m}^3\text{/s} = .01 \text{ m}^3\text{/s}$

Viscosity, $\mu = 8 \text{ poise} = \frac{8}{10} \frac{\text{Ns}}{\text{m}^2} = 0.8 \text{ N s/m}^2$ $\left(\because 1 \text{ poise} = \frac{1}{10} \text{Ns/m}^2\right)$

Kinematic Viscosity, $\nu = 6.0 \text{ stokes}$

$$= 6.0 \text{ cm}^2\text{/s} = 6.0 \times 10^{-4} \text{ m}^2\text{/s}$$

Loss of pressure head is given by equation (10.6) as $h_f = \frac{32\mu\bar{u}L}{\rho g D^2}$

$$\text{Power required} = W \times h_f \text{ watts} \qquad ...(i)$$

where W = weight of oil flowing per sec = $\rho g \times Q$

Substituting the values of W and h_f in equation (*i*),

Power required $= (\rho g \times Q) \times \frac{(32\,\mu\bar{u}L)}{\rho g D^2} \text{ watts} = \frac{Q \times 32\,\mu\bar{u}L}{D^2}$ (cancelling ρg)

But $\bar{u} = \frac{Q}{\text{Area}} = \frac{.01}{\frac{\pi}{4} D^2} = \frac{.01 \times 4}{\pi \times (.1)^2} = 1.273 \text{ m/s}$

$\therefore$ Power required $= \frac{.01 \times 32 \times 0.8 \times 1.273 \times 1000}{(.1)^2}$

$= 32588.8 \text{ W} =$ **32.588 kW. Ans.**

▶ 10.5 LAMINAR FLOW BETWEEN TWO PARALLEL STATIONARY PLATES

In this case also, the shear stress distribution, the velocity distribution across a section ; the ratio of maximum velocity to average velocity and difference of pressure head for a given length of parallel plates, are to be calculated.

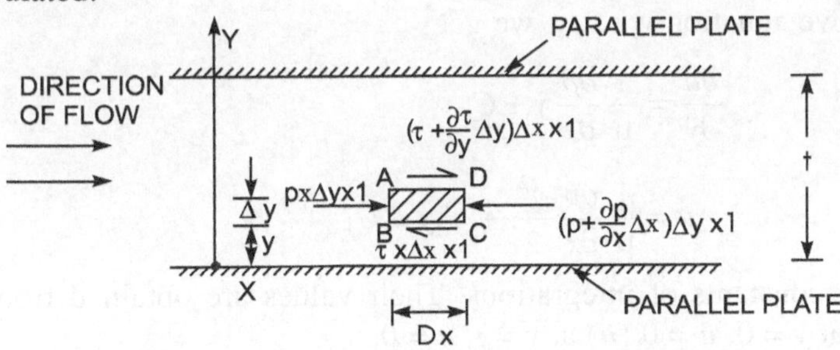

Fig. 10.6 *Viscous flow between two parallel plates.*

Consider two parallel fixed plates kept at a distance 't' apart as shown in Fig. 10.6. A viscous fluid is flowing between these two plates from left to right. Consider a fluid element of length Δx and thickness Δy at a distance y from the lower fixed plate. If p is the intensity of pressure on the face AB of the fluid element then intensity of pressure on the face CD will be $\left(p + \frac{\partial p}{\partial x} \Delta x\right)$. Let τ is the shear stress acting on the face *BC* then the shear stress on the face *AD* will be $\left(\tau+\frac{\partial\tau}{\partial y}\Delta y\right)$. If the width of the element in the direction perpendicular to the paper is unity then the forces acting on the fluid element are :

1. The pressure force, $p \times \Delta y \times 1$ on face *AB*.
2. The pressure force, $\left(p+\frac{\partial p}{\partial x}\Delta x\right)\Delta y \times 1$ on face *CD*.
3. The shear force, $\tau \times \Delta x \times 1$ on face *BC*.
4. The shear force, $\left(\tau+\frac{\partial\tau}{\partial y}\Delta y\right)\Delta x \times 1$ on face *AD*.

For steady and uniform flow, there is no acceleration and hence the resultant force in the direction of flow is zero.

$$\therefore \qquad p\Delta y \times 1-\left(p+\frac{\partial p}{\partial x}\Delta x\right)\Delta y \times 1-\tau\Delta x \times 1+\left(\tau+\frac{\partial\tau}{\partial y}\Delta y\right)\Delta x \times 1=0$$

or

$$-\frac{\partial p}{\partial x}\Delta x\Delta y+\frac{\partial\tau}{\partial x}\Delta y\Delta x=0$$

Dividing by $\Delta x\Delta y$, we get $-\frac{\partial p}{\partial x}+\frac{\partial\tau}{\partial y}=0$ or $\frac{\partial p}{\partial x}=\frac{\partial\tau}{\partial y}$ ...(10.7)

10.5.1 Velocity Distribution. To obtain the velocity distribution across a section, the value of shear stress $\tau=\mu\frac{du}{dy}$ from Newton's law of viscosity for laminar flow is substituted in equation (10.6).

$$\therefore \qquad \frac{\partial p}{\partial x}=\frac{\partial}{\partial y}\left(\mu\frac{du}{dy}\right)=\mu\frac{\partial^2 u}{dy^2}$$

$$\therefore \qquad \frac{\partial^2 u}{dy^2}=\frac{1}{\mu}\frac{\partial p}{\partial x}$$

Integrating the above equation w.r.t. y, we get

$$\frac{\partial u}{\partial y}=\frac{1}{\mu}\frac{\partial p}{\partial x}y+C_1 \qquad \left\{\because \frac{\partial p}{\partial x}\text{ is constant}\right\}$$

Integrating again

$$u=\frac{1}{\mu}\frac{\partial p}{\partial x}\frac{y^2}{2}+C_1 y+C_1 \qquad ...(10.8)$$

where C_1 and C_2 are constants of integrations. Their values are obtained from the two boundary conditions that is (*i*) at $y=0$, $u=0$ (*ii*) at $y=t$, $u=0$.

The substitution of at $y=0$, $u=0$ in equation (10.8) gives

$$0=0+C_1\times 0+C_2 \text{ or } C_2=0$$

The substitution of at $y=t$, $u=0$ in equation (10.8) gives

$$0=\frac{1}{\mu}\frac{\partial p}{\partial x}\frac{t^2}{2}+C_1\times t+0$$

$$\therefore \qquad C_1 = -\frac{1}{\mu}\frac{\partial p}{\partial x}\frac{t^2}{2\times t} = -\frac{1}{2\mu}\frac{\partial p}{\partial x}t$$

Substituting the values of C_1 and C_2 in equation (10.8)

$$u = \frac{1}{2\mu}\frac{\partial p}{\partial x}y^2 + y\left(-\frac{1}{2\mu}\frac{\partial p}{\partial x}t\right)$$

or
$$u = -\frac{1}{2\mu}\frac{\partial p}{\partial x}[ty - y^2] \qquad \text{...(10.9)}$$

In the above equation, μ, $\frac{\partial p}{\partial x}$ and t are constant. It means u varies with the square of y. Hence equation (10.9) is a equation of a parabola. Hence velocity distribution across a section of the parallel plate is parabolic. This velocity distribution is shown in Fig. 10.7 (*a*).

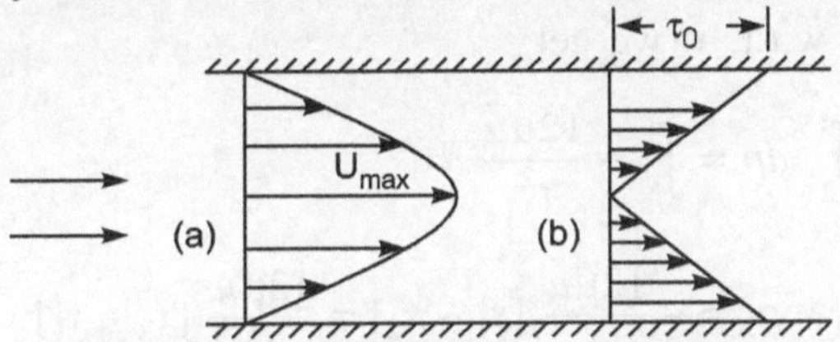

Fig. 10.7 *Velocity distribution and shear stress distribution across a section of parallel plates.*

10.5.2 Ratio of Maximum Velocity to Average Velocity. The velocity is maximum, when $y = t/2$. Substituting this value in equation (10.9), we get

$$U_{max} = -\frac{1}{2\mu}\frac{\partial p}{\partial x}\left[t \times \frac{t}{2} - \left(\frac{t}{2}\right)^2\right]$$

$$= -\frac{1}{2\mu}\frac{\partial p}{\partial x}\left[\frac{t^2}{2} - \frac{t^2}{4}\right] = -\frac{1}{2\mu}\frac{\partial p}{\partial x}\frac{t^2}{4} = -\frac{1}{8\mu}\frac{\partial p}{\partial x}t^2 \qquad \text{...(10.10)}$$

The average velocity, $\bar{u}$, is obtained by dividing the discharge (Q) across the section by the area of the section ($t \times 1$). And the discharge Q is obtained by considering the rate of flow of fluid through the strip of thickness dy and integrating it. The rate of flow through strip is

$$dQ = \text{Velocity at a distance } y \times \text{Area of strip}$$

$$= -\frac{1}{2\mu}\frac{\partial p}{\partial x}[ty - y^2] \times dy \times 1$$

$$\therefore \qquad Q = \int_0^t dQ = \int_0^t -\frac{1}{2\mu}\frac{\partial p}{\partial x}[ty - y^2]\,dy$$

$$= -\frac{1}{2\mu}\frac{\partial p}{\partial x}\left[\frac{ty^2}{2} - \frac{y^3}{3}\right]_0^t = \frac{1}{2\mu}\frac{\partial p}{\partial x}\left[\frac{t^3}{2} - \frac{t^3}{3}\right]$$

$$= -\frac{1}{2\mu}\frac{\partial p}{\partial x}\frac{t^3}{6} = -\frac{1}{12\mu}\frac{\partial p}{\partial x}t^3$$

$$\therefore \qquad \bar{u} = \frac{Q}{\text{Area}} = -\frac{\frac{1}{12\mu}\frac{\partial p}{\partial x}.t^3}{t \times 1} = -\frac{1}{12\mu}\frac{\partial p}{\partial x}t^2 \qquad \text{...(10.11)}$$

Dividing equation (10.10) by equation (10.11), we get

$$\frac{U_{max}}{\bar{u}} = \frac{-\frac{1}{8\mu}\frac{\partial p}{\partial x}t^2}{-\frac{1}{12\mu}\frac{\partial p}{\partial x}t^2} = \frac{12}{8} = \frac{3}{2} \qquad \text{...(10.12)}$$

10.5.3. Drop of Pressure Head for a given Length. From equation (10.11), we have

$$\bar{u} = -\frac{1}{12\mu}\frac{\partial p}{\partial x}t^2 \quad \text{or} \quad \frac{\partial p}{\partial x} = -\frac{12\mu\bar{u}}{t^2}$$

Integrating this equation w.r.t. x, we get

$$\int_2^1 dp = \int_2^1 -\frac{12\mu\bar{u}}{t^2}\, dx$$

or
$$p_1 - p_2 = \frac{12\mu\bar{u}}{t^2}[x_1 - x_2] = \frac{12\mu\bar{u}}{t^2}[x_2 - x_1]$$

or
$$p_1 - p_2 = \frac{12\mu\bar{u}L}{t^2} \qquad [\because x_1 - x_2 = L]$$

If h_f is the drop of pressure head, then

$$h_f = \frac{p_1 - p_2}{\rho g} = \frac{12\mu\bar{u}L}{\rho g t^2} \qquad \text{...(10.13)}$$

Fig. 10.8

10.5.4 Shear Stress Distribution. It is obtained by substituting the value of u from equation (10.9) into

$$\tau = \mu\frac{\partial u}{\partial y}$$

$$\therefore \qquad \tau = \mu\frac{\partial u}{\partial y} = u\frac{\partial}{\partial y}\left[-\frac{1}{2\mu}\frac{\partial p}{\partial x}(ty - y^2)\right] = \mu\left[-\frac{1}{2\mu}\frac{\partial p}{\partial x}(t - 2y)\right]$$

$$\tau = -\frac{1}{2}\frac{\partial p}{\partial x}[t - 2y] \qquad \text{...(10.14)}$$

In equation (10.14), $\frac{\partial p}{\partial x}$ and t are constant. Hence τ varies linearly with y. The shear stress distribution is shown in Fig. 10.7 (*b*). Shear stress is maximum, when $y = 0$ or t at the walls of the plates. Shear stress is zero, when $y = t/2$ that is at the centre line between the two plates. Max. shear stress (τ_0) is given by

$$\tau_0 = -\frac{1}{2}\frac{\partial p}{\partial x}t. \qquad \text{...(10.15)}$$

Problem 10.7 *Calculate : (a) the pressure gradient along flow, (b) the average velocity, and (c) the discharge for an oil of viscosity 0.02 Ns/m² flowing between two stationary parallel plates 1 m wide maintained 10 mm apart. The velocity midway between the plates is 2 m/s.*

(Delhi University, 1982)

Solution. Given :

Viscosity, $\mu = .02$ N s/m^2

Width, $b = 1$ m

Distance between plates, $t = 10$ mm $= .01$ m

Velocity midway between the plates, $U_{max} = 2$ m/s.

(*i*) **Pressure gradient** $\left(\frac{dp}{dx}\right)$

Using equation (10.10), $U_{max} = -\frac{1}{8\mu}\frac{dp}{dx}t^2$ or $2.0 = -\frac{1}{8 \times .02}\left(\frac{dp}{dx}\right)(.01)^2$

$$\therefore \quad \frac{dp}{dx} = -\frac{2.0 \times 8 \times .02}{.01 \times .01} = -\ \mathbf{3200\ N/m^2\ per\ m.\ Ans.}$$

(*ii*) **Average velocity** $(\bar{u})$

Using equation (10.12), $\frac{U_{max}}{\bar{u}} = \frac{3}{2}$ $\quad \therefore \quad \bar{u} = \frac{2\,U_{max}}{3} = \frac{2 \times 2}{3} = \mathbf{1.33\ m/s.\ Ans.}$

(*iii*) **Discharge** (Q) $= \text{Area of flow} \times \bar{u} = b \times t \times \bar{u} = 1 \times .01 \times 1.33 = \mathbf{.0133\ m^3/s.\ Ans.}$

Problem 10.8 *Determine (a) the pressure gradient, (b) the shear stress at the two horizontal parallel plates and (c) the discharge per metre width for the laminar flow of oil with a maximum velocity of 2 m/s between two horizontal parallel fixed plates which are 100 mm apart. Given μ = 2.4525 N s/m².*

Solution. Given :

$$U_{max} = 2 \text{ m/s}, \ t = 100 \text{ mm} = 0.1 \text{ m}, \ \mu = 2.4525 \text{ N/m}^2$$

Find (*i*) Pressure gradient, $\frac{dp}{dx}$

(*ii*) Shear stress at the wall, τ_0

(*iii*) Discharge per metre width, Q.

(*i*) **Pressure gradient,** $\frac{dp}{dx}$

Maximum velocity, U_{max}, is given by equation (10.10)

$$U_{max} = -\frac{1}{8\mu}\frac{\partial p}{\partial x}t^2$$

Substituting the values

$$\text{or} \quad 2.0 = -\frac{1}{8 \times 2.4525} \times \frac{\partial p}{\partial x} \times (.1)^2$$

$$\therefore \quad \frac{\partial p}{\partial x} = -\frac{2.0 \times 8 \times 2.4525}{.1 \times .1} = -\ \mathbf{3924\ N/m^2\ per\ m.\ Ans.}$$

(*ii*) **Shear stress at the wall,** τ_0

τ_0 is given by equation (10.15) as $\tau_0 = -\frac{1}{2}\frac{\partial p}{\partial x} \times t = -\frac{1}{2}(-3924) \times 0.1 =$ **196.2 N/m². Ans.**

(*iii*) **Discharge per metre width**, Q

$$= \text{Mean velocity} \times \text{Area}$$

$$= \frac{2}{3} U_{max} \times (t \times 1) = \frac{2}{3} \times 2.0 \times 0.1 \times 1 = \mathbf{0.133\ m^3/s.\ Ans.}$$

Problem 10.9 *An oil of viscosity 10 poise flows between two parallel fixed plates which are kept at a distance of 50 mm apart. Find the rate of flow of oil between the plates if the drop of pressure in a length of 1.2 m be 0.3 N/cm³. The width of the plates is 200 mm.*

Solution. Given :

$$\mu = 10 \text{ poise}$$

$$= \frac{10}{10} \text{ N s/m}^2 = 1 \text{ N s/m}^2 \qquad \left(\because\ 1 \text{ poise} = \frac{1}{10}\frac{\text{Ns}}{\text{m}^2}\right)$$

$$t = 50 \text{ mm} = 0.05 \text{ m}$$

$$p_1 - p_2 = 0.3 \text{ N/m}^2 = 0.3 \times 10^4 \text{ N/m}^2$$

$$L = 1.20 \text{ m}$$

Width, $B = 200 \text{ mm} = 0.20 \text{ m}.$

Find Q, rate of flow

The difference of pressure is given by equation (10.13)

$$p_1 - p_2 = \frac{12\mu\bar{u}L}{t^2}$$

Substituting the values, we get

$$0.3 \times 10^4 = 12 \times 1.0 \times \frac{\bar{u} \times 1.20}{.05 \times .05}$$

$$\therefore \qquad \bar{u} = \frac{0.3 \times 10^4 \times 1.0 \times .05 \times .05}{12 \times 1.20} = 0.52 \text{ m/s}$$

$\therefore$ Rate of flow

$$= \bar{u} \times \text{Area} = 0.52 \times (B \times t)$$

$$= 0.52 \times 0.20 \times .05 \text{ m}^3/\text{s} = .0052 \text{ m}^3/\text{s}$$

$$= 0.0052 \times 10^3 \text{ litre/s} = \mathbf{5.2\ litre/s.\ Ans.}$$

Problem 10.10 *Water at 15°C flows between two large parallel plates at a distance of 1.6 mm apart. Determine (a) the maximum velocity (b) the pressure drop per unit length and (c) the shear stress at the walls of the plates if the average velocity is 0.2 m/s. The viscosity of water at 15°C is given as 0.01 poise.* (VTU, Feb 2003)

Solution. Given :

$$t = 1.6 \text{ mm} = 1.6 \times 10^{-3} \text{ m}$$

$$= 0.0016 \text{ m}$$

$$\bar{u} = 0.2 \text{ m/sec}, \ \mu = .01 \text{ poise} = \frac{.01}{10} = 0.001 \text{ N s/m}^2$$

(*i*) **Maximum velocity**, U_{max} is given by equation (10.12) *i.e.,* $$U_{max} = \frac{3}{2}\bar{u} = 1.5 \times 0.2 = \textbf{0.3 m/s. Ans.}$$

(*ii*) **The pressure drop**, $(p_1 - p_2)$ is given by equation (10.13)

$$p_1 - p_2 = \frac{12\mu\bar{u}L}{t^2}$$

or pressure drop per unit length $= \dfrac{12\mu\bar{u}}{t^2}$

or $$\frac{\partial p}{\partial x} = 12 \times \frac{.01}{10} \times \frac{0.2}{(.0016)^2} = 937.44 \text{ N/m}^2 \text{ per m.}$$

(*iii*) **Shear stress at the walls** is given by equation (10.15)

$$\tau_0 = -\frac{1}{2}\frac{\partial p}{\partial x} \times t = \frac{1}{2} \times 937.44 \times .0016 = \textbf{0.749 N/m}^2\textbf{. Ans.}$$

Problem 10.11 *There is a horizontal crack 40 mm wide and 2.5 mm deep in a wall of thickness 100 mm. Water leaks through the crack. Find the rate of leakage of water through the crack if the difference of pressure between the two ends of the crack is 0.02943 N/cm². Take the viscosity of water equal to 0.01 poise.*

Solution. Given :

Width of crack, $b = 40 \text{ mm} = 0.04 \text{ m}$

Depth of crack, $t = 2.5 \text{ mm} = .0025 \text{ m}$

Length of crack, $L = 100 \text{ mm} = 0.1 \text{ m}$

$$p_1 - p_2 = 0.02943 \text{ N/cm}^2 = 0.02943 \times 10^4 \text{ N/m}^2 = 294.3 \text{ N/m}^2$$

$$\mu = .01 \text{ poise} = \frac{.01}{10}\frac{\text{Ns}}{\text{m}^2}$$

Find rate of leakage (Q)

$(p_1 - p_2)$ is given by equation (10.13) as

$$p_1 - p_2 = \frac{12\mu\bar{u}L}{t^2} \quad \text{or} \quad 294.3 = 12 \times \frac{.01}{10} \times \frac{\bar{u} \times 0.1}{(.0025 \times .0025)}$$

$$\therefore \quad \bar{u} = \frac{294.3 \times 10 \times .0025 \times .0025}{12 \times .01 \times 0.1} = 1.5328 \text{ m/s}$$

$\therefore$ Rate of leakage $= \bar{u} \times$ area of cross-section of crack

$= 1.538 \times (b \times t)$

$= 1.538 \times .04 \times .0025 \text{ m}^3\text{/s} = 1.538 \times 10^{-4} \text{ m}^3\text{/s}$

$= 1.538 \times 10^{-4} \times 10^3$ litre/s $=$ **0.1538 litre/s. Ans.**

Problem 10.12 *The radial clearance between a hydraulic plunger and the cylinder walls is 0.1 mm ; the length of the plunger is 300 mm and diameter 100 mm. Find the velocity of leakage and rate of leakage past the plunger at an instant when the difference of the pressure between the two ends of the plunger is 9 m of water. Take μ = 0.0127 poise.*

Solution. Given :

The flow through the clearance area will be the same as the flow between two parallel surfaces,

$$t = 0.1 \text{ mm} = 0.0001 \text{ m}$$
$$L = 300 \text{ mm} = 0.3 \text{ m}$$

Diameter, $D = 100 \text{ mm} = 0.1 \text{ m}$

Difference of presure $= \dfrac{p_1 - p_2}{\rho g} = 9$ m of water

$\therefore$ $$p_1 - p_2 = 9 \times 1000 \times 9.81 \text{ N/m}^2 = 88290 \text{ N/m}^2$$

Viscosity, $\mu = .0127 \text{ poise} = \dfrac{.0127}{10} \dfrac{\text{Ns}}{\text{m}^2}$

Find (*i*) Velocity of leakage, *i.e.*, mean velocity $\bar{u}$

(*ii*) Rate of leakage, Q

(*i*) **Velocity of leakage** ($\bar{u}$). The average velocity ($\bar{u}$) is given by equation (10.11)

$$\bar{u} = -\frac{1}{12\mu}\frac{\partial p}{\partial x}t^2$$

$$= \frac{1}{12 \times \dfrac{.0127}{10}} \times \frac{p_1 - p_2}{L} \times (.0001) \times (.0001)$$

$$= \frac{1}{12 \times .0127} \times \frac{88290}{0.3} \times (.0001) \times (.0001)$$

$= .193$ m/s $=$ **19.3 cm/s. Ans.**

(*ii*) **Rate of leakage**, Q

$$Q = \bar{u} \times \text{area of flow}$$
$$= 0.193 \times \pi D \times t \text{ m}^3\text{/s} = 0.193 \times \pi \times .1 \times .0001 \text{ m}^3\text{/s}$$
$$= 6.06 \times 10^{-6} \text{ m}^3\text{/s} = 6.06 \times 10^{-6} \times 10^3 \text{ litre/s}$$

$=$ **6.06×10^{-3} litre/s. Ans.**

▶ 10.6 KINETIC ENERGY CORRECTION AND MOMENTUM CORRECTION FACTORS

Kinetic energy correction factor is defined as the ratio of the kinetic energy of the flow per second based on actual velocity across a section to the kinetic energy of the flow per second based on average velocity across the same section. It is denoted by α. Hence mathematically,

$$\alpha = \frac{\text{K.E./sec based on actual velocity}}{\text{K.E./sec based on average velocity}} \qquad \text{...(10.16)}$$

Momentum Correction Factor. It is defined as the ratio of momentum of the flow per second based on actual velocity to the momentum of the flow per second based on average velocity across a section. It is denoted by β. Hence mathematically,

$$\beta = \frac{\text{Momentum per second based on actual velocity}}{\text{Momentum per second based on average velocity}}. \qquad \text{...(10.17)}$$

Problem 10.13 *Show that the momentum correction factor and energy correction factor for laminar flow through a circular pipe are 4/3 and 2.0 respectively.*

Solution. (*i*) **Momentum Correction Factor or β**

The velocity distribution through a circular pipe for laminar flow at any radius r is given by equation (10.3)

or $$u = \frac{1}{4\mu}\left(-\frac{\partial p}{\partial x}\right)(R^2 - r^2) \quad ...(1)$$

Consider an elementary area dA in the form of a ring at a radius r and of width dr, then

$$dA = 2\pi r\, dr$$

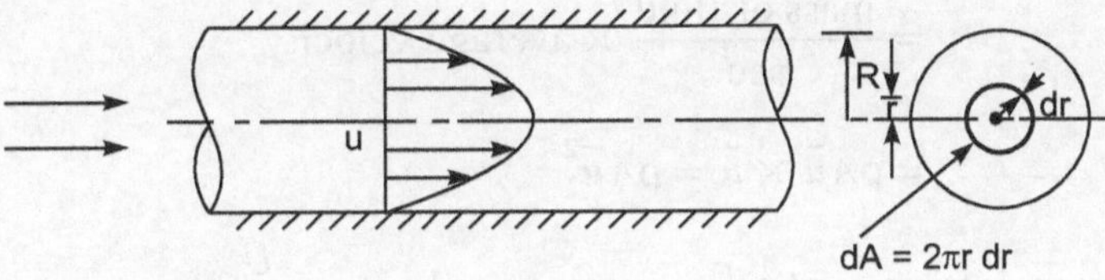

Fig. 10.9

Rate of fluid flowing through the ring

$= dQ =$ velocity × area of ring element

$= u \times 2\pi r\, dr$

Momentum of the fluid through ring per second

= mass × velocity

$= \rho \times dQ \times u = \rho \times 2\pi r\, dr \times u \times u = 2\pi\rho\, u^2 r\, dr$

∴ Total actual momentum of the fluid per second across the section

$$= \int_0^R 2\pi\rho\, u^2 r\, dr$$

Substituting the value of u from (1)

$$= 2\pi\rho \int_0^R \left[\frac{1}{4\mu}\left(\frac{-\partial p}{\partial x}\right)(R^2 - r^2)\right]^2 rdr$$

$$= 2\pi\rho\left[\frac{1}{4\mu}\left(\frac{-\partial p}{\partial x}\right)\right]^2 \int_0^R [R^2 - r^2]\, rdr$$

$$= 2\pi\rho \frac{1}{(16\mu^2)}\left(\frac{\partial p}{\partial x}\right)^2 \int_0^R (R^4 + r^4 - 2R^2r^2)\, dr$$

$$= \frac{\pi\rho}{8\mu^2}\left(\frac{\partial p}{\partial x}\right)^2 \int_0^R (R^4 r + r^5 - 2R^2 r^3)\, dr$$

$$= \frac{\pi\rho}{8\mu^2}\left(\frac{\partial p}{\partial x}\right)^2 \left[\frac{R^2 r^2}{2} + \frac{r^6}{6} - \frac{2R^2 r^4}{4}\right]_0^R = \frac{\pi\rho}{8\mu^2}\left(\frac{\partial p}{\partial x}\right)^2 \left[\frac{R^6}{2} + \frac{R^6}{6} - \frac{2R^6}{4}\right]$$

$$= \frac{\pi\rho}{8\mu^2}\left(\frac{\partial p}{\partial x}\right)^2 \frac{6R^6 + 2R^6 - 2R^6}{12}$$

$$= \frac{\pi\rho}{8\mu^2}\left(\frac{\partial p}{\partial x}\right)^2 \times \frac{R^6}{6} = \frac{\pi\rho}{48\mu^2}\left(\frac{\partial p}{\partial x}\right)^2 R^6 \qquad \text{...(2)}$$

Momentum of the fluid per second based on average velocity

$$= \frac{\text{mass of fluid}}{\text{sec}} \times \text{average velocity}$$

$$= \rho A\bar{u} \times \bar{u} = \rho A\bar{u}^2$$

where A= Area of cross-section $= \pi R^2$, $\bar{u}$ = average velocity $= \frac{U_{max}}{2}$

$$= \frac{1}{2} \times \frac{1}{4\mu}\left(-\frac{\partial p}{\partial x}\right) R^2 \qquad \left\{\because \; U_{max} = \frac{1}{4\mu}\left(-\frac{\partial p}{\partial x}\right) R^2\right\}$$

$$= \frac{1}{8\mu}\left(-\frac{\partial p}{\partial x}\right) R^2$$

$\therefore$ Momentum/sec based on average velocity

$$= \rho \times \pi R^2 \times \left[\frac{1}{8\mu}\left(-\frac{\partial p}{\partial x}\right) R^2\right]^2 = \rho \times \pi R^2 \times \frac{1}{64\mu^2}\left(-\frac{\partial p}{\mu x}\right)^2 R^4$$

$$= \frac{\rho\pi\left(-\frac{\partial p}{\partial x}\right)^2 R^6}{64\mu^2} \qquad \text{...(3)}$$

$$\therefore \qquad \beta = \frac{\text{Momentum / sec based on actual velocity}}{\text{Momentum / sec based on average velocity}}$$

$$= \frac{\frac{\pi\rho}{48\mu^2}\left(\frac{\partial p}{\partial x}\right)^2 R^6}{\frac{\pi\rho}{64\mu^2}\left(-\frac{\partial p}{\partial x}\right)^2 R^6} = \frac{64}{48} = \mathbf{\frac{4}{3}}. \textbf{ Ans.}$$

(*ii*) **Energy Correction Factor**, α. Kinetic energy of the fluid flowing through the elementary ring of radius 'r' and of width 'dr' per sec

$$= \frac{1}{2} \times \text{mass} \times u^2 = \frac{1}{2} \times \rho dQ \times u^2$$

$$= \frac{1}{2} \times \rho \times (u \times 2\pi r\, dr) \times u^2 = \frac{1}{2}\rho \times 2\pi r u^3\, dr = \pi\rho r u^3\, dr$$

$\therefore$ Total actual kinetic energy of flow per second

$$= \int_0^R \pi\rho r u^3\, dr = \int_0^R \pi\rho r \left[\frac{1}{4\mu}\left(-\frac{\partial p}{\partial x}\right)\left(R^2 - r^2\right)\right]^3 dr$$

$$= \pi\rho \times \left[\frac{1}{4\mu}\left(\frac{\partial p}{\partial \mu}\right)\right]^3 \int_0^R [R^2 - r^2]^3\, r dr$$

$$= \pi\rho \times \frac{1}{64\mu^3}\left(-\frac{\partial p}{\partial x}\right)^3 \int_0^R (R^6 - r^6 - 3R^4 r^2 + 3R^6 r^4) r\, dr$$

$$= \frac{\pi\rho}{64\mu^3}\left(-\frac{\partial p}{\partial x}\right)^3 \int_0^R (R^6 r - r^7 - 3R^4 r^3 + 3R^2 r^5)\, dr$$

$$= \frac{\pi\rho}{64\mu^3}\left(-\frac{\partial p}{\partial x}\right)^3 \left[\frac{R^6 r^2}{2} - \frac{r^8}{8} - \frac{3R^4 r^4}{4} + \frac{3R^2 r^6}{6}\right]_0^R$$

$$= \frac{\pi\rho}{64\mu^3}\left(-\frac{\partial p}{\partial x}\right)^3 \left[\frac{R^8}{2} - \frac{R^8}{8} - \frac{3R^8}{4} + \frac{3R^6}{6}\right]$$

$$= \frac{\pi\rho}{64\mu^3}\left(-\frac{\partial p}{\partial x}\right)^3 R^8 \left[\frac{12 - 3 - 18 + 12}{24}\right]$$

$$= \frac{\pi\rho}{64\mu^3}\left(-\frac{\partial p}{\partial x}\right)^3 \frac{R^8}{8} \qquad \text{...(4)}$$

Kinetic energy of the flow based on average velocity

$$= \frac{1}{2} \times \text{mass} \times \bar{u}^2 = \frac{1}{2} \times \rho A \bar{u} \times \bar{u}^2 = \frac{1}{2} \times \rho A \bar{u}^3$$

Substituting the value of $A = (\pi R^2)$

and $$\bar{u} = \frac{1}{8\mu}\left(-\frac{\partial p}{\partial x}\right) R^2$$

$\therefore$ Kinetic energy of the flow/sec

$$= \frac{1}{2} \times \rho \times \pi R^2 \times \left[\frac{1}{8\mu}\left(-\frac{\partial p}{\partial x}\right) R^2\right]^3$$

$$= \frac{1}{2} \times \rho \times \pi R^2 \times \frac{1}{64 \times 8\mu^3}\left(-\frac{\partial p}{\partial x}\right)^3 \times R^6$$

$$= \frac{\rho\pi}{128 \times 8\mu^3}\left(-\frac{\partial p}{\partial x}\right)^3 \times R^8 \quad \text{...(5)}$$

$$\therefore \quad \alpha = \frac{\text{K.E./sec based on actual velocity}}{\text{K.E./sec based on average velocity}} = \frac{\text{Equation (4)}}{\text{Equation (5)}}$$

$$= \frac{\dfrac{\pi\rho}{64\mu^3}\left(-\dfrac{\partial p}{\partial x}\right)^3 \times \dfrac{R^8}{8}}{\dfrac{\rho}{128 \times 8\mu^3}\left(-\dfrac{\partial p}{\partial x}\right)^3 \times R^8} = \frac{128 \times 8}{64 \times 8} = \textbf{2.0. Ans.}$$

▶ 10.7 POWER ABSORBED IN VISCOUS FLOW

For the lubrication of the machine parts, an oil is used. Flow of oil in bearings is an example of viscous flow. If a highly viscous oil is used for lubrication of bearings, it will offer great resistance and thus a greater power loss will take place. But if a light oil is used, a required film between the rotating part and stationary metal surface will not be possible. Hence, the wear of the two surface will take place. Hence an oil of correct viscosity should be used for lubrication. The power required to overcome the viscous resistance in the following cases will be determined :

1. Viscous resistance of Journal Bearings,
2. Viscous resistance of Foot-step Bearings,
3. Viscous resistance of Collar Bearings.

10.7.1 Viscous Resistance of Journal Bearings. Consider a shaft of diameter D rotating in a journal bearing. The clearance between the shaft and journal bearing is filled with a viscous oil. The oil film in contact with the shaft rotates as the same speed as that of shaft while the oil film in contact with journal bearing is stationary. Thus the viscous resistance will be offered by the oil to the rotating shaft.

Let N = speed of shaft in r.p.m.

t = thickness of oil film

L = length of oil film

$\therefore$ Angular speed of the shaft, $\omega = \dfrac{2\pi N}{60}$

$\therefore$ Tangential speed of the shaft= $\omega \times R$ or $V = \dfrac{2\pi N}{60} \times \dfrac{D}{2} = \dfrac{\pi D N}{60}$

The shear stress in the oil is given by, $\tau = \mu \dfrac{du}{dy}$

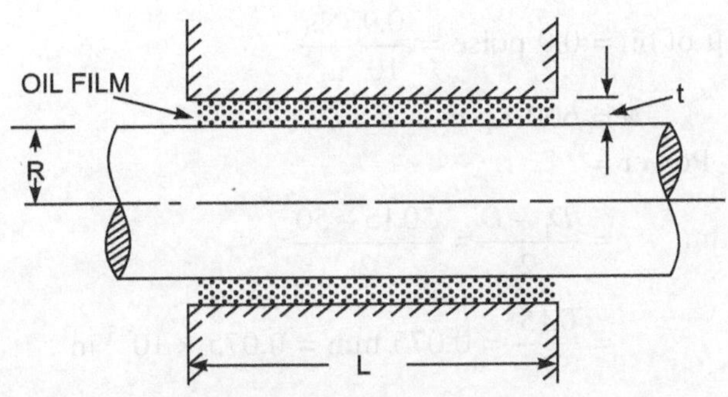

Fig. 10.10 *Journal bearing.*

As the thickness of oil film is very small, the velocity distribution in the oil film can be assumed as linear.

Hence $$\frac{du}{dy} = \frac{V - 0}{t} = \frac{V}{t} = \frac{\pi DN}{60 \times t}$$

$\therefore$ $$\tau = \mu \frac{\pi DN}{60 \times t}$$

$\therefore$ Shear force or viscous resistance = $\tau \times$ Area of surface of shaft

$$= \frac{\mu \pi DN}{60t} \times \pi DL = \frac{\mu \pi^2 D^2 NL}{60t}$$

$\therefore$ Torque required to overcome the viscous resistance,

$$T = \text{Viscous resistance} \times \frac{D}{2}$$

$$= \frac{\mu \pi^2 D^2 NL}{60t} \times \frac{D}{2} = \frac{\mu \pi^2 D^3 NL}{120t}$$

$\therefore$ Power absorbed in overcoming the viscous resistance

$$*P = \frac{2\pi NT}{60} = \frac{2\pi N}{60} \times \frac{\mu \pi^2 D^3 NL}{120t}$$

$$= \frac{\mu \pi^3 \mathbf{D^3 N^2 L}}{\mathbf{60 \times 60 \times t}} \textbf{ watts. Ans.} \quad \text{...(10.18)}$$

Problem 10.14 *A shaft having a diameter of 50 mm rotates centrally in a journal bearing having a diameter of 50.15 mm and length 100 mm. The angular space between the shaft and the bearing is filled with oil having viscosity of 0.9 poise. Determine the power absorbed in the bearing when the speed of rotation is 60 r.p.m.*

Solution. Given :

Dia. of shaft, $D = 50$ mm or .05 m

Dia. of bearing, $D_1 = 50.15$ mm or 0.05015 m

---

*Power, $P = T \times \omega = T \times \frac{2\pi N}{60} = \frac{2\pi NT}{60}$ watts $= \frac{2\pi NT}{60{,}000}$ kW.

Length, $L = 100$ mm or 0.1 m

$$\mu \text{ of oil} = 0.9 \text{ poise} = \frac{0.9}{10} \frac{\text{Ns}}{\text{m}^2}$$

$N = 600$ r.p.m.

Power = ?

$\therefore$ Thickness of oil film, $$t = \frac{D_1 - D}{2} = \frac{50.15 - 50}{2}$$

$$= \frac{0.15}{2} = 0.075 \text{ mm} = 0.075 \times 10^{-3} \text{ m}$$

Tangential speed of shaft, $$V = \frac{\pi DN}{60} = \frac{\pi \times 0.05 \times 600}{60} = 0.5 \times \pi \text{ m/s}$$

Shear stress $$\tau = \mu \frac{du}{dy} = \mu \frac{V}{t} = \frac{0.9}{10} \times \frac{0.5 \times \pi}{0.075 \times 10^{-3}} = 1883.52 \text{ N/m}^2$$

$\therefore$ Shear force (F) $$= \tau \times \text{Area} = 1883.52 \times \pi D \times L$$
$$= 1883.52 \times \pi \times .05 \times 0.1 = 29.586 \text{ N}$$

Resistance torque $$T = F \times \frac{D}{2} = 29.586 \times \frac{.05}{2} = 0.7387 \text{ Nm}$$

Power $$= \frac{2\pi NT}{60} = \frac{2\pi \times 600 \times 0.7387}{60} = \mathbf{46.41 \text{ W. Ans.}}$$

Problem 10.15 *A shaft of 100 mm, diameter rotates at 60 r.p.m. in a 200 mm long bearing. Taking that the two surfaces are uniformly separated by a distance of 0.5 mm and taking linear velocity distribution in the lubricating oil having dynamic viscosity of 4 centipoises, find the power absorbed in the bearing.* (A.M.I.E., Summer, 1976)

Solution. Given :

Dia. of shaft, $D = 100$ mm = 0.1 m

Length of bearing, $L = 200$ mm = 0.2 m

$t = 0.5$ mm = $.5 \times 10^{-3}$ m

$$\mu = 4 \text{ centipoise} = .04 \text{ poise} = \frac{0.04}{10} \frac{\text{Ns}}{\text{m}^2}$$

$N = 60$ r.p.m.

Find power (absorbed)

Using the equation (10.18), $$P = \frac{\mu \pi^3 D^3 N^2 L}{60 \times 60 \times t}$$

$$= \frac{.04}{10} \times \frac{\pi^3 \times (.1)^3 \times (60)^2 \times 0.2}{60 \times 60 \times 0.5 \times 10^{-3}} = \mathbf{4.961 \times 10^{-2} \text{ W. Ans.}}$$

Problem 10.16 *A shaft of diameter 0.35 m rotates at 200 r.p.m. inside a sleeve 100 mm long. The dynamic viscosity of lubricating oil in the 2 mm gap between sleeve and shaft is 8 poises. Calculate the power lost in the bearing.* (A.M.I.E., Summer, 1983)

Solution. Given :

Dia. of shaft, $D = 0.35$ m

Speed of shaft, $N = 200$ r.p.m.

Length of sleeve, $L = 100 \text{ mm} = 0.1 \text{ m}$

Distance between sleeve and shaft, $t = 2 \text{ mm} = 2 \times 10^{-3} \text{ m}$

Viscosity, $\mu = 8 \text{ poise} = \frac{8}{10} \frac{\text{Ns}}{\text{m}^2}$

The power lost in the bearing is given by equation (10.18) as

$$P = \frac{\mu \pi^3 D^3 N^2 L}{60 \times 60 \times t} \text{ watts}$$

$$= \frac{8}{10} \times \frac{\pi^3 \times (.35)^3 \times (200)^2 \times 0.1}{60 \times 60 \times 2 \times 10^{-3}} = \mathbf{590.8 \text{ W} = 0.59 \text{ kW. Ans.}}$$

Problem 10.17 *A sleeve, in which a shaft of diameter 75 mm, is running at 1200 r.p.m., is having a radial clearance of 0.1 mm. Calculate the torque resistance if the length of sleeve is 100 mm and the space is filled with oil of dynamic viscosity 0.96 poise.*

Solution. Given :

Dia. of shaft, $D = 75 \text{ mm} = 0.075 \text{ m}$

$N = 1200$ r.p.m.

$t = 0.1 \text{ mm} = 0.1 \times 10^{-3} \text{ m}$

Length of sleeve, $L = 100 \text{ mm} = 0.1 \text{ m}$

$\mu = 0.96 \text{ poise} = \frac{0.96}{10} \frac{\text{Ns}}{\text{m}^2}$

Tangential velocity of shaft, $V = \frac{\pi DN}{60} = \frac{\pi \times .075 \times 1200}{60} = 4.712 \text{ m/s}$

Shear stress, $\tau = \mu \frac{V}{t} = \frac{.96}{10} \times \frac{4.712}{.1 \times 10^{-3}} = 4523.5 \text{ N/m}^2$

Shear force, $F = \tau \times \pi DL$

$= 4523.5 \times \pi \times .075 \times .1 = 106.575 \text{ N}$

$\therefore$ Torque resistance $= F \times \frac{D}{2}$

$$= 106.575 \times \frac{.075}{2} = \mathbf{3.996 \text{ Nm. Ans.}}$$

Problem 10.18 *A shaft 100 mm diameter runs in a bearing of length 200 mm with a radial clearance of 0.025 mm at 30 r.p.m. Find the velocity of the oil, if the power required to overcome the viscous resistance is 183.94 watts.*

Solution. Given :

$D = 100 \text{ mm} = 0.1 \text{ m}$

$L = 200 \text{ mm} = 0.2 \text{ m}$

$t = .025 \text{ mm} = 0.025 \times 10^{-3} \text{ m}$

$N = 30$ r.p.m. ; H.P. = 0.25

Find viscosity of oil, μ.

The h.p. is given by equation (10.18) as

$$P = \frac{\mu\pi^3 D^3 N^2 L}{60 \times 60 \times t} \quad \text{or} \quad 183.94 = \frac{\mu\pi^3 \times (.1)^3 \times (30)^2 \times 0.2}{60 \times 60 \times 0.025 \times 10^{-3}}$$

$$\therefore \qquad \mu = \frac{183.94 \times 60 \times 60 \times .025 \times 10^{-3}}{\pi^3 \times .001 \times 900 \times 0.2} \frac{\text{Ns}}{\text{m}^2}$$

$$= 2.96 \frac{\text{Ns}}{\text{m}^2} = 2.96 \times 10 = \mathbf{29.6 \ poise.\ Ans.}$$

10.7.2 Viscous Resistance of Foot-Step Bearing. Fig. 10.11 shows the foot-step bearing, in which a vertical shaft is rotating. An oil film between the bottom surface of the shaft and bearing is provided, to reduce the wear and tear. The viscous resistance is offered by the oil to the shaft. In this case the radius of the surface of the shaft in contact with oil is not constant as in the case of the journal bearing. Hence, viscous resistance in foot-step bearing is calculated by considering an elementary circular ring of radius r and thickness dr as shown in Fig. 10.11.

Let N = speed of the shaft

t = thickness of oil film

R = radius of the shaft

Area of the elementary ring $= 2\pi r dr$

Now shear stress is given by $\tau = \mu \dfrac{du}{dy} = \mu \dfrac{V}{t}$

where V is the tangential velocity of shaft at radius r and is equal to

$$\omega \times r = \frac{2\pi N}{60} \times r$$

$\therefore$ Shear force on the ring $= dF = \tau \times$ area of elementary ring

$$= \mu \times \frac{2\pi N}{60} \times \frac{r}{t} \times 2\pi r \, dr = \frac{\mu}{15} \frac{\pi^2 N r^2}{t} dr$$

Fig. 10.11 *Foot-step bearing.*

$\therefore$ Torque required to overcome the viscous resistance,

$$dT = dF \times r$$

$$= \frac{\mu}{15t} \pi^2 N r^2 \, dr \times r = \frac{\mu}{15t} \pi^2 N r^3 \, dr \qquad \ldots(10.19)$$

$\therefore$ Total torque required to overcome the viscous resistance,

$$T = \int_0^R dT = \int_0^R \frac{\mu}{15t} \pi^2 N r^3 dr$$

$$= \frac{\mu}{15t} \pi^2 N \int_0^R r^3 dr = \frac{\mu}{15t} \pi^2 N \left[\frac{r^4}{4}\right]_0^R = \frac{\mu}{15t} \pi^2 N \frac{R^4}{4}$$

$$= \frac{\mu}{60t} \pi^2 N R^4 \qquad \ldots(10.19A)$$

$\therefore$ Power absorbed, $\quad P = \dfrac{2\pi N T}{60}$ watts

$$= \frac{2\pi N}{60} \times \frac{\mu}{60t} \pi^2 N R^4 = \frac{\mu\pi^2 N^2 R^4}{60 \times 30t} \qquad \ldots(10.20)$$

Problem 10.19 *Find the torque required to rotate a vertical shaft of diameter 100 mm at 750 r.p.m. The lower end of the shaft rests in a foot-step bearing. The end of the shaft and surface of the bearing are both flat and are separated by an oil film of thickness 0.5 mm. The viscosity of the oil is given 1.5 poise.* (A.M.I.E., Summer, 1977)

Solution. Given :

Dia. of shaft, $D = 100 \text{ mm} = 0.1 \text{ m}$

$\therefore$ $R = \frac{D}{2} = \frac{0.1}{2} = 0.05 \text{ m}$

$N = 750 \text{ r.p.m.}$

Thickness of oil film, $t = 0.5 \text{ mm} = 0.0005 \text{ m}$

$\mu = 1.5 \text{ poise} = \frac{1.5}{10} \frac{\text{Ns}}{\text{m}^2}$

The torque required is given by equation (10.19) or

$$T = \frac{\mu}{60t}\,\pi^2 N R^4 \text{ Nm}$$

$$= \frac{1.5}{10} \times \frac{\pi^2 \times 750 \times (.05)^4}{60 \times .0005} = \mathbf{0.2305 \text{ Nm. Ans.}}$$

Problem 10.20 *Find the power required to rotate a circular disc of diameter 200 mm at 1000 r.p.m. The circular disc has a clearance of 0.4 mm from the bottom flat plate and the clearance contains oil of viscosity 1.05 poise.*

Solution. Given :

Dia. of disc, $D = 200 \text{ mm} = 0.2 \text{ m}$

$\therefore$ $R = \frac{D}{2} = \frac{0.2}{2} = 0.1 \text{ m}$

$N = 1000 \text{ r.p.m.}$

Thickness of oil film, $t = 0.4 \text{ mm} = 0.0004 \text{ m}$

$\mu = 1.05 \text{ poise} = \frac{1.05}{10} \text{ N s/m}^2$

The power required to rotate the disc is given by equation (10.20) or

$$P = \frac{\mu \pi^3 N^2 R^4}{60 \times 30 \times t} \text{ watts}$$

$$= \frac{1.05}{10} \times \frac{\pi^3 \times 1000^2 \times (0.1)^4}{60 \times 30 \times .0004} = \mathbf{452.1 \text{ W. Ans.}}$$

10.7.3 Viscous Resistance of Collar Bearing. Fig. 10.12 shows the collar bearing, where the face of the collar is separated from bearing surface by an oil film of uniform thickness.

Let N = Speed of the shaft in r.p.m.

R_1 = Internal radius of the collar

R_2 = External radius of the collar

t = Thickness of oil film.

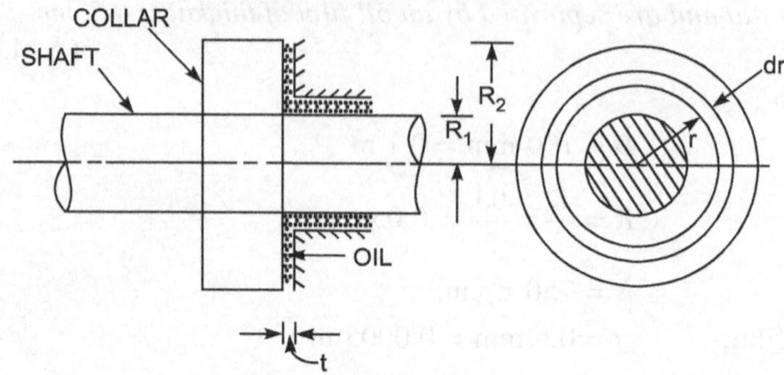

Fig. 10.12 *Collar bearing.*

Consider an elementary circular ring of radius 'r' and width dr of the bearing surface. Then the torque (dT) required to overcome the viscous resistance on the elementary circular ring is the same as given by equation (10.19A) or

$$dT = \frac{\mu}{15t} \pi^2 N r^3 \, dr$$

$\therefore$ Total torque, required to overcome the viscous resistance, on the whole collar is

$$T = \int_{R_1}^{R_2} dT = \int_{R_1}^{R_2} \frac{\mu}{15t} \pi^2 N r^3 \, dr = \frac{\mu}{15t} \pi^2 N \left[\frac{r^4}{4}\right]_{R_1}^{R_2}$$

$$= \frac{\mu}{15t \times 4} \pi^2 N [R_2^4 - R_1^4] = \frac{\mu}{60t} \pi^2 N [R_2^4 - R_1^4] \quad \text{...(10.21)}$$

$\therefore$ Power absorbed in overcoming viscous resistance

$$P = \frac{2\pi NT}{60} = \frac{2\pi N}{60} \times \frac{\mu}{60t} \pi^2 N [R_2^4 - R_1^4]$$

$$= \frac{\mu \pi^3 N^2}{60 \times 30t} [R_2^4 - R_1^4] \text{ watts.} \quad \text{...(10.22)}$$

Problem 10.21 *A collar bearing having external and internal diameters 150 mm and 100 mm respectively is used to take the thrust of a shift. An oil film of thickness 0.25 mm is maintained between the collar surface and the bearing. Find the power lost in overcoming the viscous resistance when the shaft rotates at 300 r.p.m. Take μ = 0.91 poise.*

Solution. Given :

External Dia. of collar, $D_2 = 150 \text{ mm} = 0.15 \text{ m}$

$\therefore$ $R_2 = \frac{D}{2} = \frac{.15}{2} = 0.075 \text{ m}$

Internal Dia. of collar, $D_1 = 100 \text{ mm} = 0.1 \text{ m}$

$\therefore$ $R_1 = \frac{D_1}{2} = \frac{0.1}{2} = 0.05 \text{ m}$

Thickness of oil film, $t = 0.25 \text{ mm} = 0.00025 \text{ m}$

$N = 300 \text{ r.p.m.}$

$\mu = 0.91 \text{ poise} = \frac{0.91}{10} \frac{\text{Ns}}{\text{m}^2}$

The power required is given by equation (10.22) or

$$P = \frac{\mu \pi^3 N^2}{60 \times 30 \times t} [R_3{}^4 - R_1{}^4]$$

$$= \frac{0.91}{10} \times \frac{\pi^3 \times 300^2 \times \left[.075^4 - .05^4\right]}{60 \times 30 \times .00025}$$

$$= 564314\ [.00003164 - .00000625]$$

$$= 564314 \times .00002539 = \mathbf{14.327\ W.\ Ans.}$$

Problem 10.22 *The external and internal diameters of a collar bearing are 200 mm and 150 mm respectively. Between the collar surface and the bearing, an oil film of thickness 0.25 mm and of viscosity 0.9 poise, is maintained. Find the torque and the power lost in overcoming the viscous resistance of the oil when the shaft is running at 250 r.p.m.*

Solution. Given :

$$D_2 = 200 \text{ mm} = 0.2 \text{ m}$$

$$\therefore \quad R_2 = \frac{D_2}{2} = \frac{0.2}{2} = 0.1 \text{ m}$$

$$D_1 = 150 \text{ mm} = 0.15 \text{ m}$$

$$\therefore \quad R_1 = \frac{D_1}{2} = \frac{0.15}{2} = .075 \text{ m}$$

$$t = 0.25 \text{ mm} = .00025 \text{ m}$$

$$\mu = 0.9 \text{ poise} = \frac{0.9}{10} \frac{\text{Ns}}{\text{m}^2}$$

Torque required is given by equation (10.21)

$$\therefore \quad T = \frac{\mu}{60t} \pi^2 N [R_2{}^4 - R_1{}^4] = \frac{0.9}{10} \times \frac{\pi^2 \times 250\left[0.1^4 - .075^4\right]}{60 \times 0.00025} \text{ Nm}$$

$$= 14804.4\ [\ .0001 - .00003164] = \mathbf{1.0114\ Nm.\ Ans.}$$

$\therefore$ Power lost in viscous resistance

$$= \frac{2\pi NT}{60} = \frac{2\pi \times 250 \times 1.0114}{60} = \mathbf{26.48\ W.\ Ans.}$$

▶ 10.8 LOSS OF HEAD DUE TO FRICTION IN VISCOUS FLOW

The loss of pressure head, h_f in a pipe of diameter D, in which a viscous fluid of viscosity μ is flowing with a velocity $\bar{u}$ is given by Hagen Poiseuille formula *i.e.*, by equation (10.6) as

$$h_f = \frac{32 \mu \bar{u} L}{\rho g D^2} \qquad \ldots(i)$$

where L = length of pipe

The loss of head due to friction* is given by

$$h_f = \frac{4 \,.\, f \,.\, L \,.\, V^2}{D \times 2g} = \frac{4 \,.\, f \,.\, L \,.\, \bar{u}^2}{D \times 2g} \qquad \ldots(ii)$$

$\{\because$ velocity in pipe is always average velocity $\therefore \quad V = \bar{u}\}$

where f = coefficient of friction between the pipe and fluid.

---

*For derivation, please refer to Art. 9.4

Equating (*i*) and (*ii*), we get $\frac{32\mu\bar{u}L}{\rho g D^2} = \frac{4.f.L.\bar{u}^2}{D \times 2g}$

$$f = \frac{32\mu\bar{u}L \times D \times 2g}{4.L.\bar{u}^2.\rho g.D^2} = \frac{16\mu}{\bar{u}.\rho.D} \qquad \{\because \quad \bar{u} = V\}$$

$$= 16 \times \frac{\mu}{\rho V D} = 16 \times \frac{1}{R_e}$$

where $\frac{\mu}{\rho V D} = \frac{1}{R_e}$ and R_e = Reynold number = $\frac{\rho V D}{\mu}$

$$\therefore \qquad f = \frac{16}{R_e}. \qquad \text{...(10.23)}$$

Problem 10.23 *Water is flowing through a 200 mm diameter pipe with coefficient of friction $f = 0.04$. The shear stress at a point 40 mm from the pipe axis is 0.00981 N/cm². Calculate the shear stress at the pipe wall.* (A.M.I.E., Summer, 1980)

Solution. Given :

Dia. of pipe, D = 200 mm = 0.20 m

Coefficient of friction, $f = 0.04$

Shear stress at r = 40 mm, $\tau = 0.00981$ N/cm²

Let the shear stress at pipe wall = τ_0.

First find whether the flow is viscous or not. The flow will be viscous if Reynold number R_e is less than 2000.

Using equation (10.23), we get $f = \frac{16}{R_e}$ or $.04 = \frac{16}{R_e}$

$$\therefore \qquad R_e = \frac{16}{.04} = 400$$

This means flow is viscous. The shear stress in case of viscous flow through a pipe is given by the equation (10.1) as

$$\tau = -\frac{\partial p}{\partial x}\frac{r}{2}$$

But $\frac{\partial p}{\partial x}$ is constant across a section. Across a section, there is no variation of x and there is no variation of p.

$$\therefore \qquad \tau \propto r$$

At the pipe wall, radius = 100 mm and shear stress is τ_0

$$\therefore \qquad \frac{\tau}{r} = \frac{\tau_0}{100} \quad \text{or} \quad \frac{0.00981}{40} = \frac{\tau_0}{100}$$

$$\therefore \qquad \tau_0 = \frac{100 \times 0.00981}{40} = \mathbf{0.0245\ N/cm^2.\ Ans.}$$

Problem 10.24 *A pipe of diameter 20 cm and length 10^4 m is laid at a slope of 1 in 200. An oil of sp. gr. 0.9 and viscosity 1.5 poise is pumped up at the rate of 20 litres per second. Find the head lost due to friction. Also calculate the power required to pump the oil.*

Solution. Given :

Dia. of pipe, $D = 20 \text{ cm} = 2.50 \text{ m}$

Length of pipe, $L = 10000 \text{ m}$

Slope of pipe, $i = 1 \text{ in } 200 = \dfrac{1}{200}$

Sp. gr. of oil, $S = 0.9$

$\therefore$ Density of oil, $\rho = 0.9 \times 1000 = 900 \text{ kg/m}^3$

Viscosity of oil, $\mu = 1.5 \text{ poise} = \dfrac{1.5}{10} \dfrac{\text{Ns}}{\text{m}^2}$

Discharge, $Q = 20 \text{ litres/s} = 0.02 \text{ m}^3\text{/s}$ $\{\because \ 1000 \text{ litres} = 1 \text{ m}^3\}$

$\therefore$ Velocity of flow, $\bar{u} = \dfrac{Q}{\text{Area}} = \dfrac{0.020}{\frac{\pi}{4}D^2} = \dfrac{0.020}{\frac{\pi}{4}(.2)^2} = 0.6366 \text{ m/s}$

$$\therefore \quad R_e = \text{Reynold number}$$

$$= \frac{\rho VD}{\mu} = \frac{900 \times 0.6366 \times .2}{\frac{1.5}{10}}$$

$$= \frac{900 \times .6366 \times .2 \times 10}{1.5} \qquad \{\because \ V = \bar{u} = 0.6366\}$$

$$= 763.89$$

As the Reynold number is less than 2000, the flow is viscous. The co-efficient of friction for viscous flow is given by equation (10.23) as

$$f = \frac{16}{R_e} = \frac{16}{763.89} = 0.02094$$

$\therefore$ Head lost due to friction, $h_f = \dfrac{4 \,.\, f \,.\, L \,.\, \bar{u}^2}{D \times 2g}$

$$= \frac{4 \times .02094 \times 10000 \times (.6366)^2}{0.2 \times 2 \times 9.81} \text{ m} = \textbf{86.50 m. Ans.}$$

Due to slope of pipe 1 in 200, the height through which oil is to be raised by pump

$$= \text{Slope} \times \text{Length of pipe}$$

$$= i \times L = \frac{1}{200} \times 10000 = 50 \text{ m}$$

$\therefore$ Total head against which pump is to work,

$$H = h_f + i \times L = 86.50 + 50 = 136.50 \text{ m}$$

$\therefore$ Power required to pump the oil

$$= \frac{\rho g \,.\, Q \,.\, H}{1000} = \frac{900 \times 9.81 \times 0.20 \times 136.50}{1000} = \textbf{24.1 kW. Ans.}$$

▶ 10.9 METHODS OF DETERMINATION OF CO-EFFICIENT OF VISCOSITY

The following are the experimental methods of determining the co-efficient of viscosity of a liquid :

1. Capillary tube method,
2. Falling sphere resistance method,
3. By rotating cylinder method, and
4. Orifice type viscometer.

The apparatus used for determining the viscosity of a liquid is called viscometer.

10.9.1 Capillary Tube Method. In capillary tube method, the viscosity of a liquid is calculated by measuring the pressure difference for a given length of the capillary tube. The Hagen Poiseuille law is used for calculating viscosity.

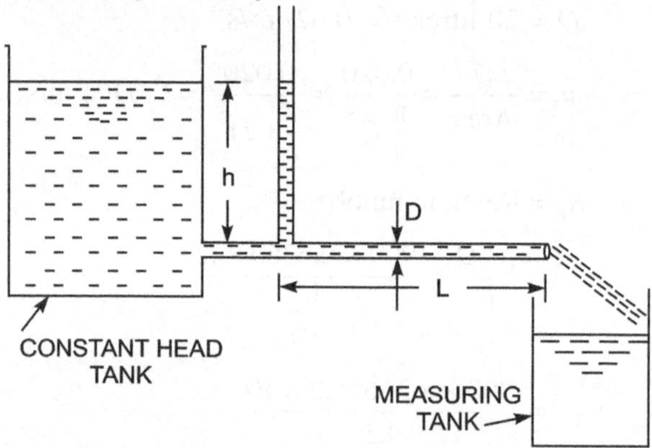

Fig. 10.13 *Capillary tube viscometer.*

Fig. 10.13 shows the capillary tube viscometer. The liquid whose viscosity is to be determined is filled in a constant head tank. The liquid is maintained at constant temperature and is allowed to pass through the capillary tube from the constant head tank. Then, the liquid is collected in a measuring tank for a given time. Then the rate of liquid collected in the tank per second is determined. The pressure head 'h' is measured at a point far away from the tank as shown in Fig. 10.13.

Then h = Difference of pressure head for length L.

The pressure at outlet is atmospheric,

Let D = Diameter of capillary tube,

L = Length of tube for which difference of pressure head is known,

ρ = Density of fluid,

and μ = Coefficient of viscosity.

Using Hagen Poiseuilli's Formula, $h = \dfrac{32\mu\bar{u}L}{\rho g D^2}$

But
$$\bar{u} = \frac{Q}{\text{Area}} = \frac{Q}{\frac{\pi}{4}D^2}$$

where Q is rate of liquid flowing through tube.

$$h = \frac{32\mu \times \dfrac{D}{\frac{\pi}{4}D^2} \times L}{\rho g D^2} = \frac{128\,\mu\, Q \,.\, L}{\pi \rho g D^4}$$

or $$\mu = \frac{\pi \rho g h \overline{D}}{128\, Q\, .\, L} \qquad \text{...(10.24)}$$

Measurement of D should be done very accurately.

10.9.2 Falling Sphere Resistance Method.

Theory. This method is based on Stoke's law, according to which the drag force, F on a small sphere moving with a constant velocity, U through a viscous fluid of viscosity, μ for viscous conditions is given by

$$F = 3\pi\mu U d \qquad \text{...}(i)$$

where d = diameter of sphere

U = velocity of sphere.

When the sphere attains a constant velocity U, the drag force is the difference of between the weight of sphere and buoyant force acting on it.

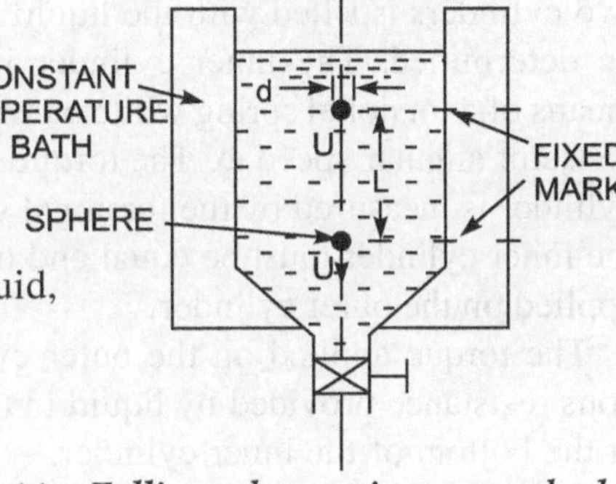

Fig. 10.14 *Falling sphere resistance method.*

Let L = distance travelled by sphere in viscous fluid,

t = time taken by sphere to cover distance l,

ρ_s = density of sphere,

ρ_f = density of fluid,

W = weight of sphere,

and F_B = buoyant force acting on sphere.

Then constant velocity of sphere, $U = \frac{L}{t}$

Weight of sphere, W = volume × density of sphere × g

$$= \frac{\pi}{6} d^3 \times \rho_s \times g \qquad \left\{\because \text{ volume of sphere} = \frac{\pi}{6} d^3\right\}$$

and buoyant force, F_B = weight of fluid displaced

= volume of liquid displaced × density of fluid × g

$$= \frac{\pi}{6} d^3 \times \rho_t \times g \qquad \{\text{volume of liquid displaced} = \text{volume of sphere}\}$$

For equilibrium,

Drag force = Weight of sphere – buoyant force

or $$F = W - F_B$$

Substituting the values of F, W and F_B, we get

$$3\pi\mu U d = \frac{\pi}{6} d^3 \times \rho_s \times g - \frac{\pi}{6} d^3 \times \rho_f \times g = \frac{\pi}{6} d^3 \times g \qquad [\rho_s - \rho_f]$$

or $$\mu = \frac{\pi}{6} \frac{d^3 \times g[\rho_s - \rho_f]}{3\pi U d} = \frac{g d^2}{18U} [\rho_s - \rho_f] \qquad \text{...(10.25)}$$

where ρ_f = Density of liquid

Hence in equation (10.25), the values of d, U, ρ_s and ρ_f are known and hence the viscosity of liquid can be determined.

Method. Thus this method consists of a tall vertical transparent cylindrical tank, which is filled with the liquid whose viscosity is to be determined. This tank is surrounded by another transparent tank to keep the temperature of the liquid in the cylindrical tank to be constant.

A spherical ball of small diameter 'd' is placed on the surface of liquid. Provision is made to release the drop this ball. After a short distance of travel, the ball attains a constant velocity. The time to travel a known vertical distance between two fixed marks on the cylindrical tank is noted to calculate the constant velocity U of the ball. Then with the known values of d, ρ_s, ρ_f the viscosity μ of the fluid is calculated by using equation (10.25).

10.9.3 Rotating Cylinder Method. This method consists of two concentric cylinders of radii R_1 and R_2 as shown in Fig. 10.15. The narrow space between the two cylinders is filled with the liquid whose viscosity is to be determined. The inner cylinder is held stationary by means of a torsional spring while outer cylinder is rotated at constant angular speed ω. The torque T acting on the inner cylinder is measured by the torsional spring. The torque on the inner cylinder must be equal and opposite to the torque applied on the outer cylinder.

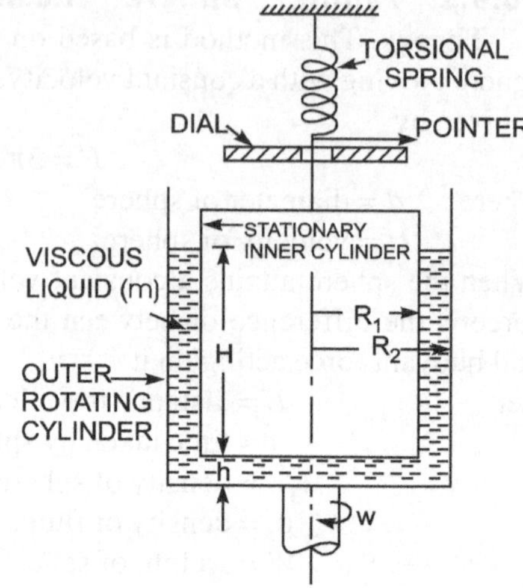

Fig. 10.15 *Rotating cylinder viscometer.*

The torque applied on the outer cylinder is due to viscous resistance provided by liquid in the annular space and at the bottom of the inner cylinder.

Let ω = angular speed of outer cylinder.

Tangential (peripheral) speed of outer cylinder

$$= \omega \times R_2$$

Tangential velocity of liquid layer in contact with outer cylinder will be equal to the tangential velocity of outer cylinder.

$\therefore$ Velocity of liquid layer with outer cylinder $= \omega \times R_2$

Velocity of liquid layer with inner cylinder = 0 {$\because$ Inner cylinder is stationary}

$\therefore$ Velocity gradient over the radial distance $(R_2 - R_1)$

$$= \frac{du}{dy} = \frac{\omega R_2 - 0}{R_2 - R_1} = \frac{\omega R_2}{R_2 - R_1}$$

$\therefore$ Shear stress (τ)

$$= \mu \frac{du}{dy} = \mu \frac{\omega R_2}{(R_2 - R_1)}$$

$\therefore$ Shear force (F) = shear stress $\times$ area of surface

$$= \tau \times 2\pi R_1 H$$

{$\because$ shear stress is acting on surface area $= 2\pi R_1 \times H$}

$$= \mu \frac{\omega R_2}{(R_2 - R_1)} \times 2\pi R_1 H$$

The torque T_1 on the inner cylinder due to shearing action of the liquid in the annular space is

$$T_1 = \text{shear force} \times \text{radius}$$

$$= \mu \frac{\omega R_2}{(R_2 - R_1)} \times 2\pi R_1 H \times R_1$$

$$= \frac{2\pi\mu\omega H R_1^2 R_2}{(R_2 - R_1)} \quad \ldots(i)$$

If the gap between the bottom of the two cylinders is 'h', then the torque applied on inner cylinder (T_2) is given by equation (10.19) as

$$T_2 = \frac{\mu}{60t}\ \pi^2\ NR^4$$

But here $R = R_1,\ t = h = \frac{\mu}{60h}\ \pi^2\ NR_1^{\ 4}$

$$\omega = \frac{2\pi N}{60} \text{ or } N = \frac{60\omega}{2\pi}$$

$$\therefore \quad T_2 = \frac{\mu}{60} \times \pi^2 \times \frac{60\omega}{2\pi} \times R_1^{\ 4} = \frac{\pi\mu\omega}{2h}\ R_1^{\ 4} \quad \text{...(ii)}$$

∴ Total torque T acting on the inner cylinder is

$$T = T_1 + T_2$$

$$= \frac{2\pi\mu\omega H\ R_1^2 R_2}{(R_2 - R_1)} + \frac{\pi\mu\omega}{4h}\ R_1^{\ 4} = 2\pi\mu R_1^{\ 2}\left[\frac{R_2 H}{R_2 - R_1} + \frac{R_1^2}{2h}\right] \times \omega$$

$$\therefore \quad \mu = \frac{2(R_2 - R_1)hT}{\pi R_1^2 \omega\left[4HhR_2 + R_1^2\ (R_2 - R_1)\right]} \quad \text{...(10.26)}$$

where T = torque measured by the strain of the torsional spring,
$R_1,\ R_2$ = radii of inner and outer cylinder,
h = clearance at the bottom of cylinders,
H = height of liquid in annular space,
μ = co-efficient of viscosity to be determined.

Hence the value of μ can be calculated from equation (10.26).

10.9.4 Orifice Type Viscometer. In this method, the time taken by a certain quantity of the liquid whose viscosity is to be determined, to flow through a short capillary tube is noted down. The co-efficient of viscosity is then obtained by comparing with the co-efficient of viscosity of a liquid whose viscosity is known or by the use conversion factors.

Viscometers such as Saybolt, Red wood or Engler are usually used. The principle for all the three viscometer is same. In the United Kingdom Redwood viscometer is used while in U.S.A., Saybolt viscometer is commonly used.

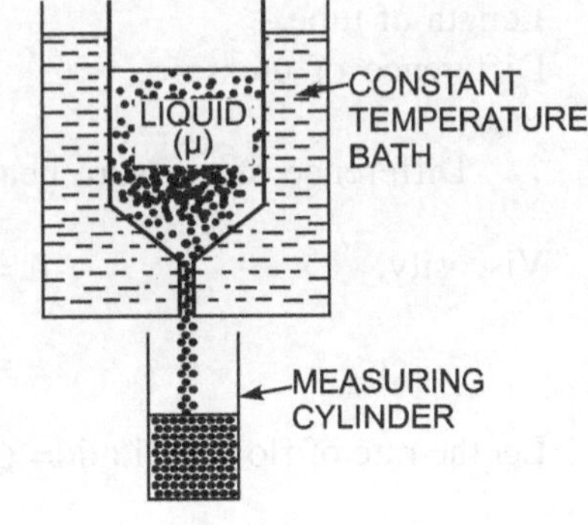

Fig. 10.16 *Saybolt viscometer.*

Fig. 10.16 shows that Saybolt viscometer, which consists of a tank at the bottom of which a short capillary tube is fitted. In this tank the liquid whose viscosity is to be determined is filled. This tank is surrounded by another tank, called constant temperature bath. The liquid is allowed to flow through capillary tube at a standard temperature. The time taken by 60 c.c. of the liquid to flow through the capillary tube is noted down. The initial height of liquid in the tank is previously adjusted to a standard height. From the time measurement, the kinematic viscosity of liquid is known from the relation,

$$\nu = At - \frac{B}{t}$$

where $A = 0.24$, $B = 190$, t = time noted in seconds, ν = kinematic viscosity in stokes.

Problem 10.25 *The viscosity of an oil of sp. gr. 0.9 is measured by a capillary tube of diameter 50 mm. The difference of pressure head between two points 2 m apart is 0.5 m of water. The mass of oil collected in a measuring tank is 60 kg in 100 seconds. Find the viscosity of oil.*

Solution. Given :

Sp. gr. of oil $= 0.9$

Dia. of capillary tube, $D = 50 \text{ mm} = 5 \text{ cm} = 0.05 \text{ m}$

Length of tube, $L = 2 \text{ m}$

Difference of pressure head, $h = 0.5 \text{ m}$

Mass of oil, $M = 60 \text{ kg}$

Time, $t = 100 \text{ s}$

Mass of oil per second $= \dfrac{60}{100} = 0.6 \text{ kg/s}$

Density of oil, $\rho = \text{sp. gr. of oil} \times 1000 = 0.9 \times 1000 = 900 \text{ kg/m}^3$

$\therefore$ Discharge, $Q = \dfrac{\text{Mass of oil / s}}{\text{Density}} = \dfrac{0.6}{900} \text{ m}^3\text{/s} = 0.000667 \text{ m}^3\text{/s}$

Using equation (10.24), we get viscosity,

$$\mu = \frac{\pi \rho g h D^4}{128\, Q\, .\, L} \qquad \text{[here } h = h_f = 0.5]$$

$$= \frac{\pi \times 900 \times 9.81 \times 0.5 \times (.05)^4}{128 \times 0.000667 \times 2.0} = 0.5075 \text{ (SI Units) N s/m}^2$$

$= 0.5075 \times 10$ poise $=$ **5.075 poise. Ans.**

Problem 10.26 *A capillary tube of diameter 2 mm and length 100 mm is used for measuring viscosity of a liquid. The difference of pressure between the two ends of the tube is 0.6867 N/cm² and the viscosity of liquid is 0.25 poise. Find the rate of flow of liquid through the tube.*

Solution. Given :

Dia. of capillary tube, $D = 2 \text{ m} = 2 \times 10^{-3} \text{ m}$

Length of tube, $L = 100 \text{ mm} = 10 \text{ cm} = 0.1 \text{ m}$

Difference of pressure, $\Delta p = 0.6867 \text{ N/cm}^2 = 0.6867 \times 10^4 \text{ N/m}^2$

$\therefore$ Difference of pressure head, $h = \dfrac{\Delta p}{\rho g} = \dfrac{0.07 \times 10^4}{\rho g}$

Viscosity, $\mu = 0.25$ poise

$= \dfrac{0.25}{10} \text{ Ns/m}^2$

Let the rate of flow of liquid $= Q$

Using equation (10.24), we get $\mu = \dfrac{\pi \rho g h D^4}{128\, .\, Q\, .\, L} = \pi \rho g \times \dfrac{\dfrac{0.6867 \times 10^4}{\rho g} \times \left(2 \times 10^{-3}\right)^4}{128 \times Q \times 0.1}$

or
$$\frac{0.25}{10} = \frac{\pi \times 0.6867 \times 10^4 \times \left(2 \times 10^{-3}\right)^4}{128 \times Q \times 0.1}$$

or $$Q = \frac{\pi \times 0.6867 \times 10^4 \times 2^4 \times 10^{-12} \times 10}{128 \times 0.1 \times 0.25} \text{ m}^3/\text{s}$$

$$= 107.86 \times 10^{-8} \text{ m}^3/\text{s} = 107.86 \times 10^{-8} \times 10^6 \text{ cm}^3/\text{s}$$

$$= 107.86 \times 10^{-2} \text{ cm}^3/\text{s} = \mathbf{1.078 \text{ cm}^3/\text{s}}. \textbf{ Ans.}$$

Problem 10.27 *A sphere of diameter 2 mm falls 150 mm in 20 seconds in a viscous liquid. The density of the sphere is 7500 kg/m³ and of liquid is 900 kg/m³. Find the co-efficient of viscosity of the liquid.*

Solution. Given :

Dia. of sphere, $d = 2 \text{ mm} = 2 \times 10^{-3} \text{ m}$

Distance travelled by sphere $= 150 \text{ mm} = 0.15 \text{ m}$

Time taken, $t = 20$ seconds

Velocity of sphere, $U = \frac{\text{Distance}}{\text{Time}} = \frac{0.15}{20} = .0075 \text{ m/s}$

Density of sphere, $\rho_s = 7500 \text{ kg/m}^3$

Density of liquid, $\rho_f = 900 \text{ kg/m}^3$

Using relation (10.25), we get $\mu = \frac{gd^2}{18U} [\rho_s - \rho_f] = \frac{9.81 \times [2 \times 10^{-3}]^2}{18 \times 0.0075} [7500 - 900]$

$$= \frac{9.81 \times 4 \times 10^{-6} \times 6600}{18 \times 0.0075} = 1.917 \frac{\text{Ns}}{\text{m}^2}$$

$$= 1.917 \times 10 = \mathbf{19.17 \text{ poise.}} \textbf{ Ans.}$$

Problem 10.28 *Find the viscosity of a liquid of sp. gr. 0.8, when a gas bubble of diameter 10 mm rises steadily through the liquid at a velocity of 1.2 cm/s. Neglect the weight of the bubble.*

Solution. Given :

Sp. gr. of liquid $= 0.8$

$\therefore$ Density of liquid, $\rho_f = 0.8 \times 1000 = 800 \text{ kg/m}^3$

Dia. of gas bubble, $D = 10 \text{ mm} = 1 \text{ cm} = 0.01 \text{ m}$

Velocity of bubble, $U = 1.2 \text{ cm/s} = .012 \text{ m/s}$

As weight of bubble is neglected and density of bubble

$$\rho_s = 0$$

Now using the relation, $\mu = \frac{gd^2}{18U} [\rho_s - \rho_f]$ which is for a falling sphere.

For a rising bubble, the relation will become as

$$\mu = \frac{gd^2}{18U} [\rho_f - \rho_s]$$

Substituting the values, we get $\mu = \frac{9.81 \times .01 \times .01}{18 \times .012} [800 - 0] \frac{\text{Ns}}{\text{m}^2} = 3.63 \frac{\text{Ns}}{\text{m}^2}$

$$= 3.63 \times 10 = \mathbf{36.3 \text{ poise.}} \textbf{ Ans.}$$

Problem 10.29 *The viscosity of a liquid is determined by rotating cylinder method, in which case the inner cylinder of diameter 20 cm is stationary. The outer cylinder of diameter 20.5 cm, contains the liquid up to a height of 30 cm. The clearance at the bottom of the two cylinders is 0.5 cm. The outer cylinder is rotated at 400 r.p.m. The torque registered on the torsion meter attached to the inner cylinder is 5.886 Nm. Find the viscosity of fluid.*

Solution. Given :

Dia. of inner cylinder, $D_1 = 20$ cm

∴ Radius of inner cylinder, $R_1 = 10$ cm $= 0.1$ m

Dia. of outer cylinder, $D_2 = 20.5$ cm

∴ Radius of outer cylinder, $R_2 = \dfrac{20.5}{2} = 10.25$ cm $= .1025$ m

Height of liquid from bottom of outer cylinder = 30 cm

Clearance at the bottom of two cylinder, $h = 0.5$ cm $= .005$ m

∴ Height of inner cylinder immersed in liquid

$$= 30 - h = 30 - 0.5 = 29.5 \text{ m}$$

or $H = 29.5$ cm $= .295$ m

Speed of outer cylinder, $N = 400$ r.p.m.

∴ $$\omega = \frac{2\pi N}{60} = \frac{2 \times \pi \times 400}{60} = 41.88$$

Torque measured, $T = 5.886$ Nm

Using equation (10.26), we get $$\mu = \frac{2(R_2 - R_1) \times h \times T}{\pi R_1^2 \omega \left[4HhR_2 + R_1^2(R_2 - R_1)\right]}$$

$$= \frac{2(.1025 - 0.1) \times .005 \times 5.886}{\pi \times (.1)^2 \times 41.88\left[4 \times .295 \times .005 \times .1025 + .1^2\ (.1025 - .1)\right]}$$

$$= \frac{2 \times .0025 \times .005 \times 5.886}{\pi \times .01 \times 41.88\left[.0006047 - .000025\right]}$$

$= 0.19286$ Ns/m^2 $= 0.19286 \times 10 =$ **1.9286 poise. Ans.**

Problem 10.30 *A sphere of diameter 1 mm falls through 335 m in 100 seconds in a viscous fluid. If the relative densities of the sphere and the liquid are 7.0 and 0.96 respectively, determine the dynamic viscosity of the liquid.* (A.M.I.E., Winter 1990)

Solution. Given :

Dia. of sphere, $d = 1$ mm $= 0.001$ m

Distance travelled by sphere = 335 mm = 0.335 m

Time taken, $t = 100$ seconds

∴ Velocity of sphere, $$U = \frac{\text{Distance}}{\text{Time}} = \frac{0.335}{100} = 0.00335 \text{ m/sec}$$

Relative density of sphere = 7

∴ Density of sphere, $\rho_s = 7 \times 1000 = 7000$ kg/m^3

Relative density of liquid = 0.96

$\therefore$ Density of liquid, $\rho_f = 0.96 \times 1000 = 960$ kg/m^3

Using the relation (10.25), we get $\mu = \frac{gd^2}{18U}[\rho_s - \rho_f] = \frac{9.81 \times 0.001^2}{18 \times 0.00335}[7000 - 960]$

$$= \frac{0.00000981 \times 6040}{18 \times 0.00335} = 0.981 \text{ Ns/m}^2$$

$$= 0.981 \times 10 = \textbf{9.81 poise. Ans.}$$

Problem 10.31 *Determine the fall velocity of 0.06 mm sand particle (specific gravity = 2.65) in water at 20°C, take $\mu = 10^{-3}$ kg/ms.*

Solution. Given :

Dia. of sand particle, $d = 0.06 \text{ mm} = 0.06 \times 10^{-3}$ m

Specific gravity of sand = 2.65

$\therefore$ Density of sand, $\rho_s = 2.65 \times 1000$ kg/m^3 ($\because$ ρ for water in S.I. unit = 1000 kg/m^3)
$= 2650$ kg/m^3

Viscosity of water, $\mu^* = 10^{-3}$ kg/ms $= 10^{-3}$ Ns/m^2 $\left[\because \frac{\text{Ns}}{\text{m}^2} = \left(\text{kg} \times \frac{\text{m}}{\text{s}^2}\right) \times \frac{\text{s}}{\text{m}^2} = \frac{\text{kg}}{\text{ms}}\right]$

Density of water, $\rho_f = 1000$ kg/m^3

Sand particle is just like a sphere.

For equilibrium of sand particle,

Drag force = Weight of sand particle – buoyant force

or $F_D = W - F_B$ ...(*i*)

But $F_D = 3\pi\mu \times U \times d$, where U = Velocity of particle

$= 3\pi \times 10^{-3} \times U \times 0.06 \times 10^{-3}$ N

W = Weight of sand particle

$$= \frac{\pi}{6} \times d^3 \times \rho_s \times g = \frac{\pi}{6} \times (0.06 \times 10^{-3})^3 \times 2650 \times 9.81 \text{ N}$$

F_B = Buoyant force = Weight of water displaced

$$= \frac{\pi}{6} \times d^3 \times \rho_f \times g = \frac{\pi}{6} \times (0.06 \times 10^{-3})^3 \times 1000 \times 9.81 \text{ N}$$

Substituting the above values in equation (*i*), we get

$$3\pi \times 10^{-3} \times U \times 0.06 \times 10^{-3} = \frac{\pi}{\pi} \times (0.06 \times 10^{-3})^3 \times 2650 \times 9.81 - \frac{\pi}{6} \times (0.06 \times 10^{-3})^3 \times 1000 \times 9.81$$

Cancelling ($\pi \times 0.06 \times 10^{-3}$) throughout, we get

$$3 \times U = \frac{1}{6} \times 0.06^2 \times 10^{-3} \times 2650 \times 9.81 - \frac{1}{6} \times 0.06^2 \times 10^{-3} \times 1000 \times 9.81$$

---

*Viscosity in S.I. unit = N s/m^2. But 1 N = 1 kg × 1 m/s^2

Hence viscosity $= \left(\frac{1 \text{ kg} \times 1 \text{ m}}{\text{s}^2}\right) \times \frac{\text{s}}{\text{m}^2}$ = kg/ms. Hence kg/ms $= \frac{\text{Ns}}{\text{m}^2}$.

$$= \frac{1}{6} \times 0.06^2 \times 10^{-3} \times 9.81\ (2650 - 1000)$$

$$= \frac{1}{6} \times 0.0036 \times 10^{-3} \times 9.81 \times 1650 = 0.009712$$

$\therefore$ $U = 0.009712/3 =$ **0.00323 m/sec. Ans.**

HIGHLIGHTS

1. A flow is said to be viscous if the Reynold number is less than 2000, or the fluid flows in layers.
2. For the viscous flow through circular pipes,

(i) Shear stress ...... $\tau = -\frac{\partial p}{\partial x}\frac{r}{2}$ (ii) Velocity ......$u = -\frac{1}{4\mu}\frac{\partial p}{\partial x}[R^2 - r]$

(iii) Ratio of velocities $\frac{U_{max}}{u} = 2.0$ (iv) Loss of pressure head, $h_f = \frac{32\mu\bar{u}L}{\rho g D^2}$

where $\frac{\partial p}{\partial x}$ = pressure gradient, r = radius at any point,

R = radius of the pipe, U_{max} = maximum velocity or velocity at $r = 0$,

$\bar{u}$ = average velocity = $\frac{Q}{\pi R^2}$, μ = co-efficient of viscosity,

D = diameter of the pipe.

3. For the viscous flow between two parallel plates,

$$u = \frac{1}{2\mu}\frac{\partial p}{\partial x}(ty - y^2) \text{ ...... Velocity distribution}$$

$$\frac{U_{max}}{\bar{u}} = 1.5 \text{ ...... Ratio of maximum and average velocity}$$

$$h_f = \frac{12\mu\bar{u}L}{\rho g t^2} \text{ ...... Loss of pressure head}$$

$$\tau = -\frac{1}{2}\frac{\partial p}{\partial x}[t - 2y] \text{ ...... Shear stress distribution}$$

where t = thickness or distance between two plates,
y = distance in the vertical direction from the lower plate
τ = shear stress at any point in flow.

4. The kinetic energy correction factor α is given as

$$\alpha = \frac{\text{K.E. per second based on actual velocity}}{\text{K.E. per second based on average velocity}}$$

$= 2.0$ ...... for a circular pipe.

5. Momentum correction factor, β is given by

$$\beta = \frac{\text{Momentum per second based on actual velocity}}{\text{Momentum per second based on average velocity}}$$

$= \frac{4}{3}$ ...... for a circular pipe.

6. For the viscous resistance of Journal Bearing.

$$V = \frac{\pi DN}{60}, \frac{du}{dy} = \frac{V}{t} = \frac{\pi DN}{60t}$$

$$\tau = \frac{\mu\pi dN}{660t}, \text{ Shear force} = \frac{\mu\pi^2 D^2 NL}{60t}$$

Torque, $T = \dfrac{\mu\pi^2 D^3 NL}{120t}$ and power $= \dfrac{\mu\pi^3 D^3 N^2 L}{60 \times 60 \times t}$

where L = length of bearing, N = speed of shaft

t = clearance between the shaft and bearing.

7. For the Foot-Step Bearing, the shear force, torque and h.p. absorbed are given as :

Shear force, $F = \dfrac{\mu}{15} \dfrac{\pi^2 N}{t} \dfrac{R^3}{3}$

Torque, $T = \dfrac{\mu}{60t} \pi^2 NR^4$

Power $= \dfrac{\mu\pi^3 N^2 R^4}{60 \times 30 \times t}$

where R = radius of the shaft, N = speed of the shaft.

8. For the collar bearing the torque and power absorbed are given as

$$T = \frac{\mu}{60t} \pi^2 N [R_2{}^4 - R_1{}^4], \qquad P = \frac{\mu\pi^3 N^2}{60 \times 30t} [R_2{}^4 - R_1{}^4]$$

where R_1 = internal radius of the collar, R_2 = external radius of the collar,

t = thickness of oil film, P = power in watts.

9. For the viscous flow the co-efficient of friction is given by, $f = \dfrac{16}{R_e}$

where R_e = the Reynold number $= \dfrac{\rho VD}{\mu} = \dfrac{VD}{\nu}$.

10. The co-efficient of viscosity of a liquid is also determined experimentally by the following method :

(*i*) Capillary tube method, $\mu = \dfrac{\pi wh\, D^4}{128\, QL}$

(*ii*) Falling sphere method, $\mu = \dfrac{gd^2 \left[\rho_s - \rho_f\right]}{18U}$

(*iii*) Rotating cylinder method, $\mu = \dfrac{2\,(R_2 - R_1)hT}{\pi R_1^2 \omega \left[4HhR_2 + R_1^2\,(R_2 - R_1)\right]}$

where w = specific weight of fluid, L = length of the tube,

D = diameter of the capillary tube, Q = rate of flow of fluid through capillary tube,

d = diameter of the sphere, ρ_s = density of sphere,

ρ_f = density of fluid, U = velocity of sphere,

R_2 = radius of outer rotating cylinder, R_1 = radius of inner stationary cylinder,

T = torque.

EXERCISE 10

(A) THEORETICAL PROBLEMS

1. Define the terms : Viscosity, kinematic viscosity, velocity gradient and pressure gradient.
2. What do you mean by 'Viscous Flow'.
3. Derive an expression for the velocity distribution for viscous flow through a circular pipe. Also sketch the velocity distribution and shear stress distribution across a section of the pipe. *(VTU, Jan 2003)*
4. Prove that the maximum velocity in a circular pipe for viscous flow is equal to two times the average velocity of the flow. *(Delhi University, December 2002)*
5. Find an expression for the loss of head of a viscous fluid flowing through a circular pipe.
6. What is Hagen Poiseuille's Formula ? Derive an expression for Hagen Poiseuille's Formula. *(VTU, March 2000)*
7. Prove that the velocity distribution for viscous flow between two parallel plates when both plates are fixed across a section is parabolic in nature. Also prove that maximum velocity is equal to one and a half times the average velocity.
8. Show that the difference of pressure head for a given length of the two parallel plates which are fixed and through which viscous fluid is flowing is given by

$$h_f = \frac{12\mu\bar{u}L}{\rho g\, t^2}$$

where μ = Viscosity of fluid, $\bar{u}$ = Average velocity,
t = Distance between the two parallel plates, L = Length of the plates.
9. Define the terms : Kinetic energy correction factor and momentum correction factor.
10. Prove that for viscous flow through a circular pipe the kinetic energy correction factor is equal to 2 while momentum correction factor = $\frac{4}{3}$. *(Delhi University, 1992)*
11. A shaft is rotating in a journal bearing. The clearance between the shaft and the bearing is filled with a viscous oil. Find an expression for the power absorbed in overcoming viscous resistance.
12. Show that the value of the co-efficient of friction for viscous flow through a circular pipe is given by,

$$f = \frac{16}{R_e}, \quad \text{where} \quad R_e = \text{Reynold number.}$$

13. What are the different methods of determining the co-efficient of viscosity of a liquid ? Describe any two method in details.
14. Prove that the loss of pressure head for the viscous flow through a circular pipe is given by

$$h_f = \frac{32\mu\,\bar{u}L}{wd^2}$$

where $\bar{u}$ = Average velocity, w = Specific weight. *(Delhi University, 1987)*
15. For a laminar steady flow, prove that the pressure gradient in a direction of motion is equal to the shear gradient normal to the direction of motion.
16. Describe Reynolds experiments to demonstrate the two types of flow.
17. For the laminar flow through a circular pipe, prove that :
   (*i*) the shear stress variation across the section of the pipe is linear and
   (*ii*) the velocity variation is parabolic. *(VTU, Jan 2004; Delhi University, May 1998)*

(B) NUMERICAL PROBLEMS

1. A crude oil of viscosity 0.9 poise and sp. gr. 0.8 is flowing through a horizontal circular pipe of diameter 80 mm and of length 15 m. Calculate the difference of pressure at the two ends of the pipe, if 50 kg of the oil is collected in a tank in 15 seconds. [**Ans.** 0.559 N/cm^2]

2. A viscous flow is taking place in a pipe of diameter 100 mm. The maximum velocity is 2 m/s. Find the mean velocity and the radius at which this occurs. Also calculate the velocity at 30 mm from the wall of the pipe. [**Ans.** 1 m/s, $r = 35.35$ mm, $u = 1.68$ m/s]

3. A fluid of viscosity 0.5 poise and specific gravity 1.20 is flowing through a circular pipe of diameter 100 mm. The maximum shear stress at the pipe wall is given as 147.15 N/m^2, find : (*a*) the pressure gradient, (*b*) the average velocity, and (*c*) the Reynold number of the flow.
[**Ans.** (*a*) – 64746 N/m^2 per m, (*b*) 3.678 m/s, (*c*) 882.72]

4. Determine (*a*) the pressure gradient, (*b*) the shear stress at the two horizontal parallel plates and (*c*) the discharge per metre width for the laminar flow of oil with a maximum velocity of 1.5 m/s between two horizontal parallel fixed plates which are 80 mm apart. Take viscosity of oil as $\frac{1.962 \text{ Ns}}{\text{m}^2}$.
[**Ans.** (*a*) – 3678.7 N/m^2 per m, (*b*) 147.15 N/m^2, (*c*) .08 m^3/s]

5. Water is flowing between two large parallel plates which are 2.0 mm apart. Determine : (*a*) maximum velocity, (*b*) the pressure drop per unit length and (*c*) the shear stress at walls of the plate if the average velocity is 0.4 m/s. Take viscosity of water as 0.01 poise.
[**Ans.** (*a*) 0.6 m/s, (*b*) 1199.7 N/m^2 per m, (*c*) 1.199 N/m^3]

6. There is a horizontal crack 50 mm wide and 3 mm deep in a wall of thickness 150 mm. Water leaks through the crack. Find the rate of leakage of water through the crack if the difference of pressure between the two ends of the crack is 245.25 N/m^2. Take the viscosity of water as 0.01 poise.
[**Ans.** 183.9 cm^3/s]

7. A shaft having a diameter of 10 cm rotates centrally in a journal bearing having a diameter of 10.02 cm and length 20 cm. The annular space between the shaft and the bearing is filled with oil having viscosity of 0.8 poise. Determine the power absorbed in the bearing when the speed of rotation is 500 r.p.m. [**Ans.** 343.6 W]

8. A shaft 150 mm diameter runs in a bearing of length 300 mm, with a radial clearance of 0.04 mm at 40 r.p.m. Find the viscosity of the oil, if the power required to overcome the viscous resistance is 220.725 W.
[**Ans.** 6.32 poise]

9. Find the torque required to rotate a vertical shaft of diameter 8 cm at 800 r.p.m. The lower end of the shaft rests in a foot-step bearing. The end of the shaft and surface of the bearing are both flat and are separated by an oil film of thickness 0.075 cm. The viscosity of the oil is given as 1.2 poise.
[**Ans.** 0.0538 Nm]

10. A collar bearing having external and internal diameters 20 cm and 10 cm respectively is used to take the thrust of a shaft. An oil film of thickness 0.03 cm is maintained between the collar surface and the bearing. Find the power lost in overcoming the viscous resistance when the shaft rotates at 250 r.p.m. Take $\mu = 0.9$ poise. [**Ans.** 30.165 W]

11. Water is flowing through a 150 mm diameter pipe with a co-efficient of friction $f = .05$. The shear stress at a point 40 mm from the pipe wall is 0.01962 N/cm^2. Calculate the shear stress at the pipe wall.
[**Ans.** 0.04198 N/cm^2]

12. The viscosity of oil of sp. gr. 0.8 is measured by a capillary tube of diameter 40 mm. The difference of pressure head between two points 1.5 m apart is 0.3 m of water. The mass of oil collected in a measuring tank is 40 kg in 120 seconds. Find the viscosity of the oil. [**Ans.** 2.36 poise]

13. A capillary tube of diameter 4 mm and length 150 mm is used for measuring viscosity of a liquid. The difference of pressure between the two ends of the tube is 0.7848 N/cm^2 and the viscosity of the liquid is 0.2 poise. Find the rate of flow of liquid through the tube. [**Ans.** 16.43 cm^3/s]

14. A sphere of diameter 3 mm falls 100 mm in 1.5 seconds in a viscous liquid. The density of the sphere is 7000 kg/m^3 and of liquid is 800 kg/m^3. Find the co-efficient of viscosity of the liquid. [**Ans.** 45.61 poise]

15. The viscosity of a liquid is determined by rotating cylinder method, in which case the inner cylinder of diameter 25 cm is stationary. The outer cylinder of diameter 25.5 cm contains the liquid up to a height of 40 cm. The clearance at the bottom of the two cylinders is 0.6 cm. The outer cylinder is rotated at 300 r.p.m. The torque registered on the torsion metre attached to the inner cylinder is 4.905 Nm. Find the viscosity of liquid. **[Ans.** .77 poise]

16. Calculate : (*a*) the pressure gradient along the flow, (*b*) the average velocity, and (*c*) the discharge for an oil of viscosity 0.02 Ns/m^2 flowing between two stationary parallel plates 1 m wide maintained 10 mm apart. The velocity midway between the plates is 2.5 m/s. (*Delhi University, 1981*)
[Ans. – 4000 N/m^2 per m, 1.667 m/g, .01667 m^3/s]

17. Calculate :

(*i*) the pressure gradient along the flow,

(*ii*) the average velocity, and

(*iii*) the discharge for an oil of viscosity 0.03 N s/m^2 flowing between two stationary plates which are parallel and are at 10 mm apart. Width of plates is 2 m. The velocity midway between the plates is 2.0 m/s. (*Delhi University, 1988*)

18. A cylinder of 100 mm diameter, 0.15 m length and weighing 10 N slides axially in a vertical pipe of 104 mm dia. If the space between cylinder surface and pipe wall is filled with liquid of viscosity μ and the cylinder slides downwards at a velocity of 0.45 m/s, determine μ. (*A.M.I.E., W 1991*)
[**Hint.** $D = 100$ mm $= 0.1$, $L = 0.15$ m, $W = 10$ N, $D_p = 1.4$ mm $= 0.104$ m,
$V = 0.45$ m/s. Hence $t = (0.104 - 0.1)/2 = 0.002$ m.

$$\mu = \frac{4Wt^3}{3\pi D^3 LV} = \frac{4 \times 10 \times 0.002^3}{3\pi \times 0.1^3 \times 0.15 \times .45} = 503 \times 10^{-6} \text{ N s/m}^2 \Big]$$

19. A liquid is pumped through a 15 cm diameter and 300 m long pipe at the rate of 20 tonnes per hour. The density of liquid is 910 kg/m^3 and kinematic viscosity = 0.002 m^2/s. Determine the power required and show that the flow is viscous.
[**Hint.** $D = 15$ cm $= 0.15$ m, $L = 300$ m, $W = 20$ tonnes/hr
$= 20 \times 1000$ kgf/60×60 sec $= 5.555$ kgf/sec $= 5.555 \times 9.81$ N/s.

$$Q = \frac{W}{\rho g} = \frac{5.555 \times 9.81}{910 \times 9.81} = 0.0061 \text{ m}^3\text{/s. } V = \frac{Q}{A} = \frac{0.0061}{\frac{\pi}{4}(.15^2)}$$

$= 0.345$ m/s, $v = 0.002$ m^2/s.

Now $$R_e = \frac{\rho VD}{\mu} = \frac{V \times D}{v} = \frac{0.345 \times 0.15}{0.002} = 25.87$$

which is less than 2000. Hence flow is viscous.

$$h_f = 32\,\mu LV/\rho gD^2, \text{ where } v = \frac{\mu}{\rho} \quad \therefore \mu = v \times \rho = 0.002 \times 910 = 1.82$$

Hence, $$h_f = \frac{32 \times 1.82 \times 300 \times 0.345}{(910 \times 9.81 \times 0.15^2)} = 30$$

∴ $P = \rho g.Q.h_f/1000 = 910 \times 9.81 \times 0.0061 \times 30/1000 =$ **1.633 kW. Ans.]**

20. An oil of specific gravity 0.9 and viscosity 10 poise is flowing through a pipe of diameter 110 mm. The velocity at the centre is 2 m/s, find : (*i*) pressure gradient in the direction of flow, (*ii*) shear stress at the pipe wall ; (*iii*) Reynolds numbers, and (*iv*) velocity at a distance of 30 mm from the wall. (*Delhi University, 1996*)

[**Hint.** $\rho = 900$ kg/m^3 ; $\mu = 10$ poise $= 1$ N s/m^2 ; $D = 110$ mm $= 0.11$ m,

$$U_{max} = 2 \text{ m/s} ; \bar{u} = 1 \text{ m/s} ; U_{max} = \frac{1}{4\mu}\left(\frac{-dp}{dx}\right) R^2$$

(i) $\left(\frac{-dp}{dx}\right) = \frac{4\mu \times U_{max}}{R^2} = \frac{4 \times 1 \times 2}{0.055^2} = 2644.6 \text{ N/m}^3$;

(ii) $\tau_0 = \left(\frac{-dp}{dx}\right) \times \frac{R}{2} = 2644.6 \times \frac{0.055}{2} = 72.72 \text{ N/m}^2$;

(iii) $R_e = \frac{\rho \times \bar{u} \times D}{\mu} = \frac{900 \times 1 \times 0.11}{1} = 99$; and

(iv) $u = \frac{1}{4\mu}\left(\frac{-dp}{dx}\right)(R^2 - r^2) = \frac{1}{4 \times 1}(2644.6)(0.055^2 - 0.025^2) = 1.586$ m/s.]

21. Determine (*i*) the pressure gradient, (*ii*) the shear stress at the two horizontal plates, (*iii*) the discharge per metre width for laminar flow of oil with a maximum velocity of 2 m/s between two plates which are 150 mm apart. Given : $\mu = 2.5$ N s/m^2. (*Delhi University, December 2002*)

[**Hint.** $U_{max} = 2$ m/s, $t = 150$ mm $= 0.15$ m, $\mu = 2.5$ N s/m^2

(i) $U_{max} = -\frac{1}{8\mu}\frac{dp}{dx}t^2 \quad \therefore \quad \frac{dp}{dx} = \frac{-8\mu\, U_{max}}{t^2} = \frac{-8 \times 2.5 \times 2}{0.15^2} =$ **– 1777.77 N/m². Ans.**

(ii) $\tau_0 = -\frac{1}{2}\frac{dp}{dx} \times t = -\frac{1}{2}(-1777.77) \times 0.15 =$ **133.33 N/m². Ans.**

(iii) Q = Mean velocity × Area $= \left(\frac{2}{3}U_{max}\right) \times (t \times 1) = \left(\frac{2}{3} \times 2\right) \times (0.15 \times 1) =$ **0.2 m³/s. Ans.]**

Flow Past Immersed Bodies

▶ 11.1 INTRODUCTION

When a fluid is flowing over a stationary body, a force is exerted by the fluid on the body. Similarly, when a body is moving in a stationary fluid, a force is exerted by the fluid on the body. Also when the body and fluid both are moving at different velocities, a force is exerted by the fluid on the body. Some of the examples of the fluids flowing over stationary bodies or bodies moving in a stationary fluid are :

1. Flow of air over buildings,
2. Flow of water over bridges,
3. Submarines, ships, airplanes and automobiles moving through water or air.

▶ 11.2 FORCE EXERTED BY A FLOWING FLUID ON A STATIONARY BODY

Consider a body held stationary in a real fluid, which is flowing at a uniform velocity U as shown in Fig. 11.1.

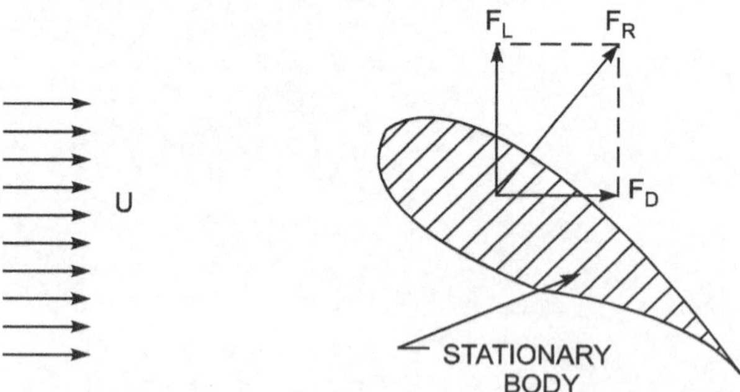

Fig. 11.1 *Force on a stationary body.*

The fluid will exert a force on the stationary body. The total force (F_R) exerted by the fluid on the body is perpendicular to the surface of the body. Thus the total force is inclined to the direction of motion. The total force can be resolved in two components, one in the direction of motion and other perpendicular to the direction of motion.

▶ 11.3 DRAG

The component of the total force (F_R) in the direction of motion is called 'drag'. This component is denoted by F_D. Thus drag is the force exerted by the fluid in the direction of motion.

▶ 11.4 LIFT

The component of the total force (F_R) in the direction perpendicular to the direction of motion is known as 'lift'. This is denoted by F_L. Thus lift is the force exerted by the fluid in the direction perpendicular to the direction of motion. Lift force occurs only when the axis of the body is inclined to the direction of fluid flow. If the axis of the body is parallel to the direction of fluid flow, lift force is zero. In that case only drag force acts.

If the fluid is assumed ideal and the body is symmetrical such as a sphere or cylinder, both the drag and lift will be zero.

▶ 11.5 EXPRESSION FOR DRAG AND LIFT

Consider an arbitrary shaped solid body placed in a real fluid, which is flowing with a uniform velocity U in a horizontal direction as shown in Fig. 11.2. Consider a small elemental area dA on the surface of the body. The forces acting on the surface area dA are :

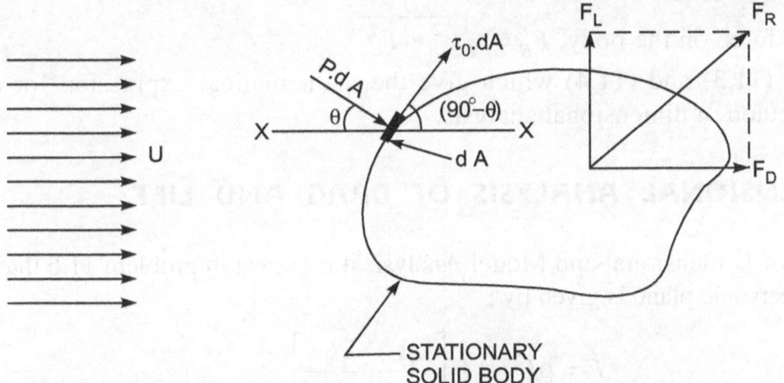

Fig. 11.2 *Drag and lift.*

1. Pressure force equal to $p \times dA$, acting perpendicular to the surface and
2. Shear force equal to $\tau_0 \times dA$, acting along the tangential direction to the surface.

Let θ = Angle made by pressure force with horizontal direction.

(*a*) **Drag Force (F_D).** The drag force on elemental area

= Force due to pressure in the direction of fluid motion + force due to shear stress in the direction of fluid motion.

$$= pdA \cos\theta + \tau_0 dA \cos(90° - \theta) = pdA \cos\theta + \tau_0 dA \sin\theta$$

∴ Total drag,

$$F_D = \text{Summation of } pdA\cos\theta + \text{Summation of } \tau_0 dA \sin\theta$$

$$= \int p\cos\theta dA + \int \tau_0 \sin\theta dA. \quad \text{...(11.1)}$$

The term $\int p \cos\theta dA$ is called the pressure drag or form drag while the term $\int \tau_0 \sin\theta dA$ is called the friction drag or skin drag or shear drag.

(*b*) **Lift Force (F_L).** The lift force an elemental area

= Force due to pressure in the direction perpendicular to the direction of motion
+ Force due to shear stress in the direction perpendicular to the direction of motion.

$= -pdA \sin\theta + \tau_0 dA \sin(90° - \theta) = -pdA \sin\theta + \tau_0 dA \cos\theta$

Negative sign is taken with pressure force as it is acting in the downward direction while shear force is acting vertically up.

$\therefore$ Total lift,

$$F_L = \int \tau_0\, dA \cos\theta - \int pdA \sin\theta$$

$$= \int \tau_0 \cos\theta dA - \int p \sin\theta dA \quad \text{...(11.2)}$$

The drag and lift for a body moving in a fluid of density ρ, at a uniform velocity U are calculated mathematically, as

$$F_D = C_D A \frac{\rho U^2}{2} \quad \text{...(11.3)}$$

$$F_L = C_L A \frac{\rho U^2}{2} \quad \text{...(11.4)}$$

where C_D = Co-efficient of drag,
C_L = Co-efficient of lift,
A = Area of the body which is the projected area of the body perpendicular to the direction of flow

Or

= Largest projected area of the immersed body.

Then resultant force on the body, $F_R = \sqrt{F_D^2 + F_L^2}$ ...(11.5)

The equations (11.3) and (11.4) which give the mathematical expression for drag and lift are derived by the method of dimensional analysis.

▶ 11.6 DIMENSIONAL ANALYSIS OF DRAG AND LIFT

In the chapter of Dimensional and Model Analysis it is shown in problem 11.6 that the force exerted by a fluid on a supersonic plane is given by :

$$F = \rho L^2 U^2 \phi \left[\frac{\mu}{\rho UL}, \frac{k}{\rho U^2}\right] \quad \text{...}(i)$$

Also in problem 11.7, it is shown that the force exerted by a fluid on a partially sub-merged body is given by :

$$F = \rho L^2 U^2 \phi \left[\frac{\mu}{\rho UL}, \frac{Lg}{U^2}\right] \quad \text{...}(ii)$$

Thus the general expression for the force exerted by a fluid (air or water) on a body (completely sub-merged or partially sub-merged) is given as

$$F = \rho L^2 U^2 \phi \left[\frac{\mu}{\rho UL}, \frac{Lg}{U^2}, \frac{k}{\rho U^2}\right] \quad \text{...(11.6)}$$

where L = Length of body, U = Velocity of body,
μ = Viscosity of fluid, ρ = Density of fluid
F = Force exerted, k = Bulk modulus of fluid,
g = Acceleration due to gravity.

If the body is completely sub-merged in the fluid, the force exerted by the fluid on the body due to gravitational effect is negligible. Hence the non-dimensional term containing '*g*' in equation (11.6), *i.e.*, $\frac{Lg}{U^2}$ is neglected. If the velocity of the body is comparable with velocity of sound, the effect due to compressibility is to be considered. But if the ratio of the velocity of the body to the velocity of the sound is less than 0.3, the force exerted by the fluid on the body due to compressibility is negligible. Hence the non-dimensional term in equation (11.6) containing K can be neglected. Then the force exerted by fluid on the body is given as

$$F = \rho L^2 U^2 \phi \left[\frac{\mu}{\rho UL}\right] = \rho L^2 U^2 \phi \left[\frac{\rho UL}{\mu}\right]$$

where $\frac{\rho UL}{\mu}$ = Reynold number = R_e

$$F = \rho L^2 U^2 \phi\, [R_e]. \qquad ...(11.7)$$

Now F is the total force exerted by the fluid on the body. The total force is having two components, one in the direction of motion called drag force and other component in the direction perpendicular to the direction of motion, called lift force.

The two components of F are expressed as

$$F_D = \frac{\rho L^2 U^2}{2} \times C_D$$

where C_D is a function of R_e and is called co-efficient of drag

$$= C_D A \frac{\rho U^2}{2} \qquad \{\because \quad L^2 = \text{Area} = A\}$$

And
$$F_L = \frac{\rho L^2 U^2}{2} \times C_L$$

where C_L is a function of R_e and is called co-efficient of lift

$$= C_L \cdot A \frac{\rho U^2}{2}.$$

Problem 11.1 *A flat plate 1.5 m × 1.5 m moves at 50 km/hour in stationary air of density 1.15 kg/m³. If the co-efficients of drag and lift are 0.15 and 0.75 respectively, determine :*

(i) The lift force, *(ii) The drag force,*
(iii) The resultant force, and
(iv) The power required to keep the plate in motion. (A.M.I.E., Winter, 1977)

Solution. Given :

Area of the plate, $A = 1.5 \times 1.5 = 2.25$ m²

Velocity of the plate, $U = 50 \text{ km/hr} = \frac{50 \times 1000}{60 \times 60} \text{ m/s} = 13.89 \text{ m/s}$

Density of air $\rho = 1.15 \text{ kg/m}^3$

Co-efficient of drag, $C_D = 0.15$

Co-efficient of lift, $C_L = 0.75$

(i) **Lift Force (F_L).** Using equation (11.4),

$$F_L = C_L A \times \frac{\rho U^2}{2} = 0.75 \times 2.25 \times \frac{1.15 \times 13.89^2}{2} \text{ N} = \mathbf{187.20 \text{ N. Ans.}}$$

(*ii*) **Drag Force (F_D).** Using equation (11.3),

$$F_D = C_D \times A \times \frac{\rho U^2}{2} = 0.15 \times 2.25 \times \frac{1.15 \times 13.89^2}{2} \text{ N} = \mathbf{37.44.\ N.\ Ans.}$$

(*iii*) **Resultant Force (F_R).** Using equation (11.5),

$$F_R = \sqrt{F_D^2 + F_L^2} = \sqrt{37.44^2 + 187.20^2} \text{ N}$$

$$= \sqrt{1400 + 35025} = \mathbf{190.85\ N.\ Ans.}$$

(*iv*) **Power Required to keep the Plate in Motion**

$$P = \frac{\text{Force in the direction of motion} \times \text{Velocity}}{1000} \text{ kW}$$

$$= \frac{F_D \times U}{1000} = \frac{37.425 \times 13.89}{1000} \text{ kW} = \mathbf{0.519\ kW.\ Ans.}$$

Problem 11.2 *Experiments were conducted in a wind tunnel with a wind speed of 50 km/hour on a flat plate of size 2 m long and 1 m wide. The density of air is 1.15 kg/m³. The co-efficients of lift and drag are 0.75 and 0.15 respectively. Determine :*

(*i*) *the lift force,* (*ii*) *the drag force,*
(*iii*) *the resultant force,* (*iv*) *direction of resultant force and*
(*v*) *power exerted by air on the plate.* (Delhi University, 1992)

Solution. Given :

Area of plate, $A = 2 \times 1 = 2 \text{ m}^2$

Velocity of air, $U = 50 \text{ kg/hr} = \frac{50 \times 1000}{60 \times 60} \text{ m/s} = 13.89 \text{ m/s}$

Density of air, $\rho = 1.15 \text{ kg/m}^3$

Value of $C_D = 0.15$ and $C_L = 0.75$

(*i*) *Lift force (F_L)*

Using equation (11.4), $F_L = C_L \times A \times \rho \times U^2/2$

$= 0.75 \times 2 \times 1.15 \times 13.89^2/2 =$ **166.404 N. Ans.**

(*ii*) *Drag force (F_D)*

Using equation (11.3), $F_D = C_D \times A \times \rho \times U^2/2$

$= 0.15 \times 2 \times 1.15 \times 13.89^2/2 =$ **33.28 N. Ans.**

(*iii*) *Resultant force (F_R)*

Using equation (11.5), $F_R = \sqrt{F_D^2 + F_L^2} = \sqrt{33.28^2 + 166.404^2} =$ **169.67 N. Ans.**

(*iv*) *The direction of resultant force* (θ)

The direction of resultant force is given by,

$$\tan\theta = \frac{F_L}{F_D} = \frac{166.38}{33.275} = 5$$

$\therefore$ $\theta = \tan^{-1} 5 =$ **78.69°. Ans.**

(*v*) *Power exerted by air on the plate*

Power = Force in the direction of motion × Velocity

$= F_D \times U$ N m/s $= 33.280 \times 13.89$ W (∵ Watt = N m/s)

= **462.26 W. Ans.**

Problem 11.3 *Find the difference in drag force exerted on a flat plate of size 2 m × 2 m when the plate is moving at a speed of 4 m/s normal to its plane in : (i) water, (ii) air of density 1.24 kg/m³. Co-efficient of drag is given as 1.15.*

Solution. Given :

Area of plate, $A = 2 \times 2 = 4 \text{ m}^2$

Velocity of plate, $U = 4$ m/s

Co-efficient of drag, $C_D = 1.15$

(*i*) Drag force when the plate is moving in water.

Using equation (11.3), $F_D = C_D \times A \times \dfrac{\rho U^2}{2}$, where ρ for water = 1000

$$= 1.15 \times 4 \times 1000 \times \frac{4^2}{2} \text{ N} = 36800 \text{ N.} \quad ...(i)$$

(*ii*) Drag force when the plate is moving in air,

$$F_D = C_D \times A \times \frac{\rho U^2}{2}, \quad \text{where } \rho \text{ for air} = 1.24$$

$$\therefore \quad F_D = 1.15 \times 4.0 \times 1.24 \times \frac{4.0^2}{2.0} \text{ N} = 45.6 \text{ N} \quad ...(ii)$$

∴ Difference in drag force = (*i*) – (*ii*)

= 36800 – 45.6 = **36754. 4 N. Ans.**

Problem 11.4 *A truck having a projected area of 6.5 square metres travelling at 70 km/hour has a total resistance of 2000 N. Of this 20 per cent is due to rolling friction and 10 per cent is due to surface friction. The rest is due to form drag. Calculate the co-efficient of form drag. Take density of air = 1.25 kg/m³.* (A.M.I.E., Winter, 1976)

Solution. Given :

Area of truck, $A = 6.5 \text{ m}^2$

Speed of truck, $U = 70 \text{ km/hr} = \dfrac{70 \times 100}{60 \times 60} = 19.44 \text{ m/s}$

Total resistance, $F_T = 2000$ N

Rolling friction resistance, $F_C = 20\%$ of total resistance $= \dfrac{20}{100} \times 2000 = 400$ N

Surface friction resistance, $F_S = 10\%$ of total resistance $= \dfrac{10}{100} \times 2000 = 200$ N

∴ Form drag, $F_D = 2000 - F_C - F_S = 2000 - 400 - 200 = 1400$ N

Using equation (11.3), $F_D = C_D \times A \times \dfrac{\rho U^2}{2}$

where if F_D = Form drag then C_D = Co-efficient of form drag

$$\therefore \quad 1400 = C_D \times 6.5 \times 1.25 \times \frac{19.44^2}{2} \quad (\rho = \text{Density of air} = 1.25 \text{ kg/m}^3)$$

$$\therefore \quad C_D = \frac{1400 \times 2}{6.5 \times 1.25 \times 19.44 \times 19.44} = \textbf{0.912. Ans.}$$

Problem 11.5 *A circular disc 3 m in diameter is held normal to a 26.4 m/s wind of density 0.0012 gm/cc. What force is required to hold it at rest ? Assume co-efficient of drag of disc = 1.1.*

(A.M.I.E., November, 1967)

Solution. Given :

Diameter of disc = 3 m

$\therefore$ Area, $A = \frac{\pi}{4} \times (3)^2 = 7.0685 \text{ m}^2$

Velocity of wind, $U = 26.4$ m/s

Density of wind, $\rho = 0.0012 \text{ gm/cm}^3 = \frac{.0012}{1000} \text{ kg/cm}^3$

$$= \frac{0.0012}{1000} \times 10^6 \frac{\text{kg}}{\text{m}^3} = 1.2 \text{ kg/m}^3$$

Co-efficient of drag, $C_D = 1.1$

The force required to hold the disc at rest is equal to the drag exerted by wind on the disc. Drag (F_D) is given by equation (11.3) as

$$F_D = C_D \times A \times \frac{\rho U^2}{2} = \frac{1.1 \times 7.0685 \times 1.2 \times 26.4^2}{2.0} = \textbf{3251.4 N. Ans.}$$

Problem 11.6 *A man weighing 90 kgf descends to the ground from an aeroplane with the help of a parachute against the resistance of air. The velocity with which the parachute, which is hemispherical in shape, comes down is 20 m/s. Find the diameter of the parachute. Assume $C_D = 0.5$ and density of air = 1.25 kg/m^3.*

Solution. Given :

Weight of man, $W = 90 \text{ kgf} = 90 \times 9.81 \text{ N} = 882.9 \text{ N}$ $(\because \; 1 \text{ kgf} = 9.81 \text{ N})$

Velocity of parachute, $U = 20$ m/s

Co-efficient of drag, $C_D = 0.5$

Density of air, $\rho = 1.25 \text{ kg/m}^3$

Let the diameter of parachute $= D$

$\therefore$ Area, $A = \frac{\pi}{4} D^2 \text{ m}^2.$

When the parachute with the man comes down with a uniform velocity, $U = 20$ m/s, the drag resistance will be equal to the weight of man, neglecting the weight of parachute. And projected area of the hemispherical parachute will be equal to $\frac{\pi}{4} D^2$.

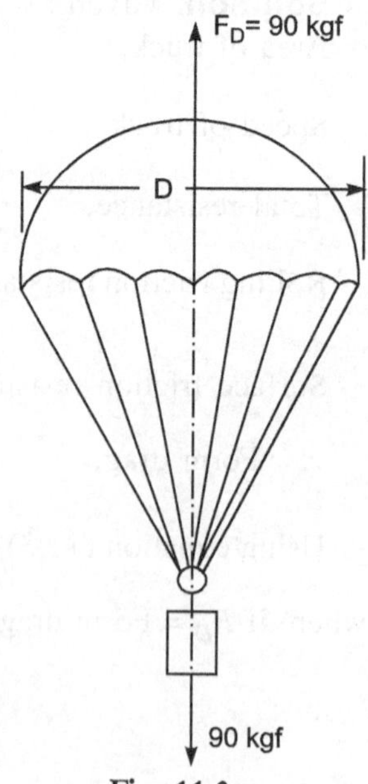

Fig. 11.3

$\therefore$ Drag, $F_D = 90 \text{ kgf} = 90 \times 9.81 = 882.9 \text{ N}$

Using equation (11.3), $F_D = C_D \times A \times \frac{\rho U^2}{2}$

$$\therefore \quad 882.9 = 0.5 \times \frac{\pi}{4} D^2 \times \frac{1.25 \times 20^2}{2}$$

$$\therefore \quad D^2 = \frac{882.9 \times 4 \times 2.0}{0.5 \times \pi \times 1.25 \times 20 \times 20} = 8.9946 \text{ m}^2$$

or $D = \sqrt{8.9946} = \textbf{2.999 m. Ans.}$

Problem 11.7 *A man weighing 981 N descends to the ground from an aeroplane with the help of a parachute against the resistance of air. The shape of the parachute is hemispherical of 2 m diameter. Find the velocity of the parachute with which it comes down. Assume $C_d = 0.5$ and ρ for air $= 0.00125$ gm/cc and $\nu = 0.015$ stoke.* (A.M.I.E., Winter, 1975)

Solution. Given :

Weight of the man, $W = 981$ N

$\therefore$ Drag force, $F_D = W = 981$ N

Diameter of the parachute, $D = 2$ m

$\therefore$ Projected area, $A = \frac{\pi}{4} D^2 = \frac{\pi}{4} \times 2^2 = \pi \text{ m}^2$

Co-efficients of drag, $C_d = 0.5$

Density for air, $\rho = 0.00125 \text{ gm/cm}^3 = \frac{.00125}{1000} \text{ kg/cm}^3$

$$= \frac{.00125}{1000} \times 10^6 \frac{\text{kg}}{\text{m}^3} = 1.25 \text{ kg/m}^3$$

Let the velocity of parachute $= U$

Using equation (11.3), $F_D = C_D \times A \times \frac{\rho U^2}{2}$ or $981 = 0.5 \times \pi \times \frac{1.25 \times U^2}{2.0}$

$$\therefore \quad U = \sqrt{\frac{981 \times 2.0}{0.5 \times \pi \times 1.25}} = \textbf{31.61 m/s. Ans.}$$

Problem 11.8 *A man descends to the ground from an aeroplane with the help of a parachute which is hemispherical having a diameter of 4 m against the resistance of air with a uniform velocity of 25 m/s. Find the weight of the man if the weight of parachute is 9.81 N. Take $C_D = 0.6$ and density of air $= 1.25$ kg/m^3.* (Osmania University, 1992)

Solution. Given :

Diameter of parachute, $D = 4$ m

$\therefore$ Projected area, $A = \frac{\pi}{4} \times D^2 = \frac{\pi}{4} \times 4^2 = 4\pi \text{ m}^2$

Velocity of parachute, $U = 25$ m/s

Weight of parachute, $W_1 = 9.81$ N

Co-efficient of drag, $C_D = 0.6$

Density of air, $\rho = 1.25 \text{ kg/m}^3$

Let the weight of man $= W_2$

Then weight of man + Weight of parachute $= W_2 + W_1 = (W_2 + 9.81)$

Hence drag force will be equal to the weight of man plus weight of parachute.

$\therefore$ Drag force, $F_D = (W_2 + 9.81)$

Using equation (11.3), we have $F_D = C_D \times A \times \frac{\rho U^2}{2}$

or $$(W_2 + 9.81) = 0.6 \times 4\pi \times \frac{1.25 \times 25^2}{2.0} = 2945.24 \text{ N}$$

$$\therefore \quad W_2 = 2945.24 - 9.81 = \textbf{2935.43 N. Ans.}$$

Problem 11.9 *Calculate the diameter of a parachute to be used for dropping an object of mass 100 kg so that the maximum terminal velocity of dropping is 5 m/s. The drag co-efficient for the parachute, which may be treated as hemispherical is 1.3. The density of air is 1.216 kg/m³.*

(A.M.I.E., Winter, 1980)

Solution. Given :

Mass of object, $M = 100$ kg

Weight of object, $W = 100 \times 9.81 = 981$ N

∴ Drag force, $F_D = 981$ N

Velocity of object, $U = 5$ m/s

Drag Co-efficient, $C_D = 1.3$

Density of air, $\rho = 1.216$ kg/m³

Let the diameter of parachute $= D$ m

∴ Projected area, $A = \frac{\pi}{4} D^2 \text{ m}^2$

Using equation (11.3), $F_D = C_D \times A \times \frac{\rho U^2}{2}$

or $$981 = 1.3 \times \frac{\pi}{4} D^2 \times \frac{1.216 \times 5^2}{2}$$

or $$D^2 = \frac{981 \times 4 \times 2}{1.3 \times \pi \times 1.216 \times 5 \times 5} = 63.21$$

∴ $D = \sqrt{63.21} =$ **7.95 m. Ans.**

Problem 11.10 *A kite 0.8 m × 0.8 m weighing 3.924 N assumes an angle, of 12° to the horizontal. The string attached to the kite makes an angle of 45° to the horizontal. The pull on the string is 24.525 N when the wind is flowing at a speed of 30 km/hour. Find the corresponding co-efficient of drag and lift. Density of air is given as 1.25 kg/m³.*

(A.M.I.E., Summer, 1984)

Solution. Given :

Projected area of kite, $A = 0.8 \times 0.8 = 0.64 \text{ m}^2$

Weight of kite, $W = 3.924$ N

Angle made by kite with horizontal, $\theta_1 = 12°$

Angle made by string with horizontal, $\theta_2 = 45°$

Pull on the string, $P = 24.525$ N

Speed of wind, $U = 30 \text{ km/hr} = \frac{30 \times 1000}{60 \times 60} \text{ m/s} = 8.333 \text{ m/s}$

Density of air, $\rho = 1.25$ kg/m³

Drag force, F_D = Force exerted by wind in the direction of motion (*i.e.*, in the *X-X* direction)

= Component of pull, *P* along *X-X*

$= P \cos 45° = 24.525 \cos 45° = 17.34$ N

And lift force, F_L = Force exerted by wind on the kite perpendicular to the direction of motion (*i.e.*, along *r-Y* direction)

= Component of *P* in vertically downward direction + Weight of kite (*W*)

$$= P \sin 45° + W = 24.525 \sin 45° + 3.924 \text{ N}$$
$$= 17.34 + 3.924 = 21.264 \text{ N.}$$

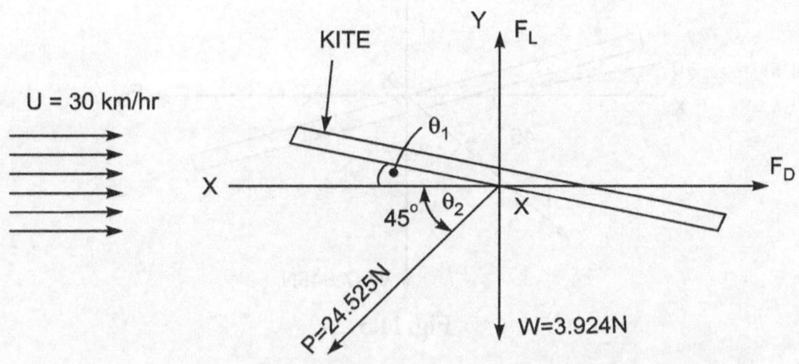

Fig. 11.4

(*i*) **Drag Co-efficient (C_D).** Using equation (11.3), we have

$$F_D = C_D \times A \times \frac{\rho U^2}{2}$$

or
$$C_D = \frac{2 \times F_D}{A\rho U^2} = \frac{2 \times 17.34}{0.64 \times 1.25 \times 8.333^2} = \mathbf{0.624.\ Ans.}$$

(*ii*) **Lift Co-efficient (C_L).** Using equation (11.4), we have

$$F_L = C_L \times A \times \frac{\rho U^2}{2}$$

or
$$C_L = \frac{2 \times F_L}{A \times \rho \times U^2} = \frac{2 \times 21.264}{0.64 \times 1.25 \times 8.333^2} = \mathbf{0.765.\ Ans.}$$

Problem 11.11 *A kite weighing 7.848 N has an effective area of 0.8 m². It is maintained in air at an angle of 10° to the horizontal. The string attached to the kite makes an angle of 45° to the horizontal and at this position the value of co-efficient of drag and lift are 0.6 and 0.8 respectively. Find the speed of the wind and the tension in the string. Take the density of air as 1.25 kg/m³.*

Solution. Given :

Weight of kite, $W = 7.848$ N

Effective area, $A = 0.8$ m²

Angle made by kite with horizontal = 10°

Angle made by string with horizontal = 45°

$$C_D = 0.6$$
$$C_L = 0.8$$

∴ Density of air, $\rho = 1.25$ kg/m³

Let the speed of the wind $= U$ m/s

and tension in string $= T$

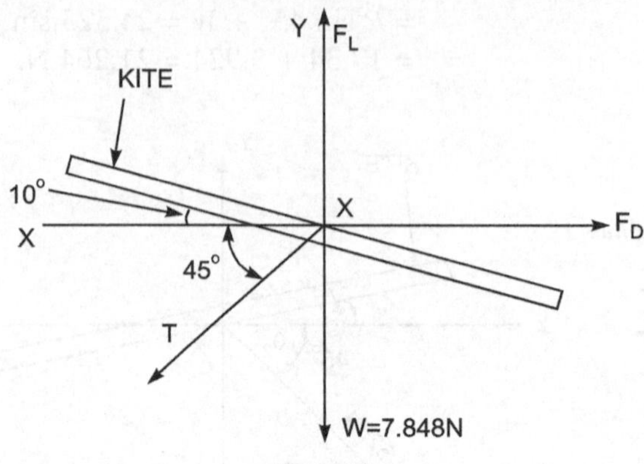

Fig. 11.5

The free body diagram for the kite is shown in Fig. 11.5.

Drag force, F_D = Component of T along X-X = $T \cos 45°$

But drag force F_D is also $= C_D \times A \times \dfrac{\rho U^2}{2} = \dfrac{0.6 \times 0.8 \times 1.25 \times U^2}{2} = 0.3\ U^2$

Equating the two values of F_D, $T \cos 45° = 0.3\ U^2$ ...(*i*)

Now lift force from Fig. 11.5, F_L = Component of T vertically downward + Weight of kite

$= T \sin 45° + 7.848$

Also lift force, $F_L = \dfrac{C_L \times A \times \rho U^2}{2} = \dfrac{0.8 \times 0.8 \times 1.25 \times U^2}{2.0} = 0.4\ U^2$

Equating the two values of F_L, $T \sin 45° + 7.848 = 0.4\ U^2$

or $T \sin 45° = .04076\ U^2 - 7.848$ ...(*ii*)

From equations (*i*) and (*ii*), as $T \cos 45° = T \sin 45°$,

$$0.3\ U^2 = 0.4\ U^2 - 7.848$$

or $7.848 = 0.4\ U^2 - 0.3\ U^2 = 0.1\ U^2$

$\therefore$ $U^2 = \dfrac{7.848}{0.1} = 78.48$ ...(*iii*)

$\therefore$ $U = \sqrt{78.48} = 8.86 \text{ m/s} = \dfrac{8.86 \times 60 \times 60}{1000}$ km/hr = **31.89 km/hr. Ans.**

Substituting the value of $U^2 = 78.48$ given by equation (*iii*) into equation (*i*), we get

$$T \cos 45° = 0.3 \times 78.48 = 23.544$$

$\therefore$ $T = \dfrac{23.544}{\cos 45°}$ = **33.3 N. Ans.**

Problem 11.12 *The air is flowing over a cylinder of diameter 50 mm and infinite length with a velocity of 0.1 m/s. Find the total drag, shear drag and pressure drag on 1 m length of the cylinder if the total drag co-efficient is equal to 1.5 and shear drag co-efficient equal to 0.2. Take density of air = 1.25 kg/cm³.*

Solution. Given :

Diameter of cylinder, $D = 50 \text{ mm} = 0.05 \text{ m}$

Length of cylinder, $L = 1.0 \text{ m}$

$\therefore$ Projected Area, $A = L \times D = 1 \times .05 = 0.05 \text{ m}^2$

Velocity of air, $U = 0.1 \text{ m/s}$

Total drag co-efficient, $C_{DT} = 1.5$

Shear drag co-efficient, $C_{DB} = 0.2$

Density of air, $\rho = 1.25 \text{ kg/m}^3$

Total drag is given by, $$F_D = C_{DT} \times A \times \frac{\rho U^2}{2}$$

$$= 1.5 \times .05 \times \frac{1.25 \times (0.1)^2}{2} = \mathbf{0.000468\ N.\ Ans.}$$

Shear drag is given by, $$F_{DS} = C_{DS} \times A \times \frac{\rho U^2}{2}$$

$$= 0.2 \times .05 \times \frac{1.25 \times (0.1)^2}{2} = \mathbf{0.0000625\ N.\ Ans.}$$

From equation (11.1), Total drag, F_D = Pressure drag + Shear drag

$\therefore$ Pressure drag, = Total drag – Shear drag

= 0.000468 – 0.0000625 = **0.0004055 N. Ans.**

Problem 11.13 *A body of length 2.0 m has a projected area 1.5 m² normal to the direction of its motion. The body is moving through water, which is having viscosity = 0.01 poise. Find the drag on the body if it has a drag co-efficient 0.5 for a Reynold number of 8 × 10⁶.*

Solution. Given :

Length of body, $L = 2.0 \text{ m}$

Projected Area, $A = 1.5 \text{ m}^2$

Viscosity of water, $\mu = 0.01 \text{ poise} = \frac{0.01}{10} = 0.001 \frac{\text{Ns}}{\text{m}^2}$

Drag co-efficients, $C_d = 0.5$

Reynold number, $R_e = 8 \times 10^6$

Let the drag force on body $= F_D$

First find the velocity with which body is moving in water. It is calculated from the given Reynold number.

$$R_e = \frac{\rho U L}{\mu}, \text{ where } \rho \text{ for water} = 1000$$

or $$8 \times 10^6 = \frac{1000 \times U \times 2.0}{0.001} = 2 \times 10^6\ U$$

$\therefore$ $$U = \frac{8 \times 10^6}{2 \times 10^6} = 4.0 \text{ m/s}$$

Using equation (11.3), $F_D = C_d \times A \times \frac{\rho U^2}{2} = 0.5 \times 1.5 \times 1000 \times \frac{4.0^2}{2}$ = **6049 N. Ans.**

Problem 11.14 *A sub-marine which may be supposed to approximate a cylinder 4 m in diameter and 20 m long travels sub-merged at 1.3 m/s in sea-water. Find the drag exerted on it, if the drag co-efficient for Reynold number greater than 10^5 may be taken as 0.75. The density of sea-water is given as 1035 kg/m^3 and kinematic viscosity as .015 stokes.*

Solution. Given :

Dia. of cylinder, $D = 4$ m

Length of cylinder, $L = 20$ m

Velocity of cylinder, $U = 1.3$ m/s

Density of sea-water, $\rho = 1035$ kg/m^3

Kinematic viscosity $\nu = 0.015$ stokes $= .015$ cm^2/s $= .015 \times 10^{-4}$ m^2/s

Let the drag force $= F_D$

Reynold number, $R_e = \frac{U \times D}{\nu} = \frac{1.3 \times 4.0}{.015 \times 10^{-4}} = 3.466 \times 10^6$

Since $R_e > 10^5$, hence $C_D = 0.75$

Drag force is given by equation (11.3) as

$$F_D = C_D \times A \times \frac{\rho U^2}{2}$$

where A = projected area of cylinder $= L \times D = 20 \times 4.0 = 80.0$ m^2

$\therefore$ $F_D = 0.75 \times 80 \times \frac{1035 \times 1.3^2}{2.0}$ kgf = **52472.2 N. Ans.**

Problem 11.15 *A jet plane which weighs 29.43 kN and having a wing area of 20 m^2 flies at a velocity of 950 km/hour, when the engine delivers 7357.5 kW power. 65% of the power is used to overcome the drag resistance of the wing. Calculate the co-efficients of lift and drag for the wing. The density of the atmospheric air is 1.21 kg/m^3.* (A.M.I.E., Summer, 1988)

Solution. Given :

Weight of plane, $W = 29.43$ kN $= 29.43 \times 1000$ N $= 29430$ N

Wing area, $A = 20$ m^2

Speed of plane, $U = 950$ km/hr $= \frac{950 \times 1000}{60 \times 60} = 263.88$ m/s

Engine power, $P = 7357.5$ kW

Power used to overcome drag resistance = 65% or 7357.5 $= \frac{65}{100} \times 7357.5 = 4782.375$ kW

$\therefore$ Density of air, $\rho = 1.21$ kg/m^3

Let C_D = Co-efficient of drag and C_L = Co-efficient of lift.

Now power used in kW to overcome drag resistance $= \frac{F_D \times U}{1000}$ or $4782.375 = \frac{F_D \times 263.88}{1000}$

$$\therefore \qquad F_D = \frac{4782.375 \times 1000}{263.88}$$

But from equation (11.3), we have $F_D = C_D \, . \, A \, . \, \dfrac{\rho U^2}{2}$

$$\therefore \qquad \frac{4782.375 \times 1000}{263.88} = C_D \times 20 \times 1.21 \times \frac{263.88^2}{2}$$

$$\therefore \qquad C_D = \frac{4782.375 \times 1000 \times 2}{20 \times 1.21 \times 263.88^3} = \mathbf{0.0215. \ Ans.}$$

The lift force should be equal to weight of the plane

$$\therefore \qquad F_L = W = 29430 \text{ N}$$

But $\qquad F_L = C_L \, . \, A \, . \, \dfrac{\rho U^2}{2} \quad$ or $\quad 29430 = C_L \times 20 \times 1.21 \times \dfrac{263.88^2}{2}$

$$\therefore \qquad C_L = \frac{29430 \times 2}{20 \times 1.21 \times 263.88^2} = \mathbf{0.0349. \ Ans.}$$

▶ 11.7 PRESSURE DRAG AND FRICTION DRAG

The total drag on a body is given by the equation (11.1) as

Total drag, $\qquad F_D = \int p \cos \theta dA + \int \tau_0 \sin \theta dA \qquad$ ...(*i*)

where $\int p \cos \theta dA$ = Pressure drag or form drag, and

$\int \tau_0 \sin \theta dA$ = Friction drag or skin drag or shear drag

The relative contribution of the pressure drag and friction drag to the total drag depends on :

(*i*) Shape of the immersed body,

(*ii*) Position of the body immersed in the fluid, and

(*iii*) Fluid characteristics.

Consider the flow of a fluid over a flat plate when the plate is placed parallel to the direction of the flow as shown in Fig. 11.6. In this cos θ, which is the angle made by pressure with the direction motion, will be 90°. Thus the term $\int p \cos \theta dA$ will be zero and hence total drag will be equal to friction drag (or shear drag). If the plate is placed perpendicular to the flow as shown in Fig. 11.7, the angle θ, made by the pressure with the direction of motion will be zero. Hence the term $\int \tau_0 \sin \theta dA$ will become equal to zero and hence total drag will be due to the pressure difference between the upstream and downstream side of the plate. If the plate is held at an angle with the direction of flow, both the terms $\int p \cos \theta dA$ and $\int \tau_0 \sin \theta dA$ will exist and total drag will be equal to the sum of pressure drag and friction drag.

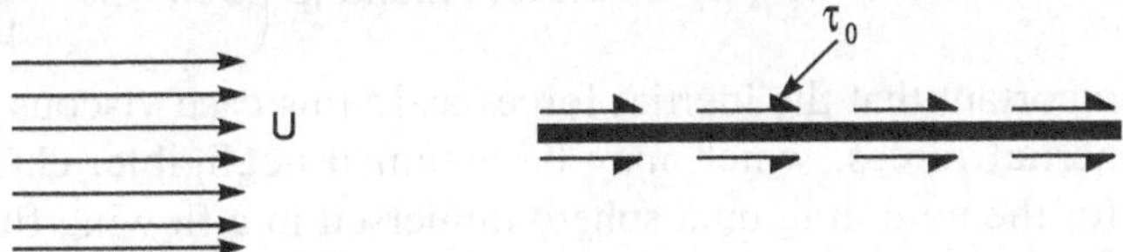

Fig. 11.6 *Flat plate parallel to flow.*

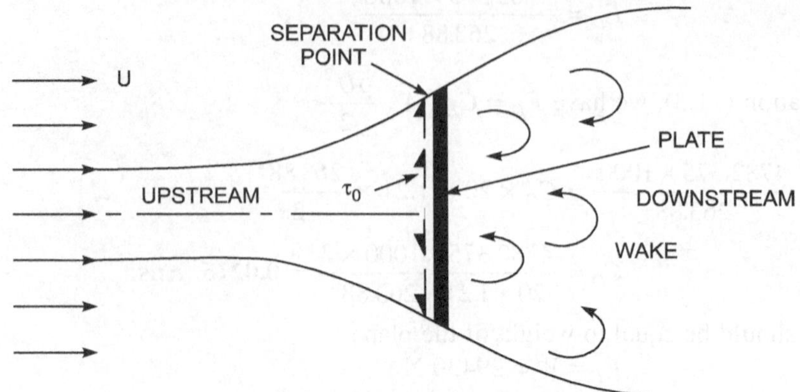

Fig. 11.7 *Flat plate perpendicular to flow.*

11.7.1 Stream-lined Body. A stream-lined body is defined as that body whose surface coincides with the stream-lines, when the body is placed in a flow. In that case the separation of flow will take place only at the trailing edge (or rearmost part of the body). Though the boundary layer will start at the leading edge, will become turbulent from laminar, yet it does not separate upto the rearmost part of the body in the case of stream-lined body. Thus behind a stream-lined body, wake formation zone will be very small and consequently the pressure drag will be very small. Then the total drag on the stream-lined body will be due to friction (shear) only. A body may be stream-lined :

1. at low velocities but may not be so at higher velocities.
2. when placed in a particular position in the flow but may not be so when placed in another position.

11.7.2 Bluff Body. A bluff body is defined as that body whose surface does not coincide with the streamlines, when placed in a flow. Then the flow is separated from the surface of the body much ahead of its trailing edge with the result of a very large wake formation zone. Then the drag due to pressure will be very large as compared to the drag due friction on the body. Thus the bodies of such a shape in which the pressure drag is very large as compared to friction drag are called bluff bodies.

11.7.3 Drag on a Sphere. Consider the flow of a real fluid past a sphere.

Let U = Velocity of the flow of fluid over sphere,
D = Diameter of sphere,
ρ = Mass density of fluid, and
μ = Viscosity of fluid.

If the Reynold number of the flow is very small less than 0.2 $\left(i.e., R_e = \dfrac{UD\rho}{\mu} < 0.2 \right)$, the viscous forces are much more important that the inertial forces as in this case viscous forces are much more predominate than the inertial forces, which may be assumed negligible. G.G. Stokes developed a mathematical equation for the total drag on a sphere immersed in a flowing fluid for which Reynold number is upto 0.2, so that inertia forces may be assumed negligible. According to his solution, total drag is

$$F_D = 3\pi\mu DU. \qquad \text{...(11.8)}$$

He further observed that out of the total drag given by equation (11.8), two-third is contributed by skin friction and the remaining one-third by pressure difference. Thus

Skin friction drag, $F_{D_f} = \frac{2}{3} F_D = \frac{2}{3} \times 3\pi\mu DU = 2\pi\mu DU$

and pressure drag, $F_{D_p} = \frac{1}{3} F_D = \frac{1}{3} \times 3\pi\mu DU = 2\pi\mu DU.$

(*i*) **Expression for C_d for Sphere for Reynold Number less than 0.2.** From equation (11.3), the total drag is given by

$$F_D = C_D \times A \times \frac{\rho U^2}{2}$$

For sphere, $F_D = 3\pi\mu DU$

$$A = \text{Projected area of the sphere} = \frac{\pi}{4} D^2$$

$$\therefore \quad 3\pi\mu DU = C_D \times \frac{\pi}{4} D^2 \frac{\rho U^2}{2}$$

$$C_D = \frac{3\pi\mu DU}{\frac{\pi}{4} D^2 \times \frac{\rho U^2}{2}} = \frac{24\mu}{\rho UD} = \frac{24}{R_e} \qquad \text{...(11.9)} \quad \left(\because \frac{\mu}{\rho DU} = R_e\right)$$

Equation (11.9) is called **'Stoke's law'.**

(*ii*) **Value of C_D for Sphere when R_e is between 0.2 and 5.** With the increase of Reynold number, the inertia forces increase and must be taken into account. When R_e lies between 0.2 and 5, Oseen, a Swedish physicist, improved Stoke's law as

$$C_D = \frac{24}{R_e}\left[1 + \frac{3}{16 R_e}\right] \qquad \text{...(11.10)}$$

Equation (11.10) is called Oseen formulae and is valid for R_e between 0.2 and 5.

(*iii*) **Value of C_D for R_e from 5.0 to 1000.** The drag co-efficient for the Reynold of number from 5 to 1000 is equal to 0.4.

(*iv*) **Value of C_D for R_e 1000 to 100,000.** In this range, C_D is independent of the Reynold number and its value is approximately equal to 0.5.

(*v*) **Value of C_D for R_e more than 10^5.** The value of C_D is approximately equal to 0.2 for the Reynold number more than 10^5.

Problem 11.16 *Calculate the weight of a ball of diameter 80 mm which is just supported in a vertical air stream which is flowing at a velocity of 7 m/s. The density of air is given as 1.25 kg/m³. The kinematic viscosity of air = 1.5 stokes.*

Solution. Given :

Dia. of ball, $D = 80 \text{ mm} = 0.08 \text{ m}$

Velocity of air, $U = 7 \text{ m/s}$

Density of air, $\rho = 1.25 \text{ kg/m}^3$

Kinematic viscosity, $\nu = 1.5 \text{ stokes} = 1.5 \times 10^{-4} \text{ m}^2/\text{s}$

Reynold number, $R_e = \frac{U \times D}{\nu} = \frac{7 \times .08}{1.5 \times 10^{-4}} = 0.373 \times 10^{-4} = 3730.$

Thus the value of R_e lies between 1000 and 100,000 and hence $C_D = 0.5$.

When the ball is supported in a vertical air stream, the weight of ball is equal to the drag force as shown in Fig. 11.8.

But drag force, $F_D = C_D \times A \times \frac{\rho U^2}{2}$

where A = projected area of ball

$$= \frac{\pi}{4} D^2 = \frac{\pi}{4} (0.08)^2 = 0.005026 \text{ m}^2$$

$\therefore$ Drag force, $F_D = 0.5 \times 0.005026 \times \frac{1.25 \times 7^2}{2}$ N

$= 0.07696$ N

$\therefore$ Weight of ball $= F_D =$ **0.07696 N. Ans.**

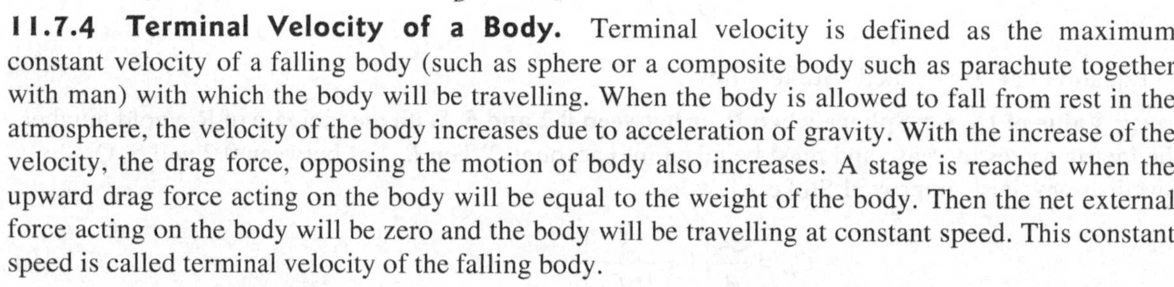

Fig. 11.8

11.7.4 Terminal Velocity of a Body. Terminal velocity is defined as the maximum constant velocity of a falling body (such as sphere or a composite body such as parachute together with man) with which the body will be travelling. When the body is allowed to fall from rest in the atmosphere, the velocity of the body increases due to acceleration of gravity. With the increase of the velocity, the drag force, opposing the motion of body also increases. A stage is reached when the upward drag force acting on the body will be equal to the weight of the body. Then the net external force acting on the body will be zero and the body will be travelling at constant speed. This constant speed is called terminal velocity of the falling body.

If the body drops in a fluid, at the instant it has acquired terminal velocity the net force acting on the body will be zero. The forces acting on the body at this state will be :

1. Weight of body (W), acting downward,
2. Drag force (F_D), acting vertically upward, and
3. Buoyant force (F_B), acting vertically up.

The net force on the body should be zero, *i.e.*, $W = F_D + F_B$ ...(11.10A)

Problem 11.17 *A metallic sphere of sp. gr. 7.0 falls in an oil of density 800 kg/m³. The diameter of the sphere is 8 mm and it attains a terminal velocity of 40 mm/s. Find the viscosity of the oil in poise.*

Solution. Given :

Sp. gr. of metallic sphere $= 7.0$

$\therefore$ Density of metallic sphere, $\rho_s = 7 \times 1000 = 7000$ kg/m^3

Density of oil, $\rho_o = 800$ kg/m^3

Dia. of sphere, $D = 8$ mm $= 8 \times 10^{-3}$ m

Terminal velocity, $U = 40$ mm/s $= .04$ m/s

Let the viscosity of oil $= \mu$

Weight of sphere, $W = \rho_s \times g \times$ Volume of sphere

$$= 7000 \times 9.81 \times \frac{\pi}{6} D^3$$

$$= 7000 \times 9.81 \times \frac{\pi}{6} \times (8 \times 10^{-3})^3 = 18.4 \times 10^{-3} = 0.0184 \text{ N}$$

Buoyant force on sphere, F_B = Density of oil × g × Volume of sphere

$$= 800 \times 9.81 \times \frac{\pi}{6} (8 \times 10^{-3})^3 \text{ N} = 0.002103 \text{ N.}$$

Drag force, F_D on the sphere is given by equation (11.8) as

$$F_D = 3\pi\mu DU = 3\pi\mu \times 8 \times 10^{-3} \times .04 = .003015\ \mu$$

Using equation (11.11), $W = F_D + F_B$

or $0.0184 = .003015\ \mu + 0.002103$

or $.003015\ \mu = 0.0184 - 0.002103 = 0.016297$

or $\mu = \dfrac{0.016297}{.003015} = 5.4 \dfrac{\text{Ns}}{\text{m}^2} = 5.4 \times 10 =$ **54.0 poise. Ans.**

The expression for drag given by equation (11.10 *A*) is valid only up to Reynold number less than 0.2. Hence it is necessary to calculate Reynold number for the flow.

∴ Reynold number, $R_e = \dfrac{\rho UD}{\mu}$, where ρ for oil = 800 kg/m^3

∴ $R_e = 800 \times \dfrac{.04 \times 8 \times 10^{-3}}{5.4} = 0.0474$

Hence $R_e < 0.2$ and so the expression $F_D = 3\pi\mu DU$ is valid.

Problem 11.18 *A spherical steel ball of diameter 40 mm and of density 8500 kg/m^3 is dropped in large mass of water. The co-efficient of drag of the ball in water is given as 0.45. Find the terminal velocity of the ball in water. If the ball is dropped in air, find the increase in terminal velocity of ball. Take the density of air = 1.25 kg/m^3 and C_D = 0.1.*

Solution. Given :

Diameter of steel ball, D = 40 mm = 0.04 m

Density of ball, ρ_s = 8500 kg/m^3

C_D for ball in water = 0.45

Let the terminal velocity in water = U_1

The forces acting on the spherical ball are :

1. Weight, W = Density of ball × g × Volume of spherical ball

$$= \rho_s \times g \times \frac{\pi}{6} D^3 = 8500 \times 9.81 \times \frac{\pi}{6} (.04)^3 = \mathbf{2.794\ N.}$$

2. Buoyant force, F_B = Density of water × g × Volume of ball

$$= 1000 \times 9.81 \times \frac{\pi}{6} (0.04)^3 = 0.3286 \text{ N.}$$

3. Drag force, $F_D = C_D \times A \times \dfrac{\rho U^2}{2}$

where A = projected area = $\frac{\pi}{4} D^2 = \frac{\pi}{4} (.04)^2$, $\rho = 1000$ for water

$$F_D = 0.45 \times \frac{\pi}{4} \times (.04)^2 \times 1000 \times \frac{U_1^2}{2} = 0.2825\ U_1^2$$

Using equation (11.10A), we get $W = F_D + F_B$

or $$2.794 = 0.2825\ U_1^2 + 0.3286$$

or $$U_1^2 = \frac{2.794 - 0.3286}{0.2825} = 8.725$$

$\therefore$ $$U_1 = \sqrt{8.725} = \mathbf{2.953\ m/s.\ Ans.}$$

When ball is dropped in air. Let the terminal velocity = U_2

Weight, $W = 2.794$

Buoyant force, F_B = Density of air × g × Volume of ball

$$= 1.25 \times 9.81 \times \frac{\pi}{6} (.04)^3 = 0.000411 \text{ N}$$

Drag force, $F_D = C_D \times A \times \frac{\rho U^2}{2}$, where ρ for air = 1.25

$$F_D = 0.1 \times \frac{\pi}{4} (.04)^2 \times 1.25\ \frac{U_2^2}{2} = 0.0000785\ U_2^2.$$

The buoyant force in air is 0.000411, while weight of the ball is 2.794 N. Hence buoyant force is negligible.

$\therefore$ For equilibrium of the ball in air, F_D = Weight of ball

or $$0.0000785\ U_2^2 = 2.794 \text{ or } U_2 = \sqrt{\frac{2.794}{0.0000785}} = 188.67 \text{ m/s}$$

$\therefore$ Increase in terminal velocity in air = $U_2 - U_1$ = 188.67 – 2.9533 = **185.717 m/s. Ans.**

Problem 11.19 *A metallic ball of diameter 2×10^{-3} m drops in a fluid of sp. gr. 0.95 and viscosity 15 poise. The density of the metallic ball is 12000 kg/m^3. Find :*

(i) The drag force exerted by fluid on metallic ball,

(ii) The pressure drag and skin friction drag,

(iii) The terminal velocity of ball in fluid.

Solution. Given :

Diameter of metallic ball, $D = 2 \times 10^{-3}$ m

Sp. gr. of fluid, $S_0 = 0.95$

$\therefore$ Density of fluid, $\rho_0 = 0.95 \times 1000 = 950$ kg/m^3

Viscosity of fluid, $\mu = 15 \text{ poise} = \frac{15}{10} = 1.5 \frac{\text{Ns}}{\text{m}^2}$

Density of ball, $\rho_s = 12000$ kg/m^3

The forces acting on the ball are :

Weight of ball, W = Density of ball × g × Volume of ball

$$= 12000 \times 9.81 \times \frac{\pi}{6} D^3$$

$$= 12000 \times 9.81 \times \frac{\pi}{6} \times (2 \times 10^{-3})^3 \text{ N} = 0.000493 \text{ N}$$

Buoyant force, F_B = Density of fluid × g × Volume of ball

$$= 950 \times 9.81 \times \frac{\pi}{6} (2 \times 10^{-3})^3 \text{ N} = 0.000039 \text{ N}$$

When the metallic ball reaches the terminal velocity, equation (11.10A) is applicable.

$\therefore$ $W = F_D + F_B$ or $F_D = W - F_B = 0.000493 - 0.000039$

$= $ **0.000454 N. Ans.**

(i) Drag force, $F_D = 0.000454$ N

(ii) Pressure drag $= \frac{1}{3} F_D = \frac{1}{3} \times 0.000454 =$ **0.0001513 N. Ans.**

Skin friction drag $= \frac{2}{3} \times F_D = \frac{2}{3} \times 0.000454 =$ **0.0003028 N. Ans.**

(iii) Let the terminal velocity $= U$

Then drag force (F_D) is given by equation (11.8) as $F_D = 3\pi\mu DU$

But $F_D = 0.000454$ N

Equating the two values of F_D, we have

$$3\pi\mu DU = 0.000454 \quad \text{or} \quad 3\pi \times \frac{15}{10} \times 2 \times 10^{-3} \times U = 0.000454$$

or

$$U = \frac{10 \times 0.000454}{3\pi \times 15 \times 2 \times 10^{-3}} = \textbf{0.016 m/s. Ans.}$$

Let us check for Reynold number, R_e

$$R_e = \frac{\rho UD}{\mu}, \quad \text{where } \rho = 950 \text{ kg/m}^3$$

$$= 950 \times \frac{0.016 \times 2 \times 10^{-3}}{1.5} = 0.02$$

Hence the Reynold number is less than 0.2 and so the expression $F_D = 3\pi\mu DU$ for calculating terminal velocity is valid.

Problem 11.20 *Determine the velocity of fall of rain drops of a 30 × 10^{-3} cm diameter, density 0.0012 gm/cm³ and kinematic viscosity 0.15 cm²/s.* (A.M.I.E., Winter, 1983)

Solution. Given :

Diameter of rain drops, $D = 30 \times 10^{-3}$ cm

Density of rain drops, $\rho = 0.0012$ gm/cm³

Kinematic viscosity, $\nu = 0.15$ cm²/s

Using the relation, $\nu = \frac{\mu}{\rho}$ or $0.15 = \frac{\mu}{.0012}$

$$\therefore \quad \mu = 0.15 \times .0012 = 0.00018 \frac{\text{gm}}{\text{cm sec}}$$

Now weight of rain drop $= \rho \times g \times$ Volume of rain drop

$$= \rho \times g \times \frac{\pi}{6} D^3 \qquad (\because \text{ Rain drop is a sphere})$$

Drag force, F_D, on rain drop is given by equation (11.8) as $F_D = 3\pi\mu DU$

When rain drop is falling with a uniform velocity U, the drag force must be equal to the weight of rain drop. Hence equating these two values, we get

Weight of rain drop = Drag force

or $$\rho \times g \times \frac{\pi}{6} D^3 = 3\pi\mu DU \quad \text{or} \quad U = \frac{\rho \times g \times \frac{\pi}{6} \times D^3}{3\pi\mu D} = \frac{\rho g D^2}{18\mu}$$

$$= \frac{0.0012 \times 981 \times (30 \times 10^{-3})^2}{18 \times .00018} = \mathbf{0.327\ cm/s.\ Ans.}$$

Let us check for Reynold number, R_e

$$R_e = \frac{\rho UD}{\mu} = \frac{UD}{\nu} = \frac{0.327 \times 30 \times 10^{-3}}{0.15} = 0.0654$$

As the Reynold number is less than 0.2, the expression $F_D = 3\pi\mu DU$ is valid.

11.7.5 Drag on a Cylinder. Consider a real fluid flowing over a circular cylinder of diameter D and length L when the cylinder is placed in the fluid such that its length is perpendicular to the direction of flow. If the Reynolds number of the flow is less than 0.2 $\left(i.e., \frac{U \times d}{\nu} < 0.2\right)$, the inertia force is negligibly small as compared to viscous force and hence the flow pattern about the cylinder will be symmetrical. As the Reynold number is increased, inertia forces increase and hence they must be taken into consideration for analysis of flow over cylinder. With the increase of the Reynold number, the flow pattern becomes unsymmetrical with respect to an axis perpendicular to the direction of flow. The drag force, *i.e.*, the force exerted by the flowing fluid on the cylinder in the direction of flow depends upon the Reynold number of the flow. From experiments, it has been observed that :

(*i*) When Reynold number (R_e) < 1, the drag force is directly proportional to velocity and hence the drag co-efficient (C_D) is inversely proportional to Reynold number.

(*ii*) With the increase of the Reynold number from 1 to 2000, the drag co-efficient decreases and reaches a minimum value of 0.95 at $R_e = 2000$.

(*iii*) With the further increase of the Reynold number from 2000 to 3×10^4, the co-efficient of drag increases and attains maximum value of 1.2 at $R_e = 3 \times 10^4$.

(*iv*) The value of co-efficient of drag decreases if the Reynold number is increased from 3×10^4 to 3×10^5. At $R_e = 3 \times 10^5$, the value of $C_D = 0.3$.

(*v*) If the Reynold number is increased beyond 3×10^6, the value of C_D increases and it becomes equal to 0.7 in the end.

▶ 11.8 BOUNDARY LAYER CONCEPT

When a real fluid flows past a solid body or a solid wall, the fluid particles adhere to the boundary and condition of no slip occurs. This means that the velocity of fluid close to the boundary will be same as that of the boundary. If the boundary is stationary, the velocity of fluid at the boundary will be zero. Farther away from the boundary, the velocity will be higher and as a result of this variation of velocity, the velocity gradient $\frac{du}{dy}$ will exist. The velocity of fluid increases from zero velocity on the stationary boundary to free-stream velocity (U) of the fluid in the direction normal to the boundary. This variation of velocity from zero to free-stream velocity in the direction normal to the boundary takes place in a narrow region in the vicinity of solid boundary. This narrow region of the fluid is called boundary layer. The theory dealing with boundary layer flows is called boundary layer theory.

According to boundary layer theory, the flow of fluid in the neighbourhood of the solid boundary may be divided into two regions as shown in Fig. 11.9.

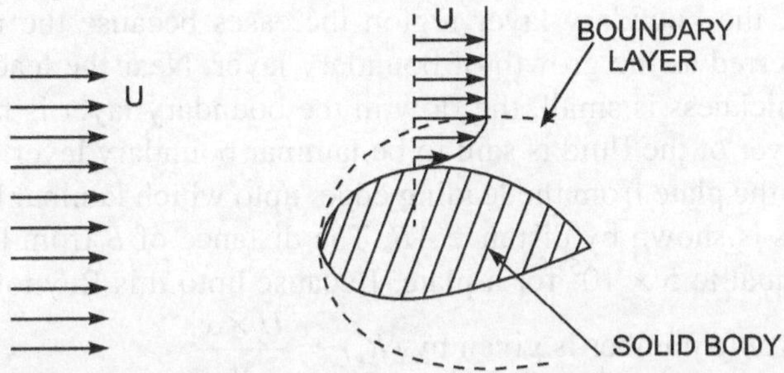

Fig. 11.9 *Flow over solid body.*

1. A very thin layer of the fluid, called the boundary layer, in the immediate neighbourhood of the solid boundary, where the variation of velocity from zero at the solid boundary to free stream velocity in the direction normal to the boundary takes place. In this region, the velocity gradient $\frac{du}{dy}$ exists and hence the fluid exerts a shear stress on the wall in the direction of motion. The value of shear stress is given by

$$\tau = \mu \frac{du}{dy}.$$

2. The remaining fluid, which is outside the boundary layer. The velocity outside the boundary layer is constant and equal to free-stream velocity. As there is no variation of velocity in this region, the velocity gradient $\frac{du}{dy}$ becomes zero. As a result of this the shear stress is zero.

▶ 11.9 DEFINITIONS

11.9.1 Laminar Boundary Layer. For defining the boundary layer (*i.e.*, laminar boundary layer or turbulent boundary layer) consider the flow of a fluid, having free-stream velocity (U), over a smooth thin plate which is flat and placed parallel to the direction for free stream of fluid as shown in Fig. 11.10. Let us consider the flow with zero pressure gradient on one side of the plate, which is stationary.

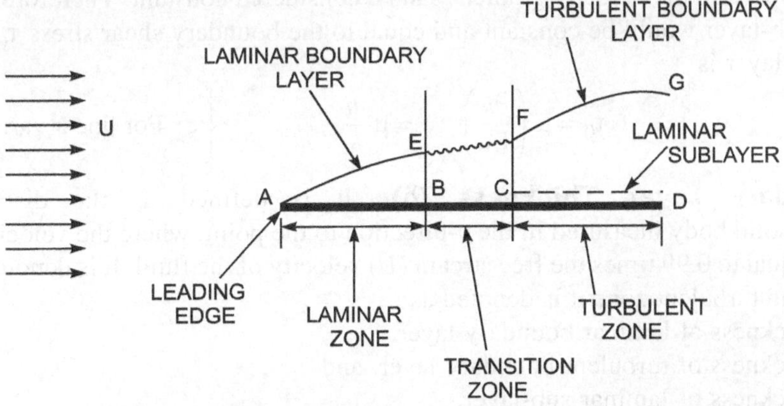

Fig. 11.10 *Flow over a plate.*

The velocity of fluid on the surface of the plate should be equal to the velocity of the plate. But plate is stationary and hence velocity of fluid on the surface of the plate is zero. But at a distance away from the plate, the fluid is having certain velocity. Thus a velocity gradient is set up in the fluid near the surface of the plate. This velocity gradient develops shear resistance, which retards the fluid. Thus the fluid with a uniform free stream velocity (U) is retarded in the vicinity of the solid surface of the plate and the boundary layer region begins at the sharp leading edge. At subsequent points downstream the leading edge, the boundary layer region increases because the retarded fluid is further retarded. This is also referred as the growth of boundary layer. Near the leading edge of the surface of the plate, where the thickness is small, the flow in the boundary layer is laminar though the main flow is turbulent. This layer of the fluid is said to be laminar boundary layer. This is shown by AE in Fig. 11.10. The length of the plate from the leading edge, upto which laminar boundary layer exists, is called laminar zone. This is shown by distance AB. The distance of B from leading edge is obtained from Reynold number equal to 5×10^5 for a plate. Because upto this Reynold number the boundary layer is laminar. The Reynold number is given by $(R_e)_x = \dfrac{U \times x}{\nu}$

where x = Distance from leading edge,
U = Free-stream velocity of fluid,
ν = Kinematic viscosity of fluid,

Hence for laminar boundary layer, we have $5 \times 10^5 = \dfrac{U \times x}{\nu}$ ...(11.11)

If the values of U and ν are known, x or the distance from the leading edge upto which laminar boundary layer exists can be calculated.

11.9.2 Turbulent Boundary Layer. If the length of the plate is more than the distance x, calculated from equation (11.11), the thickness of boundary layer will go on increasing in the down-stream direction. Then the laminar boundary layer becomes unstable and motion of fluid within it, is disturbed and irregular which leads to a transition from laminar to turbulent boundary layer. This short length over which the boundary layer flow changes from laminar to turbulent is called transition zone. This is shown by distance BC in Fig. 11.10. Further downstream the transition zone, the boundary layer is turbulent and continues to grow in thickness. This layer of boundary is called turbulent boundary layer, which is shown by the portion FG in Fig. 13.10.

11.9.3 Laminar Sub-layer. This is the region in the turbulent boundary layer zone, adjacent to the solid surface of the plate as shown in Fig. 11.10. In this zone, the velocity variation is influenced only by viscous effects. Though the velocity distribution would be a parabolic curve in the laminar sub-layer zone, but in view of the very small thickness we can reasonably assume that velocity variation is linear and so the velocity gradient can be considered constant. Therefore, the shear stress in the laminar sub-layer would be constant and equal to the boundary shear stress τ_0. Thus the shear stress in the sub-layer is

$$\tau_0 = \mu \left(\frac{\partial u}{\partial y}\right)_{y=0} = \mu \frac{u}{y} \qquad \left\{\because \text{ For linear variation, } \frac{\partial u}{\partial y} = \frac{u}{y}\right\}$$

11.9.4 Boundary Layer Thickness (δ). It is defined as the distance from the boundary of the solid body measured in the y-direction to the point, where the velocity of the fluid is approximately equal to 0.99 times the free stream (U) velocity of the fluid. It is denoted by the symbol δ. For laminar and turbulent zone it is denoted as :

1. δ_{lam} = Thickness of laminar boundary layer,
2. δ_{tur} = Thickness of turbulent boundary layer, and
3. δ' = Thickness of laminar sub-layer.

11.9.5 Displacement Thickness (δ^*). It is defined as the distance, measured perpendicular to the boundary of the solid body, by which the boundary should be displaced to compensate for the reduction in flow rate on account of boundary layer formation. It is denoted by δ^*. It is also defined as :

"The distance, perpendicular to the boundary, by which the free stream is displaced due to the formation of boundary layer".

Expression for δ^*.

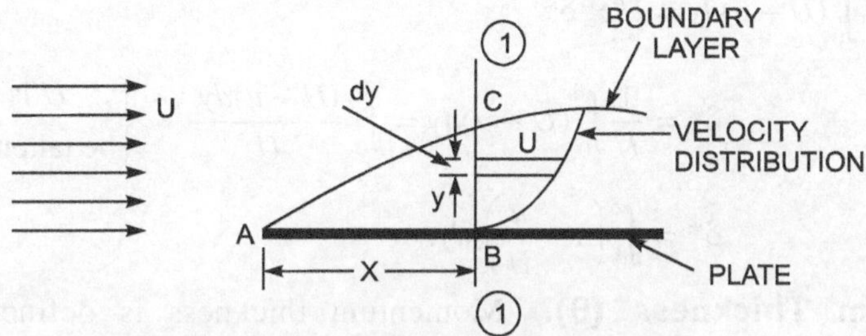

Fig. 11.11 *Displacement thickness.*

Consider the flow of a fluid having free-stream velocity equal to U over a thin smooth plate as shown in Fig. 11.11. At a distance x from the leading edge consider a section 1-1. The velocity of fluid at B is zero and at C, which lies on the boundary layer, is U. Thus velocity varies from zero at B to U at C, where BC is equal to the thickness of boundary layer *i.e.*,

Distance $BC = \delta$

At the section 1-1, consider an elemental strip.

Let y = distance of elemental strip from the plate,

dy = thickness of the elemental strip,

u = velocity of fluid at the elemental strip,

b = width of plate.

Then area of elemental strip, $dA = b \times dy$

Mass of fluid per second flowing through elemental strip

$$= \rho \times \text{Velocity} \times \text{Area of elemental strip}$$

$$= \rho u \times dA = \rho u \times b \times dy \qquad \ldots(i)$$

If there had been no plate, then the fluid would have been flowing with a constant velocity equal to free-stream velocity (U) at the section 1-1. Then mass of fluid per second flowing through elemental strip would have been

$$= \rho \times \text{Velocity} \times \text{Area} = \rho \times U \times b \times dy \qquad \ldots(ii)$$

As U is more than u, hence due to the presence of the plate and consequently due to the formation of the boundary layer, there will be a reduction in mass flowing per second through the elemental strip.

This reduction in mass/sec flowing through elemental strip

= mass/sec given by equation (*ii*) – mass/sec given by equation (*i*)

$$= \rho U b dy - \rho u b dy = \rho b(U - u)dy$$

$\therefore$ Total reduction in mass of fluid/s flowing through BC due to plate

$$= \int_0^\delta \rho b(U - u)dy = \rho b \int_0^\delta (U - u)dy \qquad \ldots(iii)$$

{if fluid is incompressible}

Let the plate is displaced by a distance δ^* and velocity of flow for the distance δ^* is equal to the free-stream velocity (*i.e.*, U). Loss of the mass of the fluid/sec flowing through the distance δ^*

$$= \rho \times \text{Velocity} \times \text{Area}$$

$$= \rho \times U \times \delta^* \times b \qquad \{\because \text{ Area} = \delta^* \times b\} \quad ...(iv)$$

Equating equations (*iii*) and (*iv*), we get

$$\rho b \int_0^\delta (U - u)dy = \rho \times U \times \delta^* b$$

Cancelling ρb to both sides, we have

$$\int_0^\delta (U - u)dy = U \times \delta^*$$

or

$$\delta^* = \frac{1}{U}\int_0^\delta (U - u)dy = \int_0^\delta \frac{(U - u)dy}{U} \qquad \left\{\begin{matrix}\because U \text{ is constant and can} \\ \text{be taken inside the integral}\end{matrix}\right\}$$

$\therefore$

$$\delta^* = \int_0^\delta \left(1 - \frac{u}{U}\right) dy. \qquad ...(11.12)$$

11.9.6 Momentum Thickness (θ). Momentum thickness is defined as the distance, measured perpendicular to the boundary of the solid body, by which the boundary should be displaced to compensate for the reduction in **momentum** of the flowing fluid on account of boundary layer formation. It is denoted by θ.

Consider the flow over a plate as shown in Fig. 11.11. Consider the section 1-1 at a distance x from leading edge. Take an elemental strip at a distance y from the plate having thickness (dy). The mass of fluid flowing per second through this elemental strip is given by equation (*i*) and is equal to $\rho ubdy$.

Momentum of this fluid = Mass × Velocity = $(\rho ubdy)u$

Momentum of this fluid in the absence of boundary layer = $(\rho ubdy)U$

$\therefore$ Loss of momentum through elemental strip = $(\rho ubdy)U - (\rho ubdy) \times u = \rho bu(U - u)dy$

$\therefore$ Total loss of momentum/sec through $BC = \int_0^\delta \rho bu(U - u)dy$ ...(11.13)

Let θ = distance by which plate is displaced when the fluid is flowing with a constant velocity U

$\therefore$ Loss of momentum/sec of fluid flowing through distance θ with a velocity U

$$= \text{Mass of fluid through } \theta \times \text{velocity}$$

$$= (\rho \times \text{area} \times \text{velocity}) \times \text{velocity}$$

$$= [\rho \times \theta \times b \times U] \times U \qquad \{\because \text{ Area} = \theta \times b\}$$

$$= \rho\theta b U^2 \qquad ...(11.14)$$

Equating equations (11.14) and (11.13), we have

$$\rho\theta b U^2 = \int_0^\delta \rho bu(U - u)dy = \rho b \int_0^\delta u(U - u)dy \qquad \{\text{If fluid is assumed incompressible}\}$$

or

$$\theta U^2 = \int_0^\delta u(U - u)dy \qquad \{\text{cancelling } \rho b \text{ to both sides}\}$$

or

$$\theta = \frac{1}{U}\int_0^\delta u(U - u)dy = \int_0^\delta \frac{u(U - u)dy}{U^2}$$

$\therefore$

$$\theta = \int_0^\delta \frac{u}{U}\left[1 - \frac{u}{U}\right] dy. \qquad ...(11.15)$$

11.9.7 Energy Thickness (δ^{**}). It is defined as the distance, measured perpendicular to the boundary of the solid body, by which the boundary should be displaced to compensate for the reduction in *kinetic energy* of the flowing fluid on account of boundary layer formation. It is denoted by δ^{**}.

Consider the flow over the plate as shown in Fig. 11.11 having section 1-1 at a distance x from leading edge. The mass of fluid flowing per second through the elemental strip of thickness 'dy' at a distance y from the plate as given by equation (i) = $\rho ubdy$

Kinetic energy of this fluid $= \frac{1}{2}\ m \times \text{velocity}^2 = \frac{1}{2}\ (\rho ubdy)\ u^2$

Kinetic energy of this fluid in the absence of boundary layer

$$= \frac{1}{2}\ (\rho ubdy)U^2$$

∴ Loss of K.E. through elemental strip

$$= \frac{1}{2}\ (\rho ubdy)U^2 - \frac{1}{2}\ (\rho ubdy)\ u^2 = \frac{1}{2}\ \rho ub\ [U^2 - u^2]\ dy$$

∴ Total loss of K.E. of fluid passing through BC

$$= \int_0^\delta \frac{1}{2}\ \rho ub\ [U^2 - u^2]\ dy = \frac{1}{2}\rho b \int_0^\delta u\ (U^2 - u^2)\ dy$$

{If fluid is considered incompressible}

Let δ^{**} = distance by which the plate is displaced to compensate for the reduction in K.E.

∴ Loss of K.E. through δ^{**} of fluid flowing with velocity U

$$= \frac{1}{2}\ (\text{mass})\ \text{velocity}^2 = \frac{1}{2}\ (\rho \times \text{area} \times \text{velocity})\ \text{velocity}^2$$

$$= \frac{1}{2}\ (\rho \times b \times \delta^{**} \times U)U^2 \qquad \{\because \quad \text{Area} = b \times \delta^{**}\}$$

$$= \frac{1}{2}\ \rho b \delta^{**} U^3$$

Equating the two losses of K.E., we get

$$\frac{1}{2}\ \rho b \delta^{**} U^3 = \frac{1}{2}\rho b \int_0^\delta u\ (U^2 - u^2)\ dy$$

or

$$\delta^{**} = \frac{1}{U^3} \int_0^\delta u\ (U^2 - u^2)\ dy$$

∴

$$\delta^{**} = \int_0^\delta \frac{u}{U}\left[1 - \frac{u^2}{U^2}\right] dy. \qquad ...(11.16)$$

Problem 11.21 *Find the displacement thickness, the momentum thickness and energy thickness for the velocity distribution in the boundary layer given by* $\frac{u}{U} = \frac{y}{\delta}$, *where u is the velocity at a distance y from the plate and* $u = U$ *at* $y = \delta$, *where* δ = *boundary layer thickness. Also calculate the value of* δ^*/θ.

Solution. Given :

Velocity distribution $\quad \frac{u}{U} = \frac{y}{\delta}$

(i) Displacement thickness δ^* is given by equation (11.12),

$$\delta^* = \int_0^\delta \left(1 - \frac{u}{U}\right) dy = \int_0^\delta \left(1 - \frac{y}{\delta}\right) dy \qquad \left\{\because \frac{u}{U} = \frac{y}{\delta}\right\}$$

$$= \left[y - \frac{y^2}{2\delta}\right]_0^\delta \qquad \{\delta \text{ is constant across a section}\}$$

$$= \delta - \frac{\delta^2}{2\delta} = \delta - \frac{\delta}{2} = \frac{\delta}{2}. \textbf{ Ans.}$$

(*ii*) Momentum thickness, θ is given by the equation (11.15),

$$\theta = \int_0^\delta \frac{u}{U}\left(1 - \frac{u}{U}\right) dy$$

Substituting the value of $\frac{u}{U} = \frac{y}{\delta}$,

$$\theta = \int_0^\delta \frac{y}{\delta}\left(1 - \frac{y}{\delta}\right) dy = \int_0^\delta \left(\frac{y}{\delta} - \frac{y^2}{\delta^2}\right) dy$$

$$= \left[\frac{y^2}{2\delta} - \frac{y^3}{3\delta^2}\right]_0^\delta = \frac{\delta^2}{2\delta} - \frac{\delta^2}{3\delta^2} = \frac{\delta}{2} - \frac{\delta}{3} = \frac{3\delta - 2\delta}{6} = \frac{\delta}{6}. \textbf{ Ans.}$$

(*iii*) Energy thickness δ** is given by equation (11.16), as

$$\delta^{**} = \int_0^\delta \frac{u}{U}\left[1 - \frac{u^2}{U^2}\right] dy = \int_0^\delta \frac{y}{\delta}\left[1 - \frac{y^2}{\delta^2}\right] dy \qquad \left\{\because \frac{u}{U} = \frac{y}{\delta}\right\}$$

$$= \int_0^\delta \left[\frac{y}{\delta} - \frac{y^3}{\delta^3}\right] dy = \left[\frac{y^2}{2\delta} - \frac{y^4}{4\delta^3}\right]_0^\delta = \frac{\delta^2}{2\delta} - \frac{\delta^4}{4\delta^3}$$

$$= \frac{\delta}{2} - \frac{\delta}{4} = \frac{2\delta - \delta}{4} = \frac{\delta}{4}. \textbf{ Ans.}$$

(*iv*)

$$\frac{\delta^*}{\theta} = \frac{\left(\frac{\delta}{2}\right)}{\left(\frac{\delta}{6}\right)} = \frac{\delta}{2} \times \frac{6}{\delta} = \textbf{3.0. Ans.}$$

Problem 11.22 *Find the displacement thickness, the momentum thickness and energy thickness for the velocity distribution in the boundary layer given by* $\frac{u}{U} = 2\left(\frac{y}{\delta}\right) - \left(\frac{y}{\delta}\right)^2$.

Solution. Given :

Velocity distribution $\qquad \frac{u}{U} = 2\left(\frac{y}{\delta}\right) - \left(\frac{y}{\delta}\right)^2$

(*i*) Displacement thickness δ* is given by equation (11.12),

$$\delta^* = \int_0^\delta \left(1 - \frac{u}{U}\right) dy$$

Substituting the value of $\frac{u}{U} = 2\left(\frac{y}{\delta}\right) - \left(\frac{y}{\delta}\right)^2$, we have

$$\delta^* = \int_0^\delta \left\{1 - \left[2\left(\frac{y}{\delta}\right) - \left(\frac{y}{\delta}\right)^2\right]\right\} dy$$

$$= \int_0^\delta \left\{1 - 2\left(\frac{y}{\delta}\right) + \left(\frac{y}{\delta}\right)^2\right\} dy = \left[y - \frac{2y^2}{2\delta} + \frac{y^3}{3\delta^2}\right]_0^\delta$$

$$= \delta - \frac{\delta^2}{\delta} + \frac{\delta^3}{3\delta^2} = \delta - \delta + \frac{\delta}{3} = \mathbf{\frac{\delta}{3}}. \quad \textbf{Ans.}$$

(*ii*) Momentum thickness θ, is given by equation (11.15),

$$\theta = \int_0^\delta \frac{u}{U}\left\{1 - \frac{u}{U}\right\} dy = \int_0^\delta \left(\frac{2y}{\delta} - \frac{y^2}{\delta^2}\right)\left[1 - \left(\frac{2y}{\delta} - \frac{y^2}{\delta^2}\right)\right] dy$$

$$= \int_0^\delta \left[\frac{2y}{\delta} - \frac{y^2}{\delta^2}\right]\left[1 - \frac{2y}{\delta} + \frac{y^2}{\delta^2}\right] dy$$

$$= \int_0^\delta \left[\frac{2y}{\delta} - \frac{4y^2}{\delta^2} + \frac{2y^3}{\delta^3} - \frac{y^2}{\delta^2} + \frac{2y^3}{\delta^3} - \frac{y^4}{\delta^4}\right] dy$$

$$= \int_0^\delta \left[\frac{2y}{\delta} - \frac{5y^2}{\delta^2} + \frac{4y^3}{\delta^3} - \frac{y^4}{\delta^4}\right] dy = \left[\frac{2y^2}{2\delta} - \frac{5y^3}{3\delta^2} + \frac{4y^4}{4\delta^3} - \frac{y^5}{5\delta^4}\right]_0^\delta$$

$$= \left[\frac{\delta^2}{\delta} - \frac{5\delta^3}{3\delta^2} + \frac{\delta^4}{\delta^3} - \frac{y^5}{5\delta^4}\right] = \delta - \frac{5\delta}{3} + \delta - \frac{\delta}{5}$$

$$= \frac{15\delta - 25\delta + 15\delta - 3\delta}{15} = \frac{30\delta - 28\delta}{15} = \mathbf{\frac{2\delta}{15}}. \quad \textbf{Ans.}$$

(*iii*) Energy thickness δ** is given by equation (11.16),

$$\delta^{**} = \int_0^\delta \frac{u}{U}\left[1 - \frac{u^2}{U^2}\right] dy = \int_0^\delta \left(\frac{2y}{\delta} - \frac{y^2}{\delta^2}\right)\left(1 - \left[\frac{2y}{\delta} - \frac{y^2}{\delta^2}\right]^2\right) dy$$

$$= \int_0^\delta \left(\frac{2y}{\delta} - \frac{y^2}{\delta^2}\right)\left(1 - \left[\frac{4y^2}{\delta^2} + \frac{y^4}{\delta^4} - \frac{4y^3}{\delta^3}\right]\right) dy$$

$$= \int_0^\delta \left(\frac{2y}{\delta} - \frac{y^2}{\delta^2}\right)\left(1 - \frac{4y^2}{\delta^2} - \frac{y^4}{\delta^4} + \frac{4y^3}{\delta^3}\right) dy$$

$$= \int_0^\delta \left(\frac{2y}{\delta} - \frac{8y^2}{\delta^3} - \frac{2y^5}{\delta^5} + \frac{8y^4}{\delta^4} - \frac{y^2}{\delta^2} + \frac{4y^4}{\delta^4} + \frac{y^6}{\delta^6} - \frac{4y^5}{\delta^5}\right) dy$$

$$= \int_0^\delta \left[\frac{2y}{\delta} - \frac{y^2}{\delta^2} - \frac{8y^3}{\delta^3} + \frac{12y^4}{\delta^4} - \frac{6y^5}{\delta^5} + \frac{y^6}{\delta^6}\right] dy$$

$$= \left[\frac{2y^2}{2\delta} - \frac{y^3}{3\delta^2} - \frac{8y^4}{4\delta^3} + \frac{12y^5}{5\delta^4} - \frac{6y^6}{6\delta^5} + \frac{y^7}{7\delta^6}\right]_0^\delta$$

$$= \frac{\delta^2}{\delta} - \frac{\delta^3}{3\delta^2} - \frac{2\delta^4}{\delta^3} + \frac{12\delta^5}{5\delta^4} - \frac{\delta^6}{\delta^5} + \frac{\delta^7}{7\delta^6} = \delta - \frac{\delta}{3} - 2\delta + \frac{12}{5}\delta - \delta + \frac{\delta}{7}$$

$$= -2\delta - \frac{\delta}{3} + \frac{12}{5}\delta + \frac{8}{7} = \frac{-210\delta - 35\delta + 252\delta + 15\delta}{105}$$

$$= \frac{-245\delta + 267\delta}{105} = \frac{\mathbf{22\delta}}{\mathbf{105}}. \quad \textbf{Ans.}$$

▶ 11.10. DRAG FORCE ON A FLAT PLATE DUE TO BOUNDARY LAYER

Consider the flow of a fluid having free stream velocity equal to U, over a thin plate as shown in Fig. 11.12. The drag force on the plate can be determined if the velocity profile near the plate is known. Consider a small length Δx of the plate at a distance of x from the leading edge as shown in Fig. 11.12 (a). The enlarged view of the small length of the plate is shown in Fig. 11.12 (b).

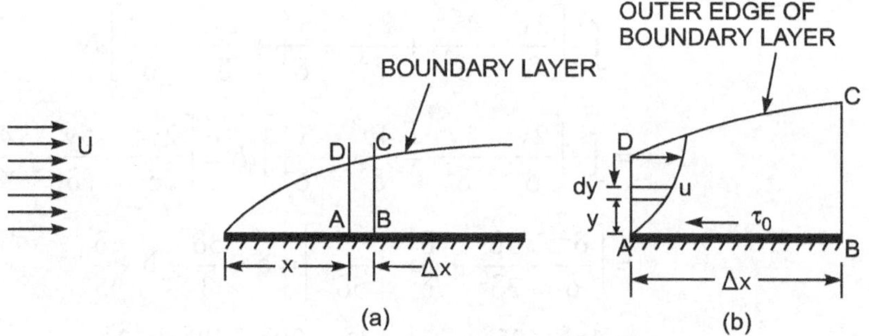

Fig. 11.12 *Drag force on a plate due to boundary layer.*

The shear stress τ_0 is given by $\tau_0 = \mu\left(\frac{du}{dy}\right)_{y=0}$, where $\left(\frac{du}{dy}\right)_{y=0}$ is the velocity distribution near the plate at $y = 0$.

Then drag force or shear force on a small distance Δx is given by

$$\Delta F_D = \text{shear stress} \times \text{area}$$
$$= \tau_0 \times \Delta x \times b \qquad \text{...(11.17)} \quad \{\text{Taking width of plate} = b\}$$

where ΔF_D = drag force on distance Δx

The drag force ΔF_D must also be equal to the rate of change of momentum over the distance Δx.

Consider the flow over the small distance Δx. Let $ABCD$ is the control volume of the fluid over the distance Δx as shown in Fig. 11.12 (b). The edge DC represents the outer edge of the boundary layer.

Let u = velocity at any point within the boundary layer

b = width of plate

Then mass rate of flow entering through the side AD

$$= \int_0^\delta \rho \times \text{velocity} \times \text{area of strip of thickness } dy$$

$$= \int_0^\delta \rho \times u \times b \times dy \qquad \{\because \text{ Area of strip} = b \times dy\}$$

$$= \int_0^\delta \rho u b dy$$

Mass rate of flow leaving the side BC

$$= \text{mass through } AD + \frac{\partial}{\partial x}(\text{mass through } AD) \times \Delta x$$

$$= \int_0^\delta \rho u b dy \, \frac{\partial}{\partial x}\left[\int_0^\delta (\rho u b dy)\right] \times \Delta x$$

From continuity equation for a steady incompressible fluid flow, we have

Mass rate of flow entering AD + mass rate of flow entering DC

= mass rate of flow leaving BC

$\because$ Mass rate of flow entering DC = mass rate of flow through BC – mass rate of flow through AD

$$= \int_0^\delta \rho u b dy + \frac{\partial}{\partial x}\left[\int_0^\delta \rho u b dy\right] \times \Delta x - \int_0^\delta \rho u b dy$$

$$= \frac{\partial}{\partial x}\left[\int_0^\delta \rho u b dy\right] \times \Delta x$$

The fluid is entering through side DC with a uniform velocity U.

Now let us calculate momentum flux through control volume.

Momentum flux entering through AD

$$= \int_0^\delta \text{momentum flux through strip of thickness } dy$$

$$= \int_0^\delta \text{mass through strip} \times \text{velocity} = \int_0^\delta (\rho u b dy) \times u = \int_0^\delta \rho u^2 b dy$$

Momentum flux leaving the side $BC = \int_0^\delta \rho u^2 b dy + \frac{\partial}{\partial x}\left[\int_0^\delta \rho u^2 b dy\right] \times \Delta x$

Momentum flux entering the side DC = mass rate through DC × velocity

$$= \frac{\partial}{\partial x}\left[\int_0^\delta \rho u b dy\right] \times \Delta x \times U \qquad (\because \text{ Velocity} = U)$$

$$= \frac{\partial}{\partial x}\left[\int_0^\delta \rho u U b dy\right] \times \Delta x$$

As U is constant and so it can be taken inside the differential and integral.

$\therefore$ Rate of change of momentum of the control volume

= Momentum flux through BC – Momentum flux through AD

– momentum flux through DC

$$= \int_0^\delta \rho u^2 b dy \frac{\partial}{\partial x}\left[\int_0^\delta \rho u^2 b dy\right] \times \Delta x - \int_0^\delta \rho u^2 b dy - \frac{\partial}{\partial x}\left[\int_0^\delta \rho u U b dy\right] \times \Delta x$$

$$= \frac{\partial}{\partial x}\left[\int_0^\delta \rho u^2 b dy\right] \times \Delta x - \frac{\partial}{\partial x}\left[\int_0^\delta \rho u U b dy\right] \times \Delta x$$

$$= \frac{\partial}{\partial x}\left[\int_0^\delta \rho u^2 b dy - \int_0^\delta \rho u U b dy\right] \times \Delta x$$

$$= \frac{\partial}{\partial x}\left[\int_0^\delta \left(\rho u^2 b - \rho u U b\right) dy\right] \times \Delta x$$

$$= \frac{\partial}{\partial x}\left[\rho b \int_0^\delta \left(u^2 - uU\right) dy\right] \times \Delta x$$

{For incompressible fluid ρ is constant}

$$= \rho b \frac{\partial}{\partial x}\left[\int_0^\delta \left(u^2 - uU\right) dy\right] \times \Delta x \qquad ...(11.18)$$

Now the rate of change of momentum on the control volume $ABCD$ must be equal to the total force on the control volume in the same direction according to the momentum principle. But for a flat plate $\frac{\partial p}{\partial x} = 0$, which means there is no external pressure force on the control volume. Also the force on the side DC is negligible as the velocity is constant and velocity gradient is zero approximately. The only external force acting on the control volume is the shear force acting on the side AB in the direction from B to A as shown in Fig. 11.12 (b). The value of this force is given by equation (11.17) as

$$\Delta F_D = \tau_0 \times \Delta x \times b$$

∴ Total external force in the direction of rate of change of momentum

$$= -\tau_0 \times \Delta x \times b \qquad ...(11.19)$$

According to momentum principle, the two values given by equations (11.19) and (11.18) should be the same.

$$\therefore \qquad -\tau_0 \times \Delta x \times b = \rho b \frac{\partial}{\partial x}\left[\int_0^\delta \left(u^2 - uU\right) dy\right] \times \Delta x$$

Cancelling $\Delta x \times b$, to both sides, we have

$$-\tau_0 = \rho \frac{\partial}{\partial x}\left[\int_0^\delta \left(u^2 - uU\right) dy\right]$$

or

$$\tau_0 = -\rho \frac{\partial}{\partial x}\left[\int_0^\delta \left(u^2 - uU\right) dy\right] = \rho \frac{\partial}{\partial x}\left[\int_0^\delta \left(uU - u^2\right) dy\right]$$

$$= \rho \frac{\partial}{\partial x}\left[\int_0^\delta U^2\left(\frac{u}{U} - \frac{u^2}{U^2}\right) dy\right] = \rho U^2 \frac{\partial}{\partial x}\left[\int_0^\delta \frac{u}{U}\left[1 - \frac{u}{U}\right] dy\right]$$

or

$$\frac{\tau_0}{\rho U^2} = \frac{\partial}{\partial x}\left[\int_0^\delta \frac{u}{U}\left[1 - \frac{u}{U}\right] dy\right] \qquad ...(11.20)$$

In equation (11.20), the expression $\int_0^\delta \frac{u}{U}\left[1 - \frac{u}{U}\right] dy$ is equal to momentum thickness θ. Hence equation (11.20) is also written as

$$\frac{\tau_0}{\rho U^2} = \frac{\partial \theta}{\partial x} \qquad ...(11.21)$$

Equation (11.21) is known as **Von Karman momentum integral equation** for boundary layer flows.

This is applied to :

1. Laminar boundary layers,
2. Transition boundary layers, and
3. Turbulent boundary layer flows.

For a given velocity profile in laminar zone, transition zone or turbulent zone of a boundary layer, the shear stress τ_0 is obtained from equation (11.20) or (11.21). Then drag force on a small distance Δx of the plate is obtained from equation (11.17) as

$$\Delta F_D = \tau_0 \times \Delta x \times b$$

Then total drag on the plate of length L on one side is

$$F_D = \int \Delta F_D = \int_0^L \tau_0 \times b \times dx \quad \{\text{change } \Delta x = dx\}. \qquad ...(11.22)$$

11.10.1 Local Co-efficient of Drag [C_D*]. It is defined as the ratio of the shear stress τ_0 to the quantity $\frac{1}{2}\rho U^2$. It is denoted by C_D*

Hence
$$C_D{}^* = \frac{\tau_0}{\frac{1}{2}\rho U^2}. \qquad ...(11.23)$$

11.10.2 Average Co-efficient of Drag [C_D]. It is defined as the ratio of the total drag force to the quantity $\frac{1}{2}\rho A U^2$. It is also called co-efficient of drag and is denoted by C_D.

Hence
$$C_D = \frac{F_D}{\frac{1}{2}\rho A U^2} \qquad ...(11.24)$$

where A = Area of the surface (or plate)

U = Free stream velocity

ρ = Mass density of fluid.

11.10.3 Boundary Conditions for the Velocity Profiles. The followings are the boundary conditions which must be satisfied by any velocity profile, whether it is in laminar boundary layer zone, or in turbulent boundary layer zone :

1. At $y = 0$, $u = 0$ and $\frac{du}{dy}$ has some finite value
2. At $y = \delta$, $u = U$
3. At $y = \delta$, $\frac{du}{dy} = 0$.

Problem 11.23 *For the velocity profile for laminar boundary layer flows given as*

$$\frac{u}{U} = 2(y/\delta) - (y/\delta)^2$$

find an expression for boundary layer thickness (δ), shear stress (τ_0) and co-efficient of drag (C_D) in terms of Reynold number.

Solution. Given :

(*i*) **The velocity distribution** $\frac{u}{U} = 2\left(\frac{y}{\delta}\right) - \left(\frac{y}{\delta}\right)^2$ ...(*i*)

Substituting this value of $\frac{u}{U}$ in equation (11.20), we get

$$\frac{\tau_0}{\rho U^2} = \frac{\partial}{\partial x}\left[\int_0^\delta \frac{u}{U}\left(1 - \frac{u}{U}\right) dy\right] = \frac{\partial}{\partial x}\left[\int_0^\delta \left[\frac{2y}{\delta} - \frac{y^2}{\delta^2}\right]\left[1 - \left(\frac{2y}{\delta} - \frac{y^2}{\delta^2}\right)\right] dy\right]$$

$$= \frac{\partial}{\partial x}\left[\int_0^\delta \left[\frac{2y}{\delta} - \frac{y^2}{\delta^2}\right]\left[1 - \frac{2y}{\delta} + \frac{y^2}{\delta^2}\right] dy\right]$$

$$= \frac{\partial}{\partial x}\left[\int_0^\delta \left[\frac{2y}{\delta} - \frac{4y^2}{\delta^2} + \frac{2y^3}{\delta^3} - \frac{y^2}{\delta^2} + \frac{2y^3}{\delta^3} - \frac{y^4}{\delta^4}\right] dy\right]$$

$$= \frac{\partial}{\partial x}\int_0^\delta \left[\frac{2y}{\delta} - \frac{5y^2}{\delta^2} + \frac{4y^3}{\delta^3} - \frac{y^4}{\delta^4}\right] dy = \frac{\partial}{\partial x}\left[\frac{2y^2}{2\delta} - \frac{5 \times y^3}{3\delta^2} + \frac{4y^4}{4\delta^3} - \frac{y^5}{5\delta^4}\right]_0^\delta$$

$$= \frac{\partial}{\partial x}\left[\frac{\delta^2}{\delta} - \frac{5}{3}\frac{\delta^3}{\delta^2} + \frac{\delta^4}{\delta^3} - \frac{\delta^5}{5\delta^4}\right] = \frac{\partial}{\partial x}\left[\delta - \frac{5}{3}\delta + \delta - \frac{\delta}{5}\right]$$

$$= \frac{\partial}{\partial x}\left[\frac{15\delta - 25\delta + 15\delta - 3\delta}{15}\right] = \frac{\partial}{\partial x}\left[\frac{30\delta - 28\delta}{15}\right] = \frac{\partial}{\delta x}\left[\frac{2\delta}{15}\right] = \frac{2}{15}\frac{\partial}{\partial x}[\delta]$$

$\therefore$ $$\tau_0 = \rho U^2 \times \frac{2}{15}\frac{\partial}{\partial x}[\delta] = \frac{2}{15}\rho U^2 \frac{\partial[\delta]}{\partial x} \quad \text{...(11.25)}$$

The shear stress at the boundary in laminar flow is also given by Newton's law of viscosity as

$$\tau_0 = \mu\left(\frac{du}{dy}\right)_{y=0} \quad \text{...}(ii)$$

But from equation (*i*), $$u = U\left[\frac{2y}{\delta} - \frac{y^2}{\delta^2}\right]$$

$\therefore$ $$\frac{du}{dy} = U\left[\frac{2}{\delta} - \frac{2y}{\delta^2}\right] \quad \{\because U \text{ is constant}\}$$

$\therefore$ $$\left(\frac{du}{dy}\right)_{y=0} = U\left[\frac{2}{\delta} - \frac{2 \times (0)}{\delta^2}\right] = \frac{2U}{\delta}$$

Substituting this value in (*ii*), we get

$$\tau_0 = \mu \times \frac{2U}{\delta} = \frac{2\mu U}{\delta} \qquad ...(iii)$$

Equating the two values of τ_0 given by equation (11.25) and (*iii*)

$$\frac{2}{15}\rho U^2 \frac{\partial}{\partial x}[\delta] = \frac{2\mu U}{\delta}$$

or
$$\frac{\delta\partial}{\partial x}[\delta] = \frac{15\mu U}{\rho U^2} = \frac{15\mu}{\rho U} \quad \text{or} \quad \delta\partial[\delta] = \frac{15\mu}{\rho U}\partial x$$

As the boundary layer thickness (δ) is a function of x only.

Hence partial derivative can be changed to total derivative

$$\therefore \qquad \delta d[\delta] = \frac{15\mu x}{\rho U}dx$$

On integration, we get
$$\frac{\delta^2}{2} = \frac{15\mu}{\rho U}x + C \qquad \left\{\frac{\mu}{\rho U} \text{ is constant}\right\}$$

$$x = 0,\ \delta = 0 \text{ and hence } C = 0$$

$$\therefore \qquad \frac{\delta^2}{2} = \frac{15\mu x}{\rho U}$$

$$\therefore \qquad \delta = \sqrt{\frac{2 \times 15\mu x}{\rho U}} = \sqrt{\frac{30\mu x}{\rho U}} = 5.48\sqrt{\frac{\mu x}{\rho U}} \qquad ...(11.26)$$

$$= 5.48\sqrt{\frac{\mu x \times x}{\rho U \times x}} = 5.48\sqrt{\frac{x^2}{R_{e_x}}} \qquad \left\{\because \ R_{e_x} = \frac{\rho U x}{\mu}\right\}$$

$$= 5.48\frac{x}{\sqrt{R_{e_x}}} \qquad ...(11.27)$$

In equation (11.26), μ, ρ and U are constant and hence it is clear from this equation that thickness of laminar boundary layer is proportional to the square root of the distance from the leading edge. Equation (11.27) gives the thickness of laminar boundary layer in terms of Reynold number.

(*ii*) **Shear stress (τ_0) in terms of Reynold number**

From equation (*iii*), we have $\tau_0 = \dfrac{2\mu U}{\delta}$

Substituting the value of δ from equation (11.27), in the above equation, we get

$$\tau_0 = \frac{2\mu U}{5.48\dfrac{x}{\sqrt{R_{e_x}}}} = \frac{2\mu U\sqrt{R_{e_x}}}{5.48x} = 0.365\frac{\mu U}{x}\sqrt{R_{e_x}}$$

(*iii*) **Co-efficient of Drag (C_D)**

From equation (11.24), we have $C_D = \dfrac{F_D}{\frac{1}{2}\rho A U^2}$

where F_D is given by equation (11.22) as

$$F_D = \int_0^L \tau_0 \times b \times dx = \int_0^L 0.365 \frac{\mu U}{x} \sqrt{R_{e_x}} \times b \times dx$$

$$= 0.365 \int_0^L \frac{\mu U}{x} \sqrt{\frac{\rho U x}{\mu}} \times b \times dx \qquad \left\{\because R_{e_x} = \frac{\rho U x}{\mu}\right\}$$

$$= 0.365 \int_0^L \mu U \sqrt{\frac{\rho U}{\mu}} \times \frac{1}{\sqrt{x}} \times b \times dx$$

$$= 0.365\, \mu U \sqrt{\frac{\rho U}{\mu}} \times b \int_0^L x^{1/2}\, dx = 0.365\, \mu U \sqrt{\frac{\rho U}{\mu}} \times b \times \left[\frac{x^{1/2}}{\frac{1}{2}}\right]_0^L$$

$$= 0.365 \times 2\mu U \sqrt{\frac{\rho U}{\mu}} \times b \times \sqrt{L}$$

$$= 0.73\, b\mu U \sqrt{\frac{\rho U L}{\mu}} \qquad \text{...(11.28)}$$

$$\therefore \qquad C_D = \frac{0.73\, b\mu U \sqrt{\frac{\rho U L}{\mu}}}{\frac{1}{2}\rho A U^2}$$

where A = Area of plate = Length of plate × width = $L \times b$

$$\therefore \qquad C_D = \frac{0.73\, b\mu U}{\frac{1}{2}\rho \times L \times b \times U^2} \sqrt{\frac{\rho U L}{\mu}} = \frac{1.46\,\mu}{\rho L U} \sqrt{\frac{\rho U L}{\mu}}$$

$$= \frac{1.46\sqrt{\mu}}{\sqrt{\rho L}} = 1.46 \sqrt{\frac{\mu}{\rho U L}} = \frac{1.46}{\sqrt{R_{e_L}}} \text{ ...(11.29)} \qquad \left\{\because \frac{\mu}{\rho U L} = \frac{1}{\sqrt{R_{e_L}}}\right\}$$

Problem 11.24 *For the velocity profile given in problem 11.23, find the thickness of boundary layer at the end of the plate and the drag force on one side of a plate 1 m long and 0.8 m wide when placed in water flowing with a velocity of 150 mm per second. Calculate the value of co-efficient of drag also. Take μ for water = 0.01 poise.*

Solution. Given :

Length of plate, $L = 1$ m

Width of plate, $b = 0.8$ m

Velocity of fluid (water), $U = 150$ mm/s $= 0.15$ m/s

μ for water $= 0.01$ poise $= \dfrac{0.01}{10}\dfrac{\text{Ns}}{\text{m}^2} = 0.001\ \dfrac{\text{Ns}}{\text{m}^2}$

Reynold number at the end of the plate *i.e.*, at a distance of 1 m from leading edge is given by

$$R_{e_L} = \frac{\rho UL}{\mu} = 1000 \times \frac{0.15 \times 1.0}{.001} \qquad (\because \ \rho = 1000)$$

$$= \frac{1000 \times .15 \times 1.0}{0.001} = 150000$$

(*i*) As laminar boundary layer exists upto Reynold number $= 2 \times 10^5$. Hence this is the case of laminar boundary layer. Thickness of boundary layer at $x = 1.0$ m is given by equation (11.27) as

$$\delta = 5.48 \frac{x}{\sqrt{R_{e_x}}} = \frac{5.48 \times 1.0}{\sqrt{150000}} = 0.01415 \text{ m} = \textbf{14.15 mm. Ans.}$$

(*ii*) Drag force on one side of the plate is given by equation (11.28)

$$F_D = 0.73\, b\mu U \sqrt{\frac{\rho UL}{\mu}}$$

$$= 0.73 \times 0.8 \times 0.001 \times 0.15 \times \sqrt{150000} \qquad \left\{\because \ \frac{\rho UL}{\mu} = R_{e_L}\right\}$$

$$= \textbf{0.0338 N. Ans.}$$

(*iii*) Co-efficient of drag, C_D is given by equation (11.29) as

$$C_D = \frac{1.46}{\sqrt{R_{e_L}}} = \frac{1.46}{\sqrt{150000}} = \textbf{.00376. Ans.}$$

Problem 11.25 *For the velocity profile for laminar boundary layer* $\dfrac{u}{U} = \dfrac{3}{2}\left(\dfrac{y}{\delta}\right) - \dfrac{1}{2}\left(\dfrac{y}{\delta}\right)^2$. *Determine the boundary layer thickness, shear stress, drag force and co-efficient of drag in terms of Reynold number.*

Solution. Given :

Velocity distribution, $\dfrac{u}{U} = \dfrac{3}{2}\left(\dfrac{y}{\delta}\right) - \dfrac{1}{2}\left(\dfrac{y}{\delta}\right)^3$

Using equation (11.20), we have $\dfrac{\tau_0}{\rho U^2} = \dfrac{\partial}{\partial x}\left[\displaystyle\int_0^\delta \frac{u}{U}\left(1 - \frac{u}{U}\right) dy\right]$

Substituting the value of $\dfrac{u}{U} = \dfrac{3}{2}\left(\dfrac{y}{\delta}\right) - \dfrac{1}{2}\left(\dfrac{y}{\delta}\right)^3$ in the above equation

$$\frac{\tau_0}{\rho U^2} = \frac{\partial}{\partial x}\left[\int_0^\delta \left[\frac{3}{2}\left(\frac{y}{\delta}\right) - \frac{1}{2}\left(\frac{y}{\delta}\right)^3\right]\left[1 - \left\{\frac{3}{2}\left(\frac{y}{\delta}\right) - \frac{1}{2}\left(\frac{y}{\delta}\right)^3\right\}\right] dy\right]$$

$$= \frac{\partial}{\partial x}\left[\int_0^\delta \left(\frac{3y}{2\delta} - \frac{y^3}{2\delta^3}\right)\left(1 - \frac{3y}{2\delta} + \frac{y^3}{2\delta^3}\right) dy\right]$$

$$= \frac{\partial}{\partial x}\left[\int_0^\delta \left(\frac{3y}{2\delta} - \frac{9y^2}{4\delta^2} + \frac{3y^4}{2\delta^4} - \frac{y^3}{2\delta^3} + \frac{3y^4}{4\delta^4} - \frac{y^6}{4\delta^6}\right) dy\right]$$

$$= \frac{\partial}{\partial x}\left[\frac{3y^2}{2\times 2\delta} - \frac{9y^3}{3\times 4\delta^2} + \frac{3y^5}{5\times 4\delta^4} - \frac{y^4}{4\times 2\delta^3} + \frac{3y^5}{5\times 4\delta^4} - \frac{y^2}{7\times 4\delta^6}\right]_0^\delta$$

$$= \frac{\partial}{\partial x}\left[\frac{3\delta^2}{4\delta} - \frac{3\delta^3}{4\delta^2} + \frac{3}{20}\frac{\delta^5}{\delta^4} - \frac{1}{8}\frac{\delta^4}{\delta^3} + \frac{3}{20}\frac{\delta^5}{\delta^4} - \frac{1}{28}\frac{\delta^7}{\delta^6}\right]$$

$$= \frac{\partial}{\partial x}\left[\frac{3}{4}\delta - \frac{3}{4}\delta + \frac{3}{20}\delta - \frac{1}{8}\delta + \frac{3}{20}\delta - \frac{1}{28}\delta\right]$$

$$= \frac{\partial}{\partial x}\left[\frac{6}{20}\delta - \frac{1}{8}\delta - \frac{1}{28}\delta\right] = \frac{\partial\delta}{\partial x}\left[\frac{84 - 35 - 10}{280}\right] = \frac{39}{280}\frac{\partial\delta}{\partial x}$$

$$\tau_0 = \rho U^2 \times \frac{39}{280}\frac{\partial x}{\partial x} = \frac{39}{280}\rho U^2 \frac{\partial\delta}{\partial x} \qquad \text{...(11.30)}$$

Also the shear stress τ_0 is given by $\tau_0 = \mu\left(\frac{du}{dy}\right)_{y=0}$, where $u = U\left[\frac{3}{2}\frac{y}{\delta} - \frac{y^3}{2\delta^3}\right]$

$$\therefore \qquad \frac{du}{dy} = U\left[\frac{3}{2\delta} - \frac{3y^2}{2\delta^3}\right]$$

Hence $$\left(\frac{du}{dy}\right)_{y=0} = U\left[\frac{3}{2\delta} - \frac{3}{2\delta^3}\times 0\right] = \frac{3U}{2\delta}$$

$$\therefore \qquad \tau_0 = \mu\left(\frac{du}{dy}\right)_{y=0} = \mu\,\frac{3U}{2\delta} = \frac{3}{2}\frac{\mu U}{\delta} \qquad \text{...(11.31)}$$

Equating the two values of τ_0 given by equations (11.30) and (11.31)

$$\frac{39}{280}\rho U^2 \frac{\partial\delta}{\partial x} = \frac{3}{2}\frac{\mu U}{\delta}$$

$$\therefore \qquad \delta\partial\delta = \frac{3}{2}\mu U \times \frac{280}{39} \times \frac{1}{\rho U^2}\partial x = \frac{420}{39}\frac{\mu}{\rho U}\partial x$$

Integrating, we get $$\frac{\delta^2}{2} = \frac{420}{39}\frac{\mu}{\rho U}x + C$$

where $x = 0$, $\delta = 0$, $\therefore\ C = 0$

$$\therefore \qquad \frac{\delta^2}{2} = \frac{420}{39}\cdot\frac{\mu}{\rho U}x$$

or $$\delta = \sqrt{\frac{420 \times 2}{39}\frac{\mu}{\rho U}x} = 4.64\sqrt{\frac{\mu x}{\rho U}} = 4.64\sqrt{\frac{\mu x \times x}{\rho U x}}$$

$$= 4.64\sqrt{\frac{\mu}{\rho U x}}\,x = \frac{4.64\,x}{\sqrt{R_{e_x}}} \qquad \text{...(11.32)}$$

(*i*) **Shear Stress τ_0.** Substituting the value of δ from equation (11.32) into equation (11.31), we get

$$\tau_0 = \frac{3}{2}\frac{\mu U}{\dfrac{4.64\,x}{\sqrt{R_{e_x}}}} = \frac{3}{9.28}\frac{\mu U\sqrt{R_{e_x}}}{x} = 0.323\,\frac{\mu U}{x}\sqrt{R_{e_x}}$$

(*ii*) **Drag force (F_D)**

Using equation (11.22), we get the drag force as

$$F_D = \int_0^L \tau_0 \times b \times dx = \int_0^L 0.323\,\frac{\mu U}{x}\sqrt{R_{e_x}} \times b \times dx$$

$$= 0.323\int_0^L \frac{\mu U}{x}\sqrt{\frac{\rho U x}{\mu}} \times b \times dx = 0.323\,\mu U\sqrt{\frac{\rho U}{\mu}} \times b\int_0^L \frac{1}{\sqrt{x}}dx$$

$$= 0.323\,\mu U\sqrt{\frac{\rho U}{\mu}} \times b\int_0^L x^{-1/2}\,dx$$

$$= 0.323\,\mu U\sqrt{\frac{\rho U}{\mu}} \times b\left[\frac{x^{1/2}}{\frac{1}{2}}\right]_0^L = 0.323 \times 2\,\mu U\sqrt{\frac{\rho U}{\mu}} \times b\,[\sqrt{L}]$$

$$= 0.646\,\mu U\sqrt{\frac{\rho U L}{\mu}} \times b \qquad \text{...(11.33)}$$

(*iii*) **Drag Co-efficient (C_D).** Using equation (11.24), we get the value of C_D as

$$C_D = \frac{F_D}{\frac{1}{2}\rho A U^2}, \text{ where } A = b \times L$$

$$= \frac{0.646\,\mu U\sqrt{\dfrac{\rho U L}{\mu}} \times b}{\frac{1}{2}\rho \times b \times L \times U^2} = 0.646 \times 2 \times \frac{\mu}{\rho U L} \times \sqrt{\frac{\rho U L}{\mu}} = \frac{1.292}{\sqrt{\dfrac{\rho U L}{\mu}}}$$

$$= \frac{1.292}{\sqrt{R_{e_L}}}. \qquad \left\{ \because \sqrt{\frac{\rho UL}{\mu}} = \sqrt{R_{e_L}} \right\} \qquad ...(11.34)$$

Problem 11.26 *For the velocity profile for laminar boundary layer*

$$\frac{u}{U} = 2\,(y/\delta) - 2(y/\delta)^3 + (y/\delta)^4$$

obtain an expression for boundary layer thickness, shear stress, drag force on one side of the plate and co-efficient of drag in term of Reynold number.

Solution. Given :

(*i*) **The velocity profile,** $\quad \frac{\mathbf{u}}{\mathbf{U}} = \frac{\mathbf{2y}}{\delta} - \frac{\mathbf{2y^3}}{\delta^3} + \frac{\mathbf{y^4}}{\delta^4}$

Using equation (11.20), we have

$$\frac{\tau_0}{\rho U^2} = \frac{\partial}{\partial x}\left[\int_0^{\delta} \frac{u}{U}\left(1 - \frac{u}{U}\right) dy\right]$$

Substituting the given velocity profile in the above equation

$$\frac{\tau_0}{\rho U^2} = \frac{\partial}{\partial x}\left[\int_0^{\delta}\left(\frac{2y}{\delta} - \frac{2y^3}{\delta^3} - \frac{y^4}{\delta^4}\right)\left(1 - \left\{\frac{2y}{\delta} - \frac{2y^3}{\delta^3} + \frac{y^4}{\delta^4}\right\}\right) dy\right]$$

$$= \frac{\partial}{\partial x}\left[\int_0^{\delta}\left(\frac{2y}{\delta} - \frac{2y^3}{\delta^3} + \frac{y^4}{\delta^4}\right)\left(1 - \frac{2y}{\delta} + \frac{2y^3}{\delta^3} - \frac{y^4}{\delta^4}\right) dy\right]$$

$$= \frac{\partial}{\partial x}\left[\int_0^{\delta}\left(\frac{2y}{\delta} - \frac{4\delta^4}{\delta^2} + \frac{4y^2}{\delta^4} - \frac{2y^5}{\delta^5} + \frac{2y^3}{\delta^3} + \frac{4y^4}{\delta^4} - \frac{4y^6}{\delta^6} + \frac{2y^7}{\delta^7} + \frac{y^4}{\delta^4} - \frac{2y^5}{\delta^5} + \frac{2y^7}{\delta^7} - \frac{y^8}{\delta^8}\right) dy\right]$$

$$= \frac{\partial}{\partial x}\left[\int_0^{\delta}\left(\frac{2y}{\delta} - \frac{4y^2}{\delta^2} - \frac{2y^3}{\delta^3} + \frac{9y^4}{\delta^4} - \frac{4y^5}{\delta^5} - \frac{4y^6}{\delta^6} + \frac{4y^7}{\delta^7} - \frac{y^8}{\delta^8}\right) dy\right]$$

$$= \frac{\partial}{\partial x}\left[\frac{2y^2}{2\delta} - \frac{4y^3}{3\delta^2} - \frac{2y^4}{4\delta^3} + \frac{9y^5}{5\delta^4} - \frac{9y^6}{6\delta^5} - \frac{4y^7}{7\delta^6} + \frac{4y^8}{8\delta^7} - \frac{y^9}{\delta^8}\right]_0^{\delta}$$

$$= \frac{\partial}{\partial x}\left[\delta - \frac{4}{3}\delta + \frac{9}{5}\delta - \frac{2}{3}\delta - \frac{4}{7}\delta - \frac{1}{9}\delta\right]$$

$$= \frac{\partial}{\partial x}\left[\frac{315 - 420 + 63 \times 9 - 210 - 45 \times 4 - 35}{315}\right]\delta$$

$$= \frac{\partial}{\partial x}\left[\frac{315 - 420 + 567 - 210 - 180 - 35}{315}\right]\delta$$

$$= \frac{\partial}{\partial x}\left[\frac{882 - 845}{815}\right]\delta = \frac{\partial}{\partial x}\left[\frac{37}{315}\right]\delta = \frac{37}{315}\frac{\partial \delta}{\partial x}$$

$$\therefore \quad \tau_0 = \frac{37}{315} \rho U^2 \frac{\partial \delta}{\partial x} \qquad \text{...(11.35)}$$

Also shear stress is given by Newton's law of viscosity as

$$\tau_0 = \mu \left(\frac{\partial u}{\partial y} \right)_{y=0}$$

where $u = U \left[\frac{2y}{\delta} - \frac{2y^2}{\delta^2} + \frac{y^4}{\delta^4} \right]$

$$\therefore \quad \left(\frac{du}{dy} \right) = U \left[\frac{2}{\delta} - \frac{4y}{\delta^2} - \frac{4y^3}{\delta^4} \right]$$

$$\therefore \quad \left(\frac{\partial u}{\partial y} \right)_{y=0} = U \left[\frac{2}{\delta} - \frac{4}{\delta^2}(0) - \frac{4}{\delta^4}(0) \right] = \frac{2U}{\delta}$$

$$\therefore \quad \tau_0 = \mu \left(\frac{\partial u}{\partial y} \right)_{y=0} = \mu \times \frac{2U}{\delta} = \frac{2U\mu}{\delta} \qquad \text{...(11.36)}$$

Equating the two values of τ_0 given by equations (11.35) and (11.36)

$$\frac{37}{315} \rho U^2 \frac{\partial \delta}{\partial x} = \frac{2U\mu}{\delta} \quad \text{or} \quad \delta \partial \delta = \frac{315}{37} \times \frac{2U\mu}{\rho U^2} \partial x = \frac{630}{37} \frac{\mu}{\rho U} \partial x$$

On integration, we get $\frac{\delta^2}{2} = \frac{630}{37} \frac{\mu}{\rho U} x + C$, where C = Constant of integration

At $x = 0$, $\delta = 0$ and hence $C = 0$

$$\therefore \quad \frac{\delta^2}{2} = \frac{630}{37} \frac{\mu}{\rho U} x$$

$$\therefore \quad \delta = \sqrt{\frac{630 \times 2}{37} \frac{\mu}{\rho U}} \, x = 5.84 \sqrt{\frac{\mu x}{\rho U}}$$

$$= 5.84 \sqrt{\frac{\mu x \times x}{\rho U x}} = 5.84 \sqrt{\frac{\mu}{\rho U x}} \times x = \frac{5.84 x}{\sqrt{R_{e_x}}} \qquad \text{...(11.37)}$$

(*ii*) **Shear Stress (τ_0).** Substituting the value of δ from (11.37) into (11.36)

$$\tau_0 = \frac{2U\mu}{\delta} = \frac{2U\mu}{\frac{5.84x}{\sqrt{R_{e_x}}}} = \frac{2U\mu}{5.84x} \sqrt{R_{e_x}} = 0.34 \frac{U\mu}{x} \sqrt{R_{e_x}} \, .$$

(*iii*) **Drag Force (F_D)** on one side of the plate :
Using equation (11.22), we get

$$F_D = \int_0^L \tau_0 \times b \times dx = \int_0^L 0.34 \frac{U\mu}{x} \sqrt{R_{e_x}} \times b \times dx = \int_0^L 0.34 \frac{U\mu}{\mu} \frac{\sqrt{\rho U \mu}}{\mu} bdx$$

$$= 0.34 \, U\mu \sqrt{\frac{\rho U}{\mu}} \times b \int_0^L x^{-1/2} \, dx = 0.34 \, U\mu \sqrt{\frac{\rho U}{\mu}} \times b \times \left[\frac{x^{1/2}}{1/2} \right]_0^L$$

$$= 0.34 \times 2U\mu \sqrt{\frac{\rho U}{\mu}}\, b\sqrt{L} = 0.68\, b\mu U \sqrt{\frac{\rho UL}{\mu}} \quad ...(11.38)$$

(*iv*) **Drag Co-efficient (C_D)**

Using equation (11.24), $C_D = \dfrac{F_D}{\frac{1}{2}\rho A U^2}$, where $A = b \times L$

$$= \frac{0.68 \times b \times \mu U \times \sqrt{\frac{\rho UL}{\mu}}}{\frac{1}{2}\rho \times b \times L \times U^2} = 0.68 \times 2\, \frac{\mu}{\rho UL} \times \sqrt{\frac{\rho UL}{\mu}} = 1.36 \times \frac{1}{\sqrt{\frac{\rho UL}{\mu}}}$$

$$= 1.36 \times \frac{1}{\sqrt{R_{e_L}}}. \quad ...(11.39)$$

Problem 11.27 *For the velocity profile for laminar boundary flow* $\dfrac{u}{U} = \sin\left(\dfrac{\pi}{2}\dfrac{y}{\delta}\right)$. *Obtain an expression for boundary layer thickness, shear stress, drag force on one side of the plate and co-efficient of drag in terms of Reynold number.*

Solution. (*i*) **The velocity profile is** $\dfrac{\mathbf{u}}{\mathbf{U}} = \sin\left(\dfrac{\pi}{2}\dfrac{\mathbf{y}}{\delta}\right)$.

Substituting this value in equation (11.20), we have

$$\frac{\tau_0}{\rho U^2} = \frac{\partial}{\partial x}\left[\int_0^\delta \frac{u}{U}\left(1 - \frac{u}{U}\right) dy\right] = \frac{\partial}{\partial x}\left[\int_0^\delta \sin\left(\frac{\pi}{2}\frac{y}{\delta}\right)\left[1 - \sin\left(\frac{\pi}{2}\cdot\frac{y}{\delta}\right)\right] dy\right]$$

$$= \frac{\partial}{\partial x}\left[\int_0^\delta \left[\sin\left(\frac{\pi}{2}\frac{y}{\delta}\right) - \sin^2\left(\frac{\pi}{2}\frac{y}{\delta}\right)\right] dy\right]$$

$$= \frac{\partial}{\partial x}\left[\left[\frac{-\cos\frac{\pi y}{2\delta}}{\frac{\pi}{2\delta}}\right] - \left[\frac{\frac{\pi y}{2\delta} \times \frac{1}{2}}{\frac{\pi}{2\delta}} - \frac{\sin 2\left(\frac{\pi}{2}\frac{y}{\delta}\right)}{4 \times \frac{\pi}{2\delta}}\right]\right]_0^\delta$$

$$\left\{\because \int \sin^2\left(\frac{\pi}{2}\frac{y}{\delta}\right) dy = \frac{\frac{\pi y}{2\delta} \times \frac{1}{2}}{\frac{\pi}{2\delta}} - \frac{\sin 2\left(\frac{\pi}{2}\frac{y}{\delta}\right)}{4 \times \frac{\pi}{2\delta}}\right\}$$

$$\therefore \quad \frac{\tau_0}{\rho U^2} = \frac{\partial}{\partial x}\left[\left(\frac{-\cos\frac{\pi}{2}\frac{\delta}{\delta}}{\frac{\pi}{2\delta}} + \frac{\cos\frac{\pi}{2} \times \frac{0}{\delta}}{\frac{\pi}{2\delta}}\right) - \left[\frac{\frac{\pi}{2}\frac{\delta}{\delta} \times \frac{1}{2}}{\frac{\pi}{2\delta}} - 0\right]\right]$$

$$= \frac{\partial}{\partial x}\left[\left(0 + \frac{1}{\frac{\pi}{2\delta}}\right) - \frac{\left(\frac{\pi}{4}\right)}{\frac{\pi}{2\delta}}\right] = \frac{\partial}{\partial x}\left[\frac{2\delta}{\pi} - \frac{\pi}{4} \times \frac{2\delta}{\pi}\right]$$

$$= \frac{\partial}{\partial x}\left[\frac{2\delta}{\pi} - \frac{\delta}{2}\right] = \frac{\partial}{\partial x}\left[\frac{4-\pi}{2\pi}\right]\delta = \left(\frac{4-\pi}{2\pi}\right)\frac{\partial\delta}{\partial x}$$

$\therefore$
$$\tau_0 = \left(\frac{4-\pi}{2\pi}\right)\rho U^2 \frac{\partial\delta}{\partial x} \qquad \text{...(11.40)}$$

τ_0 is also equal $= \mu \left(\frac{d\mu}{dy}\right)_{\text{at } y=0}$

But
$$u = U \sin\left(\frac{\pi}{2}\frac{y}{\delta}\right)$$

$\therefore$
$$\left(\frac{du}{dy}\right) = U \cos\left(\frac{\pi}{2}\frac{y}{\delta}\right) \times \frac{\pi}{2\delta}$$

$$\left(\frac{du}{dy}\right)_{y=0} = U \times \frac{\pi}{2\delta} \cos\left(\frac{\pi}{2} \times \frac{0}{\delta}\right) = \frac{U\pi}{2\delta}$$

$\therefore$
$$\tau_0 = \mu \left(\frac{\partial\mu}{\partial y}\right)_{y=0} = \frac{\mu U\pi}{2\delta} \qquad \text{...(11.41)}$$

Equating the two values τ_0 given by equations (11.40) and (11.41)

$$\left(\frac{4-\pi}{2\pi}\right)\rho U^2 \frac{\partial\delta}{\partial x} = \frac{\mu U\pi}{2\delta} \quad \text{or} \quad \delta\partial\delta = \frac{\mu U\pi}{3} \times \frac{2\pi}{4-\pi} \times \frac{1}{\rho U^2}\partial x$$

$\therefore$
$$\delta\partial\delta = \frac{\pi^2}{(4-\pi)}\frac{\mu U}{\rho U^2} \,.\, \partial x = 11.4975 \frac{\mu}{\rho U}\partial x$$

Integrating, we get
$$\frac{\delta^2}{2} = 11.4975 \frac{\mu}{\rho U} x + C$$

At $x = 0$, $\delta = 0$ and hence $C = 0$

$\therefore$
$$\frac{\delta^2}{2} = 11.4975 \frac{\mu}{\rho U} x$$

$\therefore$
$$\delta = \sqrt{2 \times 11.4975 \frac{\mu}{\rho U} x} = 4.795 \sqrt{\frac{\mu}{\rho U}}\, x$$

$$= 4.795 \sqrt{\frac{\mu}{\rho U x}} = 4.795 \sqrt{\frac{\mu}{\rho U x}} \times x$$

$$= \frac{4.795\, x}{\sqrt{R_{e_x}}} \qquad \text{...(11.42)}$$

(*ii*) **Shear Stress (τ_0)**

From equation (11.41),
$$\tau_0 = \frac{\mu U \pi}{2\delta} = \frac{\mu U \pi}{\dfrac{2 \times 4.795\, x}{\sqrt{R_{e_x}}}} = \frac{\mu U \pi \sqrt{R_{e_x}}}{2 \times 4.795\, x}$$

$$= \frac{\pi}{2 \times 4.795} \frac{\mu U}{x} \sqrt{R_{e_x}} = 0.327 \frac{\mu U}{x} \sqrt{R_{e_x}} .$$

(*iii*) **Drag force (F_D)** on one side of the plate is given by equation (11.22)

$$F_D = \int_0^L \tau_0 \times b \times dx = \int_0^L 0.327 \frac{\mu U}{x} \sqrt{R_{e_x}} \times b \times dx = 0.327\, \mu U \times b \int_0^L \frac{1}{x} \sqrt{\frac{\rho U x}{\mu}}\, dx$$

$$= 0.327\, \mu U \times b \times \sqrt{\frac{\rho U}{\mu}} \int_0^L x^{-1/2}\, dx = 0.327\, \mu U \times b \times \sqrt{\frac{\rho U}{\mu}} \left[\frac{x^{1/2}}{\frac{1}{2}} \right]_0^L$$

$$= 0.327 \times 2 \times \mu U \times b \sqrt{\frac{\rho U}{\mu}} \times \sqrt{L}$$

$$= 0.655 \times \mu U \times b \times \sqrt{\frac{\rho U L}{\mu}} \qquad \text{...(11.43)}$$

(*iv*) **Co-efficient of drag, C_D** is given by equation (11.24),

$$C_D = \frac{F_D}{\frac{1}{2} \rho A U^2}, \text{ where } A = b \times L$$

$\therefore$
$$C_D = \frac{0.655 \times \mu U \times b \times \sqrt{\dfrac{\rho U L}{\mu}}}{\frac{1}{2} \rho U^2 \times b \times L} = 0.655 \times 2 \times \frac{\mu}{\rho U L} \times \sqrt{\frac{\rho U L}{\mu}}$$

$$= 1.31 \times \frac{1}{\sqrt{\dfrac{\rho U L}{\mu}}} = \frac{1.31}{\sqrt{R_{e_L}}} \qquad \text{...(11.44)}$$

Note. $\int \sin^2 x ds = \left(\frac{x}{2} - \frac{\sin 2x}{4} \right)$ is used.

Table 11.1 shows the values of boundary layer thickness and co-efficients of drag in terms of Reynold number for various velocity distributions.

Table 11.1

	Velocity Distribution	δ	C_D
1.	$\frac{u}{U} = 2\left(\frac{y}{\delta}\right) - \left(\frac{y}{\delta}\right)^2$	$5.48\, x/\sqrt{R_{e_x}}$	$1.46/\sqrt{R_{e_L}}$
2.	$\frac{u}{U} = \frac{3}{2}\left(\frac{y}{\delta}\right) - \frac{1}{2}\left(\frac{y}{\delta}\right)^3$	$4.64\, x/\sqrt{R_{e_x}}$	$1.292/\sqrt{R_{e_L}}$
3.	$\frac{u}{U} = 2\left(\frac{y}{\delta}\right) - 2\left(\frac{y}{\delta}\right)^3 + \left(\frac{y}{\delta}\right)^4$	$5.84\, x/\sqrt{R_{e_x}}$	$1.37/\sqrt{R_{e_L}}$
4.	$\frac{u}{U} = \sin\left(\frac{\pi}{2}\frac{y}{\delta}\right)$	$4.79\, x/\sqrt{R_{e_x}}$	$1.31/\sqrt{R_{e_L}}$
5.	Blasius's Solution	$4.91\, x/\sqrt{R_{e_x}}$	$1.328/\sqrt{R_{e_L}}$

Problem 11.28 *For the velocity profile in laminar boundary layer as,*

$$\frac{u}{U} = \frac{3}{2}\left(\frac{y}{\delta}\right) - \frac{1}{2}\left(\frac{y}{\delta}\right)^3$$

find the thickness of the boundary layer and the shear stress 1.5 m from the leading edge of a plate. The plate is 2 m long and 1.4 m wide and is placed in water which is moving with a velocity of 200 mm per second. Find the total drag force on the plate if μ for water = .01 poise.

(Delhi University, 1986)

Solution. Given :

Velocity profile is $\frac{u}{U} = \frac{3}{2}\left(\frac{y}{\delta}\right) - \frac{1}{2}\left(\frac{y}{\delta}\right)^3$

Distance of x from leading edge, $x = 1.5$ m

Length of plate, $L = 2$ m

Width of plate, $b = 1.4$ m

Velocity of plate, $U = 200$ mm/s $= 0.2$ m/s

Viscosity of water, $\mu = 0.01 \text{ poise} = \frac{0.01}{10} = 0.001 \text{ Ns/m}^2$

For the given velocity profile, thickness of boundary layer is given by equation (11.32) as

$$\delta = \frac{4.64\, x}{\sqrt{R_{e_x}}} \qquad \left[\text{Here } R_{e_x} = \frac{\rho U x}{\mu} = 1000 \times \frac{0.2 \times 1.5}{0.001} = 300000\right]$$

$$\delta = \frac{4.64 \times 1.5}{\sqrt{300000}} = 0.0127 \text{ m} = \textbf{12.7 mm. Ans.}$$

Shear stress (τ_0) is given by $\tau_0 = 0.323\, \frac{\mu U}{x} \sqrt{R_{e_x}}$

$$= 0.323 \times 0.001 \times \frac{0.2}{1.5} \times \sqrt{300000} = \textbf{0.0235 N/m}^2\textbf{. Ans.}$$

Drag Force (F_D) on one side of the plate is given by (11.33) as

$$F_D = 0.646\,\mu U \sqrt{\frac{\rho UL}{\mu}} \times b$$

$$= 0.646 \times 0.001 \times 0.2 \times \sqrt{1000 \times \frac{0.2 \times 2.0}{0.001}} \times 1.4$$

$$= .646 \times 0.001 \times 0.2 \times \sqrt{400000} \times 1.4 = 0.1138 \text{ N}$$

∴ Total drag force = Drag force on both sides of the plate

= 2 × 0.1138 = **0.2276 N. Ans.**

Problem 11.29 *Air is flowing over a smooth plate with a velocity of 10 m/s. The length of the plate is 1.2 m and width 0.8 m. If laminar boundary layer exists up to a value of $R_e = 2 \times 10^5$, find the maximum distance from the leading edge upto which laminar boundary layer exists. Find the maximum thickness of laminar boundary layer if the velocity profile is given by*

$$\frac{u}{U} = 2\left(\frac{y}{\delta}\right) - \left(\frac{y}{\delta}\right)^2$$

Take kinematic viscosity for air = 0.15 stokes.

Solution. Given :

Velocity of air, $U = 10$ m/s

Length of plate, $L = 1.2$ m

Width of plate, $b = 0.8$ m

Reynold number upto which laminar boundary exists = 2×10^5

ν for air = 0.15 stokes = 0.15×10^{-4} m²/s

Reynold number $R_{e_x} = \dfrac{\rho Ux}{\mu} = \dfrac{Ux}{\nu}$

If $R_{e_x} = 2 \times 10^5$, then x denotes the distance from leading edge upto which laminar boundary layer exists

$$\therefore \quad 2 \times 10^5 = \frac{10 \times x}{0.15 \times 10^{-4}}$$

$$\therefore \quad x = \frac{2 \times 10^5 \times 0.15 \times 10^{-4}}{10} = 0.30 \text{ m} = \textbf{300 mm. Ans.}$$

Maximum thickness of the laminar boundary for the velocity profile, $\dfrac{u}{U} = 2\left(\dfrac{y}{\delta}\right) - \left(\dfrac{y}{\delta}\right)^2$ is given by equation (11.27) as

$$\delta = \frac{5.48 \times x}{\sqrt{R_{e_x}}} = \frac{5.48 \times 0.30}{\sqrt{2 \times 10^5}} = 0.00367 \text{ m} = \textbf{3.67 mm. Ans.}$$

Problem 11.30 *Air is flowing over a flat plate 500 mm long and 600 mm wide with a velocity of 4 m/s. The kinematic viscosity of air is given as 0.15×10^{-4} m²/s. Find (i) the boundary layer thickness at the end of the plate, (ii) Shear stress at 200 mm from the leading edge and (iii) drag force* *on one side of the plate. Take the velocity profile over the plate as* $\frac{u}{U} = \sin\left(\frac{\pi}{2} \cdot \frac{y}{\delta}\right)$ *and density of air 1.24 kg/m³.*

Solution. Given :

Length of plate, $L = 500 \text{ mm} = 0.5 \text{ m}$

Width of plate, $b = 600 \text{ mm} = 0.6 \text{ m}$

Velocity of air, $U = 4 \text{ m/s}$

Kinematic viscosity, $\nu = 0.15 \times 10^{-4} \text{ m}^2\text{/s}$

∴ Mass density, $\rho = 1.24 \text{ kg/m}^3$

For the velocity profile $\frac{u}{U} = \sin\left(\frac{\pi}{2}\frac{y}{\delta}\right)$, we have

(*i*) Boundary layer thickness at the end of the plate means value of δ at $x = 0.5$ m. First find Reynold number.

$$R_{e_x} = \frac{\rho U x}{\mu} = \frac{Ux}{\nu} = \frac{4 \times 0.5}{0.15 \times 10^{-4}} = 1.33 \times 10^5.$$

Hence boundary layer is laminar over the entire length of the plate as Reynold number at the end of the plate is 1.33×10^5.

∴ δ at $x = 0.5$ m for the given velocity profile is given by equation (11.42) as

$$\delta = \frac{4.795x}{\sqrt{R_{e_x}}} = \frac{4.795 \times 0.5}{\sqrt{1.33 \times 10^5}} = 0.00656 \text{ m} = \textbf{6.56 mm.} \quad \textbf{Ans.}$$

(*ii*) Shear stress at any distance from leading edge is given by $\tau_0 = 0.327 \frac{\mu U}{x}\sqrt{R_{e_x}}$

At $x = 200 \text{ mm} = 0.2$ m, $R_{e_x} = \frac{U \times x}{\nu} = \frac{4 \times 0.2}{0.15 \times 10^{-4}} = 53333$

$$\therefore \quad \tau_0 = \frac{0.327 \times \mu \times 4 \times \sqrt{53333}}{0.2}$$

But $\mu = \nu \times \rho \qquad \left\{\because \nu = \frac{\mu}{\rho}, \ \therefore \mu = \nu \times \rho\right\}$

$$= 0.15 \times 10^{-4} \times 1.24 = 0.186 \times 10^{-4}$$

$$\therefore \quad \tau_0 = \frac{0.327 \times 0.186 \times 10^{-4} \times 4 \times \sqrt{53333}}{0.2} = \textbf{0.02805 N/m}^2\textbf{. Ans.}$$

(*iii*) Drag force on one side of the plate is given by equation (11.43)

$$F_D = 0.655 \times \mu U \times b \times \sqrt{\frac{\rho U L}{\mu}}$$

$$= 0.655 \times 0.186 \times 10^{-4} \times 4.0 \times 0.6 \times \sqrt{\frac{UL}{\nu}} \qquad \left\{\because \nu = \frac{\mu}{\rho}\right\}$$

$$= 0.29234 \times 10^{-4} \times \sqrt{\frac{4 \times 0.5}{.15 \times 10^{-4}}} = \mathbf{0.01086\ N.\ Ans.}$$

Problem 11.31 *A thin plate is moving in still atmospheric air at a velocity of 5 m/s. The length of the plate is 0.6 m and width 0.5 m. Calculate (i) the thickness of the boundary layer at the end of the plate, and (ii) drag force on one side of the plate. Take density of air as 1.24 kg/m³ and kinematic viscosity 0.15 stokes.*

Solution. Given :

Velocity of plate, $U = 5$ m/s

Length of plate, $L = 0.6$ m

Width of plate, $b = 0.5$ m

Density of air, $\rho = 1.24$ kg/m³

Kinematic viscosity, $\nu = 0.15$ stokes $= 0.15 \times 10^{-4}$ m²/s

Reynold number, $R_e = \frac{UL}{\nu} = \frac{5 \times 0.6}{0.15 \times 10^{-4}} = 200000.$

As R_e is less than 5×10^5, hence boundary layer is laminar over the entire length of the plate.

(*i*) Thickness of boundary layer at the end of the plate by Blasius's solution is

$$\delta = \frac{4.91x}{\sqrt{R_{e_x}}} = \frac{4.91\,L}{\sqrt{R_{e_1}}} = \frac{4.91 \times 0.6}{\sqrt{200000}} = .00658 \text{ m} = \mathbf{6.58\ mm.\ Ans.}$$

(*ii*) Drag force on one side of the plate is given by equation (11.24) as

$$C_D = \frac{F_D}{\frac{1}{2}\rho A U^2}$$

$$\therefore \quad F_D = \frac{1}{2}\rho\, A U^2 \times C_D$$

where C_D from Blasius's solution, $C_D = \frac{1.328}{\sqrt{R_{e_L}}} = \frac{1.328}{\sqrt{200000}} = 0.002969 \simeq .00297$

$$\therefore \quad F_D = \frac{1}{2} \times 1.24 \times 0.6 \times 0.5 \times 5^2 \times .002970 \quad \{\because A = L \times b = 0.6 \times 0.5\}$$

$$= \mathbf{0.01373\ N.\ Ans.}$$

Note. If no velocity profile is given in the numerical problem but boundary layer is laminar, then Blasius's solution is used.

Problem 11.32 *A plate of 600 mm length and 400 mm wide is immersed in a fluid of sp. gr. 0.9 and kinematic viscosity (ν =) 10⁻⁴ m²/s. The fluid is moving with a velocity of 6 m/s. Determine (i) boundary layer thickness, (ii) shear stress at the end of the plate, and (iii) drag force on one side of the plate.*

Solution. As no velocity profile is given in the above problem, hence Blasius's solution will be used.

Given : length of plate, $L = 600$ mm $= 0.60$ m

Width of plate, $b = 400$ mm $= 0.40$ m

Sp. gr. of fluid, $S = 0.9$

∴ Density, $\rho = 0.9 \times 1000 = 900$ kg/m^3

Velocity of fluid, $U = 6$ m/s

Kinematic viscosity, $\nu = 10^{-4}$ m^2/s

Reynold number, $R_{e_L} = \dfrac{U \times L}{\nu} = \dfrac{6 \times 0.6}{10^{-4}} = 3.6 \times 10^4.$

As R_{e_L} is less than 5×10^5, hence boundary layer is laminar over the entire length of the plate.

(*i*) Thickness of boundary layer at the end of the plate from Blasius's solution is

$$\delta = \frac{4.91\, x}{\sqrt{R_{e_x}}}, \text{ where } x = 0.6 \text{ m and } R_{e_x} = 3.6 \times 10^4$$

$$= \frac{4.91 \times 0.6}{\sqrt{3.6 \times 10^4}} = 0.0155 \text{ m} = \mathbf{15.5\ mm.\ Ans.}$$

(*ii*) Shear stress at the end of the plate is

$$\tau_0 = 0.332\, \frac{\rho U^2}{\sqrt{R_{e_L}}} = \frac{0.332 \times 900 \times 6^2}{\sqrt{3.6 \times 10^4}} = \mathbf{56.6\ N/m^2.\ Ans.}$$

(*iii*) Drag force (F_D) on one side of the plate is given by

$$F_D = \frac{1}{2} \rho\, A U^2 \times C_D$$

where C_D from Blasius's solution is $C_D = \dfrac{1.328}{\sqrt{R_{e_L}}} = \dfrac{1.328}{\sqrt{3.6 \times 10^4}} = 0.00699$

∴
$$F_D = \frac{1}{2} \rho\, A U^2 \times C_D$$

$$= \frac{1}{2} \times 900 \times 0.6 \times 0.4 \times 6^2 \times .00699 \quad \{\because A = L \times b = 0.6 \times .4\}$$

$$= \mathbf{26.78\ N.\ Ans.}$$

▶ 11.11 TURBULENT BOUNDARY LAYER ON A FLAT PLATE

The thickness of the boundary layer, drag force on one side of the plate and co-efficient of drag due to turbulent boundary layer on a smooth plate at zero pressure gradient are determined as in case of laminar boundary layer provided the velocity profile is known. Blasius on the basis of experiments give the following velocity profile for turbulent boundary layer

$$\frac{u}{U} = \left(\frac{y}{\delta}\right)^n \qquad \text{...(11.45)}$$

where $n = \dfrac{1}{7}$ for $R_e < 10^7$ but more than 5×10^5

∴
$$\frac{u}{U} = \left(\frac{y}{\delta}\right)^{1/7} \qquad \text{...(11.46)}$$

Equation (11.46) is not applicable very near the boundary, where the thin laminar sub-layer of thickness δ' exists. Here velocity distribution is influenced only by viscous effects.

The value of τ_0 for flat plate is taken as $\tau_0 = 0.0225\ \rho\ U^2 \left(\frac{\mu}{\rho\delta U}\right)^{1/4}$ ...(11.47)

Problem 11.33 *For the velocity profile for turbulent boundary layer $\frac{u}{U} = \left(\frac{y}{\delta}\right)^{1/7}$, obtain an expression for boundary layer thickness, shear stress, drag force on one side of the plate and co-efficient of drag in terms of Reynold number. Given the shear stress (τ_0) for turbulent boundary layer as*

$$\tau_0 = 0.0225\ \rho\ U^2 \left(\frac{\mu}{\rho U\delta}\right)^{1/4}.$$

Solution. Given : $\frac{u}{U} = \left(\frac{y}{\delta}\right)^{1/7}$

(*i*) **Substituting this value in Von Karman momentum integral equation (11.20),**

$$\frac{\tau_0}{\rho U^2} = \frac{\partial}{\partial x}\left[\int_0^\delta \frac{u}{U}\left(1-\frac{u}{U}\right)dy\right]$$

$$= \frac{\partial}{\partial x}\left[\int_0^\delta \left(\frac{y}{\delta}\right)^{1/7}\left[1-\left(\frac{y}{\delta}\right)^{1/7}\right]dy\right] = \frac{\partial}{\partial x}\left[\int_0^\delta \left(\frac{y^{1/7}}{\delta^{1/7}} - \frac{y^{2/7}}{\delta^{2/7}}\right)dy\right]$$

$$= \frac{\partial}{\partial x}\left[\frac{y^{1.7+1}}{\left(\frac{1}{7}+1\right)\delta^{1/2}} - \frac{y^{2.7+1}}{\left(\frac{2}{7}+1\right)\delta^{2/7}}\right]_0^2$$

$$= \frac{\partial}{\partial x}\left[\frac{7}{8}\frac{y^{8/7}}{\delta^{1/2}} - \frac{7}{9}\frac{y^{9/7}}{\delta^{2/7}}\right]_0^\delta = \frac{\partial}{\partial x}\left[\frac{7}{8}\frac{\delta^{8/7}}{\delta^{1/7}} - \frac{7}{9}\frac{\delta^{9/7}}{\delta^{2/7}}\right]$$

$$= \frac{\partial}{\partial x}\left[\frac{7}{8}\delta - \frac{7}{9}\delta\right] = \frac{\partial}{\partial x}\left[\frac{63-56}{72}\right]\delta = \frac{\partial}{\partial x}\left[\frac{7}{72}\right]\delta = \frac{7}{72}\frac{\partial\delta}{\partial x}$$

In the above expression, the integration limits should be from δ' to δ. But as the laminar sub-layer is very thin that is δ' is very small. Hence the limits of integration are taken from 0 to δ.

Now $$\tau_0 = \frac{7}{72}\rho U^2 \frac{\partial\delta}{\partial x} \qquad ...(11.48)$$

But the value of τ_0 for turbulent boundary layer is given,

$$\tau_0 = 0.0225\ \rho U^2 \left(\frac{\mu}{\rho U\delta}\right)^{1/4} \qquad ...(11.49)$$

Equating the two values of τ_0 given by equations (11.48) and (11.49), we have

$$\frac{7}{72}\rho U^2 \frac{\partial\delta}{\partial x} = .0225\ \rho U^2 \left(\frac{\mu}{\rho U\delta}\right)^{1/4}$$

or $$\frac{7}{72}\frac{\partial\delta}{\partial x} = .0225\left(\frac{\mu}{\rho U}\right)^{1/4} \times \frac{1}{\delta^{1/4}} \qquad \{\text{cancelling } \rho U^2\}$$

or $$\delta^{1/4}\partial\delta = .0225 \times \frac{72}{7} \times \left(\frac{\mu}{\rho U}\right)^{1/4} \partial x = 0.2314\left(\frac{\mu}{\rho U}\right)^{1/4} \partial x.$$

Integrating, we get $$\frac{\delta^{1/4+1}}{\left(\frac{1}{4}+1\right)} = 0.2314\left(\frac{\mu}{\rho U}\right)^{1/4} x + C$$

or $$\frac{4}{5} \times \delta^{5/4} = 0.2314\left(\frac{\mu}{\rho U}\right)^{1/4} x + C$$

where C is constant of integration.

To determine the value of C, assume turbulent boundary layer starts from the leading edge, though in actual practice the turbulent boundary layer starts after the transition from laminar boundary layer. The laminar layer exists for a very short distance and hence this assumption will not affect the subsequent analysis.

Hence at $x = 0$, $\delta = 0$ and so $C = 0$

$$\therefore \quad \frac{4}{5}\delta^{5/4} = 0.2314\left(\frac{\mu}{\rho U}\right)^{1/4} x \text{ or } \delta^{5/4} = \frac{0.2314 \times 5}{4}\left(\frac{\mu}{\rho U}\right)^{1/4} x$$

or $$\delta = \left[\frac{0.2314 \times 5}{4}\left(\frac{\mu}{\rho U}\right)^{1/4} x\right]^{4/5} = \left(\frac{0.2314 \times 5}{4}\right)^{4/5}\left(\frac{\mu}{\rho U}\right)^{1/5} x^{4/5}$$

$$= 0.37\left(\frac{\mu}{\rho U}\right)^{1/5} x^{4/5} \quad \text{...(11.50)}$$

$$= 0.37\left(\frac{\mu}{\rho U x}\right)^{1/5} x^{1/5} \times x^{4/5} = 0.37\left(\frac{1}{R_{e_x}}\right)^{1/5} \times x = \frac{0.37x}{(R_{e_x})^{1/5}} \quad \text{...(11.51)}$$

From equation (11.50), it is clear that δ varies as $x^{4/5}$ in turbulent boundary layer while in case of laminar boundary layer δ varies as $\sqrt{x}$.

(*ii*) **Shear Stress (τ_0)** at any point from leading edge is given by equation (11.50) as

$$\tau_0 = 0.225\,\rho U^2\left(\frac{\mu}{\rho U \delta}\right)^{1/4}$$

Substituting the value of δ from equation (11.50), we have

$$\tau_0 = 0.0225\,\rho U^2\left(\frac{\mu}{\rho U \times 0.37 \times \left(\frac{\mu}{\rho U}\right)^{1/5} \times x^{4/5}}\right)^{1/4}$$

$$= \frac{.0225 \times 2}{2}\rho U^2\left(\frac{\mu^{4/5}}{0.37 \times (\rho U)^{4/5} \times x^{4/5}}\right)^{1/4}$$

$$= .0225 \times 2 \times \frac{\rho U^2}{2} \times \frac{1}{(0.37)^{1/4}} \left(\frac{\mu}{\rho U x}\right)^{1/5}$$

$$= 0.0577 \times \frac{\rho U^2}{2} \left(\frac{\mu}{\rho U x}\right)^{1/5} \quad \text{...(11.52)}$$

(*iii*) **Drag force (F_D)** on one side of the plate is

$$F_D = \int_0^L \tau_0 \times b \times dx = \int_0^L 0.0577 \times \frac{\rho U^2}{2} \times \left(\frac{\mu}{\rho U}\right)^{1/5} \frac{1}{x^{1/5}} \times b \times dx$$

$$= 0.0577 \times \frac{\rho U^2}{2} \times \left(\frac{\mu}{\rho U}\right)^{1/5} \times b \int_0^L x^{-1/5}\, dx$$

$$= .0577 \times \frac{\rho U^2}{2} \times \left(\frac{\mu}{\rho U}\right)^{1/5} \times b \times \left[\frac{x^{4/5}}{4/5}\right]_0^L$$

$$= .0577 \times \frac{5}{4} \times \frac{\rho U^2}{2} \left(\frac{\mu}{\rho U}\right)^{1/5} \times b \times L^{4/5}$$

$$= 0.072 \times \frac{\rho U^2}{2} \times \left(\frac{\mu}{\rho U}\right)^{1/5} \times b \times L^{4/5}$$

(*iv*) **Drag co-efficient, C_D** is given by

$$C_D = \frac{F_D}{\frac{1}{2}\rho A U^2}, \text{ where } A = L \times b$$

$$= \frac{.072 \times \frac{\rho U^2}{2} \times \left(\frac{\mu}{\rho U}\right)^{1/5} \times b \times L^{4/5}}{\frac{\rho U^2}{2} \times b \times L}$$

$$= 0.072 \times \left(\frac{\mu}{\rho U}\right)^{1/5} \cdot \frac{1}{L^{1/5}} = 0.072 \left(\frac{\mu}{\rho U L}\right)^{1/5}$$

$$= \frac{.072}{R_{e_L}^{1/5}} \quad \text{...(11.53)} \quad \left\{\because\ R_{e_L} = \frac{\rho U L}{\mu}\right\}$$

This is valid for $R_{e_L} > 5 \times 10^5$ but less than 10^7.

▶ 11.12 SEPARATION OF BOUNDARY LAYER

When a solid body is immersed in a flowing fluid, a thin layer of fluid called the boundary layer is formed adjacent to the solid body. In this thin layer of fluid, the velocity varies from zero to free stream velocity in the direction normal to the solid body. Along the length of the solid body, the thickness of the boundary layer increases. The fluid layer adjacent to the solid surface has to do work against surface friction at the expense of its kinetic energy. This loss of the kinetic energy is recovered from the immediate fluid layer in contact with the layer adjacent to solid surface through momentum exchange process. Thus the velocity of the layer goes on decreasing. Along the length of the solid body, at a certain point a stage may come when the boundary layer may not be able to keep sticking to the solid body if it cannot provide kinetic energy to overcome the resistance offered by the solid body. In other words, the boundary layer will be separated from the surface. This phenomenon is called the boundary layer separation. The point on the body at which the boundary layer is on the verge of separation from the surface is called point of separation.

11.12.1 Effect of Pressure Gradient on Boundary Layer Separation. The effect of pressure gradient $\left(\frac{dp}{dx}\right)$ on boundary layer separation can be explained by considering the flow over a curved surface *ABCSD* as shown in Fig. 11.13. In the region *ABC* of the curved surface, the area of flow decreases and hence velocity increases. This means that flow gets accelerated in this region. Due to the increase of the velocity, the pressure decreases in the direction of the flow and hence pressure gradient $\frac{dp}{dx}$ is negative in this region. As long as $\frac{dp}{dx} < 0$, the entire boundary layer moves forward as shown in Fig. 11.13.

Region *CSD* of the curved surface. The pressure is minimum at the point *C*. Along the region *CSD* of the curved surface, the area of flow increases and hence velocity of flow along the direction of fluid decreases. Due to decrease of velocity, the pressure increases in the direction of flow and hence pressure gradient $\frac{dp}{dx}$ is positive or $\frac{dp}{dx} < 0$. Thus in the region *CSD*, the pressure gradient is positive and velocity of fluid layers along the direction of flow decreases. As explained in the Art. 11.13, the velocity of the layer adjacent to the solid surface along the length of the solid surface goes on decreasing as the kinetic energy of the layer is used to overcome the frictional resistance of the surface. Thus the combined effect of positive pressure gradient and surface resistance reduce the momentum of the fluid is unable to the surface. A stage comes, when the momentum of the fluid is unable to overcome the surface resistance and the boundary layer starts separating from the surface at the point *S*. Downstream the point *S*, the flow is taking place in reverse direction and the velocity gradient becomes negative.

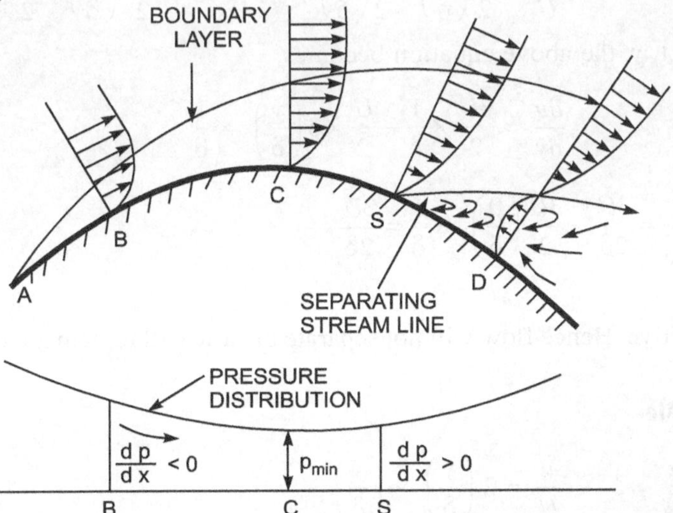

Fig. 11.13 *Effect of pressure gradient on boundary layer separation.*

Thus the positive pressure gradient helps in separating the boundary layer.

11.12.2 Location of Separation Point. The separation point S is determined from the condition,

$$\left(\frac{\partial u}{\partial y}\right)_{y=0} = 0 \qquad \text{...(11.54)}$$

For a given velocity profile, it can be determined whether the boundary layer has separated, or on the verge of separation or will not separate from the following conditions :

1. If $\left(\frac{\partial u}{\partial y}\right)_{y=0}$ is negative ..... the flow has separated.

2. If $\left(\frac{\partial u}{\partial y}\right)_{y=0}$ = 0 ... the flow is on the verge of separation.

3. If $\left(\frac{\partial u}{\partial y}\right)_{y=0}$ is positive ..... the flow will not separate or flow will remain attached with the surface.

Problem 11.34 *For the following velocity profiles, determine whether the flow has separated or on the verge of separated or will attach with the surface :*

(*i*) $\frac{u}{U} = \frac{3}{2}\left(\frac{y}{\delta}\right) - \frac{1}{2}\left(\frac{y}{\delta}\right)^2$, (*ii*) $\frac{u}{U} = 2\left(\frac{y}{\delta}\right)^2 - \left(\frac{y}{\delta}\right)^3$,

(*iii*) $\frac{u}{U} = -2\left(\frac{y}{\delta}\right) + \left(\frac{y}{\delta}\right)^2$.

Solution. Given :

1st Velocity Profile

$$\frac{u}{U} = \frac{3}{2}\left(\frac{y}{\delta}\right) = \frac{1}{2}\left(\frac{y}{\delta}\right)^3 \quad \text{or} \quad u = \frac{3U}{2}\left(\frac{y}{\delta}\right) - \frac{U}{2}\left(\frac{y}{\delta}\right)^3$$

Differentiating w.r.t. y, the above equation becomes,

$$\frac{\partial u}{\partial y} = \frac{3U}{2} \times \frac{1}{\delta} - \frac{U}{2} \times 3\left(\frac{y}{\delta}\right)^2 \times \frac{1}{\delta}$$

At $y = 0$, $\left(\frac{\partial u}{\partial y}\right)_{y=0} = \frac{3U}{2\delta} - \frac{3U}{2}\left(\frac{0}{\delta}\right)^2 \times \frac{1}{\delta} = \frac{3U}{2\delta}$.

As $\left(\frac{\partial u}{\partial y}\right)_{y=0}$ is positive. Hence flow will not separate or flow will remain attached with the surface.

2nd Velocity Profile

$$\frac{u}{U} = 2\left(\frac{y}{\delta}\right)^2 - \left(\frac{y}{\delta}\right)^2$$

$$\therefore \qquad u = 2U\left(\frac{y}{\delta}\right)^2 - U\left(\frac{y}{\delta}\right)^2$$

$$\therefore \qquad \frac{\partial u}{\partial y} = 2U \times 2\left(\frac{y}{\delta}\right) \times \frac{1}{\delta} - U \times 3\left(\frac{y}{\delta}\right)^2 \times \frac{1}{\delta}$$

at $y = 0$, $\left(\frac{\partial u}{\partial y}\right)_{y=0} = 2U \times 2\left(\frac{0}{\delta}\right) \times \frac{1}{\delta} - U \times 3\left(\frac{0}{\delta}\right)^2 \times \frac{1}{\delta} = 0$

As $\left(\frac{\partial u}{\partial y}\right)_{y=0} = 0$, the flow is on the verge of separation. **Ans.**

3rd Velocity Profile

$$\frac{u}{U} = -2\left(\frac{y}{\delta}\right) + \left(\frac{y}{\delta}\right)^2$$

$$\therefore \qquad u = -2U\left(\frac{y}{\delta}\right) + U\left(\frac{y}{\delta}\right)^2$$

$$\therefore \qquad \frac{\partial u}{\partial y} = -2U\left(\frac{1}{\delta}\right) + 2U\left(\frac{y}{\delta}\right) \times \frac{1}{\delta}$$

at $y = 0$, $\left(\frac{\partial u}{\partial y}\right)_{y=0} = -\frac{2U}{\delta} + 2U\left(\frac{0}{\delta}\right) \times \frac{1}{\delta} = -\frac{2U}{\delta}$

As $\left(\frac{\partial u}{\partial y}\right)_{y=0}$ is negative the flow has separated. **Ans.**

11.12.3 Methods of Preventing the Separation of Boundary Layer. When the boundary layer separates from the surface as shown in Fig. 11.13 at point *S*, a certain portion adjacent to the surface has a back flow and eddies are continuously formed in this region and hence continuous loss of energy takes place. Thus separation of boundary layer is undesirable and attempts should be made to avoid separation by various methods. The following are the methods for preventing the separation of boundary layer :

1. Suction of the slow moving fluid by a suction slot.
2. Supplying additional energy from a blower.
3. Providing a bypass in the slotted wing.
4. Rotating boundary in the direction of flow.
5. Providing small divergence in a diffuser.
6. Providing guide-blades in a bend.
7. Providing a trip-wire ring in the laminar region for the flow over a sphere.

HIGHLIGHTS

1. The force exerted by a fluid on a solid body immersed in the fluid in the direction of motion is called drag force while the force perpendicular to the direction of motion, on the body is known as lift force.
2. The mathematical expression for the drag and lift force are,

$$F_L = C_D A \frac{\rho U^2}{2} \; ; F_L = C_L A \frac{\rho U^2}{2}$$

where C_D = Co-efficient of drag, C_L = Co-efficient of lift,
A = Projected area of the body, ρ = Density of fluid,
U = Free-stream velocity of fluid.

3. The resultant force exerted by fluid on solid body is $F_R = \sqrt{F_D^2 + F_L^2}$.
4. Total drag on a body is the sum of pressure drag and friction drag.
5. A body whose surface coincides with the stream-lines, when the body is placed in a flow is called stream-lined body. If the surface of the body does not coincide with the stream-lines, the body is called bluff body.
6. The drag on a sphere for Reynolds number less than 0.2 is given by $F_D = 3\pi\mu DU$

Out of this total drag, Skin friction drag $= \frac{2}{3} \times 3\pi\mu DU = 2\pi\mu DU$

and pressure drag $= \frac{1}{3} \times 3\pi\mu DU = \pi\mu DU.$

7. Values of C_D for sphere for different Reynold number is

$$C_D = \frac{24}{R_e} \text{ ... when } R_e < 0.2$$

$$= \frac{24}{R_e}\left[1 + \frac{3}{16R_e}\right] \text{ ..... when } 0.2 < R_e < 5.0$$

$$= 0.4 \text{ ..... when } R_e \text{ lies between 5 and 1000}$$

$$= 0.5 \text{ ..... when } R_e \text{ lies between 100 and 100000}$$

$$= 0.2 \text{..... when } R_e > 10^5.$$

8. Terminal velocity is defined as the maximum constant velocity of a falling body with which it will travel. At the terminal velocity, the weight of the body is equal to the drag force plus the buoyant force. Hence $W = F_D + F_B$.

Boundary layer

9. When a solid body is immersed in a flowing fluid, there is a narrow region of the fluid in the neighbourhood of the solid body, where the velocity of fluid varies from zero to free-stream velocity. This narrow region of fluid is called boundary layer.
10. The boundary layer is called laminar boundary layer if the Reynold number of the flow defined as

$$R_e = \frac{U \times x}{\nu} \text{ is less than } 5 \times 10^5$$

where U = Free-stream velocity of flow, x = Distance from leading edge,
and ν = Kinematic viscosity of fluid.

11. If the Reynold number is more than 5×10^5, the boundary layer is called turbulent boundary layer.
12. The distance from the surface of the solid body in the direction perpendicular to flow, where the velocity of fluid is approximately equal to 0.99 times the free-stream velocity is called boundary layer thickness and is denoted by δ. For different zones, δ is represented as

$\delta_{lam.}$ = Thickness of laminar boundary layer
$\delta_{tur.}$ = Thickness of turbulent boundary layer
δ' = Thickness of laminar sub-layer.

13. Displacement thickness (δ^*) is given by $\delta^* = \int_0^\delta \left(1 - \frac{u}{U}\right) dy$.

14. Momentum thickness (θ) is given by $\theta = \int_0^\delta \frac{u}{U}\left[1 - \frac{u}{U}\right] dy$.

15. Energy thickness (δ^{**}) is given by $\delta^{**} = \int_0^\delta \frac{u}{U}\left[1 - \frac{u^2}{U^2}\right] dy$.

16. Von Karman momentum integral equation is given as $\frac{\tau_0}{\rho U^2} = \frac{\partial \theta}{\partial x}$

where $\theta = \int_0^d \frac{u}{U}\left[1 - \frac{u}{U}\right] dy$, τ_0 = shear stress at surface.

This equation is applicable to laminar, transition and turbulent boundary layer flows.

17. Thickness of laminar boundary layer and co-efficient of drag from Blasius's solution is given as

$$\delta = \frac{4.91\,x}{\sqrt{R_{e_x}}}$$

where R_{e_x} = Reynold number, $C_D = \frac{1.328}{\sqrt{R_{e_L}}}$

18. Velocity profile for turbulent boundary layer is $\frac{u}{U} = \left(\frac{y}{\delta}\right)^{1/7}$

This equation is not valid very near the boundary, where laminar sub-layer exists.

19. If the pressure gradient is positive, the boundary layer separates from the surface and back flow and eddies formation take place due to which a great loss of energy occur.

20. The conditions for separation, attached flow and detached flow are :

(*i*) $\left(\frac{\partial u}{\partial y}\right)_{y=0} = 0$ ..... condition for separation (*ii*) $\left(\frac{\partial u}{\partial y}\right)_{y=0}$ = positive ..... condition for attached flow

(*iii*) $\left(\frac{\partial u}{\partial y}\right)_{y=0}$ = negative ..... condition for detached flow.

EXERCISE 11

(A) THEORETICAL PROBLEMS

1. Define the terms : drag and lift.
2. What do you understand by : Total drag on a body, resultant force on a body, co-efficient of drag and co-efficient of lift.
3. Differentiate between (*i*) stream-lines body and bluff body, (*ii*) Friction drag and pressure drag.
4. (*a*) What is the expression for the drag on a sphere, when the Reynolds number of the flow is up to 0.2 ? Hence prove that the co-efficient of drag for sphere for this range of the Reynold number is given by

$$C_D = \frac{24}{R_e}, \text{ where } R_e = \text{Reynold number.}$$

(*b*) Draw C_D verses R_e diagram for a sphere and explain why C_D suddenly drops at $R_e = 3 \times 10^3$.

(*c*) Draw pressure distribution diagrams in dimensionless form for flow past sphere when fluid has no viscosity, when $R_e = 10^4$ and when $R_e = 10^6$. (*A.M.I.E., Winter, 1991*)

5. What do you mean by 'Terminal velocity of a body' ? What is the relation between the weight of the body, drag force on the body and buoyant force when the body has acquired terminal velocity ?

6. Explain the terms : (*i*) Friction drag, (*ii*) Pressure drag and profile drag. (*A.M.I.E., Winter, 1988*)

7. (*a*) How are drag and lift forces caused on a body immersed in a moving fluid. (*University of Kerala, November, 1997*)

(*b*) What is the drag force on a sphere in the stoke range ?

Boundary layer

8. What do you understand by the terms boundary layer, and boundary layer theory ?

9. Define : laminar boundary layer, turbulent, boundary layer, laminar sub-layer and boundary layer thickness. (*Delhi University, 1987*)

10. Define displacement thickness. Derive an expression for the displacement thickness.

11. Prove that the momentum thickness and energy thickness for boundary layer flows are given by

$$\theta = \int_0^\delta \frac{u}{U}\left[1 - \frac{u}{U}\right] dy \quad \text{and} \quad \delta^{**} = \int_0^\delta \frac{u}{U}\left[1 - \frac{u^2}{U}\right] dy.$$

(*Osmania University, 1992*)

12. Obtain an expression for the boundary shear stress in terms of momentum thickness.

13. Obtain Von Karman momentum integral equation.

14. What are the boundary conditions that must be satisfied by a given velocity profile in laminar boundary layer flows ?

15. How will you find the drag on a flat plate due to laminar and turbulent boundary layers ?

16. What do you mean by separation of boundary layer ? What is the effect of pressure gradient on boundary layer separation ?

17. How will you determine whether a boundary layer flow is attached flow, detached flow or on the verge of separation ?

18. What are the different methods of preventing the separation of boundary layers ?

19. What is meant by boundary layer ? Why does it increase with distance from the upstream edge ? (*A.M.I.E., Summer, 1984*)

20. Define the terms : boundary layer, boundary layer thickness, drag, lift and momentum thickness. (*Delhi University, 1982, 96*)

21. What do you mean by boundary layer separation ? What is the effect of pressure gradient on boundary layer separation ? (*R.G.P.V., Bhopal S, 2001*)

(B) NUMERICAL PROBLEMS

1. A flat plate 2 m × 2 m moves at 40 km/hour in stationary air of density 1.25 kg/m^3. If the co-efficient of drag and lift are 0.2 and 0.8 respectively, find : (*i*) the lift force (*ii*) the drag force, (*iii*) the resultant force and (*iv*) the power required to keep the plate in motion.

[**Ans.** (*i*) 246.86 N, (*ii*) 61.715 N, (*iii*) 254.4 N, (*iv*) 0.684 kW]

2. Find the drag force difference on a flat plate of size 1.5 m × 1.5 m when the plate is moving at a speed of 5 m/s normal to its plate first in water and second in air of density 1.24 kg/m^3. Co-efficient of drag is given as 1.10. [**Ans.** 30899 N]

3. A truck having a projected area of 12 square metres travelling at 60 km/hr has a total resistance of 2943 N. Of this 25% is due to rolling friction and 15% is due to surface fraction. The rest is due to form drag. Calculate the coefficient of form drag if the density of air = 1.25 kg/m^3. [**Ans.** 0.847]

4. A circular disc 4 m in diameter is held normal to 30 m/s wind of density 1.25 kg/m^3. If the co-efficient of drag of disc = 1.1, what force is required to hold the disc at rest ? [**Ans.** 7775.4 N]

5. Find the diameter of a parachute with which a man of mass 80 kg descends to the ground from an aeroplane against the resistance of air, with a velocity of 25 m/s. Take C_d = 0.5 and density of air = 1.25 kg/m^3. [**Ans.** 2.26 m]

6. A man descends to the ground with the help of a parachute from an aeroplane against the resistance of air with a uniform velocity of 10 m/s. The parachute is hemispherical in shape and is having diameter is 5 m. Find the weight of man if C_d = 0.5 and density of air = 1.25 kg/m^3. [**Ans.** 613.52 N]

7. A kite 60 cm × 60 cm weighing 2.943 N assumes an angle of 10° to the horizontal. The string attached to the kite makes an angle of 45° to the horizontal. If the pull on the string is 29.43 N when the wind is flowing at a speed of 40 km/hr. Find the corresponding co-efficient to drag and lift. Density of air is given as 1.25 kg/m^3. [**Ans.** C_D = 0.7489, C_L = .8548]

8. The air is flowing over a cylinder of diameter 100 mm and of infinite length with a velocity of 150 mm/s. Find the total drag, shear drag and pressure drag on 1 m length of the cylinder if the total drag co-efficient = 1.5 m and shear drag co-efficient = 0.25. The density of air is given as = 1.25 kg/m^3. [**Ans.** 0.00211 N, 0.000351 N, 0.001756 N]

9. A body of length 2.5 m has a projected area 1.8 m^2 normal to the direction of its motion. The body is moving through water with a velocity such that the Reynold number = 6 × 10^6 and the drag co-efficient = 0.5. Find the drag on the body. Take viscosity of water = 0.01 poise. [**Ans.** 2592 N]

10. Calculate the weight of a ball of diameter 50 mm which is just supported in a vertical air stream which is flowing at a velocity of 10 m/s. The density of air = 1.25 kg/m^3 and kinematic viscosity = 1.5 stokes. [**Ans.** 0.0613 N]

11. A metallic sphere of sp. gr. 8.0 falls in an oil of density 800 kg/m^3. The diameter of the sphere is 10 mm and it attains a terminal velocity of 50 mm/s. Find the viscosity of the oil in poise. [**Ans.** 78.48 poise]

12. A metallic ball of diameter 5 mm drops in a fluid of sp. gr. 0.8 and viscosity 30 poise. The specific gravity of the metallic ball, is 9.0. Find : (*i*) the drag force exerted by fluid on metallic ball, (*ii*) the pressure drag and skin friction drag, and (*iii*) terminal velocity of ball in fluid.
[**Ans.** (*i*) 0.005264 N, (*ii*) 0.001754 N, 0.003527 N, (*iii*) 3.7 cm/s]

13. Experiments were conducted in a wind tunnel with a wind speed of 50 km/hour on a flat plate of size 2 m long and 1 m wide. The density of air is 1.15 kg/m^3. The plate is kept at such an angle that co-efficients of lift and drag are 0.75 and 0.15 respectively. Determine : (*i*) lift force, (*ii*), drag force, (*iii*) resultant force, (*iv*) its direction and (*v*) power exerted by the air stream on the plate. (*A.M.I.E., Winter, 1987*)
[**Ans.** (*i*) 166.4 N (*ii*) 33.28 N (*iii*) 169.64 N (*iv*) 78.7° (*v*) 0.461 kW]

Boundary layer

14. (*a*) Find the ratios of displacement thickness to momentum thickness and momentum thickness to energy thickness for the velocity distribution in the boundary layer given by

$$\frac{u}{U} = 2\,(y/\delta) - (y/\delta)^2$$

where u = Velocity in boundary layer at a distance y
U = Free-stream velocity
δ = Boundary layer thickness [**Ans.** 2.5, 7/11]

(*b*) Find the displacement thickness, the momentum thickness and energy thickness for the velocity distribution in the boundary layer given by

$$\frac{u}{U} = 2\left(\frac{y}{\delta}\right) - \left(\frac{y}{\delta}\right)^2.$$

(*Delhi University, December, 2002*)

15. For the velocity profile in laminar boundary layer given as $\frac{u}{U} = \frac{3}{2}(y/\delta) - \frac{1}{2}(y/\delta)^3$, find the thickness of the boundary layer and shear stress 1.8 m from the leading edge of a plate. The plate is 2.5 m long and 1.5 m wide and is placed in water which is moving with a velocity of 15 cm per second. Find the drag on one side of the plate if the viscosity of water = 0.01 poise. [**Ans.** 1.6 cm, 0.0014 N/m^2, 0.0889 N]

16. Air is flowing over a smooth plate with a velocity of 8 m/s. The length of the plate is 1.5 m and width 1 m. If the laminar boundary exists upto a value of Reynold number = 5 × 10^5, find the maximum distance from the leading edge upto which laminar boundary layer exists. Find the maximum thickness of laminar boundary layer if the velocity profile is given by

$$\frac{u}{U} = (y/\delta) - (y/\delta)^2.$$ Take ν for air = 0.15 stokes. **[Ans.** 0.9375 m, 7.26 mm]

17. If in the problem 16, the velocity profile over the plate is given as $\frac{u}{U} = \sin\left(\frac{\pi}{2} \times \frac{y}{\delta}\right)$ and density of air as 1.24 kg/m^3, find : (*i*) maximum thickness of the laminar boundary layer, (*ii*) shear stress at 20 cm from the leading edge and (*iii*) drag force on one side of the plate assuming the laminar boundary layer over the entire length of the plate. **[Ans.** (*i*) 0.635 cm, (*ii*) 0.099 N/m^2, (*iii*) 0.0871 N]

18. A thin plate is moving in still atmospheric air at a velocity of 4 m/s. The length of the plate is 0.5 m and width 0.4 m. Calculate the (*i*) thickness of the boundary layer at the end of the plate and (*ii*) drag force on one side of the plate. Take density of air as 1.25 kg/m^3 and kinematic viscosity 0.15 stokes.
[Ans. (*i*) 0.672 cm (*ii*) 0.00728 N]

19. Find the frictional drag on one side of the plate 200 mm wide and 500 mm long placed longitudinally in a steam of crude oil (specific gravity = 0.925, kinematic viscosity = 0.9 stoke) flowing with undisturbed velocity of 5 m/s. Also find the thickness of boundary layer and the shear stress at the trailing edge of the plate. (*A.M.I.E., May, 1975*) **[Ans.** 9.34 N, 14.75 mm]

20. A smooth flat plate of length 5 m and width 2 m is moving with a velocity of 4 m/s in stationary air of density as 1.25 kg/m^3 and kinematic viscosity 1.5 × 10^{-5} m^2/s. Determine thickness of the boundary layer at the trailing edge of the smooth plate. Find the total drag on one side of the plate assuming that the boundary layer is turbulent from the very beginning. **[Ans.** 110 mm, 0.43 N]

21. Water is flowing over a thin smooth plate of length 4.5 m and width 2.5 m at a velocity of 0.9 m/s. If the boundary layer flow changes from laminar to turbulent at a Reynold number 5 × 10^5, find (*i*) the distance from leading edge upto, which boundary layer is laminar, (*ii*) thickness of the boundary layer at the transition point, and (*iii*) the drag force on-one side of the plate. Take viscosity of water as 0.01 poise. **[Ans.** (*i*) 555 mm (*ii*) 3.85 mm (*iii*) 13.75 N]

22. For the velocity profile given below, state whether the boundary layer has separated or on the verge of separation or will remain attached with the surface :

(*i*) $\frac{u}{U} = 2\,(y/\delta) - (y/\delta)^2$ (*ii*) $\frac{u}{U} = -2\,(y/\delta) + \frac{1}{2}\,(y/\delta)^3$ and

(*iii*) $\frac{u}{U} = \frac{3}{2}\,(y/\delta)^2 + \frac{1}{2}(y/\delta)^3$.

[Ans. (*i*) Remain attached (*ii*) has separated (*iii*) on the verge of separation]

23. Oil with a free-stream velocity of 1.5 m/s flow over a thin plate 1.4 m wide and 2.2 m long. Calculate the boundary layer thickness and the shear stress at the trailing end point and determine the total surface resistance of the plate. Take specific gravity of oil as 0.80 and kinematic viscosity as 0.1 stoke.
[Ans. 1.88 cm, 1.04 N/cm^2, 12.8 N]

24. (*a*) For the velocity profile $\frac{u}{U} = \frac{3}{2}\left(\frac{y}{\delta}\right) - \frac{1}{2}\left(\frac{y}{\delta}\right)^2$. Calculate the co-efficient of drag in terms of Reynolds number.

(*b*) A thin smooth plate of 0.3 m width and 1.0 m length moves at 4 m/s viscosity in still atmospheric air of density 1.20 kg/m^3 and kinematic viscosity of 1.49 × 10^{-5} m^2/s. Calculate the drag force on the plate.
(*Delhi University, 1982*) **[Ans.** 0.00716 N]

25. Find the displacement thickness, the momentum thickness and energy thickness for the velocity distribution in the boundary layer given by,

$$\frac{u}{U} = 2\left(\frac{y}{\delta}\right) - \left(\frac{y}{\delta}\right)^2,$$ where δ = boundary layer thickness.

(*Delhi University, 1996*) $\left[\textbf{Ans. } \frac{\delta}{3}; \frac{2}{15}\delta; \frac{22}{105}\delta\right]$

Compressible Flow

▶ 12.1 INTRODUCTION

Compressible flow is defined as that flow in which the density of the fluid does not remain constant during flow. This means that the density changes from point to point in compressible flow. But in case of incompressible flow, the density of the fluid is assumed to be constant. In the previous chapters, the fluid was assumed incompressible, and the basic equations such as equation of continuity, Bernoulli's equation and impulse momentum equations were derived on the assumption that fluid is incompressible. This assumption is true for flow of liquids, which are incompressible fluids. But in case of flow of fluids, such as

(*i*) flow of gases through orifices and nozzles,

(*ii*) flow of gases in machines such as compressors, and

(*iii*) projectiles and airplanes flying at high altitude with high velocities, the density of the fluid changes during the flow. The change in density of a fluid is accompanied by the changes in pressure and temperature and hence the thermodynamic behaviour of the fluids will have to be taken into account.

▶ 12.2 THERMODYNAMIC RELATIONS

The thermodynamic relations have been discussed in Chapter 1, which are as follows :

12.2.1 Equation of State. Equation of state is defined as the equation which gives the relationship between the pressure, temperature and specific volume of a gas. For a perfect gas the equation of state is

$$p\forall = RT \qquad \text{...(12.1)}$$

where p = Absolute pressure in kgf/m² or N/m²

$\forall$ = Specific volume or volume per unit mass

T = Absolute temperature = 273 + $t°$ (centigrade)

R = Gas constant in kgf-m/kg °K or (J/kg K)

= 29.2 kgf-m/kg °K or 287 J/kg K for air.

In equation (12.1), $\forall$ is the specific volume which is the reciprocal of density or

$$\forall = \frac{1}{\rho}.$$

Substituting this value of ∀ in equation (12.1), we get

$$\frac{p}{\rho} = RT \qquad ...(12.2)$$

Note. In the equation of state given by equation (12.2), the dimensions of p, ρ and R should be used with care. The following points must be remembered :

1. If the value of R is given as 29.2 kgf-m/kg °K for air, the corresponding value of p and ρ should be taken in kgf/m^2 and kg/m^3. The mass rate of flow of the gas will be in kg/sec.
2. If the value of R is given as 287 J/kg K, the corresponding value of p and ρ should be in N/m^2 and kg/m^3. The mass rate of flow will be in kg/sec.

Value of $\frac{p}{\rho}$ in Bernoulli's Equation*. (*i*) If the value of p is taken in N/m^2, the corresponding value of ρ is in kg/m^3. And as mentioned above (point number 2), the value of R should be 287 J/kg K.

(*ii*) If the value of p is taken in kgf/m^2 in Bernoulli's equation, the corresponding value of ρ should be in msl/m^3. But as mentioned in point number 1, if the value of R is taken 29.2, the corresponding values of p and ρ are in kgf/m^2 and kg/m^3. Hence the mass density in equation of state is in kg/m^3 while in Bernoulli's equation it is in msl/m^3. The density calculated from equation of state must be converted into msl/m^3.

Note. It is better to use pressure in N/m^2, density kg/m^3 and value of R = 287 J/kg K. The value of density calculated from equation of state will be in the same dimensions as used in Bernoulli's equation.

12.2.2 Expansion and Compression of Perfect Gas. When the expansion or compression of a perfect gas takes place, the pressure, temperature and density are changed. The change in pressure, temperature and density of a gas is brought about by the two processes which are known as

1. Isothermal process, and 2. Adiabatic process.

1. Isothermal Process. This is the process in which a gas is compressed or expanded while the temperature is kept constant. The gas obeys Boyle's law, according to which we have

$$p\forall = \text{Constant}, \text{ where } \forall = \text{Specific volume}$$

or

$$\frac{p}{\rho} = \text{Constant} \qquad \left(\because \forall = \frac{1}{\rho}\right) \quad ...(12.3)$$

2. Adiabatic Process. If the compression or expansion of a gas takes place in such a way that the gas neither gives heat, nor takes heat from its surrounding, then the process is said to be adiabatic. According to this process,

$$p\forall^k = \text{Constant}$$

where k = Ratio of the specific heat at constant pressure to the specific heat at constant volume

$$= \frac{C_p}{C_v} = 1.4 \text{ for air.}$$

The above relation is also written as $\frac{p}{\rho^k}$ = Constant. ...(12.4)

In the adiabatic process is reversible (or frictionless), it is known as isentropic process. And if the pressure and density are related in such a way that k is not equal to $\frac{C_p}{C_v}$ but equal to some positive value then the process is known as polytropic. According to which

$$\frac{p}{\rho^n} = \text{Constant} \qquad ...(12.5)$$

where $n \neq k$ but equal to some positive constant.

* Please, refer to equations (12.10) and (12.11).

▶ 12.3 BASIC EQUATIONS OF COMPRESSIBLE FLOW

The basic equations of the compressible flows are

1. Continuity Equation,
2. Bernoulli's Equation or Energy Equation,
3. Momentum Equation,
4. Equation of state.

12.3.1 Continuity Equation. This is based on law of conservation of mass which states that matter cannot be created nor destroyed. Or in other words, the matter or mass is constant. For one-dimensional steady flow, the mass per second = ρAV

where ρ = Mass density, A = Area of cross-section, V = Velocity

As mass or mass per second is constant according to law of conservation of mass. Hence

$$\rho AV = \text{Constant.} \qquad \text{...(12.6)}$$

Differentiating equation (12.6), $d(\rho AV) = 0$ or $\rho d(AV) + AVd\rho = 0$

or $\rho[AdV + VdA] + AVd\rho = 0$ or $\rho AdV + \rho VdA + AVd\rho = 0$

Dividing by ρAV, we get
$$\frac{dV}{V} + \frac{dA}{A} + \frac{d\rho}{\rho} = 0. \qquad \text{...(12.7)}$$

Equation (12.7) is also known as continuity equation in differential form.

12.3.2 Bernoulli's Equation. Bernoulli's equation has been derived for incompressible fluids in Chapter 7. The same procedure is followed. The flow of a fluid particle along a stream-line in the direction of S is considered. The resultant force on the fluid particle in the direction of S is equated to the mass of the fluid particle and its acceleration. As the flow of compressible fluid is steady, the same Euler's equation as given by equation (7.3) is obtained as

$$\frac{dp}{\rho} + VdV + gdZ = 0 \qquad \text{...(12.8)}$$

Integrating the above equation, we get

$$\int \frac{dp}{\rho} + \int VdV + \int gdZ = \text{Constant}$$

or
$$\int \frac{dp}{\rho} + \frac{V^2}{2} + gZ = \text{Constant} \qquad \text{...(12.9)}$$

In case of incompressible flow, the density ρ is constant and hence integration of $\frac{dp}{\rho}$ is equal to $\frac{p}{\rho}$. But in case of compressible flow, the density ρ is not constant. Hence ρ cannot be taken outside the integration sign. With the change of ρ, the pressure p also changes for compressible fluids. This change of ρ and p takes place according to equations (12.3) or (12.4) depending upon the type of process during compressible flow. The value of ρ from these equations in terms of p is obtained and is substituted in $\int \frac{dp}{\rho}$ and then the integration is done. The Bernoulli's equation will be different for isothermal process and for adiabatic process.

(A) Bernoulli's Equation for Isothermal Process. For isothermal process, the relation between pressure (p) and density (ρ) is given by equation (12.3) as

$$\frac{p}{\rho} = \text{Constant} = C_1 \text{ (say)} \qquad ...(i)$$

$$\therefore \qquad \rho = \frac{p}{C_1}$$

Hence $$\int \frac{dp}{\rho} = \int \frac{dp}{p/C_1} = \int \frac{C_1 dp}{p} = C_1 \int \frac{dp}{p} \qquad (\because C_1 \text{ is constant})$$

$$= C_1 \log_e p = \frac{p}{\rho} \log_e p \qquad \left(\because \; C_1 = \frac{p}{\rho} \text{ from equation } (i)\right)$$

Substituting the value $\int \frac{dp}{\rho}$ in equation (12.9), we get

$$\frac{p}{\rho} \log_e p + \frac{V^2}{2} + gZ = \text{Constant}$$

Dividing by 'g', $$\frac{p}{\rho g} \log_e p + \frac{V^2}{2g} + Z = \text{Constant}. \qquad ...(12.10)$$

Equation (12.10) is the Bernoulli's equation for compressible flow undergoing isothermal process. For the two points 1 and 2, this equation is written as

$$\frac{p_1}{\rho_1 g} \log_e p_1 + \frac{V_1^2}{2g} + Z_1 = \frac{p_2}{\rho_2 g} \log_e p_2 + \frac{V_2^2}{2g} + Z_2 \qquad ...(12.11)$$

(B) Bernoulli's Equation for Adiabatic Process. For the adiabatic process, the relation between pressure (p) and density (ρ) is given by equation (12.4) as

$$\frac{p}{\rho^k} = \text{Constant} = \text{say } C_2 \qquad ...(ii)$$

$$\therefore \qquad \rho^k = \frac{p}{C_2} \quad \text{or} \quad \rho = \left(\frac{p}{C_2}\right)^{1/k}$$

Hence $$\int \frac{dp}{\rho} = \int \frac{dp}{\left(\frac{p}{C_2}\right)^{1/k}} = \int \frac{C_2^{1/k}}{p^{1/k}} dp = C_2^{1/k} \int \frac{1}{p^{1/k}} dp$$

$$= C_2^{1/k} \int p^{-1/k} dp = C_2^{1/k} \frac{p^{\left(-\frac{1}{k}+1\right)}}{\left(-\frac{1}{k}+1\right)}$$

$$= \frac{C_2^{1/k} p^{\left(\frac{k-1}{k}\right)}}{\left(\frac{k-1}{k}\right)} = \left(\frac{k}{k-1}\right) C_2^{1/k} p^{\left(\frac{k-1}{k}\right)}$$

$$= \left(\frac{k}{k-1}\right)\left(\frac{p}{\rho^k}\right)^{1/k} p^{\left(\frac{1-k}{k}\right)} \qquad \left(\because \; C_2^{1/k} = \frac{p}{\rho^k} \text{ from } (ii)\right)$$

$$= \left(\frac{k}{k-1}\right)\frac{p^{1/k}}{\rho^{k\times 1/k}}\, p^{\left(\frac{k-1}{k}\right)} = \left(\frac{k}{k-1}\right)\frac{p^{\frac{1}{k}+\frac{k-1}{k}}}{\rho} = \left(\frac{k}{k-1}\right)\frac{p}{\rho}$$

Substituting the value of $\int \frac{dp}{\rho} = \left(\frac{k}{k-1}\right)\frac{p}{\rho}$ in equation (12.9), we get

$$\left(\frac{k}{k-1}\right)\frac{p}{\rho} + \frac{V^2}{2} + gZ = \text{Constant}$$

Dividing by 'g'

$$\left(\frac{k}{k-1}\right)\frac{p}{\rho g} + \frac{V^2}{2g} + Z = \text{Constant.} \quad \text{...(12.12)}$$

Equation (12.12) is the Bernoulli's equation for compressible flow undergoing adiabatic process. For the two points 1 and 2, this equation is written as

$$\left(\frac{k}{k-1}\right)\frac{p_1}{\rho_1 g} + \frac{V_1^2}{2g} + Z_1 = \left(\frac{k}{k-1}\right)\frac{p_2}{\rho_2 g} + \frac{V_2^2}{2g} + Z_2 \quad \text{...(12.13)}$$

Problem 12.1 *A gas is flowing through a horizontal pipe at a temperature of 4°C. The dimaeter of the pipe is 8 cm and at a section 1-1 in this pipe, the pressure is 30.3 N/cm² (gauge). The diameter of the pipe changes from 8 cm to 4 cm at the section 2-2, where pressure is 20.3 N/cm² (gauge). Find the velocities of the gas at these sections assuming an isothermal process. Take R = 287.14 Nm/kg K, and atmospheric pressure = 10 N/cm².*

Solution. Given :

For the section 1-1,

Temperature, $t_1 = 4°C$

∴ Absolute temperature, $T_1 = 4 + 273 = 277$ K

Diameter pipe, $D_1 = 8$ cm $= 0.08$ m

∴ Area of pipe, $A_1 = \frac{\pi}{4} D_1^2 = \frac{\pi}{4}(.08)^2 = .005026 \text{ m}^2$

Pressure, $p_1 = 30.3 \text{ N/cm}^2$ (gauge)

$= 30.3 + 10 = 40.3 \text{ N/cm}^2 \text{ (absolute)} = 40.3 \times 10^4 \text{ N/m}^2$ (abs.)

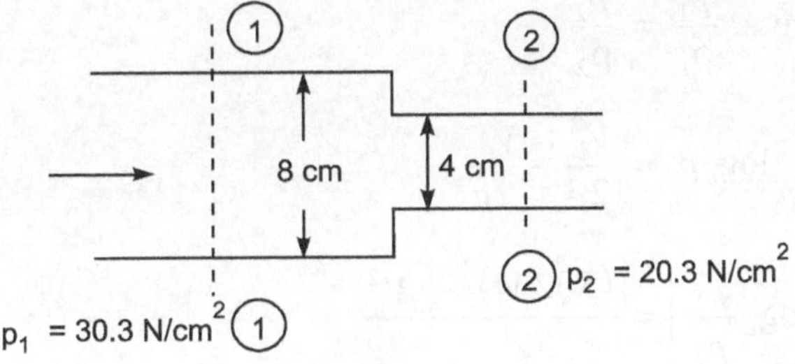

Fig. 12.1

For the section 2-2,

Diameter of pipe, $D_2 = 4$ cm $= .04$ m

∴ Area, $A_2 = \frac{\pi}{4}(.04)^2 = .0012565 \text{ m}^2$

Pressure, $p_2 = 20.3 + 10 = 30.3 \text{ N/cm}^2 \text{ (abs.)} = 30.3 \times 10^4 \text{ N/m}^2 \text{ (abs.)}$

Gas constant, $R = 287.14$ N-m/kg K

Ratio of specific heat, $k = 1.4$.

Applying continuity equation at (1) and (2), we get

$$\rho_1 A_1 V_1 = \rho_2 A_2 V_2$$

or
$$\frac{V_2}{V_1} = \frac{\rho_1 A_1}{\rho_2 A_2} = \frac{\rho_1 \times .005026}{\rho_2 \times .0012565} = 4 \times \frac{\rho_1}{\rho_2} \quad \text{...}(i)$$

For isothermal process using equation (12.3),

$$\frac{p_1}{\rho_1} = \frac{p_2}{\rho_2} \text{ or } \frac{\rho_1}{\rho_2} = \frac{p_1}{p_2} = \frac{40.3 \times 10^4}{30.3 \times 10^4} = 1.33$$

Substituting the value of $\frac{\rho_1}{\rho_2} = 1.33$ in equation (*i*), we get

$$\frac{V_2}{V_1} = 4 \times 1.33 = 5.32$$

$$\therefore \quad V_2 = 5.32\ V_1 \quad \text{...}(ii)$$

Applying Bernoulli's equation at sections 1-1 and 2-2 for isothermal process which is given by equation (12.11), we get

$$\frac{p_1}{\rho_1 g} \log_e p_1 + \frac{V_1^2}{2g} + Z_1 = \frac{p_2}{\rho_2 g} \log_e p_2 + \frac{V_2^2}{2g} + Z_2$$

For horizontal pipe, $Z_1 = Z_2$

$$\therefore \quad \frac{p_1}{\rho_2 g} \log_e p_1 + \frac{V_1^2}{2g} = \frac{p_2}{\rho_2 g} \log_e p_2 + \frac{V_2^2}{2g}$$

or
$$\frac{p_1}{\rho_1 g} \log_e p_1 - \frac{p_2}{\rho_2 g} \log_e p_2 = \frac{V_2^2}{2g} - \frac{V_1^2}{2g}$$

But for isothermal process, $\frac{p_1}{\rho_1} = \frac{p_2}{\rho_2}$

$$\therefore \quad \frac{p_1}{\rho_1 g} \log_e p_1 - \frac{p_1}{\rho_1 g} \log_e p_2 = \frac{V_2^2}{2g} - \frac{V_1^2}{2g}$$

or
$$\frac{p_1}{\rho_1 g}\left[\log_e \frac{p_1}{p_2}\right] = \frac{(5.32\ V_1)^2}{2g} - \frac{V_1^2}{2g} \quad (\because \text{ From } (ii),\ V_2 = 5.32\ V_1)$$

or
$$\frac{p_1}{\rho_1 g} \log_e\left(\frac{40.3 \times 10^4}{30.3 \times 10^4}\right) = \frac{V_1^2}{2g}(5.32^2 - 1) = 27.30\ \frac{V_1^2}{2g}$$

or
$$\frac{p_1}{\rho_1 g} \log_e 1.33 = 27.30\ \frac{V_1^2}{2g}$$

or $$\frac{p_1}{\rho_1 g} \times 0.285 = 27.30 \frac{V_1^2}{2g}$$

or $$\frac{p_1}{\rho_1} = \frac{27.30}{2 \times .285} V_1^2 = 47.894\ V_1^2 \quad ...(iii)$$

Now from equation of state, *i.e.*, from equation (12.2), we have

$$\frac{p}{\rho} = RT \text{ or at section 1, } \frac{p_1}{\rho_1} = RT_1$$

or $$\frac{p_1}{\rho_1} = RT_1 = 287.14 \times 277 = 79537.4$$

Substituting this value of $\frac{p_1}{\rho_1} = 79537.4$ in equation (*iii*), we get $79537.4 = 47.894\ V_1^2$

$\therefore$ $$V_1 = \sqrt{\frac{79537.4}{47.894}} = \mathbf{40.75\ m/s.\ Ans.}$$

From equation (*ii*), $V_2 = 5.32 \times V_1 = 5.32 \times 40.75 =$ **216.79 m/s. Ans.**

Problem 12.2 *A gas is flowing through a horizontal pipe which is having area of cross-section as 40 cm², where pressure is 40 N/cm² (gauge) and temperature 15°C. At another section the area of cross-section is 20 cm² and pressure is 30 N/cm² (gauge). If the mass rate of flow of gas through the pipe is 0.5 kg/s, find the velocities of the gas at these sections, assuming an isothermal change. Take R = 292 N-m/kg°K, and atmospheric pressure = 10 N/cm².*

Solution. Given :

Section 1		Section 2	
Area,	$A_1 = 40\ cm^2 = 40 \times 10^{-4}\ m^2$	Area,	$A_2 = 20\ cm^2 = 20 \times 10^{-4}\ m^2$
Pressure,	$p_1 = 40\ N/cm^2$ (gauge)	Pressure,	$p_2 = 30\ N/cm^2$ (gauge)
	$= 40 + 10 = 50\ N/cm^2$ (abs.)		$= 30 + 10 = 40\ N/cm^2$ (abs.)
	$= 50 \times 10^4\ N/m^2$		$= 40 \times 10^4\ N/m^2$
Temperature,	$t_1 = 15°C$		
$\therefore$	$T_1 = 15 + 273 = 288°K$		

Mass rate of flow $= 0.5$ kg/s.

Gas constant, $R = 292$ N-m/kg°K

From equation of state, *i.e.*, equation (12.2), $\frac{p_1}{\rho_1} = RT_1$

$\therefore$ $$\rho_1 = \frac{p_1}{RT_1} = \frac{50 \times 10^4}{292 \times 288} \frac{kg}{m^3} = 5.945 \frac{kg}{m^3}$$

Mass rate of flow is given by $= \rho_1 A_1 V_1$ or $0.5 = 5.945 \times 40 \times 10^{-4} \times V_1$

$\therefore$ $$V_1 = \frac{0.5}{5.945 \times 40 \times 10^{-4}} = 21.02\ m/s$$

For isothermal process, temperature is constant and hence temperature at section 2 is also 288°K.

$$\therefore \qquad T_2 = 288°\text{K}$$

Using equation (12.2), we get $\frac{p_2}{\rho_2} = RT_2$

$$\therefore \qquad \rho_2 = \frac{p_2}{RT_2} = \frac{40 \times 10^4}{292 \times 288} = 4.756 \text{ kg/m}^3$$

Now mass rate of flow $= \rho_2 A_2 V_2$

$$\therefore \qquad 0.5 = 4.756 \times 20 \times 10^{-4} \times V_2$$

$$\therefore \qquad V_2 = \frac{0.5}{4.756 \times 20 \times 10^{-4}} = \mathbf{52.565 \text{ m/s. Ans.}}$$

Problem 12.3 *A gas with a velocity of 300 m/s is flowing through a horizontal pipe at a section where pressure is 6×10^4 N/m² (absolute) and temperature 40°C. The pipe changes in diameter and at this section the pressure is 9×10^4 N/m². Find the velocity of the gas at this section if the flow of the gas is adiabatic.*

Take R= 287 J/kg K and k = 1.4.

Solution. Given :

Section 1	Section 2
$V_1 = 300$ m/s	$p_2 = 9 \times 10^4$ N/m²
$p_1 = 6 \times 10^4$ N/m²	V_2 = velocity at section 2
$t_1 = 40°$C	$R = 287$ J/kg K
$\therefore \quad T_1 = 273 + 40 = 313$ K	

Adiabatic flow, $\quad k = 1.4$

Applying Bernoulli's equation at sections 1 and 2, given by equation (12.13), we get

$$\left(\frac{k}{k-1}\right)\frac{p_1}{\rho_1 g} + \frac{V_1^2}{2g} = \left(\frac{k}{k-1}\right)\frac{p_2}{\rho_2 g} + \frac{V_2^2}{2g} \qquad (\because \quad Z_1 = Z_2)$$

or

$$\left(\frac{k}{k-1}\right)\left[\frac{p_1}{\rho_1 g} - \frac{p_2}{\rho_2 g}\right] = \frac{V_2^2}{2g} - \frac{V_1^2}{2g}$$

Dividing by 'g', we get $\left(\frac{k}{k-1}\right)\left[\frac{p_1}{\rho_1} - \frac{p_2}{\rho_2}\right] = \frac{V_2^2}{2} - \frac{V_1^2}{2}$

or

$$\left(\frac{1.4}{1.4 - 1.0}\right)\frac{p_1}{\rho_1}\left[1 - \frac{p_2}{\rho_2} \times \frac{\rho_1}{p_1}\right] = \frac{V_2^2}{2} - \frac{V_1^2}{2} \qquad \ldots(i)$$

For adiabatic flow, using equation (12.4), we get

$$\frac{p_1}{\rho_1^k} = \frac{p_2}{\rho_2^k} \quad \text{or} \quad \frac{p_1}{p_2} = \left(\frac{\rho_1}{\rho_2}\right)^k \quad \text{or} \quad \left(\frac{\rho_1}{\rho_2}\right) = \left(\frac{p_1}{p_2}\right)^{1/k}$$

Substituting the value of $\frac{\rho_1}{\rho_2}$ in equation (*i*), we get

$$\frac{1.4}{0.4}\frac{p_1}{\rho_1}\left[1-\frac{p_2}{p_1}\times\left(\frac{p_1}{p_2}\right)^{1/k}\right]=\frac{V_2^2}{2}-\frac{V_1^2}{2}$$

or $$3.5\frac{p_1}{\rho_1}\left[1-\left(\frac{p_2}{p_1}\right)\times\left(\frac{p_2}{p_1}\right)^{-1/k}\right]=\frac{V_2^2}{2}-\frac{V_1^2}{2}$$

or $$3.5\frac{p_1}{\rho_1}\left[1-\left(\frac{p_2}{p_1}\right)^{1-\frac{1}{k}}\right]=\frac{V_2^2}{2}-\frac{V_1^2}{2} \text{ or } 3.5\frac{p_1}{\rho_1}\left[1-\left(\frac{p_2}{p_1}\right)^{\frac{k-1}{k}}\right]=\frac{V_2^2}{2}-\frac{V_1^2}{2}$$

Substituting the value of p_2 and p_1, we get

$$3.5\frac{p_1}{\rho_1}\left[1-\left(\frac{9\times10^4}{6\times10^4}\right)^{\frac{1.4-1}{1.4}}\right]=\frac{V_2^2}{2}-\frac{V_1^2}{2}$$

or $$3.5\frac{p_1}{\rho_1}\left[1-1.5^{2/7}\right]=\frac{V_2^2}{2}-\frac{V_1^2}{2} \text{ or } 3.5\frac{p_1}{\rho_1}\left[1-1.5^{.2857}\right]=\frac{V_2^2}{2}-\frac{V_1^2}{2}$$

or $$-0.4298\frac{p_1}{\rho_1}=\frac{V_2^2}{2}-\frac{V_1^2}{2}=\frac{V_2^2}{2}-\frac{300^2}{2}$$

or $$-.4298\frac{p_1}{\rho_1}=\frac{V_2^2}{2}-45000 \quad \text{...(ii)}$$

From equation of state, we have $\frac{p_1}{\rho_1}=RT_1=287\times313=89831$

Substituting the value of $\frac{p_1}{\rho_1}=89831$ in equation (*ii*),

$$-0.4298\times89831=\frac{V_2^2}{2}-45000$$

or $$\frac{V_2^2}{2}=45000-.4298\times89831=6390.6$$

$\therefore$ $$V_2=\sqrt{2\times6390.6}=\mathbf{113.0\ m/s.\ Ans.}$$

12.3.3 Momentum Equations. The momentum per second of a flowing fluid (or momentum flux) is equal to the product of mass per second and the velocity of the flow. Mathematically, the momentum per second of a flowing fluid (compressible or incompressible) is

$$=\rho AV\times V, \quad \text{where } \rho AV = \text{Mass per second.}$$

The term ρAV is constant at every section of flow due to continuity equation. This means the momentum per second at any section is equal to the product of a constant quantity and the velocity. This also implies that momentum per second is independent of compressible effect. Hence the momentum equation for incompressible and compressible fluid is the same. The momentum equation for compressible fluid for any direction may be expressed as,

Net force in the direction of S= Rate of change of momentum in the direction of S

$$= \text{Mass per second [change of velocity]}$$

$$= \rho AV[V_2 - V_1] \quad \text{...(12.14)}$$

where V_2 = Final velocity in the direction of S,

V_1 = Initial velocity in the direction of S.

▶12.4 VELOCITY OF SOUND OR PRESSURE WAVE IN A FLUID

The disturbance in a solid, liquid or gas is transmitted from one point to the other. The velocity with which the disturbance is transmitted depends upon the distance between the molecules of the medium. In case of solids, molecules are closely packed and hence the disturbance is transmitted instantaneously. In case of liquids and gases (or fluids) the molecules are relatively apart. The disturbance will be transmitted from one molecule to the next molecule. But in case of fluids, there is some distance between two adjacent molecules. Hence each molecule will have to travel a certain distance before it can transmit the disturbance. Thus the velocity of disturbance in case of fluids will be less than the velocity of the disturbance in solids.

The distance between the molecules is related with the density, which in turn depends upon pressure in case of fluids. Hence the velocity of disturbance depends upon the changes of pressure and density of the fluid.

12.4.1 Expression for Velocity of Sound Wave in a Fluid. The disturbance creates the pressure waves in a fluid. These pressure waves travel with a velocity of sound waves in all directions. But for the sake of simplicity, one-dimensional case will be considered.

Fig. 12.2 shows the model for one-dimensional propagation of the pressure waves. It is a right long pipe of uniform cross-sectional area, fitted with a piston. Let the pipe is filled with a compressible fluid, which is at rest initially. The piston is moved towards right and a disturbance is created in the fluid. This disturbance is in the form of pressure wave, which travels in the fluid with a velocity of sound wave.

Let A = Cross-sectional area of the pipe

V = Velocity of piston

p = Pressure of the fluid in pipe before the movement of the piston

ρ = Density of fluid before the movement of the piston

dt = A small interval of time with which piston is moved

C = Velocity of pressure wave or sound wave travelling in the fluid

Distance travelled by the piston in time dt

$$= \text{Velocity of piston} \times dt = V \times dt$$

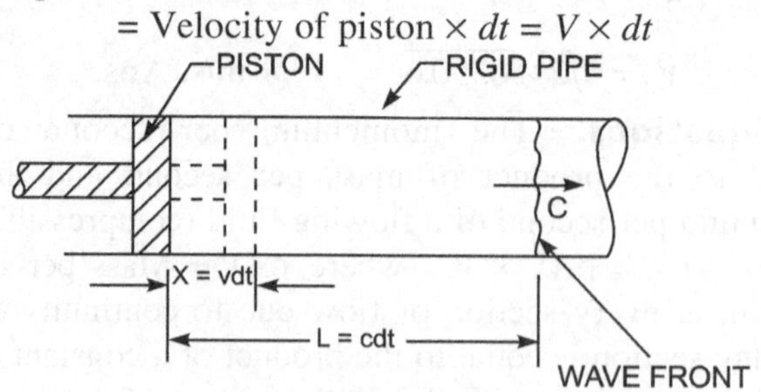

Fig. 12.2 *Propagation of pressure wave.*

Distance travelled by the pressure waves in time dt

$$= \text{Velocity of pressure wave} \times dt = C \times dt$$

As the value of C will be very large, hence $C \times dt$ will be more than $V \times dt$. For the time interval 'dt', the pressure wave has travelled a distance L and piston has moved through x. Thus in the length of the tube equal to $(L - x)$, the fluid will be compressed. Due to compression of the fluid, the pressure and density of the fluid will change.

Let $p + dp$ = Pressure after compression

$\rho + d\rho$ = Density after compression or the density of fluid in the length $(L - x)$

Now mass of fluid for a length 'L' before compression

$$= \rho \times \text{Volume of fluid upto length } L$$

$$= \rho \times A \times L = \rho \times A \times C \times dt \qquad (\because \; L = Cdt) \quad ...(i)$$

Mass of fluid after compression for length $(L - x)$

$$= \text{Density after compression} \times \text{Area} \times \text{Length}$$

$$= (\rho + d\rho)\, A \times (L - x)$$

$$= (\rho + d\rho)\, A \times (Cdt - Vdt) \qquad (\because \; L = Cdt,\; x = Vdt) \quad ...(ii)$$

From the continuity equation, we have

Mass of fluid before compression

$$= \text{Mass of fluid after compression}$$

$$\therefore \quad \rho ACdt = (\rho + d\rho)\, A \times (Cdt - Vdt) \text{ or } \rho ACdt = (\rho + d\rho)\, A \times dt\, (C - V)$$

Dividing by $A \times dt$, $\quad \rho C = (\rho + d\rho)(C - V) = \rho C - \rho V + Cd\rho - Vd\rho$

$$\therefore \quad Cd\rho = \rho C - \rho C + \rho V + Vd\rho = \rho V + Cd\rho. \quad ...(iii)$$

But the velocity of the piston, V, is very small as compared to the velocity of the pressure wave C. Also the value of $d\rho$ is very small. Hence the term $(V \times d\rho)$ will be very-very small and can be neglected. Hence equation (*iii*) becomes,

$$Cd\rho = \rho V \quad ...(iv)$$

Now when the piston is moved with a velocity V for time dt, the fluid which is at rest initially will move with a velocity equal to the velocity of the piston. Also the pressure of the fluid with increase from p to $p + dp$ due to the movement of the piston. Hence applying the impulse momentum equation, we get

Net force on the fluid = Rate of change of momentum

or $\quad (p + dp)\, A - p \times A$ = Mass per second [change of velocity of fluid]

or
$$dp \times A = \frac{\text{Total mass}}{\text{Time}} [V - 0] = \frac{\rho AL}{dt} [V - 0]$$

$$= \frac{\rho ACdt}{dt} [V - 0] \qquad (\because \; L = Cdt)$$

$$= \rho AC\, [V - 0] = \rho ACV \quad \text{or} \quad dp = \frac{\rho ACV}{A} = \rho CV$$

or $$C = \frac{dp}{\rho V} \qquad \text{...}(v)$$

Multiplying equations (*iv*) and (*v*), we get

$$C^2 d\rho = \rho V \times \frac{dp}{\rho V} = dp$$

$$C^2 = \frac{dp}{d\rho}$$

$$\therefore \qquad C = \sqrt{\frac{dp}{d\rho}} \qquad \text{...(12.15)}$$

Hence equation (12.15) gives the velocity of sound wave which is the square root of the ratio of change of pressure to the change of density of a fluid due to disturbance.

12.4.2 Velocity of Sound in Terms of Bulk Modulus. Bulk modulus K is defined as

$$K = \frac{\text{Increase in pressure}}{\frac{\text{Decrease in volume}}{\text{Original volume}}} = \frac{dp}{-\left(\frac{d\forall}{\forall}\right)} \qquad \text{...}(vi)$$

where $d\forall$ = Decrease in volume, $\forall$ = Original volume.

Negative sign is taken as with the increase of pressure, volume decreases.

Now we know mass of a fluid is constant. Hence

$$\rho \times \text{Volume} = \text{Constant} \qquad (\because \ \text{Mass} = \rho \times \text{Volume})$$

or $$\rho \times \forall = \text{Constant}$$

Differentiating the above equation (ρ and $\forall$ are variables),

$$\rho d\forall + \forall d\rho = 0 \quad \text{or} \quad \rho d\forall = -\forall d\rho \text{ or } \frac{d\forall}{\forall} = \frac{d\rho}{\rho}$$

Substituting the value $\left(-\frac{d\forall}{\forall}\right)$ in equation (*vi*), we get

$$K = \frac{dp}{\frac{d\rho}{\rho}} = \rho\frac{dp}{d\rho} \text{ or } \frac{dp}{d\rho} = \frac{K}{\rho}$$

From equation (12.15), the velocity of sound wave is

$$C = \sqrt{\frac{dp}{d\rho}} = \sqrt{\frac{K}{\rho}} \qquad \text{...(12.16)}$$

Equation (12.16) gives the velocity of sound wave in terms of bulk modulus and density. This equation is applicable for liquids and gases.

12.4.3 Velocity of Sound for Isothermal Process. Isothermal process is given by the equation (12.3), as

$$\frac{p}{\rho} = \text{Constant or } p\rho^{-1} = \text{Constant}$$

Differentiating the above equation (p and ρ both are variable),

$$p(-1)\,\rho^{-2}\,d\rho + \rho^{-1}\,dp = 0$$

Dividing by ρ^{-1}, we get $-\,p\rho^{-1}\,d\rho + dp = 0$ or $\dfrac{-p}{\rho}\,d\rho + dp = 0$

$$\therefore \qquad dp = \frac{p}{\rho}\,d\rho \text{ or } \frac{dp}{d\rho} = \frac{p}{\rho} = RT \qquad \left(\because \text{ From equation of state } \frac{p}{\rho} = RT\right)$$

Substituting the value of $\dfrac{dp}{d\rho}$ in equation (12.15), $C = \sqrt{\dfrac{p}{\rho}} = \sqrt{RT}$ ...(12.17)

12.4.4 Velocity of Sound for Adiabatic Process. Adiabatic process is given by the equation (12.4), as

$$\frac{p}{\rho^k} = \text{Constant or } p\rho^{-k} = \text{Constant}$$

Differentiating the above equation, we get

$$p(-k)\,\rho^{-k-1}\,d\rho + \rho^{-k}\,dp = 0$$

Dividing by ρ^{-k}, we get $-\,pk\rho^{-1}\,d\rho + dp = 0$ or $dp = \dfrac{pk}{\rho}\,d\rho$

$$\therefore \qquad \frac{dp}{d\rho} = \frac{p}{\rho}\,k = RTk \qquad \left(\because \frac{p}{\rho} = RT\right)$$

$$= kRT$$

Substituting the value of $\dfrac{dp}{d\rho}$ in equation (12.15), we get $C = \sqrt{kRT}$. ...(12.18)

Note 1. For the propagation of the minor disturbances through air, the process is assumed to be adiabatic. The velocity of the disturbances (pressure waves) through air is very high and hence there is no time for any appreciable heat transfer.

2. Isothermal process is considered for the calculation of the velocity of the sound waves (or pressure waves) only when it is given in the numerical problem that process is isothermal. If no process is mentioned, it is assumed to be adiabatic.

▶12.5 MACH NUMBER

In Chapter 6, Mach number has been defined as the square root of the ratio of the inertia force of a flowing fluid to the elastic force.

$$\therefore \quad \text{Mach number} \qquad = M = \sqrt{\frac{\text{Inertia force}}{\text{Elastic force}}} = \sqrt{\frac{\rho A V^2}{KA}}$$

$$= \sqrt{\frac{V^2}{K/\rho}} = \frac{V}{\sqrt{K/\rho}} = \frac{V}{C} \qquad \left[\because \sqrt{\frac{K}{\rho}} = C \text{ from equation (12.16)}\right]$$

Thus Mach number $= M$

$$= \frac{\text{Velocity of fluid or body moving in fluid}}{\text{Velocity of sound in the fluid}}$$

$$= \frac{V}{C}. \qquad \text{...(12.19)}$$

For the compressible fluid flow, Mach number is an important non-dimensional parameter. On the basis of the Mach number, the flow is defined as :

1. Sub-sonic flow, 2. Sonic flow, and 3. Super-sonic flow.

1. Sub-sonic Flow. A flow is said sub-sonic flow if the Mach number is less than 1.0 (or $M < 1$) which means the velocity of flow is less than the velocity of sound wave (or $V < C$).

2. Sonic Flow. A flow is said to be sonic flow if the Mach number (M) is equal to 1.0. This means that when the velocity of flow V is equal to the velocity of sound C, the flow is said sonic flow.

3. Super-sonic Flow. A flow is said to be super-sonic flow if the Mach number is greater than 1.0 (or $M > 1$). This means that when velocity of flow V is greater than the velocity of sound wave, the flow is said to be super-sonic flow.

Problem 12.4 *Find the sonic velocity for the following fluids :*

(i) Crude oil of sp. gr. 0.8 and bulk modulus 153036 N/cm².

(ii) Mercury having a bulk modulus of 2648700 N/cm².

(A.M.I.E., Winter 1975 ; Delhi University, 1982)

Solution. Given :

(*i*) For Crude oil, sp. gr. = 0.8

∴ Density, $\rho = 0.8 \times 1000 = 800 \text{ kg/m}^3$

Bulk modulus, $K = 153036 \text{ N/cm}^2 = 153036 \times 10^4 \text{ N/m}^2$

Using equation (12.16) for sonic velocity, as

$$C = \sqrt{\frac{K}{\rho}} = \sqrt{\frac{153036 \times 10^4}{800}} = 1383.09 \simeq \mathbf{1383 \text{ m/s. Ans.}}$$

(*ii*) For Mercury, sp. gr. = 13.6

∴ Density of Mercury, $\rho = 13.6 \times 1000 \text{ kg/m}^3$

Bulk modulus, $K = 2648700 \text{ N/cm}^2 = 2648700 \times 10^4 \text{ N/m}^2$

The sonic velocity, C is given by $C = \sqrt{\frac{K}{\rho}} = \sqrt{\frac{2648700 \times 10^4}{13.6 \times 1000}} = \mathbf{1395.55 \text{ m/s. Ans.}}$

Problem 12.5 *Find the speed of the sound wave in air at sea-level where the pressure and temperature are 10.1043 N/cm² (abs.) and 15°C respectively. Take $R = 287$ J/kg K and $k = 1.4$.*

Solution. Given :

Pressure, $p = 10.1043 \text{ N/cm}^2 = 10.1043 \times 10^4 \text{ N/m}^2$

Temperature, $t = 15°C$

$\therefore$ $T = 273 + 15 = 288 \text{ K}, R = 287 \text{ J/kg K}, k = 1.4$

For adiabatic process, the velocity of sound is given by equation (12.18), as

$$C = \sqrt{kRT} = \sqrt{1.4 \times 287 \times 288} = \mathbf{340.17 \text{ m/s. Ans.}}$$

Problem 12.6 *Calculate the Mach number at a point on a jet propelled aircraft, which is flying at 1100 km/hour at sea-level where air temperature is 20°C. Take k = 1.4 and R = 287 J/kg K.*

Solution. Given :

Speed of aircraft, $V = 1100 \text{ km/hour} = \dfrac{1100 \times 1000}{60 \times 60} = 305.55 \text{ m/s}$

Temperature, $t = 20°C$

$\therefore$ $T = 273 + 20 = 293, k = 1.4, R = 287 \text{ J/kg K}$

Using equation (12.18), the velocity of sound is

$$C = \sqrt{kRT} = \sqrt{1.4 \times 287 \times 293} = 343.11 \text{ m/s}$$

Mach number is given by equation (12.19) as

$$M = \frac{V}{C} = \frac{305.55}{343.11} = \mathbf{0.89. \text{ Ans.}}$$

Problem 12.7 *An aeroplane is flying at an height of 15 km where the temperature is – 50°C. The speed of the plane is corresponding to M = 2.0. Assuming k = 1.4 and R = 287 J/kg K, find the speed of the plane.*

Solution. Given :

Height of the plane, $Z = 15 \text{ km}$ (Extra Data)

Temperature, $t = -50°C$

$\therefore$ $T = -50 + 273 = 223°C$

Mach number, $M = 2.0, k = 1.4, R = 287 \text{ J/kg K.}$

Using equation (12.18), we get the velocity of sound as

$$C = \sqrt{kRT} = \sqrt{1.4 \times 287 \times 223} = 299.33 \text{ m/s}$$

Using equation (12.19), we have $M = \dfrac{V}{C}$ or $2.0 = \dfrac{V}{299.33}$

$\therefore$ $V = 2.0 \times 299.33 = 598.66 \text{ m/s}$

$$= \frac{598.66 \times 60 \times 60}{1000} = \mathbf{2155.17 \text{ km/hour. Ans.}}$$

▶12.6 PROPAGATION OF PRESSURE WAVES (OR DISTURBANCES) IN A COMPRESSIBLE FLUID

Whenever any disturbance is produced in a compressible fluid, the disturbance is propagated in all direction with a velocity of sound (*i.e.*, equal to C). The nature of propagation of the disturbance depends upon the Mach number. Let us consider a small projectile moving from left to right in a straight line in a stationary fluid. Due to the movement of the projectile, the disturbances will be created in the fluid. This disturbance will be moving in all directions with a velocity C.

Hence let V = Velocity of the projectile,

C = Velocity of pressure wave or disturbance created in the fluid.

Let us find the nature of propagation of the disturbance for different Mach numbers.

1st Case : When M < 1. When Mach number is less than 1.0, the flow is called sub-sonic flow. For $M < 1$ means $\frac{V}{C} < 1$ or $V < C$. To find the nature of propagation for this case, let V = One unit and $C = 2$ unit, so that $\frac{V}{C} = \frac{1}{2}$ which is less than 1.0. Let the projectile is at A and is moving towards right. Let in 4 seconds the projectile reaches to the position B. At A, the point 4 is also marked. The position of the projectile after 1 sec, 2 sec, 3 sec and 4 sec along the lines are shown by the points 3, 2, 1 and B respectively. The projectile moves from A to B in 4 seconds and hence the distance $AB = 4 \times V = 4 \times 1 = 4$ units. The disturbance created at A in 4 seconds will move a distance $= 4C = 4 \times 2 = 8$ units in all directions. Hence taking A as centre and radius equal to 8 units, a circle is drawn. This circle gives the position of disturbance after 4 seconds. When the projectile is at point 3, it will reach B in three seconds and distance $3B = 3 \times V = 3 \times 1 = 3$ units. But the disturbance created at point 3 in three seconds will move a distance having a radius $= 3 \times C = 3 \times 2 = 6$ units. Similarly at point 2, the disturbance will have a radius $= 2 \times C = 4$ units and at point 1, the disturbance will have a radius $= 1 \times C = 1 \times 2 = 2$ units. This is shown in Fig. 12.3 (*a*).

As in this case $V < C$, the pressure wave is always ahead of the projectile and point B is inside the sphere of radius 8 units.

2nd Case : When M = 1. When $M = 1$, the flow is known as sonic flow. In this case, the disturbance always travels with the projectile as shown in Fig. 12.3 (*b*). Let $V = 1$ unit, and $C = 1$ unit so that $M = \frac{V}{C} = \frac{1}{1} = 1.0$. Let the projectile moves from A to B in 4 seconds. The disturbance created at A in 4 seconds will move

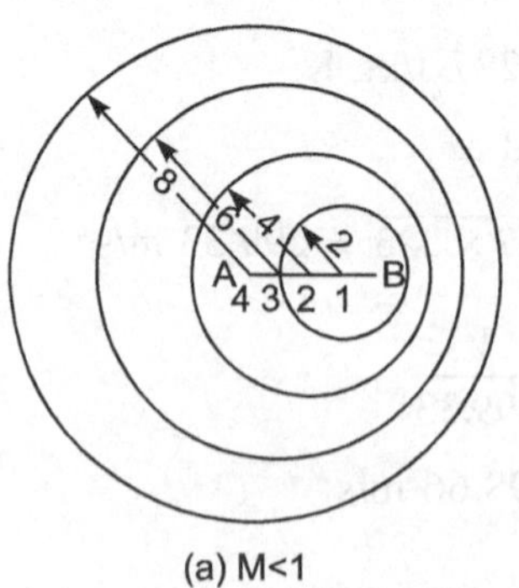

(a) M<1

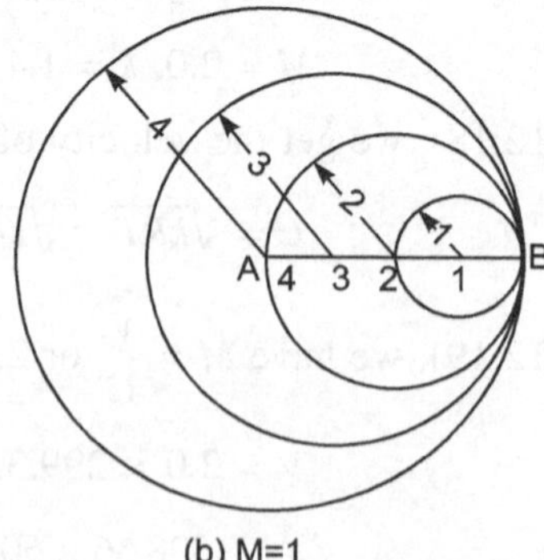

(b) M=1

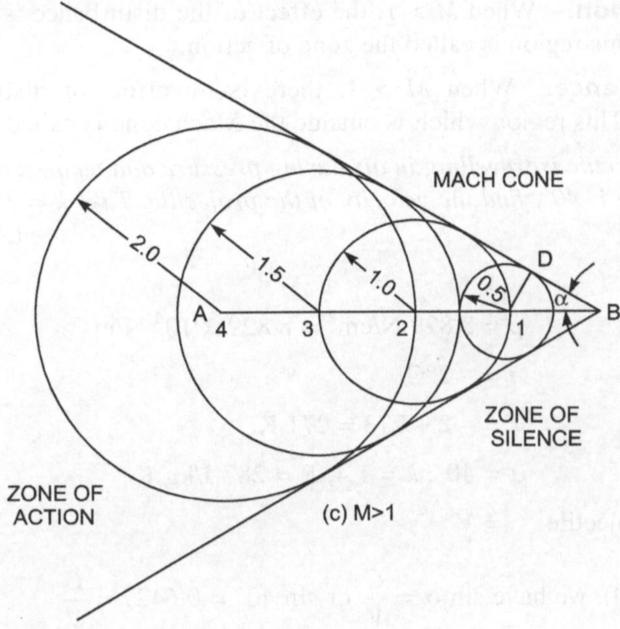

Fig. 12.3 *Propagation of disturbance for different Mach numbers.*

a distance having radius = 4 × C = 4 × 1 = 4 units in all directions. The projectile from point 3 will move to position B in three seconds. The disturbance created at point 3, will move a distance having radius = 3 × C = 3 × 1 = 3 units in all directions in three seconds. Similarly at the point 2 and point 1, the disturbance created at these points will move a distance having radius 2 and 1 in all directions respectively.

3rd Case : When M > 1. When $M > 1$, the flow is known as supersonic flow. Let $V = 1$ units and $C = 0.5$ unit so that $M = \frac{V}{C} = \frac{1}{0.5} = 2.0$, which is greater than unity. Let the projectile moves from A to B in 4 seconds. The distance travelled by the projectile in 4 seconds = 4 × V = 4 × 1 = 4 units. Hence, take AB = 4 units. The disturbance created at A will move in all directions and in 4 seconds, the radius of disturbance will be equal to 4 × C = 4 × 0.5 = 2 units. Hence taking A as centre draw a circle with radius equal to 2 units. After one second from A, the projectile will be at point 3 and distance $A3 = V \times 1 = 1 \times 1 = 1$ unit. The projectile form point 3 will reach point B in three seconds. Hence the disturbance created at point 3 will move in all directions and in three seconds, the radius of disturbance from point 3 will be equal to 3 × C = 3 × 0.5 = 1.5 units. Similarly the radius of disturbance at point 2 and 1 will be 2 × C = 2 × 0.5 = 1 unit and 1 × C = 1 × 0.5 = 0.5 unit respectively as shown in Fig. 12.3 (*c*). In this case the sphere of propagation of disturbance always lags behind the projectile. If we draw a tangent to the different circles which represent the propagated spherical waves on both sides, we shall get a cone with vertex at B. This cone is known as **Mach Cone.**

12.6.1 Mach Angle. This is defined as the half of the angle of the Mach cone. In Fig. 12.3 (*c*), angle α is known as Mach angle. In the Δ1BD of Fig. 12.3 (*c*), the distance 1B = Velocity of projectile = V, the distance 1D = Velocity of sound wave = C. Hence we have

$$\sin \alpha = \frac{ID}{IB} = \frac{C}{V} = \frac{1}{V/C} = \frac{1}{M} \qquad \text{...(12.20)}$$

12.6.2 Zone of Action. When $M > 1$, the effect of the disturbance is felt only in the region inside the Mach cone. This region is called the zone of action.

12.6.3 Zone of Silence. When $M > 1$, there is no effect of disturbance in the region outside the Mach cone. This region which is outside the Mach cone is called zone of silence.

Problem 12.8 *A projectile is travelling in air having pressure and temperature as 8.829 N/cm² and – 2°C. If the Mach angle is 40°, find the velocity of the projectile. Take k = 1.4 and R = 287 J/kg K.*

(A.M.I.E., Winter, 1975)

Solution. Given :

Pressure of air, $p = 8.829 \text{ N/cm}^2 = 8.829 \times 10^4 \text{ N/m}^2$

Temperature of air, $t = -2°\text{C}$

$\therefore$ $T = -2 + 273 = 271 \text{ K}$

Mach angle, $\alpha = 40°$, $k = 1.4$, $R = 287 \text{ J/kg K}$

Let the velocity of projectile $= V$

Using equation (12.20), we have $\sin \alpha = \dfrac{C}{V}$ or $\sin 40° = 0.6427 = \dfrac{C}{V}$

The velocity of sound, C is given by equation (12.18) as

$$C = \sqrt{kRT} = \sqrt{1.4 \times 287 \times 271} = 329.98 \text{ m/s} \simeq 330 \text{ m/s}$$

$$\therefore \quad \sin 40° = 0.6427 = \frac{C}{V} = \frac{330}{V}$$

$$\therefore \quad V = \frac{330}{0.6427} = \textbf{513 m/s. Ans.}$$

Problem 12.9 *A projectile travels in air of pressure 10.1043 N/cm² at 10°C at a speed of 1500 km/hour. Find the Mach number and the Mach angle. Take k = 1.4 and R = 287 J/kg K.*

Solution. Given :

Pressure, $p = 10.1043 \text{ N/cm}^2 = 10.1043 \times 10^4 \text{ N/cm}^2$

Temperature, $t = 10°\text{C}$

$\therefore$ $T = 10 + 273 = 283°\text{K}$

Speed of projectile, $V = 1500 \text{ km/hour} = \dfrac{1500 \times 1000}{60 \times 60} \text{ m/s} = 416.67 \text{ m/s}$

$k = 1.4$, $R = 287 \text{ J/kg K}$

For adiabatic process, the velocity of sound is given by

$$C = \sqrt{kRT} = \sqrt{1.4 \times 287 \times 283} = 337.20 \text{ m/s}$$

$$\therefore \quad \text{Mach number,} \quad M = \frac{V}{C} = \frac{416.67}{337.20} = \textbf{1.235. Ans.}$$

∴ Mach angle is obtained from equation (12.20) as

$$\sin\alpha = \frac{C}{V} = \frac{1}{M} = \frac{1}{1.235} = 0.8097$$

∴ Mach angle, $\alpha = \sin^{-1} 0.8097 =$ **54.06°. Ans.**

Problem 12.10 *Find the velocity of bullet fired in standard air if the Mach angle is 30°. Take R = 287.14 J/kg K and k = 1.4 for air. Assume temperature as 15°C.*

(V.T.U. 2000 ; A.M.I.E., Winter, 1974)

Solution. Given :

Mach angle $\alpha = 30°$

$R = 287.14$ J/kg K

$k = 1.4$

Temperature, $t = 15°C$

∴ $T = 15 + 273 = 288°K$

Velocity of sound is given by equation (12.18) as

$$C = \sqrt{kRT} = \sqrt{1.4 \times 287.14 \times 288} = 340.25 \text{ m/s}$$

Using the relation, $\sin\alpha = \frac{C}{V}$ given by equation (12.20)

$$\sin 30° = \frac{340.25}{V}$$

∴ $$V = \frac{340.25}{\sin 20} = \textbf{680.50 m/s. Ans.}$$

Problem 12.10(A) *An air plane is flying at an altitude of 15 km where the temperature is – 50°C. The speed of the plane corresponds to Mach Number of 1.6. Assuming k = 1.4 and R = 287 J/kg K for air, find the speed of the plane and Mach angle α.*

Solution. Given :

Height of plane, $H = 15$ km $= 15 \times 1000 = 15000$ m

Temperature, $t = -50°C$ ∴ $T = -50 + 273 = 223°K$

Mach number, $M = 1.6$, $k = 1.4$ and $R = 287$ J/kg K

Find : (*i*) Speed of plane (V)

(*ii*) Mach angle, α

Velocity of sound wave is given as

$$C = \sqrt{kRT} = \sqrt{1.4 \times 287 \times 223} = 229.33 \text{ m/s}$$

(*i*) *Mach angle,* α

Using the relation for Mach angle, we get

$$\sin\alpha = \frac{C}{V} = \frac{1}{\left(\frac{V}{C}\right)} = \frac{1}{M} = \frac{1}{1.6} = 0.625$$

∴ $\alpha = \sin^{-1} 0.625 =$ **38.68°. Ans.**

(ii) Speed of plane, V

We know,
$$M = \frac{V}{C}$$

or
$$1.6 = \frac{V}{299.33}$$

$\therefore$
$$V = 1.6 \times 299.33 = 478.928 \text{ m/s}$$

$$= \frac{478.928}{1000} \times 3600 = \mathbf{1724.14 \text{ km/hr. Ans.}}$$

▶12.7 STAGNATION PROPERTIES

When a fluid is flowing past an immersed body, and at a point on the body if the resultant velocity becomes zero, the values of pressure, temperature and density at that point are called stagnation properties. The point is called the **stagnation point.** The values of pressure, density and temperature are called stagnation pressure, stagnation density and stagnation temperature respectively. They are denoted by p_s, ρ_s and T_s respectively.

12.7.1 Expression for Stagnation Pressure (p_s). Consider a compressible fluid flowing past an immersed body under frictionless adiabatic conditions as in Fig. 12.4. Consider two points 1 and 2 on a stream-line as shown in Fig. 12.4.

Let p_1 = Pressure of compressible fluid at point 1,

V_1 = Velocity of fluid at 1, and

ρ_1 = Density of fluid at 1,

p_2, V_2, ρ_2 = Corresponding values of pressure, velocity and density at point 2.

Applying Bernoulli's equation for adiabatic flow given by equation (12.13) at point 1 and 2, we get

$$\left(\frac{k}{k-1}\right)\frac{p_1}{\rho_1 g} + \frac{V_1^2}{2g} + z_1 = \left(\frac{k}{k-1}\right)\frac{p_2}{\rho_2 g} + \frac{V_2^2}{2g} + z_2$$

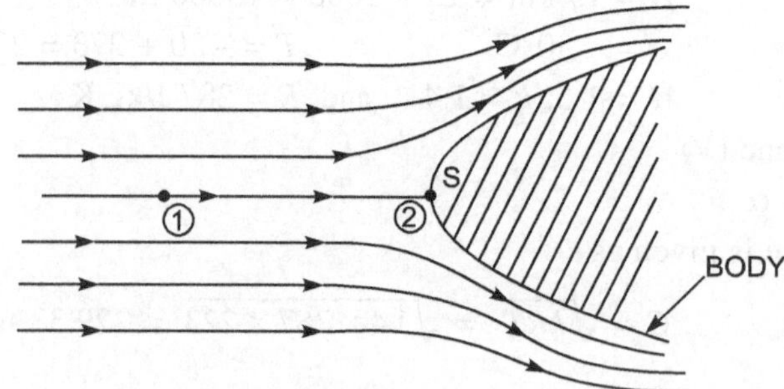

Fig. 12.4 *Stagnation properties.*

But
$$z_1 = z_2$$

$\therefore$
$$\left(\frac{k}{k-1}\right)\frac{p_1}{\rho_1 g} + \frac{V_1^2}{2g} = \left(\frac{k}{k-1}\right)\frac{p_2}{\rho_2 g} + \frac{V_2^2}{2g}$$

or $$\left(\frac{k}{k-1}\right)\frac{p_1}{\rho_1}+\frac{V_1^2}{2}=\left(\frac{k}{k-1}\right)\frac{p_2}{\rho_2}+\frac{V_2^2}{2} \qquad \left(\text{Cancelling } \frac{1}{g}\right)$$

Point 2 is a stagnation point. Hence velocity will become zero at stagnation point and pressure and density will be denoted by p_s and ρ_s.

$\therefore$ $$V_2 = 0,\ p_2 = p_s \text{ and } \rho_2 = \rho_s$$

Substituting these values in the above Bernoulli's equation,

$$\left(\frac{k}{k-1}\right)\frac{p_1}{\rho_1}+\frac{V_1^2}{2}=\left(\frac{k}{k-1}\right)\frac{p_s}{\rho_s}+O$$

or $$\left(\frac{k}{k-1}\right)\left[\frac{p_1}{\rho_1}-\frac{p_s}{\rho_s}\right]=-\frac{V_1^2}{2} \text{ or } \left(\frac{k}{k-1}\right)\frac{p_1}{\rho_1}\left[1-\frac{p_s}{\rho_s}\times\frac{\rho_1}{p_1}\right]=-\frac{V_1^2}{2}$$

or $$\left(\frac{k}{k-1}\right)\frac{p_1}{\rho_1}\left[1-\frac{p_s}{p_1}\times\frac{\rho_1}{\rho_s}\right]=-\frac{V_1^2}{2} \qquad \text{...}(i)$$

But for adiabatic process from equation (12.4), we have

$$\frac{p}{\rho^k}=\text{constant or } \frac{p_1}{\rho_1^k}=\frac{p_s}{\rho_s^k} \text{ or } \frac{p_1}{p_s}=\frac{\rho_1^k}{\rho_s^k} \text{ or } \left(\frac{\rho_1}{\rho_s}\right)=\left(\frac{p_1}{p_s}\right)^{\frac{1}{k}} \qquad \text{...}(ii)$$

Substituting the value of $\frac{\rho_1}{\rho_s}$ in equation (i), we get

$$\left(\frac{k}{k-1}\right)\frac{p_1}{\rho_1}\left[1-\frac{p_s}{p_1}\times\left(\frac{p_1}{p_s}\right)^{\frac{1}{k}}\right]=-\frac{V_1^2}{2}$$

or $$\left(\frac{k}{k-1}\right)\frac{p_1}{\rho_1}\left[1-\frac{p_s}{p_1}\times\left(\frac{p_s}{p_1}\right)^{-\frac{1}{k}}\right]=-\frac{V_1^2}{2}$$

or $$\left(\frac{k}{k-1}\right)\frac{p_1}{\rho_1}\left[1-\left(\frac{p_s}{p_1}\right)^{1-\frac{1}{k}}\right]=-\frac{V_1^2}{2}$$

or $$\left[1-\left(\frac{p_s}{p_1}\right)^{\frac{k-1}{k}}\right]=-\frac{V_1^2}{2}\times\left(\frac{k-1}{k}\right)\frac{\rho_1}{p_1}$$

or $$1+\frac{V_1^2}{2}\left(\frac{k-1}{k}\right)\frac{\rho_1}{p_1}=\left(\frac{p_s}{p_1}\right)^{\frac{k-1}{k}} \qquad \text{...}(iii)$$

Now for adiabatic process, the velocity of sound is given by equation (12.18) as

$$C = \sqrt{kRT} = \sqrt{k\frac{p}{\rho}} \qquad \left(\because \frac{p}{\rho} = RT\right)$$

For the point 1,
$$C_1 = \sqrt{k\frac{p_1}{\rho_1}} \text{ or } C_1^{\ 2} = k\frac{p_1}{\rho_1}$$

Substituting the value of $\frac{k\,p_1}{\rho_1} = C_1^{\ 2}$ in equation (*iii*)

$$1 + \frac{V_1^2}{2}(k-1) \times \frac{1}{C_1^2} = \left(\frac{p_s}{p_1}\right)^{\left(\frac{k-1}{k}\right)} \text{ or } 1 + \frac{V_1^2}{2C_1^2} \times (k-1) = \left(\frac{p_s}{p_1}\right)^{\left(\frac{k-1}{k}\right)}$$

or
$$1 + \frac{M_1^2}{2}(k-1) = \left(\frac{p_s}{p_1}\right)^{\left(\frac{k-1}{k}\right)} \qquad \left(\because \frac{V_1^2}{C_1^2} = M_1^2\right)$$

or
$$\left(\frac{p_s}{p_1}\right)^{\frac{k-1}{k}} = \left[1 + \frac{(k-1)}{2}M_1^2\right] \text{ or } \frac{p_s}{p_1} = \left[1 + \frac{k-1}{2}M_1^2\right]^{\left(\frac{k-1}{k}\right)} \qquad ...(iv)$$

$\therefore$
$$p_s = p_1\left[1 + \frac{k-1}{2}M_1^2\right]^{\left(\frac{k}{k-1}\right)} \qquad ...(12.21)$$

Equation (12.21) gives the value of stagnation pressure.

In equation (12.21), for $M < 1$, the term $\frac{k-1}{2}M_1^{\ 2}$ will be less than 1 and hence the R.H.S. of this equation can be expressed by Binomial theorem as

$$p_s = p_1\left[1 + \left(\frac{k}{k-1}\right) \times \left(\frac{k-1}{2}.M_1^2\right) + \left(\frac{k}{k-1}\right)\left(\frac{k}{k-1} - 1\right)\left(\frac{k-1}{2}.M_1^2\right)^2 \Big/ 2!\right.$$
$$\left. + \left(\frac{k}{k-1}\right)\left(\frac{k}{k-1} - 1\right)\left(\frac{k}{k-1} - 2\right)\left(\frac{k-1}{2}.M_1^2\right)^3 \Big/ 3! + ...\right]$$

$$= p_1\left[1 + \frac{k}{2}M_1^2 + \frac{k}{8}M_1^4 + \frac{k(2-k)}{48}M_1^6 + ...\right]$$

$$= p_1 + p_1\left[\frac{k}{2}M_1^2 + \frac{k}{8}M_1^4 + \frac{k(2-k)}{48}M_1^6 + ...\right]$$

or
$$\frac{p_s - p_1}{p_1} = \left[\frac{k}{2}M_1^2 + \frac{k}{8}M_1^4 + \frac{k(2-k)}{48}M_1^6 + ...\right]$$

$$= \frac{k}{2} M_1^2 \left[1 + \frac{1}{4} M_1^2 + \frac{(2-k)}{24} M_1^4 + \ldots\ldots \right]$$

$$= \frac{k}{2} \frac{V_1^2}{C_1^2} \left[1 + \frac{1}{4} M_1^2 + \frac{(2-k)}{24} M_1^4 + \ldots\ldots \right]$$

$$= \frac{k}{2} \frac{V_1^2}{\left(k \frac{p_1}{\rho_1} \right)} \left[1 + \frac{1}{4} M_1^2 + \frac{(2-k)}{24} M_1^4 + \ldots\ldots \right] \qquad \left(\because C_1 = \sqrt{\frac{kp_1}{\rho_1}} \right)$$

$$= \frac{1}{2} . \frac{\rho_1 V_1^2}{p_1} \left[1 + \frac{1}{4} M_1^2 + \frac{(2-k)}{24} M_1^4 + \ldots\ldots \right]$$

or
$$(p_s - p_1) = \frac{1}{2} . \rho_1 . V_1^2 \left[1 + \frac{1}{4} M_1^2 + \frac{(2-k)}{24} M_1^4 + \ldots\ldots \right] \qquad \ldots(12.21A)$$

From the above equation, it is clear that when the approaching velocity V_1 is small compared with the velocity of sound wave C_1, then M_1 will be very small and the term $1 + \frac{1}{4} {M_1}^2 + \frac{(2-k)}{24} {M_1}^4 + \ldots\ldots$ will be nearly equal to 1.

Hence equation (12.21*A*) becomes as

$$\therefore \qquad p_s - p_1 = \frac{1}{2} . \rho_1 \times {V_1}^2 \text{ or } p_s = p_1 + \frac{1}{2} \rho_1 {V_1}^2$$

But when approaching velocity becomes high then M_1 is not small and equation (12.21*A*) is expressed as

$$\frac{(p_s - p_1)}{\frac{1}{4} \times \rho_1 \times V_1^2} = 1 + \frac{1}{4} M_1^2 + \frac{(2-k)}{24} {M_1}^2 + \ldots\ldots \qquad \ldots(12.21B)$$

12.7.2 Expression for Stagnation Density (ρ_s). From equation (*ii*), we have

$$\left(\frac{\rho_1}{\rho_s} \right) = \left(\frac{p_1}{p_s} \right)^{\frac{1}{k}} \text{ or } \left(\frac{\rho_s}{\rho_1} \right) = \left(\frac{p_s}{p_1} \right)^{\frac{1}{k}} \qquad \text{(Taking reciprocal)}$$

$$\therefore \qquad \rho_s = \rho_1 \left[\frac{p_s}{p_1} \right]^{\frac{1}{k}}$$

Substituting the value of $\left(\frac{p_s}{p_1} \right)$ from equation (*iv*),

$$\rho_s = \rho_1 \left[\left(1 + \frac{k-1}{2} M_1^2 \right)^{\frac{k}{k-1}} \right]^{\frac{1}{k}} \text{ or } \rho_s = \rho_1 \left[1 + \frac{k-1}{2} M_1^2 \right]^{\frac{1}{k-1}} \qquad \ldots(12.22)$$

12.7.3 Expression for Stagnation Temperature (T_s). Equation of state is given by equation (12.2) as $\frac{p}{\rho} = RT$

For the stagnation point, we have equation of state as $\frac{p_s}{\rho_s} = RT$

$$\therefore \quad T_s = \frac{1}{R_s}\frac{p_s}{\rho_s}$$

Substituting the value of p_s and ρ_s from equations (12.21) and (12.22), we have

$$T_s = \frac{1}{R}\frac{p_1\left[1+\left(\frac{k-1}{2}\right)M_1^2\right]^{\left(\frac{k}{k-1}\right)}}{\rho_1\left[1+\left(\frac{k-1}{2}\right)M_1^2\right]^{\frac{1}{k-1}}}$$

$$= \frac{1}{R}\frac{p_1}{\rho_1}\left[1+\left(\frac{k-1}{2}\right)M_1^2\right]^{\left(\frac{k}{k-1}\right)-\left(\frac{1}{k-1}\right)}$$

$$= \frac{1}{R}\frac{p_1}{\rho_1}\left[1+\left(\frac{k-1}{2}\right)M_1^2\right]^{\left(\frac{k-1}{k-1}\right)} = \frac{p_1}{\rho_1 R}\left[1+\left(\frac{k-1}{2}\right)M_1^2\right]$$

$$= T_1\left[1+\left(\frac{k-1}{2}\right)M_1^2\right] \qquad \left(\because \frac{p_1}{\rho_1} = RT_1\right) \qquad \text{...(12.23)}$$

Problem 12.11 *Find the Mach number when an aeroplane is flying at 1100 km/hour through still air having a pressure of 7 N/cm² and temperature – 5°C. Wind velocity may be taken as zero. Take R = 287.14 J/kg K. Calculate the pressure, temperature and density of air at stagnation point on the nose of the plane. Take k = 1.4.* (A.M.I.E., May, 1975)

Solution. Given :

Speed of aeroplane, $V = 1100 \text{ km/hr} = \frac{1100 \times 1100}{60 \times 60} = 305.55 \text{ m/s}$

Pressure of air, $p_1 = 7 \text{ N/cm}^2 = 7 \times 10^4 \text{ N/m}^2$

Temperature, $t_1 = -5°\text{C}$

$\therefore$ $T_1 = -5 + 273 = 268°\text{K}$

$R = 287.14 \text{ J/kg K}$

$k = 1.4$

Using relation $C = \sqrt{kRT}$ for velocity of sound for adiabatic process, we have

$$C_1 = \sqrt{1.4 \times 287.14 \times 268} = 328.2 \text{ m/s}$$

$\therefore$ Mach number, $$M_1 = \frac{V_1}{C_1} = \frac{305.55}{328.20} = 0.9309 \simeq \textbf{0.931. Ans.}$$

Stagnation pressure, p_s. Using equation (12.21) for stagnation pressure,

$$p_s = p_1\left[1+\frac{k-1}{2}M_1^2\right]^{\frac{k}{k-1}}$$

$$= 7.0 \times 10^4\left[1+\frac{1.4-1.0}{2.0}\times(.931)^2\right]^{\frac{1.4}{1.4-1.0}}$$

$$= 7.0 \times 10^4\,[1 + .1733]^{\frac{1.4}{4}}$$

$$= 7.0 \times 10^4\,[1.1733]^{3.5} = 12.24 \times 10^4 \text{ N/m}^2 = \textbf{12.24 N/cm}^2\textbf{. Ans.}$$

Stagnation Temperature, T_s. Using equation (12.23) for stagnation temperature,

$$T_s = T_1\left[1+\frac{k-1}{2}M_1^2\right]$$

$$= 268\left[1+\frac{1.4-1.0}{2.0}\times(.931)^2\right] = 268\,[1.1733] = 314.44°\text{K}$$

$\therefore$ $$t_s = T_s - 273 = 314.43 - 273 = \textbf{41.44°C. Ans.}$$

Stagnation Density, ρ_s. Using equation of state (12.20) for stagnation density, $\frac{p_s}{\rho_s} = RT_s$

$\therefore$ $$\rho_s = \frac{p_s}{RT_s} \qquad ...(i)$$

In equation (*i*) given above, if R is taken as 287.14 J/kg K, then pressure should be taken in N/m^2 so that the value of ρ is in kg/m^3. Hence p_s = 12.24 × 10^4 N/m^2 and T_s = 314.44 K.

$\therefore$ $$\rho_s = \frac{12.24\times10^4}{287.14\times314.44} = \textbf{1.355 kg/m}^3\textbf{. Ans.}$$

Problem 12.12 *Calculate the stagnation pressure, temperature and density on the stagnation point on the nose of a plane, which is flying at 800 km/hour through still air having a pressure 8.0 N/cm^2 (abs.) and temperature – 10°C. Take R = 287 J/kg K and k = 1.4.*

Solution. Given :

Speed of plane, $$V = 800 \text{ km/hour} = \frac{800\times1000}{60\times60} = 222.22 \text{ m/s}$$

Pressure of air, $$p_1 = 8.0 \text{ N/cm}^2 = 8.0 \times 10^4 \text{ N/m}^2$$

Temperature, $$t_1 = -\,10°\text{C}$$

$\therefore$ $$T_1 = -\,10 + 273 = 263°\text{K}$$

$$R = 287 \text{ J/kg°K}$$

$$k = 1.4$$

For adiabatic flow, the velocity of sound is given by

$$C = \sqrt{kRT} = \sqrt{1.4 \times 287 \times 263} = 325.07 \text{ m/s}$$

$\therefore$ Mach number, $$M = \frac{V}{C} = \frac{222.22}{325.07} = 0.683.$$

This Mach number is the local Mach number and hence equal to M_1.

$$\therefore \qquad M_1 = 0.683$$

Using equation (12.21) for stagnation pressure,

$$p_s = p_1 \left[1 + \frac{k-1}{2} M_1^2\right]^{\left(\frac{k}{k-1}\right)} = 8.0 \times 10^4 \left[1 + \frac{1.4 - 1.0}{2.0} \times (.683)^2\right]^{\left(\frac{1.4}{1.4 - 1.0}\right)}$$

$$= 8.0 \times 10^4 \, [1.0933]^{3.5} = 10.93 \times 10^4 \text{ N/m}^2 = \mathbf{10.93 \text{ N/cm}^2. \text{ Ans.}}$$

Using equation (12.23) for stagnation temperature,

$$T_s = T_1 \left(1 + \frac{k-1}{2} M_1^2\right) = 263 \left(1 + \frac{1.4 - 1.0}{2.0} \times (.683)^2\right)$$

$$= 263 \, [1.0933] = 287.5 \text{ K}$$

$$\therefore \qquad t_s = T_s - 273 = 287.5 - 723 = \mathbf{14.5°C. \text{ Ans.}}$$

Using equation (12.2), $$\frac{p}{\rho} = RT$$

For stagnation point, $$\frac{p_s}{\rho_s} = RT_s \qquad \therefore \quad \rho_s = \frac{p_s}{RT_s}$$

As $R = 287$ J/kg K, the value of p_s should be taken in N/m^2 so that the value of ρ_s is obtained in kg/m^3.

$$\therefore \qquad p_s = 10.93 \times 10^4 \text{ N/m}^2$$

$\therefore$ Stagnation density, $$\rho_s = \frac{10.93 \times 10^4}{287 \times 287.5} = \mathbf{1.324 \text{ kg/m}^3. \text{ Ans.}}$$

▶12.8 AREA VELOCITY RELATIONSHIP FOR COMPRESSIBLE FLOW

The area velocity relationship for incompressible fluid is given by the continuity equation as

$$A \times V = \text{Constant.}$$

From the above equation, it is clear that with the increase of area, velocity decreases. But in case of compressible fluid, the continuity equation is given by, ρAV = Constant. ...(*i*)

From this relation, it is clear that with the change of area, both the velocity and density are affected. Hence to find the relation between area and velocity for compressible fluid we proceed as given below.

Differentiating equation (*i*), we get

$$\rho d\,(AV) + AVd\rho = 0 \quad \text{or} \quad \rho \, [AdV + VdA] + AVd\rho = 0$$

or $\rho AdV + \rho VdA + AVd\rho = 0.$

Dividing by ρAV, we get $\frac{dV}{V} + \frac{dA}{A} + \frac{d\rho}{\rho} = 0$ ...(*ii*)

The Euler's equation for compressible fluid is given by equation (12.8), as

$$\frac{dp}{\rho} + VdV + gdz = 0$$

Neglecting the Z term, the above equation is written as $\frac{dp}{\rho} + VdV = 0.$

This equation can also be written as $\frac{dp}{\rho} \times \frac{d\rho}{d\rho} + VdV = 0$ (Dividing and multiplying by $d\rho$)

or $$\frac{dp}{d\rho} \times \frac{d\rho}{\rho} + VdV = 0$$

But $\frac{dp}{d\rho} = C^2$ from equation (12.15)

Hence above equation becomes as

$$C^2 \frac{d\rho}{\rho} + VdV = 0 \quad \text{or} \quad C^2 \frac{d\rho}{\rho} = -VdV \quad \text{or} \quad \frac{d\rho}{\rho} = -\frac{VdV}{C^2}$$

Substituting the value of $\frac{d\rho}{\rho}$ in equation (*ii*), we get

$$\frac{dV}{V} + \frac{dV}{A} - \frac{VdV}{C^2} = 0 \quad \text{or} \quad \frac{dA}{A} = \frac{VdV}{C^2} - \frac{dV}{V} = \frac{dV}{V}\left[\frac{V^2}{C^2} - 1\right]$$

$$\therefore \quad \frac{dA}{A} = \frac{dV}{V}\left[M^2 - 1\right] \quad \text{...(12.24)}$$

Equation (12.24) gives the relationship between change of area with change of velocity for different Mach numbers. The followings are the important conclusions :

(*i*) For $M < 1$, the flow is sub-sonic and the right-hand side of the equation (12.24) is negative as $(M^2 - 1)$ is negative for the values of $M < 1$. Hence $\frac{dA}{A} > 0$, $\frac{dV}{V} < 0$. This means that with the increase of area, the velocity decreases and *vice-versa.*

(*ii*) For $M > 1$, the flow is super-sonic. The value of $(M^2 - 1)$ will be positive and hence right-hand side of equation (12.24) will be positive. Hence $\frac{dA}{A} > 0$ and also $\frac{dV}{V} > 0$. This means that with the increase of area, velocity also increases.

(*iii*) For $M = 1$, the flow is called sonic flow. The value of $(M^2 - 1)$ is zero. Hence right-hand side of the equation (12.24) will be zero. Hence $\frac{dA}{A} = 0$. This means area is constant.

▶ 12.9 PITOT-STATIC TUBE IN A COMPRESSIBLE FLOW

The pitot-static tube, when used for determining the velocity at any point in a compressible fluid, gives only the difference between the stagnation head and static head. From this difference, the velocity of the incompressible fluid at that point is obtained from the relation

$$V = \sqrt{2gh}, \text{ where } h = \text{Difference in two heads.}$$

But when the pitot-static tube is used for finding velocity at any point in a compressible fluid, the actual pressure difference shown by the gauges of the pitot-tube should be multiplied by a factor, for obtaining correct velocity at that point. The value of the factor depends upon the Mach number of the flow. Let us find an expression for the correction factor for *sub-sonic* flow.

At a point in pitot-static tube, the pressure becomes stagnation pressure, denoted by p_s. The expression for stagnation pressure, p_s is given by equation (12.21), as

$$p_s = p_1\left[1 + \frac{k-1}{2} M_1^2\right]^{\frac{k}{k-1}} \quad ...(i)$$

where p_1 = Pressure of fluid far away from stagnation point,
M_1 = Mach number at point 1, far away from stagnation point.

For $M < 1$, the term $\frac{k-1}{2} M_1{}^2$ will be less than 1 and hence the right-hand side of the equation (*i*) can be expanded by Binomial theorem* as

$$p_s = \left[1 + \frac{k-1}{2} M_1^2 \times \frac{k}{k-1} \times \frac{\left(\frac{k}{k-1}\right)\left(\frac{k}{k-1} - 1\right)}{2!} \times \left(\frac{k-1}{2} M_1^2\right)^2 + \frac{\left(\frac{k}{k-1}\right)\left(\frac{k}{k-1} - 1\right)\left(\frac{k}{k-1} - 2\right)\left(\frac{k-1}{2} M_1^2\right)^3}{3!} + ...\right]$$

$$= p_1\left[1 + \frac{k}{2} M_1^2 + \frac{k}{8} M_1^4 + \frac{k(2-k)}{48} M_1^6 + ...\right]$$

$$= p_1 + p_1\left[\frac{k}{2} M_1^2 + \frac{k}{8} M_1^4 + \frac{k(2-k)}{48} M_1^6 + ...\right]$$

$$p_s - p_1 = p_1 \times \frac{k}{2} M_1^2\left[1 + \frac{M_1^2}{4} + \frac{(2-k)}{24} M_1^4 + ...\right] \quad ...(ii)$$

But $\quad M_1{}^2 = \frac{V_1^2}{C_1^2}$, where $C_1{}^2 = \frac{kp_1}{\rho_1}$

* Binomial Theorem $(1+x)^x = 1 + nx + \frac{n(n-1)x^2}{2!} + \frac{n(n-1)(n-2)x^3}{3!} + \frac{n(n-1)(n-2)(n-3)x^4}{4!}$ .....

$$= \frac{V_1^2}{\frac{kp_1}{\rho_1}} = \frac{V_1^2 \rho_1}{kp_1}$$

Substituting the value of M_1^2 in equation (*ii*), we get

$$p_s - p_1 = p_1 \times \frac{k}{2} \times \frac{V_1^2 \rho_1}{kp_1}\left[1 + \frac{M_1^2}{4} + \frac{(2-k)}{24} M_1^4 + ...\right]$$

$$= \frac{\rho_1 V_1^2}{2}\left[1 + \frac{M_1^2}{4} + \frac{(2-k)}{24} M_1^4 + ...\right]$$

The term $\left[1 + \frac{M_1^2}{4} + \frac{(2-k)}{24} M_1^4 + ...\right]$ is known as **Compressibility Correction Factor.** And $\frac{\rho_1 V_1^2}{2}$ is the reading of the pitot-static tube. Thus the reading of the pitot- tube must be multiplied by a correction factor given below for correct value of velocity measured by the pitot-tube.

Compressibility Correction Factor, C.C.F. = $\left[1 + \frac{M_1^2}{4} + \frac{2-k}{24} M_1^4 + ...\right]$ ...(12.25)

Problem 12.13 *Calculate the numerical factor by which the actual pressure difference shown by the gauge of a pitot-static tube must be multiplied to allow for compressibility when the value of the Mach number is 0.9. Take k = 1.4.*

Solution. Given :

Mach number, $M_1 = 0.9$

$k = 1.4$

Using equation (12.25), Compressibility Correction Factor is

$$\text{C.C.F.} = \left[1 + \frac{M_1^2}{4} + \frac{2-k}{24} M_1^4 + ...\right] = 1 + \frac{.9^2}{4} + \frac{2-1.4}{24}(.9)^4 + ...$$

$= 1.0 + .2025 + .0164 + ... =$ **1.2189. Ans.**

∴ Numerical factor by which the actual pressure difference is to be multiplied = 1.2189.

HIGHLIGHTS

1. A flow, in which density does not remain constant during flow, is called compressible flow.
2. Equation of state is given by, $\frac{p}{\rho} = RT$

   where p = Absolute pressure in kgf/m^2 or N/m^2
   T = Absolute temperature = 273 + t°C
   R = Gas constant in J/kg K or m/kg
   ρ = Density of gas.

   If the value of p is taken in N/m^2, R is J/kg K and T in K, the value of density is given in kg/m^3

3. The pressure density of a gas are related as

$$\frac{p}{\rho} = \text{Constant.... of isothermal process}$$

$$\frac{p}{\rho^k} = \text{Constant....for adiabatic process}$$

where k = Ratio of specific heats = 1.4 for air.

4. Continuity equation for compressible flow is given as ρAV = Constant.

And in differential form, continuity equation is $\frac{d\rho}{\rho} + \frac{dA}{A} + \frac{dV}{V} = 0$.

5. Bernoulli's equation for compressible fluids is given as

$$\frac{p}{\rho g} \log_e p + \frac{V^2}{2g} + Z = \text{Constant......for isothermal process.}$$

$$\frac{k}{(k-1)} \frac{p}{\rho\phi g} + \frac{V^2}{2g} + Z = \text{Constant......for adiabatic process.}$$

6. Velocity of sound wave is given by

$$C = \sqrt{\frac{dp}{d\rho}} = \sqrt{\frac{K}{\rho}} \quad \text{......in term of Bulk modulus}$$

$$C = \sqrt{\frac{p}{\rho}} = \sqrt{RT} \quad \text{......for isothermal process}$$

$$= \sqrt{\frac{kp}{\rho}} = \sqrt{kRT} \quad \text{......for adiabatic process.}$$

7. Mach number, M is given as $M = \frac{V}{C}$

If $M < 1$ ......flow is sub-sonic flow,
$M > 1$ ......flow is super-sonic flow,
$M = 1$ ......flow is sonic flow.

8. In sub-sonic flow, the disturbance always moves ahead of the projectile. In sonic flow, the disturbance moves along the projectile while in super-sonic flow, the projectile always moves ahead of the disturbance.

9. Mach angle is given by $\sin \alpha = \frac{C}{V} = \frac{1}{M}$.

10. The pressure, temperature and density at a point where velocity is zero are called stagnation pressure, temperature and stagnation density. Their values are given as

$$p_s = p_1 \left[1 + \frac{k-1}{2} M_1^2\right]^{\frac{k}{k-1}} ; \rho_1 = \rho_1 \left[1 + \frac{k-1}{2} M_1^2\right]^{\frac{1}{k-1}}$$

$$T_s = T_1 \left[1 + \frac{k-1}{2} M_1^2\right] \text{ also } = \frac{1}{R} \frac{p_s}{\rho_1}.$$

11. Area velocity relationship for compressible fluid is given as $\frac{dA}{A} = \frac{dV}{V}[M^2 - 1]$

If $M < 1, \frac{dV}{V} < 0$ and $\frac{dA}{V} > 0$ which means with the increase of area, velocity decreases.

If $M > 1, \frac{dV}{V} > 0$ and $\frac{dA}{A} > 0$ which means with the increase of area, velocity also increases.

If $M = 1$, $\frac{dA}{A} = 0$ means area is constant.

12. The compressibility correction factor is given by C.C.F = $\left[2 + \frac{M_1^2}{4} + \frac{2-k}{24} M_1^2 + ...\right]$

EXERCISE 12

(A) THEORETICAL PROBLEMS

1. Define compressible and incompressible flow.
2. What is the relation between pressure and density of a compressible fluid for
   (*a*) isothermal process and (*b*) adiabatic process ?
3. Derive the continuity equation for one-dimensional compressible flow in differential form. *(VTU, Feb., Aug. 2002)*
4. State the Bernoulli's theorem for compressible flow. Derive an expression for Bernoulli's equation when the process is (*i*) isothermal and (*ii*) adiabatic.
5. Write an expression for momentum equation for compressible fluid.
   (*a*) isothermal process and (*b*) adiabatic process ?
6. (*a*) Obtain an expression for velocity of the sound wave in a compressible fluid in terms of change of pressure and change of density.
   (*b*) Show that the velocity of propagation of the pressure wave in a compressible fluid is given by $C = \sqrt{E/\rho}$, where E is volume modulus of elasticity of fluid. *(A.M.I.E., Summer 1991)*
   (*c*) Show that velocity of propagation of elastic wave in an adiabatic medium is given by $C = \sqrt{kRT}$ starting from fundamentals. *(VTU, March 2001, Feb, 2003)*
7. Prove that the velocity of sound wave in a compressible fluid is given by $C = \sqrt{\frac{k}{\rho}}$
   where k = Bulk modulus of fluid, ρ = Density of fluid.
   (VTU, Aug. 2002, Jan 2004; Delhi University, 1982)
8. Derive an expression for the velocity of sound wave for compressible fluid when the process is assumed as (*i*) isothermal and (*ii*) adiabatic. *(A.M.I.E., Summer, 1987)*
9. Define Mach number. What is the significance of Mach number in compressible fluid flows ?
10. Define the terms: Sub-sonic flow, super-sonic flow, sonic flow, Mach angle and Mach cone.
11. (*a*) Show by means of diagrams the nature of propagation of disturbance in compressible flow when Mach number is less than one, is equal to one and is more than one. *(Delhi University, Dec. 2002)*
    (*b*) Sketch the nature of propagation of disturbance in compressible flow when Mach number is more than one and hence define Mach angle and Mach cone. *(VTU, Jan. 2004)*
12. What do you understand by stagnation pressure? Obtain an expression for stagnation pressure of a compressible fluid in terms of approaching Mach number and pressure.

13. Prove that stagnation temperature and stagnation density are given as

$$T_s = T_1\left[1+\frac{k-1}{2}M_1^2\right] \text{ and } \rho_s = \rho_1\left[1+\frac{k-1}{2}M_1^2\right]^{\frac{1}{k-1}}$$

14. Derive an expression for area velocity relationship for a compressible fluid in the form.

$$\frac{dA}{A} = \frac{dV}{V}[M^3 - 1]$$ *(Delhi University, Dec. 2002)*

15. What do you mean by compressibility correction factor of ? Find an expression for compressibility factor.

16. Derive an expression for velocity of sound for an adiabatic process. *(A.M.I.E., Summer, 1987)*

17. What do you mean by sub-sonic, sonic and super-sonic flows? *(A.M.I.E., Summer, 1988)*

18. For frictionless adiabatic flow(*i.e.*, isentropic flow) show that the stagnation pressure at a given point is given by

$$\frac{p_s}{p_1} = 1+\frac{1}{4}M_0^2 + \frac{(2-k)}{24}M_0^4 + \ldots$$

where p_s = stagnation pressure, p_0 = pressure in ambient flow and $M_0 = U_0/\sqrt{E/\rho}$.

(A.M.I.E., Winter, 1991)

19. Differentiate between isentropic and adiabatic processes. *(A.M.I.E., Winter, 1990)*

20. (*a*) Derive an expression for the velocity of sound waves moving in a compressible fluid.
(*b*) Define and explain the terms :
Mach number, Froude number, Reynolds number, Mach cone and Mach angle.

(Delhi University, 1996)

21. State the Bernoulli's theorem for compressible flow. Derive an expression for Bernoulli's equation when the process is adiabatic. *(R.G.P.V., Summer, 2002)*

22. Differentiate between :
(*i*) Supersonic flow and sub-sonic flow,
(*ii*) Adiabatic flow and isothermal flow. *(VTU, March 2001)*

(B) NUMERICAL PROBLEMS

1. A gas is flowing through a horizontal pipe of cross-sectional area of 30 cm^2. At a point the pressure is 30 N per cm^2 (gauge) and temperature 20°C. At another section the area of cross-section is 15 cm^2 and pressure is 25 N/cm^2 gauge. If the mass rate of flow of gas is 0.15 kg/s, find the velocities of the gas at these two sections, assuming an isothermal change. Take R = 287 J/kg K, and atmospheric pressure 10 N/cm^2.

[**Ans.** V_1 = 10.71 m/s ; V_2 = 24.5 m/s]

2. A gas with a velocity of 350 m/s is flowing through a horizontal pipe at a section where pressure is 8 N/cm^2 (absolute) and temperature 30°C. The pipe changes in diameter and at this section the pressure is 12 N/cm^2 (absolute). Find the velocity of the gas at this section if the flow of the gas is adiabatic. Take R = 287 J/kg K and k = 1.4. [**Ans.** 218.63 m/s]

3. Find the speed of the sound wave in air at sea-level where the pressure and temperature are 9.81 N/cm^2 (abs.) and 20°C respectively. Take R = 287 J/kg K and k = 1.4. [**Ans.** 343.11 m/s]

4. Calculate the Mach number at a point on a jet propelled aircraft which is flying at 900 km/hour at sea-level where air temperature is 15°C. Take k = 1.4 and R = 287 J/kg K. [**Ans.** 0.735]

5. An aeroplane is flying at an height of 20 km, where the temperature is – 40°C. The speed of the plane is corresponding to M = 1.8. Assuming k = 1.4 and R = 287 J/kg K, find the speed of the plane.

[**Ans.** 1982.66 m/hr]

6. A projectile is travelling in air having pressure and temperature as 8.829 N/cm^2 and – 5°C. If the Mach angle is 30°, find the velocity of the projectile. Take k = 1.4 and R = 287 J/kg K. [**Ans.** 656.30 m/s]

7. A projectile travels in air of pressure 8.829 N/cm^2 at –10°C at a speed of 1200 km/hour. Find the Mach number and the Mach angle. Take k = 1.4 and R = 287 J/kg K. [**Ans.** 1.025, $\theta = 77.2°$]

8. Find the Mach number when an aeroplane is flying at 900 km/hour through still air having a pressure of 8.0 N/cm^2 and temperature –15°C. Take k = 1.4 and R = 287 J/kg K. Calculate the pressure, temperature and density of air at the stagnation point on the nose of the plane. *(Delhi University, 1986)* [**Ans.** 0.776, 11.9 N/cm^2, 16.06°C, 1.434 kg/m^3]

9. Calculate the numerical factor by which the actual pressure difference shown by the gauge of a pitot-tube should be multiplied to allow for compressibility when the value of the Mach number is 0.7. Take k = 1.4. [**Ans.** 1.1285] *(A.M.I.E., Summer, 1987)*

10. Find the Mach number when an aeroplane is flying at 1000 km/hour through still air having pressure of 7 N/cm^2 and temperature of –5°C. Take R = 287.14 J/kg K. Calculate the pressure and temperature of air at stagnation point. Take k = 1.4. *(Delhi University, Dec. 2002)*

[**Hint.** $V = 1000 \text{ km/hour} = \dfrac{1000 \times 1000}{60 \times 60} = 277.77 \text{ m/s}$; $p_1 = 7 \text{ N/m}^2 = 7 \times 10^4 \text{ N/m}^2$, $t_1 = -5°\text{C}$

$$\therefore \quad T_1 = -5 + 273 = 268° \text{ K} ; R = 287.14 \text{ J/kg K}, k = 1.4$$

$$\text{Now} \quad C_1 = \sqrt{kRT_1} = \sqrt{1.4 \times 287.14 \times 268} = 328.2 \text{ m/s} \qquad \therefore M = M_1 = \frac{V}{C} = \frac{V_1}{C_1} = \frac{277.77}{328.2}$$

$$\therefore \quad M_1 = 0.846 \qquad \therefore p_s = p_1\left[1 + \frac{k-1}{2} M_1^2\right]^{\frac{k}{k-1}} = 7 \times 10^4 \left[1 + \frac{1.4-1}{2} \times (0.846)^2\right]^{\frac{1.4}{1.4-1}}$$

$$= 7 \times 10^4 \left[1 + 0.2 \times 0.846^2\right]^{\frac{1.4}{0.4}} = 11.18 \times 10^4 \text{ N/m}^2 = \mathbf{11.18 \text{ N/cm}^2. \text{ Ans.}}$$

$$T_s = T_1\left[1 + \frac{k-1}{2} M_1^2\right] = 268\left[1 + \frac{1.4-1}{2} \times (0.846)^2\right] = 306.38° \text{ K} = 306.38 - 273 = \mathbf{33.38°\text{C. Ans.}}]$$